To convert from	to	Multiply by
Moment	dyne-centimeter	newton-meter (N-m)	$1.000\ 000\ (10^{-7})$*
	pound-foot	newton-meter (N-m)	1.355 818
Power	Btu/hour	watt (W)	$2.930\ 711\ (10^{-1})$
	erg/second	watt (W)	$1.000\ 000\ (10^{-7})$*
	foot-pound/second	watt (W)	1.355 818
	horsepower (550 ft–lbs)	watt (W)	$7.456\ 999\ (10^{2})$
Pressure and Stress	atmosphere	pascal (Pa)	$1.013\ 25\ (10^{5})$
	bar	pascal (Pa)	$1.000\ 000\ (10^{5})$*
	centimeter of mercury (0°C)	pascal (Pa)	$1.333\ 22\ (10^{3})$
	dyne/centimeter^2	pascal (Pa)	$1.000\ 000\ (10^{-1})$*
	pound/foot^2	pascal (Pa)	$4.788\ 026\ (10^{1})$
	pound/inch^2 (psi)	pascal (Pa)	$6.894\ 757\ (10^{3})$
Speed	foot/second	meter/second (m/s)	$3.048\ 000\ (10^{-1})$*
	inch/second	meter/second (m/s)	$2.540\ 000\ (10^{-2})$*
	kilometer/hour	meter/second (m/s)	$2.777\ 778\ (10^{-1})$
	knot	meter/second (m/s)	$5.144\ 444\ (10^{-1})$
	mile/hour	meter/second (m/s)	$4.470\ 400\ (10^{-1})$*
Volume	foot^3	meter^3 (m^3)	$2.831\ 685\ (10^{-2})$
	gallon (U.S. liquid)	meter^3 (m^3)	$3.785\ 412\ (10^{-3})$
	inch^3	meter^3 (m^3)	$1.638\ 706\ (10^{-5})$
	liter	meter^3 (m^3)	$1.000\ 000\ (10^{-3})$*
	quart (U.S. liquid)	meter^3 (m^3)	$9.463\ 529\ (10^{-4})$

*Denotes an exact quantity.

STATICS

JERRY H. GINSBERG
JOSEPH GENIN
School of Mechanical Engineering
Purdue University

JOHN WILEY & SONS
New York Santa Barbara London
Sydney Toronto

STATICS

This book was printed and bound by Kingsport Press.
It was set in Times Roman by Graphic Arts Composition, Inc.
The copyeditor was Lynn Lackenbach.
Kenneth R. Ekkens supervised production.

Cover and text design by Edward A. Butler.

Library of Congress Cataloging in Publication Data:

Ginsberg, Jerry H. 1944-
Statics.

1. Statics. I. Genin, Joseph, 1932- joint author. II. Title.
TA351.G56 620.1'053 76-55753
ISBN 0-471-29607-4

Printed in the United States of America

10 9 8 7 6 5 4 3 2 1

PREFACE

Statics is the first course in the engineering mechanics sequence. Its primary function is to help the student develop an understanding of the physical laws governing the response of engineering systems to forces. Beyond that, the study of statics enhances the student's reasoning power as applied to the field of engineering, in that the student learns to solve problems logically, using the concept of mathematical models for physical systems. A successful statics course is a strong motivation for further study of engineering.

This text is written in modular form. The modules are broader in content than the conventional grouping by chapters, being largely self-contained in order to minimize the amount of cross-referencing necessary to develop the material. This approach was also chosen because it is our desire to let the book communicate directly to the student, with a minimum of amplification and clarification required of the instructor.

Our treatment of the subject matter is founded on the recognition that the learning experience in statics is a gradual process, as emphasized by the dual nature of the topics forming a course in engineering statics. One set of topics—such as the treatments of force resultants, the laws of equilibrium, and the virtual work and energy principles—deals primarily with the development of the methodology of solving general problems. The other set of topics—such as the treatments of structural analysis, distributed forces, and the laws of friction—deals with the applications of the fundamental laws to specific systems and situations.

The modules that focus on the methodology associated with general principles are presented in a novel way. To increase comprehension of the basic principles and techniques we precede each derivation with a discussion of why the principle is needed. The derivations are then keynoted by remarks regarding common systems that illustrate the implications of the derived concepts, as well as critical comments regarding its usefulness. This is then followed by one or more solved examples that directly illustrate how the derived principle or technique is employed.

After a broad body of basic concepts has been developed in a methodology module, the question of their synthesis into a consistent procedure for solving general problems is addressed. This begins with the presentation of a set of sequential steps detailing the multiple operations necessary for the solution of general problems. These steps are merely a logical sequence to follow. (They are certainly not the only possible sequence.) With these steps, where appropriate, we indicate places where errors are commonly made. By presenting a systematic approach to problem solving we hope to enhance the student's senses of logic and deductive reasoning. Thus, the steps are not intended for memorization. Instead, as a student gains proficiency in problem solving, the procedures will be performed intuitively.

The steps are then employed in a series of three or more solved problems—called illustrative problems—whose solution requires the synthesis of the concepts previously developed. The illustrative problems are cross-referenced directly to the steps for problem solving. This allows the student to isolate a particular aspect of the solution that may prove troublesome. Numerous homework problems are presented after both examples and illustrative problems; in general, those following the latter are broader in scope.

The sole difference between the approach outlined above and the approach for the applications modules is that steps for solving problems are not presented in the latter. Emphasis in the discussion and solutions of examples in the applications modules is placed on the development of a logical approach using the basic methodology, and each group of solved examples is followed by a broad range of homework problems.

The development of physical understanding in mechanics is addressed in the solved problems, as well as in the formal text material. Where appropriate, in these problems care is taken to discuss the qualitative aspects of the solutions. Also, the mathematics presented does not overwhelm the physical aspects of the problem; each solution is implemented with the aid of only that level of mathematics appropriate to solve the problem at hand. Hopefully, this overall philosophy will give the student better insight as to how an engineer thinks.

Note from the Contents that the conventional ordering of the subject matter has been retained. Nevertheless, the viewpoint and treatment of many topics are not contained in other texts. In particular, the systematization of the methodology for treating virtual work and energy methods is entirely new and accessible to students with a broad range of backgrounds.

In addition to its organization, another feature of the book is the flexibility it affords the instructor. It was written to provide the greatest opportunity for adjustments and accommodations in instruction, in order to communicate clearly with students having a broad range of individual experience. The text is equally suitable for courses using innovative instructional approaches, such as self-paced and self-taught courses, as well as for those using the more conventional methods.

Note also that this book does not attempt to address the organizational problems involved in self-paced and self-taught courses. However, it is possible for instructors wishing to utilize such an approach to employ the book as a framework that is to be supplemented by material addressing the problem of the interaction between the student and the instructor. The self-contained aspect of the modules make the text especially valuable for such courses.

A final technical matter for consideration here is the question of physical units. We strongly believe that the SI system of metric units is the best for engineering. Hopefully, by the time this is read, the SI system will have been adopted universally. However, recognizing the transitory nature of this problem, and noting the large number of physical systems that have already been built according to the U.S.-British system of units, the numerical problems have been divided approximately into a two to one proportion between SI and British units. This proportion allows for a sufficient number of problems for courses using only one system of units.

We gratefully acknowledge the encouragement of our colleagues at Purdue University, whose helpful comments proved invaluable. We also thank the faculty of the École Nationale Supérieure d'Électricité et de Mécanique at Nancy, France, particularly Professor Michel Lucius. By providing an academic environment for Professor Ginsberg during his sabbatical leave, they greatly expedited the completion of this work. We are indebted to Rona Ginsberg for her excellent editorial comments, as well as for typing the manuscript. Finally, we note the knowledgeable assistance of the staff of John Wiley, especially our editor, Thurman R. Poston, in all phases of the production process.

JERRY H. GINSBERG

JOSEPH GENIN

CONTENTS

STATICS

MODULE I
FUNDAMENTAL CONCEPTS

A. MECHANICS

Many times when touching an object we wonder how it has been affected physically. Thoughts of this kind are basic to the field of *mechanics,* where the general concern is the study of the effect of *forces* acting on a body.

To a certain extent the concept of a force is intuitive, because forces are not actual objects. In spite of this, *force* does have a basis in our fundamental sensing system, especially in the sense of touch. For instance, the sensations our brain processes when our fingertips contact another body are a measure of how hard and in which direction we push or pull on the object. In mechanics this is referred to as the external force exerted on the object.

The study of mechanics is divided into two broad categories, *statics* and *dynamics*. *Statics* is the study of physical systems that remain at rest under the action of a set of forces. *Dynamics* is the study of physical systems in motion. This text is devoted to the study of *statics*.

A key phrase used above is ''physical system.'' By it we mean a collection of basic atomic elements that combine to form bodies. In the field of mechanics, for the convenience of problem solving, we tend to categorize these collections of elements into three broad groupings: particles, rigid bodies, and deformable bodies.

PARTICLE A body whose dimensions are negligible is said to be a particle. It follows, then, that a particle occupies only a single point in space.
RIGID BODY A body occupying more than one point in space is said to be rigid if all of the constituent elements of matter within the body are always at fixed distances from each other.
DEFORMABLE BODY A body is said to be deformable if its constituent elements of matter experience changes in their distances from each other that are significant to the problem being investigated.

Clearly there is a certain amount of ambiguity in these definitions. For instance, is the group of molecules that compose a gas a system of particles or is it a deformable body? Another basic question we could ask is whether any body can correctly be considered to be rigid; for all real materials deform when forces are exerted upon them.

There are no absolute answers to these questions, for the *model* of the system we form depends on what knowledge we wish to gain about the response of the system. This leads to the next topic, which is the modeling process.

B. THE MODELING PROCESS

The general approach in the static analysis of a physical system is to consider first the nature of the forces acting on the system and the type of information we seek from the subsequent analysis. On the basis of those considerations we construct a conceptual model of the system. That is, we consciously consider its components to be either particles, rigid bodies, or deformable bodies. Then, by applying the laws of mechanics we determine the interrelationships among the forces acting on the system. An analysis of these forces should provide us with the information we desire. This general approach to the subject of statics can be summarized in the conceptual diagram shown in Figure 1.

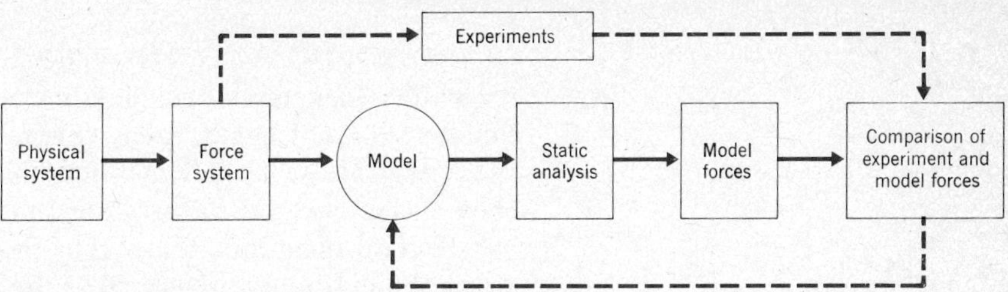

Figure 1

The dashed paths in Figure 1 illustrate a key feature of the modeling process. That is, to be sure of the validity of the model initially chosen to represent the physical system, it is necessary that the model display any relevant phenomena that are experimentally observed. If not, the model must be improved. Hence, modeling is an educated trial-and-error procedure. That is, the modeling process is, in part, an art based on prior experience and physical intuition. On the other hand, the modeling process is also a science based on a knowledge of the function that the system is designed to perform, and of the analytical methods available for studying the phenomena exhibited by the system.

As an example, consider an object in the shape of a sphere. If this sphere is a small ball bearing resting on a flat level surface, as a first attempt one would tend to model it as a particle occupying a single point in space. We shall see in later modules that this model would result in a successful analysis.

Suppose that we now step on this ball bearing, placing the ball bearing at the edge of the sole of our shoe. Our later studies will show that considering the ball bearing as a particle leaves us unable to explain why it tends to slide out from under foot when we press on it harder and harder. In this case the nature of the forces acting on the ball bearing changes drastically from the case where it was merely resting on the ground. Modeling the ball bearing as a rigid body, as opposed to a single particle, would allow us to resolve this problem.

Continuing with this example, suppose that instead of stepping on a steel ball bearing, we were stepping on a tennis ball. Needless to say, the shape of the ball would be altered considerably as we stepped on it.

Clearly, the model of a rigid body should then be discarded in favor of the model of a deformable one.

From the foregoing examples we see that the model chosen to represent the physical system is a crucial element in the solution process. In this text, a first course on the statics of physical systems, we focus on the particle and rigid body models. The relationship between the forces acting on a body and its deformation are considered in the subject of mechanics of materials, which builds on the techniques we will develop here.

C. VECTOR QUANTITIES—FORCES

A quantity that can be completely characterized by its numerical value is called a *scalar*. Length, area, volume, weight, mass, temperature, time, and energy are scalars. Scalar quantities are merely numbers, and as such obey all the laws of ordinary algebra.

Several quantities that occur in mechanics require a description in terms of their direction, as well as the numerical value of their magnitude. Such quantities behave as *vectors*. Typical vectors are force, velocity, position, displacement, and acceleration. Because vectors have both *magnitude* and *direction,* they have a unique algebra.

Pictorially, a vector may be represented by a directed line segment, that is, by a straight arrow. The length of the arrow, to any convenient scale, represents the magnitude of the vector, and the direction of the arrow is the direction of the vector. Thus, in Figure 2 we show a vector quantity $\bar{A}$.

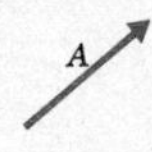

Figure 2

In this book we will denote vector quantities by placing a bar over the symbol adopted for the vector. Thus, the vector in Figure 2 is $\bar{A}$. In contrast, when we wish to refer to the magnitude of a vector, such as the magnitude of $\bar{A}$, we will do so by writing either $|\bar{A}|$ or simply A. Furthermore, because of the significant role of forces in mechanics, an arrow representing a force will be represented more boldly in diagrams than arrows representing other types of vector quantities.

For some vector quantities, a specification of their magnitude and direction describes them completely; that is, their action is not confined to, or associated with, a unique line in space. Such quantities are said to be *free vectors*. An example of a free vector is the statement that a person walked two kilometers (2 km) in a southeasterly direction, which is a description of the displacement resulting from a motion. Velocity and acceleration are other free vectors.

In contrast, when describing a force we must provide some specification of where the force is applied. To see this, consider the horizontal force $\bar{F}$ applied to corner A of the table shown in Figure 3. If the force $\bar{F}$ is sufficient to move the table, in addition to pushing the table forward it will cause the table to rotate counterclockwise as viewed looking downward.

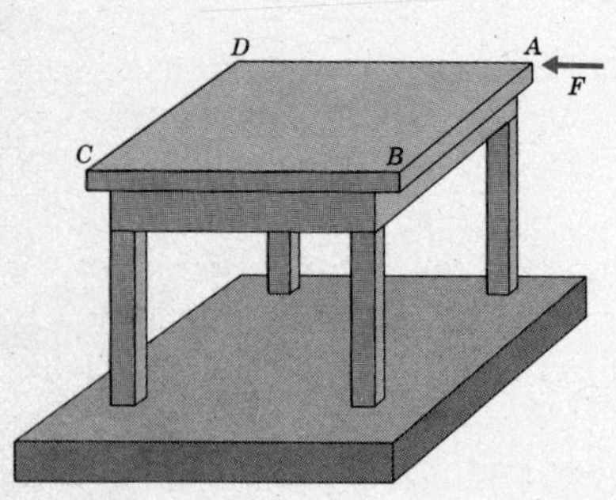

Figure 3

(Try it.) On the other hand, if a force having the same magnitude and direction as $\bar{F}$ is applied at corner B (as opposed to applying $\bar{F}$ at corner A), the result is a clockwise rotation. Hence, we can conclude that in this case we also need to specify the point of application of the force. The vectors described in the foregoing example are *fixed vectors,* for each had a unique point of application.

At times we may regard a force as a *sliding vector,* one that acts anywhere along a unique line in space. In such cases we describe the force by saying that it is collinear with a specific straight line, called its *line of action,* and then giving its magnitude and sense. The actual point of application of the force along its line of action is then considered to be immaterial to the problem. For example, in Figure 3 if we choose to regard $\bar{F}$ as a sliding vector, we can imagine it being applied at point A *or* D and we would get the same resulting motion for the table.

To gain some insight into the seemingly ambiguous cases of fixed vectors and sliding vectors, consider a stalled car that is to be moved with a tow truck. As shown in Figure 4a, the truck may pull on the front bumper with a horizontal force $\bar{F}_A$. Alternatively, as shown in Figure 4b, it may push on the rear bumper with a horizontal force $\bar{F}_B$.

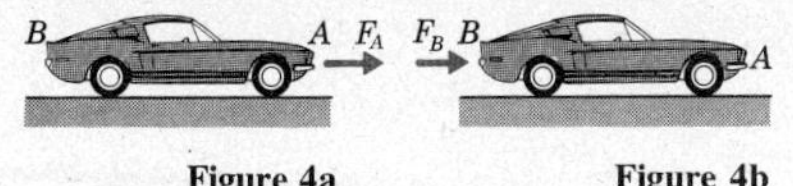

Figure 4a Figure 4b

In this case the forces $\bar{F}_A$ and $\bar{F}_B$ are fixed vectors, having specific points of application, although they share the same line of action and have the same sense. If these forces also have the same magnitude, they will produce the same motion of the automobile. Hence, if the resulting motion was what we wanted to know, we could also treat the force applied to either bumper as a sliding vector.

Suppose now that we consider what happens if the tow truck that is attempting to move the stalled vehicle exerts too large a force. If $\bar{F}_A$ is too large, the front bumper will be torn off, whereas if $\bar{F}_B$ is too large, the rear bumper will be crushed. To account for what happened, we would have to consider the specific point of application of the force. In other words, the force could then be regarded only as a fixed vector.

These observations are a result of the *principle of transmissibility,* which states that

> The effect on a rigid body, in its entirety, of a force external to the body depends only on the magnitude, direction, and line of action of the force.

When we consider questions such as the relationship between the external forces acting on a rigid body, or the motion of a rigid body resulting from a set of external forces, we may apply the principle of transmissibility. In such cases the forces are regarded as sliding vectors. However, if we are concerned with the effect of forces on one portion of an assembly of rigid bodies, or on one region within a single rigid body, we may *not* invoke the principle of transmissibility. The forces should then be regarded as fixed vectors.

The principle of transmissibility is a corollary of a theorem developed in the study of the dynamics of rigid bodies. Because of its importance, we will regard it as an axiom in statics.

Thus far we have discussed forces without really considering how they are generated. There are two broad categories of forces: those resulting from the *contact* of two bodies, and those associated with a *force field* created by the interaction of two bodies. Typical of the latter are force fields associated with gravity and electromagnetism. We will have more to say about gravity later in this module.

An important type of contact force is the one exerted by a lightweight cable fastened to a body, as shown in Figure 5. Pictorially, in Figure 5 we see a cable (or string) fastened to an arbitrary body at a point A with a force $\bar{F}$ applied at its opposite end B. The line of action of $\bar{F}$ is line AB connecting the ends of the cable, with $\bar{F}$ pulling on the end of the cable. The magnitude $|\bar{F}|$ of this force is called the *tension* in the cable because it is transmitted uniformly along the length of the cable. That is, if we ask what force is applied on the region to the left of the dashed line in Figure 5 by the portion of the cable to its right, we would find the same force $\bar{F}$.

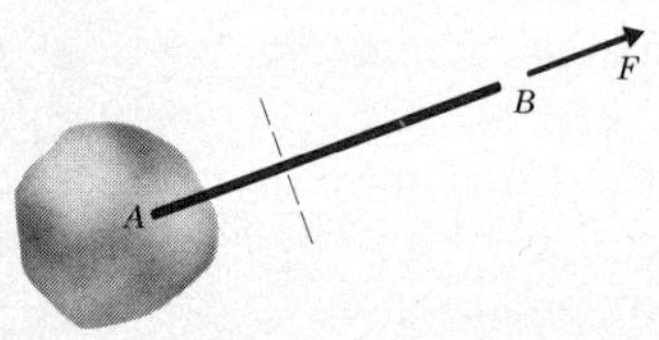

Figure 5

A more subtle feature of $\bar{F}$ is that it can pull on end B with any magnitude up to the breaking force of the cable, but it cannot push on it. When we try to push on end B, the cable becomes slack.

The foregoing discussion contains a *modeling assumption*. The model we considered was for a cable that is inextensible. Cables made of any material will elongate due to a tension force. A clear example of this is a rubber band. We will refer to cables that undergo a significant amount of stretching as *elastic cables*. Defining the elongation, Δ, as the increase in the length of the cable due to an applied force, we may plot the tension as a function of elongation. Two examples are shown in Figure 6.

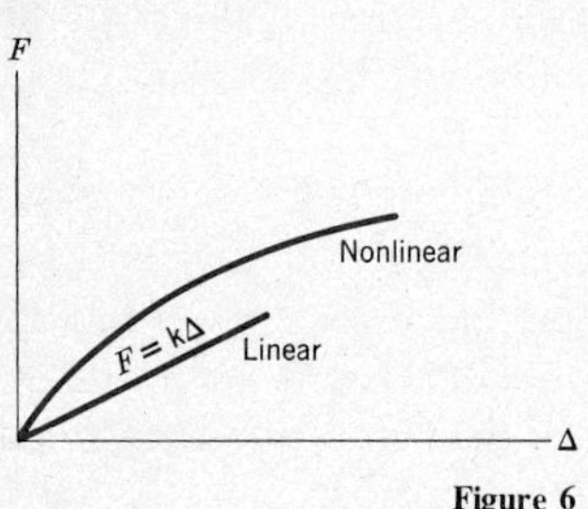

Figure 6

If the plot is a curved line, we have a nonlinear elastic cable. The straight-line plot represents a linear elastic cable, for which we can write

$$\boxed{F = k\Delta} \tag{1}$$

The constant k is called the *stiffness* of the cable, because it tells us how hard the cable pulls for a given elongation.

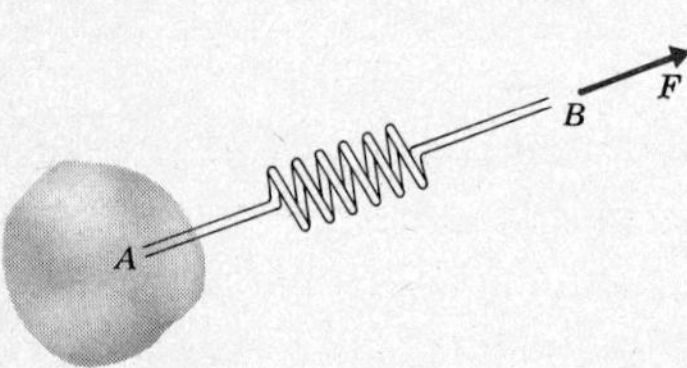

Figure 7

The case of a coil spring, such as that in Figure 7, is identical to that of an elastic cable, with one important exception; the force $\bar{F}$ may be reversed from the direction shown, resulting in the spring pushing on the body with a force whose line of action is AB. The magnitude of $\bar{F}$ is then called the *compressive force*. Equation (1) is also valid for a linear spring, except that now negative Δ corresponds to the distance the spring is shortened, and negative F denotes compression.

D. VECTOR ALGEBRA

We have thus far established that force, being a vector quantity, is specified by its magnitude and direction. For the moment we shall leave aside the question of the point of application of a force. We now wish to define some basic algebraic operations: (1) multiplication of a vector by a scalar, (2) addition of vectors, and (3) subtraction of vectors. The more subtle operations of vector algebra will be developed in later modules in situations where they are naturally motivated.

1 Multiplication by a Scalar

The expression "double the force" means that the magnitude of the force is increased by a factor of 2, but the direction of the force is unchanged. Mathematically, we state this by letting q be a scalar number and $\bar{F}$ be an arbitrary force. The resulting vector $q\bar{F}$ is defined to be parallel to $\bar{F}$. If q is positive, then $q\bar{F}$ is in the same sense as $\bar{F}$, whereas when q is negative $q\bar{F}$ is in the opposite sense from $\bar{F}$. The (positive) magnitude of the resulting vector is defined to be

$$|q\bar{F}| = |q|\,|\bar{F}|$$

This is illustrated in Figure 8.

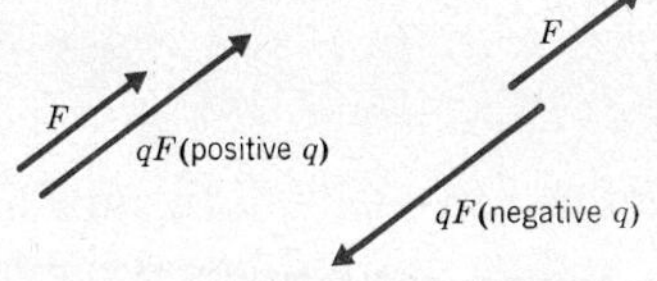

Figure 8

2 Addition of Vectors

The basic rule for adding any two vectors $\bar{A}$ and $\bar{B}$ is the parallelogram law. The vector sum $\bar{C} \equiv \bar{A} + \bar{B}$, which is called the *resultant,* is obtained by drawing the vectors $\bar{A}$ and $\bar{B}$ with their tails coincident. This forms two sides of a parallelogram. The resultant $\bar{C}$ is then the diagonal vector of this parallelgram whose tail is coincident with those of $\bar{A}$ and $\bar{B}$, as depicted in Figure 9.

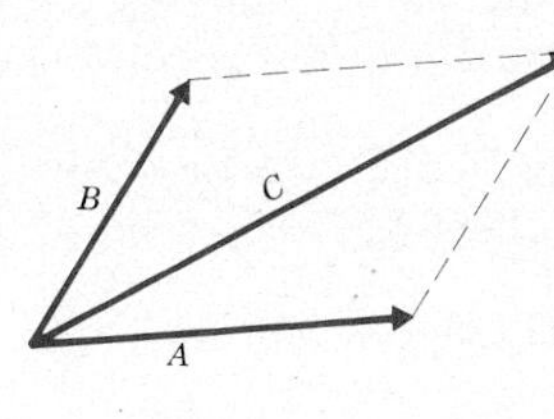

Figure 9

Notice that the order of addition is not important, so $\bar{A} + \bar{B} \equiv \bar{B} + \bar{A}$. This equality suggests an alternative rule for adding vectors. To do so we form a triangle by placing the tail of one vector at the head of the other. The resultant $\bar{C} = \bar{A} + \bar{B}$ is then the third side of the triangle, extending from the tail of the first vector to the head of the second. This is depicted in Figure 10 for the same vectors as were used in Figure 9.

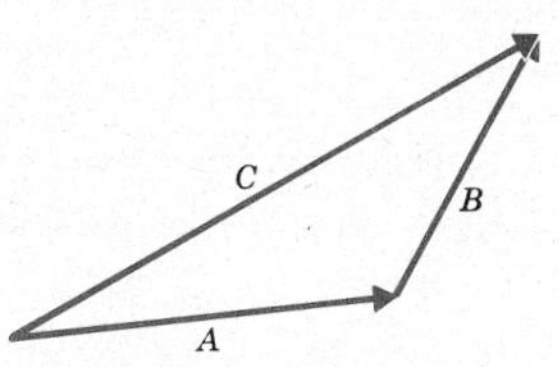

Figure 10

3 Subtraction of Vectors

The rule for subtraction of vectors is a corollary of the addition rule. This

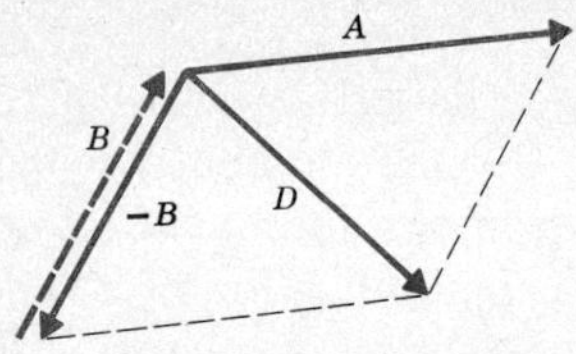

Figure 11

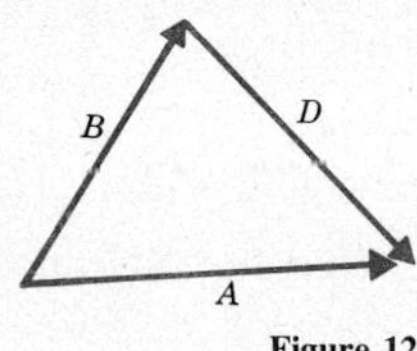

Figure 12

follows because we may write $\bar{D} = \bar{A} - \bar{B} \equiv \bar{A} + (-\bar{B})$. Then, according to the rule for multiplication by a scalar, the vector $-\bar{B} \equiv (-1)\bar{B}$ is a vector having the same magnitude as $\bar{B}$, but opposite direction. For the vectors $\bar{A}$ and $\bar{B}$ of Figure 9, the construction of the difference (subtraction) of vectors, according to the parallelogram law, is shown in Figure 11.

Noting that the difference $\bar{D}$ forms the minor diagonal in the parallelogram of Figure 9, we see that an alternative rule for constructing $\bar{D}$ is to place the two vectors $\bar{A}$ and $\bar{B}$ so that their tails coincide. The difference then goes from the head of the negative vector to the head of the positive one, as shown in Figure 12. This figure also demonstrates that $\bar{B} + \bar{D} \equiv \bar{B} + (\bar{A} - \bar{B}) = \bar{A}$, as it should.

The foregoing vector operations are illustrated in the following example.

EXAMPLE 1

Given two vectors $\bar{A}$ and $\bar{B}$, where $|\bar{A}| = 27$ directed east and $|\bar{B}|$ is directed 40° north of east, determine (a) $\bar{A} + \bar{B}$, (b) $\bar{B} - \bar{A}$, (c) $\bar{A} - 5\bar{B}$.

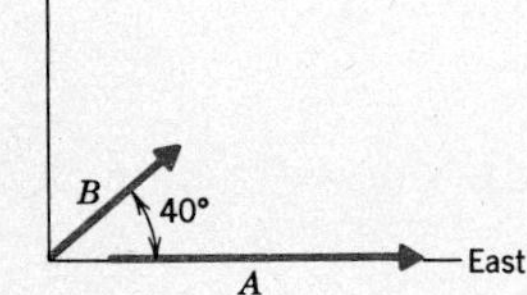

Solution

For the purpose of drawing the vectors let us consider a horizontal line to the right to define the easterly direction and an upward line to be north. The given vectors are then as shown in the sketch.

Part a

Using the triangle construction, we form the sum by moving the tail of $\bar{B}$ to the head of $\bar{A}$, giving the resultant $\bar{C}$ shown in the second sketch.

From the law of cosines we have

$$|\bar{C}|^2 = |\bar{A}|^2 + |\bar{B}|^2 - 2|\bar{A}||\bar{B}| \cos \theta$$

$$= (27)^2 + (15)^2 - 2(27)(15) \cos 140°$$

$$|\bar{C}| = 39.68$$

Then, from the law of sines we have

$$\frac{\sin \alpha}{|\bar{B}|} = \frac{\sin \theta}{|\bar{C}|}$$

$$\sin \alpha = \frac{15}{39.68} \sin 140°$$

$$\alpha = 14.06°$$

Hence

$$\bar{C} = \bar{A} + \bar{B} = 39.68 \text{ directed } 14.06° \text{ north of east}$$

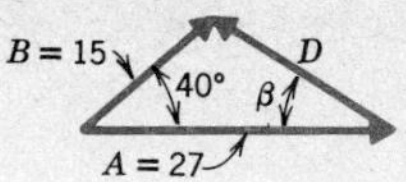

Part b

The triangle for evaluating $\bar{D} = \bar{B} - \bar{A}$ is formed by placing the vectors tail to tail. Then $\bar{D}$ goes from the head of $\bar{A}$ (the negative vector) to the head of $\bar{B}$, as shown in the sketch.

The law of cosines now gives

$$|\bar{D}|^2 = |\bar{A}|^2 + |\bar{B}|^2 - 2|\bar{A}||\bar{B}| \cos 40°$$
$$= (27)^2 + (15)^2 - 2(27)(15) \cos 40°$$
$$|\bar{D}| = 18.26$$

From the law of sines we then find

$$\frac{\sin \beta}{|\bar{B}|} = \frac{\sin 40°}{|\bar{D}|}$$
$$\sin \beta = \frac{15}{18.26} \sin 40°$$
$$\beta = 31.87°$$

Hence

$$\bar{D} = \bar{B} - \bar{A} = 18.26 \text{ directed } 31.87° \text{ north of west}$$

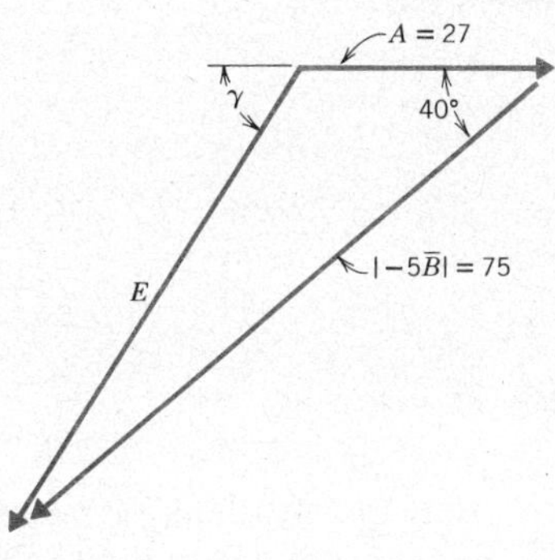

Part c

We may regard $\bar{A} - 5\bar{B}$ to be either $\bar{A} + (-5\bar{B})$ or $\bar{A} - (5\bar{B})$. Let us consider it as the former. Multiplying $\bar{B}$ by -5, we obtain a vector of magnitude 75 directed opposite of $\bar{B}$, that is, directed 40° south of west. Thus, the triangle construction for $\bar{E} = \bar{A} + (-5\bar{B})$ is as shown in the sketch.

The law of cosines gives

$$|\bar{E}|^2 = |\bar{A}|^2 + |-5\bar{B}|^2 - 2|\bar{A}||-5\bar{B}| \cos 40°$$
$$= (27)^2 + (75)^2 - 2(27)(75) \cos 40°$$
$$|\bar{E}| = 57.02$$

whereas the law of sines gives

$$\frac{\sin(180° - \gamma)}{|-5\bar{B}|} = \frac{\sin 40°}{|\bar{E}|}$$
$$\sin(180° - \gamma) = \frac{75}{57.02} \sin 40°$$
$$180° - \gamma = 57.72,\ 122.28°$$

From the computation of $|\bar{E}|$ we know that $|-5\bar{B}|$ is the longest side in the triangle. It then follows that the corresponding vertex angle, $180° - \gamma$, must be the largest interior angle. Hence, choosing $180° - \gamma = 122.28$, we find

$$\gamma = 57.72°$$

so

$$\bar{E} = \bar{A} - 5\bar{B} = 57.02 \text{ directed } 57.72° \text{ south of west}$$

HOMEWORK PROBLEMS

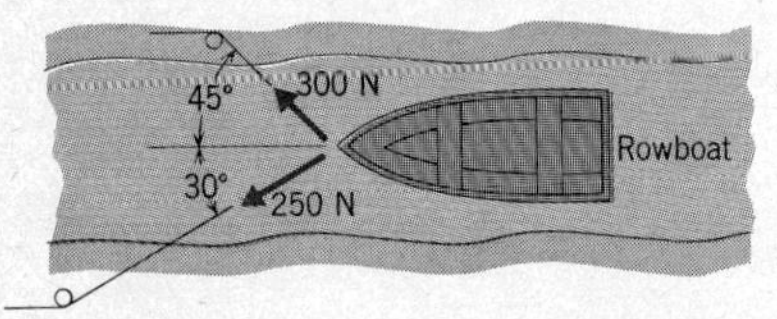

Prob. I.1

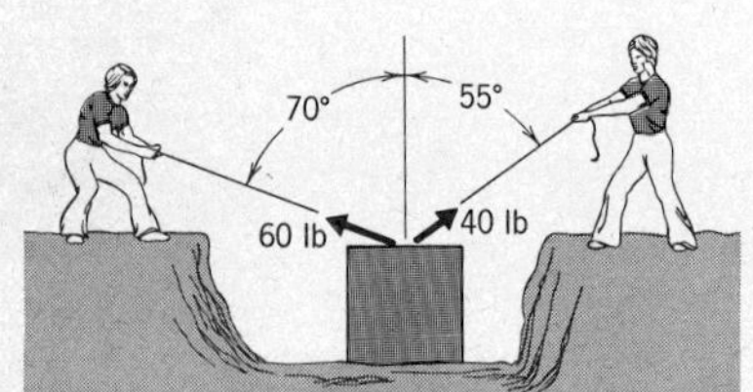
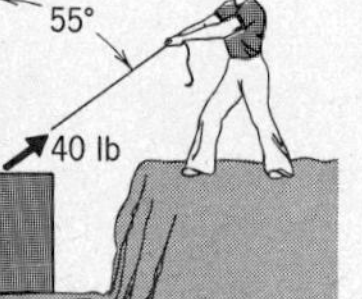

Prob. I.2

I.1 and I.2 Determine the magnitude and direction of the resultant of the two forces shown (N = newton).

In Problems I.3 through I.7: $\bar{A} = 10$ due north, $\bar{B} = 13$ at 45° north of east, $\bar{C} = 27$ at 30° south of east, $\bar{D} = 18$ at 15° west of north.

I.3 Find (a) $\bar{A} + \bar{B}$, (b) $\bar{A} - \bar{B}$, and (c) $3\bar{A} - 2\bar{B}$.

I.4 Find $\bar{A} + \bar{B} - \bar{C}$.

I.5 Find $\bar{A} + \bar{B} + \bar{C} + \bar{D}$.

I.6 Find $\bar{A} + \bar{B} - \bar{C} - \bar{D}$.

I.7 Find the vector $\bar{E}$ for which $\bar{A} + \bar{B} - \bar{C} + \bar{E} = \bar{0}$.

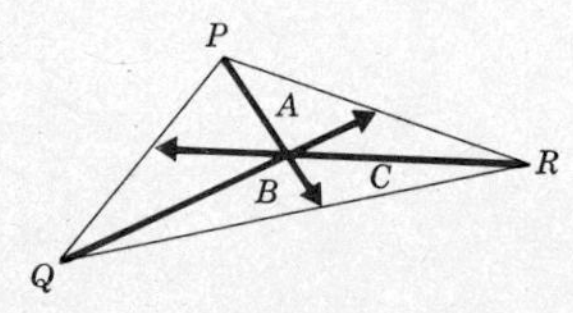

Prob. I.8

I.8 The vectors $\bar{A}$, $\bar{B}$, and $\bar{C}$ shown represent the medians of triangle *PQR*, each being drawn from a vertex to the midpoint of the opposite side. Prove that $\bar{A} + \bar{B} + \bar{C} = \bar{0}$. *Hint:* Represent each side of the triangle as a vector and describe $\bar{A}$, $\bar{B}$, and $\bar{C}$ is terms of these vectors.

E. NEWTONIAN MECHANICS

The laws of motion stated by Sir Isaac Newton (1666–1727) in his historic work *Principia* (1687) are the foundation for the study of engineering mechanics. Here we state his laws in modern language.

FIRST LAW A particle will remain at rest or move with constant speed along a straight line, unless it is acted upon by a resultant force.
SECOND LAW When a resultant force is exerted upon a particle, the acceleration of that particle is parallel to the direction of the force and the

magnitude of the acceleration is proportional to the magnitude of the force.

THIRD LAW Each force exerted upon a body is the result of an interaction with another body, the forces of action and reaction being equal in magnitude, opposite in direction, and collinear.

Our interest in statics lies primarily in the first and third laws; we leave the study of accelerating systems for dynamics. However, a system that is at rest or moving with constant velocity is a special case of an accelerating system, so the first law is really included in the second law. It is stated separately in order to emphasize that the state of motion will be unchanged unless there is a resultant force acting on a system. The third law helps us to define what forces are acting on a body.

In this text on statics, we will use the second law later in this module to discuss the force of gravity and to establish a consistent system of units for describing physical quantities. The second law may be reworded to state that the resultant force $\bar{F}$ acting on a particle is proportional to the acceleration $\bar{a}$ of the particle. As you probably know, the constant of proportionality in this relation is the mass m of the particle. Thus, we have the familiar equation

$$\boxed{\bar{F} = m\bar{a}} \tag{2}$$

A remarkable aspect of Newton's laws is that, as stated, they apply only to a particle. Part of the study of mechanics is devoted to extending these laws to deal with rigid and deformable bodies.

F. SYSTEMS OF UNITS

The quantities appearing in Newton's second law, $\bar{F} = m\bar{a}$, involve measurements of length, time, force, and mass, which we denote for brevity as L, T, F, and M, respectively. The units chosen for the measurement of these four dimensional quantities cannot be defined independently. They must obey the *law of dimensional homogeneity,* which requires that the physical units of all terms in an equation be identical. Stated in colloquial language, "apples cannot be equated to oranges."

In terms of the four basic measurements, the dimensions of acceleration are L/T^2. Dimensional homogeneity of equation (2) requires that

$$F = M\frac{L}{T^2}$$

In other words, we are free to choose the units of three of the four quantities, L, T, F, and M. The fourth must be derived. Historically, the units by which length and time are measured are well defined, and, through the efforts of international standards committees, universally accepted.

This leaves us with the choice of defining either a standard unit of mass or of force, and then deriving the units for the undefined quantity. When the unit of mass is defined according to some standard, the system is called *absolute,* whereas a system in which the unit of force is defined is called *gravitational*. This terminology originates from the relationship between the weight W and the acceleration g of a particle that is falling freely in a vacuum at the surface of the earth. According to Newton's second law, this relationship is

$$\boxed{W = mg} \tag{3}$$

In using equation (3) we must be aware that the values of W and g depend on the location of the particle with respect to the surface of the earth, and equally important on the choice of measurement system. Presently, the two systems that are in wide usage are the metric absolute system and the British gravitational system. In terms of the units of these systems, for most problems we will use the value $g = 9.806$ meters/second^2 (m/s^2) (metric) or $g = 32.17$ feet/second^2 (ft/s^2) (British) as average values. These values are looked at in considerable detail in the dynamics text.

In Table 1 we summarize the commonly used systems of units and indicate how the derived units are obtained. The basic units in each system are defined with respect to standard bodies or with respect to physical phenomena. For example, a kilogram is the mass of a particular bar of platinum alloy. On the other hand, a meter is defined as 1,650,763.73 wavelengths in a vacuum of the orange-red line of krypton 86. The derived quantities in the table are the starred values.

Table 1 System of units

UNIT/SYSTEM	LENGTH	TIME	MASS	FORCE
Metric absolute (SI)	Meter (m)	Second (s)	Kilogram (kg)	Newton (N)* = kg-m/s^2
Metric gravitational	Meter (m)	Second (s)	Metric slug* = kg-s^2/m	Kilogram (kg)
British absolute	Foot (ft)	Second (s)	Pound (lb)	Poundal (pdl)* = lb-ft/s^2
British gravitational	Foot (ft)	Second (s)	Slug* = lb-s^2/ft	Pound (lb)

The choice for systems of units presented in Table 1 is further complicated by the fact that certain multiples and submultiples of the basic units in each system are given their own names, such as the mile, which is 5280 feet, or the dyne, which is one hundred-thousandth of a newton. To simplify this situation the SI (Standard International) system is now being adopted worldwide.

Essentially, the SI system is the metric absolute system with only one unit being used to describe each type of quantity. The SI units for length, time, mass, and force are meters, seconds, kilograms, and newtons, respectively. Then, if one desires, decimal multiples and submultiples of the basic units may be indicated by the appropriate prefixes. Thus, in the SI system length dimensions of 0.2 meters (m) and 200 millimeters (mm) are identical and correct. On the other hand, $2(10^{-3})$ newtons (N) or equivalently 2 millinewtons (mN) is the correct description for a force whose magnitude is 200 dynes. Lists of the SI units and of the preferred prefixes may be found in Appendix A.

Many of the examples and problems in this text will be presented in SI units. Unfortunately, this system is not yet universally used in the United States, where many engineers, as well as nontechnical people, still use the British gravitational system. We will therefore present examples and problems that are posed in either system of units. In general, we will solve problems in the same system of units as that used for the given information, and we shall not convert between different systems of units. Obviously, the units of any physical quantity can be changed with the aid of conversion factors. These conversion factors are numbers that are physically unit values, but not numerically, for example, 25.40 mm = 1 in. A brief set of factors are given in Appendix A.

Conversion factors can be used consecutively. For instance, to change from a speed in miles per hour to a speed in meters per second, we can compute

$$1\,\frac{\text{mile}}{\text{hr}}\left(\frac{1\text{ hr}}{3600\text{ s}}\right)\left(\frac{5280\text{ ft}}{1\text{ mile}}\right)\left(\frac{12\text{ in.}}{1\text{ ft}}\right)\left(\frac{25.40\text{ mm}}{1\text{ in.}}\right)\left(\frac{1\text{ m}}{1000\text{ mm}}\right) = 0.4470\text{ m/s}$$

Remember, the units of all quantities in an equation must be consistent in order to satisfy the law of dimensional homogeneity.

As practice for obtaining a consistent set of units, and also to gain an intuitive understanding of the magnitude of British units relative to SI units, we present the following example.

EXAMPLE 2

Using the facts that 1 in. equals 25.40 mm and that a body weighing 1 lb on the surface of the earth has a mass of 0.4536 kg, determine the factors for converting the British gravitational units of pounds force, slugs, mass, and density (slugs per cubic foot) into their SI equivalents.

Solution

The conversion factors between pounds and newtons is determined by using $W = mg$. Let us compute the weight in newtons of a 0.4536-kg body. Not knowing where the body was weighed, we use the average value $g =$ 9.806 m/s^2 to obtain

$$W = (0.4536 \text{ kg})(9.806 \text{ m/s}^2) = 4.448 \text{ N}$$

In other words,

$$1 \text{ lb} = 4.448 \text{ N}$$

Then, since 1 lb-s^2/ft is 1 slug, we have

$$1 \text{ slug} = \left(1 \frac{\text{lb-s}^2}{\text{ft}}\right)\left(\frac{4.448 \text{ N}}{1 \text{ lb}}\right)\left(\frac{1 \text{ ft}}{12 \text{ in.}}\right)\left(\frac{1 \text{ in.}}{25.4 \text{ mm}}\right)\left(\frac{1000 \text{ mm}}{1 \text{ m}}\right)$$

$$= 14.593 \frac{\text{N-s}^2}{\text{m}} = 14.593 \text{ kg}$$

To convert the units of density, we use

$$1 \frac{\text{slug}}{\text{ft}^3} = \left(1 \frac{\text{slug}}{\text{ft}^3}\right)\left(\frac{14.593 \text{ kg}}{\text{slug}}\right)\left(\frac{1 \text{ ft}}{12 \text{ in.}}\right)^3\left[\left(\frac{1 \text{ in.}}{25.40 \text{ mm}}\right)\left(\frac{1000 \text{ mm}}{1 \text{ m}}\right)\right]^3$$

$$= 515.3 \text{ kg/m}^3$$

The differences between the values obtained here and those in the table in Appendix A are attributable to the fact that we were only using four significant figures for our calculations. The limitation of four significant figures follows from our choice for the value of g.

HOMEWORK PROBLEMS

In the following problems, t is time, x is distance (length), v is speed (length/time), a is the magnitude of acceleration (length/time^2), and m is mass.

I.9 Derive conversion factors for changing the following British gravitational units to their SI equivalents. (a) area: inch^2, (b) volume: foot^3, (c) force: ounce, (d) pressure: pounds/inch^2, (e) pressure: pounds/foot^2, (f) speed: foot/second, (g) speed: in./hr, (h) acceleration: miles/hour^2.

I.10 In each of the following formulas, c_1, c_2, and so on, denote constants and θ is an angle in radians. Determine the units of these constants if the formula is to be dimensionally correct. (a) $a = c_1 v^2/x$, (b) $\frac{1}{2}mv^2 = \frac{1}{2}c_1 x^2$, (c) $x = c_1 x_0 + c_2 v_0 + c_3 t^2$, (d) θ (degrees) $= c_1\theta$ (radians), (e) $d\theta/dt = c_1 + c_2 t$, (f) $t = c_1\sqrt{x}$, (g) $mc_1{}^2\, d^2\theta/dt^2 = Fx$.

I.11 At a certain large university known for its athletic prowess, the following new set of physical units are in use. The basic length unit is the *touchdown* which is 100 yards, the basic time unit is the *dash* which is 9 s, and the basic force unit is the *golf ball,* which is 1.620 oz. Determine the conversion factors between these units and their SI equivalents.

I.12 At the university in Problem I.11 the unit of mass is the *basket,* 1 basket being 1 golf ball-dash²/touchdown. Determine the number of kilograms in one basket.

I.13 When a linear coil spring in the suspension system of an automobile is compressed by 2.5 in., it exerts a compressive force of 375 lb. Determine the stiffness of the spring in British gravitational and in SI units.

I.14 When a certain linear spring has a length of 180 mm the tension within it is 150 N. For a length of 160 mm the compressive force within the spring is 100 N. Determine the unstretched length and stiffness of the spring (a) in SI units, (b) in British gravitational units.

I.15 A widely employed fluid mechanics equation for the pressure drop along a pipe of length l and diameter D is

$$P_1 - P_2 = f\frac{l}{D}\rho\frac{v^2}{2}$$

where P denotes pressure, ρ is the density of the fluid, v is its velocity, and f is the resistance coefficient, which is given by the expression

$$\sqrt{f} = 2.30\log_{10}\left(\frac{Dv}{\nu}\sqrt{f}\right) - 0.91$$

Determine the dimensions of ν, the kinematic viscosity.

I.16 A restaurant buys 1000 lb of coffee at a standard locality and also a spring scale calibrated at that locality. Determine what the coffee weighs on the spring scale in a locality where (a) $g = 32.11$ ft/s², (b) $g = 9.791$ m/s².

I.17 The work done by a force is Fx and the power developed by this force is Fv. In the SI system the basic unit of work and energy is joules (J),

1 J being the work done by a 1-N force in moving a body a distance of 1 m. The basic SI unit for power is a watt (W), 1 W being the power developed by a 1-N force when it moves a body at a speed of 1 m/s. Determine what (a) 1 J is in terms of meters, kilograms, and seconds, (b) 1 W is in terms of meters, kilograms, and seconds, and (c) 1 W is in terms of joules and seconds.

I.18 The power developed by a force is Fv. In the SI system the basic unit of power is a watt, where a watt is as defined in Problem I.17. In the British gravitational system the basic unit of power is a horsepower (hp), 1 hp being the power developed by a 550-lb force when it moves a body at a speed of 1 ft/s. Determine (a) how many kilowatts there are in 1 hp, and (b) how many horsepower there are in 1 kW.

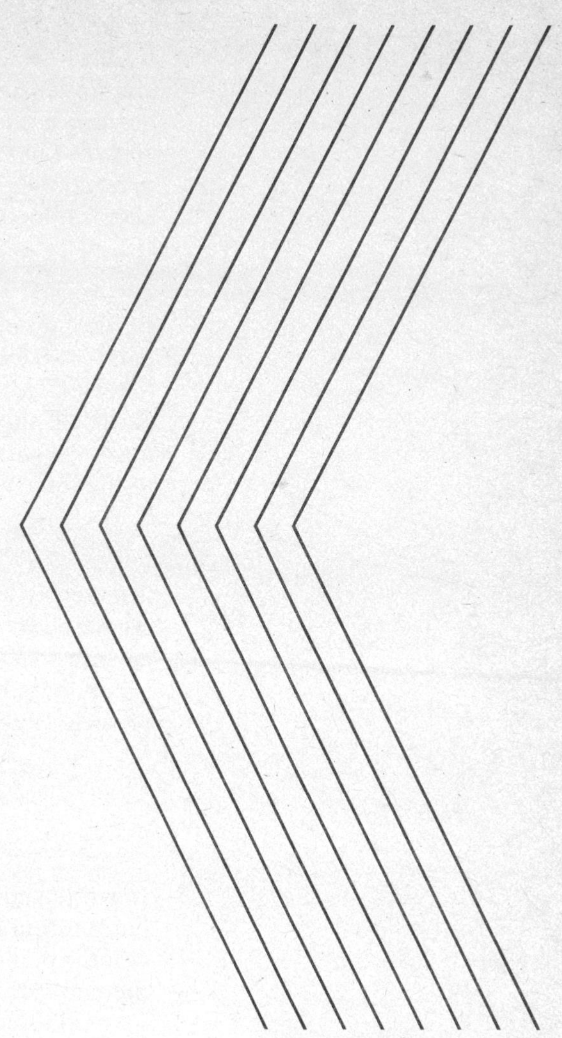

MODULE II
CONCURRENT FORCE SYSTEMS AND EQUILIBRIUM OF A PARTICLE

In this and the following module we shall establish the basic laws governing the forces acting on bodies that are in a state of static equilibrium. The concepts, techniques, and methods in this development are central to all engineering studies concerning the way in which physical systems respond to the forces acting on them. It is natural that in the initial phase, presented in this module, we consider the case of a particle, for it is the simplest model that can be used to represent a physical body.

A. BASIC CONCEPTS

By definition, a particle is a body that is considered to occupy a single point in space. Hence, the forces acting on a particle must be *concurrent* (intersect) at the particle. Consider the two forces $\bar{F}_1$ and $\bar{F}_2$ acting on the particle P shown in Figure 1. Forces $\bar{F}_1$ and $\bar{F}_2$, being vectors, add according to the parallelogram law. Their total effect on the particle is equivalent to the force $\bar{R}$ in Figure 1, where

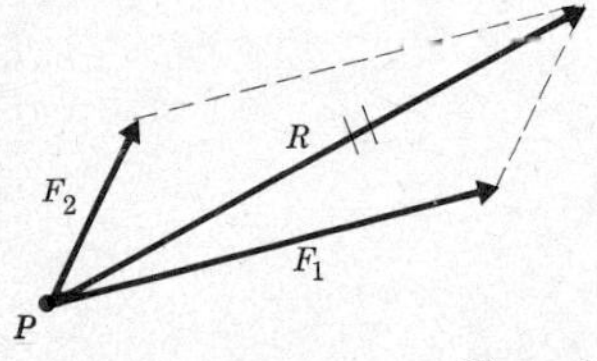

Figure 1

$$\bar{R} = \bar{F}_1 + \bar{F}_2$$

The vector $\bar{R}$ is called the *resultant* of $\bar{F}_1$ and $\bar{F}_2$. Note that we use a double slash on the arrow representing the resultant to distinguish it from an actual force acting on the particle.

If there were more than two forces acting on the particle the resultant of such a *system of forces* would be

$$\boxed{\bar{R} = \sum_{i=1}^{N} \bar{F}_i} \tag{1}$$

In writing equation (1) we made use of the fact that the sum of vectors is independent of the order in which they are added. Further, it is apparent that the point of application of the resultant force is the point occupied by the particle.

Let us consider a number of ways in which the resultant force vector can be obtained. Clearly, as was done in Figure 1, we can use the parallelogram law to add the vectors consecutively. For example, in the case of three forces, we could first obtain the vector $\bar{R}_1 = \bar{F}_1 + \bar{F}_2$ and then find the total resultant according to

$$\bar{R} = (\bar{F}_1 + \bar{F}_2) + \bar{F}_3 = \bar{R}_1 + \bar{F}_3$$

This is depicted in Figure 2a on the following page.

Another view of the process for determining the resultant is obtained from the alternative rule for the addition of vector quantities. In this method the individual vectors in the sum are moved in space, with their orientations constant, until the tail of each vector in the sum coincides

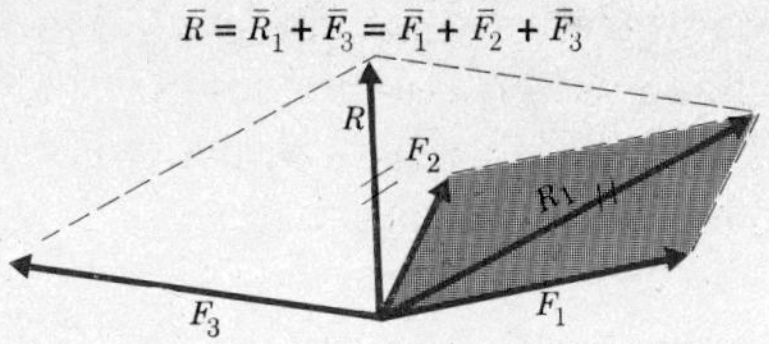

Figure 2a

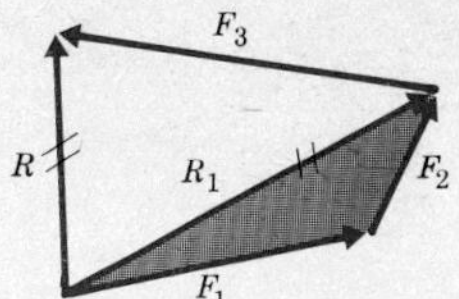

Figure 2b

with the tip of the previous vector, as illustrated in Figure 2b. The resultant $\bar{R}$ then extends from the tail of the first vector ($\bar{F}_1$) to the tip of the last vector ($\bar{F}_3$). Note that in this approach it is not necessary to first determine the intermediate resultant force $\bar{R}_1$. In general, then, the resultant of a system of forces acting on a particle is the closing vector of a polygon whose other sides are composed of all the forces in the system.

Two major difficulties arise when using the approaches of Figure 2 to determine resultant vectors. Clearly, it is difficult to construct the appropriate diagrams for a system of three-dimensional vectors. Equally important, even in the case of planar force systems, the determination of the resultant usually involves the geometry and trigonometry of scalene (general) triangles.

A graphical solution in which the vector polygon is constructed to scale circumvents the need to use trigonometry. However, this approach places unnecessary limitations on the accuracy of the solution, and still does not directly address the question of three-dimensional force systems.

Another procedure, which we shall use throughout the text, is to represent vectors by their components with respect to a set of mutually orthogonal (i.e., perpendicular) axes. These axes form an XYZ coordinate system which we refer to as the *frame of reference*. This component approach for treating vectors has the significant advantage of simplifying the geometry and trigonometry to that of right triangles, while simultaneously providing a consistent approach for both planar and three-dimensional problems.

B. COMPONENTS OF A VECTOR

Before we can use the component approach for vector operations, such as addition, it is necessary for us to develop the ability to represent vectors in terms of their components with respect to an *arbitrarily chosen* rectangular Cartesian coordinate system XYZ. A typical situation is depicted in Figure 3 where we wish to determine the components of a vector $\bar{A}$. The origin O of the coordinate system is chosen, for convenience, to coincide with the tail of the vector.

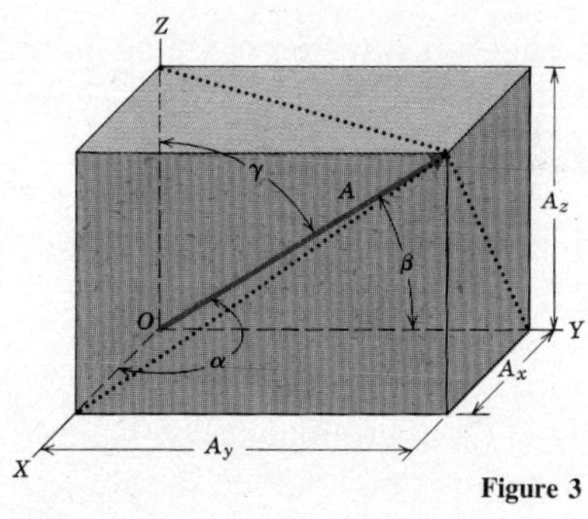

Figure 3

The dotted lines in Figure 3 are constructed to intersect the tip of the vector $\bar{A}$ and to intersect each coordinate axis at a right angle. From the intersections of these perpendiculars with the axes, we can form a rectangular parallelepiped, as depicted in the figure. The lengths A_X, A_Y, and A_Z are defined as the components of the vector $\bar{A}$ with respect to the X, Y, and Z coordinate axes, respectively. Because $\bar{A}$ is the diagonal of the parallelepiped, we may write

$$|\bar{A}| = \sqrt{A_X^2 + A_Y^2 + A_Z^2} \tag{2}$$

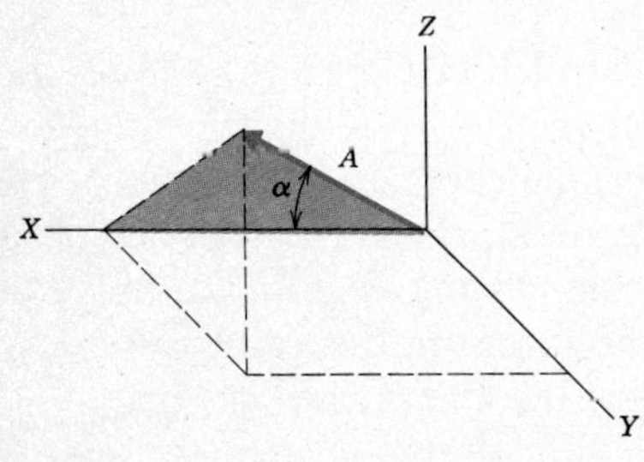

Figure 4

If we passed a plane through the vector $\bar{A}$ and the X axis, the angle that would appear between the two lines is α. This is depicted by the shaded plane in Figure 4. Clearly, $A_X = |\bar{A}| \cos \alpha$. More generally, in terms of the angles between the vector $\bar{A}$ and the three coordinate axes, we may write

$$A_X = |\bar{A}| \cos \alpha \qquad A_Y = |\bar{A}| \cos \beta \qquad A_Z = |\bar{A}| \cos \gamma \tag{3}$$

The cosines of the *direction angles* α, β, and γ are called the *direction cosines*. Placing equations (3) into equation (2) yields

$$\cos^2 \alpha + \cos^2 \beta + \cos^2 \gamma \equiv 1$$

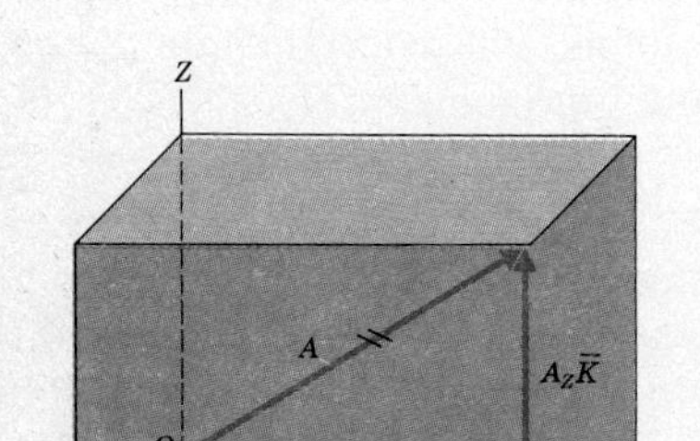

Figure 5

A way of interpreting this identity is to note that only two of the three direction angles can be independent quantities.

From the foregoing we see that the properties of a free vector are completely described by its components with respect to a specific coordinate system. The coordinate system itself is specified by introducing a set of unit vectors (vectors of magnitude one) $\bar{I}$, $\bar{J}$, and $\bar{K}$ which are oriented in the directions of the positive X, Y, and Z axes, respectively. It then follows, as shown in Figure 5, that $A_X\bar{I}$ is a vector parallel to the X axis whose magnitude is the component A_X. Figure 5 also shows that the sum of the three vectors $A_X\bar{I}$, $A_Y\bar{J}$, and $A_Z\bar{K}$ is the vector $\bar{A}$. Hence

$$\boxed{\bar{A} = A_X\bar{I} + A_Y\bar{J} + A_Z\bar{K}} \tag{4}$$

From equation (4) we observe that the unit vectors have no physical units. The physical units of the vector are the same as those of its components.

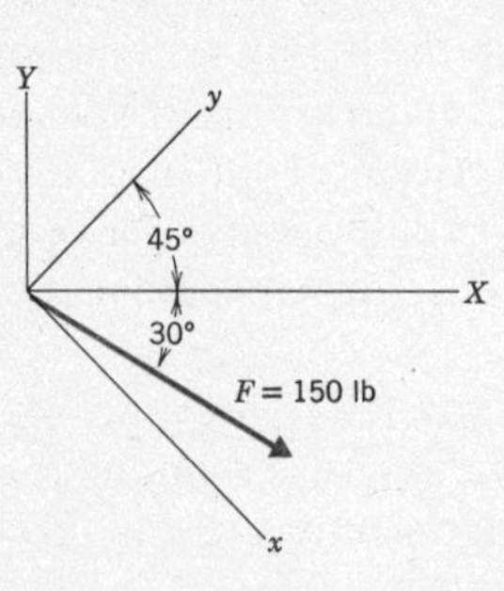

EXAMPLE 1

Represent the force $\bar{F}$ shown in the sketch in terms of its components with respect to the two coordinate systems XYZ and xyz.

Solution

The force lies in the plane of the paper, so the direction angle between $\bar{F}$ and the Z axis is 90°. Because cos 90° ≡ 0, there is no component perpen-

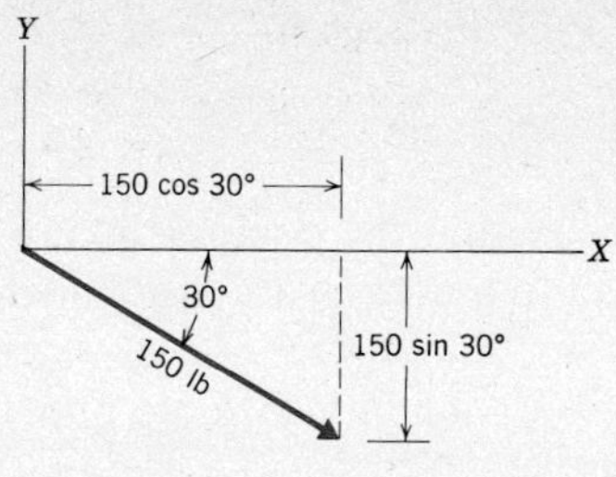

dicular to the paper. A sketch of the components of $\bar{F}$ with respect to the X and Y axes shows that

$$F_X = 150 \cos 30° = 129.90 \text{ lb}$$

$$F_Y = -150 \sin 30° = -75.00 \text{ lb}$$

Note that F_Y is negative because it is along the negative Y axis.

Equation (4) then gives

$$\bar{F} = 129.90\bar{I} - 75.00\bar{J} \text{ lb}$$

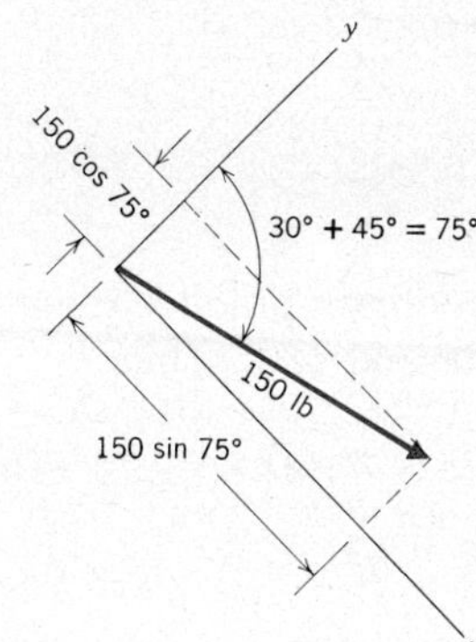

We now follow the same approach for the components with respect to xyz, beginning with a sketch.

From the sketch,

$$F_x = 150 \sin 75° = 144.89 \text{ lb}$$

$$F_y = 150 \cos 75° = 38.82 \text{ lb}$$

Thus

$$\bar{F} = 144.89\bar{i} + 38.82\bar{j} \text{ lb}$$

where $\bar{i}, \bar{j}, \bar{k}$ are the unit vectors for the xyz axes.

This result illustrates the obvious fact that the components of a vector have meaning only when the coordinate system is specified. Using equation (2) we can easily verify that both expressions represent the same vector; the differences in the component values result from the fact that the $\bar{I}$ and $\bar{J}$ unit vectors refer to the XYZ system, whereas the $\bar{i}$ and $\bar{j}$ unit vectors refer to the xyz system.

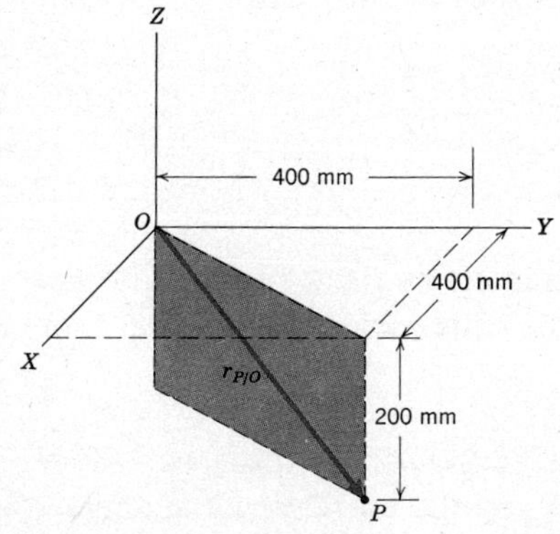

EXAMPLE 2

Determine the vector that locates a point P whose rectangular Cartesian coordinates are (400, 400, −200) mm with respect to the origin O. What are the direction angles of this vector?

Solution

A sketch shows that the desired vector is $\bar{r}_{P/O}$, which is the position of point P with respect to point O (the notation should be read as r of P with respect to O).

We see from the sketch that the components of the vector $\bar{r}_{P/O}$ in the X and Y directions are both 400 mm, whereas the Z component is −200 mm; the negative sign arises because $\bar{r}_{P/O}$ projects onto the negative Z

axis. Hence we write the vector as

$$\bar{r}_{P/O} = 400\bar{I} + 400\bar{J} - 200\bar{K} \text{ mm}$$

This expression illustrates the general result that the components of $\bar{r}_{P/O}$ are the coordinates of point P with respect to a coordinate system whose origin is point O.

We now find the direction cosines by calculating $|\bar{r}_{P/O}|$ and then applying equations (3), as follows.

$$|\bar{r}_{P/O}| = \sqrt{(400)^2 + (400)^2 + (200)^2} = 600 \text{ mm}$$

$$\cos\alpha = \frac{(r_{P/O})_X}{|\bar{r}_{P/O}|} = \frac{400}{600} = 0.6667$$

$$\cos\beta = \frac{(r_{P/O})_Y}{|\bar{r}_{P/O}|} = \frac{400}{600} = 0.6667$$

$$\cos\gamma = \frac{(r_{P/O})_Z}{|\bar{r}_{P/O}|} = \frac{-200}{600} = -0.3333$$

In order to eliminate any ambiguity, it is common practice to consider the direction angles to have values between 0° and 180°. Thus

$$\alpha = \beta = \cos^{-1}(0.6667) = 48.19°$$

$$\gamma = \cos^{-1}(-0.3333) = 109.47°$$

Recalling the definition of the direction angles, can you explain the significance of γ being larger than 90°?

The preceding examples show that the description of a vector quantity according to equation (4) fully describes the vector, provided we have a sketch of the coordinate system. Hence, when solving problems it is acceptable to leave a solution for a vector quantity in the form of equation (4), unless the magnitude and direction angles of the vector are specifically requested.

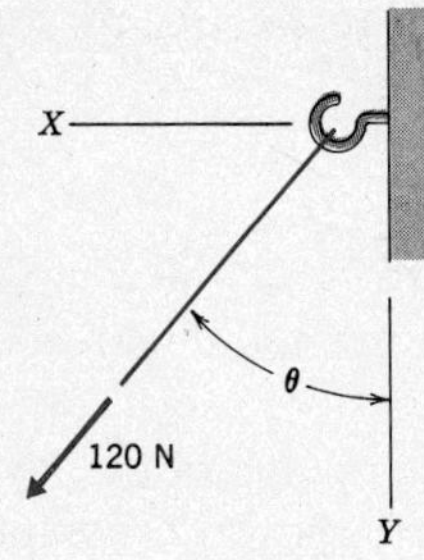

Prob. II.1

HOMEWORK PROBLEMS

II.1 The cable pulls on the eyebolt with a tensile force of 120 N. If θ = 25°, determine the components of the force with respect to the XYZ coordinate system.

II.2 In Problem II.1 the X component of the 120-N force is −90 N. Determine the Y component of the force and the corresponding angle θ.

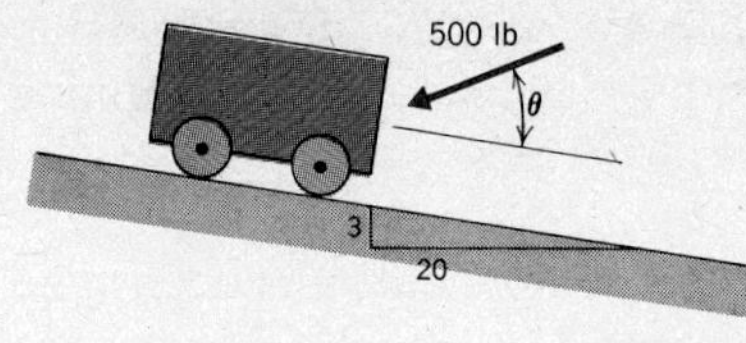

Prob. II.4

II.3 In Problem II.1 the X component of the 120-N force is twice as large as the Y component. Determine these components and the corresponding angle θ.

II.4 The cart is pushed up the 15% grade by the 500-lb force. If $\theta = 30°$, determine the components of this force (a) parallel and perpendicular to the hill, (b) horizontally and vertically.

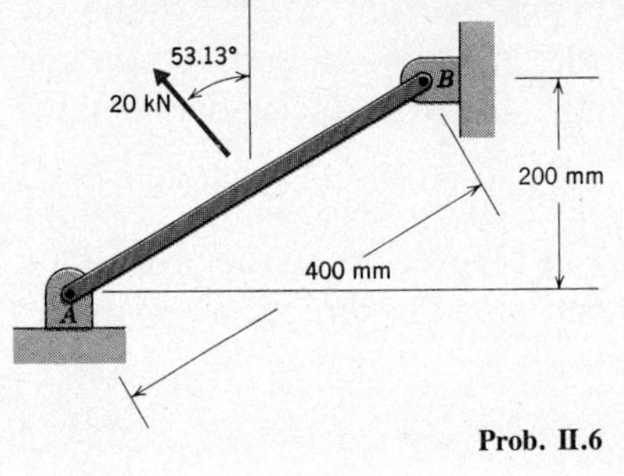

Prob. II.6

II.5 In Problem II.4 the horizontal component of the 500-lb force is 400 lb. Determine (a) the angle θ, (b) the vertical component of the force, (c) the components of the force parallel and perpendicular to the hill.

II.6 The 20-kN force is applied to the bar AB as shown. Determine the components of this force (a) horizontally and vertically, (b) parallel and perpendicular to the longitudinal axis of the bar.

II.7 A spring whose unstretched length is 2 ft and whose stiffness is 200 lb/ft is tied between points A and B. Determine the horizontal and vertical components of the force the spring exerts (a) on point A, (b) on point B.

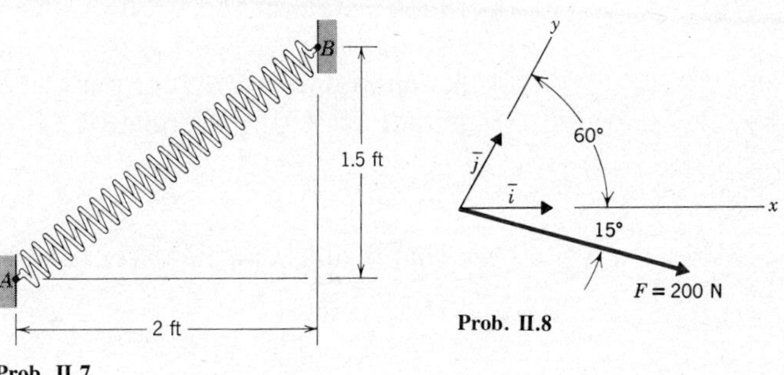

Prob. II.7

Prob. II.8

II.8 The x and y axes shown form nonorthogonal Cartesian coordinates in the plane. It is desired to write the force $\bar{F}$ in the form $\bar{F} = F_x \bar{i} + F_y \bar{j}$. Determine the values of F_x and F_y. Are these components the perpendicular projections onto the corresponding axes?

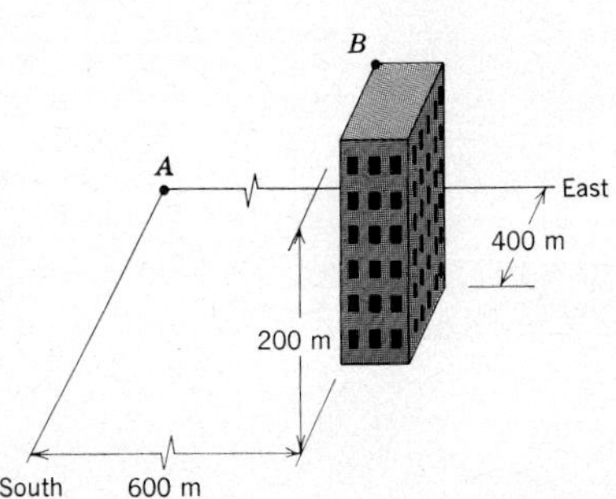

Prob. II.9

II.9 An observer on the ground at point A is looking at the top of a 200-m tall building at point B. (a) Determine the components of the position of point B with respect to point A and write the result as a vector. (b) Determine the direction cosines of this position vector with respect to the southerly and easterly directions.

II.10 A force $\bar{F}$ has a magnitude of 4 tons and the direction angles be-

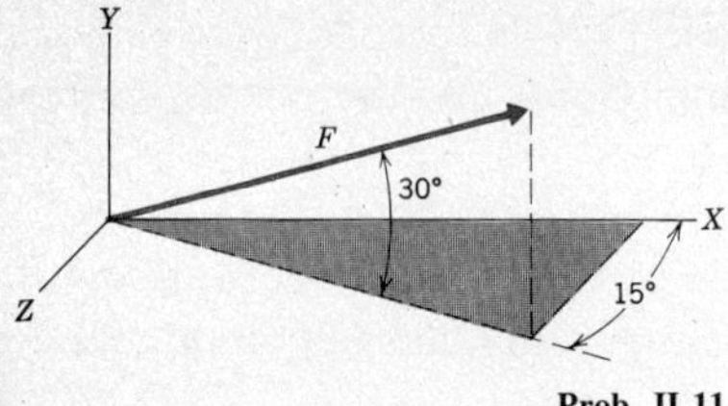

Prob. II.11

tween this force and the Y and Z axes are $\beta = 45°$, $\gamma = 120°$. Determine the components of this force and write the result as a vector. Also, determine the direction angle between the force and the X axis.

II.11 Write the force $\bar{F}$ shown in terms of its components and determine the direction angles for this force with respect to the XYZ reference frame.

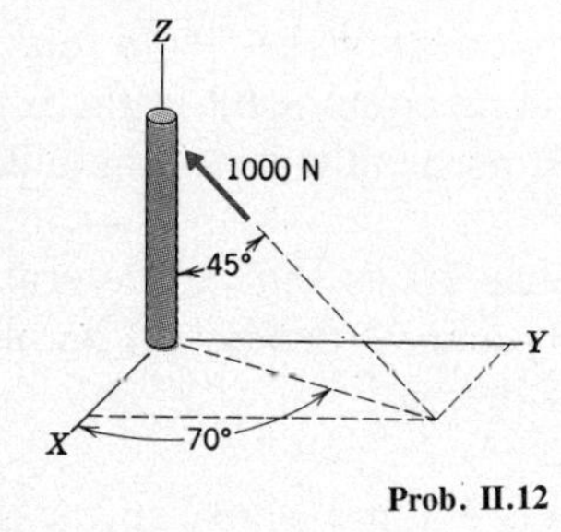

Prob. II.12

II.12 A 1000-N force acts on the vertical pole shown. (a) Determine the components of the force with respect to the XYZ coordinate system and write the force as a vector. (b) Determine the direction angles between the force and the XYZ coordinate axes.

II.13 The X, Y, and Z components of a 140-N force are in the proportion of 3:-2:6. Determine these components and also the direction angles between the force and the coordinate axes.

C. GENERAL UNIT VECTORS

An asset of the component representation of vectors is that the vector operations take on the form of algebraic equations. For instance, if we wish to multiply a vector $\bar{A}$ by a scalar number c, by definition the result of this operation is a vector parallel to $\bar{A}$ whose magnitude is c times the magnitude of $\bar{A}$. From equation (4) it then follows that

$$c\bar{A} = cA_X\bar{I} + cA_Y\bar{J} + cA_Z\bar{K} \tag{5}$$

This result is particularly useful in the situation where we must de-

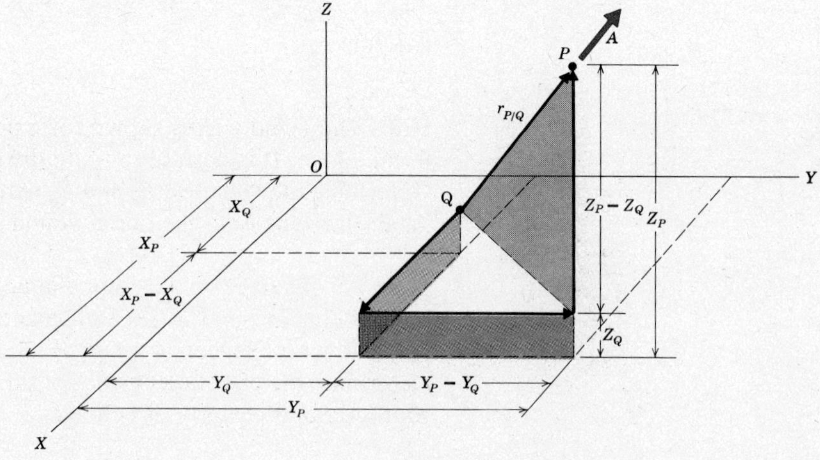

Figure 6

scribe a vector that is parallel to a specific line. This situation is depicted in Figure 6 on the preceding page, where a vector $\bar{A}$, having known magnitude $|\bar{A}|$, is formed such that it is parallel to the line from point Q to point P, and thus parallel to $\bar{r}_{P/Q}$.

Denoting the coordinates of the end points of $\bar{r}_{P/Q}$ as (X_P, Y_P, Z_P) and (X_Q, Y_Q, Z_Q), we see from the figure that the components of $\bar{r}_{P/Q}$ can be written as the differences in the coordinates; that is

$$\bar{r}_{P/Q} = (X_P - X_Q)\bar{I} + (Y_P - Y_Q)\bar{J} + (Z_P - Z_Q)\bar{K} \tag{6}$$

Once $\bar{r}_{P/Q}$ is known, we may obtain a unit vector $\bar{e}_{P/Q}$ in the direction of $\bar{r}_{P/Q}$ (and hence $\bar{A}$) by writing

$$|\bar{r}_{P/Q}|\ \bar{e}_{P/Q} = \bar{r}_{P/Q}$$

so

$$\bar{e}_{P/Q} = \frac{\bar{r}_{P/Q}}{|\bar{r}_{P/Q}|}$$

Then, to obtain $\bar{A}$ we need only multiply the unit vector by the magnitude of $\bar{A}$, which gives

$$\boxed{\bar{A} = |\bar{A}|\bar{e}_{P/Q} = |\bar{A}|\frac{\bar{r}_{P/Q}}{|\bar{r}_{P/Q}|}} \tag{7}$$

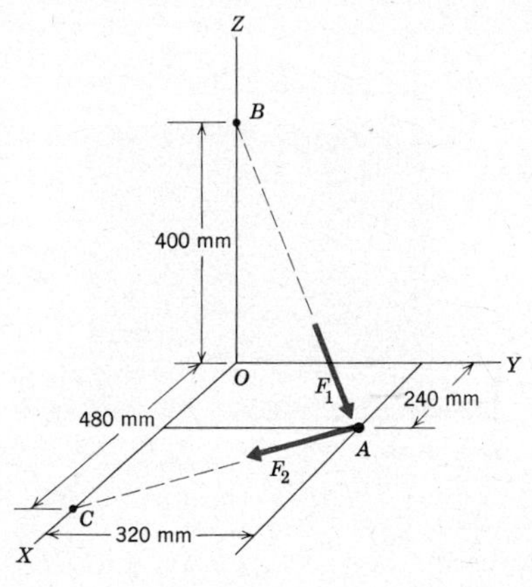

EXAMPLE 3

The magnitudes of the forces $\bar{F}_1$ and $\bar{F}_2$ shown in the sketch are 1.20 and 0.80 kN, respectively. Express these forces in terms of their components.

Solution

The forces $\bar{F}_1$ and $\bar{F}_2$ are directed along the lines from point B to point A, and point A to point C, respectively. A method for constructing the necessary position vectors $\bar{r}_{A/B}$ and $\bar{r}_{C/A}$ is to picture ourselves moving from the tail to the tip of each vector. For instance, to move from point B to point A we must go 400 mm in the negative Z direction, 240 mm in the positive X direction, and 320 mm in the positive Y direction, so that

$$\bar{r}_{A/B} = 240\bar{I} + 320\bar{J} - 400\bar{K} \text{ mm}$$

Similarly,

$$\bar{r}_{C/A} = 240\bar{I} - 320\bar{J} \text{ mm}$$

From equation (7) we then form

$$\bar{F}_1 = |\bar{F}_1|\bar{e}_{A/B}$$

$$= (1.2)\frac{\bar{r}_{A/B}}{|\bar{r}_{A/B}|} = (1.2)\frac{240\bar{I} + 320\bar{J} - 400\bar{K}}{\sqrt{(240)^2 + (320)^2 + (400)^2}}$$

$$= 0.3091\bar{I} + 0.6788\bar{J} - 0.8485\bar{K} \text{ kN}$$

$$\bar{F}_2 = (0.8)\frac{\bar{r}_{C/A}}{|\bar{r}_{C/A}|} = (0.8)\frac{240\bar{I} - 320\bar{J}}{\sqrt{(240)^2 + (320)^2}}$$

$$= 0.480\bar{I} - 0.640\bar{J} \text{ kN}$$

As an aside, note that $\bar{F}_2$, being a force in the XY plane, could have been obtained just as easily by determining the angle between the line AC and one of the coordinate axes and then using the method of Example 1. However, this method is awkward to employ in the case of the three-dimensional force $\bar{F}_1$.

HOMEWORK PROBLEMS

II.14 The piston is pushed by the force $\bar{F}$. Write an expression for the unit vector in the direction of $\bar{F}$ in terms of the X and Y axes shown.

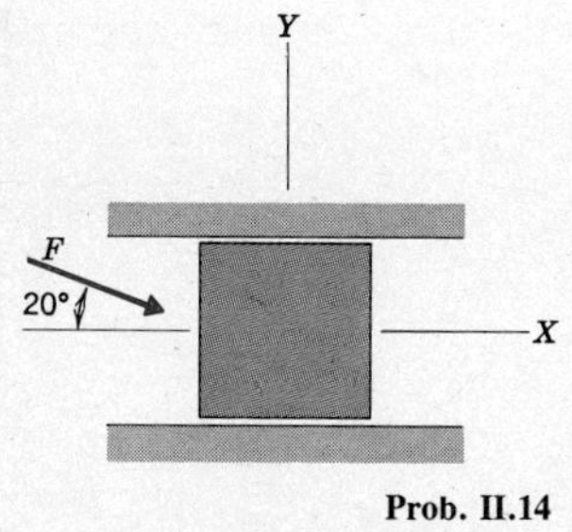

Prob. II.14

II.15 The structure shown is pulled by cable CD. If the tensile force within the cable is 8000 N, determine the components with respect to the XYZ axes of the force exerted by the cable on the structure at point D. Also, determine the unit vector parallel to this force.

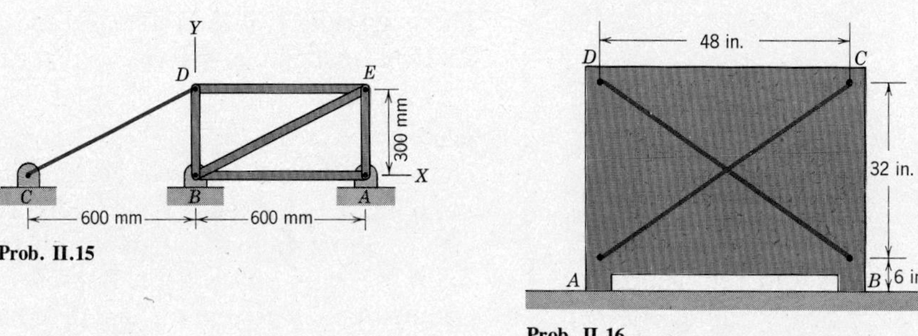

Prob. II.15

Prob. II.16

II.16 The cabinet shown is braced by tensioned crosswires AC and BD. The tensile forces in both cables are 40 lb. Determine the X and Y components of the unit vectors associated with the forces exerted by the cables on the cabinet at the four corners.

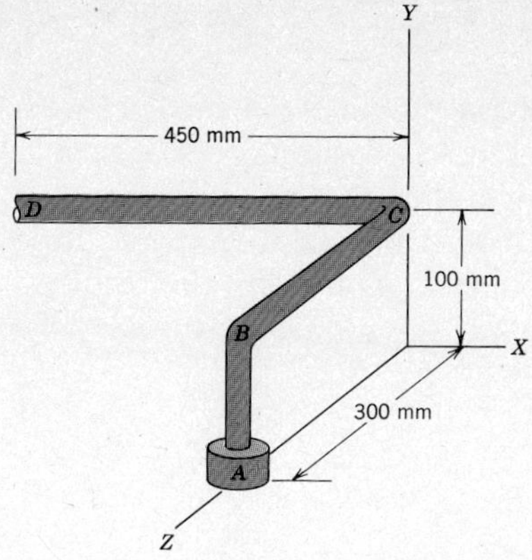

Prob. II.17

II.17 Determine the unit vectors extending from point A to points B, C, and D of the bent bar shown.

II.18 The rectangular plate is supported in part by cables AB and CD. If the tensile force in each cable is 150 lb, determine the components of the forces each cable exerts on the plate.

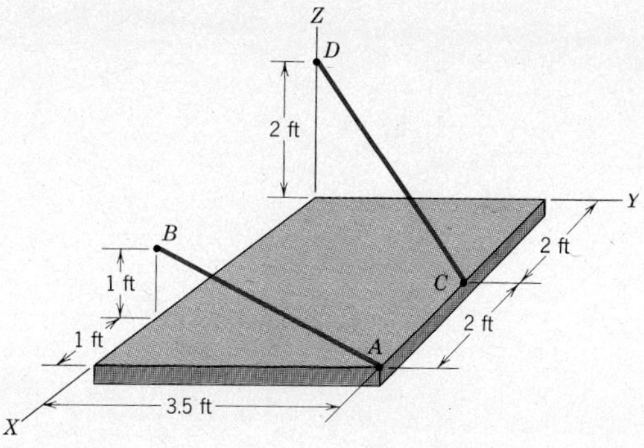

Prob. II.18

II.19 The package is being pulled at point A by cable AB. For what tensile force in the cable will the X component of the force exerted by the cable on the package be 200 N?

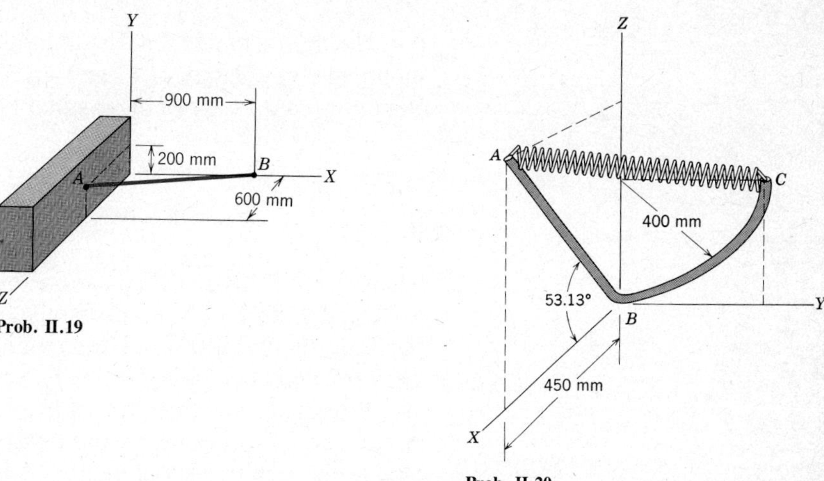

Prob. II.19

Prob. II.20

II.20 A spring whose stiffness is 400 N/m and whose unstretched length is 400 mm is stretched between ends A and C of bent rod ABC. Determine the force exerted by the spring on end C of the rod.

D. DETERMINATION OF THE RESULTANT OF CONCURRENT FORCES

It was shown in equation (1) that the resultant of a set of concurrent forces acting on a particle is the sum of the forces. Let us consider Figure 7, where two forces $\bar{F}_1$ and $\bar{F}_2$ are concurrent at the origin O, with the aim of evaluating the resultant of these forces. Toward this end we move $\bar{F}_2$ so as to align the vectors "head to tail." This allows us to form the resultant according to the polygon rule for addition.

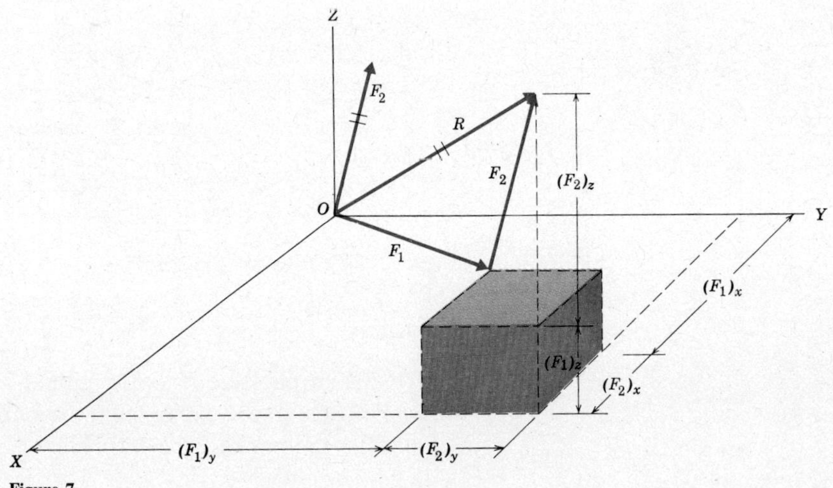

Figure 7

The figure shows that the components of $\bar{R}$ are the sum of the corresponding components of $\bar{F}_1$ and $\bar{F}_2$. This conclusion can be extended (by induction) to the case of the resultant of several forces, with the result that

$$\boxed{\bar{R} = \Sigma F_X \bar{I} + \Sigma F_Y \bar{J} + \Sigma F_Z \bar{K}} \tag{8}$$

where ΣF_X, ΣF_Y, and ΣF_Z denote the sum of the force components in the X, Y, and Z directions, respectively. Also, remember that because $\bar{R}$ is the resultant of the system of forces acting on the particle at point O, the vector $\bar{R}$ must be applied at *the point of concurrency* of the force system.

Clearly, although equation (8) was derived by considering the resultant of a set of forces, it is valid for any set of vectors. Thus, each component of the sum of a set of vectors is the sum of the corresponding components of the individual vectors.

The rule for the difference of two vectors $\bar{A}$ and $\bar{B}$ is a corollary to the foregoing development. For example, given the vector $\bar{B}$ we may write

$$-\bar{B} \equiv (-1)\bar{B} = -B_X\bar{I} - B_Y\bar{J} - B_Z\bar{K}$$

Thus, the difference between $\bar{A}$ and $\bar{B}$ is

$$\bar{A} - \bar{B} = \bar{A} + (-\bar{B}) = (A_X - B_X)\bar{I} + (A_Y - B_Y)\bar{J} + (A_Z - B_Z)\bar{K} \tag{9}$$

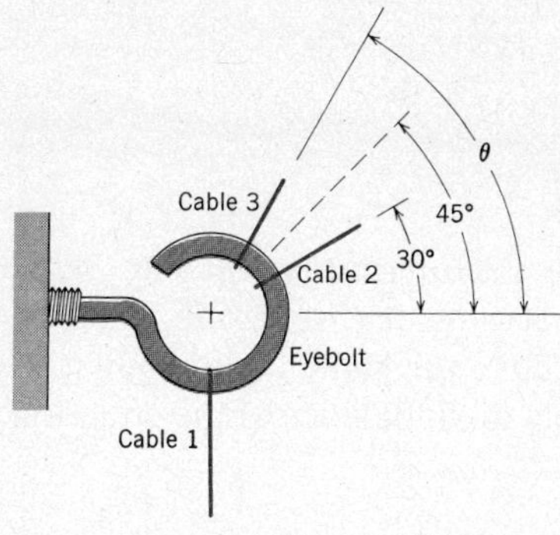

EXAMPLE 4

Three cables exert a concurrent set of forces on the eyebolt to which they are attached. Cables 1 and 2 pull on the eyebolt with forces of 6 and 8 kN, respectively. It is desired that the force exerted by cable 3 be the minimum value that causes the resultant of the three forces to be along the indicated dashed line. Determine the magnitude of this force and the corresponding value of the angle θ.

Solution

Recalling the discussion in Module I, we know that cables can exert only pulling forces on the objects to which they are attached; that is, cables can act only in tension. From the given information we know the tensile forces of cables 1 and 2. Further, we let $\bar{F}_3$ denote the tensile force in cable 3. Each of these forces is along the corresponding cable, as illustrated in the sketch.

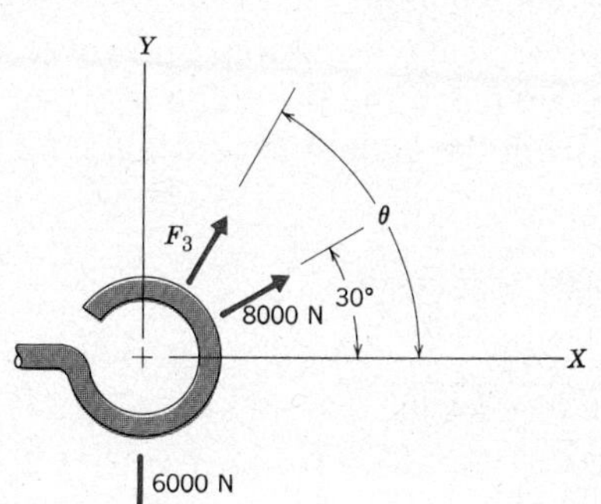

Also shown in the sketch is the coordinate system to be used for expressing components. Note that we chose one axis, the X axis, horizontal because all angles are referred to this line. The forces are all coplanar, so there are no components in the Z direction.

To determine the components of the forces we write each vector as the product of its magnitude times its unit vector. Cable 1 pulls downward, so

$$\bar{F}_1 = 6000(-\bar{J}) = -6000\bar{J} \text{ N}$$

whereas cable 2 is 30° above the X axis, so

$$\bar{F}_2 = 8000(\cos 30° \, \bar{I} + \sin 30° \, \bar{J})$$
$$= 6928\bar{I} + 4000\bar{J} \text{ N}$$

We do not as yet know the magnitude of $\bar{F}_3$ or the angle θ, hence for $\bar{F}_3$ we write

$$\bar{F}_3 = F_3(\cos\theta \, \bar{I} + \sin\theta \, \bar{J})$$

Before adding these forces to equate the sum to the resultant, let us use the given information to express vectorially the requirement that $\bar{R}$ must be parallel to the indicated 45° line. This means that

$$\bar{R} = R(\cos 45° \, \bar{I} + \sin 45° \, \bar{J})$$
$$= 0.7071R(\bar{I} + \bar{J})$$

Note that in writing this expression we have considered $\bar{R}$ to be directed outward from the origin along the dashed line. The sense of the resultant will be indicated by the sign of R.

In terms of the force vectors of the system, equation (1) gives

$$\bar{R} = \bar{F}_1 + \bar{F}_2 + \bar{F}_3$$

$$0.7071R(\bar{I} + \bar{J}) = (-6000\bar{J}) + (6928\bar{I} + 4000\bar{J}) + F_3(\cos\theta\,\bar{I} + \sin\theta\,\bar{J})$$

$$= (6928 + F_3\cos\theta)\bar{I} + (-2000 + F_3\sin\theta)J$$

This equation expresses the fact that the vector on the left side of the equality sign must be the same as the vector on the right side. Because two vectors are identical only when they have the same components, we must equate the corresponding components on both sides of the equation. Thus

$$\bar{I}\text{ components:}\qquad 0.7071R = 6928 + F_3\cos\theta$$

$$\bar{J}\text{ components:}\qquad 0.7071R = -2000 + F_3\sin\theta$$

There are three unknowns in these two scalar equations; however, we want the smallest value of F_3. Hence, we must find the value of θ that minimizes F_3. To do this we eliminate R from the equations and solve for F_3 in terms of θ, as follows.

$$6928 + F_3\cos\theta = -2000 + F_3\sin\theta$$

$$F_3 = \frac{8928}{(\sin\theta - \cos\theta)}$$

For F_3 to be a minimum, $dF_3/d\theta = 0$, hence

$$\frac{dF_3}{d\theta} = 0 = -\frac{8928}{(\sin\theta - \cos\theta)^2}(\cos\theta + \sin\theta)$$

from which we get

$$\cos\theta + \sin\theta = 0 \quad \text{so} \quad \tan\theta = -1 \qquad \theta = 135°$$

$$F_3 = \frac{8928}{\sin(135°) - \cos(135°)} = 6313\text{ N}$$

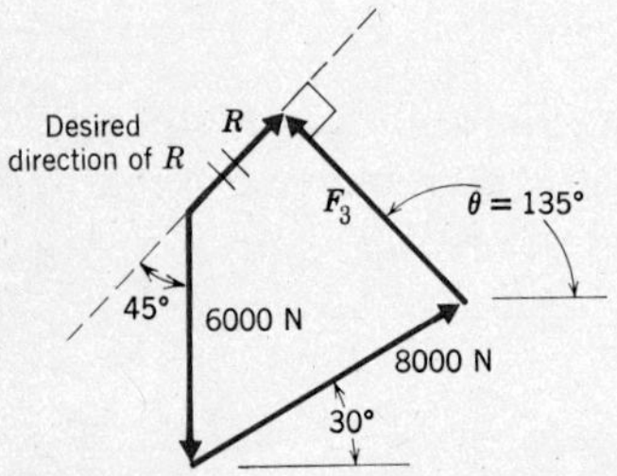

How do we know that this value of θ, which gives $dF_3/d\theta = 0$, represents the minimum value of F_3, and not the maximum? To answer this question notice that the vector $\bar{F}_3$ that we obtained is perpendicular to the desired direction of $\bar{R}$. That this should be so becomes apparent when we consider the addition of the forces as shown in the diagram. Clearly the shortest length of $\bar{F}_3$ is obtained when $\bar{F}_3$ is perpendicular to the resultant force.

Desired direction of R
R
F_3
$\theta = 135°$
45°
6000 N
8000 N
30°

EXAMPLE 5

A radio antenna is supported by three guy wires. The tensile force in cables *AB, AC,* and *AD* are 6 kilopounds (kips), 5 kips, and 8 kips, respectively. Determine the resultant force exerted on the antenna by these cables.

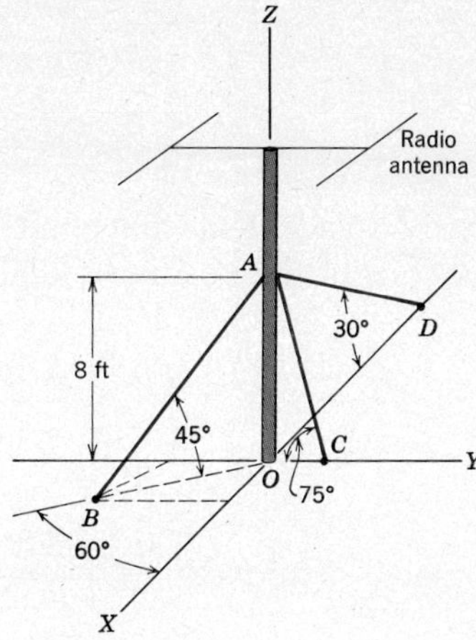

Solution

The first step is to draw a sketch showing the forces applied by the cables upon the mast; recall that cables have a pulling effect.

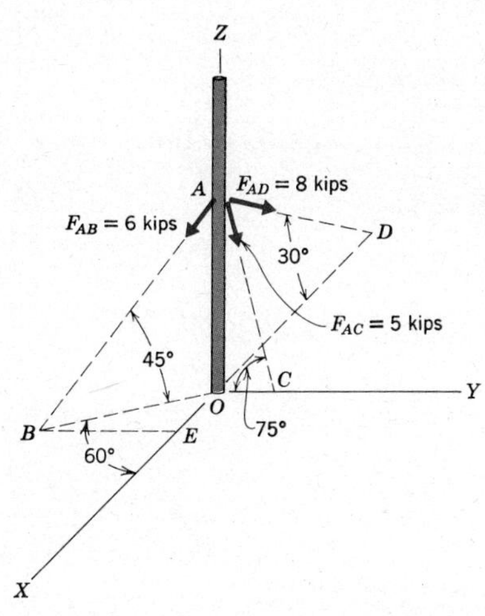

Let us first write the vector expressions for $\bar{F}_{AC}$ and $\bar{F}_{AD}$, because they have the simplifying feature of coinciding with the coordinate planes. From the sketch we see that the components of $\bar{F}_{AD}$ are oriented along the negative *X* and negative *Z* axes. Then, because the angle between the line *AD* and the *X* axis is 30°, it follows that

$$\bar{F}_{AD} = 8(-\cos 30°\, \bar{I} - \sin 30°\, \bar{K})$$
$$= -\,6.928\bar{I} - 4.0\bar{K} \text{ kips}$$

In a similar manner we find that

$$\bar{F}_{AC} = 5(\cos 75°\, \bar{J} - \sin 75°\, \bar{K})$$
$$= 1.292\bar{J} - 4.830\bar{K} \text{ kips}$$

We express $\bar{F}_{AB}$ using the unit vector oriented from point *A* to point *B*, $\bar{e}_{B/A}$. From equation 7 we have

$$\bar{F}_{AB} = 6\bar{e}_{B/A} = 6\frac{\bar{r}_{B/A}}{|\bar{r}_{B/A}|} \text{ kips}$$

The XYZ components of $\bar{r}_{B/A}$ are the lengths of the lines OE, EB, and AO, respectively (with the appropriate signs). To determine these lengths we note that points B, O, and A form a 45° right triangle, so that $BO = AO = 8$ ft. Now, because triangle OEB is also a right triangle, it follows that

$$EB = BO \sin 60° = 6.928 \text{ ft}$$

$$OE = BO \cos 60° = 4.00 \text{ ft}$$

Noting that a displacement from point A to point B requires displacements in the directions of the positive X and negative Y and Z axes, we have

$$\bar{r}_{B/A} = OE\bar{I} - EB\bar{J} - AO\bar{K} = 4\bar{I} - 6.928\bar{J} - 8\bar{K}$$

Thus

$$\bar{F}_{AB} = 6 \frac{4\bar{I} - 6.928\bar{J} - 8\bar{K}}{\sqrt{4^2 + (6.928)^2 + 8^2}}$$

$$= 2.121\bar{I} - 3.674\bar{J} - 4.234\bar{K} \text{ kips}$$

Finally, we determine the resultant by adding the corresponding components of the forces; that is,

$$\bar{R} = \bar{F}_{AB} + \bar{F}_{AC} + \bar{F}_{AD} = (2.121 + 0 - 6.928)\bar{I}$$
$$+ (-3.674 + 1.294 + 0)\bar{J} + (-4.234 - 4.830 - 4.0)\bar{K}$$
$$= -4.807\bar{I} - 2.380\bar{J} - 13.604\bar{K} \text{ kips}$$

HOMEWORK PROBLEMS

II.21-II.23 Determine the magnitude and direction of the resultant of the two forces shown.

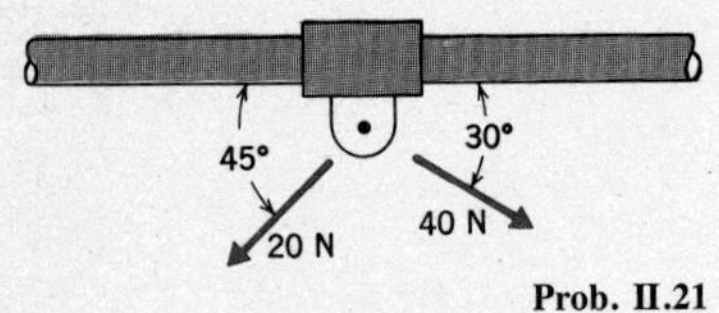

Prob. II.21

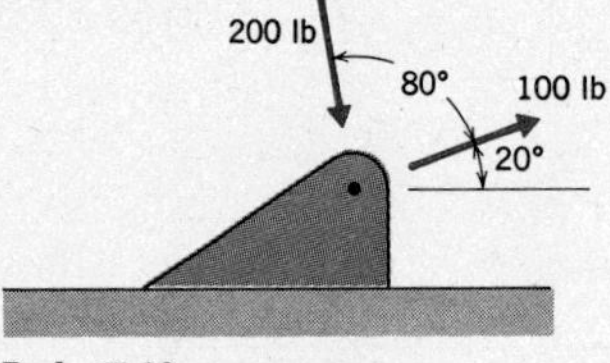

Prob. II.22

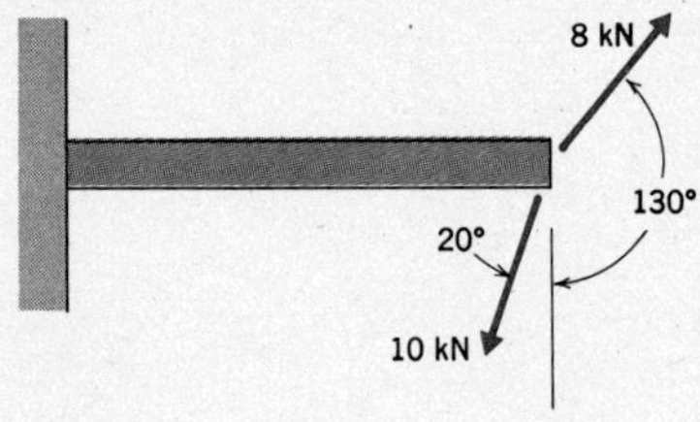

Prob. II.23

II.24 Two forces act at point C on the beam. Knowing that $F = 10$ tons and $\alpha = 40°$, determine the magnitude and angle relative to the beam of the resultant of these forces.

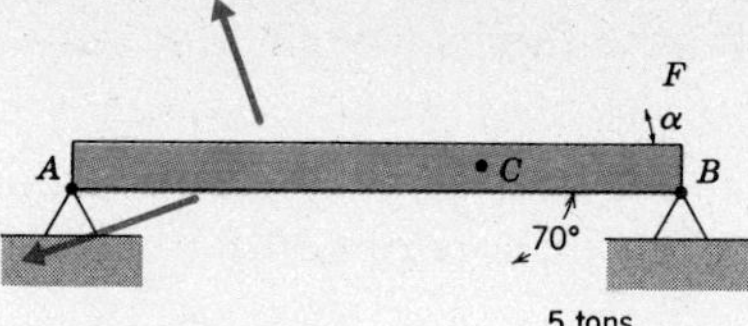

Probs. II.24 and II.25

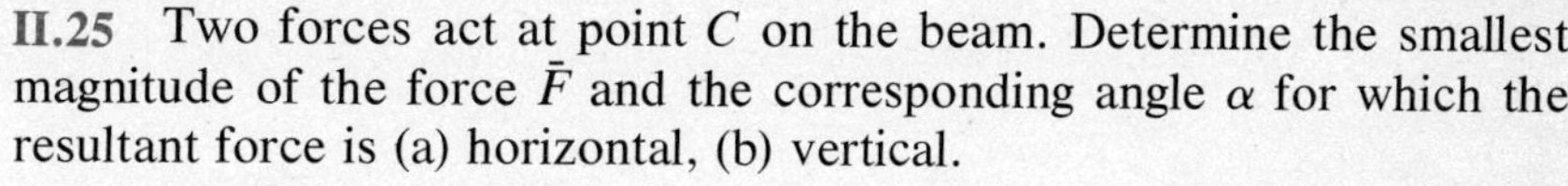

II.25 Two forces act at point C on the beam. Determine the smallest magnitude of the force $\bar{F}$ and the corresponding angle α for which the resultant force is (a) horizontal, (b) vertical.

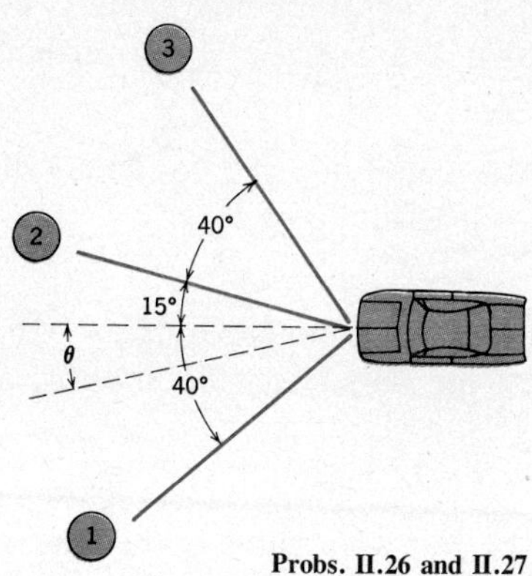

Probs. II.26 and II.27

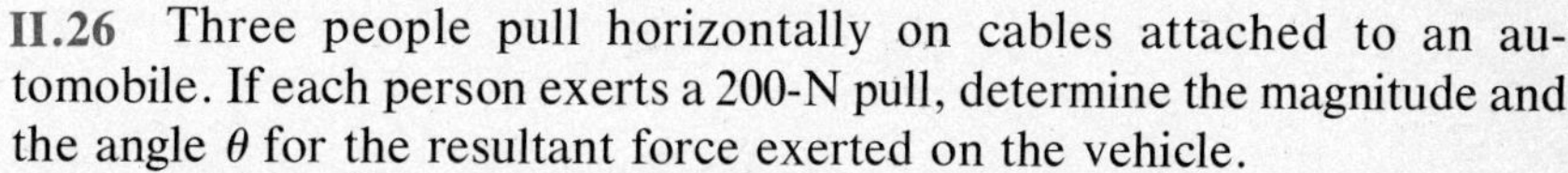

II.26 Three people pull horizontally on cables attached to an automobile. If each person exerts a 200-N pull, determine the magnitude and the angle θ for the resultant force exerted on the vehicle.

II.27 Cables 2 and 3 are pulled by 200-N forces. It is desired that the resultant of the forces exerted by the three cables be oriented along the dashed line indicated by the angle θ. Determine the required pull of cable 1 to have (a) $\theta = 0$, (b) $\theta = 20°$.

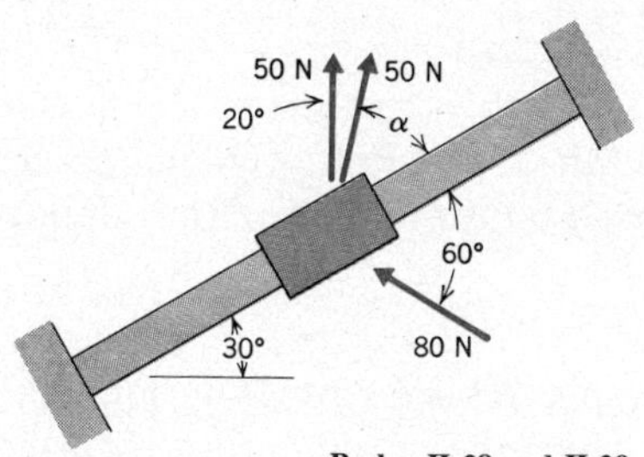

Probs. II.28 and II.29

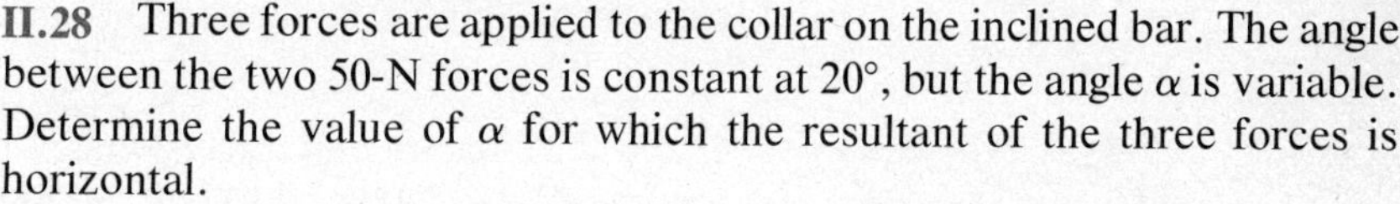

II.28 Three forces are applied to the collar on the inclined bar. The angle between the two 50-N forces is constant at 20°, but the angle α is variable. Determine the value of α for which the resultant of the three forces is horizontal.

II.29 Determine the required value of α for the resultant of the three forces to be parallel to the inclined bar.

II.30 Joint E of the bridge structure is shown isolated at the right. Knowing that the magnitudes of forces $\bar{F}_1$, $\bar{F}_2$, and $\bar{F}_3$ are 4, 5, and 6 kN, respectively, determine the magnitude of $\bar{F}_4$ such that the resultant of these forces at joint E is vertical. Note that the members are bars, not cables, so they can sustain compressive forces.

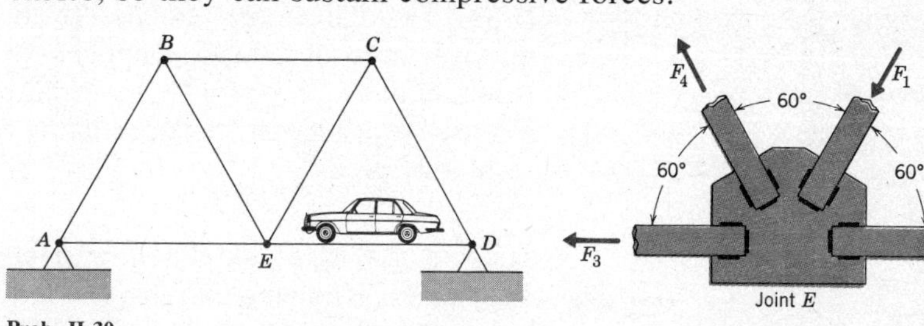

Prob. II.30

II.31 In Problem II.30 the magnitudes of $\bar{F}_3$ and $\bar{F}_1$ are 2 kN and 3 kN, respectively. Determine the magnitudes of $\bar{F}_2$ and $\bar{F}_4$ for which the resultant of these forces at joint E is zero.

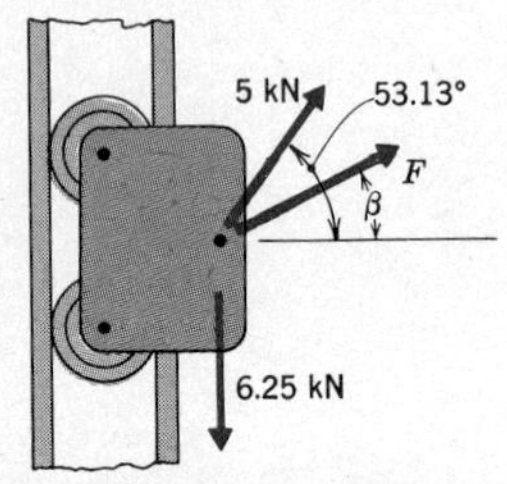

Probs. II.32, II.33 and II.34

II.32 It is desired that the resultant of the three forces acting on the roller guide be horizontal. If $F = 8$ kN, determine the corresponding angle β.

II.33 Find the smallest value of F and the corresponding angle β for which the resultant of the three forces shown is (a) horizontal, (b) vertical.

II.34 Knowing that $\beta = 30°$, determine the magnitude of $\bar{F}$ for which the resultant of the force system shown has a magnitude of 5 kN.

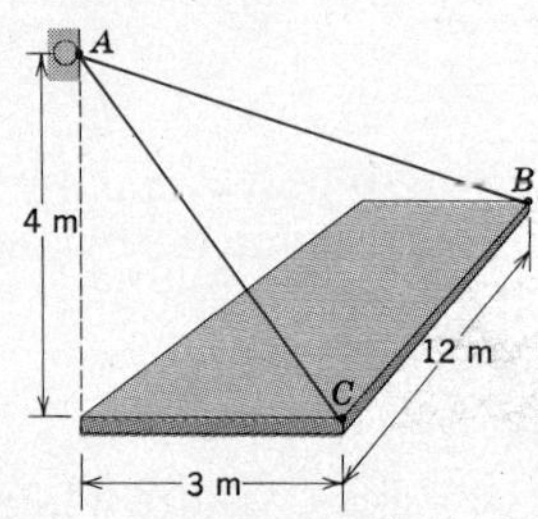

Probs. II.35 and II.36

II.35 The tensile force in cable AB is 2 kN, and that in cable AC is 3 kN. Determine the resultant force exerted by the cables on the support at point A.

II.36 The tensile force in cable AB is 2 kN. For what value of the tension force in cable AC will the resultant force exerted by the cables on the support at point A have a magnitude of 4 kN?

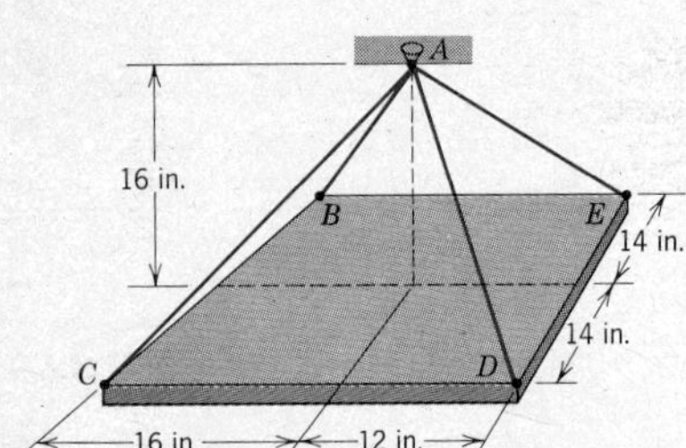

Prob. II.37

II.37 The square plate is suspended by four cables as shown. The tensile forces in cables AB and AC are 90 lb, and those in cables AD and AE are 60 lb. Determine the resultant force exerted by the cables on the support at point A.

II.38 In Example 5 the tensile force in cable AD is 8 kips. Determine the tensile forces in the other two cables for which the resultant force exerted by the cables on the antenna is vertical. Determine the magnitude of the resultant force in this case.

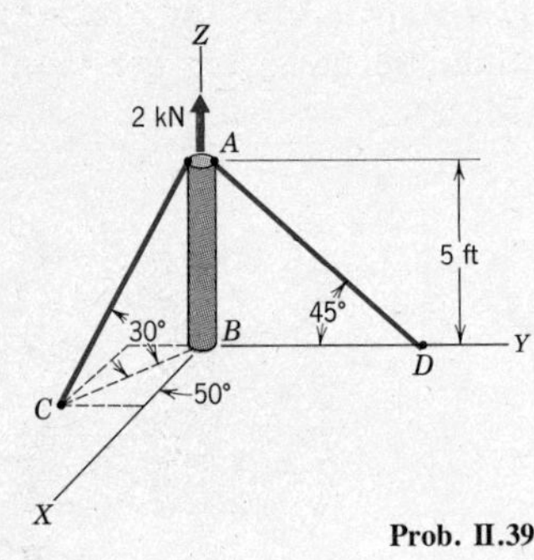

Prob. II.39

II.39 Two cables are attached to the vertical pole AB to steady it. An upward force of 2 kips is applied to the pole with the effect that the resultant of the three forces acting at point A on the pole is parallel to the X axis. Determine the tensions in the two cables and the magnitude of the resultant.

II.40 A 200-kg block is held in position on the 36.87° inclined ramp by two cables. Knowing that the tensile force in cable AB is 2500 N and that in cable AC is 3000 N, determine the resultant force exerted by the two cables and the force of gravity on the block.

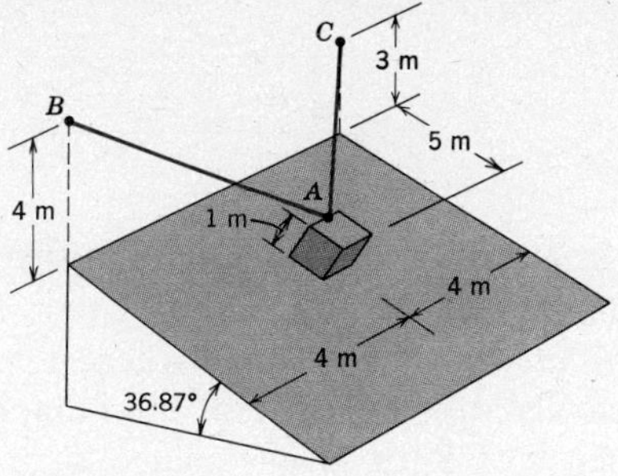

Prob. II.40

II.41 In Problem II.40 the resultant force exerted by the two cables and the gravitational force on the 200-kg block is normal to the 36.87° plane. Determine the tensile force in each cable.

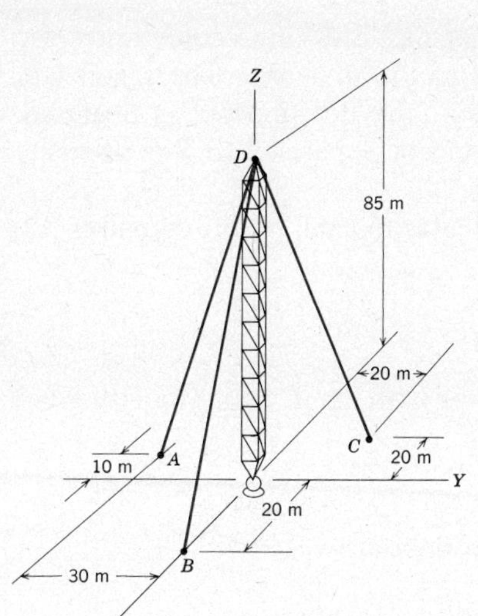

Prob. II.42

II.42 In order to hold the television tower erect, the resultant loading exerted on end D must be 20 kN in the negative Z direction. Determine the tensile force in each cable required to accomplish this.

II.43 Solve Problem II.42 for the case where an additional force of 1 kN is applied to the tower at point D in the positive Y direction.

E. DOT (SCALAR) PRODUCTS

In the preceding section we confined our efforts to determining the components of a force parallel to each of the three axes of a system of rectangular Cartesian coordinates. Essentially, this was done by multiplying the magnitude of the vector by the cosine of the angle it made with the axis. This operation can be put in a more general form in order to facilitate the determination of the components of a vector in a direction that is not a coordinate axis. The operation we will derive to accomplish this is called the *dot product* of two vectors.

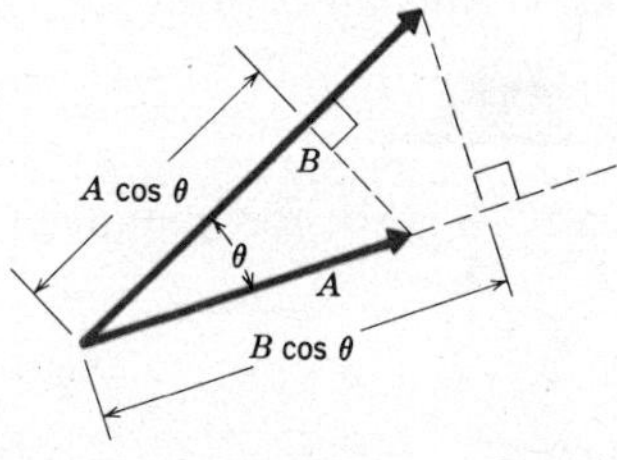

Figure 8

Let us consider the two intersecting vectors $\bar{A}$ and $\bar{B}$ shown in Figure 8. (They have been aligned to make their tails coincide.) The length $|\bar{A}|$ $\cos\theta$ is the component of $\bar{A}$ parallel to $\bar{B}$, and $|\bar{B}|\cos\theta$ is the component of $\bar{B}$ parallel to $\bar{A}$, where θ is the angle between the two vectors. Obviously, the scalar number $|A|(|B|\cos\theta)$ is the same as the scalar number $|\bar{B}|(|\bar{A}|\cos\theta)$. We write this product symbolically as $\bar{A}\cdot\bar{B}$, which should be read as $\bar{A}$ dot $\bar{B}$. Thus, the definition of a dot (or scalar) product is

$$\boxed{\bar{A} \cdot \bar{B} = \bar{B} \cdot \bar{A} \equiv |\bar{A}||\bar{B}| \cos \theta} \tag{10}$$

In words, the dot product is the magnitude of one vector times the component of the other vector parallel to the first one. The result is a *scalar number*. Note that when $\theta > 90°$, the dot product is negative. Physically, this means that the component of $\bar{A}$ parallel to $\bar{B}$ is opposite the sense of $\bar{B}$ (or vice versa).

When the two vectors are perpendicular to each other, equation (10) gives

$$\bar{A} \cdot \bar{B} = |\bar{A}||\bar{B}| \cos 90° = 0$$

If we calculate the dot product of a vector with itself, equation (10) gives

$$\bar{A} \cdot \bar{A} = |\bar{A}||\bar{A}| \cos 0 = |\bar{A}|^2$$

Two useful properties of the dot product are

$$p(\bar{A} \cdot \bar{B}) = (p\bar{A}) \cdot \bar{B} = \bar{A} \cdot (p\bar{B})$$
$$\bar{A} \cdot (\bar{B} + \bar{C}) = \bar{A} \cdot \bar{B} + \bar{A} \cdot \bar{C} \tag{11}$$

Both identities in equations (11) are easily proven by use of the definition of the dot product, equation (10).

Let us now turn to the subject of the determination of a dot product in terms of the components of the vectors. This is accomplished by first calculating the dot products of the unit vectors. For example, using equation (10),

$$\bar{I} \cdot \bar{I} = (1)(1) \cos 0° = 1$$

Hence, for the unit vectors of a rectangular Cartesian coordinate system,

$$\bar{I} \cdot \bar{I} = \bar{J} \cdot \bar{J} = \bar{K} \cdot \bar{K} = 1 \tag{12a}$$

Any other combination of unit vectors leads to zero, for instance,

$$\bar{I} \cdot \bar{J} = (1)(1) \cos 90° = 0$$

Therefore,

$$\bar{I} \cdot \bar{J} = \bar{J} \cdot \bar{I} = \bar{J} \cdot \bar{K} = \bar{K} \cdot \bar{J} = \bar{K} \cdot \bar{I} = \bar{I} \cdot \bar{K} = 0 \tag{12b}$$

We may now compute a dot product of two arbitrary vectors in terms of their components. To do this we use equations (11) to write

$$\begin{aligned}\bar{A}\cdot\bar{B} &= (A_X\bar{I} + A_Y\bar{J} + A_Z\bar{K})\cdot\bar{B}\\ &= A_X\bar{I}\cdot(B_X\bar{I} + B_Y\bar{J} + B_Z\bar{K})\\ &\quad + A_Y\bar{I}\cdot(B_X\bar{I} + B_Y\bar{J} + B_Z\bar{K})\\ &\quad + A_Z\bar{K}\cdot(B_X\bar{I} + B_Y\bar{J} + B_Z\bar{K})\end{aligned}$$

Then, in view of equations (12), this becomes

$$\bar{A}\cdot\bar{B} = A_XB_X + A_YB_Y + A_ZB_Z \tag{13}$$

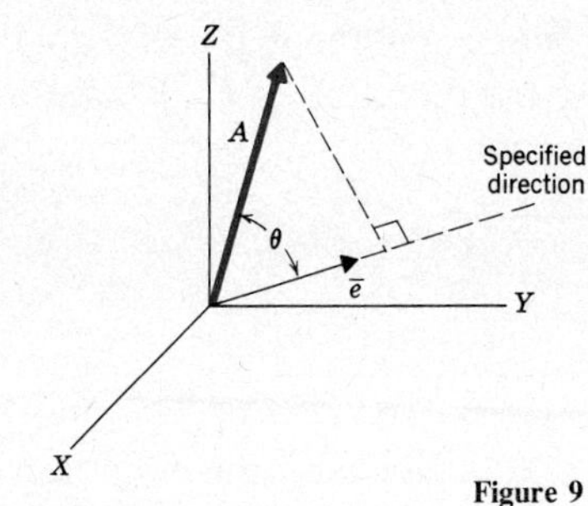

Figure 9

Let us now return to our original reason for introducing the concept of a dot product: the determination of the component of a force parallel to a specific line that is not a coordinate axis. Let $\bar{e}$ be a unit vector parallel to the specified line, as is shown in Figure 9. We desire the component of the vector $\bar{A}$ in the specified direction. Using equation (10) we have

$$\bar{A}\cdot\bar{e} = |\bar{A}||\bar{e}|\cos\theta = |\bar{A}|\cos\theta \tag{14}$$

Clearly, this is the desired component because θ is, by definition, the angle between $\bar{A}$ and the unit vector $\bar{e}$.

EXAMPLE 6

It is common practice in describing the resultant force acting on the surface of a body to refer to the component normal to the surface as a tension force if it pulls on the surface or as a compression force if it pushes on the surface. Also, the component parallel to the surface is called the shearing force. Determine the magnitudes of the tension (or compression) and shear forces acting on the shaded rectangular area shown in the sketch.

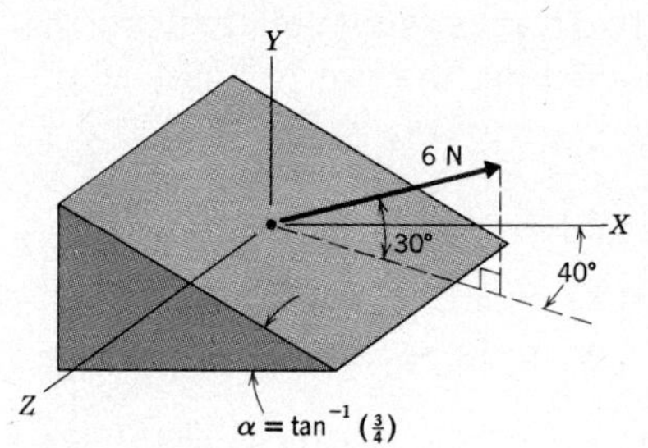

Solution

We first express the 6-kN force in terms of its components. Note in the given sketch that the projection of the force onto the XZ plane is 6(cos 30°) and that the angle between this projection and the X axis is 40°. Thus

$$\begin{aligned}\bar{F} &= 6\cos 30^\circ(\cos 40^\circ\,\bar{I} + \sin 40^\circ\,\bar{K}) + 6\sin 30\,\bar{J}\\ &= 3.980\bar{I} + 3.00\bar{J} + 3.340\bar{K}\text{ kN}\end{aligned}$$

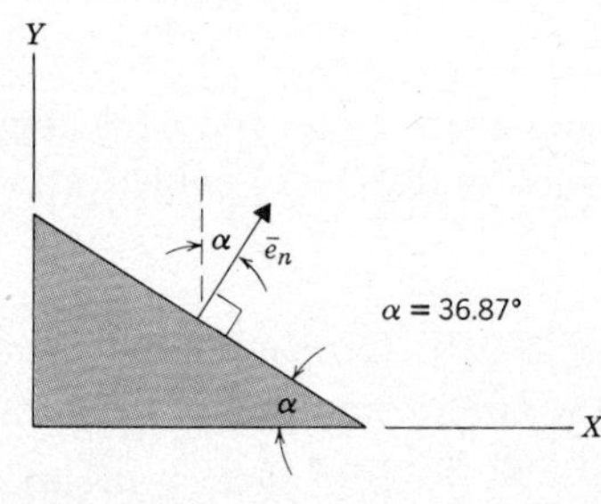

To determine the component of the force normal to the rectangular area we will need the unit vector $\bar{e}_n$ normal to the surface. In this particular case we see that $\bar{e}_n$ does not have a $\bar{K}$ component; that is, the normal vector is parallel to the XY plane, as shown in the sketch.

Because $\bar{e}_n$ is a unit vector, we have

$$\bar{e}_n = |\bar{e}_n|(\sin \alpha\, \bar{I} + \cos \alpha\, \bar{J}) = 0.60\bar{I} + 0.80\bar{J}$$

The component of the 6-kN force in the direction of $\bar{e}_n$ is found from the dot product.

$$F_n = \bar{F} \cdot \bar{e}_n = (3.980\bar{I} + 3.00\bar{J} + 3.340\bar{K}) \cdot (0.60\bar{I} + 0.80\bar{J})$$

$$= 3.980(0.60) + 3.00(0.80) + 0 = 4.788 \text{ kN}$$

Because this component is positive it means that it has the same sense as the $\bar{e}_n$ vector. Thus the component of $\bar{F}$ normal to the surface is directed outward from the surface. We conclude then that F_n is a tensile force.

Finally, the shear component F_s is the component parallel to the inclined surface, and thus perpendicular to the direction of F_n. Therefore, we can find F_s from the Pythagorean theorem.

$$F_s = \sqrt{|\bar{F}|^2 - F_n^2} = \sqrt{(6.0)^2 - (4.788)^2} = 3.616 \text{ kN}$$

HOMEWORK PROBLEMS

II.44 Prove the identities of equations (11) starting with the definition of a dot product, equation (10).

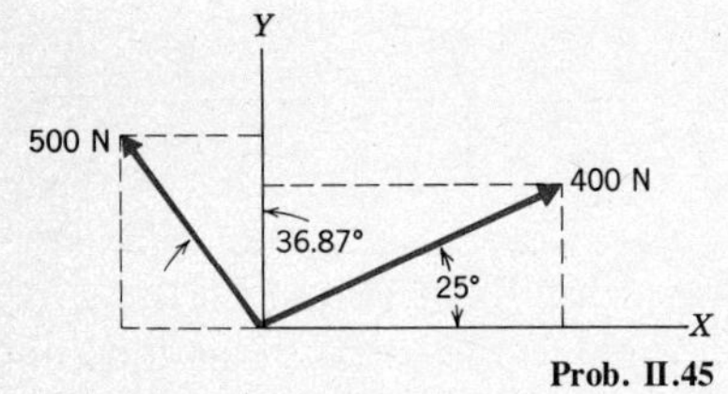

Prob. II.45

II.45 Determine the dot product of the two forces shown by (a) using the formal definition of this product, (b) calculating the dot product of the component expressions for each force.

II.46 Use the dot product of $\bar{F}_1$ and $\bar{F}_2$ to derive the trigonometric identity for $\cos(\phi_2 - \phi_1)$.

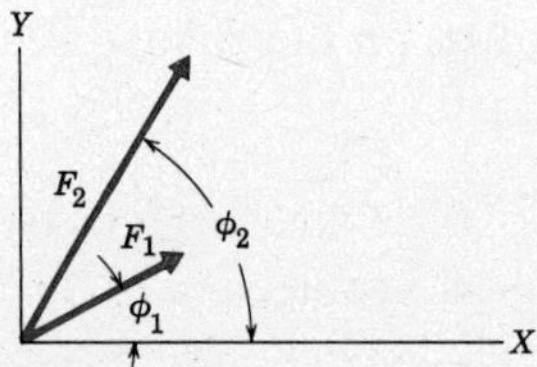

Prob. II.46

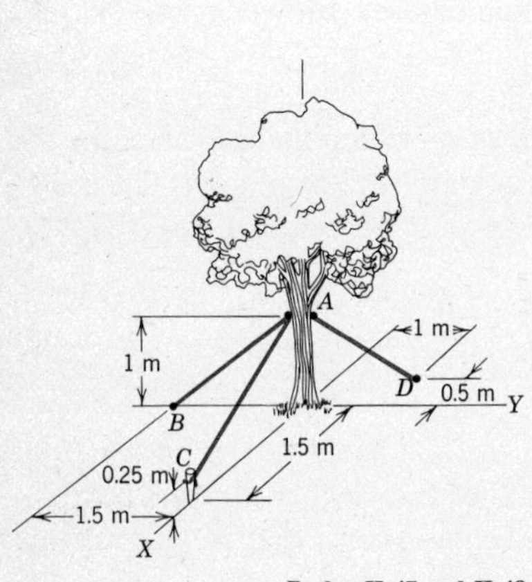

Probs. II.47 and II.48

II.47 Three cables brace a newly planted tree. Determine the angles between (a) cables AB and AC, (b) cables AB and AD, (c) cables AC and AD.

II.48 Each of the cables has a tension of 250 N. Determine the component of the resultant force exerted by these cables on the tree (a) parallel to line AB, (b) parallel to line AC, (c) parallel to line AD.

II.49 Two cables are attached to the building at point D as shown. Determine the angle between these two cables.

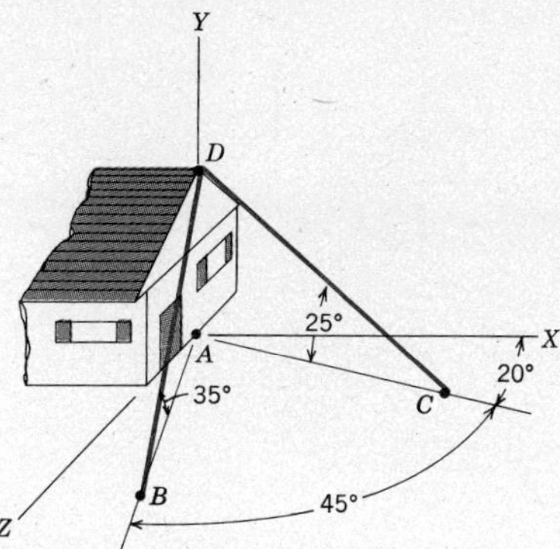

Probs. II.49 and II.50

II.50 Cable BD has a tension of 200 lb, whereas the tension in cable CD is 100 lb. Determine the component parallel to line CD of the resultant force exerted by the cables on point D.

II.51 Rectangle $ABCD$ is situated in space as shown. Determine the unit vector that is oriented from point A toward point B.

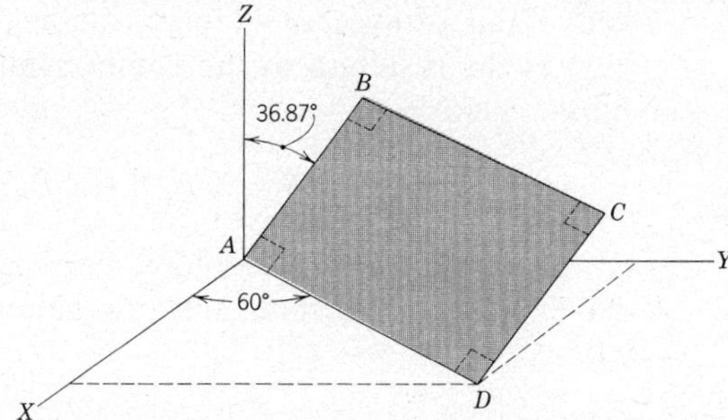

Prob. II.51

II.52 The 6-kN force in Example 6 was expressed in terms of its tension and shear components relative to the inclined surface. Express these two components as vectors, thus replacing the given force by a tension force and a shear force.

II.53-II.55 A 5000-N vertical force is applied to the surface shown at point B. (a) Determine the magnitude of the tension and shear forces produced by this force with respect to the tangent to the surface at point

B. (b) Write these tension and shear forces as vectors with respect to the *XYZ* reference frame shown.

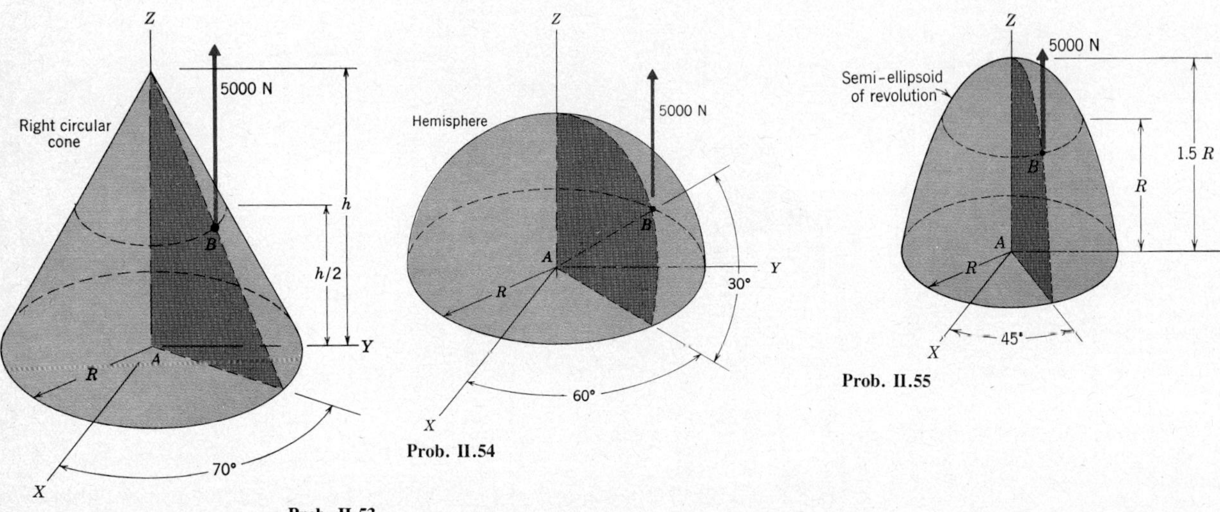

Prob. II.53

Prob. II.54

Prob. II.55

F. EQUILIBRIUM OF A PARTICLE

1 Basic Equations

The conditions for the static equilibrium of a particle are addressed explicitly in Newton's first law. It tells us that a particle will remain at rest only if the resultant of all forces acting on the particle is zero. Equation (8) relates the resultant to the components of a system of forces. Thus, for *static equilibrium*

$$\bar{R} = \Sigma F_X \bar{I} + \Sigma F_Y \bar{J} + \Sigma F_Z \bar{K} \equiv \bar{0}$$

As a zero vector can have only zero components, it follows that we have the following three scalar equations for static equilibrium of a particle.

$$\boxed{\Sigma F_X = 0 \qquad \Sigma F_Y = 0 \qquad \Sigma F_Z = 0} \tag{15}$$

In words, static equilibrium of a particle requires that the sum of all force components in each coordinate direction be zero. Equations (15) are three scalar equations that can be used to determine the *XYZ* components of a force necessary to maintain the static equilibrium of a particle.

In the special case of a planar system of forces, by definition there are no forces perpendicular to the plane. Equations (15) then reduce to the two scalar equations obtained by summing force components in the plane.

2 Free Body Diagrams

In general engineering practice, before writing the equations governing the behavior of a system, one first models the essential features of the system. This model identifies the important characteristics of the system, and equally important, serves to remove extraneous information from the problem. In mechanics, we create such models by drawing *free body diagrams*. For particle systems these diagrams isolate each particle to which forces are applied and show *every* force exerted on the particle. With the free body diagram we also sketch an XYZ coordinate reference system in order to minimize the chance for error in describing the force components, and also in order to simplify the task of checking our work.

In any problem it will be apparent that certain forces must be included in the free body diagram. Typical forces in this category are those exerted by cables, springs, and gravity (the weight force), and those that are given in the statement of the problem.

Forces that are less obvious, but equally important, are those that arise from the contact between the particle and its fixed surroundings. These contact forces are called *constraint forces*. A constraint restricts the possible movement of the particle. Although the topic of constraint forces will be considered in detail in the next module, where we study the equilibrium of rigid bodies, let us here look at a simple example of a physical system with a constraint force.

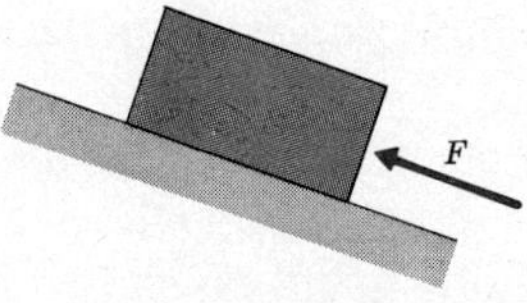

Figure 10a

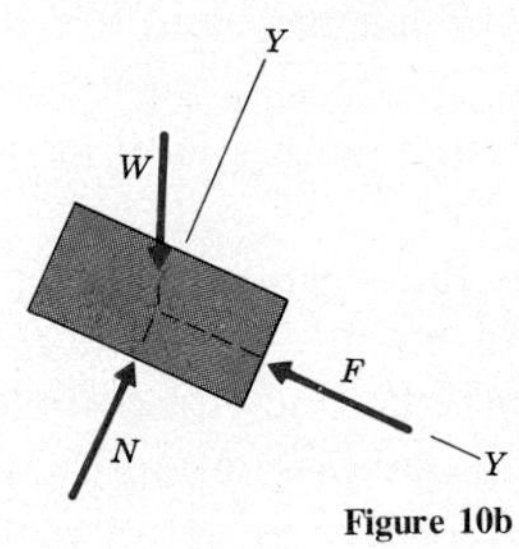

Figure 10b

Consider a box resting on a smooth (frictionless) inclined plane, as shown in Figure 10a. Clearly, the box cannot penetrate the solid surface of the incline. This is a constraint condition. The constraint condition is obtained by the constraint force, $\bar{N}$, normal to the plane, which is simply the interaction force between the box and the incline. In essence, in the free body diagram of Figure 10b we *remove* the plane and *account* for its presence by the interaction force $\bar{N}$. In addition, to complete the free body diagram, we show the given force $\bar{F}$ and the weight force $\bar{W}$. Only X and Y axes are shown in the sketch, because we have a planar system of forces. The orientation of the axes may be chosen arbitrarily. In this case they were picked to facilitate the description of the components of the forces.

In the foregoing study the box was modeled as a particle. This means that we assumed that a system of concurrent forces exists. The shortcoming of such a model is that it cannot be used to study the effect of the point of application of $\bar{F}$ to the box. However, the two planar equilibrium equations that can be written for the box ($\Sigma F_X = 0$, $\Sigma F_Y = 0$) will yield the values of F and N (in terms of W) required to have a state of static equilibrium.

It may be confusing to you that at times we can take the liberty of regarding physical systems occupying a region in space as particles. Throughout the text we will frequently make assumptions when modeling a system. These assumptions will not be loosely decided upon. Rather, they will be engineering decisions that result from our study of the various models that can be employed and the physical phenomena that these models can exhibit. Ultimately however, the verification of an analytical study, and thus of all the assumptions it contains, comes from agreement with experimental results.

3 Problem Solving

At this juncture we can deduce a series of formal steps that will enable us to attack logically many problems relating to the equilibrium of bodies acted on by a system of concurrent forces. They are *not* items to be followed by rote; memorization does not lead to understanding. Rather, these steps should be used as a guideline to assist you in developing your own analytical abilities.

1 Draw a complete free body diagram. It should exhibit all constraint forces, all gravity forces, and all external forces acting on the system. When isolating one body from a larger system, be sure to account for the forces exerted by all other parts of the system that are in contact with the body of interest.
2 Choose an *XYZ* reference frame that best fits the way in which the dimensions for the system are given. Show the axes in the free body diagram.
3 Evaluate the geometrical parameters necessary to describe all forces by their components. In planar problems quantities, such as angles, that are not obvious from the given information should be shown in the free body diagram. For three-dimensional problems the forces should be written out in component form. Here, it may be necessary to use the concept of a general unit vector.
4 Write the equations of static equilibrium corresponding to the free body diagram drawn in step 1.
5 Count the number of scalar equations that result from step 4 and the number of unknowns contained in these equations. If there are more unknowns than equations, look for another free body diagram in the original system that will give information about one or more of the unknowns. Now repeat steps 1, 3, and 4. Eventually, a solvable set of equations will be obtained.
6 Solve the equations.

ILLUSTRATIVE PROBLEM 1

The 3000-lb automobile rests on an ice-covered hill. A cable-pulley system is connected between the tow truck and the automobile. Because the friction within the pulleys is negligible, the tension within the cable is unmodified as the cable passes over the pulleys. Also, the pulleys have negligible mass. Considering the hill to be smooth and frictionless, determine the force that the tow truck must apply to the cable to hold the automobile in position.

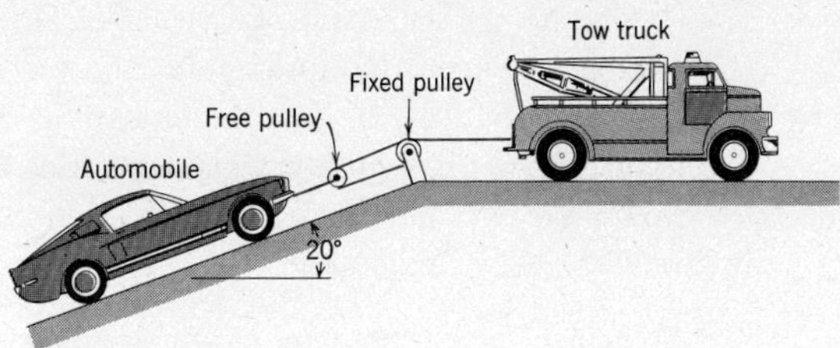

Solution

Step 1 The fact that the cable tension is constant along the length of a cable that passes over a frictionless pulley will be proven in the next module. The terminology used for referring to a pulley that is modeled as massless and frictionless is to say that it is *ideal*. Clearly, an ideal pulley is an approximation.

The automobile seems to be the focal point of the problem. Hence let us begin with a free body diagram of it. Note that the free body diagram must account for the constraint force $\bar{N}$ caused by the interaction with the hill, the weight of the car, and the force $\bar{F}$ representing the influence of the cable we cut to isolate the automobile.

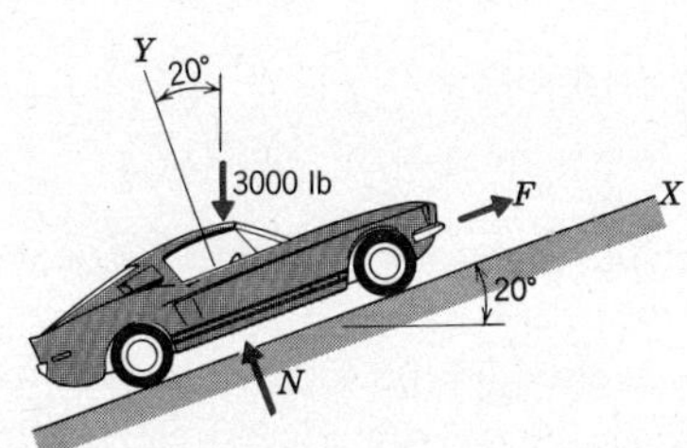

As an aside, note that if the automobile were to be modeled as a rigid body, as opposed to a particle, we would be concerned with the points of application of these forces. For example, there would be a contact force between each tire and the hill, rather than the single force $\bar{N}$, the weight force would be applied at the mass center of the vehicle, and $\bar{F}$ would have a distinct point of application.

Steps 2 and 3 The orientation of the XY axes was chosen to facilitate the writing of the equations of static equilibrium, and the necessary angles are shown in the free body diagram.

Step 4 Summing force components with respect to the axes shown in the free body diagram, the equilibrium equations are

$$\Sigma F_X = F - 3000 \sin 20° = 0$$

$$\Sigma F_Y = N - 3000 \cos 20° = 0$$

Note that the force components that act opposite to the direction of the axes were summed as negative values. Also, because this is a planar problem, force components could be summed without following the intermediate step of writing the weight force in terms of its components before summing forces.

Steps 5 and 6 The two foregoing equilibrium equations have two unknowns, F

and N, and therefore may be solved. In this problem we are interested in the cable force transmitted to the tow truck, but not the normal force $\bar{N}$. Hence, solving the first equation we obtain

$$F = 3000 \sin 20° = 1026 \text{ lb}$$

Is this the cable tension we seek? The answer is no, for $\bar{F}$ is merely the force exerted between the automobile and the free pulley. To determine the desired cable tension we note that the tension force of the cable and the force $\bar{F}$ are both applied to the free pulley, thus suggesting that we examine the equilibrium of that body. We will now repeat the procedures for studying equilibrium, using them now to study this pulley.

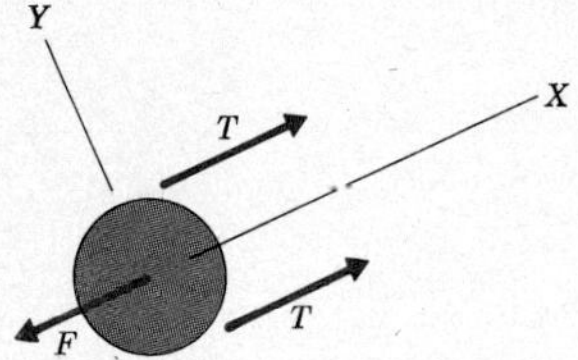

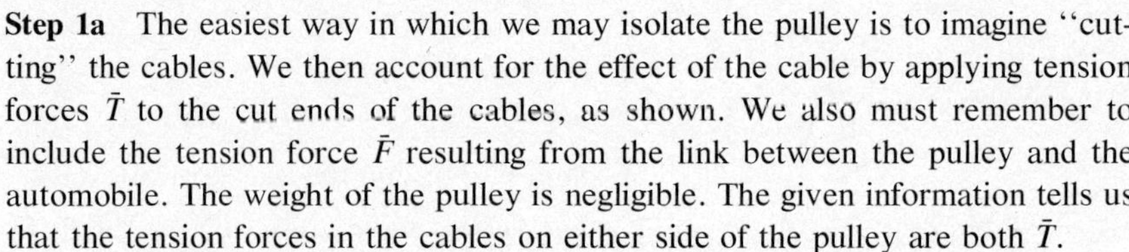

Step 1a The easiest way in which we may isolate the pulley is to imagine ''cutting'' the cables. We then account for the effect of the cable by applying tension forces $\bar{T}$ to the cut ends of the cables, as shown. We also must remember to include the tension force $\bar{F}$ resulting from the link between the pulley and the automobile. The weight of the pulley is negligible. The given information tells us that the tension forces in the cables on either side of the pulley are both $\bar{T}$.

Steps 2a and 3a Not applicable, for these steps were performed earlier.

Steps 4a-6a The only nontrivial equilibrium equation for the pulley comes from summing forces in the X direction. Thus

$$\Sigma F_X = 2T - F \equiv 2T - 1026 = 0$$

$$T = 513 \text{ lb}$$

This is the force in the cable that is attached to the tow truck. Hence we have the required solution.

Note that the magnitude of the force $|\bar{F}|$ exerted on the automobile is twice that exerted on the tow truck. In technical language we sometimes refer to the ratio of the output force of a system to the input force as a *mechanical advantage*. The mechanical advantage of the pulley system in this problem is 2.

ILLUSTRATIVE PROBLEM 2

A 500-kg crate is suspended by cables that are joined at point A. Determine the tension in each of these cables.

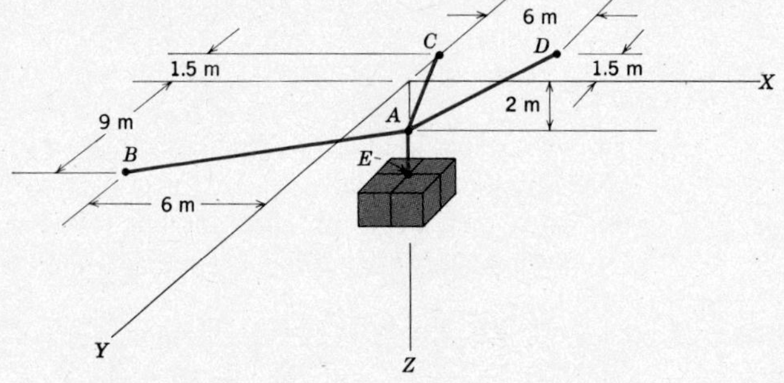

Solution

Step 1 There are two free body diagrams to be drawn here. One isolates the crate and the other isolates the focal point A at which the cables meet. The order in which they are drawn is arbitrary. Let us draw the diagram for point A first, because it will contain all the forces we seek.

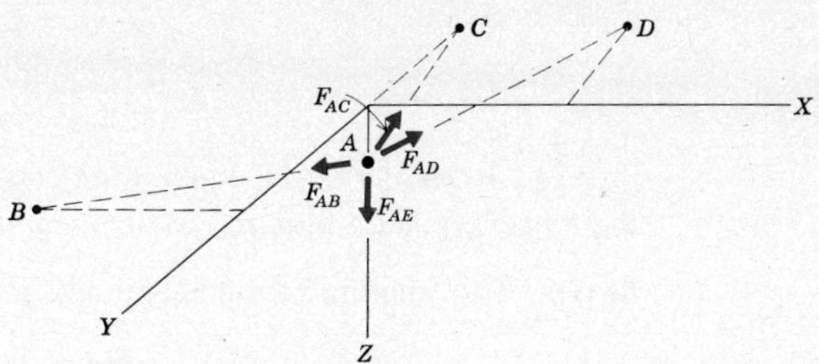

Step 2 The given XYZ coordinate system is suitable for computations.

Step 3 Because the cables are oriented in three dimensions, it is best to express the forces in component form. The forces are oriented along lines from point A to points B, C, D, and E. Hence, from equation (7) we have

$$\bar{F}_{AB} = F_{AB}\,\bar{e}_{B/A} = F_{AB}\frac{\bar{r}_{B/A}}{|\bar{r}_{B/A}|} = F_{AB}\frac{-6\bar{I} + 9\bar{J} - 2\bar{K}}{\sqrt{6^2 + 9^2 + 2^2}}$$

$$= F_{AB}(\tfrac{-6}{11}\bar{I} + \tfrac{9}{11}\bar{J} - \tfrac{2}{11}\bar{K})$$

$$\bar{F}_{AC} = F_{AC}\,\bar{e}_{C/A} = F_{AC}\frac{\bar{r}_{C/A}}{|\bar{r}_{C/A}|} = F_{AC}\frac{-1.5\bar{J} - 2\bar{K}}{\sqrt{(1.5)^2 + 2^2}}$$

$$= F_{AC}\,(-\tfrac{3}{5}\bar{J} - \tfrac{4}{5}\bar{K})$$

$$\bar{F}_{AD} = F_{AC}\frac{6\bar{I} - 1.5\bar{J} - 2\bar{K}}{\sqrt{6^2 + (1.5)^2 + 2^2}}$$

$$= F_{AD}\,(\tfrac{12}{13}\bar{I} - \tfrac{3}{13}\bar{J} - \tfrac{4}{13}\bar{K})$$

$$\bar{F}_{AE} = F_{AE}\,\bar{K}$$

Step 4 For static equilibrium the sum of the force components at point A equals zero, therefore

$$\Sigma F_X = -\tfrac{6}{11}F_{AB} + \tfrac{12}{13}F_{AD} = 0$$

$$\Sigma F_Y = \tfrac{9}{11}F_{AB} - \tfrac{3}{5}F_{AC} - \tfrac{3}{13}F_{AD} = 0$$

$$\Sigma F_Z = -\tfrac{2}{11}F_{AB} - \tfrac{4}{5}F_{AC} - \tfrac{4}{13}F_{AD} + F_{AE} = 0$$

Step 5 There are *three* equations for *four* unknowns. As mentioned earlier, the additional information comes from considering the equilibrium of the crate.

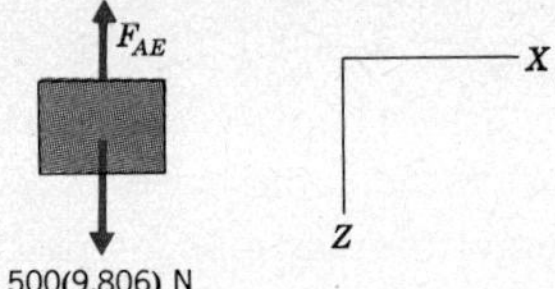

Step 1a The crate is acted upon only by cable AE and gravity, hence the free body diagram is as shown. Note that the tensile force exerted by cable AE on the crate is directed upward.

Steps 2a and 3a Not applicable.

Step 4a The only nontrivial equilibrium equation comes from summing forces in the Z direction. Thus

$$\Sigma F_Z = 4903 - F_{AE} = 0$$

Step 5a Referring to the equilibrium equations obtained in step 5, we now have the fourth equation required to obtain a solution.

Step 6 The solution of the equation just obtained is obviously

$$F_{AE} = 4903 \text{ N}$$

Placing this result into the previous set of equilibrium equations reduces the system to three simultaneous equations in the three unknown cable tensions. From the first two equations we can find F_{AD} and F_{AC} in terms of F_{AB}.

$$\begin{aligned} F_{AD} &= \tfrac{13}{12}(\tfrac{6}{11})F_{AB} = 0.5909F_{AB} \\ F_{AC} &= \tfrac{5}{3}(\tfrac{9}{11}F_{AB} - \tfrac{3}{13}F_{AD}) \\ &= 1.6667[0.8182F_{AB} - 0.2308(0.5909)F_{AB}] \\ &= 1.1364F_{AB} \end{aligned}$$

The last equation for ΣF_Z then gives

$$\tfrac{2}{11}F_{AB} + \tfrac{4}{5}(1.1364)F_{AB} + \tfrac{4}{13}(0.5909)F_{AB} = 4903$$

$$F_{AB} = 3852 \text{ N}$$

The other forces are then found to be

$$F_{AD} = 0.5909(3852) = 2276 \text{ N}$$

$$F_{AC} = 1.1364(3852) = 4377 \text{ N}$$

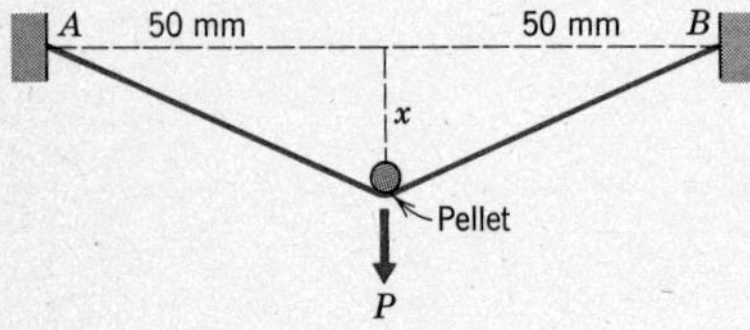

ILLUSTRATIVE PROBLEM 3

The system shown in the sketch represents a sling shot formed by tying a rubber band between points A and B. When the rubber band is deformed from its rest position (dashed line) it develops a tensile force that is proportional to its elongation; that is, $F = k\Delta$, where Δ is the elongation of the rubber band and k is the spring constant. Assuming that the rubber band has no tension when in the undeformed position, determine the force P required to hold the pellet in static equilibrium as a function of the distance x.

Solution

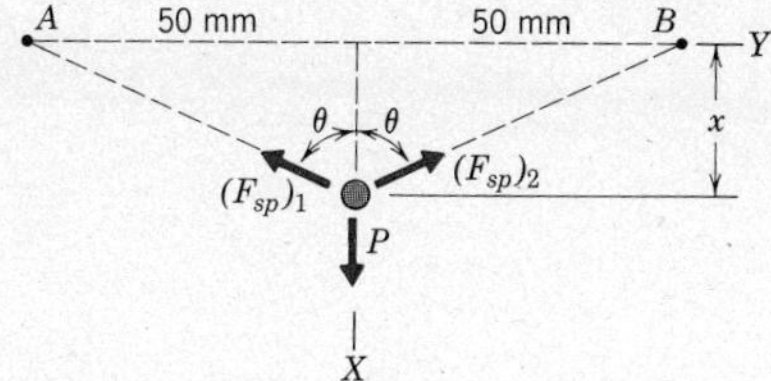

Step 1 In view of the fact that the spring forces and applied force $\bar{P}$ meet at the pellet, we draw a free body diagram of this body. (The mass of the pellet is not given, so the weight force is ignored.)

Steps 2 and 3 The X and Y axes shown in the free body diagram were selected to fit the manner in which the dimensions are given. The angle θ for the spring forces depends on the value of x, in meters, according to

$$\theta = \sin^{-1}\left(\frac{0.05}{\sqrt{(0.05)^2 + x^2}}\right) = \cos^{-1}\left(\frac{x}{\sqrt{(0.05)^2 + x^2}}\right)$$

Step 4 For equilibrium, we have

$$\Sigma F_X = P - [(F_{sp})_1 + (F_{sp})_2] \cos \theta = 0$$

$$\Sigma F_Y = [(F_{sp})_2 - (F_{sp})_1] \sin \theta = 0$$

Step 5 Clearly the second of the foregoing equilibrium equations yields $(F_{sp})_2 = (F_{sp})_1$, as one would expect from symmetry considerations. To determine the value of P in terms of x we now eliminate $(F_{sp})_1$ from the first equation. This is accomplished by using the given information that $(F_{sp})_1 = k\Delta$, where Δ is the increase in the length of the rubber band from the unstretched position. The unstretched length of the rubber band is given as 0.10 m, and the stretched length of the band is $2((0.05)^2 + x^2)^{1/2}$. Hence

$$(F_{sp})_1 = k\Delta = k(2\sqrt{(0.05)^2 + x^2} - 0.10)$$

Step 6 Using the foregoing expression for $(F_{sp})_1$ and the equation for the angle θ obtained earlier, we have

$$P = 2(F_{sp})_1 \cos \theta = 2k(2\sqrt{(0.05)^2 + x^2} - 0.10)\frac{x}{\sqrt{(0.05)^2 + x^2}}$$

$$= 4kx\left(1 - \frac{0.05}{\sqrt{0.0025 + x^2}}\right)$$

Particular attention is drawn to the fact that the force P is not proportional to the distance x, even though the springs are linear. Can you explain why this is so?

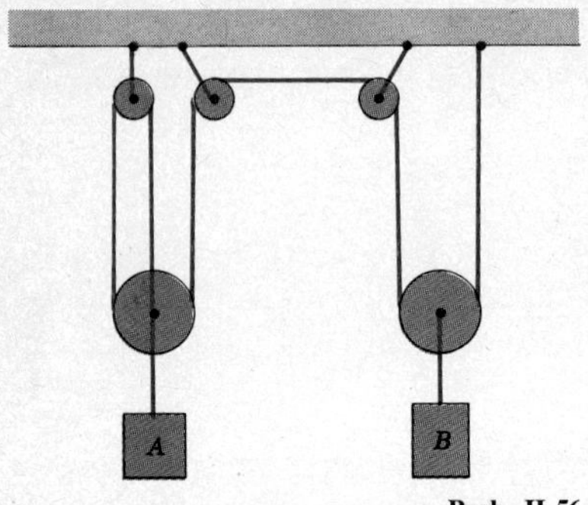

Prob. II.56

HOMEWORK PROBLEMS

II.56 Block A has a mass of 15 kg. Determine the mass of block B for static equilibrium. The pulleys are ideal.

II.57 If $W_1 = 10$ lb, determine the values of W_2 and W_3 for static equilibrium. All pulleys have ideal properties.

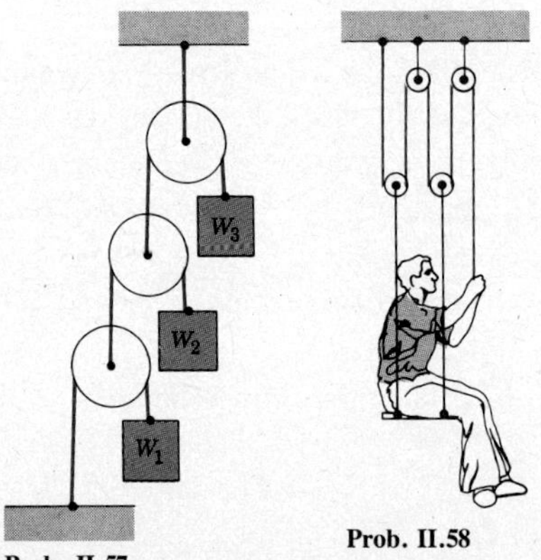

Prob. II.57

Prob. II.58

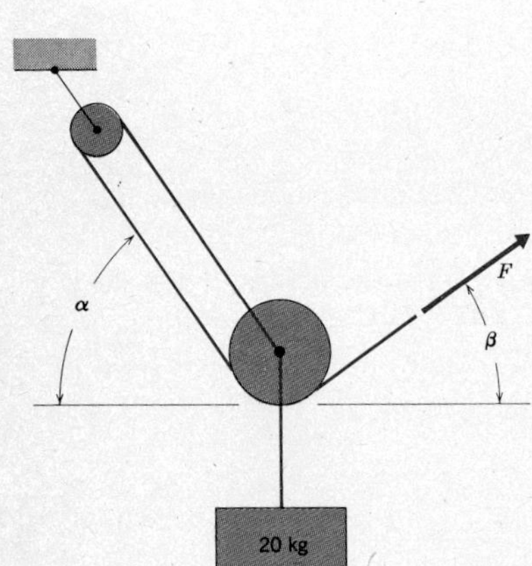

Probs. II.59 and II.60

II.58 A 70-kg sign painter sitting on a 10-kg scaffold holds the free end of the cable. The pulleys have ideal properties. Determine the tension in the cable and the reaction force exerted between the painter and the scaffold.

II.59 If $\alpha = 75°$, determine the magnitude of the force $\bar{F}$ and the angle β corresponding to static equilibrium. The pulleys are ideal.

II.60 Determine the smallest angle α for which the system will be in static equilibrium. Also find the corresponding values of the magnitude of the force F and the angle β. The pulleys are ideal.

II.61 A 50-g ball bearing rests in a smooth groove, as shown in the vertical cross section. Determine the forces acting on the bearing.

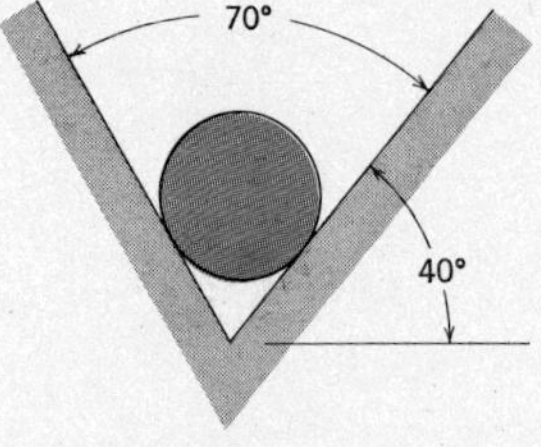

Prob. II.61

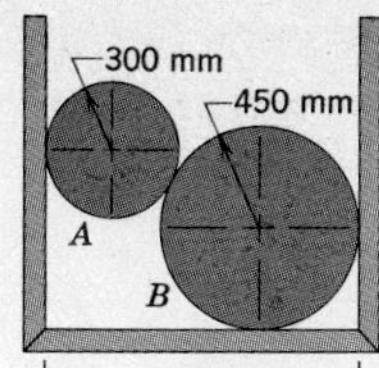

Prob. II.62

II.62 The 50-kg cylinder A and the 100-kg cylinder B are held inside the crate shown in the vertical cross-sectional view. All surfaces of contact are smooth. Determine the contact forces exerted between the cylinders, and between each cylinder and the sides of the crate.

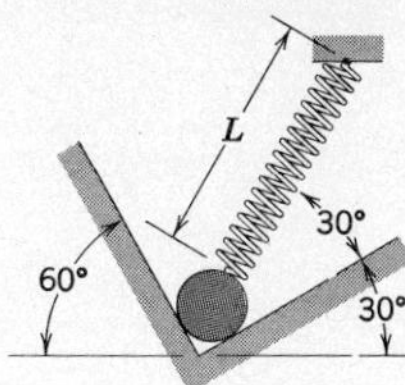

Prob. II.63

II.63 A 150-g ball bearing is attached to a spring and then brought into contact with the smooth groove whose vertical cross section is shown. The stiffness of the spring is 20 N/m and its unstretched length is 100 mm. Knowing that $L = 120$ mm, determine the reaction forces exerted by the walls of the groove on the ball bearing.

II.64 In Problem II.63, determine the maximum value of L for which the ball bearing will maintain contact with both walls of the groove.

II.65 Each end of a small-radius cylinder of mass m is connected to a spring to hold the cylinder in position against the smooth horizontal semicylinder. (Only one spring is shown in the side view.) Both springs have a stiffness k and their unstretched length is R. Relate the angle θ for static equilibrium to the other parameters.

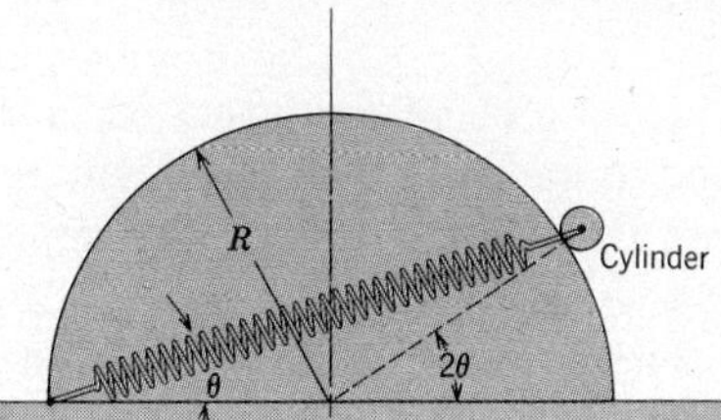

Prob. II.65

II.66-II.68 A collar on a smooth vertical rod is held in position by two springs of identical unstretched length L, but different stiffnesses k_1 and k_2. (a) Derive an expression for the distance Δ representing the downward displacement of the collar from the position where the springs are unstretched. (b) Determine the stiffness of a single equivalent spring (k_{eq}). The criterion for a spring to be equivalent is that it give the same value of Δ as obtained in part (a).

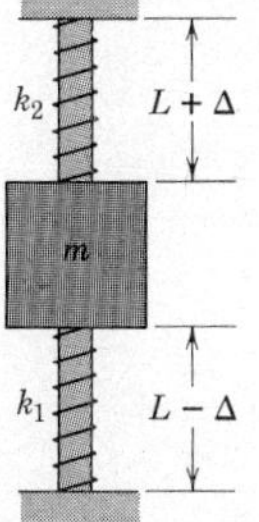

Prob. II.66

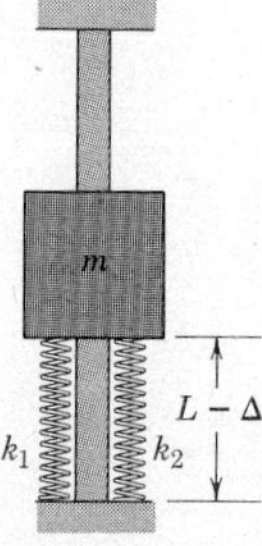

Prob. II.67

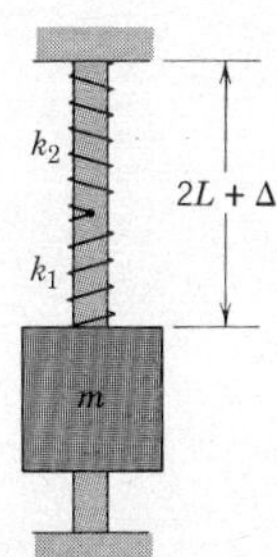

Prob. II.68

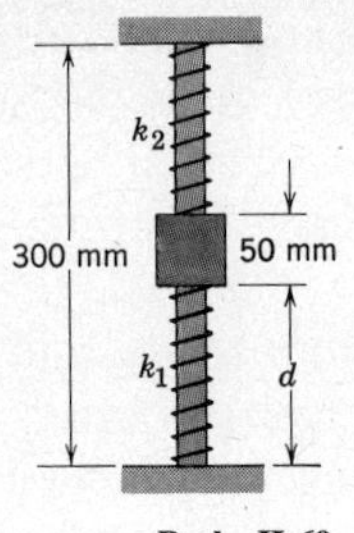

Prob. II.69

II.69 The friction between the 80-g collar and its vertical guide is negligible. The lower spring has a stiffness $k_1 = 4$ N/m and an unstretched length $L_1 = 200$ mm, whereas the corresponding properties for the upper spring are $k_2 = 3$ N/m, $L_2 = 150$ mm. Determine the distance d defining the position of static equilibrium.

II.70 Block A, whose weight is 90 lb, is supported in the position shown. Determine the weight of block B.

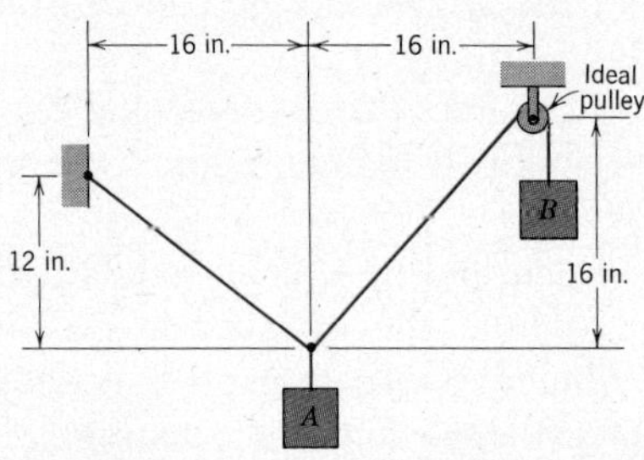

Prob. II.70

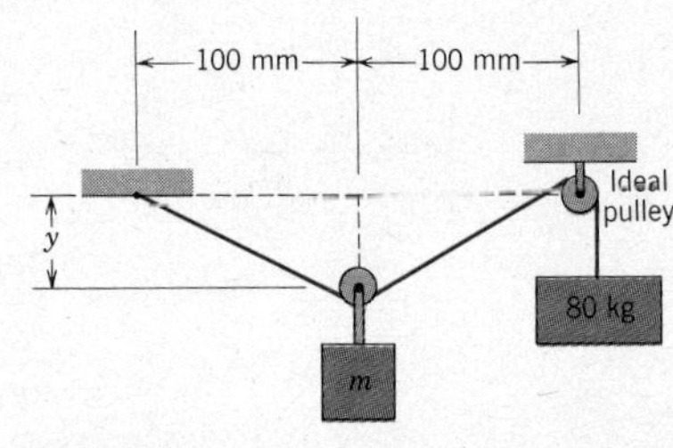

Prob. II.71

II.71 The block of unknown mass m is suspended by means of a cable attached to an 80-kg counterweight, as shown. Determine the relationship between the value of m and the distance y.

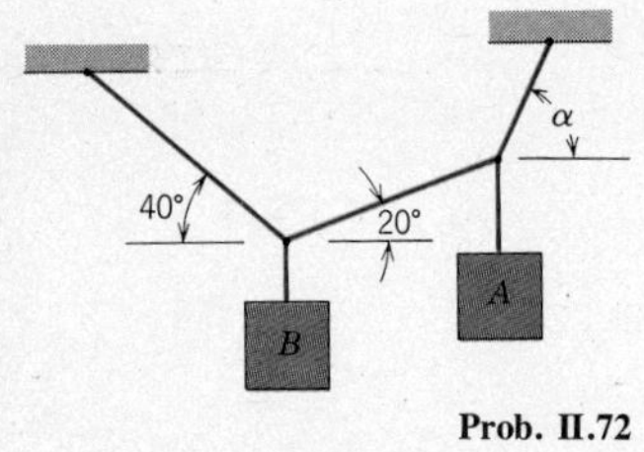

Prob. II.72

II.72 The masses of blocks A and B are 20 and 40 kg, respectively. Determine the value of α and the tension forces in the cables for static equilibrium.

II.73 Determine the tensile force in chain D. All chains are of negligible weight.

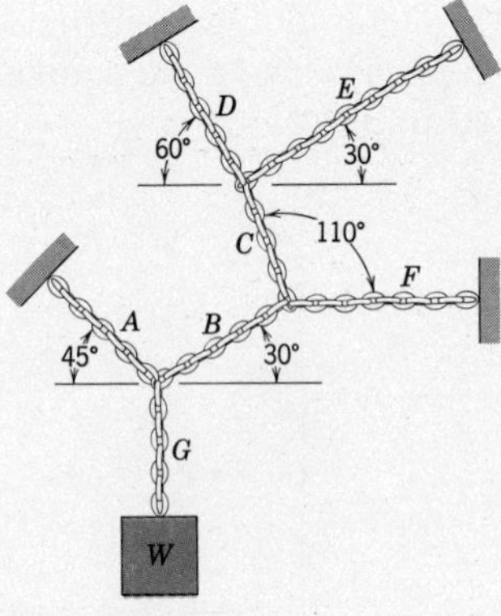

Prob. II.73

II.74 The spring has an unstretched length of 14 in. and a stiffness of 80 lb/ft. Determine the height x corresponding to static equilibrium in the position shown.

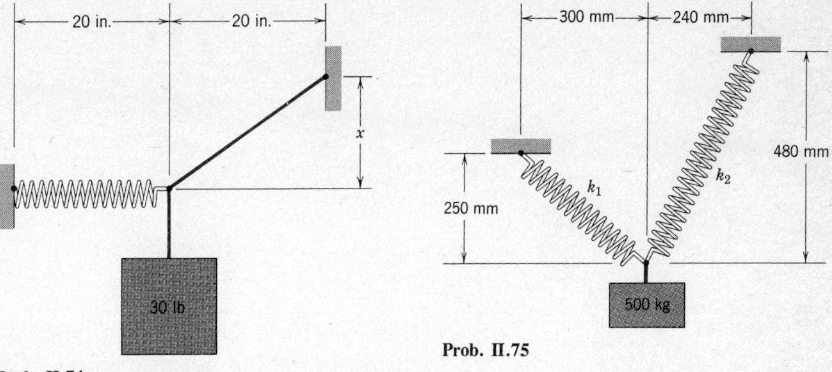

Prob. II.74

Prob. II.75

II.75 The stiffnesses of the springs are $k_1 = 20$ kN/m and $k_2 = 10$ kN/m. If the 500-kg block is suspended in static equilibrium in the position shown, determine the unstretched lengths of the springs.

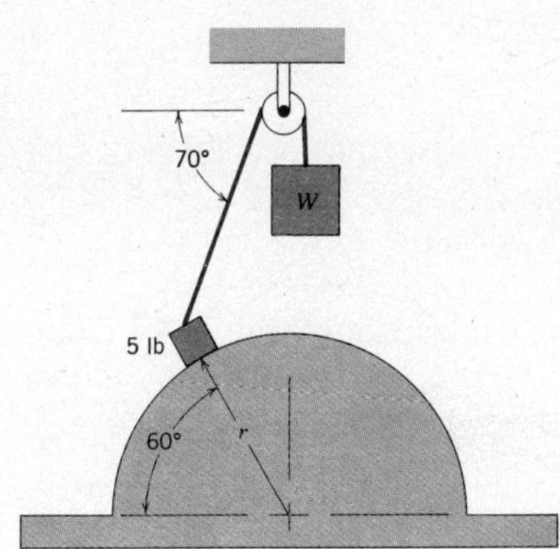

Prob. II.76

II.76 A 5-lb block is held in the position shown on the smooth semicylinder by a cable that is attached to a counterweight. Determine the weight W of the counterweight.

II.77 A 3000-lb automobile is suspended on a platform by four cables of equal length. The platform weighs 200 lb. Determine the tension in each of the cables.

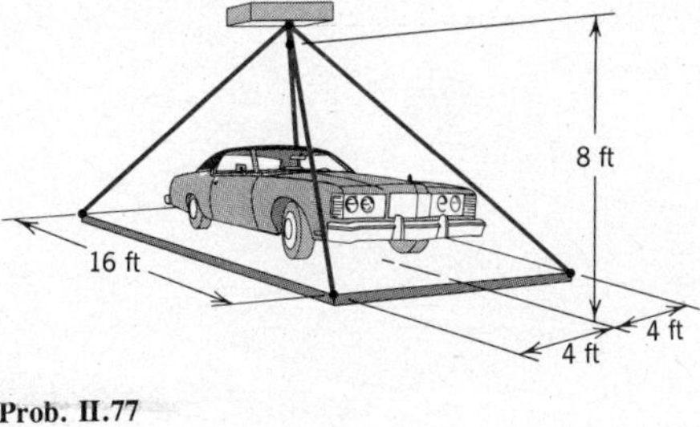

Prob. II.77

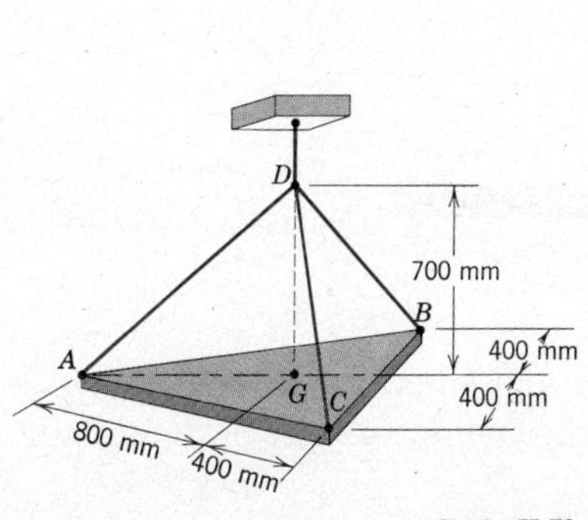

Prob. II.78

II.78 A 200-kg triangular plate is supported by three cables as shown. Determine the tension in each cable. (*Hint:* The weight of the plate may be considered to act at the centroid G of the plate.)

II.79 A 50-kg traffic light is suspended 6 m above the ground by cables from three 8-m-tall poles located as shown. Determine the tension in each cable.

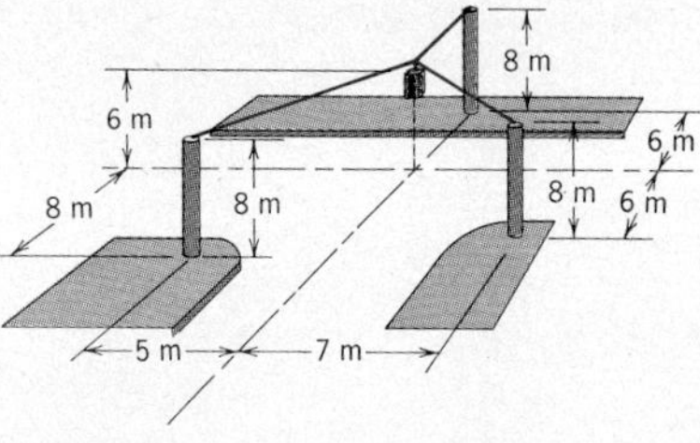

Prob. II.79

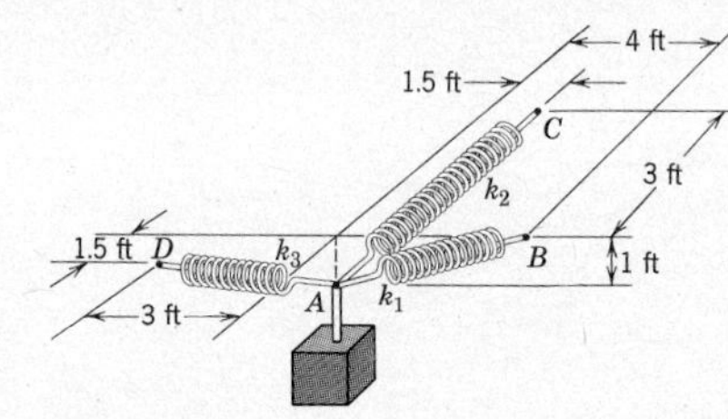

Prob. II.80

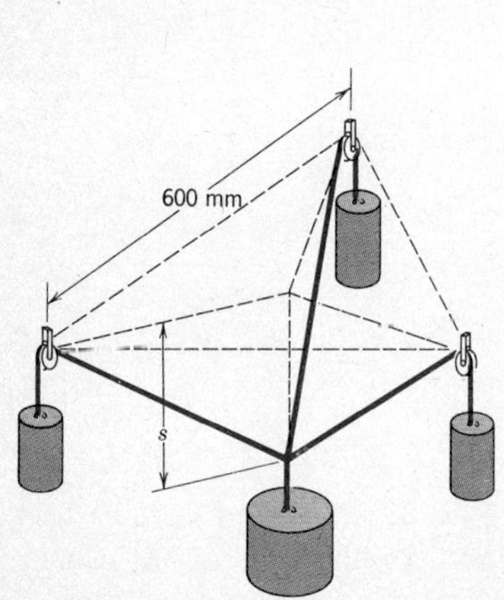

Prob. II.81

II.80 A 10-lb block is supported by three springs as shown. The stiffnesses of the springs are $k_1 = 5$ lb/in., $k_2 = 3$ lb/in., $k_3 = 4$ lb/in. Determine the unstretched length of each spring.

II.81 The 20-kg cylinder is supported by cables from three 15-kg counterweights that are equally spaced at 600 mm. Knowing that the pulleys have ideal properties, determine the dimension s representing the sag.

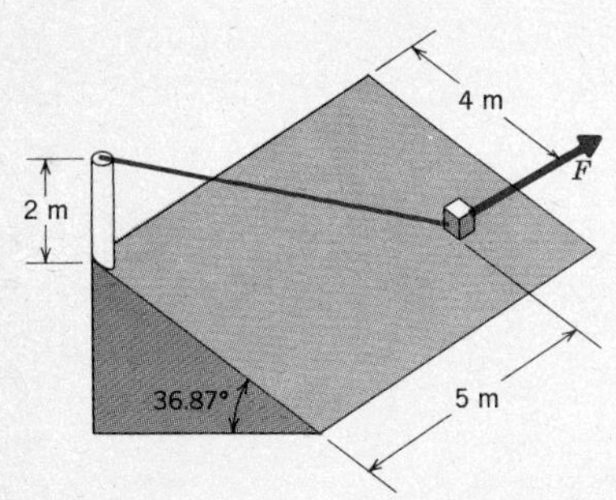

Prob. II.82

II.82 A 100-kg crate is supported on the smooth inclined plane by the cable and the horizontal force $\bar{F}$ tangent to the incline, as shown. Determine the magnitude of $\bar{F}$ and the tension in the cable.

II.83 The 3000-lb automobile on the ice-covered 20° hill is being held by cables from two tow trucks, as shown. Considering the hill to be smooth and frictionless, determine the tensile force in each cable.

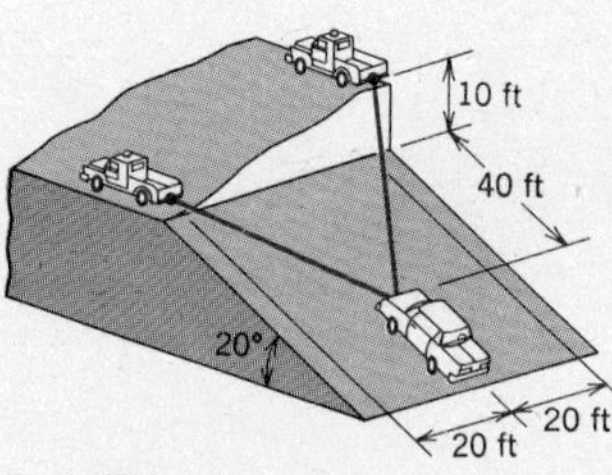

Prob. II.83

II.84 A 1-kg ball bearing is pushed into the smooth right-angle corner by the 20-N force shown, which passes through the center of the sphere.

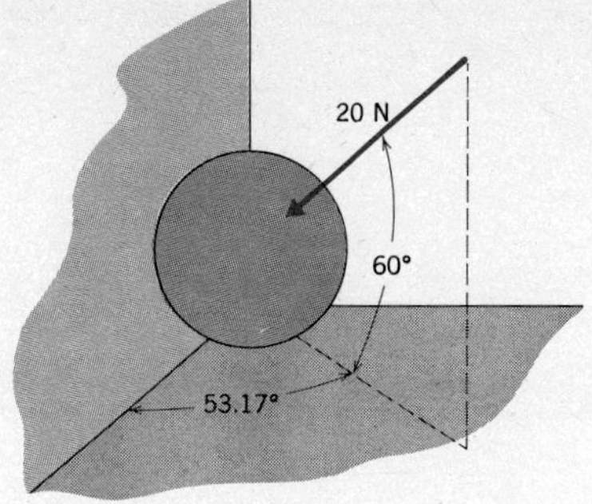

Prob. II.84

Determine the reaction forces exerted between the ball bearing and the vertical walls and horizontal floor.

II.85 The diagram shows the side view and cross section of a V-block that is holding a 200-g steel sphere. The reaction between the sphere and the fixed vertical wall is 1000 N. Determine the reaction forces exerted between the sphere and the walls of the groove, and between the sphere and the inclined movable block.

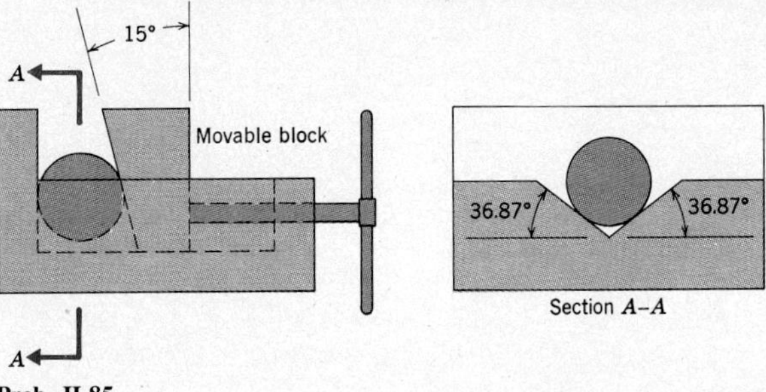

Prob. II.85

II.86 A grooved block, whose cross section is as shown, is tilted up at a 20° angle against a vertical wall, and a 1.5-kg sphere is placed in the corner. Determine the reaction forces exerted between the walls of the groove and the ball bearing. *Hint:* Use the XYZ reference frame shown whose Z axis is tangent to the groove.

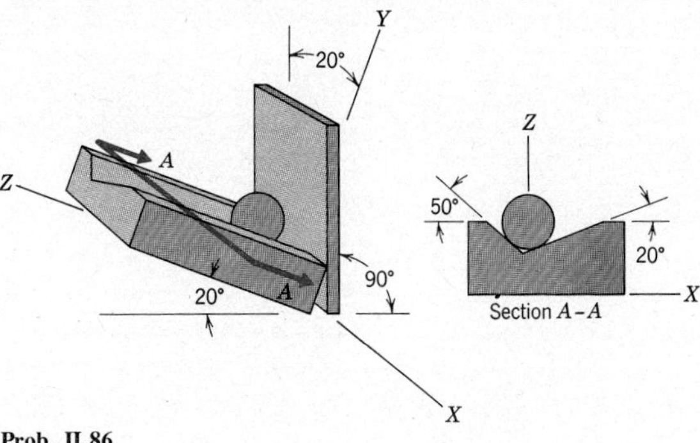

Prob. II.86

MODULE III GENERAL FORCE SYSTEMS AND EQUILIBRIUM OF RIGID BODIES

A rigid body, as compared to a particle, occupies more than a single point in space. As a result, the body will generally be subjected to a system of nonconcurrent loads having a tendency to cause the body to rotate. To describe this rotational tendency we will develop the concept of the *moment of a force*. We shall see that the moment is dependent on the line of action of a force, as well as its magnitude and direction.

The methods for determining moments, in conjunction with the methods for describing forces that were developed in the previous module, are the tools that will enable us to establish and study the equations of static equilibrium for rigid bodies. The computation of the moment of a force is addressed in the first set of topics in this module. The general topic of static equilibrium is presented in the last portion of this module. Module III, therefore, is the focal point of the study of statics.

Simply stated, a rigid body is a collection of particles that are always at a fixed distance from each other. Consider the two-particle systems shown in Figure 1.

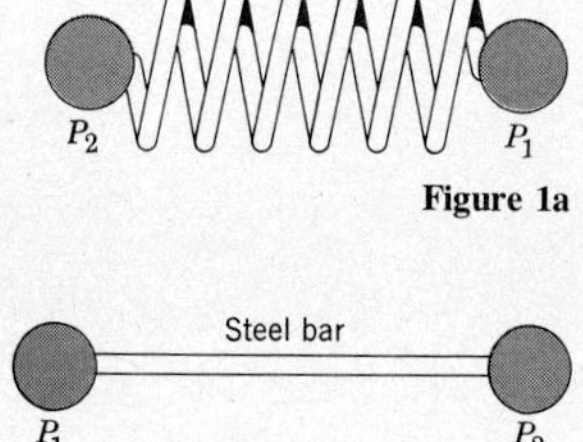

Figure 1a

Figure 1b

Figure 1a shows two particles connected by an elastic spring. If the spring is extended or compressed, the particles will no longer be at their original distance from each other; hence the system is not a rigid body. In Figure 1b, if we neglect the secondary (small) effects of the deformation of the steel bar, the particles are always at a fixed distance apart and therefore the system is termed a rigid body.

In general, the concept of a rigid body is a mathematical model, in that no real material can remain undeformed under the influence of forces. What we are saying is that for the class of problems we will study, the microscopic deformation effects will not influence the results.

Incidentally, if we did not neglect the deformation of the steel bar we would be concerned with the elastic properties of flexible bodies, which is an important topic (an integral part of the design process) that you will encounter in the study of mechanics of materials.

A. MOMENTS OF A FORCE ABOUT A LINE AND ABOUT A POINT

Let us begin this study by considering the system shown in Figure 2, where a force $\bar{F}$ is applied at point D on the perimeter of a disk mounted on a shaft. The Y axis of the coordinate system shown is chosen parallel to the shaft AB. Because $\bar{F}$ is parallel to the XY plane it may be replaced by the two vector components $\bar{F}_X$ and $\bar{F}_Y$ shown. The distance d is the perpendicular distance from the shaft to the line of action of the vector $\bar{F}_X$. In the terminology of statics, d is the *lever arm* of the force $\bar{F}_X$ about the axis AB.

Figure 2

For the situation of Figure 2, the component $\bar{F}_X$ will cause the disk to rotate about the axis of the shaft in the sense of the curved arrow. The other component $\bar{F}_Y$ causes no rotation about axis AB. Rather, it merely pulls on the disk parallel to that axis.

To quantify this rotational effect we define the concept of the *moment of a force about an axis*. Specifically, we define the moment $\bar{M}_{AB}$ of the force $\bar{F}$ about line AB such that its magnitude is the product of the component of the force $\bar{F}$ perpendicular to the line and the lever arm of the force (perpendicular distance) to the line. Thus

$$|\bar{M}_{AB}| = F_X\, d \qquad (1)$$

Dimensionally, moment carries the physical units of force times distance. Throughout this text we shall consistently give the force unit of the moment first and then the length unit. This is done in order to avoid confusion with the units of energy, which also involve the product of force and length.

In addition to having a magnitude, the moment is a vector quantity having direction. This direction is associated with the sense of the rotational effect. We could use the curved line in Figure 2 to describe this direction by saying that the moment of $\bar{F}$ about line AB is counterclockwise as viewed from point B toward point A. Such descriptions prove to be inadequate in many situations.

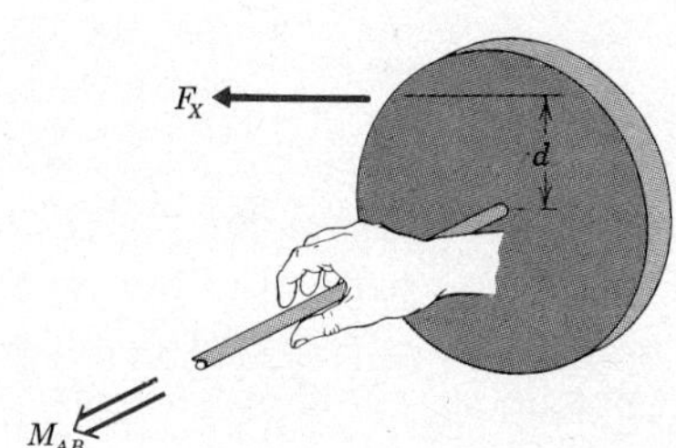

Figure 3

The accepted way to describe the moment about a line is to employ the *right-hand rule,* as depicted in Figure 3. The fingers of the right hand are curled in the sense of the rotational effect of the moment about the axis. The direction of the extended thumb is then defined as the direction of the moment. Accordingly, we now can describe the $\bar{M}_{AB}$ of Figure 2 as $|\bar{M}_{AB}|\bar{J}$, so

$$\bar{M}_{AB} = F_X\, d\, \bar{J}$$

In an effort to avoid confusion between force and moment vectors, we shall use a double line throughout the text to depict moment vectors, as is shown in Figures 2 and 3.

The concept of the moment of a force about a line will prove to be valuable in two-dimensional situations. For example, by confining our attention to the moment $\bar{M}_{AB}$ in Figure 2, in essence we were looking at the effect of $\bar{F}$ on equilibrium in the XZ plane; the force component $\bar{F}_Y$ did not have any effect in this plane. To consider the effect of $\bar{F}$ on the entire system, we form the *moment of the force about a point*.

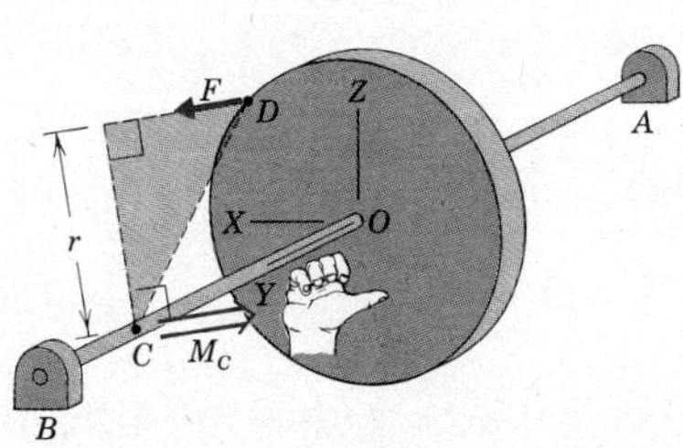

Figure 4

Consider an arbitrary point, C, on the shaft AB, as shown in Figure 4. To indicate that the desired moment is about point C, we write this quantity as $\bar{M}_C$, using a single subscript. This contrasts with a moment with double subscripts, which denotes that the moment is about an axis.

The distance r is the perpendicular distance from the line of action of the force to point C. This distance is called the *lever arm of the force about point C*. As is the case of the moment about an axis, we define the magnitude of $\bar{M}_C$ to be the product of this lever arm and the magnitude of $\bar{F}$; that is,

$$|\bar{M}_C| = |\bar{F}|r \tag{2}$$

The right-hand rule is again used to define the direction of the moment. To do this we form an analogy to the way in which we regarded the plane of the disk when defining the moment of the force about the axis of the shaft. The fingers of the right hand are curled in the sense of the rotation that would be imparted to the shaded plane formed by point C and the line of action of $\bar{F}$. The direction of $\bar{M}_C$ is then perpendicular to the shaded plane, in the sense of the extended thumb of the right hand.

The similarities in the definitions of the moment of a force about a line and the moment of a force about a point suggest that the two quantities are directly related, and they are. Later we will study this relationship. Before doing so, we must increase our ability to describe the moment of a force. For example, in Figure 4 the determination of the lever arm r and the direction of $\bar{M}_C$ can be quite complicated. To avoid cumbersome computational procedures, we will now study the remaining vectorial tool we need, the cross (vector) product.

B. THE CROSS (VECTOR) PRODUCT

The dot product of two vectors, defined in the previous module, gives a scalar value involving the component of one vector parallel to the other vector. Here, we will see that the *cross product* gives a vector result in terms of the component of one vector perpendicular to the other vector. Because of the difference in the type of result of these two multiplication operations, the dot product is sometimes called the *scalar product,* and the cross product is sometimes called the *vector product*.

Consider the two vectors $\bar{A}$ and $\bar{B}$ shown in Figure 5. Without loss of generality, let the plane of the vectors coincide with the plane of the paper. Denoting the angle between the two vectors as θ, the component of $\bar{A}$ perpendicular to $\bar{B}$ is $|\bar{A}|$ sin θ, whereas the component of $\bar{B}$ perpendicular to $\bar{A}$ is $|\bar{B}|$ sin θ, as shown. The magnitude of the cross product $\bar{A} \times \bar{B}$ is defined to be the component of $\bar{B}$ perpendicular to $\bar{A}$ multipled by the magnitude of $\bar{A}$:

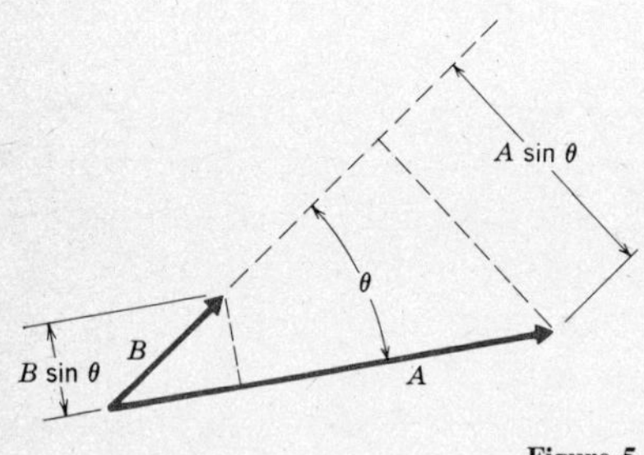

Figure 5

$$|\bar{A} \times \bar{B}| = |\bar{A}||\bar{B}| \sin\theta \tag{3}$$

Notice that $|\bar{A} \times \bar{B}| \equiv |\bar{B} \times \bar{A}|$ and that $\bar{A} \times \bar{B}$ is zero whenever $\bar{A}$ and $\bar{B}$ are parallel.

The direction of the cross product is defined to be normal to the plane formed by the two vectors. To define the sense of this normal, we again employ the right-hand rule, curling the fingers of the right hand about the common tail of the two vectors, from the first vector in the product to the second. The extended thumb is then oriented in the sense of the product.

According to this definition for the direction, $\bar{A} \times \bar{B}$ in Figure 5 is outward from the plane of the diagram, whereas $\bar{B} \times \bar{A}$ is inward. Because the order of multiplication does not affect the magnitude of the product given by equation (3), we can conclude that the two products have equal magnitude and opposite direction, so

$$\bar{B} \times \bar{A} = -\bar{A} \times \bar{B} \tag{4}$$

Primarily, we will employ the cross product operation in conjunction with a description of the vectors in terms of their components with respect to a coordinate system XYZ. To do this we must first consider the algebraic rules for this operation. In view of the fact that the order in which the vectors appear in the cross product affects the result, as shown by equation (4), we may wonder whether the other algebraic rules are different in form from those for the dot product. The answer is no. Specifically, we can prove that

$$\begin{aligned} p(\bar{A} \times \bar{B}) &= (p\bar{A}) \times \bar{B} = \bar{A} \times (p\bar{B}) \\ \bar{A} \times (\bar{B} + \bar{C}) &= (\bar{A} \times \bar{B}) + (\bar{A} \times \bar{C}) \\ (\bar{A} + \bar{B}) \times \bar{C} &= (\bar{A} \times \bar{C}) + (\bar{B} \times \bar{C}) \end{aligned} \tag{5}$$

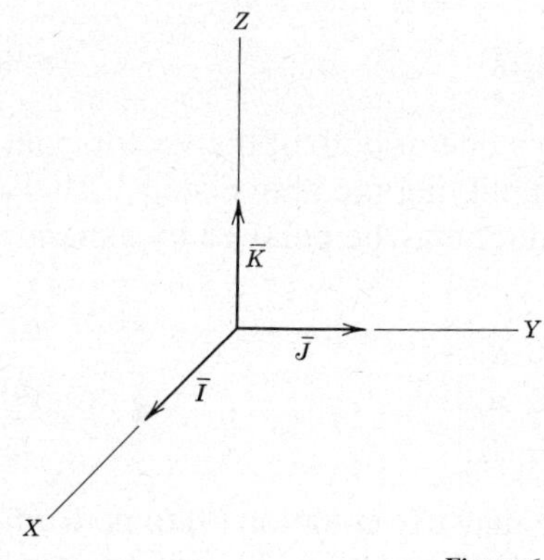

Figure 6

We begin the discussion of vector components by first considering the cross products of the unit vectors. In doing this it is essential that we use a right-handed coordinate system in order to be consistent with the use of the right-hand rule in defining the vector product. Figure 6 shows a typical coordinate system.

As each of the unit vectors is obviously parallel to itself, it follows that

$$\bar{I} \times \bar{I} = \bar{J} \times \bar{J} = \bar{K} \times \bar{K} = \bar{0} \tag{6}$$

In contrast, the angle between any two different unit vectors is 90°, so

$$\begin{aligned} |\bar{I} \times \bar{J}| &= |\bar{I}|\,|\bar{J}| \sin 90^\circ = 1 \\ &= |\bar{J} \times \bar{K}| = |\bar{K} \times \bar{I}| \end{aligned}$$

Also, the third unit vector is perpendicular to the coordinate plane of the other two unit vectors. The right-hand rule gives the appropriate sign. For example, when we curl our fingers from $\bar{J}$ to $\bar{I}$ in order to evaluate $\bar{J} \times \bar{I}$,

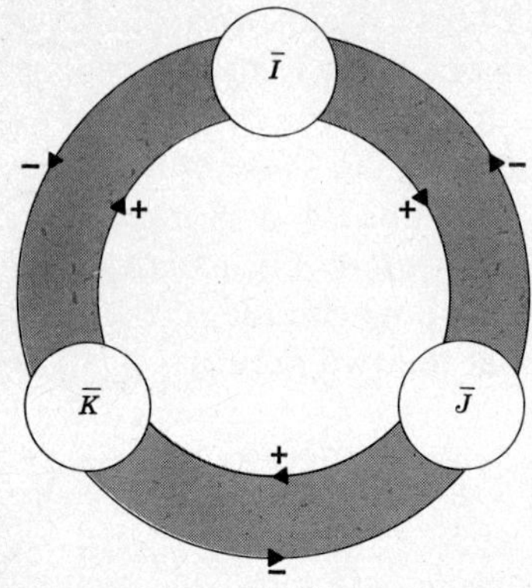

Figure 7

our thumb is oriented opposite the direction of positive $\bar{K}$. The full set of relations are

$$\begin{aligned}\bar{I} \times \bar{J} &= -\bar{J} \times \bar{I} = \bar{K} \\ \bar{J} \times \bar{K} &= -\bar{K} \times \bar{J} = \bar{I} \\ \bar{K} \times \bar{I} &= -\bar{I} \times \bar{K} = \bar{J}\end{aligned} \tag{7}$$

We will employ equations (7) frequently. As an aid in remembering them, we can consider the cross product of two unit vectors to be positive for positive alphabetical order and negative for negative alphabetical order, as depicted in Figure 7.

Finally, we find the cross product of two general vectors by expressing them in terms of their components and then employing equations (5) through (7). Thus

$$\begin{aligned}\bar{A} \times \bar{B} &= (A_X \bar{I} + A_Y \bar{J} + A_Z \bar{K}) \times \bar{B} \\ &= A_X [\bar{I} \times (B_X \bar{I} + B_Y \bar{J} + B_Z \bar{K})] \\ &\quad + A_Y [\bar{J} \times (B_X \bar{I} + B_Y \bar{J} + B_Z \bar{K})] \\ &\quad + A_Z [\bar{K} \times (B_X \bar{I} + B_Y \bar{J} + B_Z \bar{K})] \\ &= A_X (B_Y \bar{K} - B_Z \bar{J}) + A_Y (-B_X \bar{K} + B_Z \bar{I}) \\ &\quad + A_Z (B_X \bar{J} - B_Y \bar{I}) \\ \bar{A} \times \bar{B} &= (A_Y B_Z - A_Z B_Y)\bar{I} + (A_Z B_X - A_X B_Z)\bar{J} \\ &\quad + (A_X B_Y - A_Y B_X)\bar{K}\end{aligned} \tag{8}$$

It is noteworthy that in most problems one or both of the vectors will have some zero components, thereby simplifying the above steps. Alternatively, the computation of a cross product may be pursued by expanding a determinant, for example

$$\bar{A} \times \bar{B} = \begin{vmatrix} \bar{I} & \bar{J} & \bar{K} \\ A_X & A_Y & A_Z \\ B_X & B_Y & B_Z \end{vmatrix} \tag{9}$$

If you are familiar with determinants, you may use equation (9) to perform your calculations. However, the method of equation (8) will be used throughout this book.

Before we apply the vector product to the evaluation of moments, let us consider the following example, which demonstrates another use of this operation.

EXAMPLE 1

Three points A, B, and C are located in the sketch on the following page.

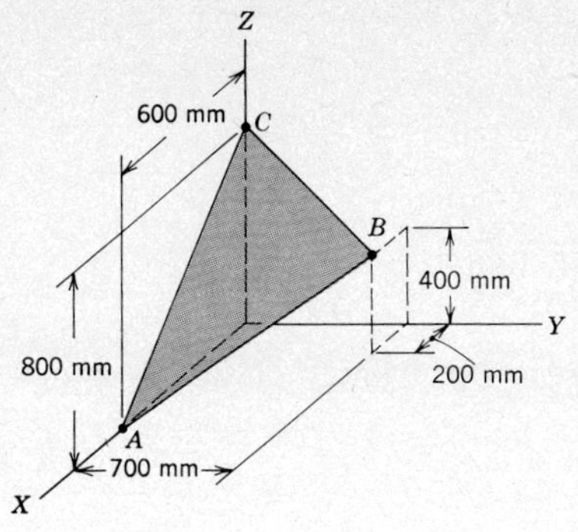

Determine the angle ϕ between the lines AB and AC, the area of triangle ABC, and a unit vector normal to the plane of this triangle.

Solution

Considering the two vectors $\bar{A}$ and $\bar{B}$ in Figure 5, we can see that the component of one of these vectors perpendicular to the other is the altitude of the triangle formed by $\bar{A}$, $\bar{B}$, and the line connecting their heads. Thus, for the vectors in Figure 5 the area is

$$\text{Area} = \tfrac{1}{2}|\bar{A} \times \bar{B}|$$

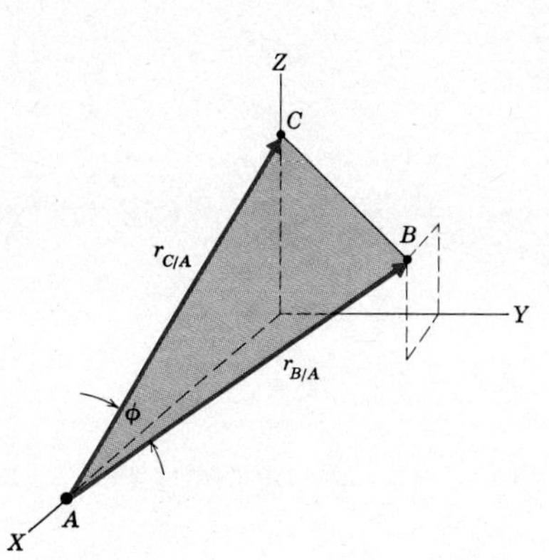

For the problem at hand, we can obtain the angle ϕ from the same cross product as that used for the area if we employ the position vectors $\bar{r}_{B/A}$ and $\bar{r}_{C/A}$ as shown in the sketch. We then form

$$\text{Area} = \tfrac{1}{2}|\bar{r}_{B/A} \times \bar{r}_{C/A}| \equiv \tfrac{1}{2}|\bar{r}_{B/A}||\bar{r}_{C/A}| \sin \phi$$

Referring to the given sketch to determine the position of the various points, we see that a displacement from point A to point B is equivalent to displacements of 700 mm in the $+Y$ direction, 400 mm in the $-X$ direction, and 400 mm in the $+Z$ direction, so that

$$\bar{r}_{B/A} = 0.70\bar{J} - 0.40\bar{I} + 0.40\bar{K} \text{ m}$$

For $\bar{r}_{C/A}$ we follow the coordinate axes from point A to point C to find

$$\bar{r}_{C/A} = -0.60\bar{I} + 0.80\bar{K} \text{ m}$$

The process of evaluating the vector product may be thought of as being similar to that for multiplying two scalar polynomials. Thus

$$\begin{aligned}
\bar{r}_{B/A} \times \bar{r}_{C/A} &= (0.7\bar{J} - 0.4\bar{I} + 0.4\bar{K}) \times (-0.6\bar{I} + 0.8\bar{K}) \\
&= -0.7(0.6)\bar{J} \times \bar{I} + 0.7(0.8)\bar{J} \times \bar{K} \\
&\quad +0.4(0.6)\bar{I} \times \bar{I} - 0.4(0.8)\bar{I} \times \bar{K} \\
&\quad -0.4(0.6)\bar{K} \times \bar{I} + 0.4(0.8)\bar{K} \times \bar{K} \\
&= 0.42\bar{K} + 0.56\bar{I} + 0.32\bar{J} - 0.24\bar{J} \\
&= 0.56\bar{I} + 0.08\bar{J} + 0.42\bar{K}
\end{aligned}$$

Incidentally, as you become more familiar with the cross product, you will find that you can skip the second of the above computational steps.

Using equation (3) we have

$$\sin \phi = \frac{|\bar{r}_{B/A} \times \bar{r}_{C/A}|}{|\bar{r}_{B/A}||\bar{r}_{C/A}|}$$

The magnitudes of these vectors are

$$|\bar{r}_{B/A}| = \sqrt{(0.7)^2 + (0.4)^2 + (0.4)^2} = 0.90 \text{ m}$$
$$|\bar{r}_{C/A}| = \sqrt{(0.6)^2 + (0.8)^2} = 1.00 \text{ m}$$
$$|\bar{r}_{B/A} \times \bar{r}_{C/A}| = \sqrt{(0.56)^2 + (0.08)^2 + (0.42)^2} = 0.7046 \text{ m}^2$$

Thus, because $|\bar{r}_{B/A} \times \bar{r}_{C/A}|$ is twice the area of triangle ABC, we have

$$\text{Area} = \tfrac{1}{2}(0.7046) = 0.3523 \text{ m}^2$$

Also, from the expression for sin ϕ, we find that

$$\sin\phi = \frac{0.7046}{0.90\,(1.00)} \qquad \phi = 51.52°$$

Finally, we divide $\bar{r}_{B/A} \times \bar{r}_{C/A}$ by its magnitude to obtain a *unit* vector normal to the plane of $\bar{r}_{B/A}$ and $\bar{r}_{C/A}$. This gives

$$\bar{e}_n = \frac{0.56\bar{I} + 0.08\bar{J} + 0.42\bar{K}}{0.7046} = 0.7948\bar{I} + 0.1135\bar{J} + 0.5961\bar{K}$$

As all of the components of $\bar{e}_n$ are positive, we can conclude that $\bar{e}_n$ is directed outward from the origin.

HOMEWORK PROBLEMS

III.1 Compute $\bar{A} \times \bar{B}$ for the following vectors: (a) $\bar{A} = 5\bar{I} - \bar{J} - \bar{K}$, $\bar{B} = -4\bar{I} + 2\bar{J} + 4\bar{K}$, (b) $\bar{A} = 2\bar{I} + \bar{J}$, $\bar{B} = 3\bar{I} - 4\bar{K}$, (c) $\bar{A} = \bar{I} + 2\bar{J} + \bar{K}$, $\bar{B} = -3\bar{I} + 2\bar{J} - \bar{K}$.

III.2 Given that $\bar{A} = A_X\bar{I} - 5\bar{J} + 2\bar{K}$, $\bar{B} = -3\bar{I} + 2\bar{J} - B_Z\bar{K}$, determine the values of A_X and B_Z for which $A \times B$ is parallel (a) to the Y axis, (b) to the X axis.

III.3 Given that $\bar{A} = \bar{I} + \bar{J} - \bar{K}$, $\bar{B} = \bar{I} - 2\bar{J} + 3\bar{K}$, and $\bar{C} = -\bar{I} + 2\bar{J} - 4\bar{K}$, compute (a) $(\bar{A} \times \bar{B}) \times \bar{C}$, (b) $\bar{A} \times (\bar{B} \times \bar{C})$.

III.4 For the vectors given in Problem III.3, compute (a) $\bar{A} \cdot \bar{B} \times \bar{C}$, (b) $\bar{C} \cdot \bar{A} \times \bar{B}$.

III.5 The product $\bar{A} \cdot \bar{B} \times \bar{C}$ of three arbitrary vectors $\bar{A}$, $\bar{B}$, and $\bar{C}$ is called the scalar triple product. Prove the following identities: (a) $\bar{A} \cdot \bar{B} \times \bar{C} \equiv \bar{A} \times \bar{B} \cdot \bar{C}$, (b) $\bar{A} \cdot \bar{B} \times \bar{C} \equiv \bar{B} \cdot \bar{C} \times \bar{A}$.

III.6 It is desired to pass a plane through point A whose unit normal vector is $0.4286\bar{I} - 0.8571\bar{J} + 0.2857\bar{K}$. Determine the values of Y_B and Z_C, thus locating where the plane intersects the corresponding coordinate axes.

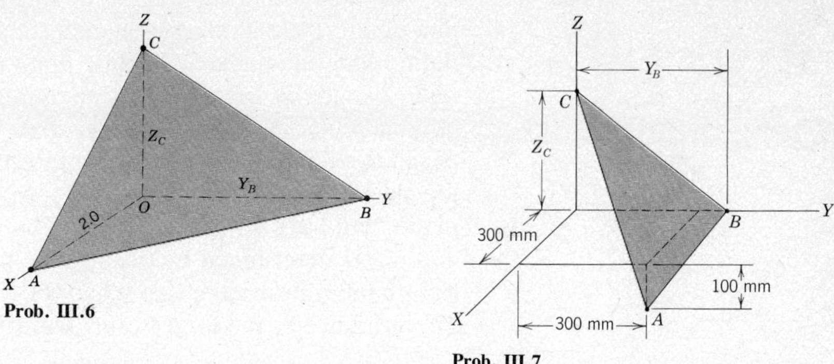

Prob. III.6

Prob. III.7

III.7 The plane formed by points A, B, and C has a unit normal vector given by $-0.4444\bar{I} + 0.1111\bar{J} + 0.8889\bar{K}$. Determine the values of Y_B and Z_C locating points B and C.

III.8 Forces $\bar{F}_1$ and $\bar{F}_2$ are both applied at point A. The force $\bar{F}_1 = 500\bar{I} - 400\bar{J} + Z\bar{K}$ N is to form a 60° angle with force $\bar{F}_2 = -200\bar{I} + 100\bar{J} + 300\bar{K}$ N. Determine (a) the value of Z, (b) the unit vector normal to the plane formed by $\bar{F}_1$ and $\bar{F}_2$.

III.9 Line AB is perpendicular to line AC. Determine the relationship between the distances X_B and Z_C, and describe the unit vector normal to the plane ABC.

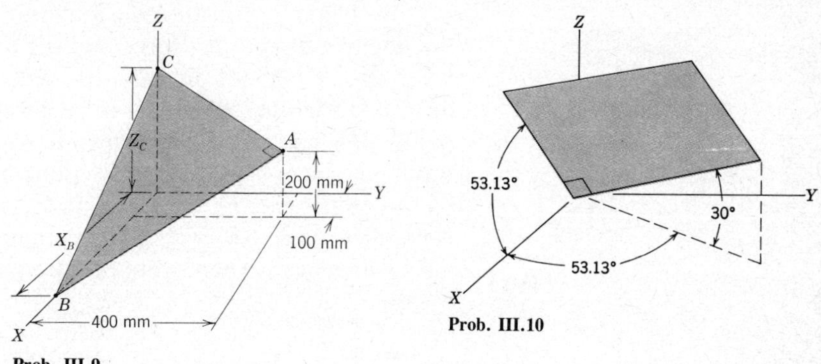

Prob. III.9

Prob. III.10

III.10 A rectangle is situated in space as shown. Determine a unit vector normal to its surface.

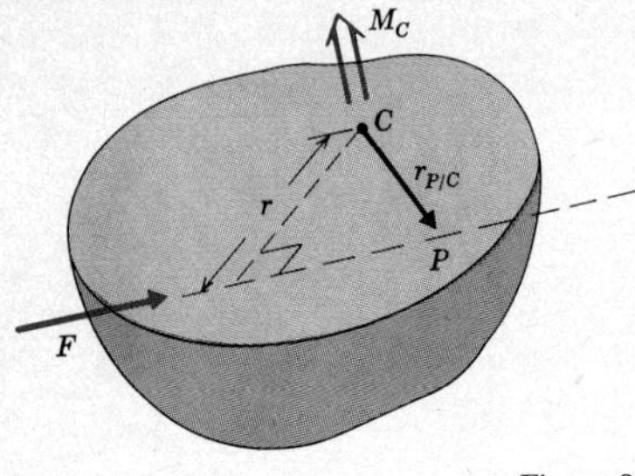

Figure 8

C. MOMENTS OF A FORCE USING CROSS PRODUCTS

Figure 8 depicts a general sitation where a force $\bar{F}$ is applied along the designated line of action. We wish to determine the moment of this force about point C, which has been chosen arbitrarily.

By definition, the magnitude of the moment $\bar{M}_C$ is $|\bar{F}|r$ in the direction of an axis that intersects point C and is normal to the plane containing the force $\bar{F}$ and point C. Letting point P be any point on the line of action of $\bar{F}$, we see in Figure 8 that the component of the position vector $\bar{r}_{P/C}$ perpendicular to $\bar{F}$ is the lever arm r. From the definition of a cross product, equation (3), it follows that the cross product of $\bar{F}$ and $\bar{r}_{P/C}$ has the magnitude of $\bar{M}_C$. Also, this cross product is perpendicular to the plane, and thus is parallel to $\bar{M}_C$. The proper order of multiplication of $\bar{F}$ and $\bar{r}_{P/C}$ is determined by placing the tail of $\bar{F}$ at point C. Then, using the right-hand rule, we see that when the thumb of the right hand is parallel to $\bar{M}_C$ in Figure 8, the fingers curl from $\bar{r}_{P/C}$ to $\bar{F}$. Thus

$$\boxed{\bar{M}_C = \bar{r}_{P/C} \times \bar{F}} \tag{10}$$

Equation (10) says that

> The moment of a force about a point C is obtained by crossing a position vector from point C to any point on the line of action of the force into the force vector.

With equation (10) as an aid for the determination of the moment about a point, we can now establish the relationship between this type of moment and the moment about a line. Consider Figure 9, where we wish to determine the moment of the force $\bar{F}$ about the line AB.

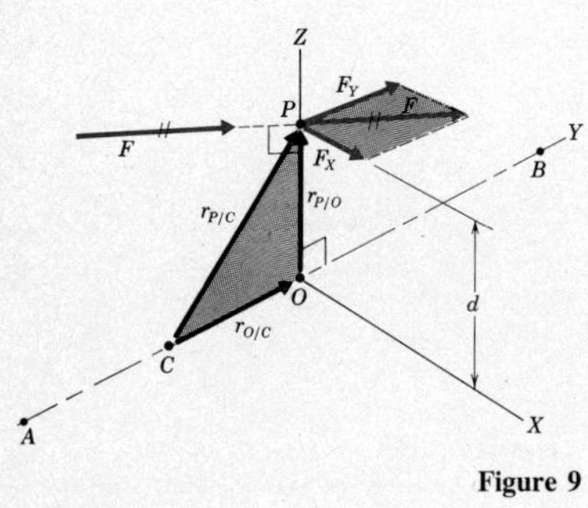

Figure 9

We begin by constructing the line OP, which is perpendicular to both the line of action of the force and the line AB. Thus, the length OP is the lever arm of $\bar{F}$ about the axis AB. We next define a convenient reference frame by orienting the Y axis to be parallel to line AB and the Z axis to be parallel to the line OP, as shown in Figure 9. Now, using the fact that a force may be applied anywhere along its line of action, we can replace $\bar{F}$ by its components $\bar{F}_X$ and $\bar{F}_Y$ ($F_Z \equiv 0$) acting at point P. The component $\bar{F}_Y$ is parallel to line AB, whereas the component $\bar{F}_X$ is perpendicular to both lines AB and OP. Then, from equation (2) for the moment about a line, we obtain

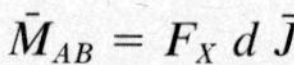

$$\bar{M}_{AB} = F_X\, d\, \bar{J}$$

Let us now consider the moment of the force $\bar{F}$ about any point C

loaded on line AB. From equation (10) we have

$$\bar{M}_C = \bar{r}_{P/C} \times \bar{F}$$

From Figure 9 we note that

$$\bar{r}_{P/C} = \bar{r}_{O/C} + \bar{r}_{P/O} = |\bar{r}_{O/C}|\bar{J} + (d)\bar{K}$$

so upon replacing $\bar{F}$ by its components we have

$$\begin{aligned} \bar{M}_C &= [|\bar{r}_{O/C}|\bar{J} + (d)\bar{K}] \times (F_X \bar{I} + F_Y \bar{J}) \\ &= -|\bar{r}_{O/C}|F_X \bar{K} + (d)F_X \bar{J} - (d)F_Y \bar{I} \end{aligned}$$

The second term in this expression is the component of $\bar{M}_C$ parallel to the Y axis (the line AB). This component is identical in magnitude and direction to the moment $\bar{M}_{AB}$. Hence, we can conclude that

$$\bar{M}_C \cdot \bar{J} = F_X\, d = |\bar{M}_{AB}|$$

In other words:

> The magnitude of the moment about a line is the component parallel to that line of the moment about any point on the line.

Finally, we express the direction of $\bar{M}_{AB}$ by again recalling equation (2), to get

$$\bar{M}_{AB} = F_X\, d\, \bar{J} = (\bar{M}_C \cdot \bar{J})\, \bar{J}$$

In Figure 9, the reference frame was intentionally chosen such that the Y axis was parallel to the line of interest, AB. To express the foregoing relationship in a situation where the Y axis of the coordinate system is *not* parallel to the axis AB, it is only necessary to replace $\bar{J}$ by a unit vector $\bar{e}_{B/A}$ parallel to line AB. Thus

$$\boxed{\bar{M}_{AB} = (\bar{M}_C \cdot \bar{e}_{B/A})\, \bar{e}_{B/A}} \qquad (11)$$

The significance of equations (10) and (11) is that they enable us to determine the moment of a force by expressing the force and position vectors in terms of the XYZ components. Before solving some examples, let us consider the general situation in Figure 10. By studying the resulting moment of the force $\bar{F}$ about point C, we can gain increased understanding of the significance of the vector operations.

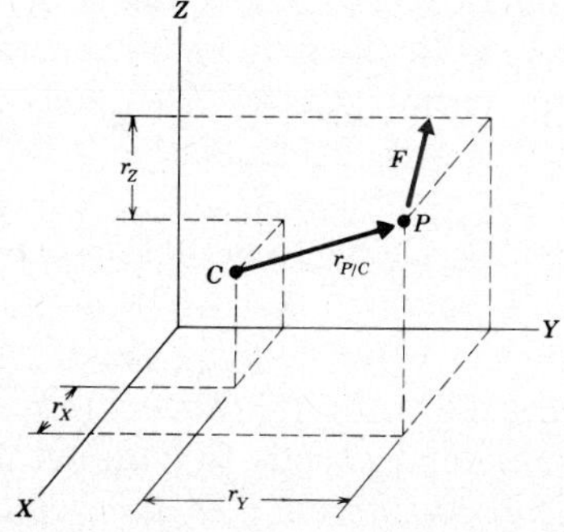

Figure 10

We compute $\bar{M}_C$ by expressing the position vector $\bar{r}_{P/C}$ and the general force $\bar{F}$ in terms of their components with respect to the XYZ coordinate system. Thus

$$\bar{r}_{P/C} = r_X\,\bar{I} + r_Y\,\bar{J} + r_Z\,\bar{K}$$

$$\bar{F} = F_X\,\bar{I} + F_Y\,\bar{J} + F_Z\,\bar{K}$$

Equation (10) then gives

$$\bar{M}_C = \bar{r}_{P/C} \times \bar{F} = (r_YF_Z - r_ZF_Y)\bar{I} + (r_ZF_X - r_XF_Z)\bar{J} + (r_XF_Y - r_YF_X)\bar{K} \quad (12)$$

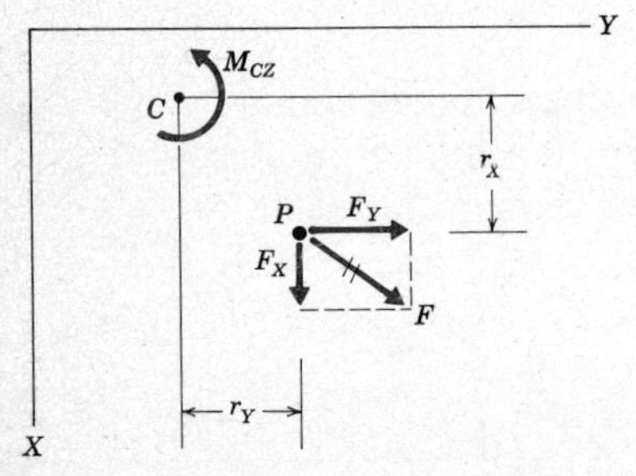

Figure 11

The meaning of the component values in equation (12) becomes apparent when we consider any one component, for example, by looking down the Z axis in Figure 10. This view is shown in Figure 11, where M_{CZ} denotes the Z component of $\bar{M}_C$. In view of the relationship between the moment about a line and the moment about a point along that line, $\bar{M}_{CZ}$ is *also the moment about the line parallel to the Z axis intersecting point C*. The moment $\bar{M}_{CZ}$ is shown as a counterclockwise curling arrow in this figure, corresponding to the way the fingers of our right hand curl according to the right-hand rule when M_{CZ} is positive.

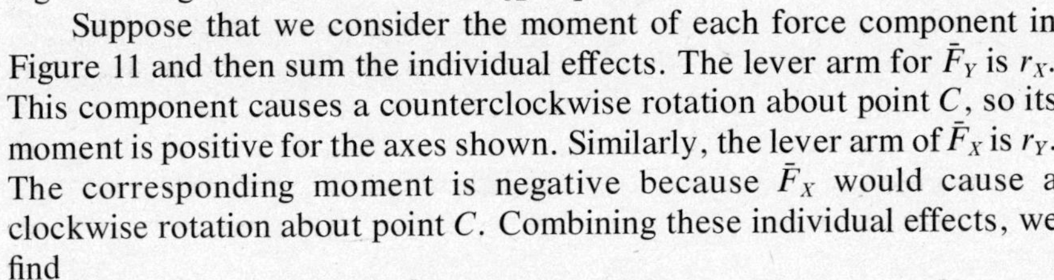

Suppose that we consider the moment of each force component in Figure 11 and then sum the individual effects. The lever arm for $\bar{F}_Y$ is r_X. This component causes a counterclockwise rotation about point C, so its moment is positive for the axes shown. Similarly, the lever arm of $\bar{F}_X$ is r_Y. The corresponding moment is negative because $\bar{F}_X$ would cause a clockwise rotation about point C. Combining these individual effects, we find

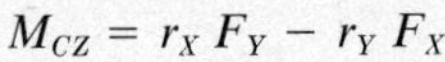

$$M_{CZ} = r_X\,F_Y - r_Y\,F_X$$

which, of course, is the same result for M_{CZ} found in equation (12) by using the cross product.

Similar arguments for the other components of $\bar{M}_C$ lead to the following conclusion. Equation (12) states that we can determine the moment about a point C by vectorially adding the moments of *each component* of a force about the lines parallel to the coordinate axes intersecting point C. This fact is known as *Varignon's theorem,* after the mathematician who formulated the result before the advent of vector algebra. This theorem provides a method for computing moments without evaluating a cross product.

When dealing with three-dimensional forces, the use of Varignon's theorem to compute the components of $\bar{M}_C$ requires that we draw three diagrams such as Figure 11, representing the views looking down each coordinate axis. This is a cumbersome procedure. For such situations we will rely on equations (10) and (11) to evaluate moments. In contrast, for problems *where the plane formed by the force and point C is parallel to a coordinate plane, the moment about C is identical to the moment about the line normal to that plane intersecting point C.* For such problems, considering the force components individually offers a definite computational advantage, as is demonstrated in the next example.

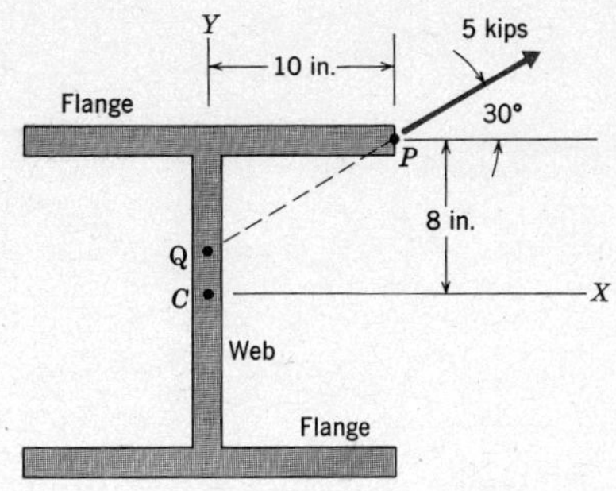

EXAMPLE 2

A 5-kip force is applied to the flange of the H beam as shown. Determine the moment of this force about the center C of the cross section by

a evaluating the lever arm of the force,
b evaluating a vector product,
c considering force components acting at the flange point P,
d considering force components acting at the web point Q.

Solution

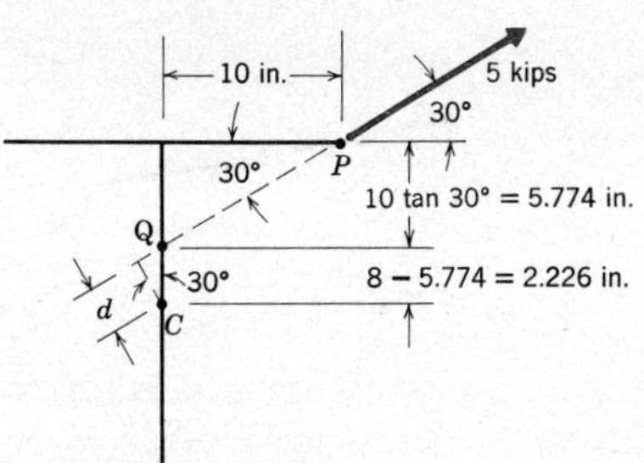

Method A

We first draw a sketch to evaluate the lever arm d. From this diagram

$$d = (2.226)\cos 30° = 1.928 \text{ in.}$$

so

$$M_C = |\bar{F}|d = 5(1.928) = 9.640 \text{ kip-in. clockwise}$$

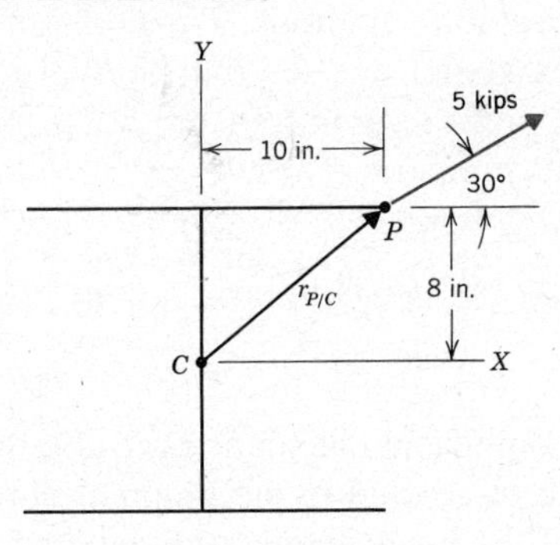

Method B

For the cross product formulation we need a position vector from point C to a point on the line of action of $\bar{F}$. Referring to the given diagram, we see that the dimensions for point P are given in terms of horizontal and vertical distances. We therefore express $\bar{r}_{P/C}$ in terms of the coordinate system shown in the sketch. From this sketch we have

$$\bar{r}_{P/C} = 10\bar{I} + 8\bar{J} \text{ in.}$$

$$\bar{F} = 5(\cos 30°\,\bar{I} + \sin 30°\,\bar{J})$$

$$= 4.330\bar{I} + 2.5\bar{J} \text{ kips}$$

$$\bar{M}_C = \bar{r}_{P/C} \times \bar{F} = (10\bar{I} + 8\bar{J}) \times (4.330\bar{I} + 2.5\bar{J})$$

$$= [10(2.5) - 8(4.330)]\bar{K}$$

$$= -9.640\bar{K} \text{ kip-in.}$$

Because $\bar{M}_C$ is in the negative Z direction, it is clockwise as viewed in the plane.

Method C

Point P is located with respect to point C in terms of horizontal and vertical distances, so we draw a sketch showing the horizontal and vertical components of the force.

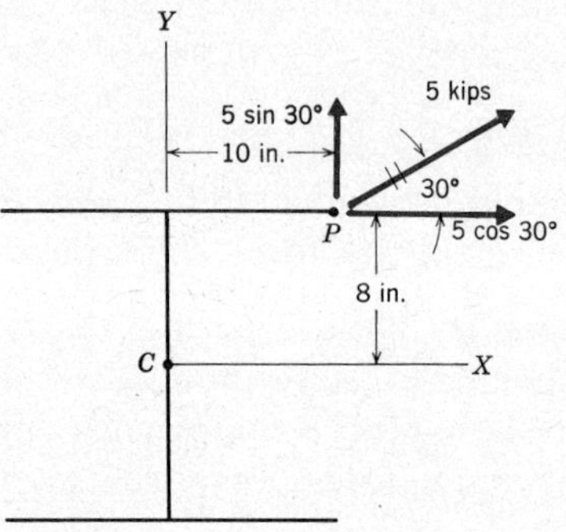

Because the Z axis is outward, counterclockwise moments are positive according to the right-hand rule. Multiplying each component by its lever arm, we then have

$$M_{CZ} = 10(5 \sin 30°) - 8(5 \cos 30°)$$

$$= -9.640 \text{ kip-in.}$$

Method D

This approach is based on the principle of transmissibility, according to which a force may be applied anywhere along its line of action. The force components are the same as in the two previous methods. The location of point Q is given in Method A. From the sketch we see that the vertical component of the force passes through point C, and that the horizontal component exerts a clockwise (negative) moment. Thus

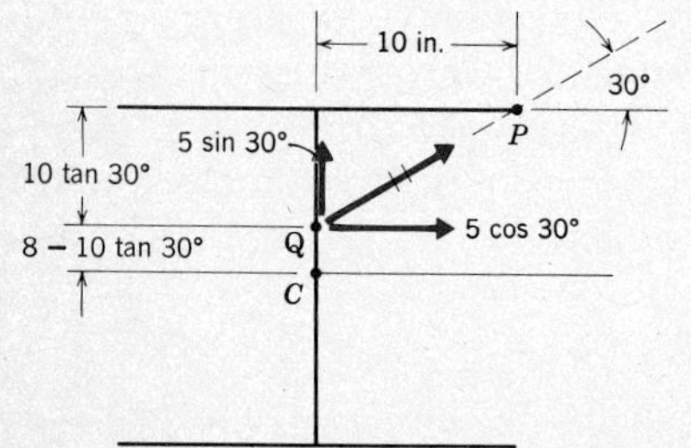

$$M_{CZ} = -(8 - 10 \tan 30°)\,(5 \cos 30°)$$

$$= -\ 9.640 \text{ kip-in.}$$

Assessing the four methods, the geometrical calculations in the first were fairly simple. However, if we were interested in the moment about some other point, such as the right tip of the lower flange, the determination of the lever arm would have been more complicated. The second and third methods make maximum use of the given geometrical information, but the third method yields the result more directly, because we do not have to write explicit expressions for the vectors. The last method is somewhat more cumbersome because the force was applied at a point whose location we had to determine. In general, for planar problems, we will employ either the first or third method for the evaluation of moments, depending on the given geometrical information.

EXAMPLE 3

A 10-kN force is applied as shown to the free end, point D, of the bent rod. Compute the moment of this force about point A and about line BC.

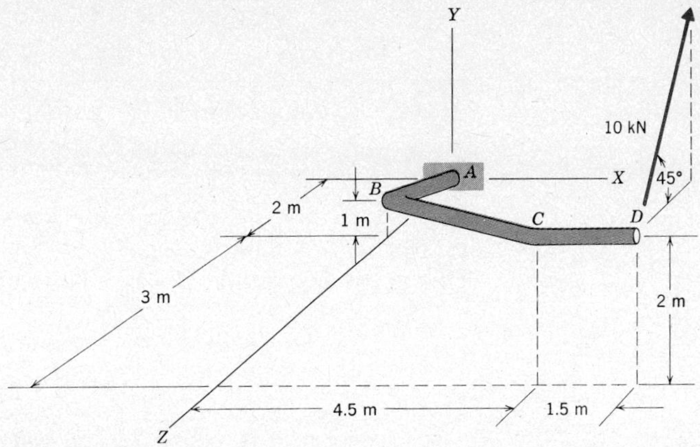

Solution

This is a three-dimensional problem. Therefore, we compute the moment about point A by constructing the position vector from this point to point D on the line of action of the force and then applying equation (10). The diagram indicates that the force is parallel to the YZ plane at an angle of 45° above the negative Z direction. Thus

$$\bar{r}_{D/A} = 5\bar{K} + 6\bar{I} + 2\bar{J} \text{ m}$$

$$\begin{aligned}\bar{F} &= 10(-\cos 45^\circ\ \bar{K} + \sin 45^\circ\ \bar{J}) \\ &= 7.071(\bar{J} - \bar{K}) \text{ kN}\end{aligned}$$

$$\begin{aligned}\bar{M}_A = \bar{r}_{D/A} \times \bar{F} &= (5\bar{K} + 6\bar{I} + 2\bar{J}) \times 7.071(\bar{J} \times \bar{K}) \\ &= 7.071[5(\bar{K} \times \bar{J}) - 5(\bar{K} \times \bar{K}) + 6(\bar{I} \times \bar{J}) \\ &\quad - 6(\bar{I} \times \bar{K}) + 2(\bar{J} \times \bar{J}) - 2(\bar{J} \times \bar{K})] \\ &= 7.071(-5\bar{I} + 6\bar{K} + 6\bar{J} - 2\bar{I}) \\ &= -49.50\bar{I} + 42.43\bar{J} + 42.43\bar{K} \text{ kN-m}\end{aligned}$$

For the second part of the problem, where the moment about line BC is required, we will first determine the moment about some point along this line and then find the moment about the line using equation (11). Let us use point C, because the position vector from point C to the force is easiest to construct. Thus

$$\begin{aligned}\bar{M}_C = \bar{r}_{D/C} \times \bar{F} &= 1.5\bar{I} \times 7.071(\bar{J} - \bar{K}) \\ &= 10.607(\bar{K} + \bar{J}) \text{ kN-m}\end{aligned}$$

The unit vector parallel to line BC is

$$\bar{e}_{C/B} = \frac{\bar{r}_{C/B}}{|\bar{r}_{C/B}|} = \frac{3\bar{K} + 4.5\bar{I} + \bar{J}}{\sqrt{(3)^2 + (4.5)^2 + (1)^2}}$$

$$= 0.8182\bar{I} + 0.1818\bar{J} + 0.5455\bar{K}$$

Hence, the component of $\bar{M}_C$ parallel to line BC is

$$\bar{M}_C \cdot \bar{e}_{C/B} = 10.607(\bar{J} + \bar{K}) \cdot (0.8182\bar{I} + 0.1818\bar{J} + 0.5455\bar{K})$$

$$= 10.607(0.1818 + 0.5455) = 7.714 \text{ kN-m}$$

This is the magnitude of $\bar{M}_{BC}$. To express its direction we write

$$\bar{M}_{BC} = (\bar{M}_C \cdot \bar{e}_{C/B})\bar{e}_{C/B} = 7.714(0.8182\bar{I} + 0.1818\bar{J} + 0.5455\bar{K})$$

$$= 6.313\bar{I} + 1.402\bar{J} + 4.208\bar{K} \text{ kN-m}$$

This vector is oriented from point B to point C. If $\bar{M}_C \cdot \bar{e}_{C/B}$ had been negative, the moment vector would then be oriented from point C to point B.

HOMEWORK PROBLEMS

III.11 A 300-lb force is applied to the rectangular plate at corner C. Determine the moment of this force about pin E (a) by computing its lever arm, (b) by replacing the force by its components acting at point C, (c) by replacing the force by its components acting at some other convenient point on its line of action.

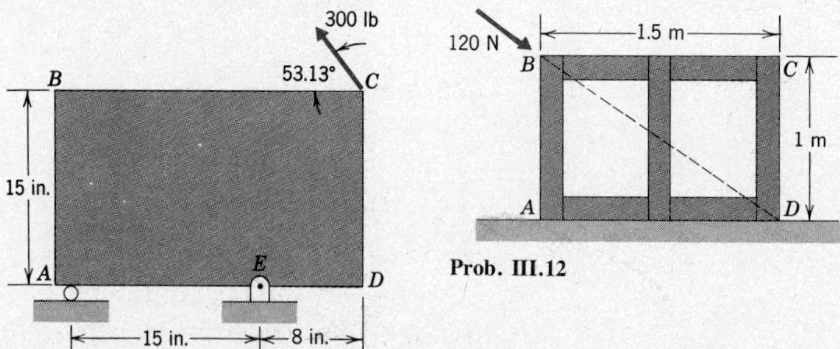

Prob. III.12

Prob. III.11

III.12 A crate resting on the ground is acted upon by the 120-N force shown. Determine the moment of this force about corner A (a) by determining the lever arm of the force, (b) by replacing the force by its components acting at point B, (c) by replacing the force by its components acting at some other convenient point on its line of action.

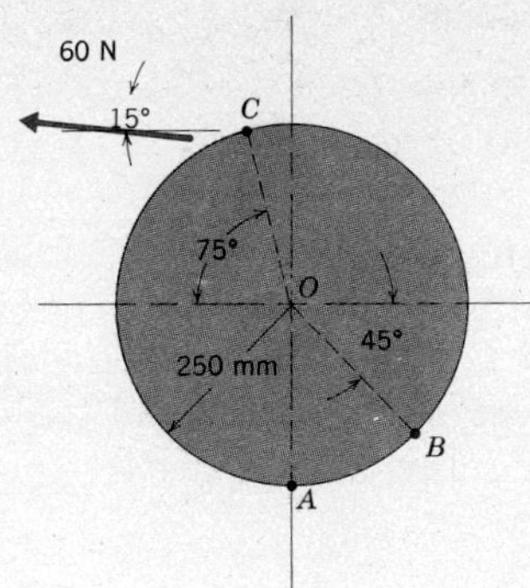

Probs. III.13, 14, and 15

III.13 The 60-N force is applied to the disk at point C. Determine the moment of the force about the center O (a) by determining the lever arm of the force, (b) by replacing the force by its X and Y components.

III.14 Using the two methods given in Problem III.13, determine the moment of the 60-N force about point A.

III.15 Using the two methods given in Problem III.13, determine the moment of the 60-N force about point B.

III.16 A 5-kN force is applied to the bent arm at point A. Determine the moment of the force about point C (a) for $\theta = 0°$, (b) for $\theta = 30°$.

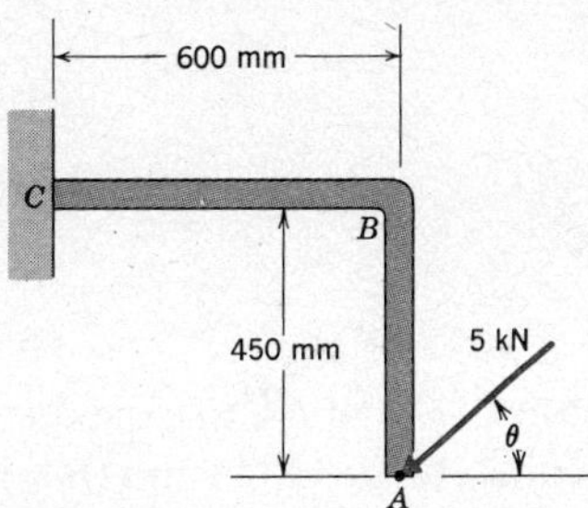

Prob. III.16

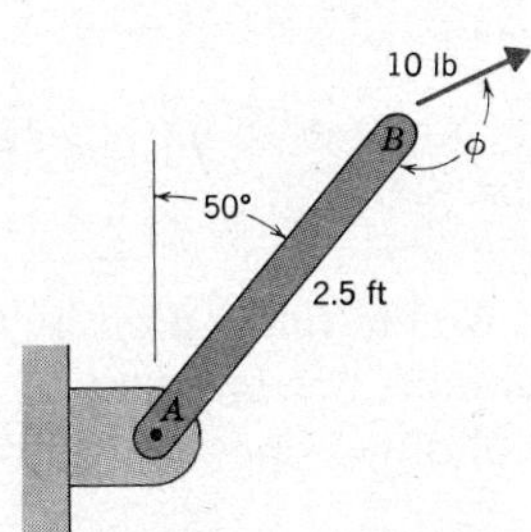

Prob. III.18

III.17 In Problem III.16, determine the angle θ such that the 5-kN force applied at point A exerts the maximum moment about point C. What is the corresponding moment?

III.18 Control bar AB is inclined at 50° from the vertical. A 10-lb force is applied to the bar at end B. Determine the moment of this force about pin A (a) if the force is horizontal, (b) if $\phi = 120°$.

III.19 In Problem III.18, determine the value of the angle ϕ for which the 10-lb force exerts the maximum moment about pin A. Also determine the corresponding moment.

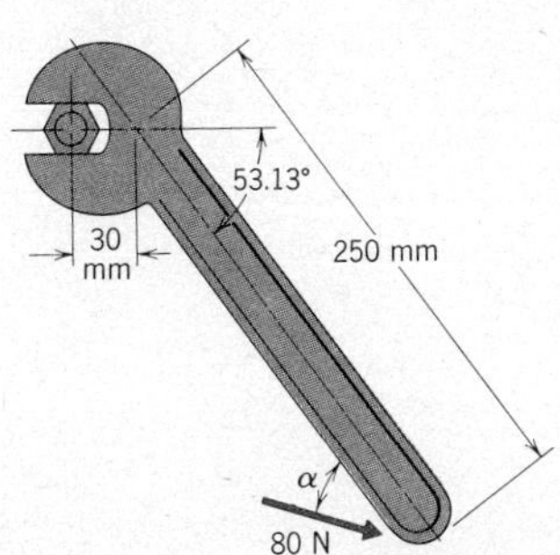

Prob. III.20

III.20 An 80-N force is applied to the special purpose wrench as shown. Determine the moment of this force about the center of the bolt when $\alpha = 75°$.

III.21 In Problem III.20, determine the angle α for which the magnitude of the moment of the 80-N force about the center of the bolt is (a) a minimum, (b) a maximum.

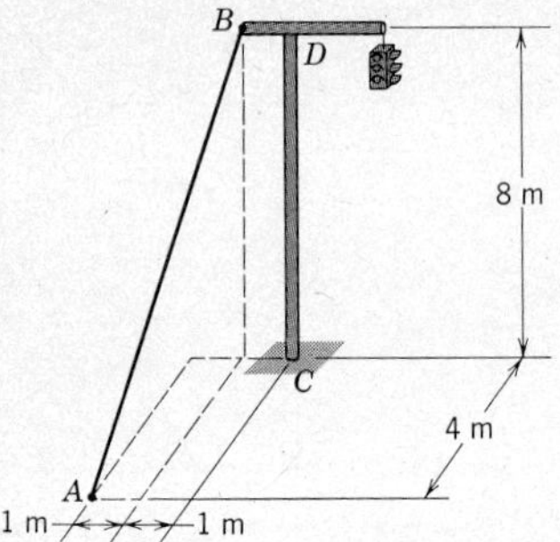
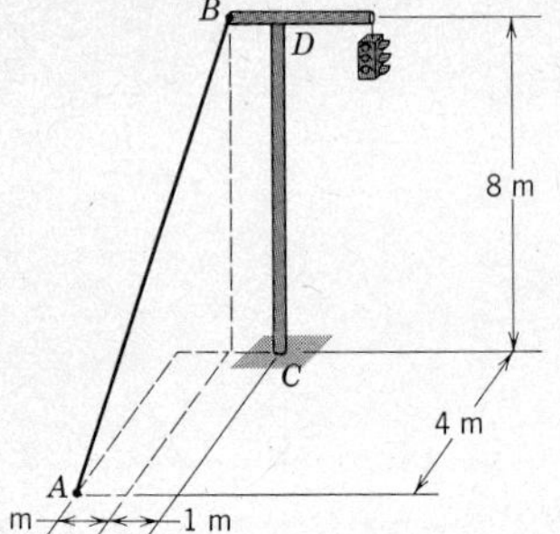

Prob. III.22

III.22 A traffic light is steadied by cable AB. If the tension in this cable is 4 kN, determine (a) the moment about the base C of the force the cable applies to the pole, (b) the moment of this force about the axis of pole CD.

III.23 A socket wrench is used to tighten a screw. A 120-N horizontal force is applied to the handle as shown. (a) Determine the moment of this force about the head of the screw. (b) What portion of this moment has the effect of tightening the screw? (c) What are the lever arms of the force about the axis of the screw and about the screw head?

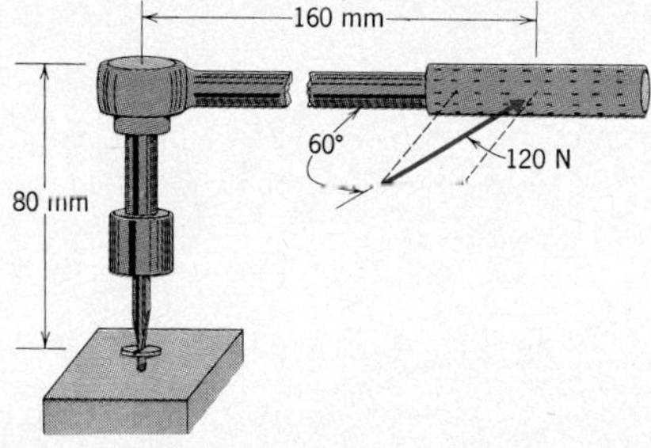

Prob. III.23

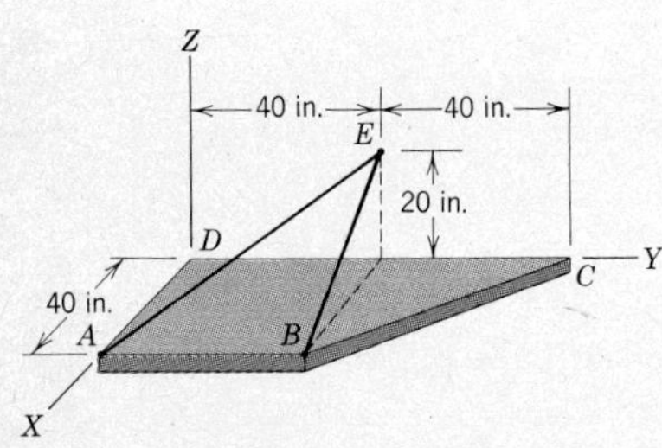

Prob. III.24

III.24 Cables AE and BE are used to support the trapezoidal plate. The tension in cable AE is 3 kips and that in cable BE is 2 kips. Determine the moment of the force that each cable exerts on the plate (a) about corner D, (b) about corner C.

III.25 In Problem III.24, determine the moment about edge BC of the force exerted by each of the cables on the plate.

III.26 The rectangular sheet of plywood is held in position by the cable. The tension in the cable is 150 N. Determine the moment of the force exerted by the cable on the plywood (a) about corner A, (b) about edge AB.

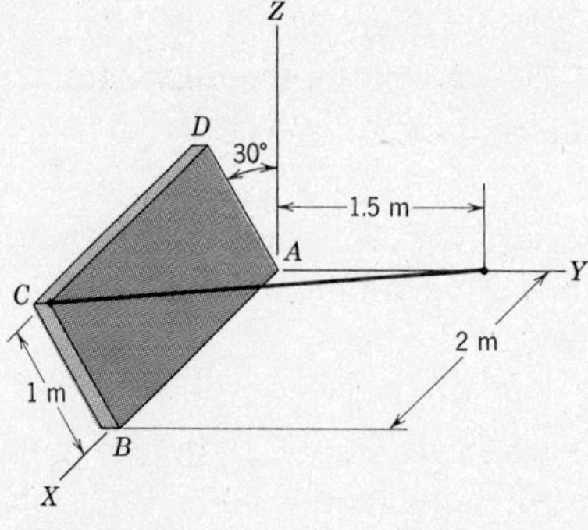

Prob. III.26

III.27 In Problem III.26, determine the moment of the force exerted by the cable on the plywood sheet (a) about corner D, (b) about edge AD.

III.28 A 1.5-kN force is applied to the pipe assembly as shown. Determine the moment of this force about each of the joints (points A, B, and C).

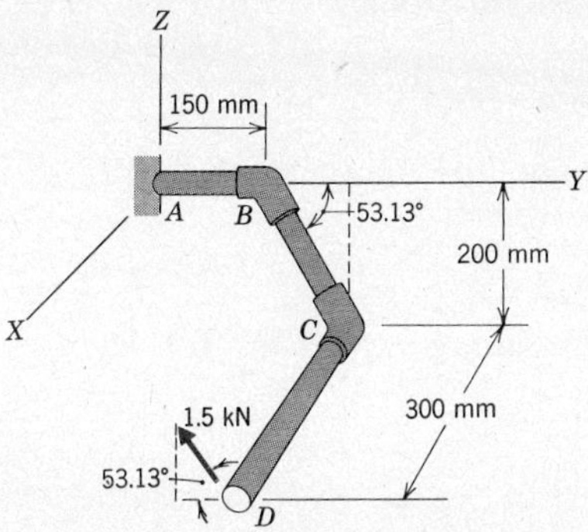

Prob. III.28

III.29 In Problem III.28, determine the moment of the 1.5-kN force about the axis of pipe segment AB and about the axis of pipe segment BC. Why is the latter value zero?

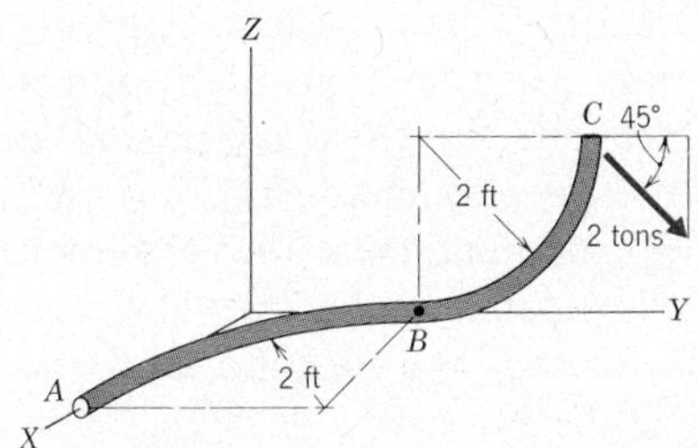

Prob. III.30

III.30 A 2-ton force is applied as shown to one end of the curved rod. Section AB coincides with the XY plane, whereas section BC coincides with the YZ plane. Determine the moment of this force (a) about the welded end A of the bar, (b) about the midpoint B.

III.31 In Problem III.30, determine the moment of the 2-ton force (a) about line AB, (b) about line BC. Why is the latter value zero?

D. FORCE COUPLES

We have now established how the magnitude, direction, and line of action of a force are characterized mathematically. With this knowledge as a foundation, the next sequence of topics we will study is how to describe the total effect of a system of forces acting on a rigid body. Our concern here is solely with the way in which a body in influenced by its surroundings. In later modules we will study the force interaction of individual elements of a rigid body.

The most fundamental system of forces is formed by two forces that are equal in magnitude and opposite in direction. Such a system is said to form a *couple*. Figure 12 depicts a couple formed by a force $\bar{F}$ applied at point A and a force $-\bar{F}$ applied at point B. For convenience the diagram is drawn in the plane of the forces.

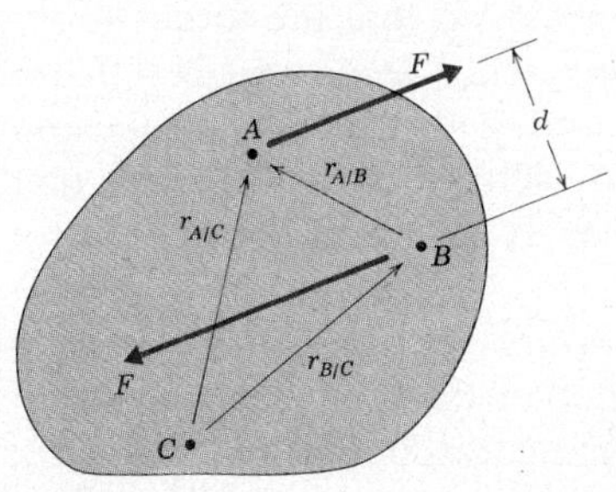

Figure 12

If the forces in this figure were collinear, we know from our study of concurrent force systems that $\bar{F}$ and $-\bar{F}$ would have a zero resultant, thereby cancelling each other. In the illustrated situation the forces still do not push or pull on the body; they only exert a twisting effect. Let us compute the total moment these two forces exert about an arbitrary point C, which need not necessarily be coplanar with the forces. Using the position vectors shown in Figure 12, we have

$$\bar{M}_C = \bar{r}_{A/C} \times \bar{F} + \bar{r}_{B/C} \times (-\bar{F}) = (\bar{r}_{A/C} - \bar{r}_{B/C}) \times \bar{F}$$

However, the diagram shows that

$$\bar{r}_{A/C} = \bar{r}_{B/C} + \bar{r}_{A/B} \quad \text{so} \quad \bar{r}_{A/B} = \bar{r}_{A/C} - \bar{r}_{B/C}$$

The foregoing expression for M_C then becomes

$$\boxed{\bar{M}_C = \bar{r}_{A/B} \times \bar{F}} \tag{13}$$

This result has several noteworthy features. First, we see that only the relative position of the points of application of the couple, as expressed by $\bar{r}_{A/B}$, affects the resultant moment. Thus, regardless of which point is selected for summing moments, a couple exerts the same moment. We call this resultant moment the torque of the couple, and denote it by $\bar{M}$. Furthermore, because the component of $\bar{r}_{A/B}$ perpendicular to $\bar{F}$ is shown in Figure 12 to be the perpendicular distance d between the lines of action of the forces forming the couple, it follows that the magnitude of the torque is given by

$$\boxed{|\bar{M}| = |\bar{F}|d} \tag{14}$$

Finally, the direction of the torque is perpendicular to the plane formed by the couple forces, in the sense obtained from the right-hand rule when the fingers are curled in the direction of the forces.

From these observations we can conclude that different couples can have equivalent effects on a rigid body provided that the forces forming the couples are in parallel planes. Equivalence then requires that the products $|\bar{F}|d$ for the couples are the same, and that their torques have the same sense. Typical examples of equivalent couples are shown on the following page in Figure 13a for a planar system and Figure 13b for a three-dimensional system.

It is important to realize that the torques of individual couples are additive as vectors. This is a consequence of the fact that the torque is simply the total moment of the forces forming the couple, and a moment is a vector quantity. One result of this statement is that we may describe the

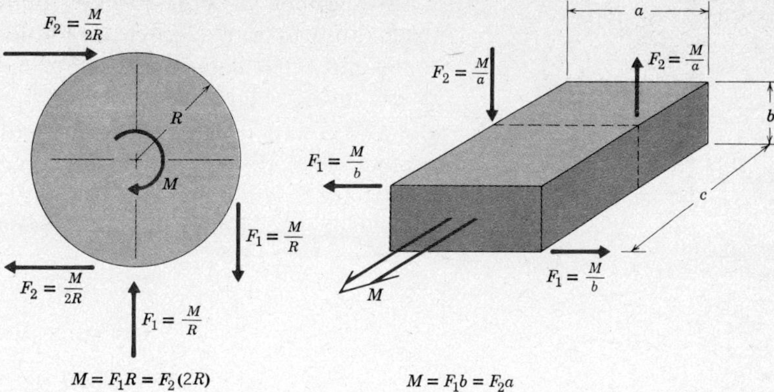

Figure 13a **Figure 13b**

torque of a couple as the sum of its components with respect to a Cartesian coordinate system; these components represent the rotational tendency about each coordinate axis.

The foregoing discussion on components of couples can be generalized as follows:

> A set of couples acting on a rigid body is equivalent to a single couple whose torque is the sum of the moments, about an arbitrary point, of all forces forming the couples.

EXAMPLE 4

Determine the smallest forces applied at ends A and B of the beam shown that are equivalent to the given loading.

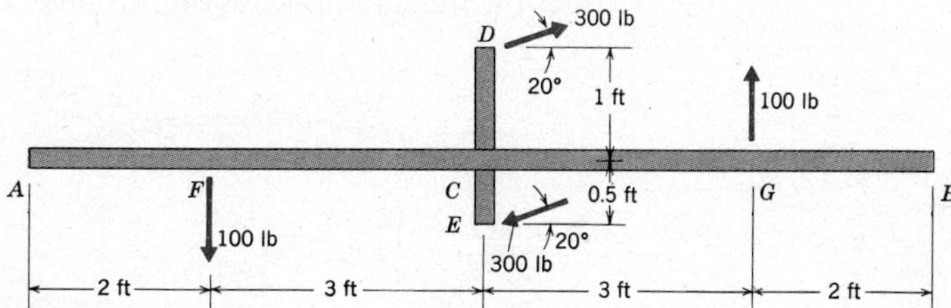

Solution

The planar force system consists of two couples formed by the 100-lb forces and the 300-lb forces. In order to find an equivalent couple we must determine the total torque of the given couples. This torque may be evaluated by summing the moments of each force about any point; the geometry of the system suggests that we use point C.

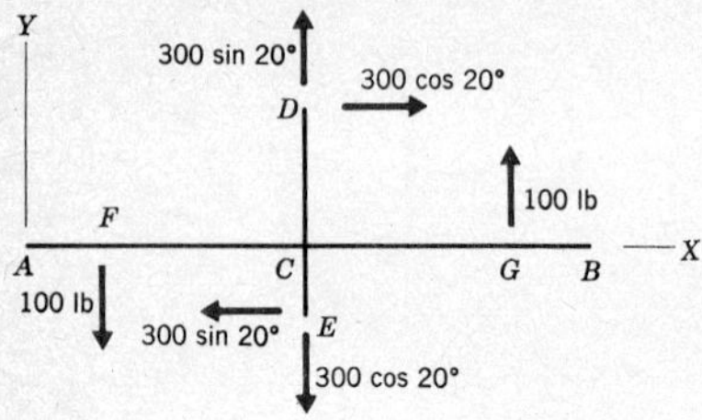

The points of application of the forces are located by horizontal and vertical dimensions. Hence we choose a coordinate system having horizontal and vertical axes. The coordinate system and force components are shown in the sketch.

The Z axis is outward for a right-hand coordinate system, so counterclockwise moments are positive according to the right-hand rule. The vertical force components at points D and E pass through point C, making their lever arms about this point zero. Hence, the resultant torque is

$$\begin{aligned} M &= -300 \cos 20°(1.0) - 300 \cos 20°(0.5) \\ &\quad + 100(3.0) + 100(3.0) \\ &= 177.14 \text{ lb-ft} \end{aligned}$$

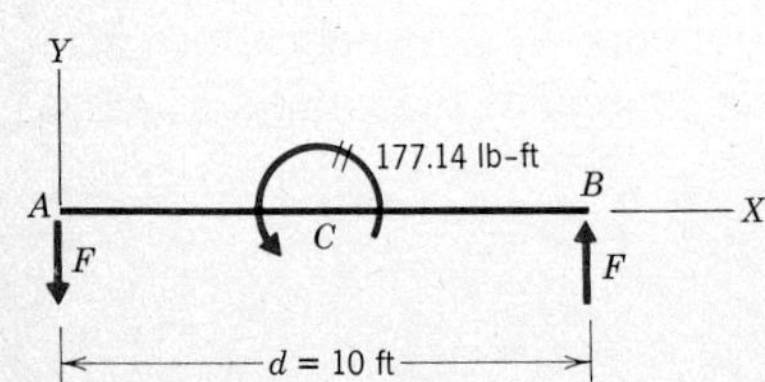

The problem now is to form a 177.14 lb-ft couple from the smallest forces $\bar{F}$ at point A and $-\bar{F}$ at point B we can find. The magnitude of the torque of this couple is $|\bar{F}|d$, so we must align $\bar{F}$ and $-\bar{F}$ to obtain the largest value for d. As shown in the sketch, this situation is obtained when $\bar{F}$ is perpendicular to the beam. Notice that the sense of the forces is determined from the requirement that they form a counterclockwise couple. From the sketch we then have

$$M = 177.14 = |\bar{F}|d = 10|\bar{F}|$$

$$|\bar{F}| = 17.714 \text{ lb}$$

$$\bar{F} = -17.714\bar{J} \text{ lb}$$

EXAMPLE 5

A bus driver applies two 60-N forces to the steering wheel as shown. What is the torque of this couple, and what portion of this torque has the effect of turning the steering wheel?

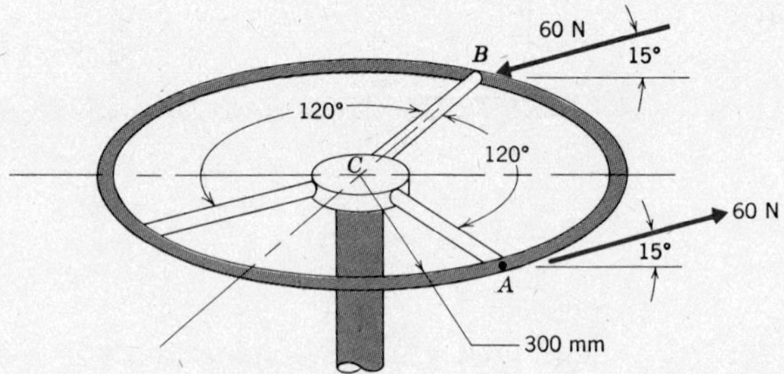

Solution

We will evaluate the torque of the couple by summing moments about the center point C, for the position vectors from this point to the points of

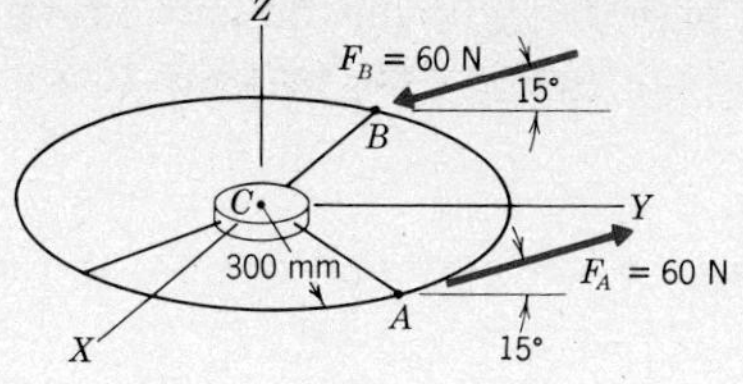

application of the forces are easily described. The XYZ reference frame shown in the sketch is chosen because points A, B, and C then all coincide with a coordinate plane, and also for the convenience it affords in describing force components.

Denoting the forces at points A and B as $\bar{F}_A$ and $\bar{F}_B$, we have

$$\bar{F}_A = -\bar{F}_B = 60(\cos 15^\circ\ \bar{J} + \sin 15^\circ\ \bar{K})$$
$$= 57.96\bar{J} + 15.53\bar{K}\ \text{N}$$

The position vectors from point C to the forces are

$$\bar{r}_{B/C} = -0.30\bar{I}\ \text{m}$$
$$\bar{r}_{A/C} = 0.30(\sin 30^\circ\ \bar{I} + \cos 30^\circ\ \bar{J})$$
$$= 0.15\bar{I} + 0.2598\bar{J}\ \text{m}$$

Hence, the torque of the couple is

$$\bar{M} = \bar{r}_{A/C} \times \bar{F}_A + \bar{r}_{B/C} \times \bar{F}_B$$
$$= (0.15\bar{I} + 0.2598\bar{J}) \times (57.96\bar{J} + 15.53\bar{K})$$
$$+ (-0.30\bar{I}) \times (-57.96\bar{J} - 15.53\bar{K})$$
$$= 0.15(57.96)\bar{K} + 0.15(15.53)(-\bar{J})$$
$$+ 0.2598(15.53)(\bar{I}) + 0.30(57.96)\bar{K}$$
$$+ 0.30(15.53)(-\bar{J})$$
$$= 4.03\bar{I} - 6.99\bar{J} + 26.08\bar{K}\ \text{N-m}$$

Only the component of the torque that is parallel to the shaft of the steering wheel has the effect of turning the wheel. This is the Z component of $\bar{M}$.

$$M_Z = 26.08\ \text{N-m}$$

Furthermore, because M_Z is positive, the driver is trying to turn the wheel counterclockwise looking down the Z axis.

HOMEWORK PROBLEMS

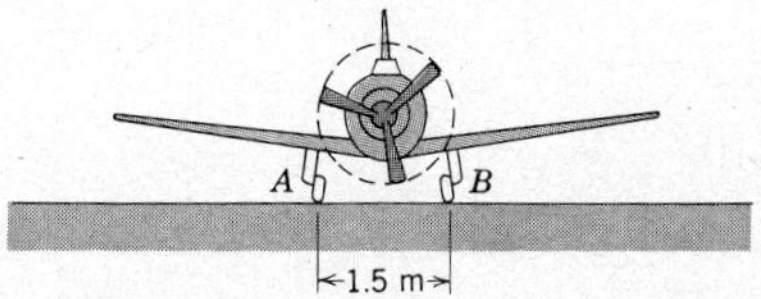

Prob. III.32

III.32 The drive shaft from the engine applies a counterclockwise torque, as viewed from the front, of 1.8 kN-m to the propeller of the aircraft shown. Determine vertical reactions between the wheels and the ground at points A and B that are equivalent to this couple.

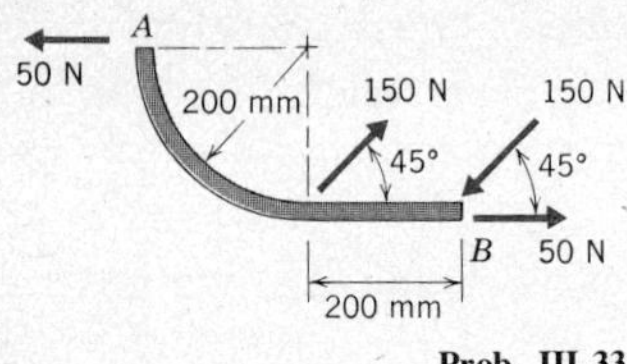

Prob. III.33

III.33 Determine the smallest forces at ends A and B of the curved bar that are equivalent to the two couples shown. Specify the magnitude and direction for these forces.

III.34 The three couples shown exert a resultant moment of 60 ft-lb on the wheel. Determine the magnitude of $\bar{F}$.

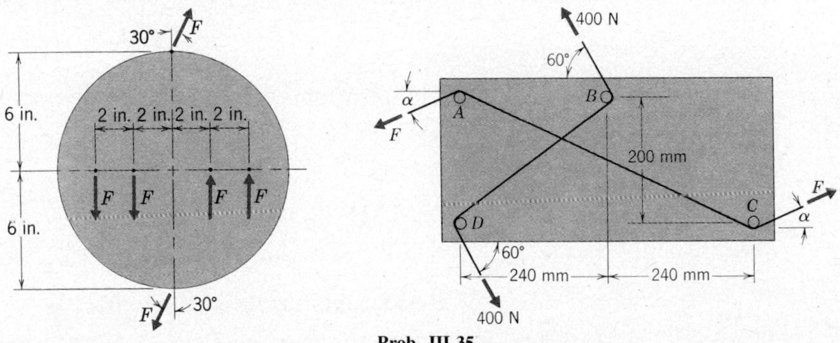

Prob. III.34

Prob. III.35

III.35 Cables are passed over smooth pegs in the block as shown. The tension in the cable passing over pegs B and D is 400 N. Knowing that the tension in the cable passing over pegs A and C is 300 N and that $\alpha = 30°$, determine the couple $\bar{M}$ that is equivalent to the force system exerted by the cables on the block.

III.36 In Problem III.35, determine the minimum value of F and the corresponding angle α for which the effect of the four cables on the block is a counterclockwise couple whose torque is 250 N-m.

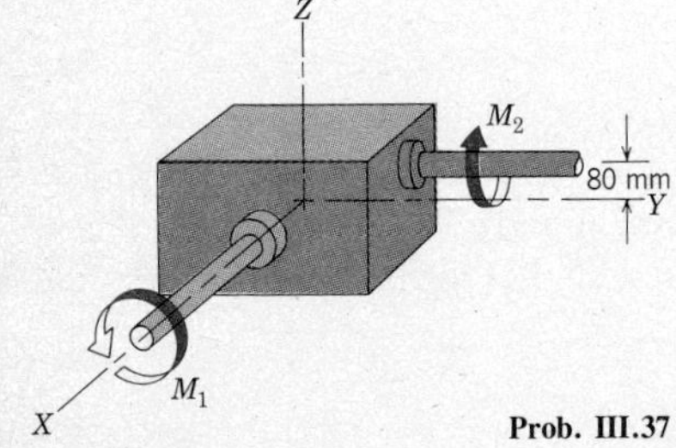

Prob. III.37

III.37 The drive shaft applies a torque $M_1 = 5$ kN-m to the gear box and the driven shaft applies a torque $M_2 = 15$ kN-m, as shown. Determine the resultant couple acting on the gear box, specifying its magnitude and its direction angles with respect to the XYZ coordinate system.

III.38 The wide flange beam shown is loaded by four forces. Determine the couple $\bar{M}$ that is equivalent to this loading.

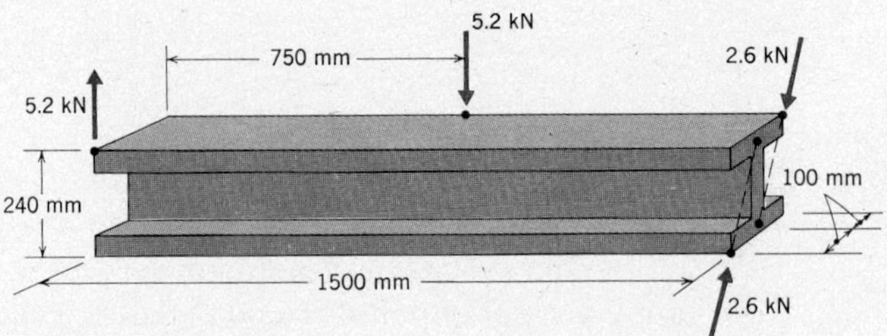

Prob. III.38

III.39 The four forces acting on the rectangular parallelepiped are equivalent to a couple $\bar{M}$. Determine the magnitude of $\bar{F}$, the angle α, and $\bar{M}$.

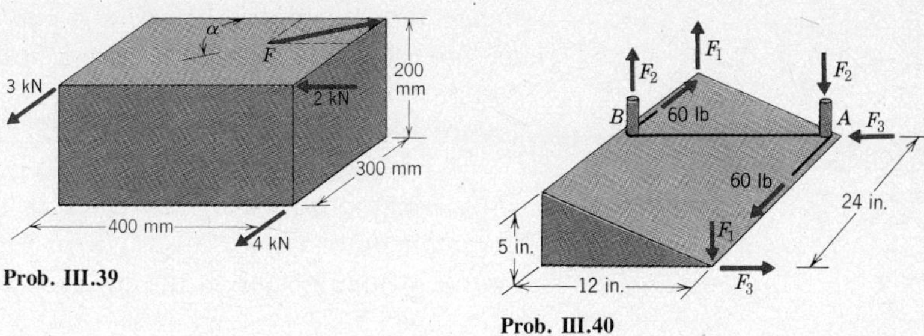

Prob. III.39

Prob. III.40

III.40 A cord is passed over the smooth pegs A and B on the triangular block. Tensile forces of 60 lb are applied at its ends. Determine the magnitude of the forces $\bar{F}_1$, $\bar{F}_2$, and $\bar{F}_3$ for which the resultant couple acting on the block is zero.

E. RESULTANT OF A GENERAL SYSTEM OF FORCES

In the preceding section we showed that the effect of a couple acting on a rigid body is a torque representing the twisting tendency of the body. Intuitively, we know that a general system of forces can also have a rotational effect, as well as having the effect of pushing or pulling a body.

Because we have already seen how to obtain the resultant of a concurrent system of forces, let us begin by investigating how noncurrent forces may be transferred to a common point of application. Consider a single force $\bar{F}$ applied to point P of a rigid body, as shown in Figure 14a. It is our desire to transfer $\bar{F}$ to the arbitrary point C.

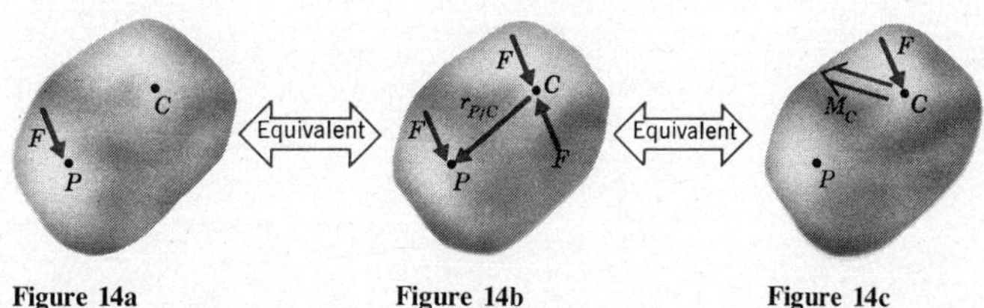

Figure 14a **Figure 14b** **Figure 14c**

To achieve the transfer we apply two forces $\bar{F}$ and $-\bar{F}$ at point C. These two forces are concurrent. Their resultant is zero, and therefore the force system in Figure 14b is equivalent to the force $\bar{F}$ in Figure 14a. However, the force $\bar{F}$ at point P and the force $-\bar{F}$ at point C form a couple

$\bar{M}_C$ which is given by

$$\bar{M}_C = \bar{r}_{P/C} \times \bar{F} \tag{15}$$

This couple is simply the moment of the force $\bar{F}$ about point C. This means that *the original force $\bar{F}$ is equivalent to the force $\bar{F}$ at point C and the couple $\bar{M}_C$, which is the moment of $\bar{F}$ about C*. This is called a *force-couple system*. The resulting equivalent force-couple system is shown in Figure 14c.

Now consider the problem of a general force system, such as the one shown in Figure 15a. Each force in the system may be transferred to a common point C, as is shown in Figure 15b, where $(\bar{M}_C)_i$ denotes the couple whose torque is the moment of force $\bar{F}_i$ about point C.

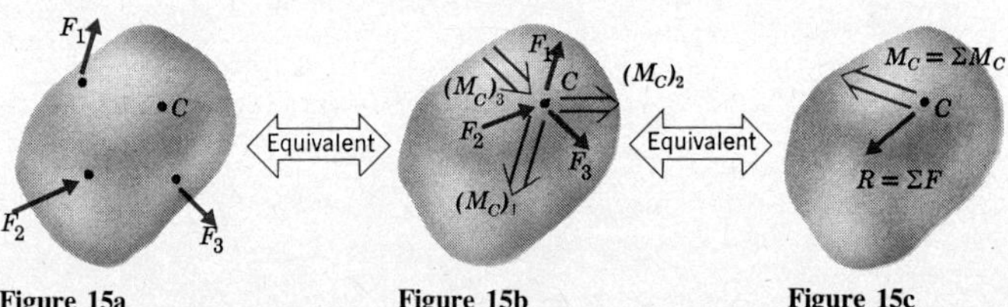

Figure 15a **Figure 15b** **Figure 15c**

The forces are now concurrent, so they may be summed to obtain their resultant $\bar{R}$. Further, because couples are additive, the individual couples $(\bar{M}_C)_i$ may be summed vectorially to obtain the resultant couple $\bar{M}_C$. The value of $\bar{M}_C$ is the sum of the moments about point C of the forces in the original system. It then follows that the original system of forces is equivalent to the force-couple system in Figure 15c. In other words:

A general system of forces is equivalent to a force $\bar{R}$ applied at any point C, and a couple $\bar{M}_C$ about point C, provided that

$$\bar{R} = \Sigma\,\bar{F} \qquad \bar{M}_C = \Sigma\,\bar{M}_C \tag{16}$$

where $\Sigma\,\bar{F}$ denotes the sum of the forces in the original system and $\Sigma\,\bar{M}_C$ denotes the sum of the moments of the forces in the original system about point C.

Clearly, the resultant force $\bar{R}$ in the equivalent force-couple system is independent of the choice for point C; but the resultant couple $\bar{M}_C$ depends on which point we choose. This observation leads to a corollary of equation (16). *Two different force systems have equivalent effects on a rigid body if the forces in each system exert the same total moment about any point, as well as having identical sums.*

EXAMPLE 6

Replace the force and couple by an equivalent force-couple system at end B of the curved beam.

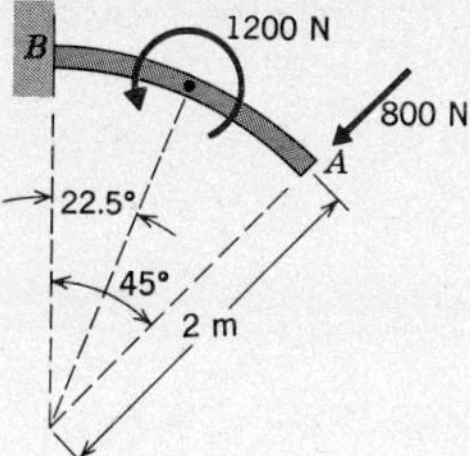

Solution

There is only one force in the given system, so the resultant $\bar{R}$ applied at point B is simply the 800-N force. The resultant couple $\bar{M}_B$ is the moment of the given force system about point B. When summing moments, we use the fact that a couple exerts the same moment about any point. Thus, the 22.5° angle is not needed in the solution. As an aid, we draw a sketch of the force system, choosing the Y axis vertical because angles are referred to this line.

The geometry is circular. Hence the lever arm of the 800-N force about point B is easily determined, as shown in the sketch. Noting that the Z axis is outward, clockwise moments are negative by the right-hand rule. Therefore,

$$M_{Bz} = -800(2 \sin 45°) + 1200 = 68.63 \text{ N-m}$$

In terms of the chosen coordinate system, the resultant force $\bar{R}$ is

$$\bar{R} = \Sigma \bar{F} = 800(-\sin 45° \, \bar{I} - \cos 45° \, \bar{J})$$

$$= -565.6(\bar{I} + \bar{J}) \text{ N}$$

The equivalent system is illustrated in the margin.

EXAMPLE 7

Determine the force-couple system at point A that is equivalent to the forces acting on the socket wrench shown.

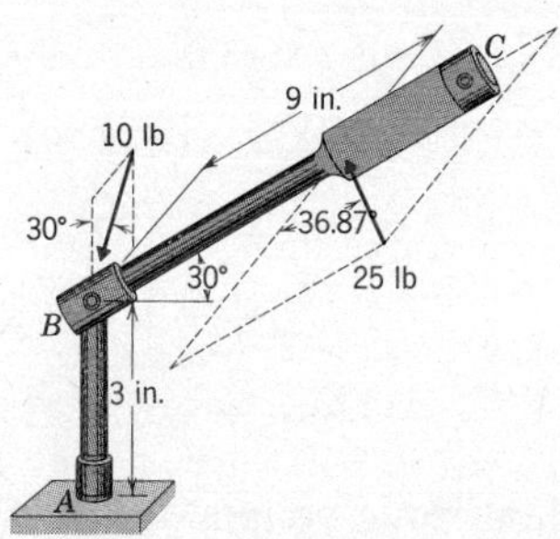

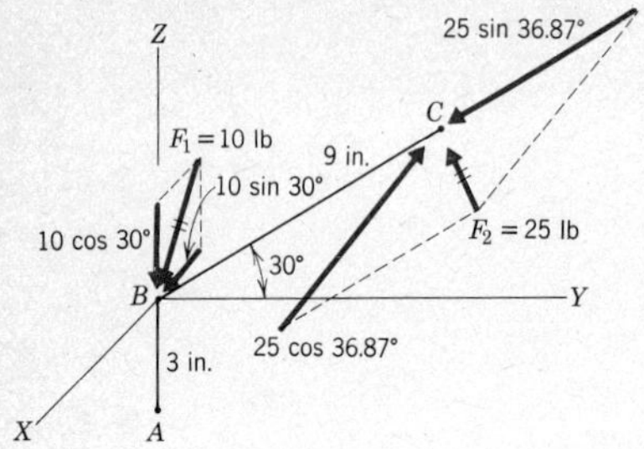

Solution

We begin with a sketch showing the forces and the XYZ reference frame, whose orientation is chosen to fit the manner in which the dimensions are given. The force components shown in the sketch are the same as those in the given diagram.

We must represent all forces by their XYZ components. The appropriate expression for $\bar{F}_1$ is easily obtained from the sketch.

$$\bar{F}_1 = (10 \sin 30° \bar{I} - 10 \cos 30° \bar{K}) = 5.0\bar{I} - 8.66\bar{K} \text{ lb}$$

The illustrated components of $\bar{F}_2$ are described with respect to bar BC. The component parallel to this bar is in the YZ plane, at an angle of 30° from the negative Y direction, whereas the component perpendicular to the bar is in the negative X direction. Hence

$$\begin{aligned}\bar{F}_2 &= (25 \sin 36.87°)(-\cos 30° \bar{J} - \sin 30° \bar{K}) - (25 \cos 36.87°)\bar{I} \\ &= -12.99\bar{J} - 7.5\bar{K} - 20.0\bar{I} \text{ lb}\end{aligned}$$

The equivalent force system we seek consists of the resultant force $\bar{R}$ applied at point A and the resultant couple $\bar{M}_A$. To find $\bar{R}$ we sum forces.

$$\begin{aligned}\bar{R} = \bar{F}_1 + \bar{F}_2 &= (5.0 - 20.0)\bar{I} - 12.99\bar{J} + (-8.66 - 7.5)\bar{K} \\ &= -15.00\bar{I} - 12.99\bar{J} - 16.16\bar{K} \text{ lb}\end{aligned}$$

To find $\bar{M}_A$ we sum moments about point A.

$$\bar{M}_A = \bar{r}_{B/A} \times \bar{F}_1 + \bar{r}_{C/A} \times \bar{F}_2$$

A displacement from point A to point B involves a movement of 3 in. in the positive Z direction, so

$$\bar{r}_{B/A} = 3.0\bar{K} \text{ in.}$$

Continuing to point C from point B requires an additional movement of 9 in. parallel to bar BC at an angle of 30° above the Y axis. Thus

$$\bar{r}_{C/A} = (3 + 9 \sin 30°)\bar{K} + 9 \cos 30° \bar{J} = 7.5\bar{K} + 7.794\bar{J} \text{ in.}$$

The moment is now found to be

$$\begin{aligned}\bar{M}_A &= (3.0\bar{K}) \times (5.0\bar{I} - 8.66\bar{K}) + (7.5\bar{K} \\ &\quad + 7.794\bar{J}) \times (-12.99\bar{J} - 7.5\bar{K} - 20.0\bar{I}) \\ &= 3.0(5.0)(\bar{J}) - 7.5(12.99)(-\bar{I}) \\ &\quad - 7.5(20.0)(\bar{J}) - 7.794(7.5)(\bar{I}) - 7.794(20.0)(-\bar{K}) \\ &= -39.0\bar{I} - 135.0\bar{J} + 155.9\bar{K} \text{ lb-in.}\end{aligned}$$

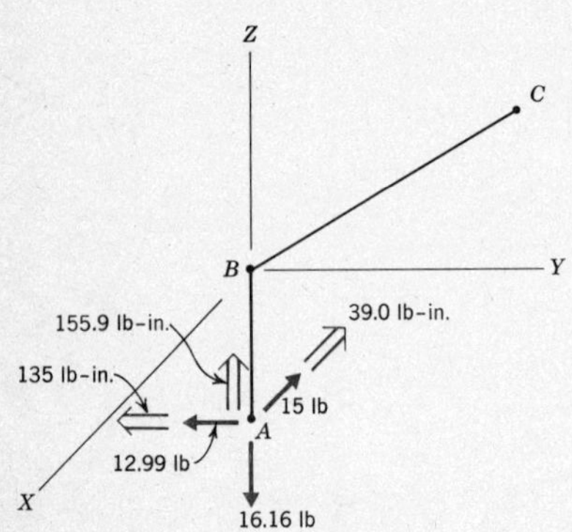

The equivalent force-couple system is sketched in the left margin.

HOMEWORK PROBLEMS

III.41 A force of 100 N is applied as shown to the tire wrench. Determine the equivalent force-couple system acting (a) at point A, (b) at the center of the wheel.

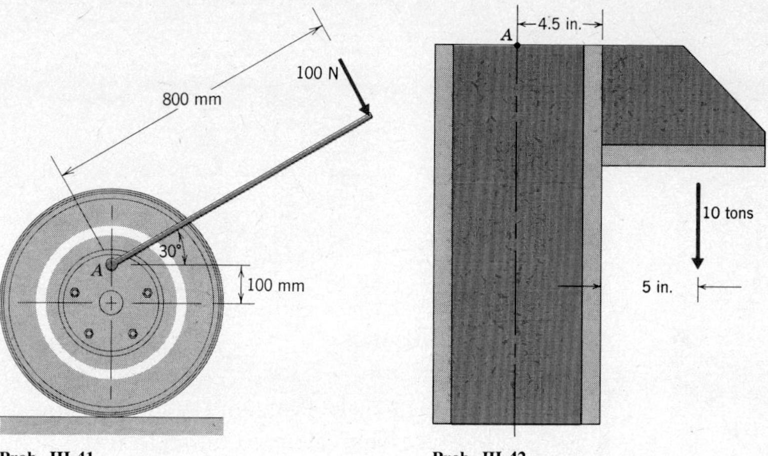

Prob. III.41

Prob. III.42

III.42 A 10-ton load is applied to the bracket, which is welded onto the I-beam. Determine the equivalent force-couple system acting at point A on the centerline of the I-beam.

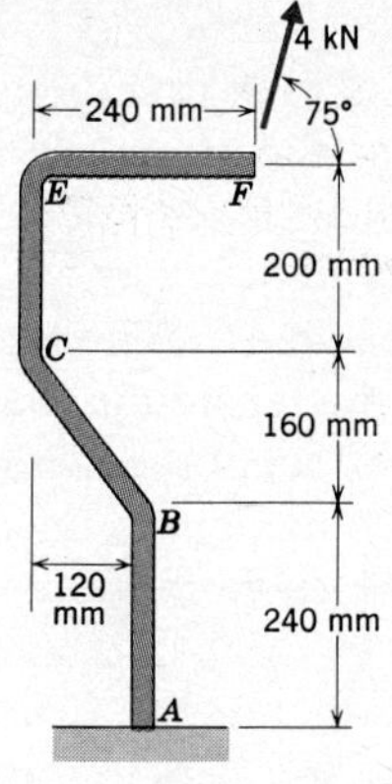

Prob. III.43

III.43 A 4-kN force is applied to the bent bar as shown. Determine an equivalent force-couple system (a) at point A, (b) at point B, (c) at point C.

III.44 A cable is passed over two ideal pulleys that are pinned to a block. The tension within the cable is 300 N. Determine an equivalent force-couple system acting (a) at pin A, (b) at pin B.

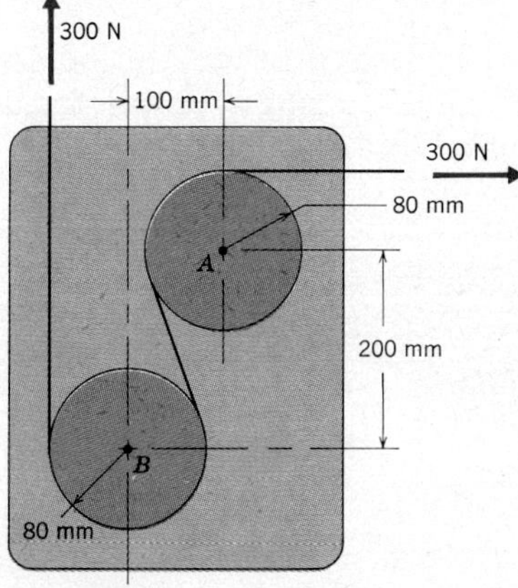

Prob. III.44

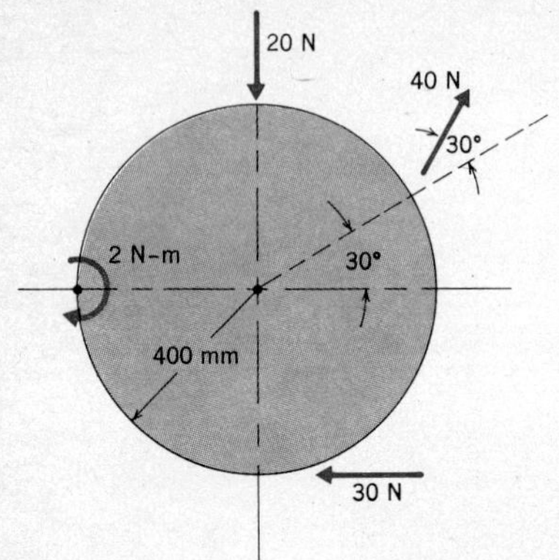

Prob. III.45

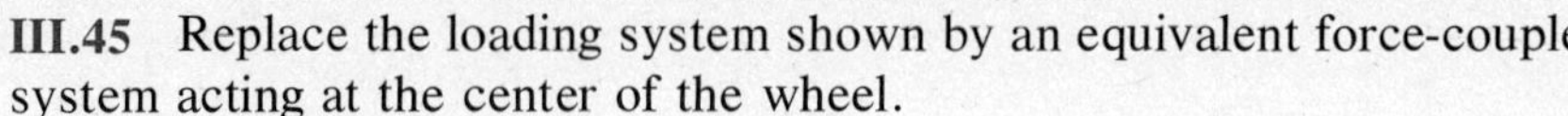

III.45 Replace the loading system shown by an equivalent force-couple system acting at the center of the wheel.

III.46 The stepped gear is subjected to two forces as shown. Knowing that $F = 200$ N, determine the force-couple system acting at the center C of the gear that is equivalent to these forces.

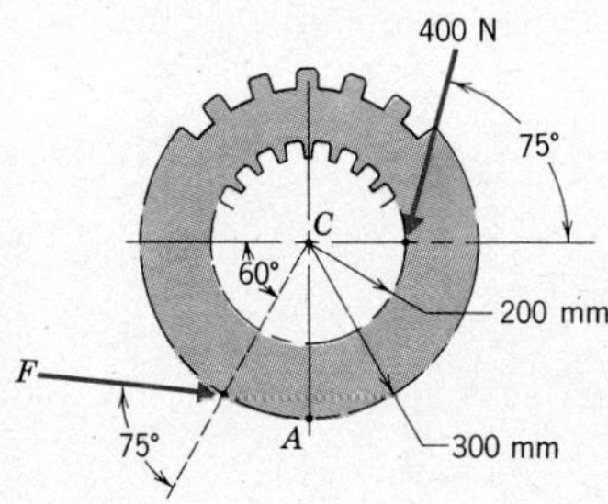

Prob. III.46

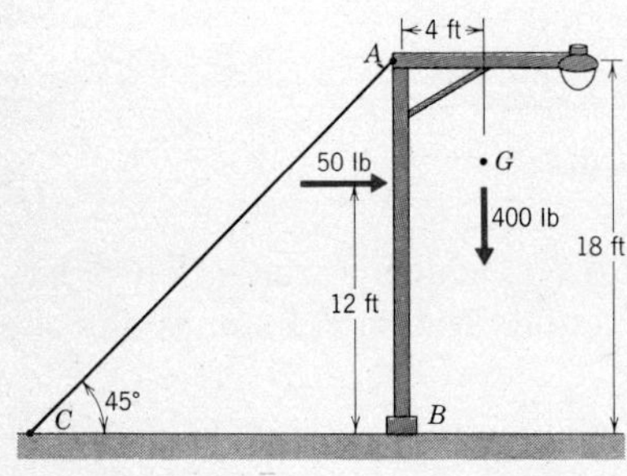

Prob. III.48

III.47 In Problem III.46, the value of F is such that the two given forces are equivalent to a single force acting at the center C. Determine this value of F and the corresponding equivalent force at C.

III.48 The street light is supported by cable AC. The total weight of the structure is 400 lb acting at point G, and the tension in the cable is 900 lb. The 50-lb horizontal force represents the effect of the wind. Determine the force-couple system at the base B that is equivalent to the combined effect on the structure of gravity, wind, and the cable tension.

III.49 In Problem III.48, the combined effect on the structure of gravity, wind, and the cable tension is equivalent to a single force acting at the base B. Determine the tension in the cable and the equivalent single force.

III.50 It is desired to replace the 400-kN force shown by two parallel forces $\bar{F}_A$ and $\bar{F}_B$ acting at points A and B on the shelf bracket. Determine the required forces $\bar{F}_A$ and $\bar{F}_B$.

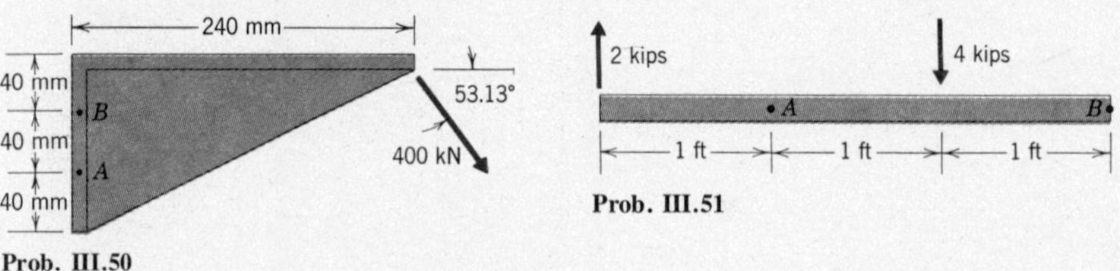

Prob. III.50

Prob. III.51

III.51 Determine parallel forces acting at points A and B that are equivalent to the vertical loads shown.

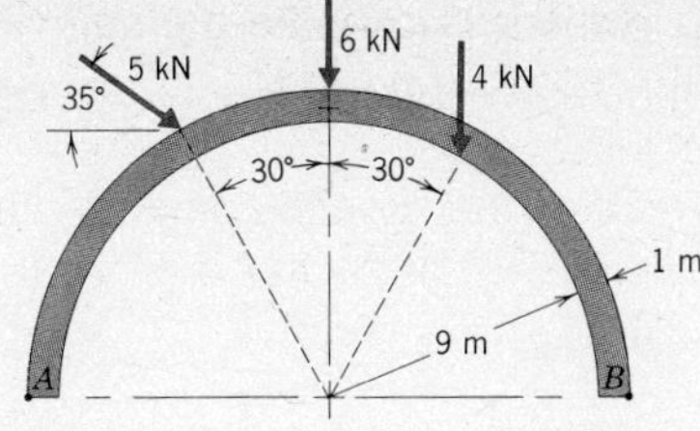

Prob. III.52

III.52 Determine the force-couple system at corner A of the circular arch that is equivalent to the system of forces shown.

III.53 In Problem III.52, it is desired to replace the system of forces shown acting on the semicircular arch by a vertical force $\bar{F}_A$ at point A and a force $\bar{F}_B$ at point B having an equivalent effect. Determine $\bar{F}_A$ and $\bar{F}_B$.

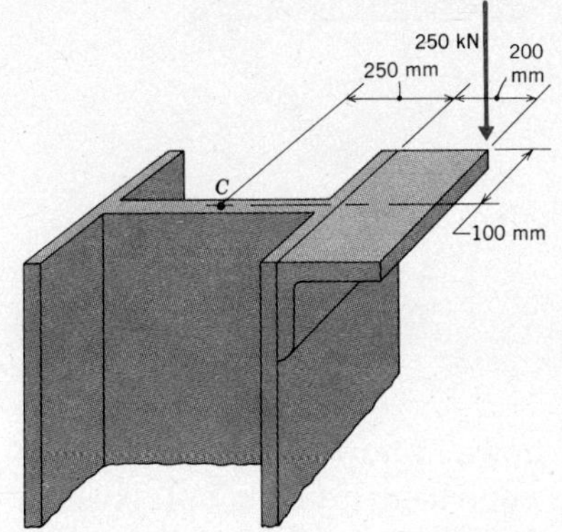

Prob. III.54

III.54 A 250-kN force is applied to the bracket, which is welded to the vertical column. Determine the equivalent force-couple system acting at point C on the centerline of the column.

III.55 A 40-lb horizontal force is applied to the wrench. Determine the equivalent force-couple system acting (a) at joint A, (b) at joint B. The pipe assembly is parallel to the vertical plane.

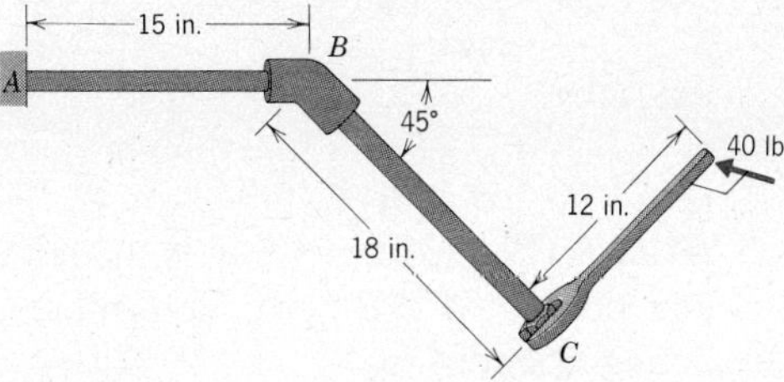

Prob. III.55

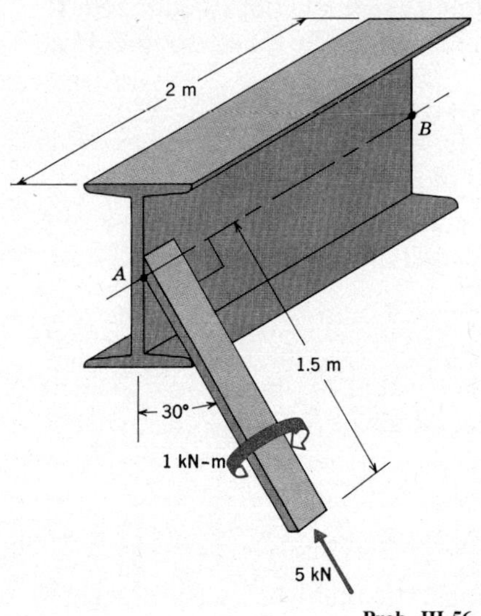

Prob. III.56

III.56 A bar is welded to the end of the I-beam. A force and a couple are applied to the end of the bar. Determine the equivalent force-couple system acting (a) at point A on the centerline of the beam, (b) at point B on the centerline of the beam.

III.57 A valve is tightened with the T-bar assembly by applying 75-N forces at points B and C as shown. (a) Replace these forces by an equivalent force-couple system acting at point A. (b) What portion of this equivalent system has the effect of turning the valve about axis AD?

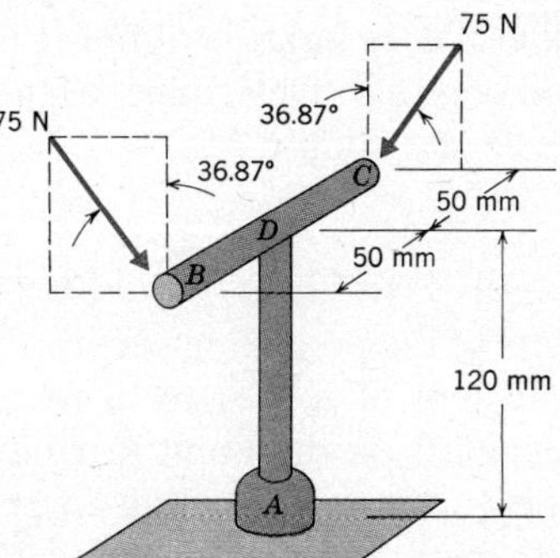

Prob. III.57

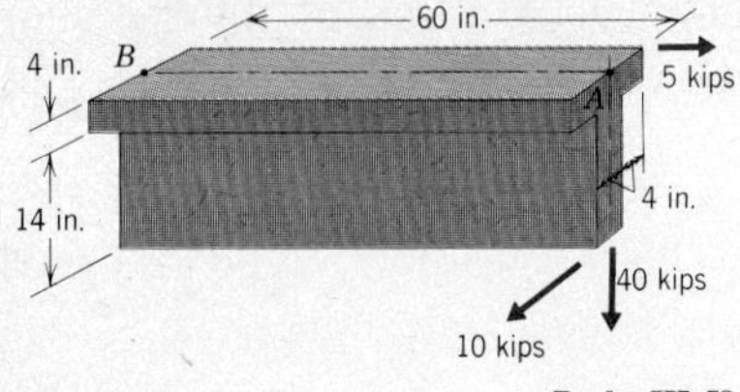

Prob. III.58

III.58 The concrete T-beam is loaded as shown. Determine the equivalent force-couple system acting (a) at point A, (b) at point B.

III.59 Replace the forces acting on the rectangular parallelepiped by an equivalent force-couple system (a) at corner A, (b) at corner B.

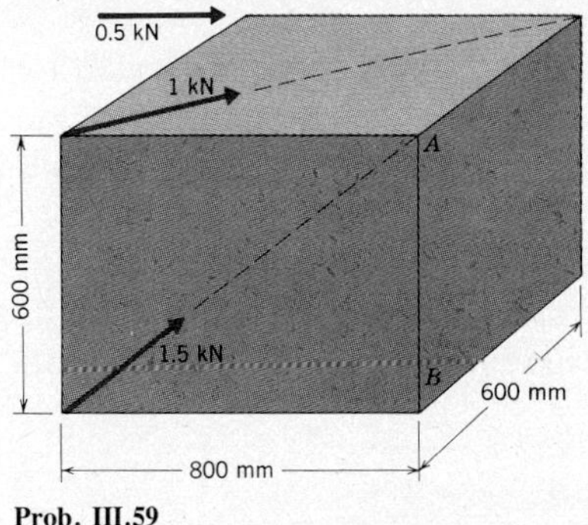

Prob. III.59

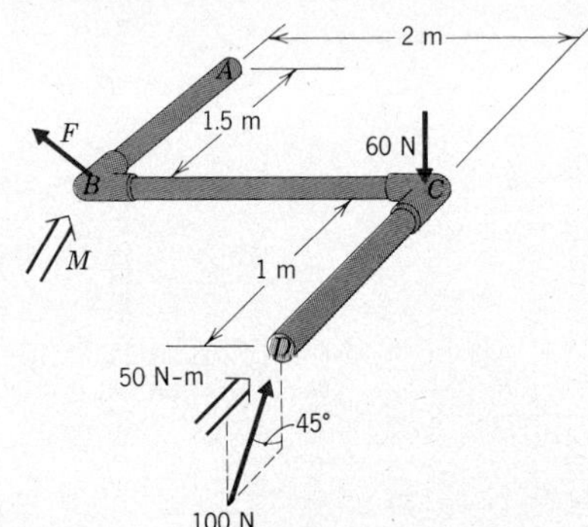

Prob. III.60

III.60 The pipe assembly is loaded as shown. Knowing that the force $\bar{F}$ and couple $\bar{M}$ are zero, determine the equivalent force-couple system acting at end A.

III.61 In Problem III.60, the force-couple system at end A equivalent to the force system shown is zero. Determine the force $\bar{F}$ and couple $\bar{M}$ at joint B.

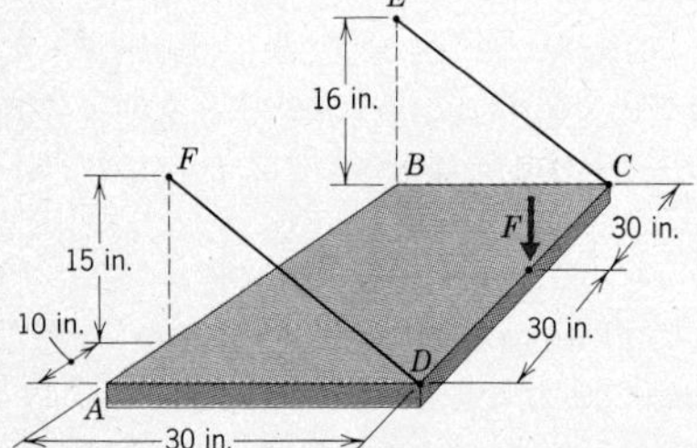

Prob. III.62

III.62 The tension in cable CE is 3400 lb and the tension in cable DF is 4200 lb. The force $\bar{F}$ is a downward load of 1000 lb. Determine the force-couple system acting at corner A that is equivalent to the effect on the plate of the cables and $\bar{F}$.

III.63 In Problem III.62, the vertical force $\bar{F}$ and the given cable tensions are equivalent to a force $\bar{R}$ and couple $\bar{M}_A$ acting at corner A. The magnitude of $\bar{F}$ is unknown, but it is known that $\bar{M}_A$ has no component parallel to edge AB. Determine (a) the magnitude of $\bar{F}$, (b) the corresponding force $\bar{R}$ and couple $\bar{M}_A$.

F. SINGLE FORCE EQUIVALENTS

The results of the preceding section demonstrate that when we represent a general system of forces by a force-couple system, the torque of the couple depends on the point to which we transfer the forces. This gives rise to the question of whether there is some point O that eliminates the couple, thereby enabling us to replace a general system of forces by a single force.

Suppose that we have determined the force $\bar{R}$ at point C and the couple $\bar{M}_C$ which are the equivalent of a system of forces. We may transfer this resultant to another point O, such as that shown in Figure 16. Both

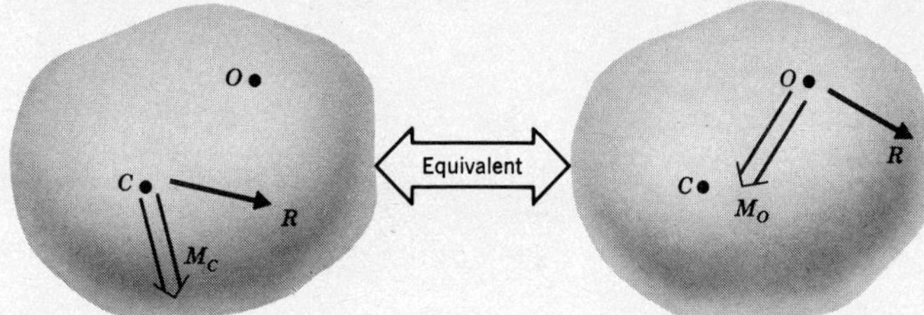

Figure 16

forces depicted in the figure are identical, so the two force-couple systems will be equivalent if they exert equal moments about any point. Choosing point C for the moment computation, we have

$$\bar{M}_C = \bar{M}_O + \bar{r}_{O/C} \times \bar{R} \tag{17}$$

Although equation (17) is the relationship between the couples for any two equivalent force-couple systems, our desire here is to determine the conditions for which $\bar{M}_O = \bar{0}$. Setting $\bar{M}_O$ equal to zero in equation (17) gives

$$\bar{M}_C = \bar{r}_{O/C} \times \bar{R} \tag{18}$$

In view of the properties of a cross product, equation (18) tells us that $\bar{M}_C$ must be perpendicular to $\bar{R}$ in order for $\bar{M}_O$ to be zero. Recalling the meaning of $\bar{M}_C$, it follows that a general system of forces may be replaced by a single force $\bar{R}$ acting at point O only when the moment of the forces in the original system about any point C is perpendicular to the resultant force $\bar{R}$. This condition is satisfied in three special cases.

1 Concurrent Forces

This case was studied in Module II. The point O of zero moment is any point along a line parallel to $\bar{R}$ which passes through the point of concurrency P, as illustrated in Figure 17. This follows because the vector $\bar{r}_{P/O}$ is parallel to $\bar{R}$, so

$$\bar{r}_{P/O} \times \bar{R} \equiv \bar{0} \equiv \bar{r}_{P/O} \times \Sigma\bar{F} \equiv \Sigma\bar{M}_O$$

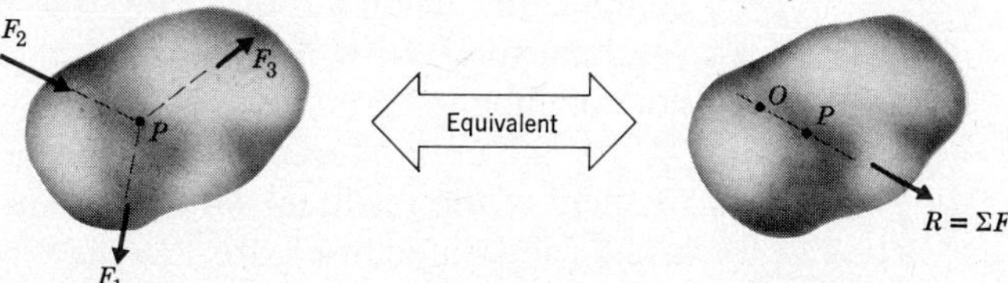

Figure 17

2 Planar Force Systems

When all forces coincide with a plane, their moment about any point in that plane is perpendicular to the plane. Consider the planar force system and its single force equivalent depicted in Figure 18.

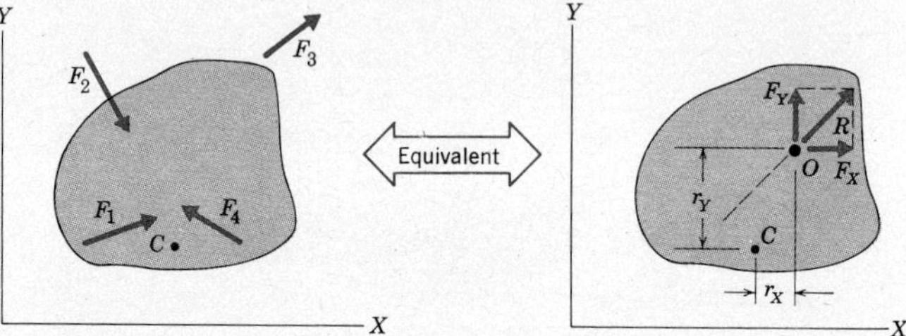

Figure 18

The components of the resultant $\bar{R}$ are the sum of the corresponding components of the original force system. To locate the point O of zero moment we sum moments about any convenient point C. Equating the moments of the forces in the two equivalent systems yields

$$\Sigma M_C = (\Sigma F_Y)r_X - (\Sigma F_X)r_Y$$

where counterclockwise moments are positive for the chosen coordinate system. This equation may be solved for the ratio of r_Y to r_X, corresponding to *any point* O along the line of action of $\bar{R}$, as indicated by the dashed line in Figure 18.

3 Parallel Forces

When all the forces are parallel to a given line, we can align one of the coordinate axes, such as the Z axis, with this line. In this case the force sum $\bar{R}$ will have only a Z component, as shown in Figure 19. A force in the

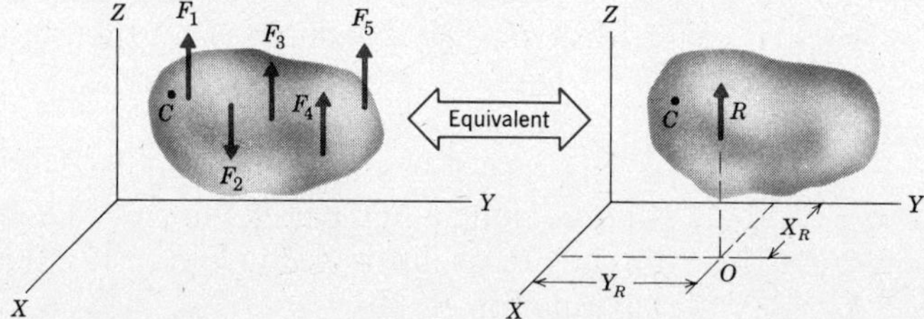

Figure 19

Z direction does not exert a moment about any line parallel to the Z axis, hence the sum of the moments about any point C will have only X and Y components. It then follows that the moment $\bar{M}_C$ is perpendicular to the resultant force $\bar{R}$. If $\bar{R}$ is nonzero, that is, if the given force system does not combine to form a couple, we may match the X and Y components of $\bar{M}_C$ for the actual force system to the corresponding components of the moment of the resultant $\bar{R}$ about point C. This will yield the values of X_R and Y_R illustrated in Figure 19. They locate where the line of action of the force resultant intersects the XY plane. Any point O along the indicated line of action of $\bar{R}$ yields $\Sigma \bar{M}_O = \bar{0}$.

4 General System of Forces

In general, there is no point C for which the torque $\bar{M}_C$ of the equivalent force-couple system is perpendicular to the resultant force $\bar{R}$. A slight simplification may be obtained by representing $\bar{M}_C$ by its components parallel and perpendicular to $\bar{R}$. We can then find a point O for which the torque $\bar{M}_O$ of the equivalent force-couple system is parallel to $\bar{R}$. Such a force system is called a *wrench*. The action of a wrench is typified by the manner in which we use a screwdriver, simultaneously pushing and twisting on the shaft on the screwdriver. The actual determination of a wrench is academic. It does little to increase understanding of the effect of the force system. Therefore, we shall not pursue the required procedure.

EXAMPLE 8

Three people exert horizontal forces on a rectangular table as shown. What single force has the same effect as these people, and where along edge BC should this force be applied?

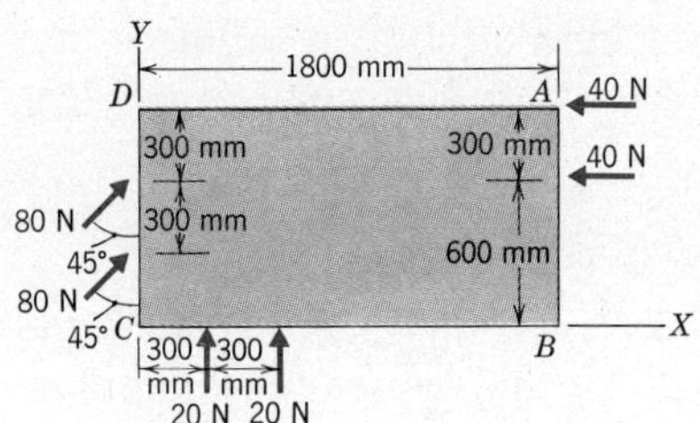

Solution

The force system is planar, so it has a single equivalent force. This force is the sum of the given forces. We therefore sum force components using the coordinate system illustrated in the sketch. Note that we could first replace each pair of parallel forces by their single equivalent, but there is no advantage in doing that.

The force sum gives

$$\bar{R} = \Sigma \bar{F} = \Sigma F_X \bar{I} + \Sigma F_Y \bar{J}$$

$$= [2(80 \cos 45°) - 2(40)]\bar{I} + [+2(80 \sin 45°) + 2(20)]\bar{J} = 33.14\bar{I} + 153.14\bar{J} \text{ N}$$

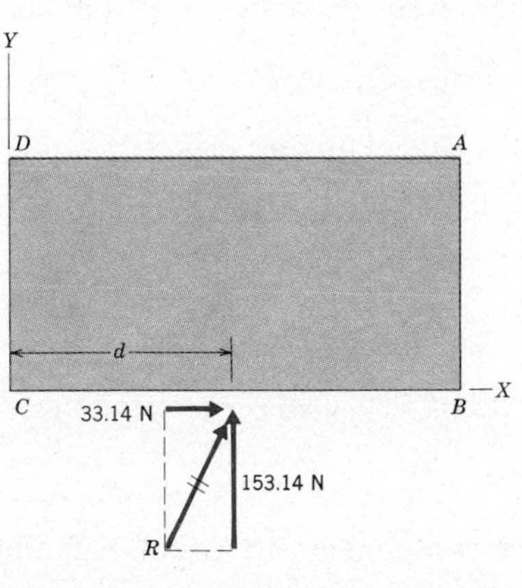

We now determine the point of application along edge BC with the aid of a sketch in which $\bar{R}$ is applied at an undetermined point along the edge. Then the moment of $\bar{R}$ about any point must equal the moment of the actual force system about that point.

The choice of the point for the moment sum is arbitrary; let us use point C. The Z axis is outward, so counterclockwise moments are positive. The lever arms of the force components are shown in the given sketch. The Y components of the 80-N forces pass through point C, so we compute the moment of the given forces to be

$$\Sigma M_C = -(80 \cos 45°)(0.3) - (80 \cos 45°)(0.6) + 20(0.3) + 20(0.6) + 40(0.6) + 40(0.9)$$

$$= 27.09 \text{ N-m}$$

The moment of the force resultant $\bar{R}$ about point C is obtained with the aid of the second sketch. Noting that the component R_X passes through point C and that the lever arm of component R_Y is the unknown distance d, we find

$$\Sigma M_C = (153.14)d$$

Equating the two expressions for ΣM_C yields

$$d = \frac{27.09}{153.14} = 0.1769 \text{ m}$$

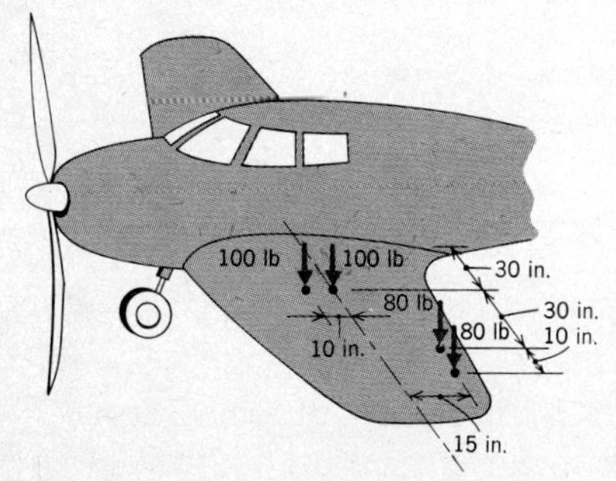

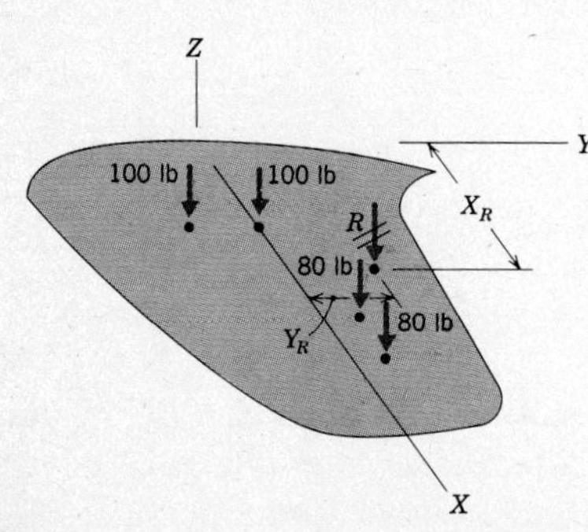

EXAMPLE 9

Two people are standing on the wing of a light aircraft, resulting in the loading shown. Determine the equivalent force and its point of application.

Solution

In order to facilitate the usage of the given dimensions and to simplify the representation of the parallel forces, we use the reference frame illustrated in the second sketch. The resultant force $\bar{R}$ is the sum of the given forces. Because these forces are all in the negative Z direction, we have

$$\bar{R} = -(80 + 80 + 100 + 100)\bar{K} = -360\bar{K} \text{ lb}$$

To determine the dimensions X_R and Y_R of the point of application of $\bar{R}$, we equate the moments about any point of the given forces and the resultant force $\bar{R}$. Using the origin of the coordinate system for the moment sum and referring to the given diagram for distances, we formulate

$$\begin{aligned}\Sigma M_C &= (30\bar{I}) \times (-100\bar{K}) + (30\bar{I} - 10\bar{J}) \times (-100\bar{K}) \\ &\quad + (60\bar{I} + 15\bar{J}) \times (-80\bar{K}) + (70\bar{I} + 15\bar{J}) \times (-80\bar{K}) \\ &= (X_R\,\bar{I} + Y_R\,\bar{J}) \times \bar{R} = (X_R\,\bar{I} + Y_R\,\bar{J}) \times (-360\bar{K}) \\ &\quad -3000(-\bar{J}) - 3000(-\bar{J}) \\ &\quad + 1000\bar{I} - 4800(-\bar{J}) - 1200\bar{I} - 5600(-\bar{J}) - 1200\bar{I} \\ &= -360\,Y_R\,\bar{I} - 360X_R(-\bar{J}) - 1400\bar{I} + 16{,}400\bar{J} \\ &= -360\,Y_R\,\bar{I} + 360X_R\,\bar{J} \text{ lb-in.}\end{aligned}$$

Because two vectors are equal only if the components are equal, equating corresponding components for the vectors on the left- and right-hand sides of the equation yields

$$\bar{I} \text{ component:} \qquad -1400 = -360Y_R$$

$$\bar{J} \text{ component:} \qquad 16{,}400 = 360X_R$$

Solving these equations, we obtain

$$X_R = 45.55 \text{ in.} \qquad Y_R = 3.89 \text{ in.}$$

HOMEWORK PROBLEMS

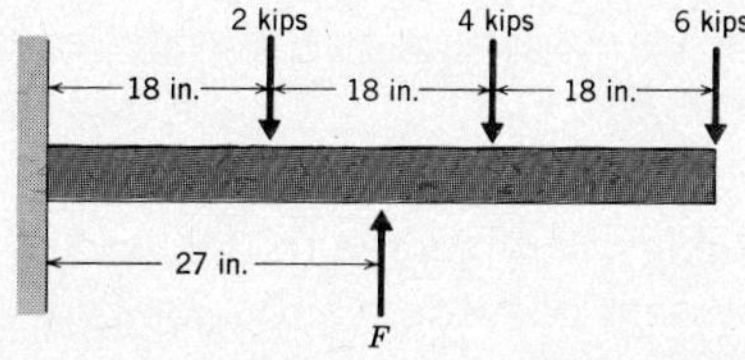

Prob. III.64

III.64 Four vertical loads are applied to the cantilever shown. Determine the resultant of this system of forces and locate where it is applied along the beam (a) when $F = 0$, (b) when $F = 6$ kips, (c) when $F = 12$ kips.

III.65 Knowing that the distance $d = 1.5$ m, determine the resultant of the loads acting on the beam shown. Specify the point of application of this resultant on the beam.

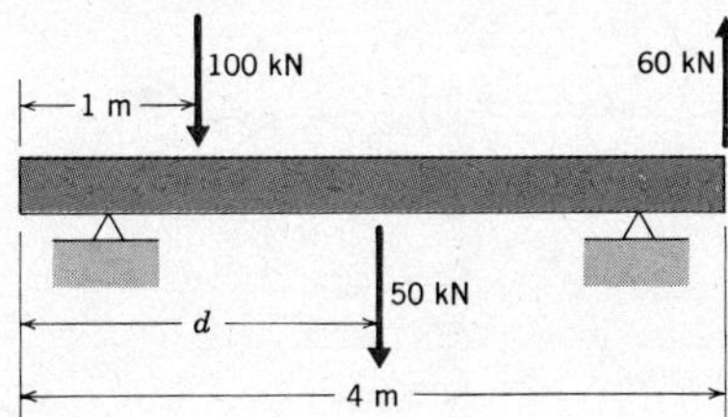

Prob. III.65

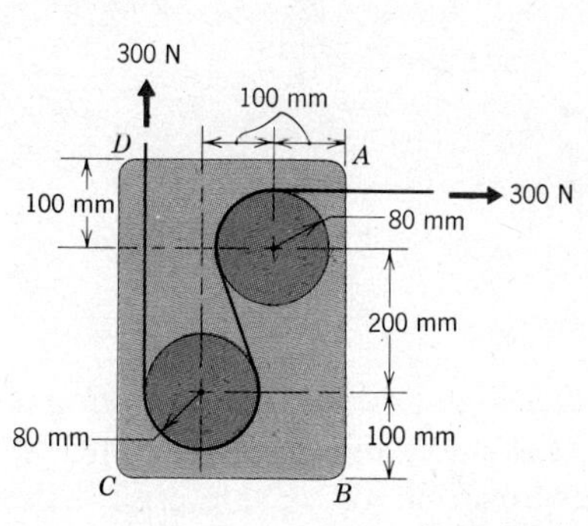

Prob. III.67

III.66 Determine the distance d for which the resultant of the three forces shown passes through the midpoint of the beam.

III.67 In a certain drive mechanism a cable is passed over the idler pulley assembly shown. The pulleys are ideal, and the tension within the pulleys is 300 N. Determine the magnitude and direction of the resultant force exerted by the cable on the assembly, and also determine the point of application of this force along edge AB.

III.68 Determine the resultant of the force system shown. Specify its magnitude, direction, and point of application.

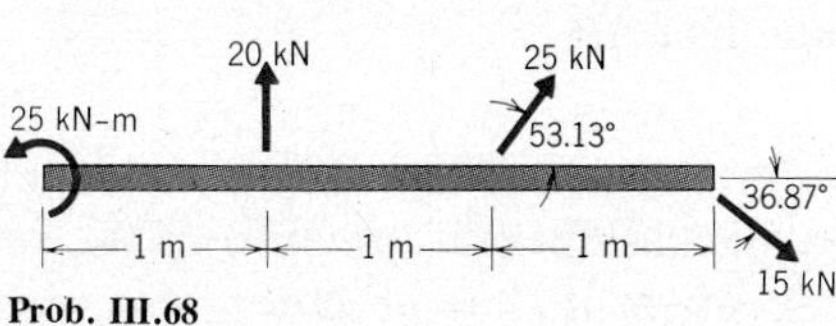

Prob. III.68

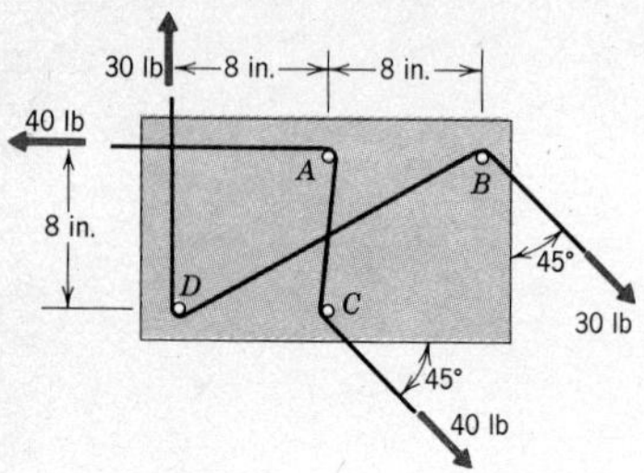

Prob. III.69

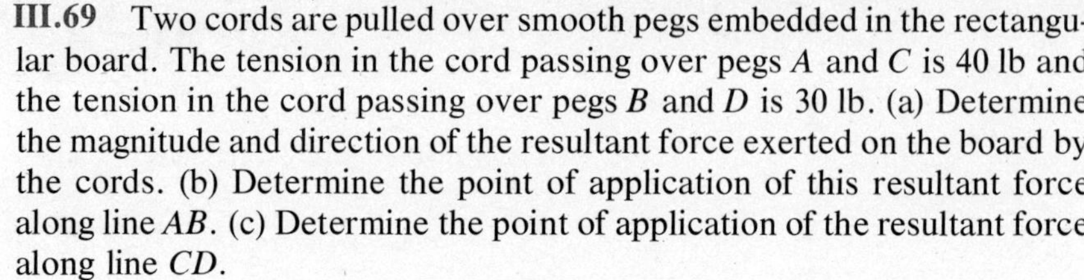

III.69 Two cords are pulled over smooth pegs embedded in the rectangular board. The tension in the cord passing over pegs A and C is 40 lb and the tension in the cord passing over pegs B and D is 30 lb. (a) Determine the magnitude and direction of the resultant force exerted on the board by the cords. (b) Determine the point of application of this resultant force along line AB. (c) Determine the point of application of the resultant force along line CD.

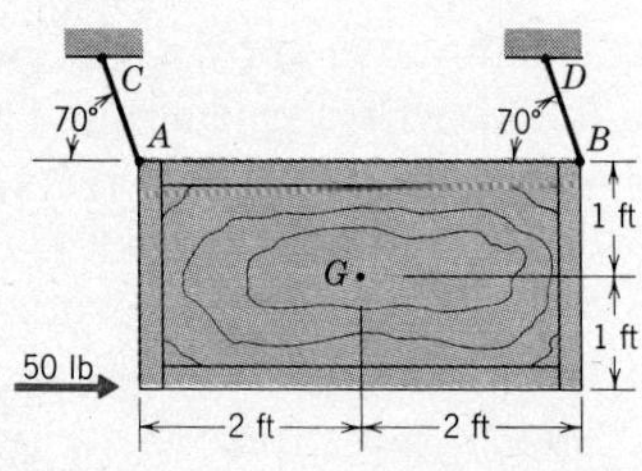

Prob. III.70

III.70 A crate supported by two cables is pushed to one side by a person who exerts the 50-lb horizontal force. The weight of the crate is 400 lb, acting at point G. The tension in cable AC is 280 lb and the tension in cable BD is 240 lb. Determine the resultant of the forces acting on the crate, and specify its point of application along edge AB.

III.71 A street lamp supported by two cables is subjected to the loading shown. Determine the resultant of this system of forces, and specify its point of application along line AB.

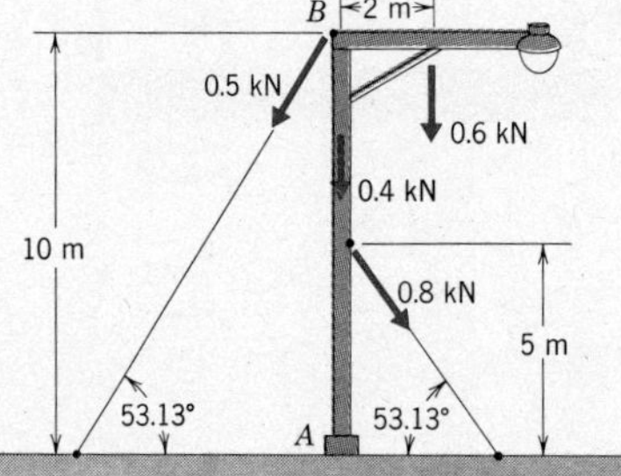

Prob. III.71

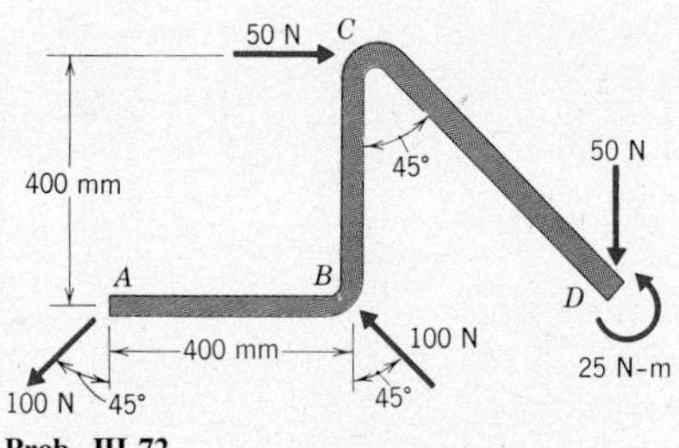

Prob. III.72

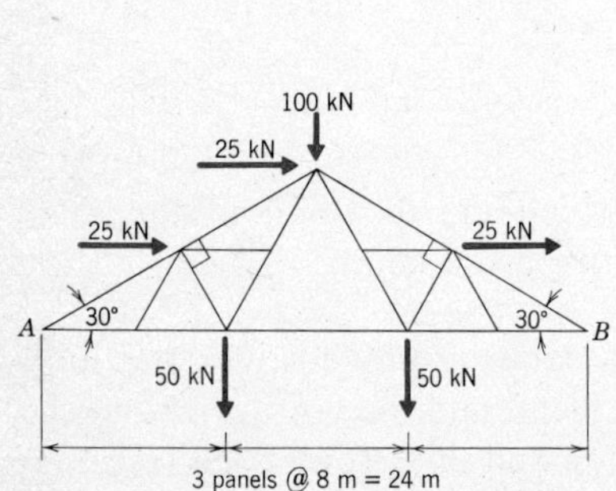

Prob. III.73

III.72 The bent bar is subjected to the loading shown. (a) Determine the magnitude and direction of the resultant of this loading. (b) Determine the point of application of the resultant force along line AB. (c) Determine the point of application of the resultant force along line BC.

III.73 The truss structure is loaded as shown. Determine the resultant of this loading, specifying its magnitude, direction, and point of application along line AB.

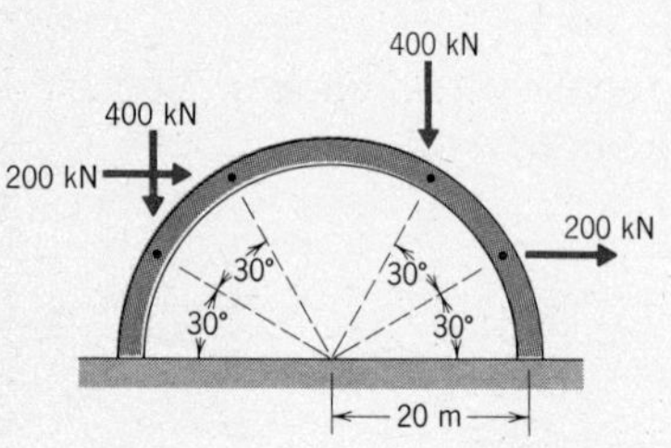

Prob. III.74

III.74 The circular arch is subjected to the loading shown. Determine the magnitude and direction of the resultant of these forces, and determine where its line of action intersects the arch.

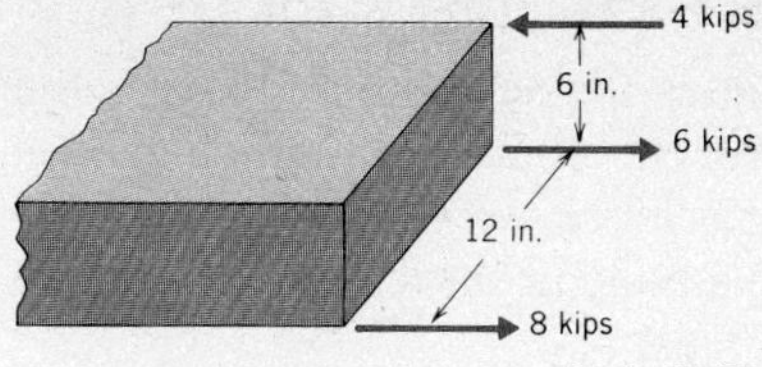

Prob. III.75

III.75 Three axial forces are exerted on the concrete beam as shown. Determine the point of application of the resultant force on the cross section of the beam.

III.76 A T-beam section is loaded as shown. Determine the resultant of these forces and its point of application on the top of the beam.

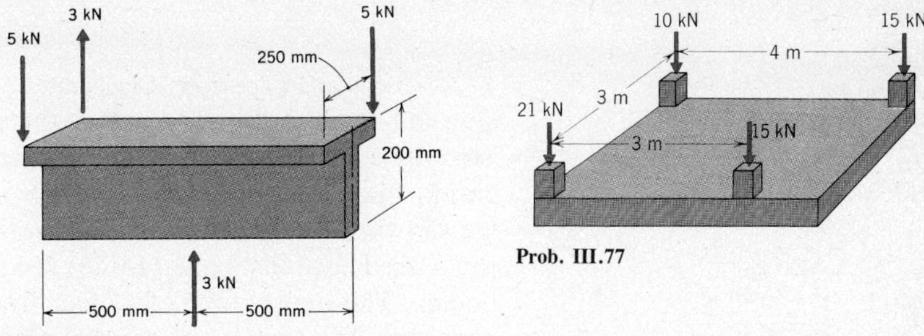

Prob. III.76

Prob. III.77

III.77 Four columns rest on the rectangular foundation slab. For the column loadings shown determine the resultant force and its point of application on the slab.

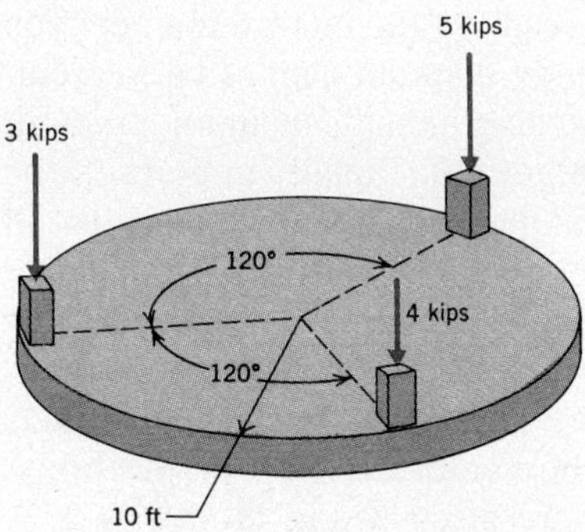

Prob. III.78

III.78 Three parallel columns equally spaced around the perimeter of the circular foundation slab carry the loads shown. Determine the resultant of this loading, and specify its point of application on the slab.

III.79 A semicircular rod embedded in a wall is loaded as shown. Determine the resultant of these forces and its point of application in the XY plane.

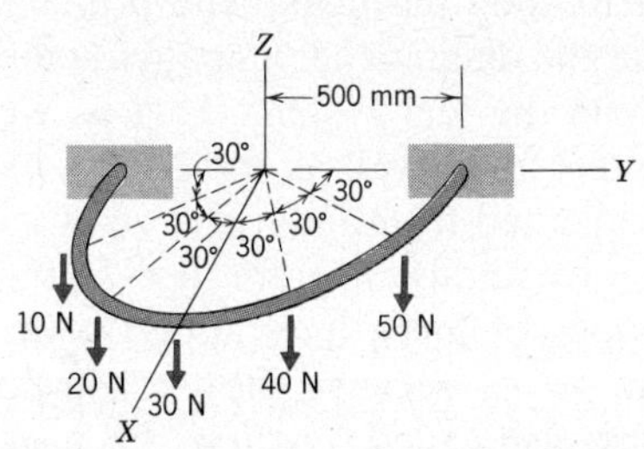

Prob. III.79

Four packages resting on the table exert the forces shown. Determine the resultant of these forces, and specify its point of application on the top of the table.

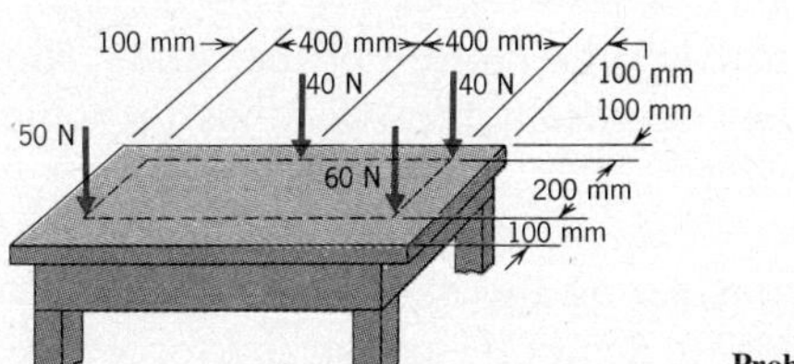

Prob. III.80

III.81 In Problem III.80, it is desired to apply to the perimeter of the table an upward force that will cause the resultant of the forces acting on the table top to be at the center of the table. Determine this force and its point of application.

G. EQUILIBRIUM OF RIGID BODIES

A fundamental application of the techniques for describing forces is the determination of the forces required to prevent a body from moving. This is the topic we will treat here. However, before we can investigate the relationship between the forces acting on a body in static equilibrium, we must learn to properly account for all forces.

1 Reactions and Free Body Diagrams

One aid in describing forces is the free body diagram. As we saw in Module II when we studied concurrent force systems, the free body diagram isolates the body of interest from its surroundings. In the diagram we show *all* forces being *exerted on the body,* bearing in mind that Newton's third law tells us that forces are the result of the interaction between bodies. The importance of a clear free body diagram cannot be overemphasized. The process of constructing the diagram aids us in understanding which parameters of the system are important. Equally important, the resulting diagram is a vital visual aid in making certain that we consider all forces during the solution process.

We will have no difficulty in depicting certain classes of forces in the diagram. Typical forces in this category are given forces, tensile forces exerted by cables, and the weight force. The latter, in particular, acts vertically downward, passing through the center of mass of the body. (The methods for determining this point are discussed in Module V.)

Free body diagrams for rigid bodies are somewhat more difficult to construct than those we encountered in treating particles. The primary reason for this difficulty lies in the fact that the diagram accompanying the statement of the problem will usually depict the real system. That is, we will be shown the body (or bodies) of interest, as well as the means by which the body is supported and thus prevented from moving.

It is part of the task of constructing a free body diagram to examine the supports in order to deduce what types of force they apply to the body. These forces are called *constraint forces* because they constrain (restrict) the motion of the body. They are also called *reactions* because they represent the way that the objects to which the body is attached react.

The characteristics of the constraint forces can be established by considering the manner in which the body is supported. In determining how to depict the constraint forces we need only two facts. First, because a force represents a pushing or pulling effect, a point in a body is prevented from moving in a specific direction by the action of a reaction force in the opposite direction. Second, because a couple represents a twisting effect, a body is prevented from rotating about a specific axis by the action of a reaction couple whose torque is parallel to the axis of rotation, twisting opposite the sense in which rotation would occur.

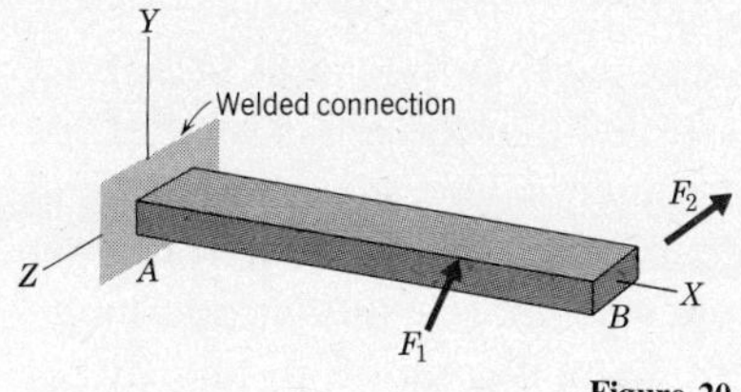

Figure 20

To see how these criteria are applied when examining the forces transmitted to a body by its supports, let us consider the welded connection between the rigid bar and the wall shown in Figure 20. A weld connection between two bodies fully joins their material. Thus, end A of the bar is not free to move in the direction of any of the coordinate axes, so there must be reaction force components in each coordinte direction. We do not know in advance the value of these reaction forces, nor even whether they are oriented in the positive or negative coordinate direction; this information depends on how the bar would tend to move under the action of the applied forces. We will denote a reaction force at a specific point by the letter corresponding to this point. Therefore, a portion of the constraint at end A consists of a reaction force $\bar{A}$ having components $\bar{A}_X$, $\bar{A}_Y$, and $\bar{A}_Z$.

To complete the determination of the nature of the constraint imposed by a welded connection, we must consider the type of rotations the end may undergo. Because the material of the bar is continuous with the material of the wall, and the wall itself is assumed to be rigid and fixed, it follows that there can be no rotation about any of the coordinate axes. This means that there must be a reaction couple $\bar{M}_A$, whose components $\bar{M}_{AX}$, $\bar{M}_{AY}$, and $\bar{M}_{AZ}$ represent the constraints necessary to prevent the rotations. We usually show the components of general reaction forces and couples in the free body diagram in order to be certain that we remember to account for each unknown component. Figure 21 depicts the free body diagram of the rigid bar.

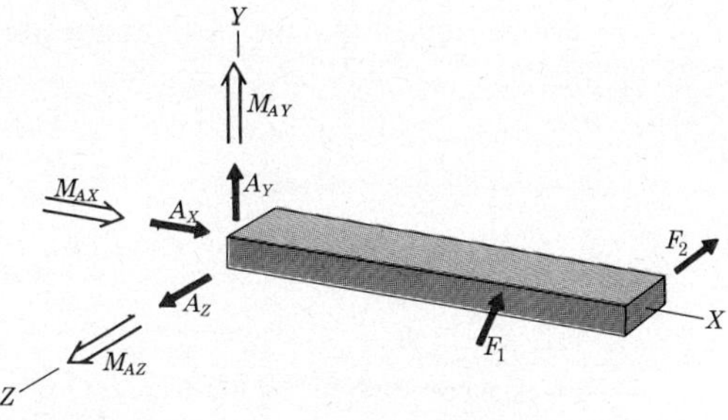

Figure 21

A welded joint represents the greatest degree of fixity (lack of freedom to move) that can be imposed by a connection. In fact, a welded support is usually described as a *fixed end*.

Whenever the type of support allows the body to move in a specific direction, or rotate, about a specific axis, there will be no reaction force, or couple, in that direction. For example, suppose that a bar is supported at one end by a pin connection, such as that shown in Figure 22a. The pin acts like a miniature shaft. The end of the bar is fully constrained from moving in any direction by the pin and the brackets. The only possible way in which the bar may move is to rotate about the axis of the pin,

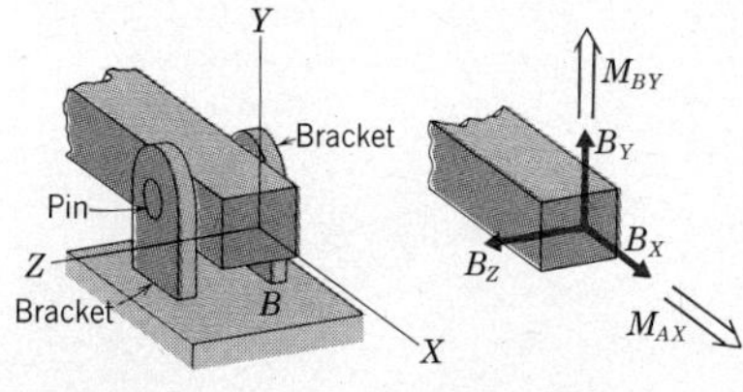

Figure 22a

Figure 22b

which is the Z axis in the figure. Thus, the reaction force and couple corresponding to this support are as shown in Figure 22b. Note that there is no couple in the Z direction, because rotation about this axis is not restricted.

The foregoing discussion implicitly assumed that the pin fit closely in the hole of the bar. Further, it was also assumed that the pin is rigid. This, of course, may not be the situation. For instance, a very common method of constructing a pin connection is to use a hinge, such as those found on doors. Anyone who has ever tried to support a door with just one hinge knows that it affords very little constraint against rotation in any direction. Thus, depending on specific knowledge of the type of construction, we may choose to retain or neglect the couples depicted in Figure 22b. We shall differentiate between the two alternatives by referring to the former condition as an ideal *pin,* and the latter one as a *hinge,* although this is not standard terminology. In a planar problem where the axis of the pin is perpendicular to the plane of interest, as shown in Figure 23a, we will be concerned only with reaction forces that are parallel to the plane and couples whose torque is perpendicular to the plane. In this case there is no difference between an ideal pin and a hinge, as depicted in Figure 23b.

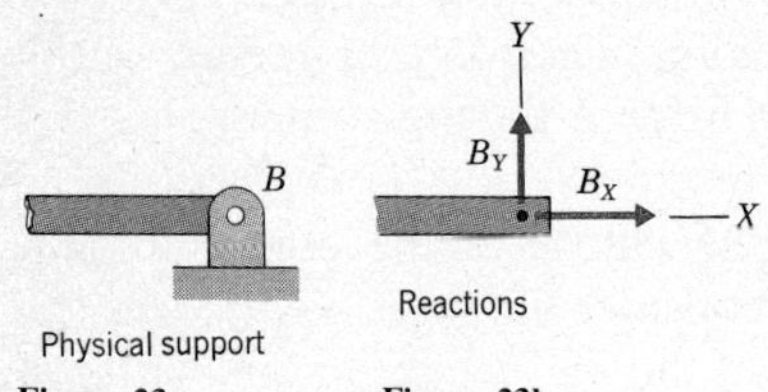

Figure 23a **Figure 23b**

As an aid for our future studies, Table 1 depicts other common methods of connecting and supporting bodies, accompanied by a description of the nature of the corresponding restriction on the motion of the body and a sketch of the reaction force and couple components. This tabulation is presented in order to supplement, but not replace, our ability to examine the means of supporting a body. This is especially so because on some occasions we will encounter a support that is not depicted in Table 1.

Table 1 Common Supports and Reactions

TYPE OF SUPPORT	MOTION RESTRICTION	REACTIONS
Cable	Attached point A cannot move outward, parallel to the cable. Free rotation.	$\bar{T}$ is a tensile force parallel to the cable.

Table 1 *(continued)*

TYPE OF SUPPORT	MOTION RESTRICTION	REACTIONS
Short link 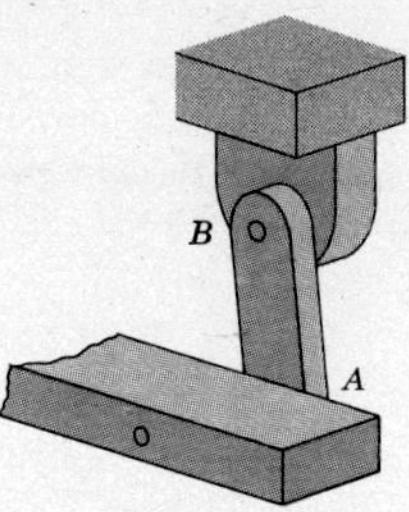	Attached point A cannot move parallel to the line AB of the link.	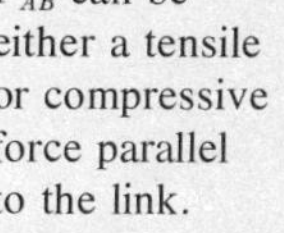 $\bar{F}_{AB}$ can be either a tensile or compressive force parallel to the link. 
Smooth surface 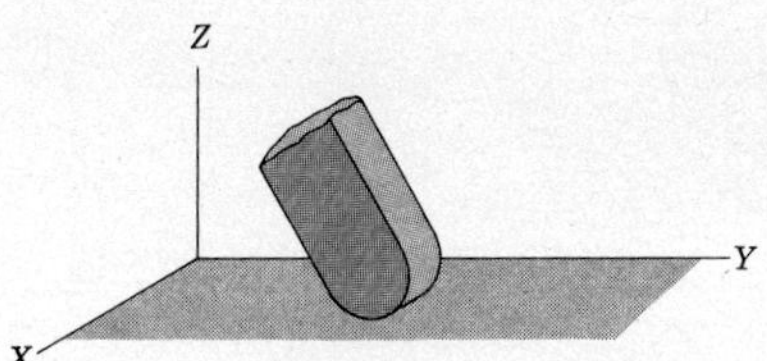	No movement that penetrates the surface. Free rotation.	$\bar{N}$ can only be normal upward 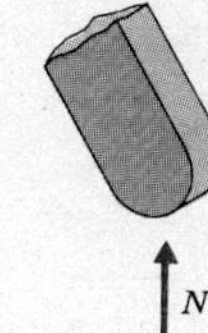
Rough surface or sharp edge 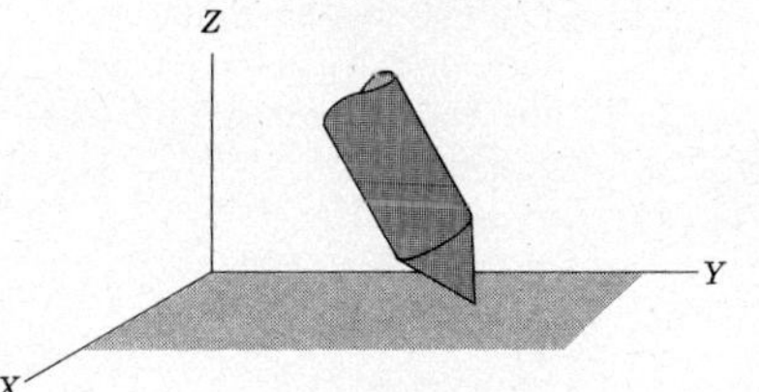	No movement that either penetrates the surface or is parallel to it. Free rotation. (For the effects of friction, see Module VII.)	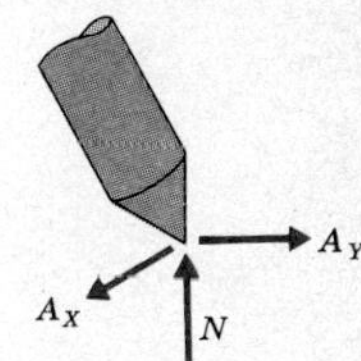
Small roller on a rough surface 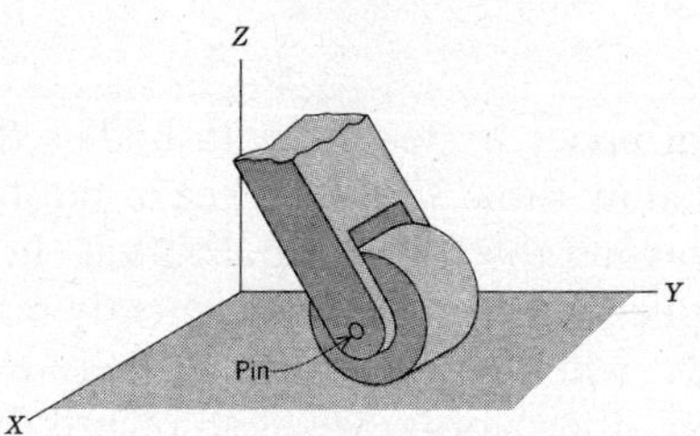	No movement that penetrates the surface or is perpendicular to the plane of the roller. Free rotation. The case of a small roller in a smooth groove is identical.	In the absence of a groove or friction, 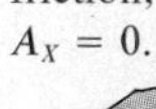 $A_X = 0$. 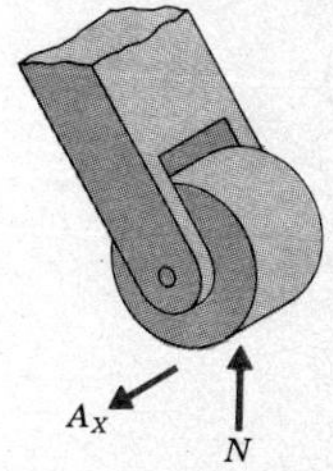

(continued)

Table 1 *(continued)*

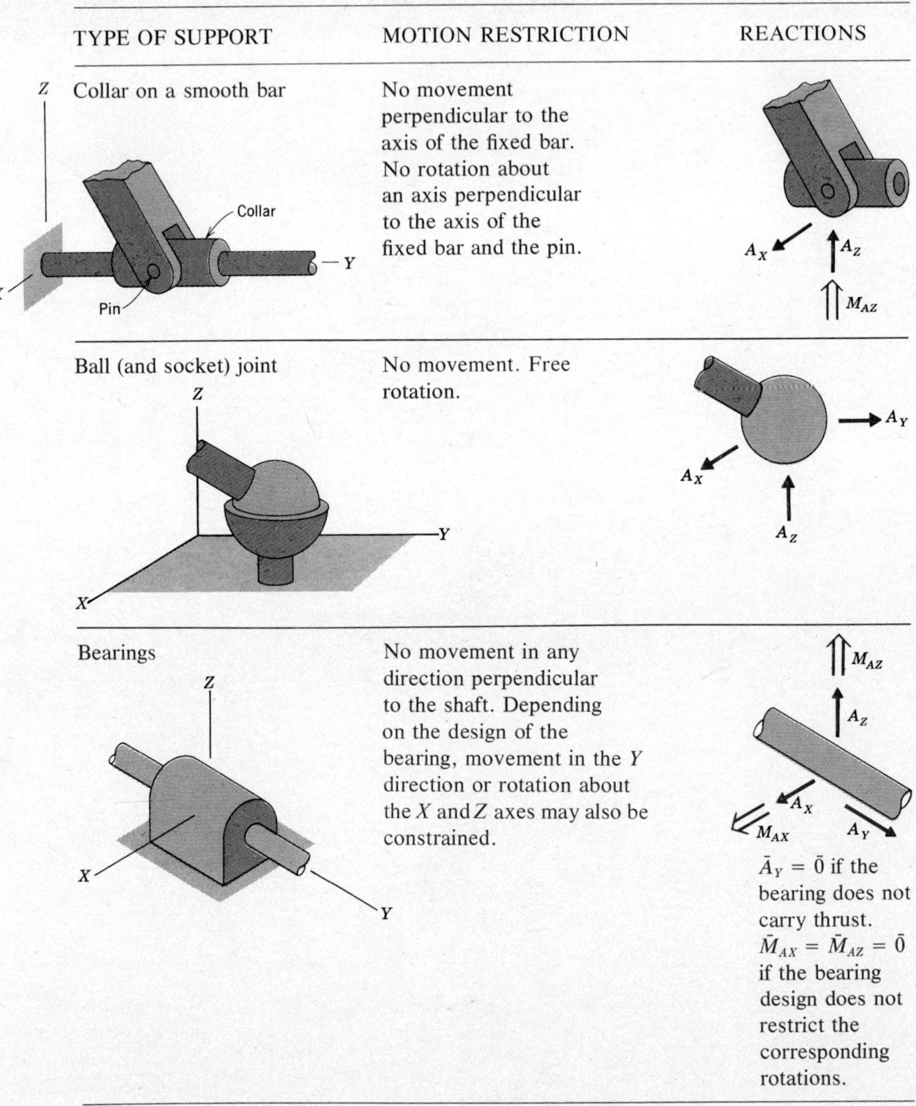

TYPE OF SUPPORT	MOTION RESTRICTION	REACTIONS
Collar on a smooth bar	No movement perpendicular to the axis of the fixed bar. No rotation about an axis perpendicular to the axis of the fixed bar and the pin.	
Ball (and socket) joint	No movement. Free rotation.	
Bearings	No movement in any direction perpendicular to the shaft. Depending on the design of the bearing, movement in the Y direction or rotation about the X and Z axes may also be constrained.	$\bar{A}_Y = \bar{0}$ if the bearing does not carry thrust. $\bar{M}_{AX} = \bar{M}_{AZ} = \bar{0}$ if the bearing design does not restrict the corresponding rotations.

A question that arises in dealing with bodies that are supported by bearings, as depicted in Table 1, is whether or not to include the couples $\bar{M}_{AX}$ and $\bar{M}_{AZ}$ in drawing the free body diagram. In that the presence of these couples complicate the problem, frequently making it insoluble, our approach in this text will be to assume that there are no couples, unless stated otherwise. In other words, we shall regard the action of a bearing and of a pin to be identical.

The following example will demonstrate how the concepts of constraints and constraint forces are implemented in drawing free body diagrams.

EXAMPLE 10

Draw free body diagrams for the systems shown. The centers of mass are denoted as points G.

a A ladder

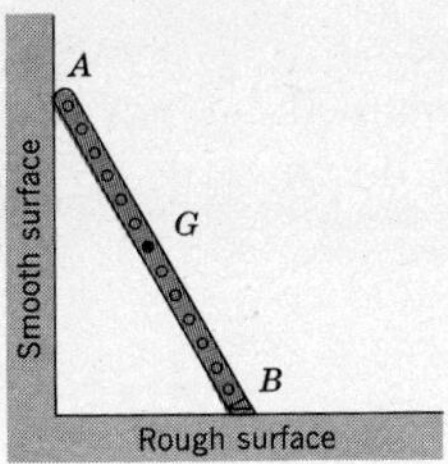

b A disk attached to a block by a cable

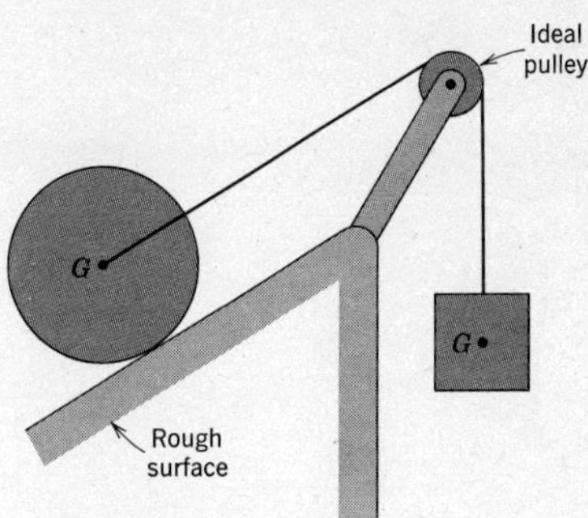

c Each of the two bars separately, and joined together

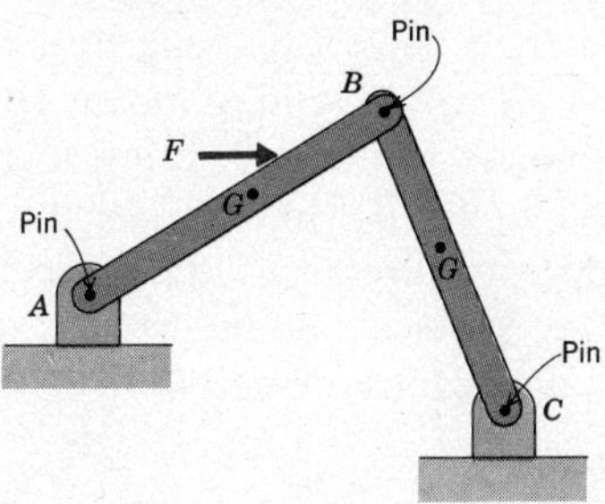

d A door suspended by two hinges

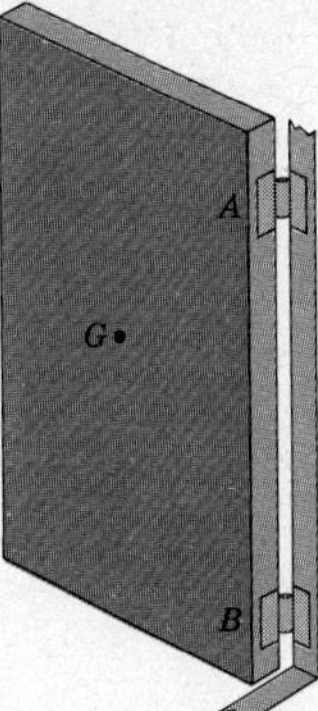

e The entire truss structure (neglect gravity)

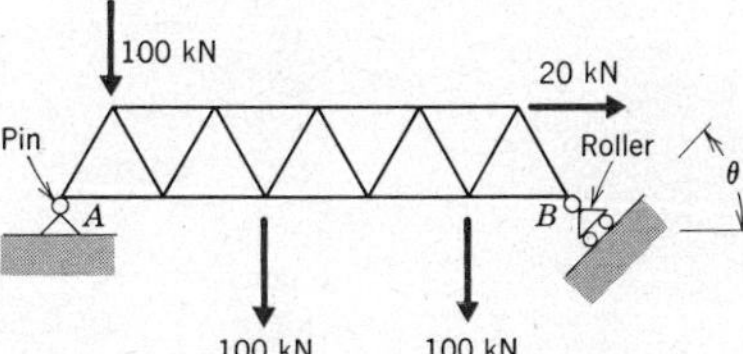

f An antenna tower (neglect gravity)

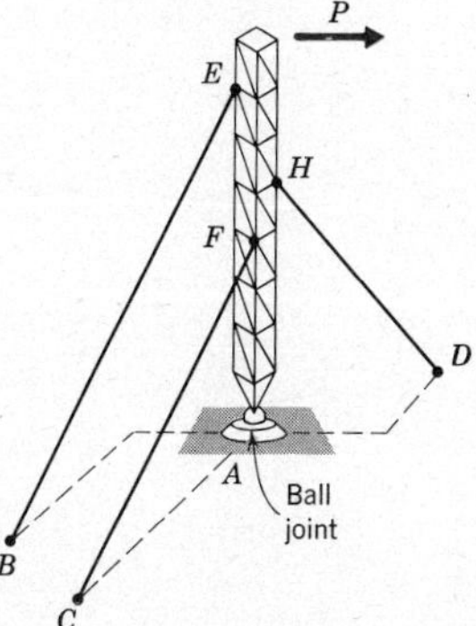

Solution

As an aid to including all forces, it is axiomatic that anything touching a body exerts a force on it.

Part a

There is a force normal to the smooth wall at point A, whereas the rough surface exerts normal and tangential forces at point B. We show the weight force acting downward, passing through point G.

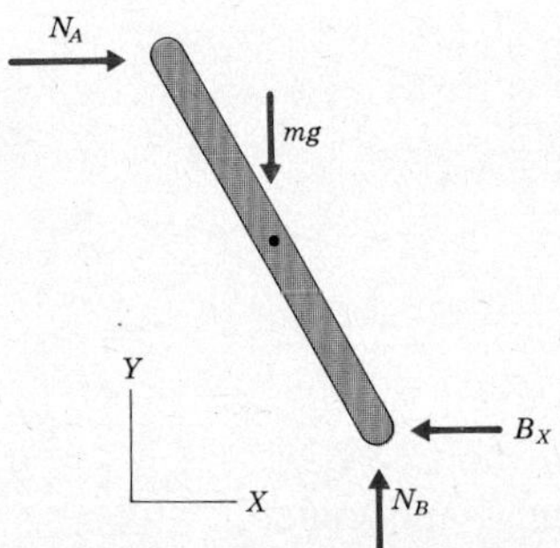

The force $\bar{B}_X$ is shown pushing to the left because it opposes the way in which point B would tend to move. However, it is also correct to show $\bar{B}_X$ in the opposite direction, acting to the right, for we would then find a negative value for B_X. In contrast, it would make no sense to show $\bar{N}_A$ and $\bar{N}_B$ in opposite directions, because the surfaces can only push on the ladder. Nevertheless, an *incorrect* choice will merely result in a negative value for the reactive force.

Part B

There are three connected bodies: the disk, the block, and the pulley. We shall draw free body diagrams of each. A useful technique to facilitate using Newton's third law is to *lay out the diagrams to resemble an assembly drawing*.

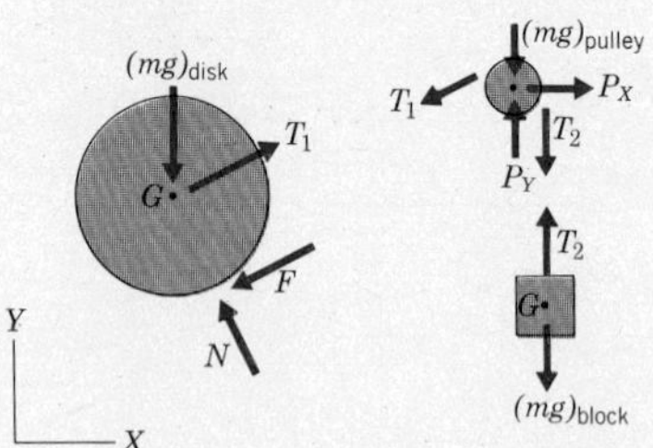

The force F is the tangential component of the reaction between the disk and the rough surface. The forces $\bar{P}_X$ and $\bar{P}_Y$ represent the reactions of the shaft on the pulley. The tensile forces T_1 and T_2 pull on the bodies they contact. We have already seen that for an ideal pulley, $T_1 = T_2$.

Part c
Each bar is pinned at both of its ends. Further, the reactions exerted between the two bars at pin B must satisfy Newton's third law of action and reaction. Thus,

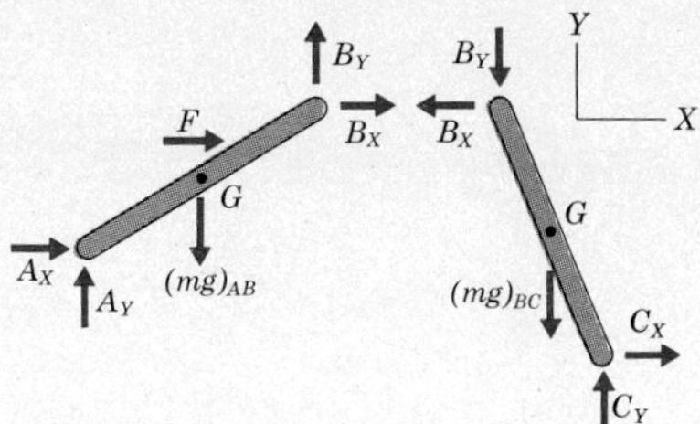

When we draw a free body of the two bars joined together, the reactions at pin B are internal to the system, so they do not appear in such a diagram. In effect, the diagram of the system is an overlay of the individual diagrams sketched above.

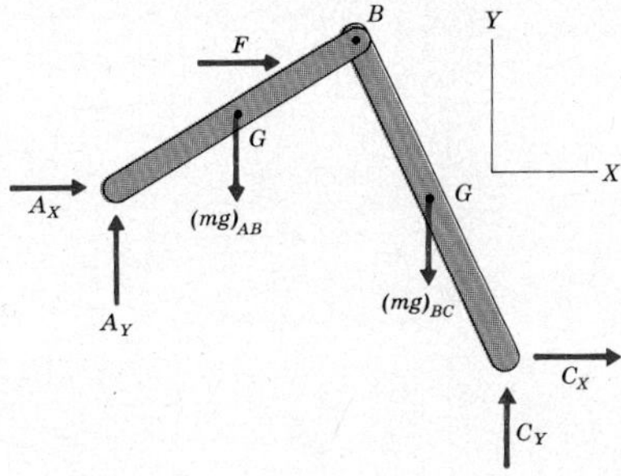

Part d
We agreed to neglect the couple reactions of hinges, hence

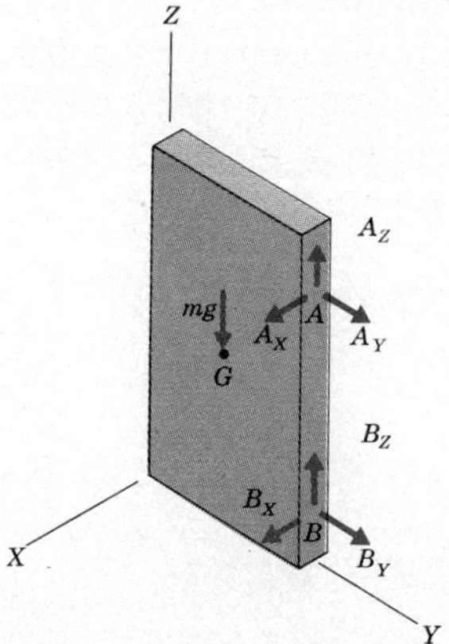

Part e

Note that the reaction of the roller is perpendicular to the inclined surface.

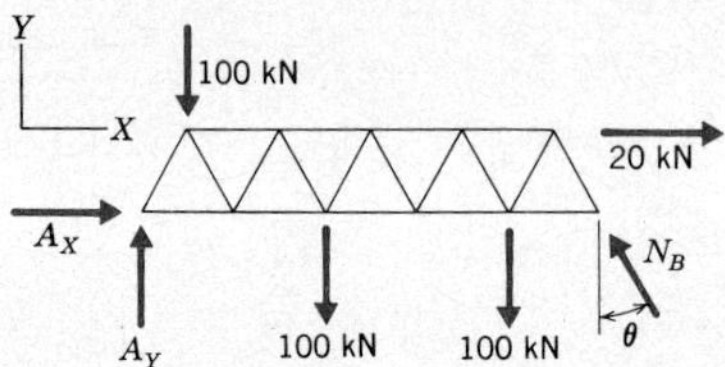

Part f

The cables exert tensile forces and the ball joint prevents movement of the base.

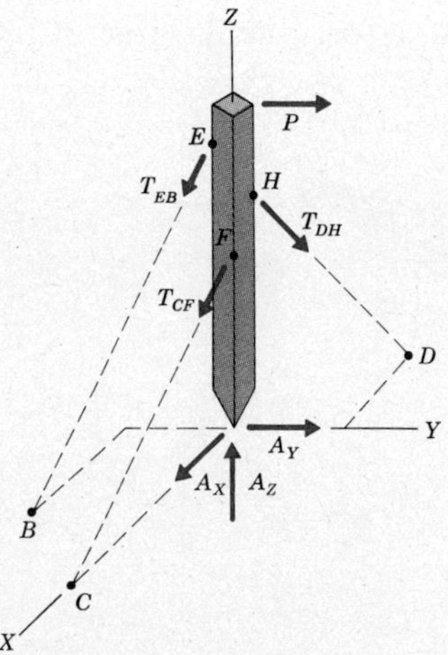

HOMEWORK PROBLEMS

Sketch free body diagrams for the systems shown. In each case the given forces are illustrated. In problems where the weight force is to be included, the location of the mass center, point G, is given. For simplicity, dimensions are omitted.

III.82 The wrecking bar

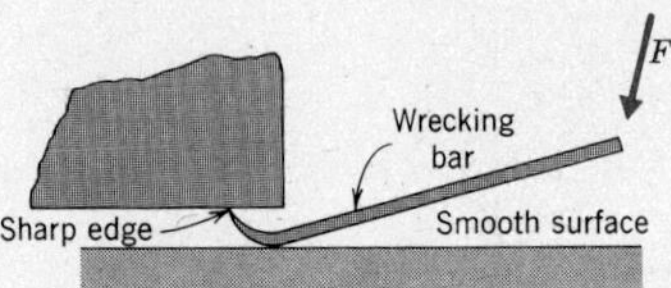

Prob. III.82

III.83 The stepped disk

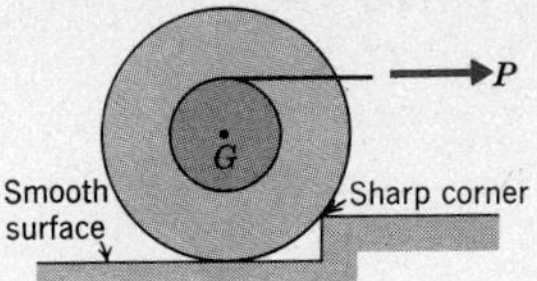

Prob. III.83

III.84 The bent bar

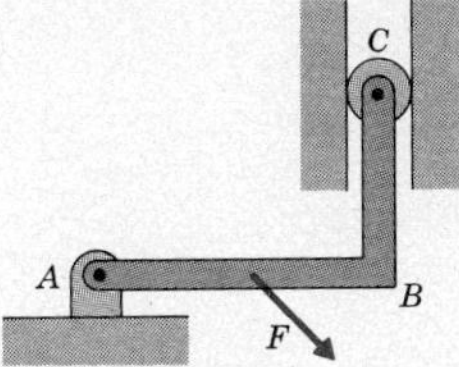

Prob. III.84

III.85 The bar *AB*

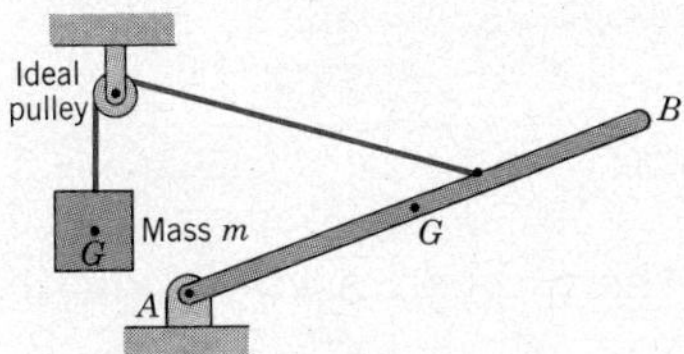

Prob. III.85

III.86 The bar *AB*

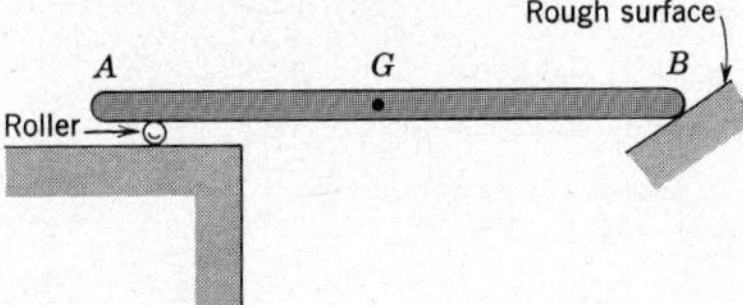

Prob. III.86

III.87 The tape guide assembly

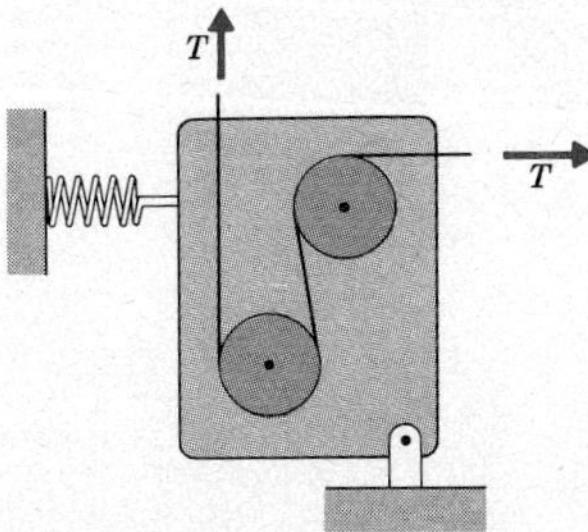

Prob. III.87

III.88 The automobile with only the brakes at the rear wheels applied

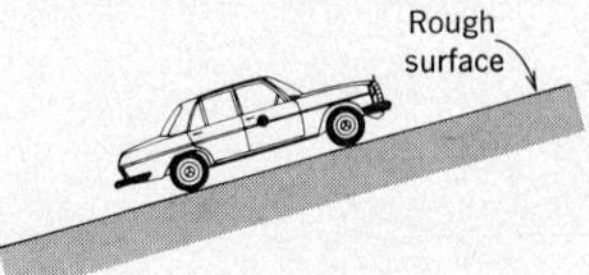

Prob. III.88

III.89 The balance arm

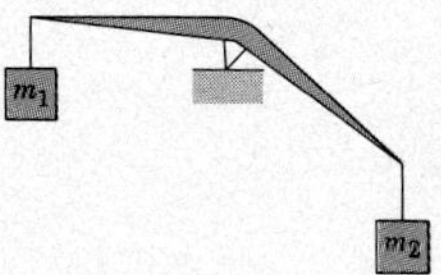

Prob. III.89

III.90 The two-beam assembly, and beams AB and BC individually

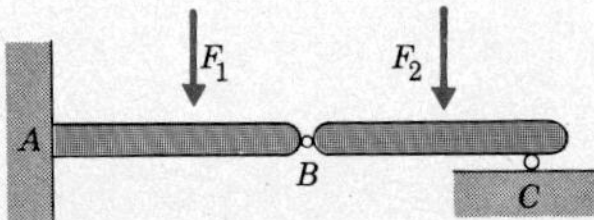

Prob. III.90

III.91 The two-bar assembly, and bars AB and CD individually

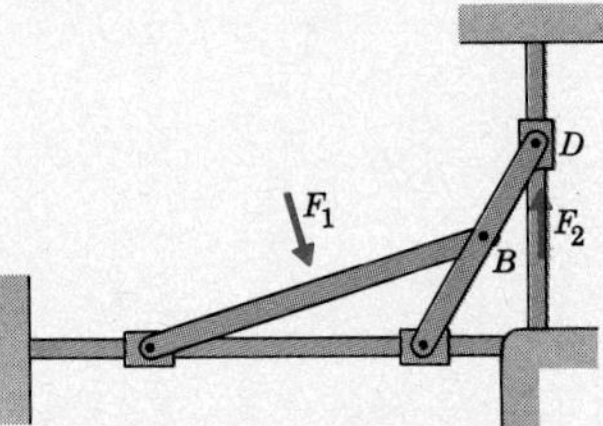

Prob. III.91

III.92 The bent bar

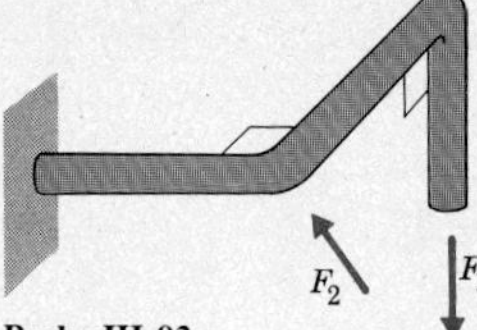

Prob. III.92

III.93 The flywheel and bar AB, each individually

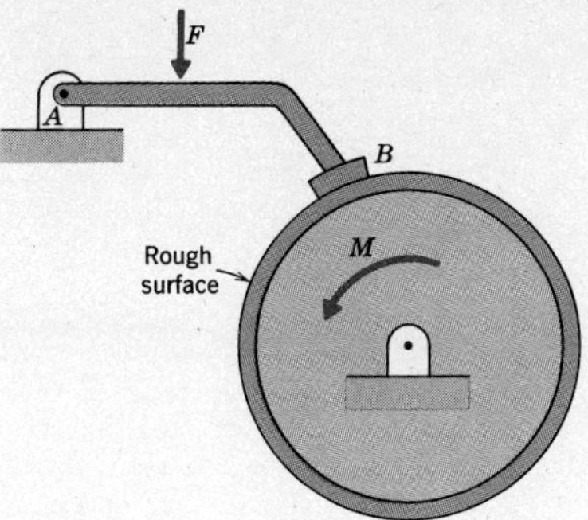

Prob. III.93

III.94 The hand drill

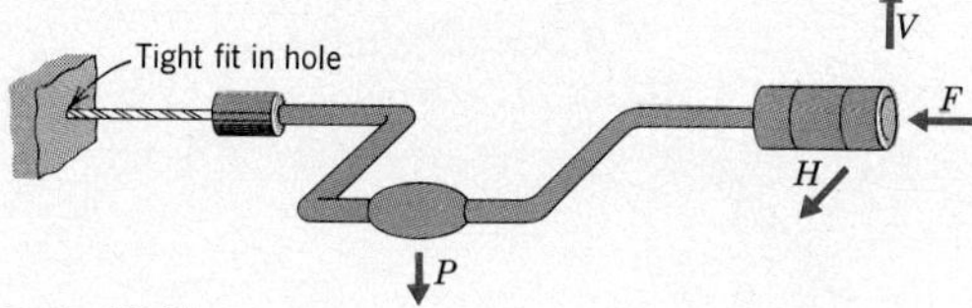

Prob. III.94

III.95 The T-bar

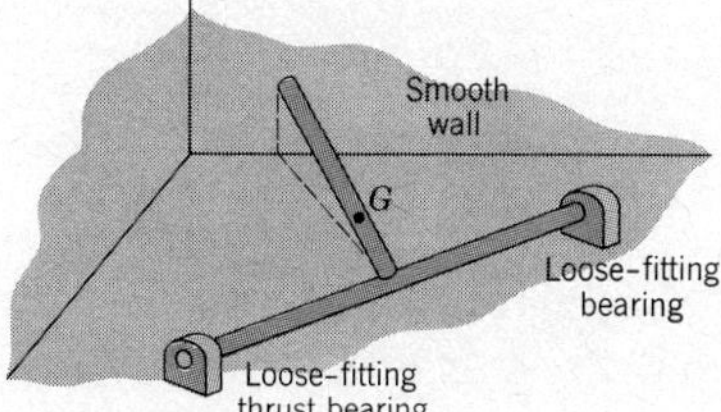

Prob. III.95

III.96 The hatch cover

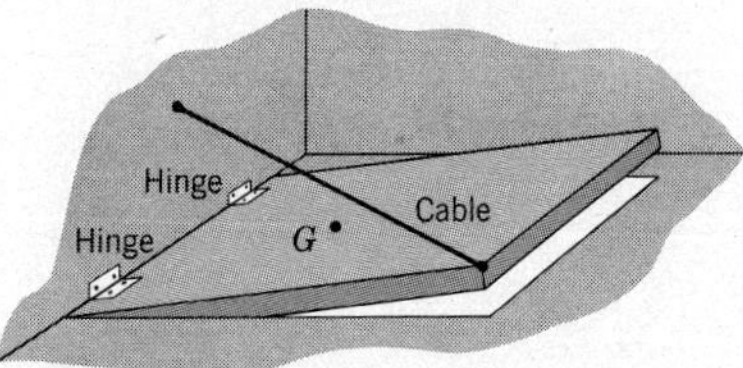

Prob. III.96

III.97 The truss structure — all connections are ball joints

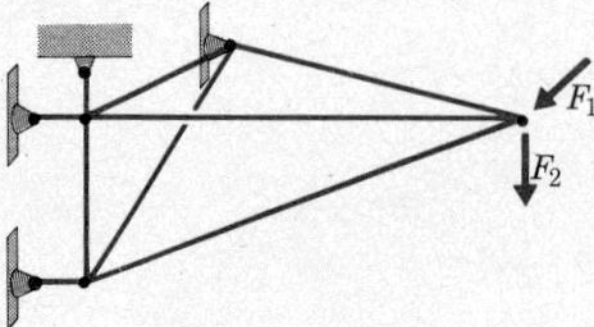

Prob. III.97

III.98 The bar assembly

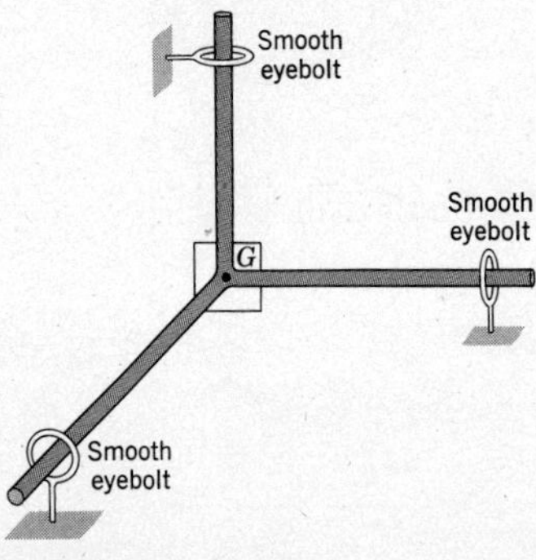

Prob. III.98

2 Conditions for Static Equilibrium

We have already seen that a general system of forces is equivalent to a force $\bar{R}$ passing through an arbitrary point C and a couple $\bar{M}_C$. The force $\bar{R}$, which is the sum of the forces in the given system, $\bar{R} = \Sigma \bar{F}_i$, describes the equivalent pushing or pulling effect on point C. The couple $\bar{M}_C$ is the sum of the moments of the forces about point C, $\bar{M}_C = \Sigma \bar{r}_{i/C} \times \bar{F}_i$. This resultant couple $\bar{M}_C$ describes the twisting effect of the force system about point C. It follows that

> The resultant force and couple should both be zero if a rigid body is to be in static equilibrium.

Notice that the equilibrium condition for concurrent force systems is a special case of the general statement.

The foregoing is an intuitive approach for determining the conditions for static equilibrium. A rigorous derivation is presented in the companion dynamics text, where mathematical proofs accounting for the effects of motion are required. The conditions of static equilibrium are obtained, as a special case, by arresting the motion of the body.

The most direct way of setting the resultant force and couple equal to zero is to actually compute the sum of the forces and the sum of the moments about any arbitrary point. We then equate the two resultant vectors to zero, thus obtaining

$$\Sigma \bar{F} = \bar{0} \qquad \Sigma \bar{M}_C = \bar{0} \tag{19a}$$

This is the approach we shall usually apply. In the case of a three-dimensional force system the force and moment sums have three components each, so equations (19a) are equivalent to the following six scalar *equations of static equilibrium*:

$$\Sigma F_X = 0 \quad \Sigma F_Y = 0 \quad \Sigma F_Z = 0$$
$$\Sigma M_{CX} = 0 \quad \Sigma M_{CY} = 0 \quad \Sigma M_{CZ} = 0 \tag{19b}$$

Because the point we choose for summing moments is arbitrary, in practice we shall choose a point to be situated along the line of action of an unknown reaction force, thereby eliminating that force from the moment equation.

The fact that we can eliminate an unknown force from the moment equation suggests that we should explore the possibility of summing moments about more than one point. Because of the frequency with which planar force systems occur, we shall first discuss such a case.

We begin by considering Figure 24, which illustrates the resultant force and couple with respect to point C for an arbitrary set of planar forces. To investigate an alternative condition for static equilibrium, let us choose another point D and resolve $\bar{R}$ into its components parallel and perpendicular to the line connecting points C and D, as shown in the figure.

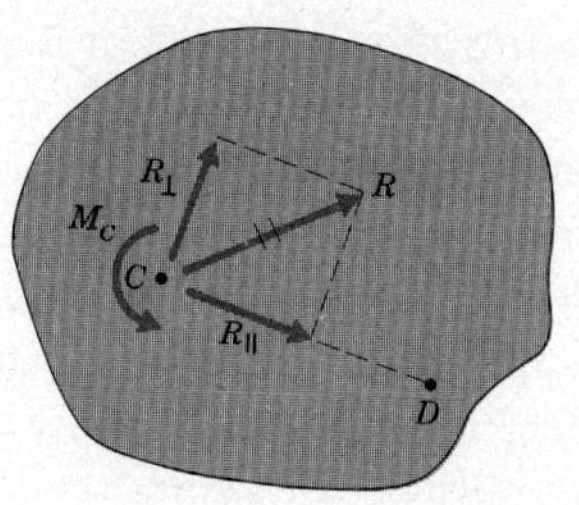

Figure 24

Consider the case where $\bar{M}_C$ has already been set to zero, leaving the nonzero force $\bar{R}$. If we also equate the moment about D of the force system to zero, this will cause $\bar{R}$ to be zero, by the definition of the moment of a force. The remaining force component $\bar{R}_\parallel$ will then vanish by equating to zero the sum of the force components in any direction other than the one that is perpendicular to line CD. Alternatively, in this case we can make $\bar{R}_\parallel$ vanish by equating the moment about a third point to zero, provided that this third point is not coincident with line CD.

Summarizing, *for planar forces the scalar equations for static equilibrium yield only three nontrivial equations*. The scalar equations may be obtained from three alternative formulations, as follows:

1 Equate to zero the moment about one point and the force sums in two directions. This is the approach of equations (19).
2 Equate to zero the moment sums about two points and the force sum in any direction that is not parallel to the line connecting the two chosen points.
3 Equate to zero the moment sums about three points that are not collinear.

We could present similar alternatives for the case of a three-dimensional system of forces. However, as we have already seen, the determination of moments in the three-dimensional case is somewhat lengthier than that for planar forces. Thus, the simplification in the resulting equilibrium equations would be offset by the additional computations required to obtain these equations. In order to have a unified approach for planar and three-dimensional forces, in both cases we will generally employ the formulation of equation (19b), where the force and moment results are equated to zero. Hence, for a planar set of forces lying in the XY plane, the equilibrium equations we shall employ are

$$\boxed{\Sigma F_X = 0, \qquad 2\, F_Y = 0, \qquad \Sigma M_{CZ} = 0} \tag{19c}$$

3 Static Indeterminacy and Partial Constraints

The static equilibrium conditions we have presented describe the relationships that must be satisfied if a body is to remain at rest. Based on these conditions, we will soon consider problems where the reactions and any other forces required to maintain the static equilibrium condition are to be determined. However, we have not yet considered whether it is possible to determine all unknown forces from the equations of statics, or indeed whether it is possible for a body to remain at rest.

In the case of a general force system there are six scalar equations of equilibrium, obtained by setting the resultant force and moment vectors to zero. In the case of a planar force system only three of these equations are nontrivial. Clearly, if there are more unknown components of the reaction forces and couples than the number of equilibrium equations available, we will not be able to evaluate the reactions. The reactions of such a body are then said to be *statically indeterminate*.

A second possibility is that there are fewer unknown components to the reactions than the number of nontrivial equations for static equilibrium. In that case there can be no set of reactions that satisfy all of the equilibrium equations. We then say that the body is *partially constrained,* because for each equilibrium equation that is unsatisfied there is a a corresponding type of motion: translational motion for a force equation and rotational motion for a moment equation. (In the text on dynamics, precise definitions are given for these motion terms.)

Clearly then, in order to be able to solve the equations of static equilibrium, and have *statically determinate* reactions, the number of unknown reaction components must be the same as the number of nontrivial equilibrium equations. However, this is merely a necessary condition, not a sufficient one. There are special situations where one of the equilibrium equations still cannot be satisfied. These situations arise when the supports of the body have not been properly arranged, with the result that motion of the body may still occur. Such bodies are termed *improperly constrained.*

To gain insight into these possibilities, let us consider the beam shown in Figure 25 along with its free body diagram. It will be shown that

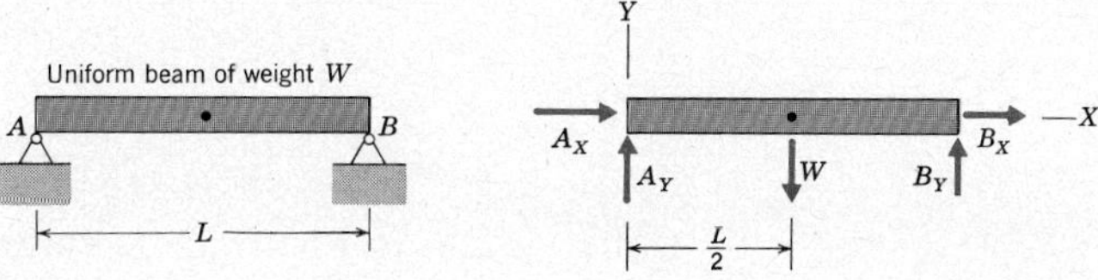

Figure 25

the beam is statically indeterminate. The pins at both ends of the bar exert unknown horizontal and vertical forces. Choosing point A for summing moments, in order to eliminate the reactions at this point from the moment equation, the equilibrium equations are

$$\Sigma F_X = A_X + B_X = 0$$

$$\Sigma F_Y = A_Y + B_Y - W = 0$$

$$\Sigma M_{AZ} = B_Y L - W\frac{L}{2} = 0$$

The solution of these equations shows that $A_Y = B_Y = W/2$. However, we can only determine that $A_X = -B_X$. It is very tempting to say that A_X and B_X are both zero, because no loading force is acting in that direction. However, we cannot prove this contention on the basis of the available equations, and indeed it is not true. Furthermore, if a horizontal force was also applied to the bar, we would have no idea of the separate values of A_X and B_X.

The condition of static indeterminancy is one situation where it is inadequate to model a system as a rigid body. The value of the indeterminate reactions can be obtained only when we consider the manner in which the body deforms, a subject that is treated in the study of mechanics of materials.

When a system is indeterminate, we can solve the basic equilibrium equations for some of the reactions in terms of the remaining ones, as was done for the system of Figure 25. The remaining reactions, which in this case was only the component B_X, are said to be *redundant* because they are not required to support the system. By modifying the supports to remove the redundant reactions, we can obtain a statically determinate system. For example, we can replace the pin at B by a roller as shown in Figure 26.

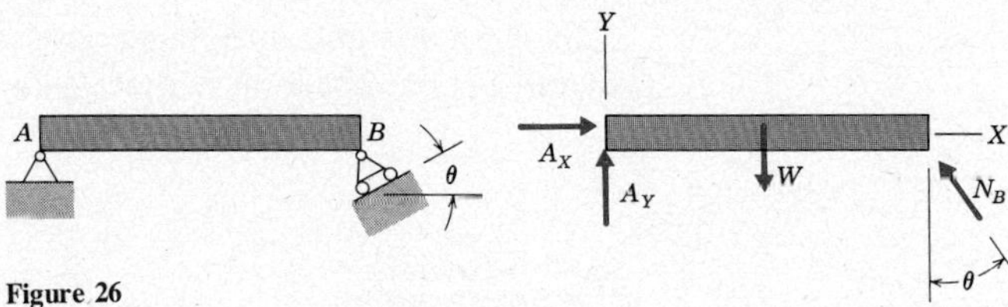

Figure 26

The system in Figure 26 is statically determinate, because all reactions may be obtained from the corresponding equilibrium equations. The solutions are $A_Y = W/2$, $A_X = W/(2 \cos \theta)$ and $N_B = (W/2) \tan \theta$. These answers are acceptable, except for the case where $\theta = 90°$, for which A_X and N_B become infinite.

Notice that when $\theta = 90°$, the reaction $\bar{N}_B$ in Figure 26 is the same as the reaction $\bar{B}_X$ in Figure 25, so the reaction that has been removed in that case is not the redundant one. The system is then capable of rigid body motion, as illustrated in Figure 27. This occurs because the roller permits point B to move in the vertical direction, which is the same direction as that in which the end would move if the bar was attached only to pin A. This particular modification of the supports causes the beam to be partially constrained.

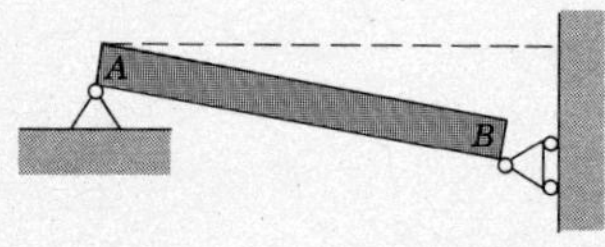

Figure 27

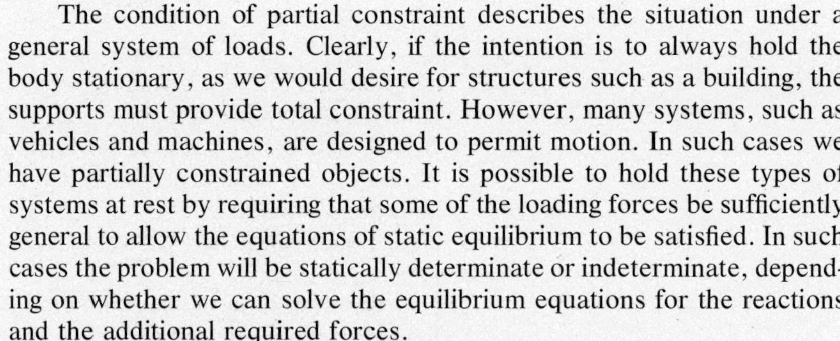

The condition of partial constraint describes the situation under a general system of loads. Clearly, if the intention is to always hold the body stationary, as we would desire for structures such as a building, the supports must provide total constraint. However, many systems, such as vehicles and machines, are designed to permit motion. In such cases we have partially constrained objects. It is possible to hold these types of systems at rest by requiring that some of the loading forces be sufficiently general to allow the equations of static equilibrium to be satisfied. In such cases the problem will be statically determinate or indeterminate, depending on whether we can solve the equilibrium equations for the reactions and the additional required forces.

To illustrate this matter, consider the beams mounted on rollers in Figures 28a and 28b. Clearly, for a general set of loads, both beams will move horizontally. Therefore, they are partially constrained. However, if the loading system is such that the sum of their components in the X direction is zero, as is the case for the loads shown, this motion will not occur. For the beam in Figure 28a we may determine the normal reactions of the two rollers by summing moments about a roller and summing forces in the Y direction. Hence, this beam is statically determinate. On the other hand, the static equilibrium equations are inadequate for the determination of the three normal reactions of the roller supports in Figure 28b. Hence, this is a condition of static indeterminancy.

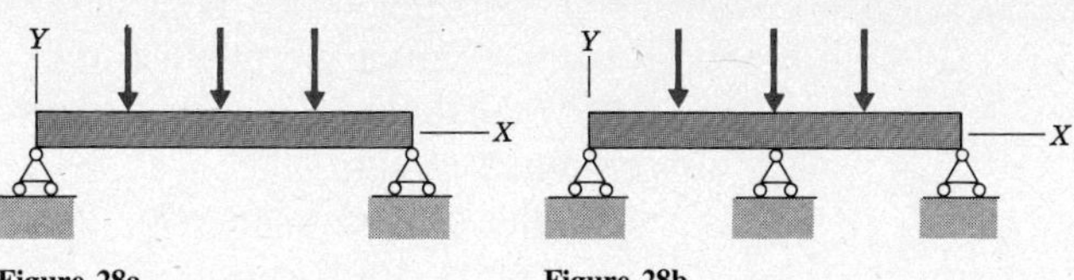

Figure 28a **Figure 28b**

Our main concern in this test is with statically determinate systems that are *properly constrained,* either because of their support conditions or because of the loads to which they are subjected. Nevertheless, you should be able to recognize situations of static indeterminancy and partial constraint on the basis of the solvability of the equilibrium equations.

4 Solving Problems in Static Equilibrium

The development of the ability to determine the forces acting on a body by applying the equations of static equilibrium is a primary goal of a course in statics. This capability is a prerequisite to the study of most of the topics in this text.

Problem solving is not an art, but rather a professional decision made by recognizing the proper keys. In what follows we present a logical way of thinking to identify the keys. It is not meant for memorization, but rather as a guide while working problems. Work enough problems *correctly* and the keys will become part of your thinking process.

1 Select the body (or bodies) to be isolated in a free body diagram (FBD) by examining where the forces of interest are applied. Draw a simple sketch of this body. Label any significant points if they are not already labeled.

2 Choose an XYZ coordinate system for the description of the components of the force and position vectors. Usually, the coordinate axes should coincide with the directions in which the dimensions of the system are given. Show this coordinate system in the sketch of step 1.

3 Complete the FBD begun in step 1 by showing all forces acting on the body. In doing this, examine each support of the body to determine what types of reactions exist. The weight force, acting at the center of mass,

should be included unless it is small compared to the other forces, or unless no information about it is given. If the dimensions describing a force do not appear in the given diagram, show the necessary dimensions in the FBD.

4 Choose a convenient point along the line of action of some unknown constraint forces and compute the sum of the moments of the forces about this point. Equate the components of the moment sum to zero.

5 Equate to zero each component of the sum of the forces appearing in the FBD. (At times, this step may be omitted if it can be seen that the force sums involve unknown forces whose values are not desired, provided that these unknown forces have not appeared in the moment equation obtained in step 4.)

6 Solve the simultaneous equations that result from steps 4 and 5. Interpret the results, for example, by checking any initial assumptions.

Note that all of the simplifications of planar force systems, such as being able to sum force components directly and being able to compute moments without actually using the cross product, may be employed.

ILLUSTRATIVE PROBLEM 1

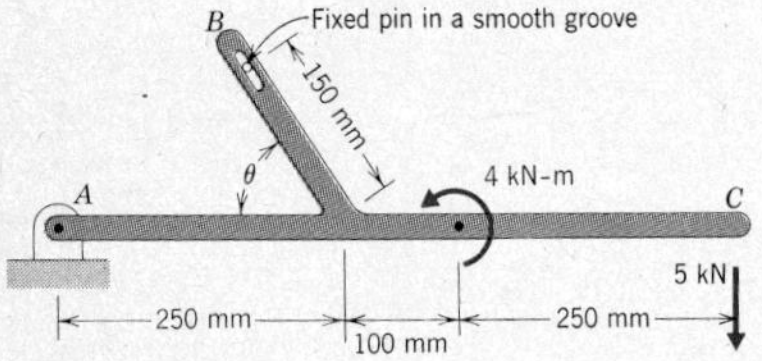

A 5-kN force and a 4-kN-m couple are applied to the forked bar. Determine the reaction forces at supports A and B when $\theta = 36.87°$.

Solution

Steps 1 and 2 We choose the forked bar for the FBD, because we wish to determine the forces required to support this body against the given set of loads. The X and Y axes for this planar problem are selected as horizontal and vertical because the dimensions are given with respect to these directions.

Step 3 The reaction at pin B is normal to the surface of the smooth groove, so we have the following FBD.

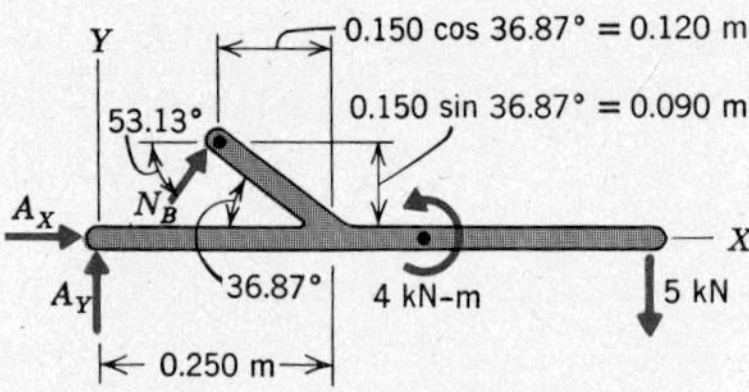

The dimensions shown in the FBD are needed to define the orientation and point of application of the normal force $\bar{N}_B$.

Step 4 We choose pin A for the moment sum in order to eliminate the unknown reactions at that pin. Thus, we have

$$\Sigma M_{AZ} = -N_B \cos 53.13°(0.090) + N_B \sin 53.13° \,(0.250 - 0.120) + 4 - 5(0.600) = 0 \text{ kN-m}$$

Step 5 The equilibrium equations obtained by summing force components in the directions of the coordinate axes are

$$\Sigma F_X = N_B \cos 53.13° + A_X = 0 \text{ kN}$$

$$\Sigma F_Y = N_B \sin 53.13° + A_Y - 5 = 0 \text{ kN}$$

Step 6 We may now solve the moment equation for N_B and substitute this result into the force equations. Therefore

$$N_B[-0.6(0.9) + 0.8(0.13)] + 1.0 = 0 \quad \text{so} \quad N_B = -20 \text{ kN}$$

$$-20(0.60) + A_X = 0 \quad \text{so} \quad A_X = 12 \; kN$$

$$-20(0.80) + A_Y - 5 = 0 \quad \text{so} \quad A_Y = 21 \text{ kN}$$

The negative value of N_B means that this force acts in the opposite direction from the one shown (guessed at) in the FBD. What does this mean in regard to which surface of the groove pin B is bearing against?

ILLUSTRATIVE PROBLEM 2

A lightweight rod is wedged between smooth pins A and B, and rests on the smooth floor. Determine all reaction forces as functions of the magnitude of the force $\bar{F}$ and the angle θ at which it is applied. Then determine the range of values of θ for which static equilibrium is possible.

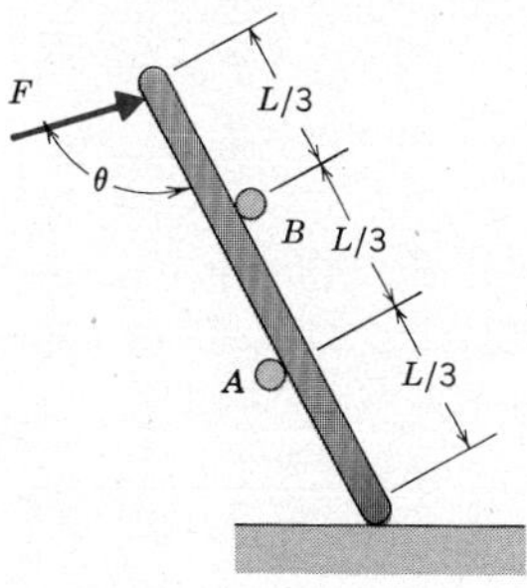

Solution

Steps 1 and 2 We isolate the bar for the FBD because all of the forces are applied

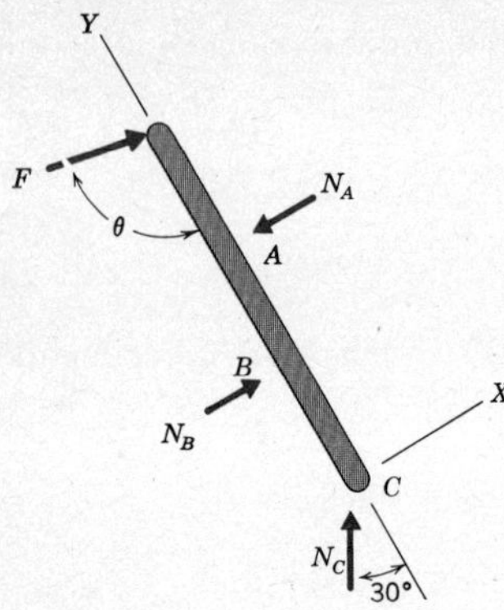

to this body. For this planar problem, we incline the Y axis parallel to the bar because the dimensions are referred to the axis of the bar.

Step 3 Because the floor and the pins are smooth, the reaction forces are normal to the surface of contact. Hence, the FBD is as shown.

Step 4 The unknown force $\bar{N}_C$ passes through point C, so the corresponding moment equilibrium equation about point C is

$$\Sigma M_{CZ} = -N_A\left(\frac{L}{3}\right) + N_B\left(\frac{2L}{3}\right) - F \sin \theta(L) = 0$$

Step 5 Equating the force sums in the X and Y direction to zero yields

$$\Sigma F_X = N_C \sin 30° + N_A - N_B + F \sin \theta = 0$$

$$\Sigma F_Y = N_C \cos 30° + F \cos \theta = 0$$

Step 6 We have three equilibrium equations for the three unknowns N_A, N_B, and N_C, which may be solved determining N_C from the second equation and using that result in the other two equations. Thus

$$N_C = -\frac{F \cos \theta}{\cos 30°} = -1.1547\, F \cos \theta$$

$$\left(-F\frac{\cos \theta}{\cos 30°}\right) \sin 30° + N_A - N_B + F \sin \theta = 0$$

$$-N_A + 2N_B - 3F \sin \theta = 0$$

The solution of the last two of these equations is

$$N_B = F(2 \sin \theta + \cos \theta \tan 30°)$$

$$= F(2 \sin \theta + 0.5774 \cos \theta)$$

$$N_A = F(\sin \theta + 2 \cos \theta \tan 30°)$$

$$= F(\sin \theta + 1.1547 \cos \theta)$$

For static equilibrium the three normal reactions we have determined may not be negative because the surfaces of contact can only press against each other. Thus, for N_C non-negative we have

$$-\cos \theta \le 0$$

which is satisfied when $90° \le \theta \le 270°$. For N_B positive or zero we have

$$2 \sin \theta + 0.5774 \cos \theta \ge 0$$

The equality sign applies when $\theta = 164.4°$ of $344.4°$. It is easily verified that the inequality is satisfied when $0 \leq \theta \leq 164.4°$ or $344.4° \leq \theta \leq 360°$. Finally, for N_A non-negative we have

$$\sin\theta + 1.1547\cos\theta \geq 0$$

The equality sign applies when $\theta = 130.9°$ or $310.9°$ and the inequality is satisfied when $0 \leq \theta \leq 130.9°$ or $310.9° \leq \theta > 360°$. Thus, the region where all reactions are positive is

$$90° \leq \theta < 130.9°$$

What happens if θ is not within this range?

ILLUSTRATIVE PROBLEM 3

The 2000-kg rectangular plate is supported by a hinge at point A, a short horizontal link at corner E, and two cables, as shown. The weight force acts at the geometric center G. Determine the tensions in the cables and the reaction force at support A.

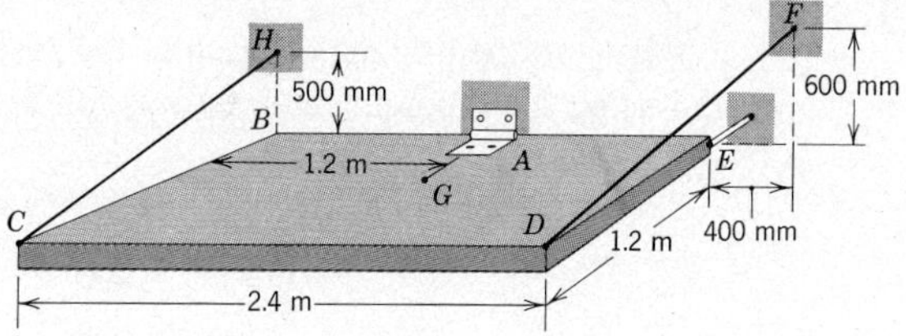

Solution

Steps 1 and 2 To determine the reaction forces required to support the plate against its own weight, we choose this body for the FBD. We select the Z axis vertical and the X and Y axis parallel to the edges of the plate.

Step 3 A hinge is considered to prevent movement of the supported point, but not prevent rotation of the body. The short link exerts a force parallel to its axis, thereby preventing movement of the body in that direction. These considerations lead to the following free body diagram.

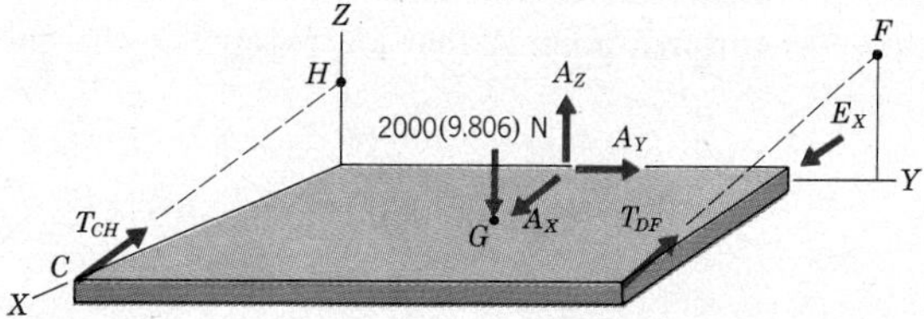

Step 4 Before considering the equilibrium equations, we shall describe the cable tensions in terms of their components. Writing each of these forces as the product of its (unknown) magnitude times its unit vector, we have

$$\bar{T}_{CH} = T_{CH} \frac{\bar{r}_{H/C}}{|\bar{r}_{H/C}|} = T_{CH} \frac{-1.2\bar{I} + 0.5\bar{K}}{\sqrt{(1.2)^2 + (0.5)^2}}$$

$$= T_{CH}(-0.9231\bar{I} + 0.3846\bar{K})$$

$$\bar{T}_{DF} = T_{DF} \frac{\bar{r}_{F/D}}{|\bar{r}_{F/D}|} = T_{DF} \frac{-1.2\bar{I} + 0.4\bar{J} + 0.6\bar{K}}{\sqrt{(1.2)^2 + (0.4)^2 + (0.6)^2}}$$

$$= T_{DF}(-0.8571\bar{I} + 0.2857\bar{J} + 0.4286\bar{K})$$

The moment equilibrium equations may now be formulated. In order to eliminate the reaction forces at hinge A, we sum moments about this point. Thus, for moment equilibrium we write

$$\Sigma \bar{M}_A = \bar{r}_{C/A} \times \bar{T}_{CH} + \bar{r}_{D/A} \times \bar{T}_{DF} + \bar{r}_{E/A} \times E_X\bar{I}$$
$$+ \bar{r}_{G/A} \times [-2(9.806)\bar{K}] = \bar{0} \text{ kN-m}$$

Referring to the original diagram of the system to determine the position vectors, and using the foregoing expressions for $\bar{T}_{CH}$ and $\bar{T}_{DF}$, yields

$$\Sigma \bar{M}_A = (1.2\bar{I} - 1.2\bar{J}) \times T_{CH}(-0.9231\bar{I} + 0.3846\bar{K})$$
$$+ (1.2\bar{I} + 1.2\bar{J}) \times T_{DF}(-0.8571\bar{I} + 0.2857\bar{J}$$
$$+ 0.4286\bar{K}) + 1.2\bar{J} \times E_X\bar{I} + 0.6\bar{I} \times [-2(9.806)\bar{K}]$$
$$= \bar{0} \text{ kN-m}$$

Carrying out the cross products gives

$$\Sigma \bar{M}_A = T_{CH}\,(-0.4615\bar{J} - 1.1077\bar{K} - 0.4615\bar{I})$$
$$+ T_{DF}(+0.3428\bar{K} - 0.5143\bar{J} + 1.0285\bar{K}$$
$$+ 0.5143\bar{I}) - 1.2E_X\bar{K} + 11.767\bar{J} = \bar{0}$$

The scalar equations are now obtained by equating the sums of like components to zero, in effect thereby summing moments about the three lines passing through point A that are parallel to the coordinate axes. Hence,

$$\Sigma M_{AX} = -0.4615T_{CH} + 0.5143T_{DF} = 0 \text{ kN-m}$$

$$\Sigma M_{AY} = -0.4615T_{CH} - 0.5143T_{DF} + 11.767 = 0 \text{ kN-m}$$

$$\Sigma M_{AZ} = -1.1077T_{CH} + (0.3428 + 1.0285)T_{DF} - 1.2E_X = 0 \text{ kN-m}$$

Step 5 The FBD shows that the reaction at hinge A has components in three directions and that the weight force is in the negative Z direction. Using the vector expressions for the cable forces derived in step 4, the sums of the force components parallel to each of the coordinate axes are

$$\Sigma F_X = -0.9231T_{CH} - 0.8571T_{DF} + A_X + E_X = 0 \text{ kN}$$

$$\Sigma F_Y = 0.2857T_{DF} + A_Y = 0$$

$$\Sigma F_Z = 0.3846T_{CH} + 0.4286T_{DF} + A_Z - 2(9.806) = 0 \text{ kN}$$

Step 6 The equilibrium equations in steps 4 and 5 may be solved by using the first two moment equations to determine T_{CH} and T_{DF}, then substituting these results into the last moment equation to find E_X. These values are then substituted into the force equations to find the forces at hinge A. This procedure yields

$$T_{CH} = 12.75 \qquad T_{DF} = 11.44 \qquad E_X = 1.30 \text{ kN}$$

$$A_X = 20.27 \qquad A_Y = 3.27 \qquad A_Z = 9.81 \text{ kN}$$

Note that the values of the cable tensions are positive, as they must be because cables can only pull on other bodies.

ILLUSTRATIVE PROBLEM 4

An ideal pulley is pinned to end A of the bent bar ABC and a cable is wrapped around the pulley as shown. A tensile force of 200 lb is applied to the free end of the cable. (a) Prove that the tensile force within the cable is constant. (b) Determine the reactions at pin B and roller C. Is the roller bearing against the upper or lower surface?

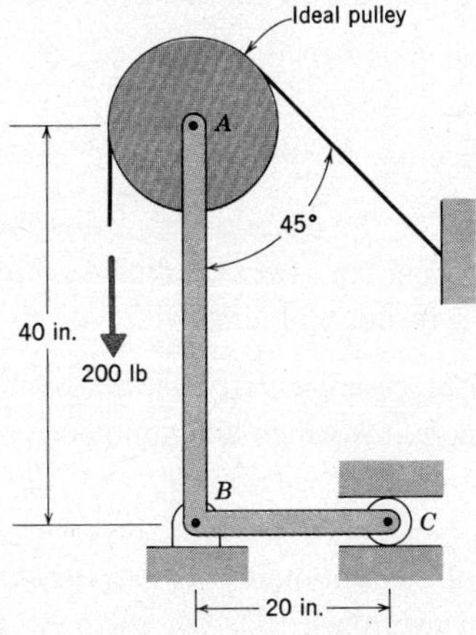

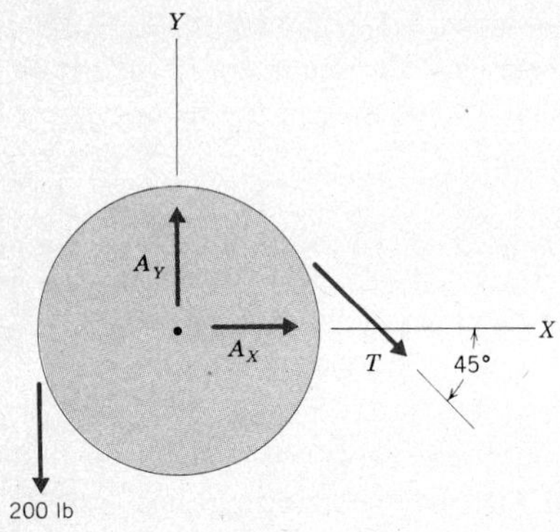

Solution

Up to now, we have accepted the fact that the tension in a cable that passes over an ideal pulley is constant. We shall now prove it.

Steps 1 a and 2a In part (a) we are concerned with the interaction between the cable and the pulley, so we isolate the pulley in a FBD. Horizontal and vertical coordinate axes in this planar problem suit the directions in which dimensions are given.

Step 3a To the left of the pulley the tension within the cable is the 200-lb force applied to its end, whereas the tension on the right is considered to be some unknown value T. The pulley is also attached to a pin at its center, so there are two reaction force components at that point, as shown.

Step 4a Choosing pin A for the moment sum in order to eliminate the reactions at this location, the moment equilibrium equation is

$$\Sigma M_{AZ} = -T(5) + 200(5) = 0 \text{ lb-in.}$$

Step 5a The force sums for the planar system of forces are

$$\Sigma F_X = T \cos 45° + A_X = 0 \text{ lb}$$

$$\Sigma F_Y = A_Y - T \sin 45° - 200 = 0 \text{ lb}$$

Step 6a The moment equation gives the desired result:

$$T = 200 \text{ lb}$$

If the tension were not constant, it would be necessary to have a couple acting on the pulley to resist the difference in the moments of the two tension forces.

The remaining force equilibrium equations give the values of A_X and A_Y, which were not requested. Nevertheless, they are

$$A_X = -141.4 \text{ lb} \qquad A_Y = 341.4 \text{ lb}$$

We will have use for these values in the following.

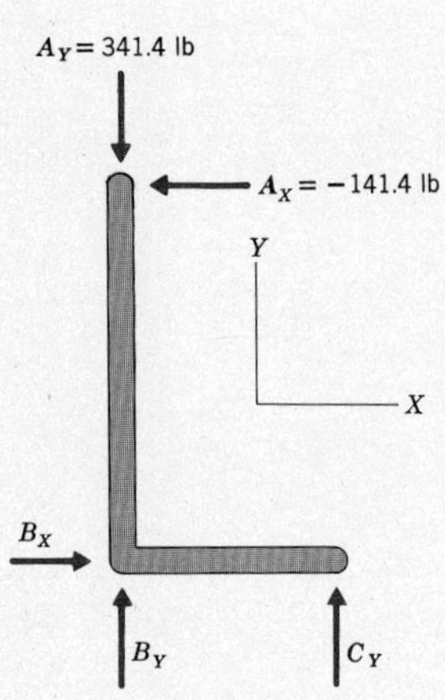

Steps 1b and 2b In order to determine the reactions at points B and C, we draw a FBD of the bent bar. The coordinate system utilized in part (a) will also suffice for this portion of the problem.

Step 3b Forces are exerted on the bent bar by the pin at point B, the roller at point C, and pin A which holds the pulley. Hence, we have the FBD shown in the left margin.

Particular attention is drawn to the way in which the reaction forces at point A are entered into the FBD. We show $\bar{A}_X$ and $\bar{A}_Y$ opposite from the directions in which they were assumed when we drew the FBD of the pulley, in accordance with Newton's third law. Then, after depicting the algebraic forces, we enter their numerical values with the same sign as they were calculated to have. Thus, we see that the pulley really pulls the bar to the right while pressing it down. This procedure is employed in order to prevent serious, and also common, sign errors. We will have more to say about this matter in the next module.

Step 4b Summing moments about pin B, we have

$$\Sigma M_{BZ} = C_Y(20) + (-141.4)(40) = 0 \text{ lb-in.}$$

Step 5b The force sums are

$$\Sigma F_X = B_X - (-141.4) = 0 \text{ lb}$$

$$\Sigma F_Y = B_Y + C_Y - 341.4 = 0 \text{ lb}$$

Step 6b The solutions of the equilibrium equations are readily found to be

$$C_Y = 282.8 \qquad C_Y = 58.6 \qquad B_X = 141.4 \text{ lb}$$

Because C_Y is positive, it acts in the same sense as that shown in the FBD. Therefore, roller C is being pushed upward, which means that it is bearing on the lower surface.

As a final comment, let us consider how we would have approached the solution of part (b) if we were allowed to consider as an established fact that the tension within a cable is unchanged when the cable passes over an ideal pulley. If we then were to isolate the assembly of bar ABC and the pulley in a FBD, we would only have the reaction forces at points B and C and the two cable tensions, as shown.

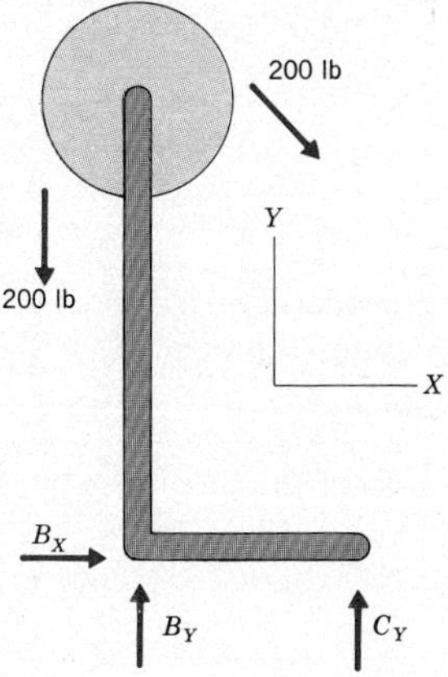

This would enable us to determine the desired reactions, without going through the intermediate step of studying the interaction between the pulley and the bent bar.

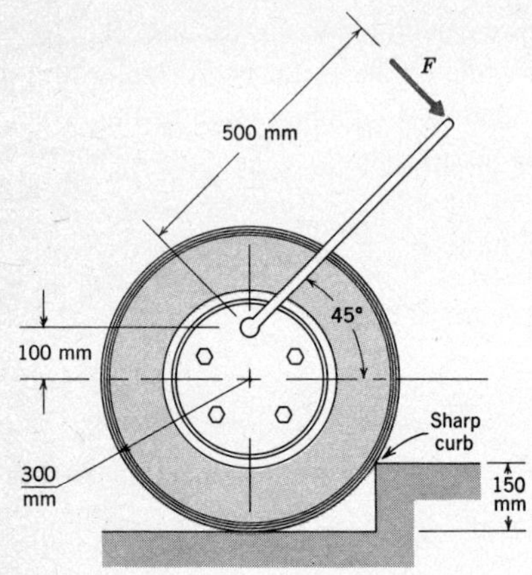

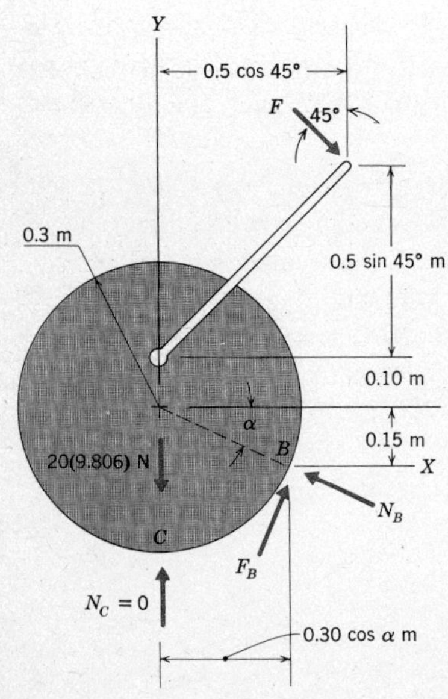

ILLUSTRATIVE PROBLEM 5

A lightweight wrench is attached to the 20-kg automobile tire as shown. Determine the magnitude of the force $\bar{F}$ that will cause the tire to begin to roll over the sharp curb whose height is 150 mm.

Solution

Steps 1 and 2 We wish to study the effect on the tire of the force $\bar{F}$ which is applied to the wrench. In order to avoid studying the interaction forces exerted between these two bodies, we draw a FBD of their assembly. Noting the way in which the dimensions of the planar system are given, we choose the X axis horizontal and the Y axis vertical.

Step 3 At the point of contact with the curb there are normal and tangential reactions because the corner is sharp. In general, there would also be a normal force exerted between the tire and the horizontal surfaces at point C, but in this problem we want to determine the condition where rolling is impending (i.e., on the verge of occurring). We therefore set this force to zero. The resulting FBD is as shown. Do not forget the weight force.

The dimensions shown locate the point of application of the given force $\bar{F}$ and the point of contact B. It is readily seen that

$$\alpha = \sin^{-1}\left(\frac{0.15}{0.30}\right) = 30°$$

Step 4 We choose point B for the moment sum in order to eliminate the unknown reaction forces $\bar{N}_B$ and $\bar{F}_B$ which are applied at this point. This yields

$$\Sigma M_{BZ} = 20(9.806)(0.30 \cos 30°) - F \cos 45°(0.15 + 0.10 + 0.50 \sin 45°) - F \sin 45°(0.50 \cos 45° - 0.30 \cos 30°) = 0 \text{ N-m}$$

Step 5 We omit consideration of force equilibrium, because the moment equation may be solved to obtain the value of F and we are not interested in the reaction components N_B and F_B.

Step 6 The solution of the moment equation is

$$F = 103.3 \text{ newtons}$$

If the magnitude of the force $\bar{F}$ exceeds this value, it will not be possible to maintain static equilibrium in the position where the wheel is almost in contact with the ground.

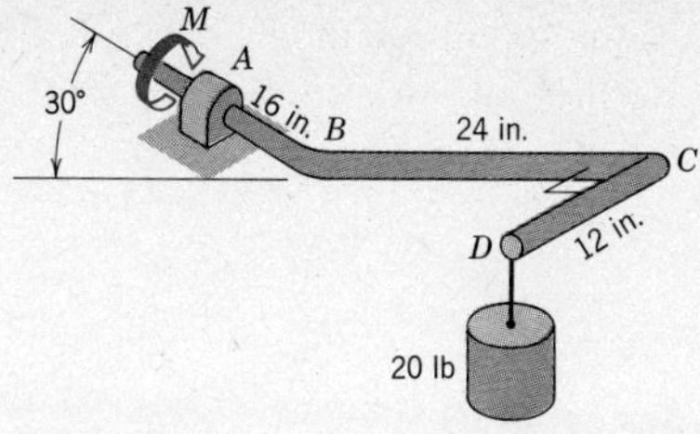

ILLUSTRATIVE PROBLEM 6

A 20-lb block is suspended from the bent bar, which is supported by a thrust bearing at end A. In the position shown, section ABC of the bar coincides with the vertical plane, and section CD is horizontal. Considering the thrust bearing to be tight fitting, it will permit rotation of the rod only about axis AB. Determine the couple $\bar{M}$ that must be exerted about this axis in order to maintain static equilibrium.

Solution

Steps 1 and 2 The problem is to determine the couple that should be applied to the bent bar to resist the loading of the suspended block, so we isolate the bar in the FBD, aligning one coordinate axis with the horizontal portion BC and choosing one of the other axes vertical.

Step 3 The reactions at bearing A consist of a force-couple system, $\bar{A}$ and $\bar{M}_A$. These are general vectors, except that $\bar{M}_A$ can have no component parallel to axis AB because the bearing does not prevent rotations about that axis. Other forces acting on the bar are the couple $\bar{M}$ parallel to axis AB, and a 20-lb force representing the tension in the cable holding the block. As can be seen, we represent the bearing reactions by their components.

Step 4 Although summing moments about point A in the conventional manner will give the value of M eventually, a shortcut is obtained by noting that we can eliminate all unknown forces and couples except for $\bar{M}$ by considering only the component of $\Sigma\, \bar{M}_A$ parallel to line AB. By definition, this is the sum of the moments about line AB. Clearly, if $\Sigma\, \bar{M}_A = 0$, it must also be true that $\Sigma\, M_{AB} = 0$.

To compute the moment of the 20-lb force about line AB, we first compute the moment of this force about any convenient point along line AB. Clearly, point B is best, because $\bar{r}_{D/B}$ is easily determined to be

$$\bar{r}_{D/B} = 24\bar{I} + 12\bar{K} \text{ in.}$$

Thus

$$\bar{M}_B = (24\bar{I} + 12\bar{K}) \times (-20\bar{J}) = 240(\bar{I} - 2\bar{K}) \text{ lb-in.}$$

We next evaluate the component of $\bar{M}_B$ parallel to line AB by calculating the dot product of $\bar{M}_B$ with a unit vector parallel to that line, directed from point A to point B. This gives

$$\begin{aligned} \bar{M}_B \cdot \bar{e}_{B/A} &= 240(\bar{I} - 2\bar{K}) \cdot (\cos 30^\circ\, \bar{I} - \sin 30^\circ\, \bar{J}) \\ &= 207.8 \text{ lb-in.} \end{aligned}$$

Finally, because the couple $\bar{M}$ is directed along the line from B to A, we have

$$\Sigma\, M_{AB} = -M + 207.8 = 0 \text{ lb-in.}$$

Step 5 We omit the computation of the force equilibrium equations, because the moment equation may be solved for M and we are not interested in any of the reactions exerted by the bearing.

Step 6 The solution of the moment equation is

$$M = 207.8 \text{ lb-in.}$$

As an aid in seeing why this method of solution was employed, you should try the alternative of solving $\Sigma \bar{M}_A = 0$.

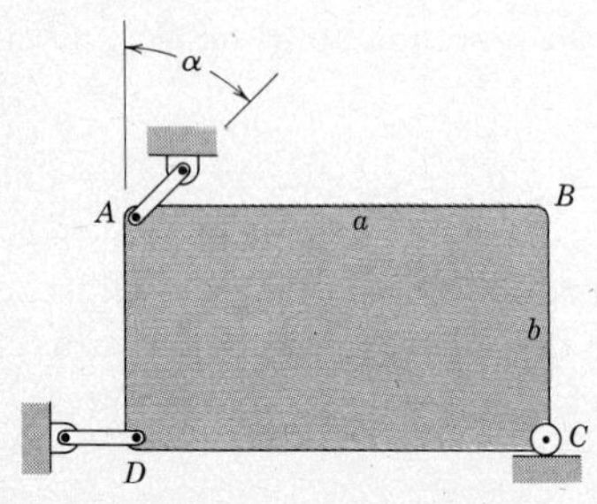

ILLUSTRATIVE PROBLEM 7

Determine the value of the angle α for the short link at point A that results in improper constraint of the rectangular plate.

Solution

Recall that the condition of partial constraint concerns equilibrium under a general set of loading forces. Hence, we will assume an arbitrary set of loads.

Steps 1 and 2 We want to study the nature of the constraints of the plate, so we draw a FBD of the plate. The given dimensions suit choosing coordinate axes parallel to the edges of the plate.

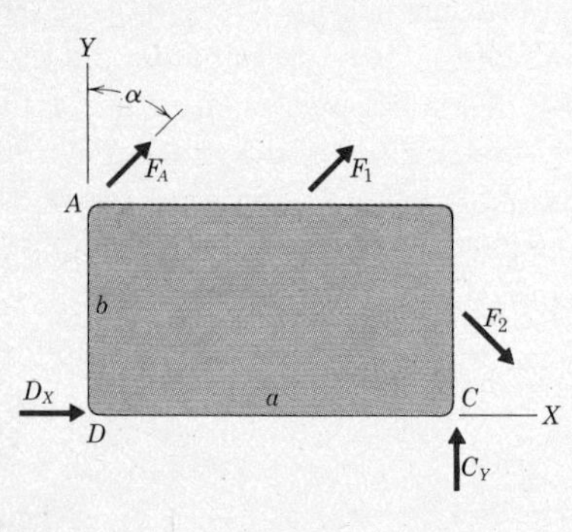

Step 3 Let two arbitrary forces $\bar{F}_1$ and $\bar{F}_2$ represent the general set of loads. The short links exert forces only along their axes, so the FBD is as shown.

Step 4 We select point C for the moment sum because the reaction forces $\bar{C}_Y$ and $\bar{D}_X$ both intersect this point. The moment equilibrium equation is then

$$\Sigma M_{CZ} = -(F_A \sin \alpha)b - (F_A \cos \alpha)a + M_1 + M_2 = 0$$

where M_1 and M_2 are the moments of $\bar{F}_1$ and F_2 respectively, about point C.

Step 5 The force equilibrium equations are

$$\Sigma F_X = D_X + F_A \sin \alpha + (F_1)_X + (F_2)_X = 0$$

$$\Sigma F_Y = C_Y + F_A \cos \alpha + (F_1)_Y + (F_2)_Y = 0$$

Step 6 The three equilibrium equations are now solved for the values of F_A, C_Y, and D_X in terms of the angle α, yielding

$$F_A = \frac{M_1 + M_2}{b \sin \alpha + a \cos \alpha}$$

$$D_X = -(F_1)_X - (F_2)_X - \frac{(M_1 + M_2) \sin \alpha}{b \sin \alpha + a \cos \alpha}$$

$$C_Y = -(F_1)_Y - (F_2)_Y - \frac{(M_1 + M_2)\cos\alpha}{b\sin\alpha + a\cos\alpha}$$

We now look for the condition of partial constraint, which is evidenced by an infinite value for at least one of the reactions. For this system, all three reactions are infinite when the denominator $b \sin \alpha + a \cos \alpha = 0$. Thus

$$\alpha = -\tan^{-1}\left(\frac{a}{b}\right)$$

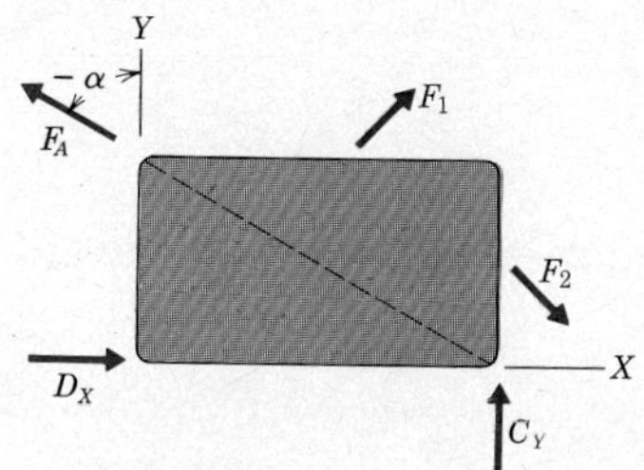

The reason that this value of α leads to partial constraint becomes obvious when we draw the FBD corresponding to this value of α, as depicted to the left. In this situation all of the reactions intersect point C, so they cannot provide a moment to counteract the moment of the loading forces about point C.

In general, there are two ways of having improper constraint of a body for the case of a planar set of forces. As happened in this problem, all of the reaction forces may be concurrent. Alternatively, as occurred for the beams in Figure 26, all of the reaction forces may be parallel.

HOMEWORK PROBLEMS

III.99 A wrecking bar wedged into a crevice is loaded at its end by a 200-N force, as shown. Considering the surface of contact at point B to be smooth, determine the forces exerted on the bar at end A.

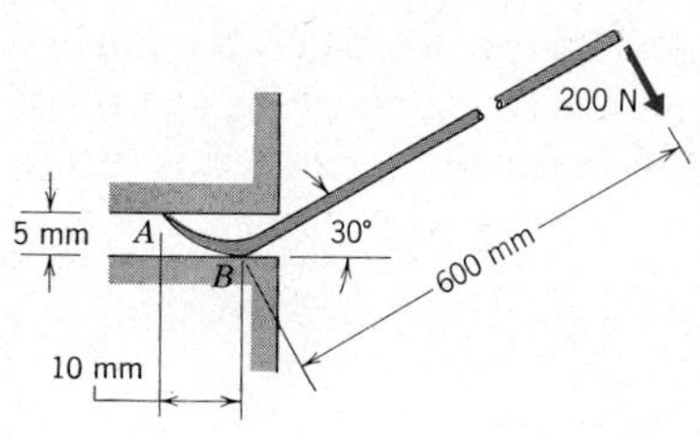

Prob. III.99

III.100 The 10-lb, 4-ft-long bar is suspended between two smooth walls as shown. The center of mass of the bar is at the midpoint G. Determine the reactions exerted by the walls on the bar.

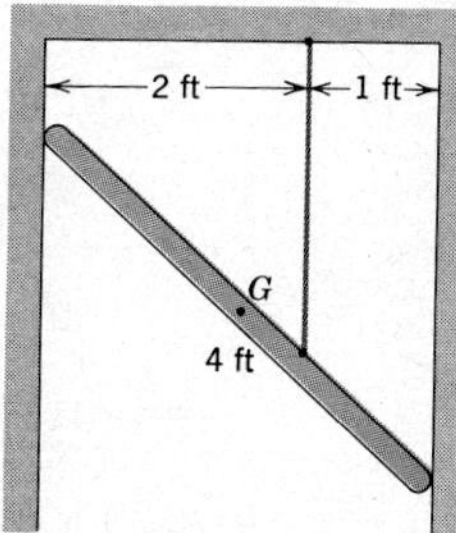

Prob. III.100

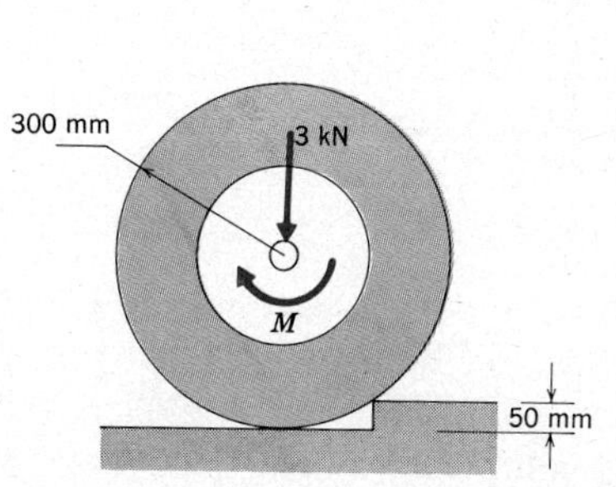

Prob. III.101

III.101 The rear axle of an automobile exerts the vertical 3-kN force and couple $\bar{M}$ on the 20-kg tire, which is at rest. Determine the magnitude of $\bar{M}$ required to cause the tire to roll over the curb.

III.102 The 1300-kg automobile is at rest on the 20° hill. The weight force acts at point G. Determine the normal force exerted between the rough ground and each tire. Compare these forces to those obtained when the vehicle is on level ground. Can the reaction forces tangent to the hill be determined? The brakes on all four wheels are applied.

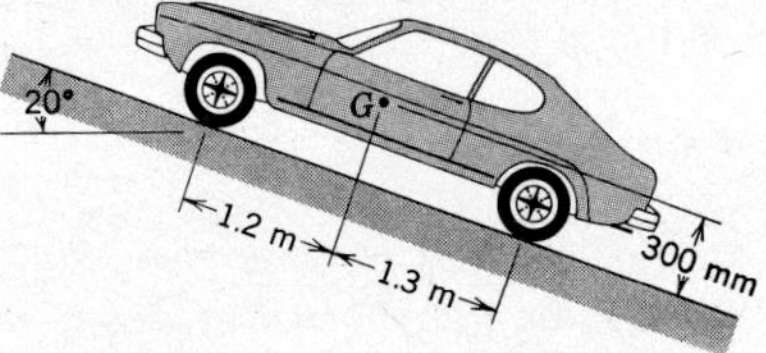

Prob. III.102

III.103 Solve Problem III.102 if the automobile is facing downward on the hill.

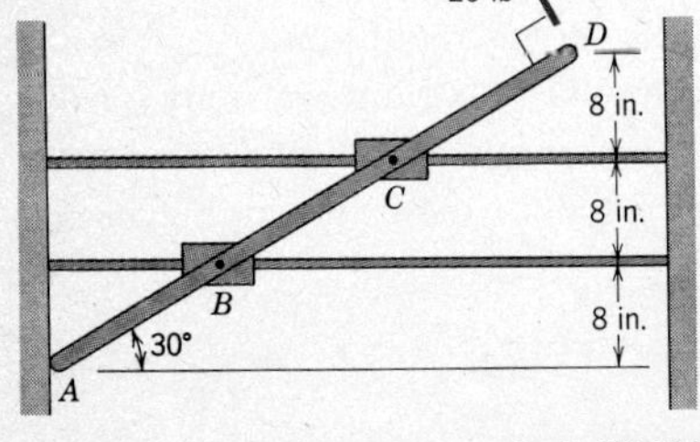

Prob. III.104

III.104 The bar AD is supported by collars B and C, which ride on smooth horizontal rods. End A of the bar rests against the smooth vertical wall. Determine the reactions resulting from the 20-lb force shown.

III.105 The symmetrical balance beam scale shown is used to compare the unknown mass m_B to the standard mass m_A. The support C consists of a frictionless knife edge. Derive an expression for m_B/m_A in terms of the angle θ. The center of mass of arm ACB coincides with point C.

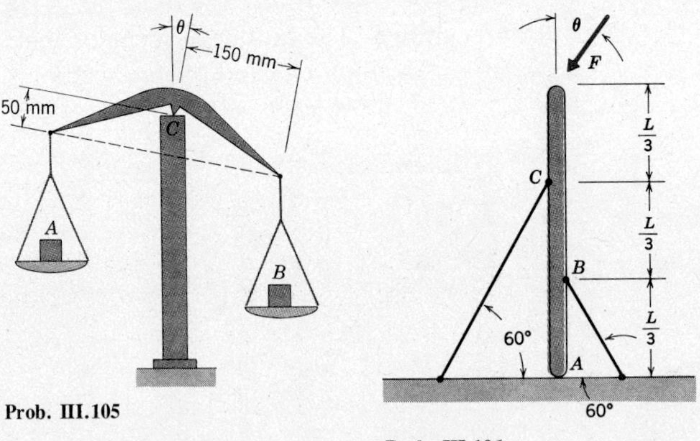

Prob. III.105

Prob. III.106

III.106 A lightweight pole, which rests on a smooth horizontal surface at end A, is supported by two cables, as shown. (a) Determine the tension in each cable as a function of the angle θ at which the force $\bar{F}$ is applied. (b) From the result of part (a), determine the allowable range of values of θ for which the pole will be in static equilibrium.

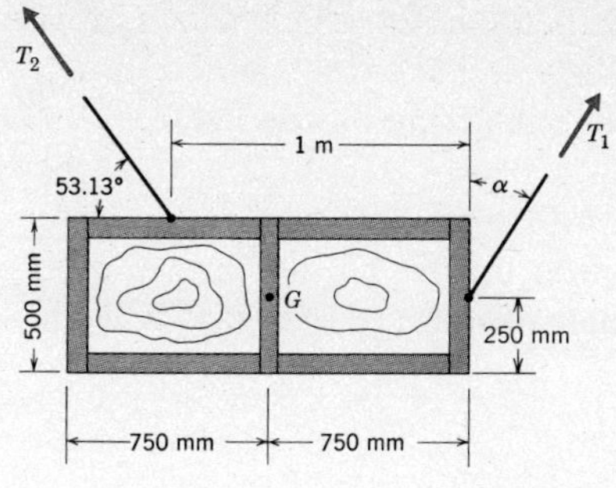

Prob. III.107

III.107 Determine the tension in the cables and the angle α required to hold the 1000-kg crate in the position shown. Point G is the center of mass.

III.108 The stiffness of the spring is 2 kN/m and its unstretched length is 400 mm. Bar AB is in equilibrium in the position shown. Determine the mass of the bar, knowing that its center of mass is at the midpoint G.

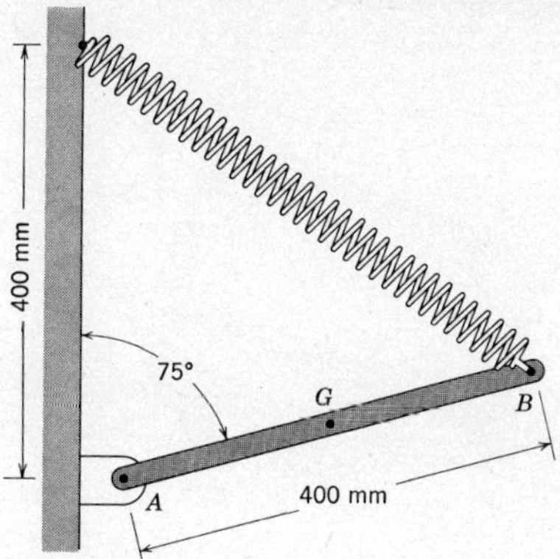

Prob. III.108

III.109 An 80-lb force and a 40 ft-lb couple are applied to the bent bar as shown. Determine the reaction at roller B. Which surface of the groove does this roller bear against?

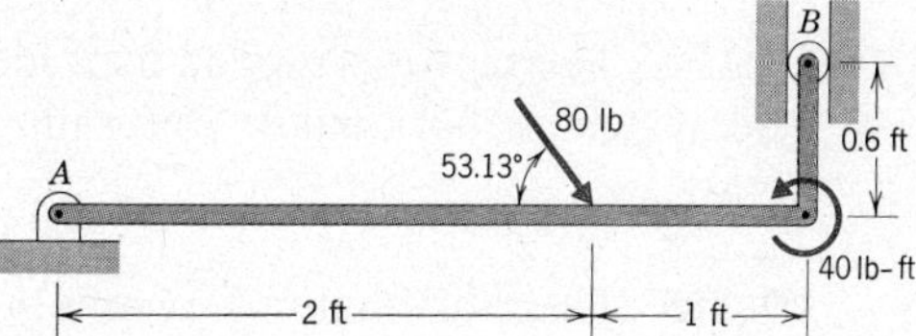

Prob. III.109

III.110 The bent bar is supported by the fixed pin A and pin B which fits into the smooth groove. Knowing that d = 750 mm, determine the reactions for the loading shown.

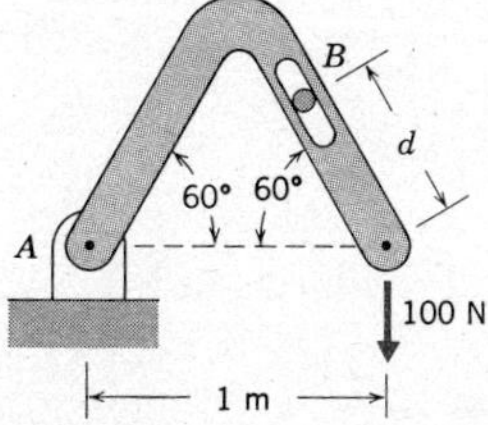

Prob. III.110

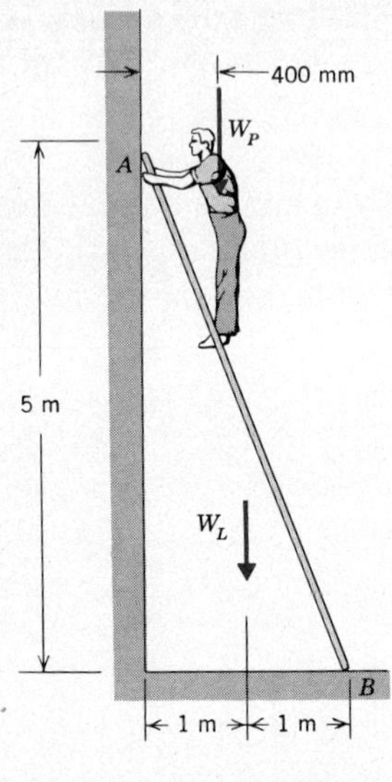

Prob. III.111

III.111 A 65-kg painter is standing on a 10-kg ladder which rests against the smooth wall and the rough ground. The weights of the painter and the ladder act as shown. Determine the reactions acting on the ladder.

III.112 Determine the horizontal force $\bar{F}$ that must be applied to corner C of the 100-kg crate to hold it in position on the hill as shown. Both surfaces of contact are smooth. The center of mass of the crate is at the geometric center G.

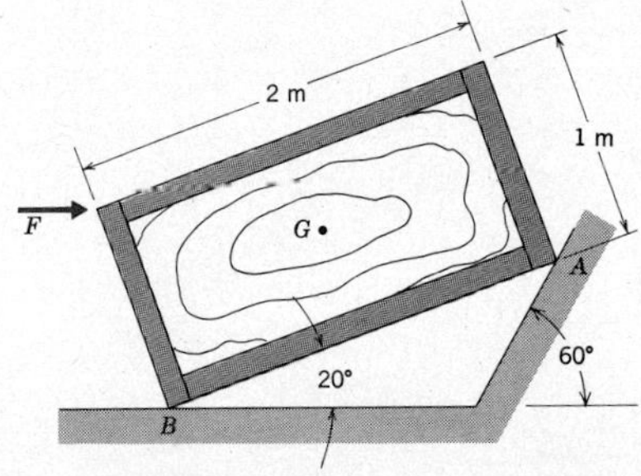

Prob. III.112

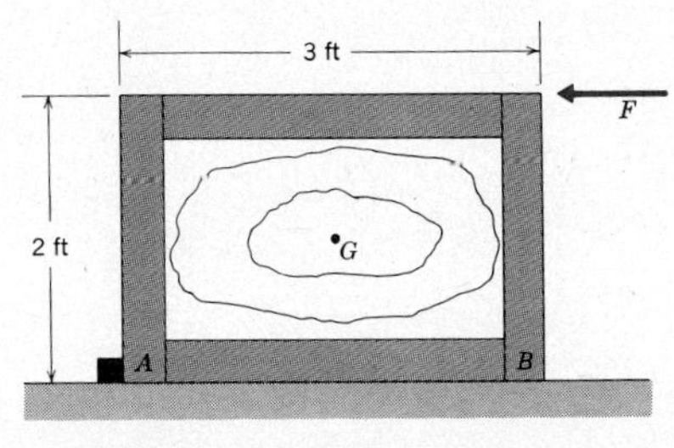

Prob. III.113

III.113 The 2-ton crate resting on the smooth horizontal surface is restrained by cleat A. Determine the maximum force $\bar{F}$ that may be exerted on the crate without causing it to begin to tip over. The center of mass of the crate is at the geometric center G.

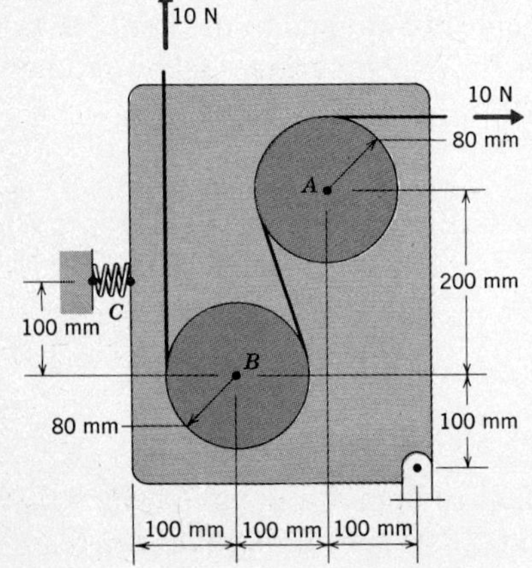

Prob. III.114

III.114 The tape is wrapped over ideal pulleys A and B in the tape guide. The system is horizontal. Determine the force exerted by the spring at point C.

III.115 The crane boom shown is supported by a pin at end A and cable CD. Determine the tension in cable CD required to hold the 10-ton crate in the position shown.

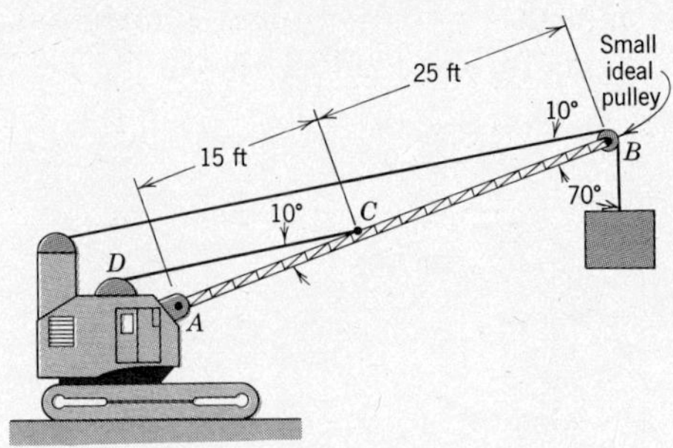

Prob. III.115

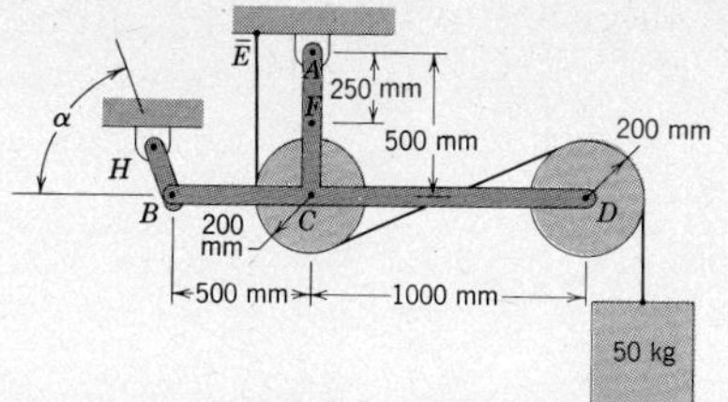

Prob. III.116

III.116 The forked bar is supported by fixed pin A and short link BH. A 50-kg block is suspended from a cable that passes over ideal pulleys C and D which are pinned to the bar, as shown. Knowing that $\alpha = 45°$, determine the reaction force exerted by the short link.

III.117 Solve Problem III.116 if, instead of being fastened at point E, the cable is fastened to the bar at point F.

III.118 The 100-lb box is supported by a knife edge at point A and by a cable that passes over the ideal pulley B. The center of mass of the box is at the geometric center G. For the loading shown, determine the tension in the cable (a) when $d = 10$ in., (b) when $d = 28$ in.

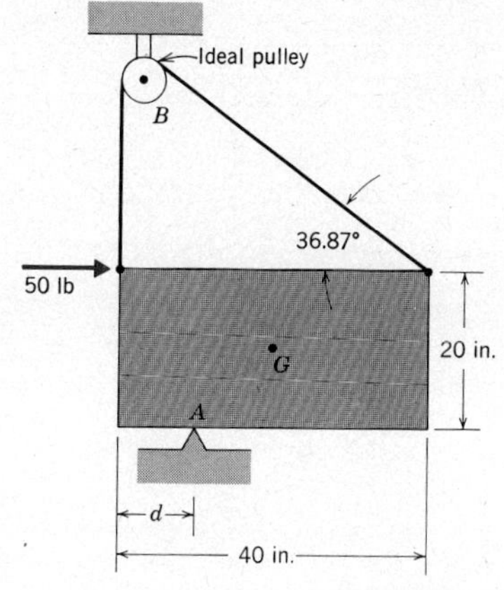

Prob. III.118

III.119 A gear box mounted on a beam is subjected to two torques, as shown. Determine the reactions at the supports of the beam.

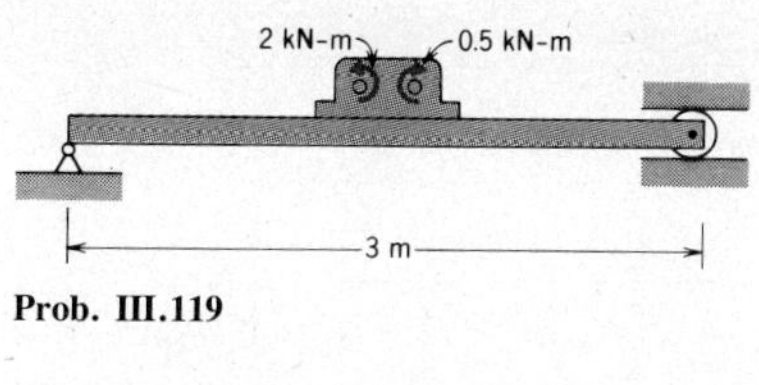

Prob. III.119

III.120–III.123 Determine the reactions required to support the beam shown.

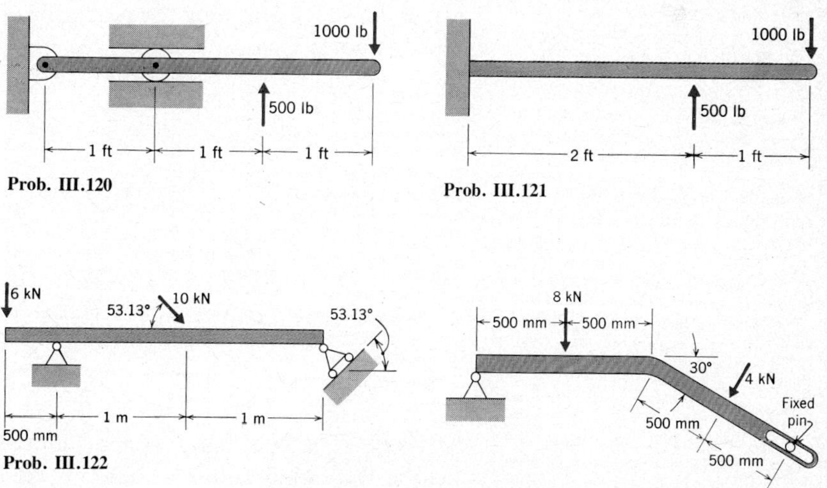

Prob. III.123

III.124–III.126 Determine the reactions required to support the structure shown.

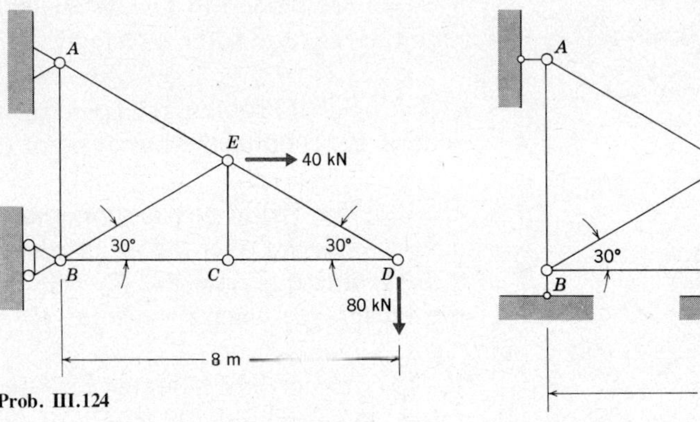

Prob. III.124

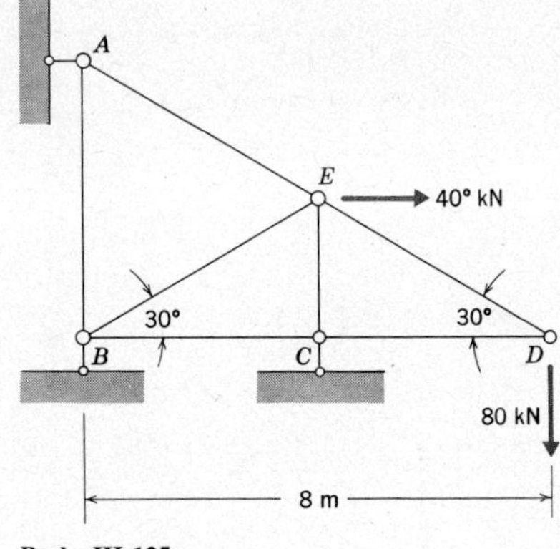

Prob. III.125

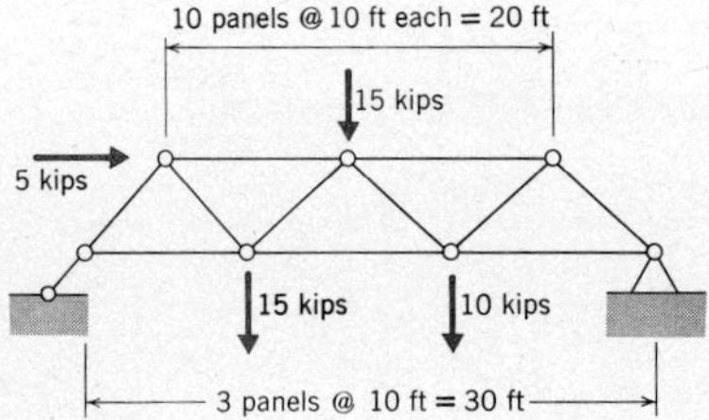

Prob. III.126

III.127 The structure shown is supported by three short links at points A, B, and C. Determine the reactions of these links when the angle for the link at point C is (a) $\alpha = 90°$, (b) $\alpha = 36.87°$.

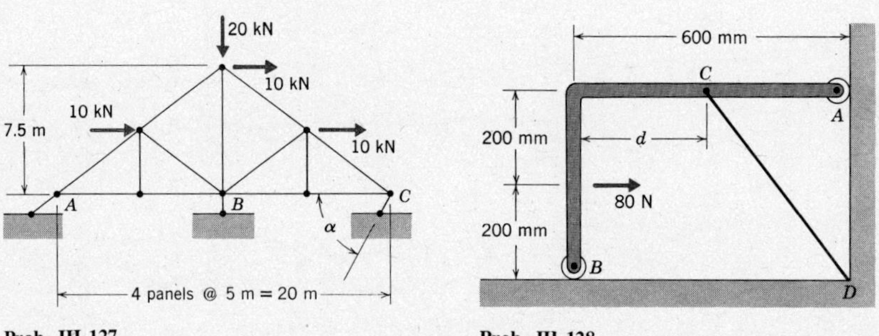

Prob. III.127

Prob. III.128

III.128 The bent bar is supported by rollers at ends A and B and by the cable. Knowing that $d = 300$ mm, determine the reactions at the rollers and the tension in the cable for the loading shown.

III.129 A tow truck exerts the horizontal force $\bar{F}$ to hold the 1500-kg automobile at rest on the hill in the position shown. The brakes are not applied. The center of mass of the automobile is point G. Determine the magnitude of the force $\bar{F}$.

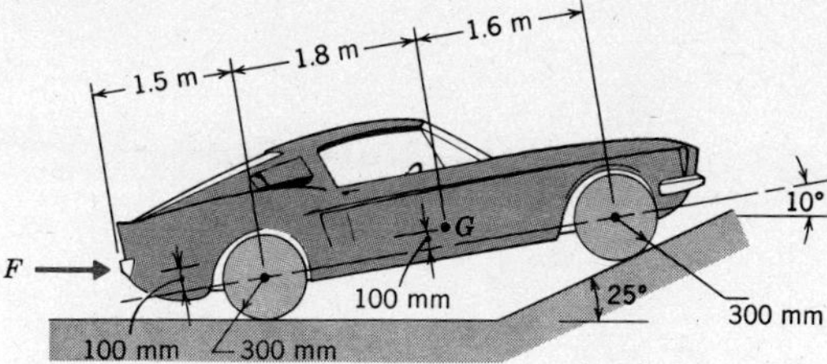

Prob. III.129

III.130 The winch supports a mass m in the position shown. Derive an expression for the required vertical force $\bar{V}$ in terms of the other parameters of the system.

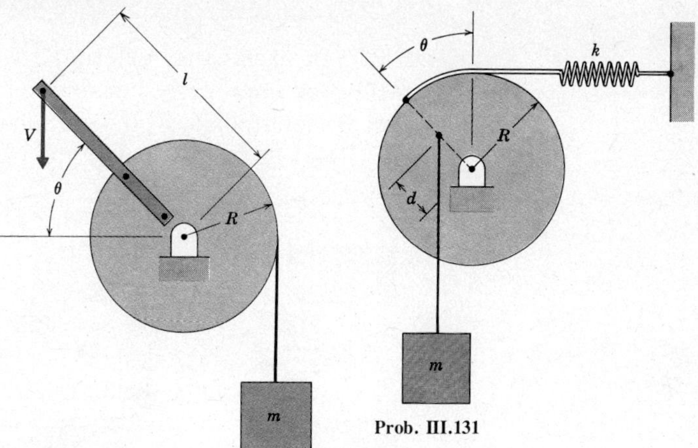

Prob. III.131

Prob. III.130

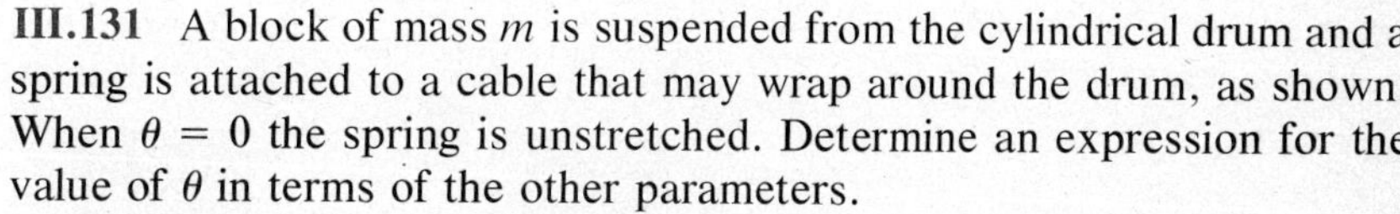

III.131 A block of mass m is suspended from the cylindrical drum and a spring is attached to a cable that may wrap around the drum, as shown. When $\theta = 0$ the spring is unstretched. Determine an expression for the value of θ in terms of the other parameters.

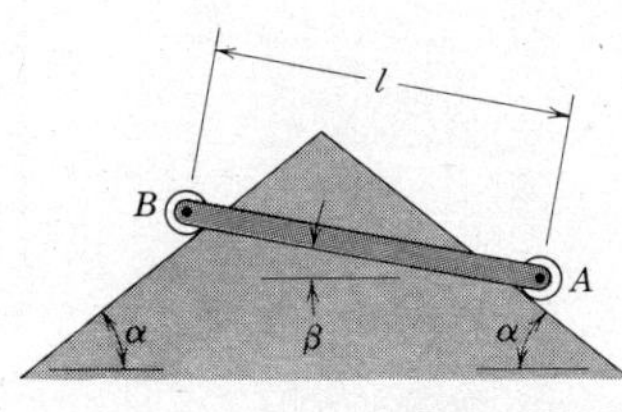

Prob. III.132

III.132 Rollers A and B of masses m_A and m_B, respectively, are interconnected by a lightweight rod of length l and placed on the smooth surface shown. Knowing that in the equilibrium position shown $\alpha = 30°$ and $\beta = 10°$, determine the ratio m_A/m_B.

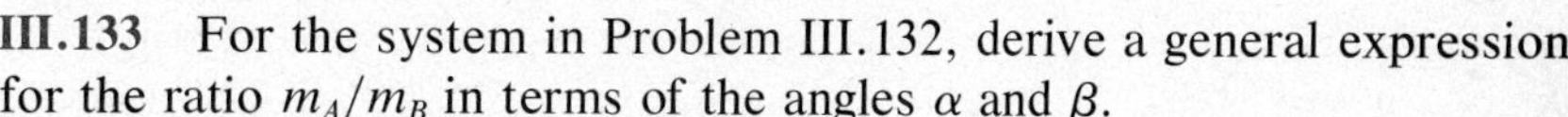

III.133 For the system in Problem III.132, derive a general expression for the ratio m_A/m_B in terms of the angles α and β.

III.134–III.136 Derive an expression for the angle θ corresponding to equilibrium for the systems shown. In each case the bar has a mass m and its center of mass is at the midpoint G. All surfaces are smooth.

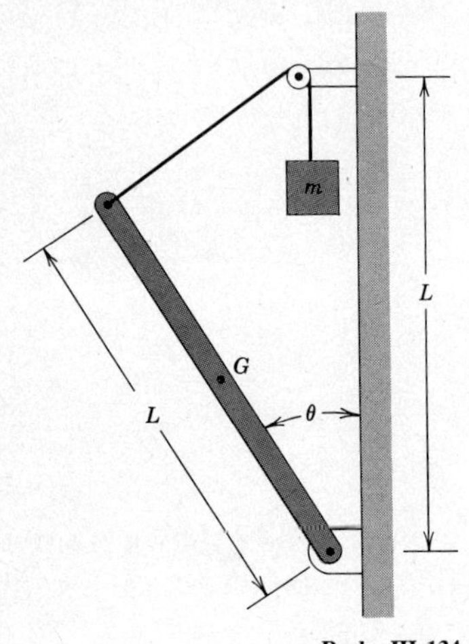

Prob. III.134

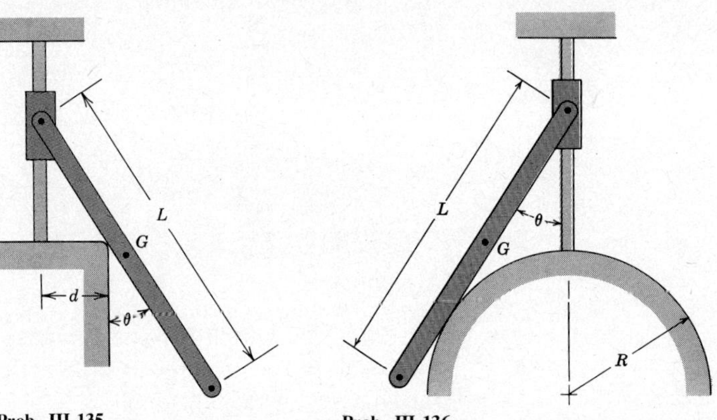

Prob. III.135 Prob. III.136

III.137 In each case, determine whether (a) the plate is completely, partially, or improperly constrained, and (b) whether the reactions are statically determinate or indeterminate for cases of complete constraint. (*Hint:* Refer to the last paragraph of Illustrative Problem 7.)

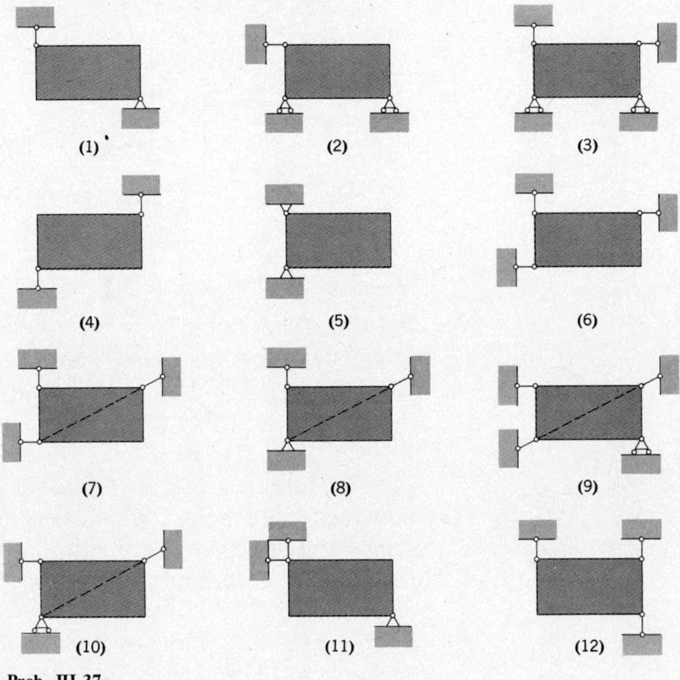

Prob. III.37

III.138 (a) For the forked bar in Illustrative Problem 1, determine the reactions at pins A and B as a function of the angle θ. (b) From the results in part (a), determine whether there are any values of θ for which the bar is improperly constrained.

For the systems in the following problems, determine the value of the dimension d that results in improper constraint of the system.

III.139 Problem III.110

III.140 Problem III.118

III.141 Problem III.128

For the systems in the following problems, determine the angle α that results in improper constraint of the system.

III.142 Problem III.116

III.143 Problem III.127

III.144 Point C on the top flange of the wide flange beam is subjected to the loads shown. Determine the reactions acting at point A on the axis of symmetry.

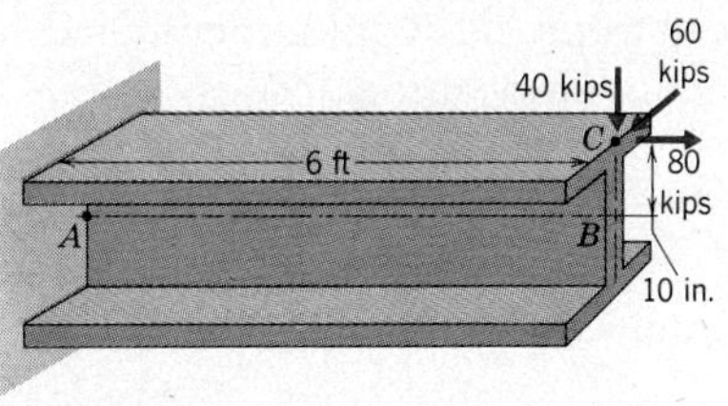

Prob. III.144

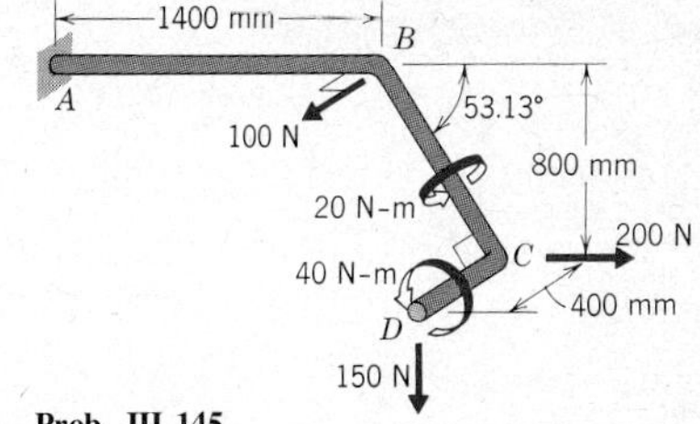

Prob. III.145

III.145 The bent bar, which is welded to the wall at end A, carries the loads shown. Determine the reactions.

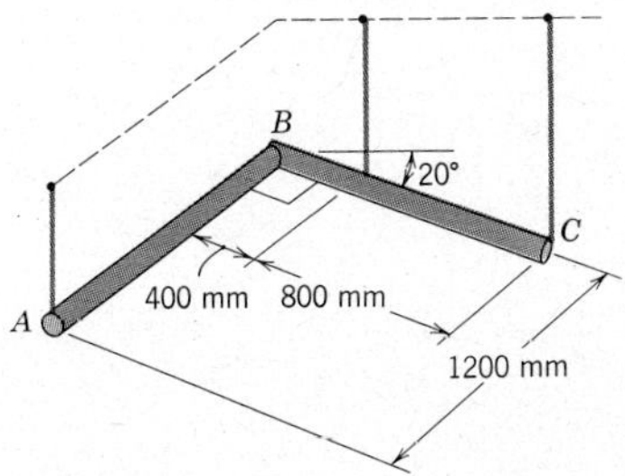

Prob. III.146

III.146 Two identical rods of 2 kg mass each are welded at right angles and suspended by vertical wires. Section AB is horizontal and section BC is at 20° below the horizontal. Determine the tension in each wire. The center of gravity of each rod is at its midpoint.

III.147 A package is placed on the three-legged circular table, whose mass is 30 kg. The reactions between the legs of the table and the ground are known to be $N_A = 100$, $N_B = 120$, $N_C = 130$ newtons. Determine the mass of the package and the values of R and θ describing its location.

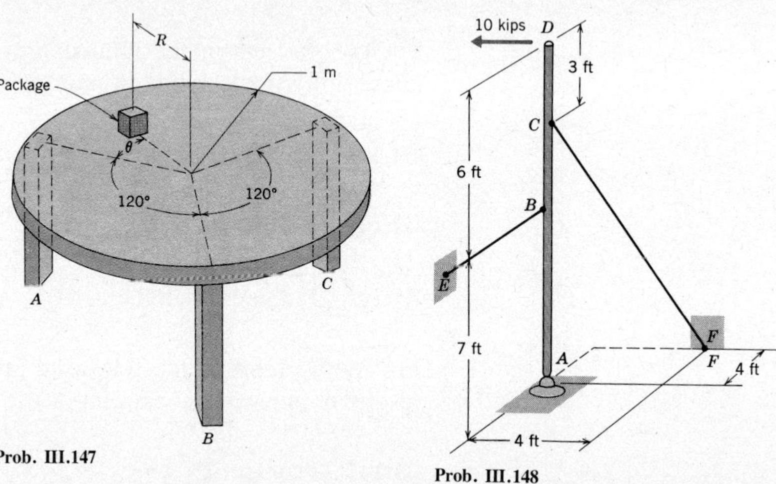

Prob. III.147

Prob. III.148

III.148 The antenna mast supported by a ball-and-socket joint at end A and two guy wires is subjected to the horizontal 10 kip load. Determine the tensions in the guy wires and the reactions at the ball-and-socket joint.

III.149 The cabinet door is supported by a rod that exerts an axial force $\bar{F}$ as shown. Knowing that the door has a mass of 10 kg and that its center of mass coincides with its geometric center, determine the magnitude of $\bar{F}$ for the open position shown.

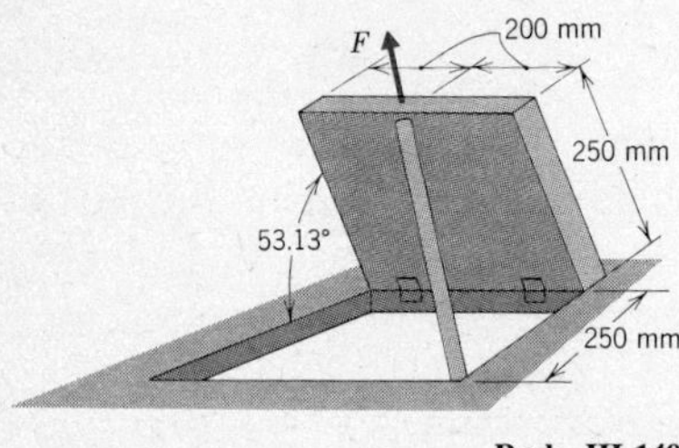

Prob. III.149

III.150 The 10-kg horizontal flagpole is supported by a ball-and-socket joint at point A and two cables. Knowing that $d = 2$ m, determine the tension in the cables. The center of mass of the boom is at the midpoint.

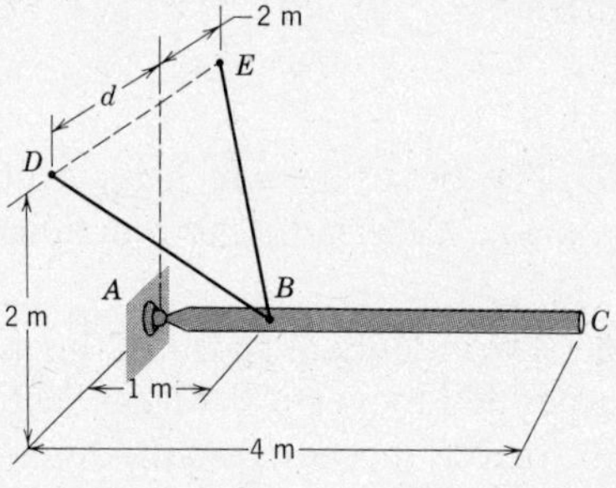

Prob. III.150

III.151 Solve Problem III.150 for $d = 1$ m.

III.152 The horizontal boom is held in position at an angle of 60° with respect to the vertical wall by the ball-and-socket joint and the two cables. Determine the reactions at the ball-and-socket joint and the tension in the cables resulting from the vertical 50-kN loads.

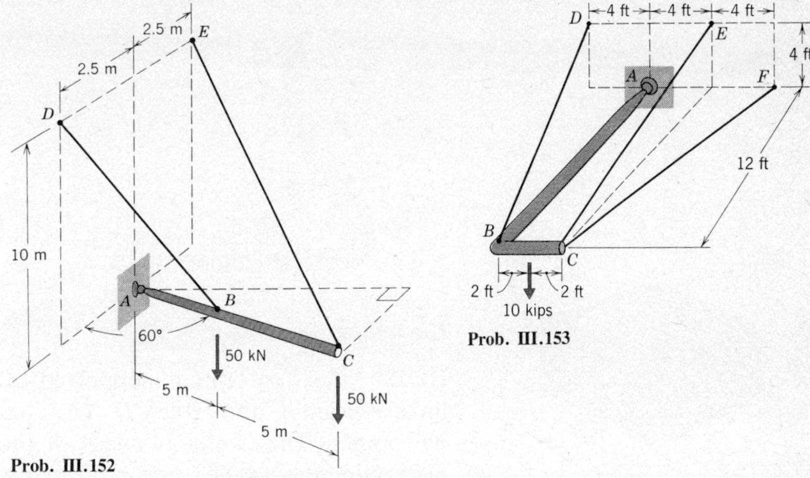

Prob. III.152

Prob. III.153

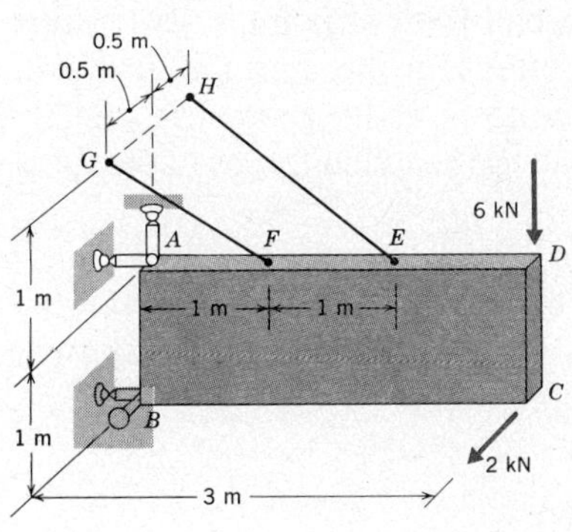

Prob. III.154

III.153 The bent bar ABD is supported by three cables and a ball-and-socket joint at end A. Determine the tension in the cables for the loading shown.

III.154 The rectangular plate is oriented vertically and supported by four short links at points A and B and by two cables. Determine the reaction forces of the links for the loading shown. The connections for the short links are ball-and-socket joints.

III.155 The bent bar is supported by three smooth eyebolts as shown. Is the system statically determinate and properly constrained?

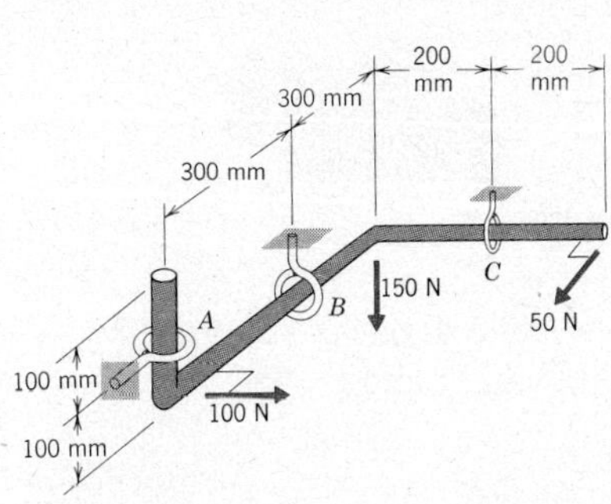

Prob. III.155

III.156 A rigid bent tube guides a flexible shaft that is transmitting a 20 lb-ft torque, as shown. The tube is supported by the smooth eyebolts at A, B, and C. Is the system properly constrained?

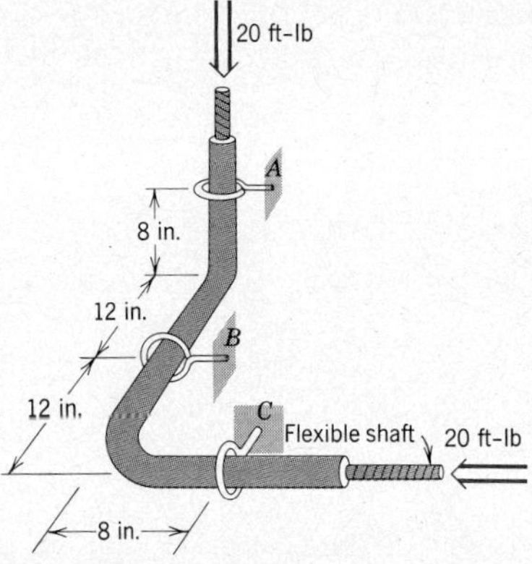

Prob. III.156

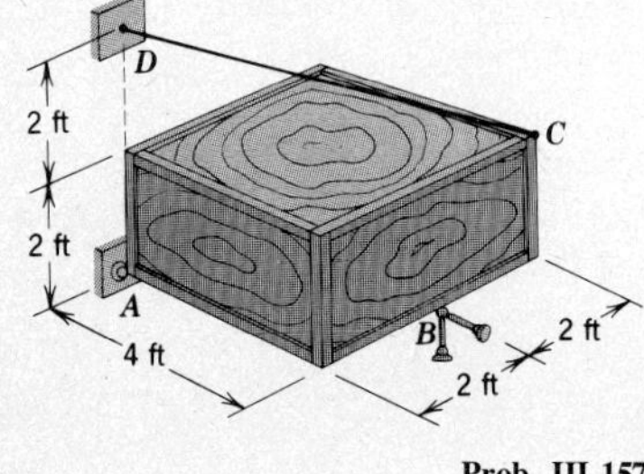

Prob. III.157

III.157 A 2-ton crate is supported by a ball joint at point A, two short links at point B, and cable CD. The connections for the short links are also ball joints. The center of mass of the crate is at its geometric center. Determine the tensile force in the cable and the reaction forces of the links at point B.

III.158 The winch is holding a 100-kg package. Determine the vertical force $\bar{F}$ and the reactions at ball bearings A and B for the position shown. The drum of the winch has a diameter of 400 mm.

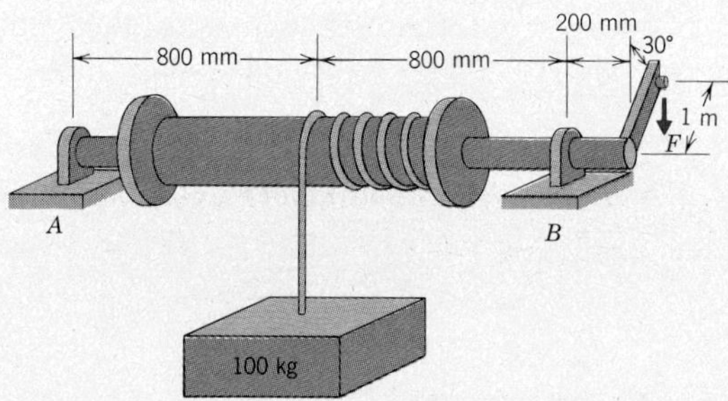

Prob. III.158

III.159 The T-bar is supported by thrust bearing A, bearing B, and the smooth wall at end D. The bar fits loosely in both bearings. Knowing that sections *AB* and *CD* each have a mass of 2 kg, and that their centers of mass are at their respective midpoints, determine the reactions exerted by each bearing and by the wall.

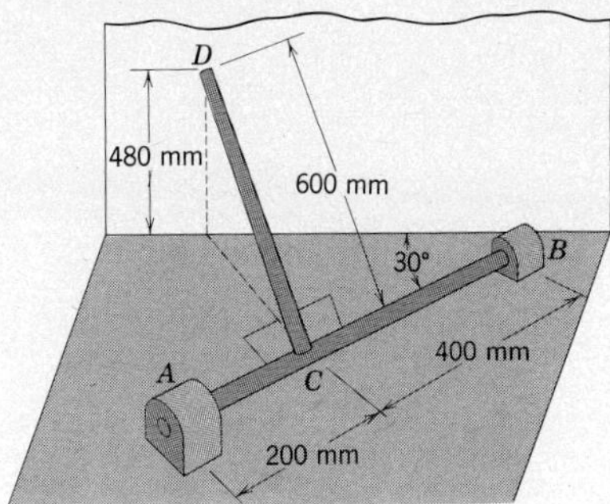

Prob. III.159

III.160 Ends *A* and *B* of the 5-kg rigid bar are attached by ball joints to lightweight collars that may slide over the smooth fixed rods shown. The center of mass of the bar is at its midpoint. Determine the horizontal force $\bar{F}$ applied to collar *A* that will result in static equilibrium in the position where $d = 1$ m. The length of the bar is 1.5 m.

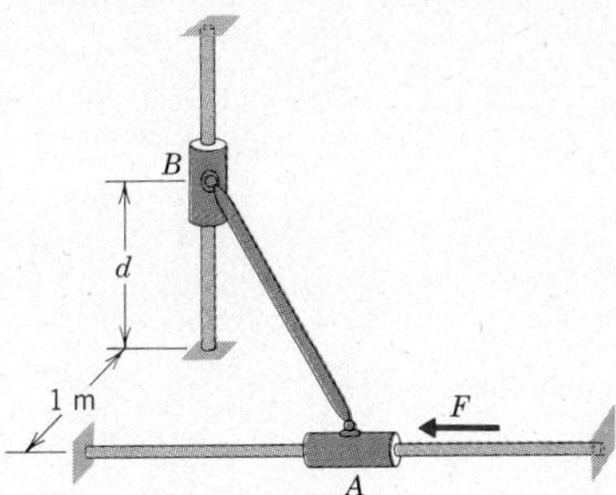

Prob. III.160

III.161 Solve Problem III.160 for the position where $d = 0.5$ m.

MODULE IV
STRUCTURES

The methods for evaluating the forces acting on a rigid body were developed in Module III. A few systems encountered there contained interconnected bodies. In those situations it was sometimes necessary to investigate each body in the system, using the law of action and reaction.

The systems we will consider in this module all share the feature of being composed of many interconnected bodies. These systems are *engineering structures,* which are designed for the purpose of safely resisting, transmitting, and, at times, modifying the various forces applied to them. The individual bodies forming a structure are called its *members*. In this module we will treat two categories of structures. They differ in the manner in which their members transfer forces.

A. TRUSSES

1 Two Force Members

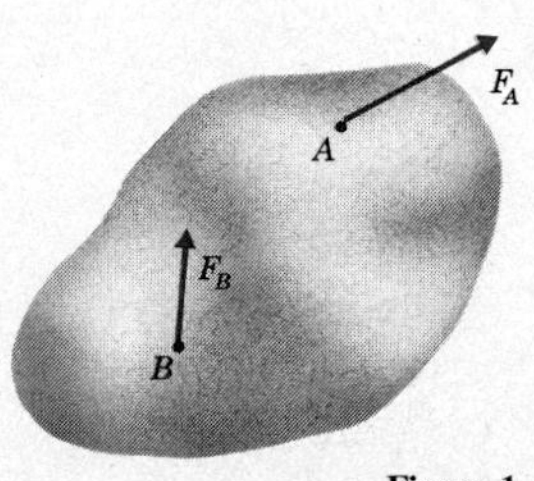

Figure 1

A primary characteristic of a truss is that is is composed *solely* of two force members, that is, members having forces applied at only two locations. To see why this feature is important we have illustrated an arbitrary two force member in Figure 1.

In this figure $\bar{F}_A$ and $\bar{F}_B$ are the forces acting at points A and B, respectively. Let us investigate the relationship that must exist between these two forces for static equilibrium of the body. Choosing point A for the moment sum, the equilibrium equations are

$$\Sigma \bar{M}_A = \bar{r}_{B/A} \times \bar{F}_A = \bar{0}$$

$$\Sigma \bar{F} = \bar{F}_A + \bar{F}_B = \bar{0} \tag{1}$$

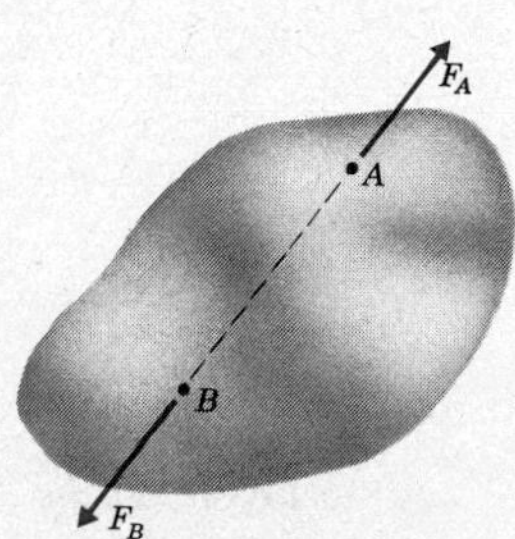

Figure 2

The force sum requires that $\bar{F}_A = -\bar{F}_B$; in other words, $\bar{F}_A$ and $\bar{F}_B$ must have equal magnitude and be oriented in opposite directions. Recalling that the cross product of two vectors is zero only when they are parallel, we see that the moment sum requires that $\bar{F}_B$ be parallel to the line connecting points A and B. Hence, for static equilibrium the forces must be directed as shown in Figure 2.

In summary, the equilibrium equations require that the forces at each end must be equal in magnitude, opposite in direction, and collinear. Clearly, it is possible for the forces to be reversed from the orientations shown.

A further restriction on the members of a truss is that they must be straight bars, such as the one illustrated in Figure 3a on the following page. The force system shown there is said to place the bar in a state of *tension,* because it tends to pull the bar apart, as if it were a cable. In fact, in this case the bar acts exactly like a cable, carrying a constant tension F_{AB} on every cross section, as shown in Figure 3a.

The difference between the bar in Figure 3a and a cable is that the forces in the bar can also have the effect of pushing the ends together.

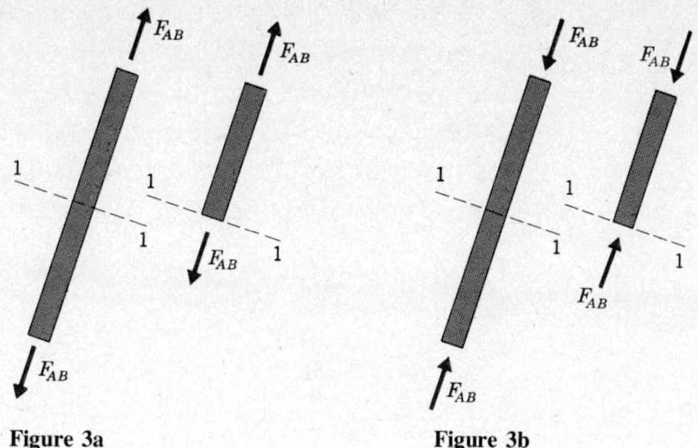

Figure 3a Figure 3b

This is the state of *compression,* illustrated in Figure 3b.

2 Characteristics of a Truss

A *truss* is a rigid framework of straight two-force members that are joined together at their ends. In order that the reactions at the ends of the members consist of forces, without couples, we model the connections of the members as pins in the case of planar trusses, or ball-and-socket joints in the case of three-dimensional trusses. The latter are usually referred to as *space trusses*. To gain some appreciation for the type of structure in which we are interested, Figure 4 shows some typical planar trusses; those with a peak are generally used to support roofs, whereas the flat ones are generally used to support bridges. The dots in each diagram indicate the pin connections.

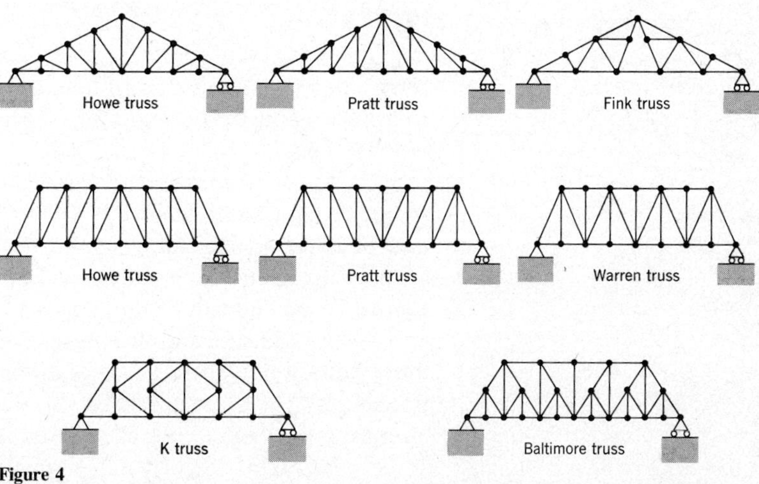

Figure 4

In an actual structure the members may be I-beams, channels, angle bars, or bars with other types of cross sections. The most common way of fastening members together is by means of welding, riveting, or bolting to gusset plates, as shown in Figure 5a. However, the model of a pin connection shown in Figure 5b is sufficiently accurate, provided that the centerlines of all members meeting at that connection are concurrent. (The proof

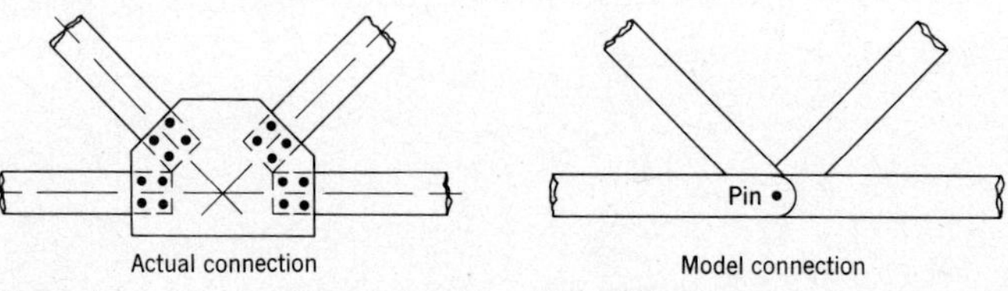

Figure 5a Figure 5b

of the validity of this approximation is beyond the scope of this book.) The terminology used to describe the model connection is to refer to it as a *joint*.

It will be noted that the members in each of the trusses illustrated in Figure 4 are interconnected to form traingles. Comparing the basic truss component in Figure 6a to the quadrilateral in Figure 6b, we see that the triangular shape is essential to the formation of a rigid truss. If we try to

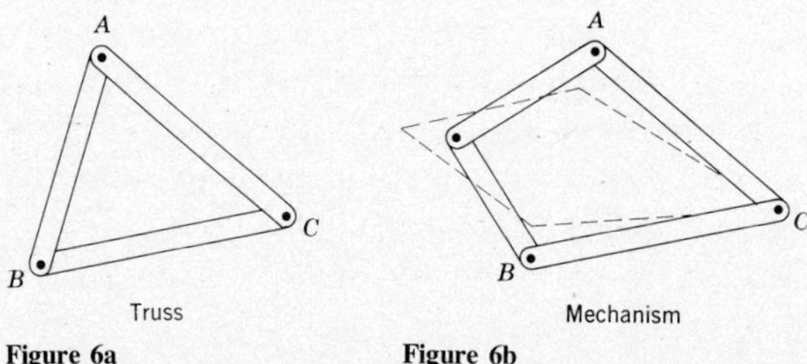

Figure 6a Figure 6b

form a polygon having more than three sides, the pin joints allow the bars to rotate relative to one, thus forming a nonrigid structure called a *mechanism*.

Using the truss of Figure 6a as a basis, we can increase the size of the structure by attaching more members. If this addition results in the formation of additional basic triangular components, we have a *simple truss,* such as the planar one shown in Figure 7a and the space truss shown in Figure 7b on the following page.

As we will see, simple trusses are statically determinate, allowing for the evaluation by the basic equilibrium equations of the forces being transmitted by each member. Simple trusses will be the focus of our attention. However, not all trusses are simple. For example, the planar

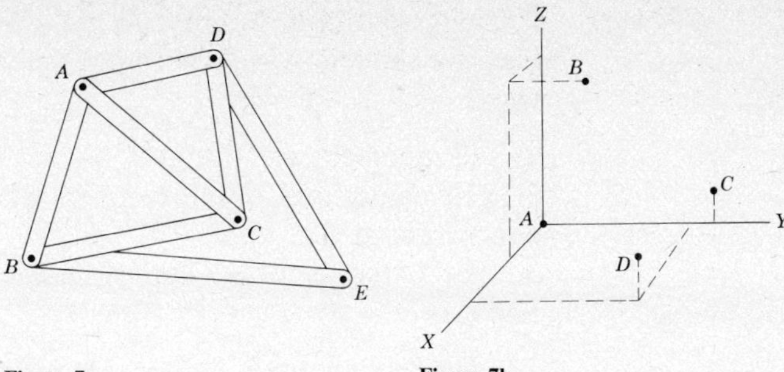

Figure 7a

Figure 7b

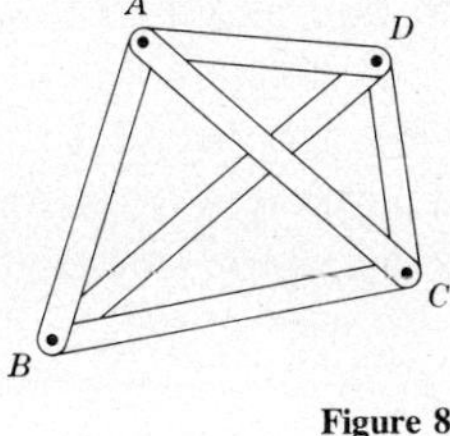

Figure 8

truss in Figure 8 was formed by adding three members to the basic truss component in Figure 6a. Not all of the members in this truss are needed to form the polygon *ABCD;* either of the diagonal members *AD* or *BC* could be removed to form a simple truss. The extra member is *redundant*. It can be shown that this structure is statically indeterminate.

For statically indeterminate trusses, in addition to the equations of static equilibrium, the evaluation of the force in each member requires that the small deformations of the bars resulting from the loading be considered. In other words, the model of a rigid body is not adequate for treating statically indeterminate structures. The subject of indeterminate trusses is treated in courses on strength of materials and structural analysis.

There is a simple relationship between the number of joints and the number of members necessary to form a simple truss. Starting from the basic triangle in Figure 6a, where there are three members and three joints, we see that for a planar truss, such as that in Figure 7a, additional triangles are formed by adding two bars for every joint added. Thus, letting m be the number of members and j be the number of joints, we find that

$$m = 2j - 3 \qquad \text{for a simple planar truss} \tag{2a}$$

In case of a simple space truss, we see that the requirement of rigid triangles leads to the formation of tetrahedrons, such as that in Figure 7b. Each additional joint now requires three additional members, so

$$m = 3j - 6 \qquad \text{for a simple space truss} \tag{2b}$$

Another characteristic of a truss pertains to the way in which it is loaded. Because the ends of the members are interconnected at the joints, it follows that, in order for there to be two force members, the external forces acting on the structure must be applied at the joints only. Any loading that is not applied at the joints requires that the member to which it is applied transfer more than just a tensile or compressive force. Hence, we have the following criteria for determining when a structure acts like a truss:

1 All members are straight.
2 The centerlines of all members that are connected at a joint must be concurrent at that joint.
3 The external forces acting on the structure must be applied at the joints.

A problem arises in considering the question of the weight of each member, because this force acts at the midpoint of the member, and thus violates the second of the above criteria. In most cases the weight force can be neglected, because it will be small compared to the other forces acting on the structure.

There are two methods that may be employed to evaluate the member forces. We shall make it a standard practice when applying either method *to evaluate first the reaction force being exerted by the supports,* so that these forces may be regarded as known when we determine the forces in the members. (There are truss problems where it is not necessary to first determine the reactions. We will discuss these cases after both methods have been fully developed.)

3 Method of Joints

To describe this method, let us consider a single triangular truss element, such as that shown in Figure 9a.

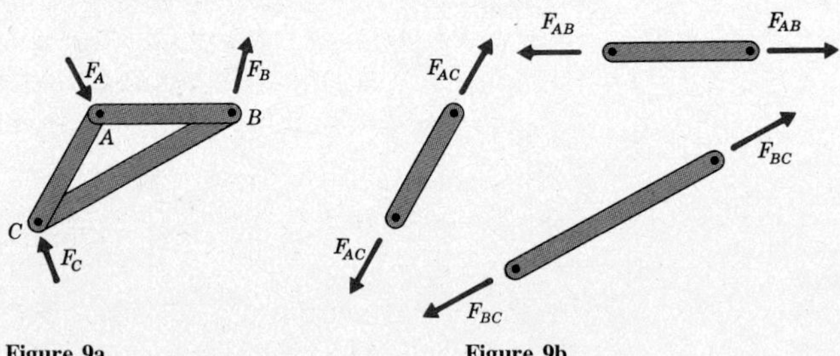

Figure 9a **Figure 9b**

The external forces $\bar{F}_A$, $\bar{F}_B$, and $\bar{F}_C$ are known. To ensure equilibrium of the structure, we now consider each member of the structure individually in a free body diagram, as shown in Figure 9b. In view of the earlier discussion, equilibrium for each member requires that its end forces be equal in magnitude, opposite in direction, and collinear. The notation we use is to denote the force in each member by subscripts corresponding to the joints at each end. Also note that in Figure 9b we begin by assuming, for the sake of convenience, that each member is in tension.

Each member is automatically in equilibrium by considering it to be in tension. To determine these axial forces we recognize that the structure is formed from pin joints, as well as straight members. The pins forming the joints must also be in equilibrium, so we draw free body diagrams of each joint; hence the name *method of joints*. Employing Newton's third law and recalling that we have assumed each member to be in tension, the force exerted by each member on a joint must pull away from that joint, toward the other end of the member. Thus, the free body diagrams of the members and joints of the truss in Figure 9a are as shown in Figure 10. Note that the external force acting at each joint must also be shown to complete the free body diagram.

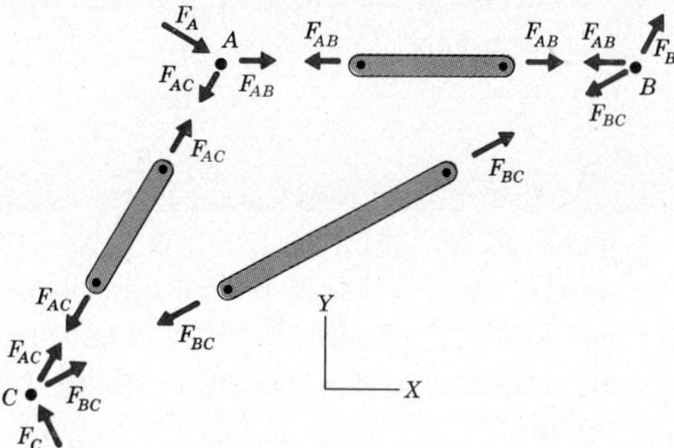

Figure 10

The unknown forces in the free body diagrams are those of the members of the truss. It is only necessary for us to consider equilibrium of the joints, for the members shown in Figure 10 are identically in equilibrium, as we have already noted. Because the forces at each joint are concurrent, each joint will be in equilibrium if the resultant force acting on it is zero. Thus, choosing a convenient coordinate system, such as that shown in Figure 10 for this planar system we have $\Sigma F_X = 0$ and $\Sigma F_Y = 0$ at each joint. In all, we have six scalar equations for the system in Figure 10.

At this juncture it might seem that we have more equations than we need, in that the only unknowns are the three member forces. This is a subtle point. Recall that earlier we decided to begin by obtaining the reactions of the truss. This procedure enables us to say that the structure as a whole is in equilibrium. However, the entire structure will also be in equilibrium if the forces acting on each joint are equilibrated. In other words, we have three excess equations, corresponding to equilibrium of the entire structure. The extra equations may be employed to check the solution, as will be seen in the examples that follow.

We have assumed that all the members are in tension. This assumption is confirmed when the magnitude of the member force is found to be positive, whereas a negative value indicates a state of compression. Clearly, it is not necessary for us to regard tension as positive, but this convention allows for an orderly interpretation of the results. It is paramount for the design of a member to know whether the force being transmitted is tensile or compressive, for the design procedure used depends entirely on this knowledge.

The foregoing discussion can be extended to treat general planar trusses, as well as space trusses. Briefly stated, we see that

> The method of joints consists of drawing free body diagrams for all joints and then solving the equilibrium equations for the concurrent force systems acting at each joint.

In the case of a planar truss, we have $\Sigma F_X = 0$ and $\Sigma F_Y = 0$ for each joint, so we obtain a total of $2j$ equilibrium equations for the j joints of a planar truss. However, the equilibrium of the entire structure is ensured by the determination of the reactions; only $2j - 3$ of the joint equations are independent. The other three equations can be used to check the results.

Hence, in a planar truss we see that in order for the member forces to be statically determinate, there must be $2j - 3$ members. Notice that this is the number of members required for a simple planar truss, as given by equation (2a). A planar truss having more than $2j - 3$ members is *internally indeterminate,* even though its reactions may be determinate. On the other hand, if there are fewer than $2j - 3$ members, the equilibrium equations for the joints cannot be satisfied. The truss is then not a rigid structure.

The foregoing are merely necessary conditions for having an internally determinate, rigid planar truss. It is possible to construct planar trusses that have j joints and $2j - 3$ members that are not rigid and statically determinate. As always, the ultimate verification of these conditions rests in the solvability of the equilibrium equations for the joints.

In the case of a space truss we have $\Sigma F_X = 0$, $\Sigma F_Y = 0$, and $\Sigma F_Z = 0$ for each joint. Thus, a space truss having j joints will have $3j$ equilibrium equations for these joints. Deducting the six checking equations corresponding to the static equilibrium of the entire structure, we see that the conditions of internal determinancy and rigidity require $3j - 6$ members. This is the same number of members as that given by equation (2b) for a simple space truss. A space truss with more than $3j - 6$ members is internally indeterminate, whereas a space truss having fewer than this number of members is not rigid.

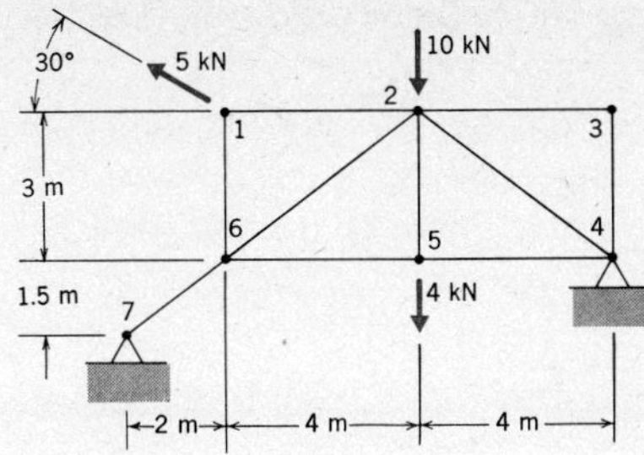

EXAMPLE 1

Determine the force in each member of the truss shown using the method of joints.

Solution

We begin by determining the reactions. Member 6-7 is a two-force member, so it can apply only an axial force to the rectangular portion of the truss. The other point of support of the truss is at pin joint 4, so we have the following free body diagram for the entire truss.

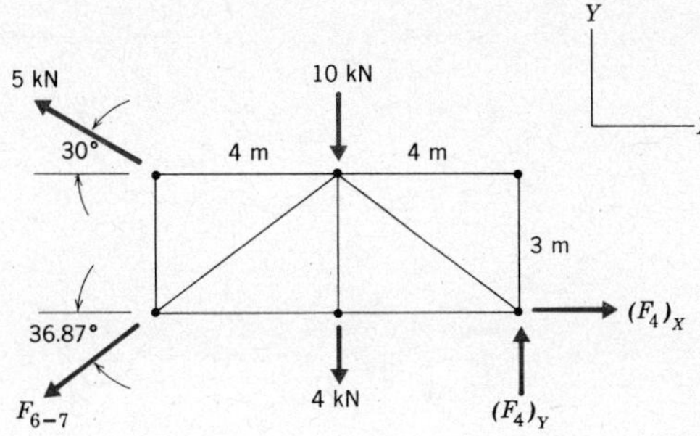

For consistency we have assumed that $\bar{F}_{6-7}$ is a tensile force. The dimensions of the truss are given in terms of horizontal and vertical distances, so the coordinate axes are aligned in these directions.

The Z axis is outward for the coordinate system shown, making counterclockwise moments positive. In order to eliminate two reaction components, we choose joint 4 for the moment sum. The equilibrium equations are then

$$\Sigma M_{4Z} = 10(4) + 4(4) + (F_{6-7} \sin 36.87°)8 - (5 \sin 30°)8 + (5 \cos 30°)\, 3 = 0 \text{ kN-m}$$

$$\Sigma F_X = -F_{6-7} \cos 36.87° + (F_4)_X - 5 \cos 30° = 0$$

$$\Sigma F_Y = -F_{6-7} \sin 36.87° + (F_4)_Y - 4 - 10 + 5 \sin 30° = 0$$

Solving the moment equation for F_{6-7} and using the force sums, we determine $\bar{F}_4$. Thus

$$F_{6-7} = -10.206 \text{ kN} \qquad (F_4)_Y = 5.376 \qquad (F_4)_X = -3.835 \text{ kN}$$

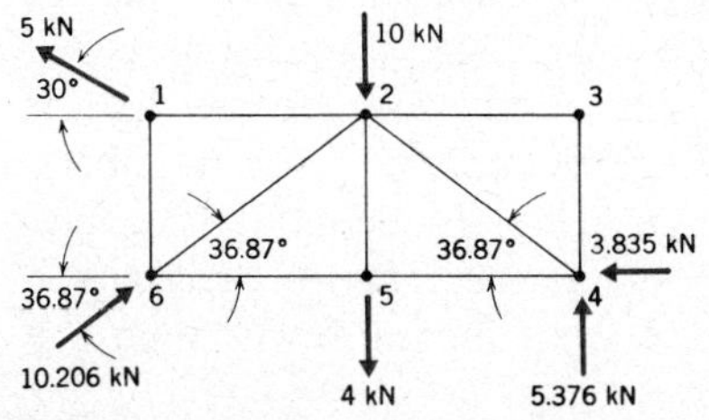

The equilibrium of each joint of the truss may now be considered. To assist in the calculations we sketch the truss showing the angles between the members and the coordinate directions, and all loadings, including the reactions, in their correct orientation.

Note that the force $(\bar{F}_4)_X$ is pushing to the left, because its value was negative. Also note that the principle of transmissibility does not apply when finding the forces in the internal members of the structure, so each external force must be shown at the joint at which it acts.

This sketch can now be used to draw free body diagrams of each joint, using the convention of considering the members to be in tension. Hence the members pull on the joint, as shown. Note that each joint is labeled in the free body diagrams.

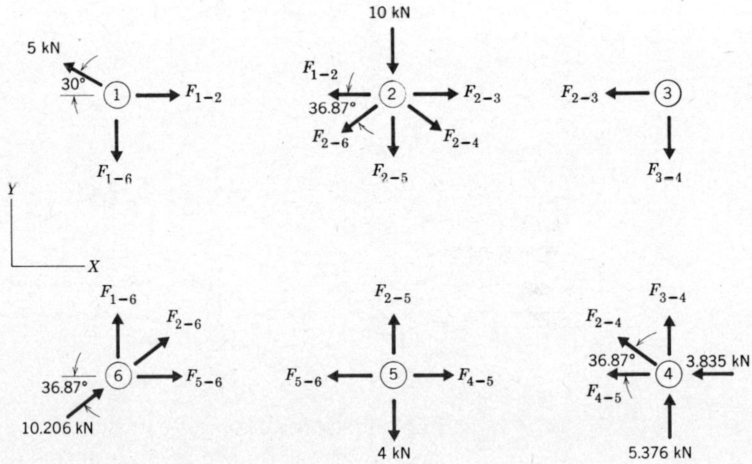

For computational purposes *it is best to consider joints connecting the fewest members first*. This allows us to solve successively for the member forces rather than dealing with a large number of simultaneous equations. In other words, in planar problems we solve the equilibrium equations at joints that contain two or less unknowns first. After referring to the free body diagrams, the solution order we choose, by joints, is 1-3-4-5-6-2.

Joint 1:

$$\Sigma F_X = F_{1-2} - 5 \cos 30^\circ = 0$$

$$\Sigma F_Y = -F_{1-6} + 5 \sin 30^\circ = 0$$

$$F_{1-2} = 4.330 \qquad F_{1-6} = 2.50 \text{ kN}$$

Joint 3:

$$\Sigma F_X = -F_{2-3} = 0 \qquad \Sigma F_Y = -F_{3-4} = 0$$

$$F_{2-3} = F_{3-4} = 0$$

Joint 4:

$$\Sigma F_X = -F_{4-5} - F_{2-4} \cos 36.87° - 3.835$$

$$\equiv -F_{4-5} - 0.8F_{2-4} - 3.835 = 0$$

$$\Sigma F_Y = F_{3-4} + F_{2-4} \sin 36.87° + 5.376$$

$$\equiv 0 + 0.6F_{2-4} + 5.376 = 0$$

$$F_{2-4} = -8.960 \qquad F_{4-5} = 3.333 \text{ kN}$$

Joint 5:

$$\Sigma F_X = F_{4-5} - F_{5-6} \equiv 3.333 - F_{5-6} = 0$$

$$\Sigma F_Y = F_{2-5} - 4 = 0$$

$$F_{5-6} = 3.333 \qquad F_{2-5} = 4 \text{ kN}$$

Joint 6:

$$\Sigma F_X = F_{5-6} + F_{2-6} \cos 36.87° + 10.206 \cos 36.87°$$

$$\equiv 3.333 + 0.8F_{2-6} + 8.165 = 0$$

$$\Sigma F_Y = F_{1-6} + F_{2-6} \sin 36.87° + 10.206 \sin 36.87°$$

$$\equiv 2.50 + 0.6F_{2-6} + 6.124 = 0$$

$$F_{2-6} = -14.373 \text{ kN}$$

Both equilibrium equations for joint 6 give the same value of F_{2-6}. This is the first check of the computations. The equilibrium equations for the remaining joint, number 2, give two more computational checks because the forces in all members are already known.

Joint 2:

$$\Sigma F_X = F_{2-3} + F_{2-4} \cos 36.87° - F_{2-6} \cos 36.87° - F_{1-2}$$

$$\equiv 0 + (-8.960)0.8 - (-14.373)0.8 - 4.330 \equiv 0$$

$$\Sigma F_Y = -(F_{2-6} + F_{2-4}) \sin 36.87° - F_{2-5} - 10$$

$$\equiv -(-14.373 - 8.960)0.6 - 4 - 10 \equiv 0$$

Because all of the check equations have been verified, we can be fairly certain that our calculations of the member forces, as well as of the reactions, are correct. The member forces that are negative indicate that the member is in compression.

When solving a problem by the method of joints, it may happen that you have chosen to consider the joints in the wrong order, so that the member forces for a certain joint cannot be found immediately (e.g., more than two unknowns at a joint in a planar problem). If that should happen, proceed to the other joints until a sufficient number of forces have been calculated to allow the joint to be reconsidered.

As a concluding remark, recall that we began the solution by considering member 6–7 as a support of the rectangular truss. This allowed us to express the reaction on the left side of the truss as the axial force in that member. Suppose we had instead considered member 6–7 to be part of the truss, and ignored the fact that this member is a two-force member. The corresponding reaction forces would have been those of pin 7, so we would have had four reaction components (two for pin 7 and two for pin 4). The equilibrium equations for the entire structure would have then been insoluble by themselves. In that case it would have been necessary to consider the equilibrium of the truss joints first, and then to use the results for the joints to reexamine the reactions. Clearly, it is better to be able to recognize two-force members that support a structure.

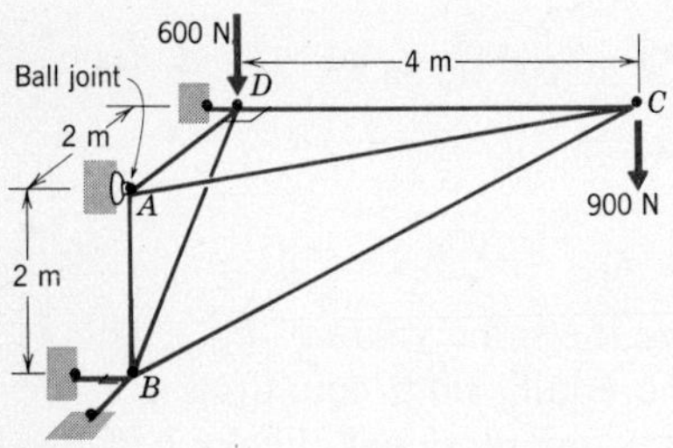

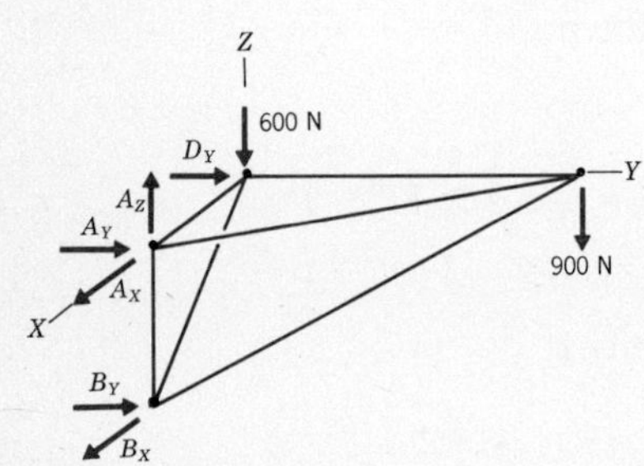

EXAMPLE 2

Using the method of joints, determine the forces in the members of the space truss shown.

Solution

We first determine the reactions by drawing a free body diagram of the truss, noting that the short links at joints B and D can only exert axial forces. Assuming that these forces are tensile, we have the free body diagram of the truss on the left. As always, the XYZ coordinate system is chosen to fit the directions in which dimensions are given.

In order to eliminate the three reaction components from the ball joint, point A is chosen for the moment sum. The equilibrium equations are then

$$\begin{aligned}\Sigma M_A &= \bar{r}_{D/A} \times (D_Y\bar{J} - 600\bar{K}) + \bar{r}_{B/A} \times (B_X\bar{I} + B_Y\bar{J}) \\ &\quad + \bar{r}_{C/A} \times (-900\bar{K}) \\ &= (-2\bar{I}) \times (D_Y\bar{J} - 600\bar{K}) + (-2\bar{K}) \times (B_X\bar{I} + B_Y\bar{J}) \\ &\quad + (-2\bar{I} + 4\bar{J}) \times (-900\bar{K}) \\ &= 2D_Y\bar{K} - 1200\bar{J} - 2B_X\bar{J} + 2B_Y\bar{I} - 1800\bar{J} - 3600\bar{I} = \bar{0}\text{ N-m}\end{aligned}$$

$$\begin{aligned}\Sigma \bar{F} &= (A_X\bar{I} + A_Y\bar{J} + A_Z\bar{K}) + (D_Y\bar{J}) + (B_X\bar{I} + B_Y\bar{J}) \\ &\quad - 900\bar{K} - 600\bar{K} = \bar{0}\text{ N}\end{aligned}$$

Equating the components of the moment sum to zero yields

$$\bar{I}\text{ component:}\qquad 2B_Y - 3600 = 0 \quad\text{so}\quad B_Y = 1800\text{ N}$$

$$\bar{J}\text{ component:}\qquad -2B_X - 3000 = 0 \quad\text{so}\quad B_X = -1500\text{ N}$$

$$\bar{K}\text{ component:}\qquad -2D_Y = 0 \quad\text{so}\quad D_Y = 0$$

Using these values of B_X, B_Y, and D_Y in equating the components of the force sum to zero yields

$$\bar{I}\text{ component:}\qquad A_X + B_X = 0 \quad\text{so}\quad A_X = 1500\text{ N}$$

$$\bar{J}\text{ component:}\qquad A_Y + D_Y + B_Y = 0 \quad\text{so}\quad A_Y = -1800\text{ N}$$

$$\bar{K}\text{ component:}\qquad A_Z - 1500 = 0 \quad\text{so}\quad A_Z = 1500\text{ N}$$

We now consider the equilibrium of the joints. The member forces are collinear with the corresponding members. To describe their directions we employ unit vectors parallel to each bar. Recalling that in our notation a unit vector extends *from* the point denoted by the second subscript *to* the point denoted by the first subscript, these unit vectors are

$$\bar{e}_{B/A} = -\bar{e}_{A/B} = \frac{\bar{r}_{B/A}}{|\bar{r}_{B/A}|} = -\bar{K}$$

$$\bar{e}_{C/A} = -\bar{e}_{A/C} = \frac{\bar{r}_{C/A}}{|\bar{r}_{C/A}|} = \frac{-2\bar{I} + 4\bar{J}}{\sqrt{2^2 + 4^2}}$$

$$= -0.4472\bar{I} + 0.8944\bar{J}$$

$$\bar{e}_{D/A} = -\bar{e}_{A/D} = \frac{\bar{r}_{D/A}}{|\bar{r}_{D/A}|} = -\bar{I}$$

$$\bar{e}_{C/B} = -\bar{e}_{B/C} = \frac{\bar{r}_{C/B}}{|\bar{r}_{C/B}|} = \frac{-2\bar{I} + 4\bar{J} + 2\bar{K}}{\sqrt{2^2 + 4^2 + 2^2}}$$

$$= -0.4082\bar{I} + 0.8165\bar{J} + 0.4082\bar{K}$$

$$\bar{e}_{D/B} = -\bar{e}_{B/D} = \frac{-2\bar{I} + 2\bar{K}}{\sqrt{2^2 + 2^2}} = -0.7071\bar{I} + 0.7071\bar{K}$$

$$\bar{e}_{D/C} = -\bar{e}_{C/D} = -\bar{J}$$

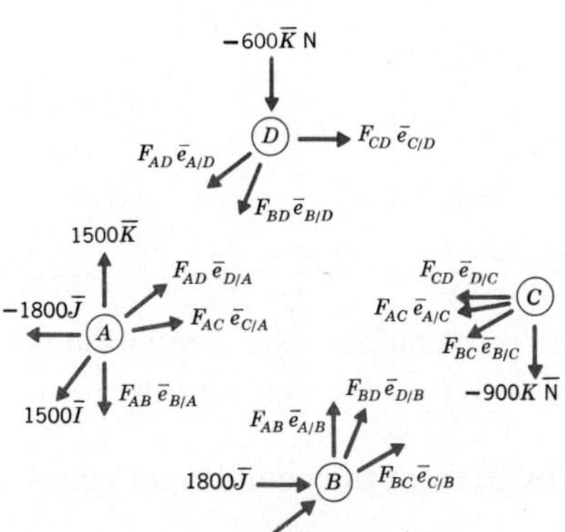

When drawing the free body diagrams for each joint, we assume that all members are in tension. Therefore, the axial force of each member at a joint is written as the product of the magnitude of the force and the unit vector from that joint to the other end of that member. For instance, the force member BC exerts on joint B is $F_{BC}\,\bar{e}_{C/B}$, whereas the force that this member exerts on joint C is $F_{BC}\,\bar{e}_{B/C}$. Thus, we have the free body diagrams on the left.

Particular attention is called to the fact that the reaction forces are entered into the free body diagram in the direction in which they were calculated to act.

In the space truss we are analyzing each joint connects only three members, so the joints may be considered in any order we desire. Starting with joint D, the equilibrium equations are

Joint D:

$$\Sigma F = F_{AD}\,\bar{e}_{A/D} + F_{BD}\,\bar{e}_{B/D} + F_{CD}\,\bar{e}_{C/D} - 600\bar{K}$$
$$\equiv F_{AD}(+\bar{I}) + F_{BD}(+0.7071\bar{I} - 0.7071\bar{K})$$
$$+ F_{CD}(+\bar{J}) - 600\bar{K} = \bar{0}$$

The following three scalar equations are found by equating the sum of corresponding components to zero.

$$\bar{I}\text{ component:} \qquad F_{AD} + 0.7071F_{BD} = 0$$
$$\bar{J}\text{ component:} \qquad F_{CD} = 0$$
$$\bar{K}\text{ component:} \qquad -0.7071F_{BD} - 600 = 0$$
$$F_{BD} = -849 \qquad F_{AD} = 600\text{ N}$$

Next we write the equilibrium equations for joint C, which are

Joint C:

$$\Sigma \bar{F} = F_{AC}\,\bar{e}_{A/C} + F_{BC}\,\bar{e}_{B/C} + F_{CD}\,\bar{e}_{D/C} - 900\bar{K}$$
$$\equiv F_{AC}(0.4472\bar{I} - 0.8944\bar{J}) + F_{BC}(0.4082\bar{I}$$
$$- 0.8165\bar{J} - 0.4082\bar{K}) + (0)(-\bar{J}) - 900\bar{K} = \bar{0}$$

Equating like components to zero yields

$$\bar{I}\text{ component:} \qquad 0.4472F_{AC} + 0.4082F_{BC} = 0$$
$$\bar{J}\text{ component:} \qquad -0.8944F_{AC} - 0.8165F_{BC} = 0$$
$$\bar{K}\text{ component:} \qquad -0.4082F_{BC} - 900 = 0$$
$$F_{BC} = -2205 \qquad F_{AC} = 2013\text{ N}$$

These values for F_{BC} and F_{AC} satisfy all three equations (the $\bar{J}$ equation is -2 times the $\bar{I}$ equation), so we have validation of the calculations from the first excess equation.

The last unknown member force is found from the equilibrium equations for joint B, which is

Joint B:

$$\Sigma \bar{F} = F_{AB}\, \bar{e}_{A/B} + F_{BC}\, \bar{e}_{C/B} + F_{BD}\, \bar{e}_{D/B} - 1500\bar{I} + 1800\bar{J}$$

$$\equiv F_{AB}\bar{K} - 2205(-0.4082\bar{I} + 0.8165\bar{J} + 0.4082\bar{K})$$

$$- 849(-0.7071\bar{I} + 0.7071\bar{K}) - 1500\bar{I} + 1800\bar{J} = \bar{0}$$

In component form this yields

$\bar{I}$ component: $\quad 900 + 600 - 1500 \equiv 0$

$\bar{J}$ component: $\quad -1800 + 1800 \equiv 0$

$\bar{K}$ component: $\quad F_{AB} - 900 - 600 = 0 \qquad F_{AB} = 1500 \text{ N}$

The first two equations are checks on the calculations. With the forces in all members known, the component equations resulting from equilibrium of joint A will be three more checks. We leave it to you to perform the necessary computations for their verification. Once again, remember that member forces that are negative indicate compression.

HOMEWORK PROBLEMS

IV.1–IV.14 Determine the forces in the members of each truss shown. State whether each member is in tension or compression.

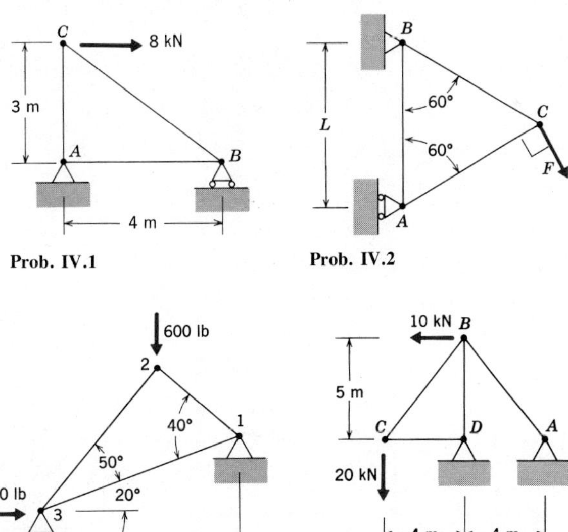

Prob. IV.1

Prob. IV.2

Prob. IV.3

Prob. IV.4

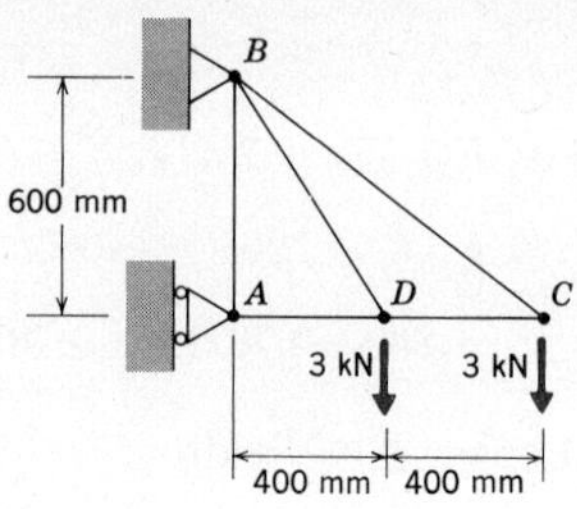

Prob. IV.5

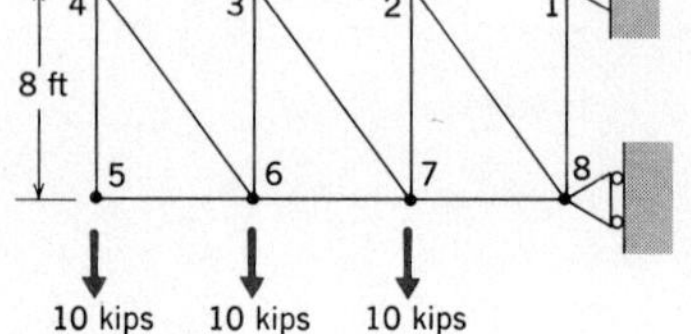

Prob. IV.6

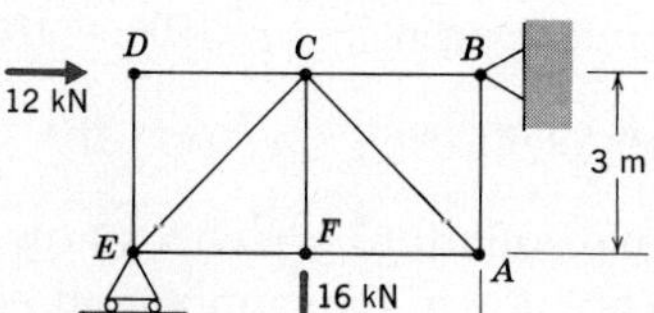

Prob. IV.7

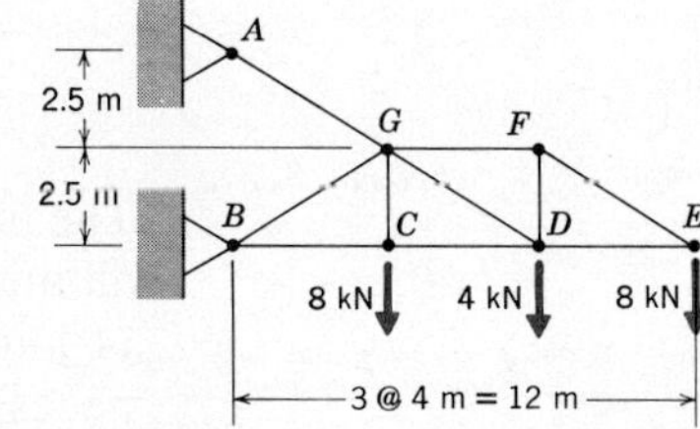

Prob. IV.8

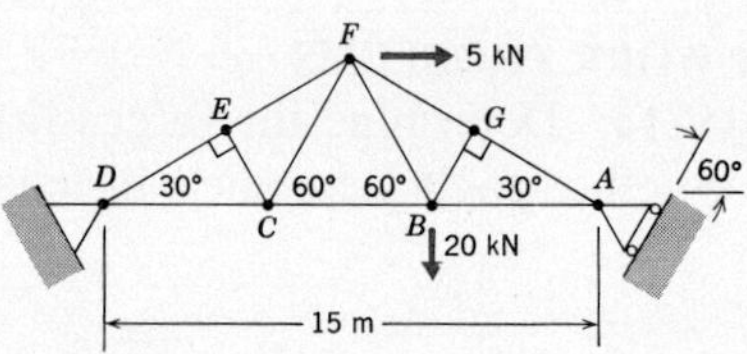
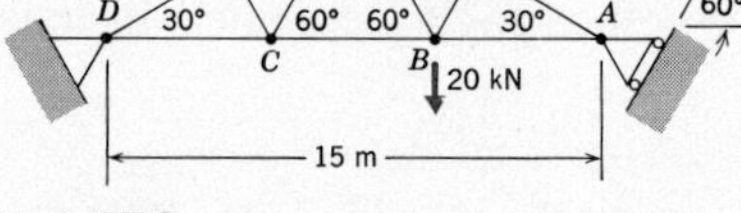

Prob. IV.9

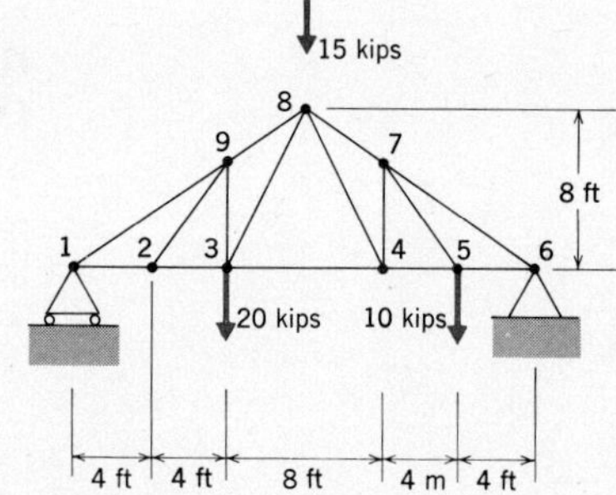

Prob. IV.10

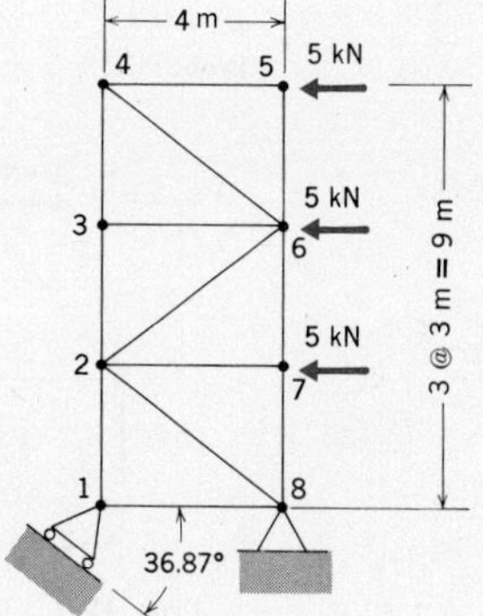

Prob. IV.11

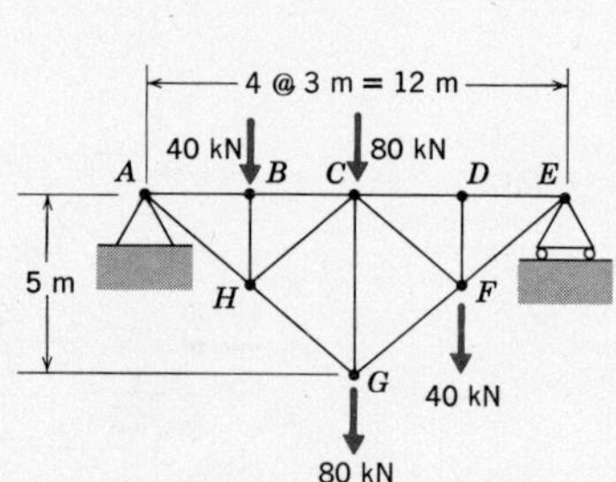

Prob. IV.12

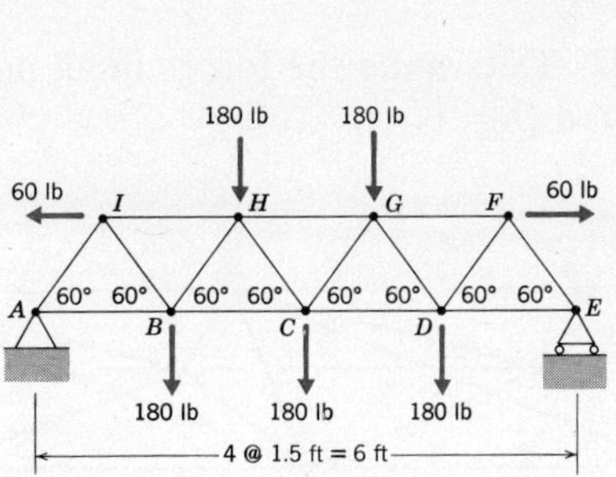

Prob. IV.13

Prob. IV.14

IV.15 Determine the forces in all members of the symmetrical space truss when it is subjected to the self-equilibrating set of loads shown. State whether the members are in tension or compression.

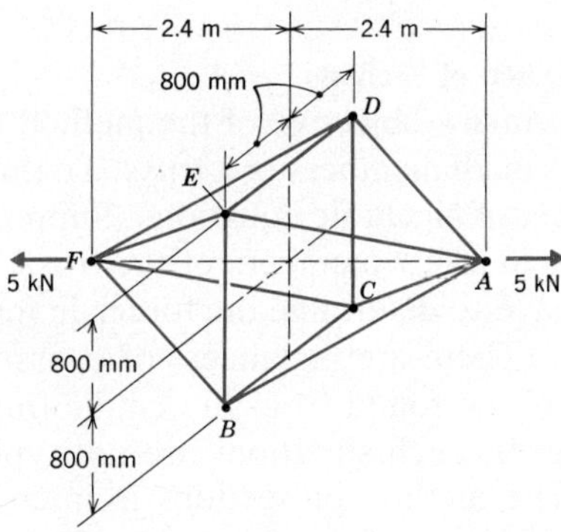

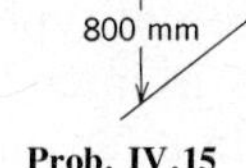

Prob. IV.15

Prob. IV.16

IV.16 Determine the forces in the members of the space truss if $P = 30$ kN and $Q = 0$. Indicate whether the members are in tension or compression.

IV.17 Solve Problem IV.16 if $P = 30$ kN and $Q = 20$ kN.

IV.18 Determine the forces in all members of the space truss if $P = 4$ kips and $Q = 0$. Indicate whether the members are in tension or compression.

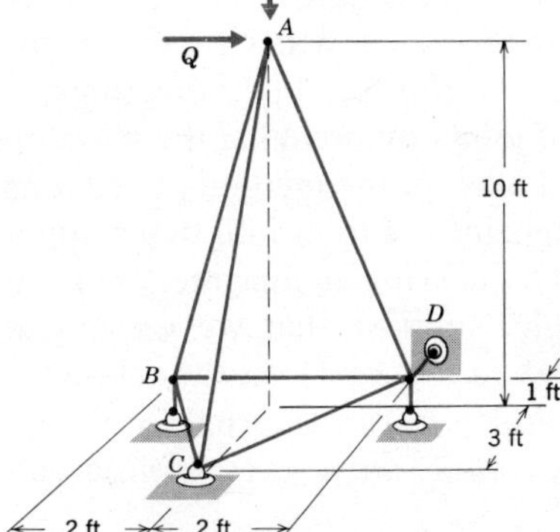

Prob. IV.18

IV.19 Solve Problem IV.18 for $P = 4$ kips and $Q = 3$ kips.

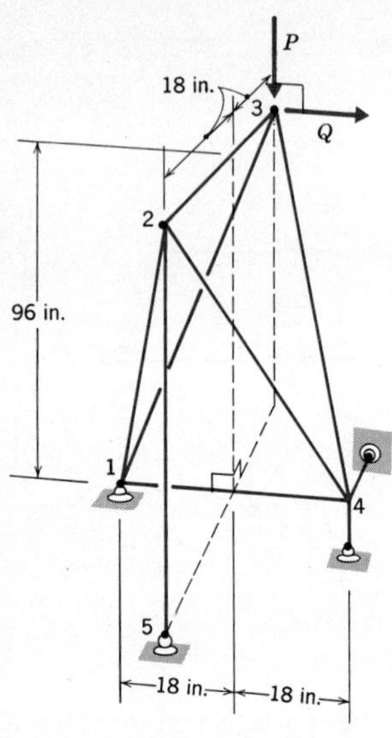

Prob. IV.20

IV.20 Determine the forces in all members of the space truss if $P = 400$ lb and $Q = 0$.

IV.21 Solve Problem IV.20 if $P = 400$ lb and $Q = 500$ lb.

IV.22 Determine the forces in all members of the space truss if $P = 12$ kN and $Q = 0$.

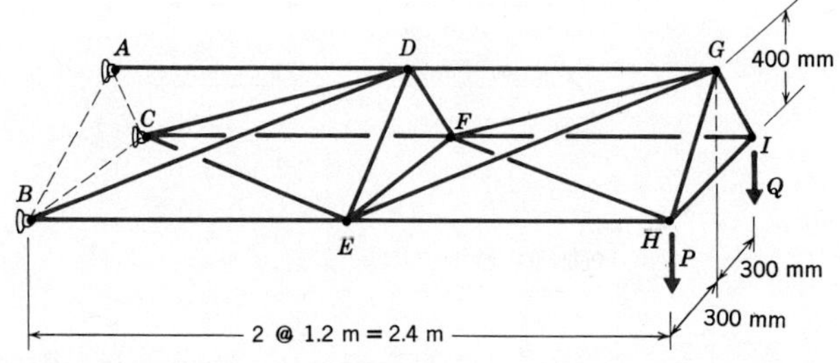

Prob. IV.22

IV.23 Solve Problem IV.22 if $P = 12$ kN and $Q = 18$ kN.

4 Method of Sections

The primary objective of the method of joints is the determination of the forces in all members of a truss. To this end it is necessary to solve a large number of algebraic equations. Suppose that we are interested only in the forces in a few members of the truss. If we use the method of joints, we will have to determine the forces in many members that are of no interest to us. (There are instances, of course, where the member force desired can still be found from the equilibrium equations for one or two joints. However, such situations are not typical.)

The method of sections is intended specifically to be used for the determination of the forces in a few members of a truss. Equally important, in learning to use the method of sections we will strengthen our ability to disassemble a structure to create useful free body diagrams.

Recall that we derived the method of joints by breaking the truss up into its individual components of bars and pins. In the method of sections we leave the structure assembled, decomposing the truss into two distinct portions. This decomposition is achieved by passing an imaginery *cutting plane* through the structure. For example, suppose that we wanted to know the force in member *CH* of the truss in Figure 11 on the following page. In order to determine $\bar{F}_{CH}$ it must appear as an external force in a free body diagram. Hence, *the cutting plane must cut through the member of interest.*

In the process of passing the cutting plane *a-a* through the structure,

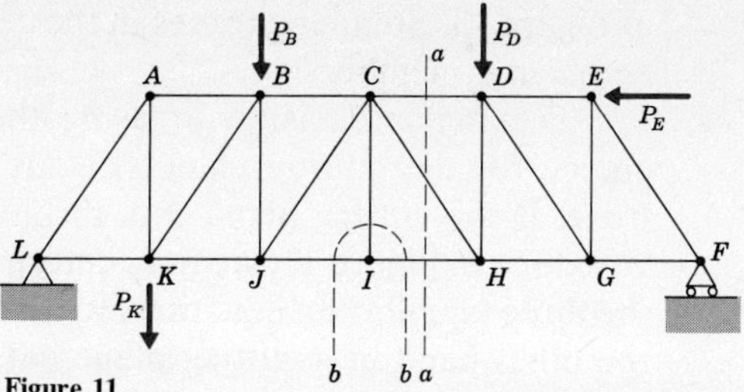

Figure 11

other members are also cut. In this manner the structure is decomposed into two parts, each of which must be in static equilibrium. The forces in the cut members are now external forces for these two truss segments. Once again, we assume the member forces to be tensile, so they pull on the isolated portions of the structure. The free body diagrams for the cut portions to the left and right of plane *a-a* are as shown in Figure 12.

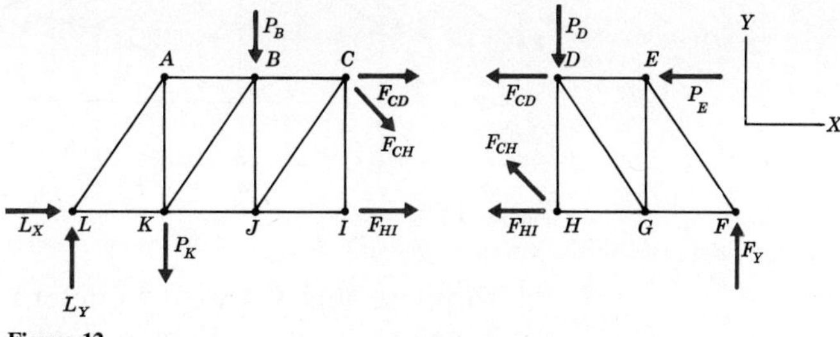

Figure 12

Recalling that our standard procedure is to determine the reaction forces first, the only unknown forces in these free body diagrams are those corresponding to the cut members: F_{CD}, F_{CH}, F_{HI}. Each isolated portion of the structure forms a rigid body carrying a planar force system, so we have three equations of static equilibrium for each body. The equations for *either* part will yield the values of the member forces being sought. Because the equilibrium of the entire structure was assured by the determination of the reactions exerted by the supports, the equilibrium equations for the remaining portion of the truss merely repeat the other equilibrium equations. The decision as to which portion of the cut truss should be equilibrated is arbitrary, although the portion having the smaller number of external forces and simpler geometry is usually the more convenient one to use for calculations.

The cutting plane need not be flat; it can be a curved plane. Also, it need not cut through an entire truss. In fact, we could even use a curved plane to isolate a joint. For example, let us determine the force in member *CI* of the truss in Figure 11. Passing plane *b*−*b* through the truss (actually around the joint), we can isolate the joint in the free body diagram shown

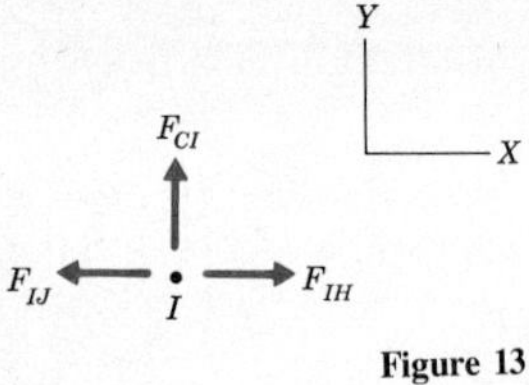

Figure 13

in Figure 13. Summing forces in the Y direction shows that member IC is a zero force member.

The major decision in using the method of sections involves the choice for the cutting plane. Let us first consider the case of a planar truss. If the cutting plane cuts through no more than three members, as was done in Figure 11, we may determine the member forces directly from the three equilibrium equations for an isolated portion of the structure. On the other hand, if a cutting plane cuts through four or more members, it will not be possible to determine all of the cut member forces, although it might be possible to obtain values for a few of the forces.

A situation where it is not possible to determine the member force of interest with one cutting plane is typified by the truss in Figure 14.

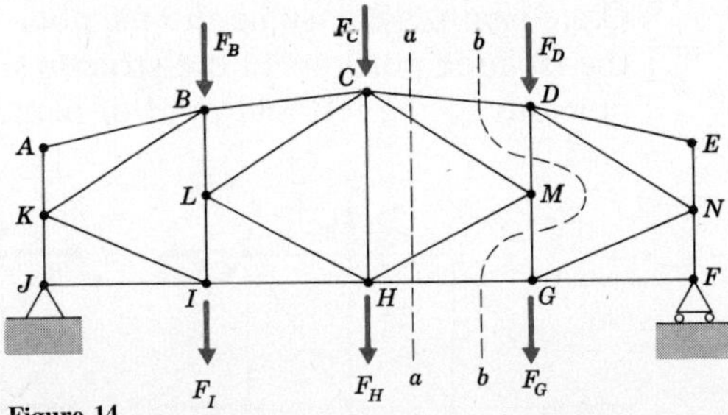

Figure 14

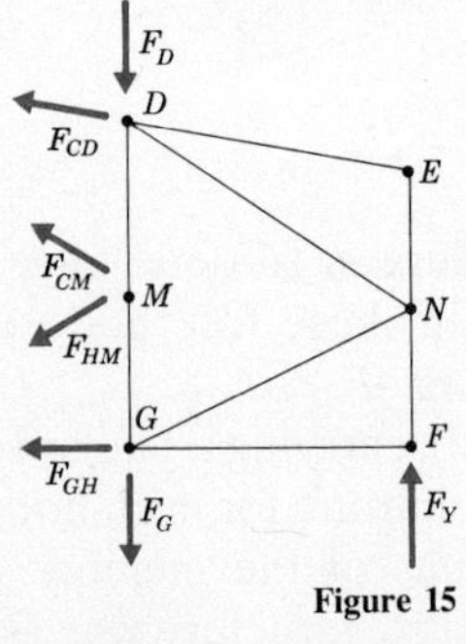

Figure 15

Suppose that we wish to determine the force in member CM. Any cutting plane, such as plane a-a, that cuts this member also cuts at least three other members. The free body diagram resulting from plane a-a is shown in Figure 15, where $\bar{F}_Y$ is the known reactive force. For the isolated section there are only three equations of static equilibrium, so the four unknowns F_{CD}, F_{CM}, F_{HM}, and F_{GH} cannot be found solely from these equations.

Let us now consider another section, formed from the curved cutting plane $b-b$ appearing in Figure 14; the corresponding free body diagram is shown in Figure 16. Once again, we cannot solve for all the unknown forces, for the free body diagram leads to three equations of static equilibrium in the four unknowns, F_{CD}, F_{DM}, F_{GM}, F_{GH}. However, by summing moments about joint G, we can determine the force in member CD. This value may then be used in the static equilibrium equations obtained from the free body diagram shown in Figure 15, which then enables us to solve for the force in member CM. From this example we see that, at times, more than one cutting plane may be necessary to determine the force in a specific member.

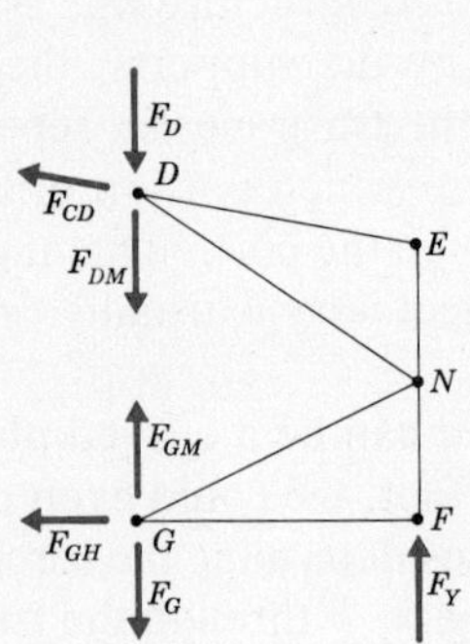

Figure 16

The method of sections may also be applied to space trusses, in which case equilibrium of the isolated portion of the structure results in six scalar equations. If the cutting plane cuts through no more than six members, we will be able to determine all of the forces in the cut members. A cutting plane intersecting seven or more members will leave us unable to determine at least some of the member forces. In the examples that follow we shall concentrate on planar trusses, because the only true difference between these trusses and three-dimensional ones involves the techniques for forming the equilibrium equations.

Finally, the discussion of the method of sections would be incomplete if we did not note that it is not universally applicable. In Figure 17 we have a simple truss that requires an excessive number of cutting planes to evaluate the force in an interior member, such as member NM. In such a case we resort to the method of joints, as we would if we wanted to know the forces in many or all of the members of any truss.

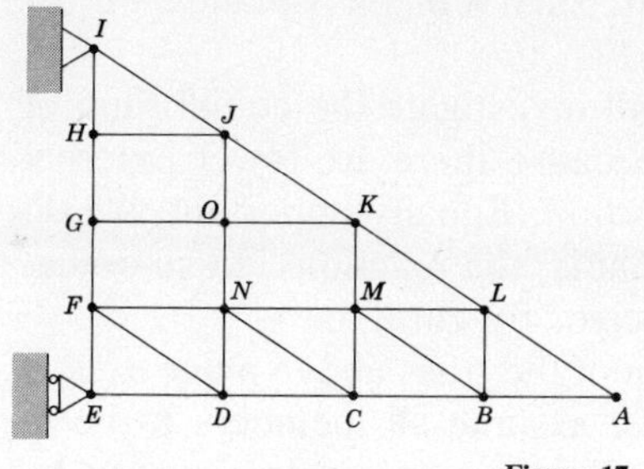

Figure 17

EXAMPLE 3

Determine the forces in members CF and FG of the truss shown.

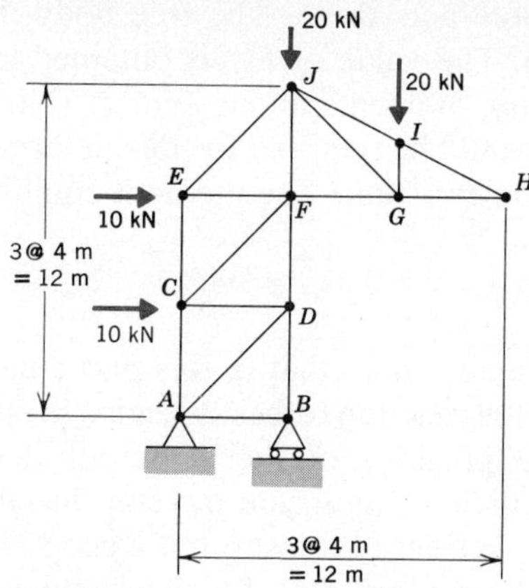

Solution

We begin with a free body diagram of the entire structure, which is used for calculating the reactions. The XYZ coordinate system is chosen horizontal and vertical, because the structure's dimensions are given in that manner.

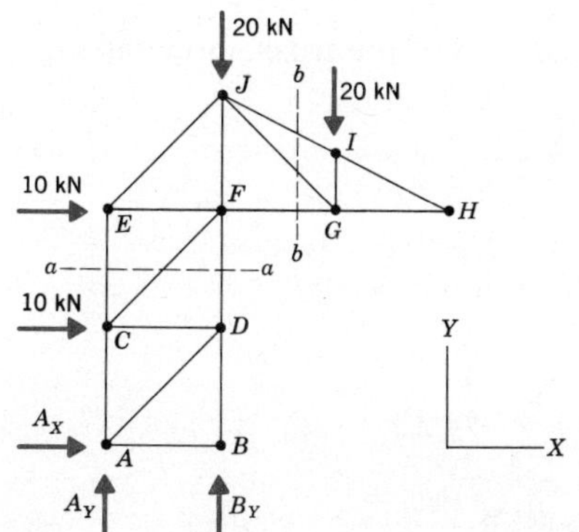

Summing forces and summing moments about point A, we find

$$\Sigma M_{AZ} \equiv 4B_Y - 10(4) - 10(8) - 20(4) - 20(8)$$

$$\equiv 4B_Y - 360 = 0$$

$$\Sigma F_X = A_X + 10 + 10 = 0$$

$$\Sigma F_Y = A_Y + B_Y - 20 - 20 = 0$$

$$B_Y = 90 \qquad A_Y = -50 \qquad A_X = -20 \text{ kN}$$

We next choose the cutting planes to be employed. If possible, the cutting planes should cut through no more than three members in the process of cutting the member of interest. Such a plane is plane $a-a$ for member CF and plane $b-b$ for member FG.

Considering the section $a-a$, we will investigate the equilibrium of the upper portion of the cut structure because there are fewer external forces acting there than on the lower portion. Equally important, we do not have to use any computed results, that is, the reactions. In so doing, we minimize the chance of carrying an error forward.

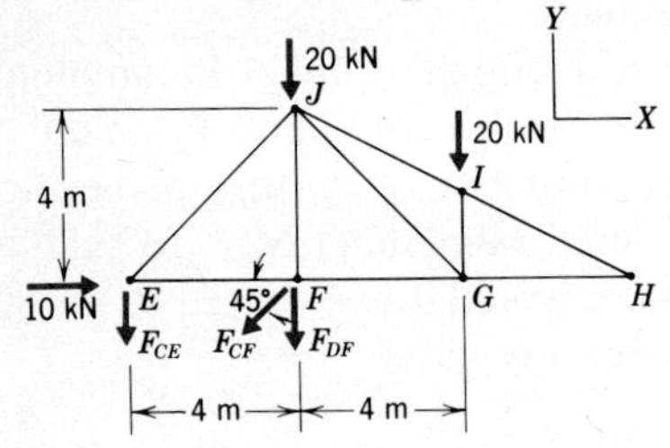

The free body diagram of the portion of the truss above plane $a-a$ is shown in the sketch. Remember that we assume all members to be in tension. The value of F_{CF} can be determined from a single equation by summing forces in the X direction. Thus

$$\Sigma F_X = -F_{CF} \cos 45^\circ + 10 = 0 \qquad F_{CF} = -14.14 \text{ kN}$$

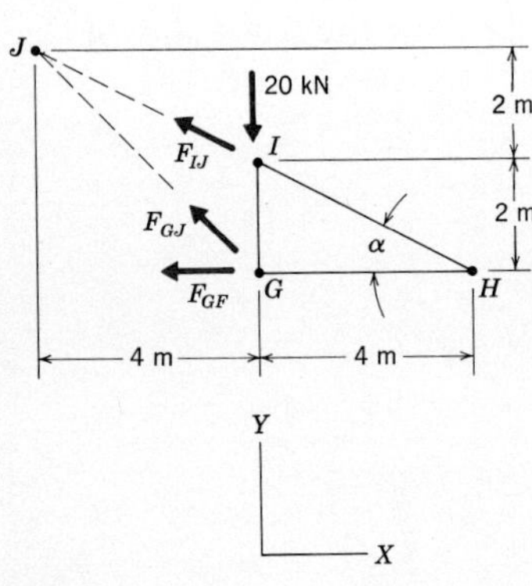

For plane $b-b$ let us consider the portion to the right, because only one force acts there. The free body diagram for the right portion is as shown. The value of F_{FG} is obtained in a single equilibrium equation by summing moments about joint J. (Note that all three equilibrium equations would be required for this determination if we chose joint I or H for the moment sum.) The moment equilibrium equation is

$$\Sigma M_{JZ} = -4F_{GF} - 20(4) = 0 \qquad F_{GF} = -20 \text{ kN}$$

Finally, note that in this particular problem it was not necessary to know the reaction forces, because the cutting planes utilized enabled us to isolate portions of the structure that did not contain the supports. Nevertheless, until you develop insight for cutting and isolating appropriate sections of trusses, it is a good idea to get into the habit of determining the reactions first, for their values will generally be needed.

EXAMPLE 4

Determine the tension in members 4–8 and 7–8 of the truss shown using the method of sections.

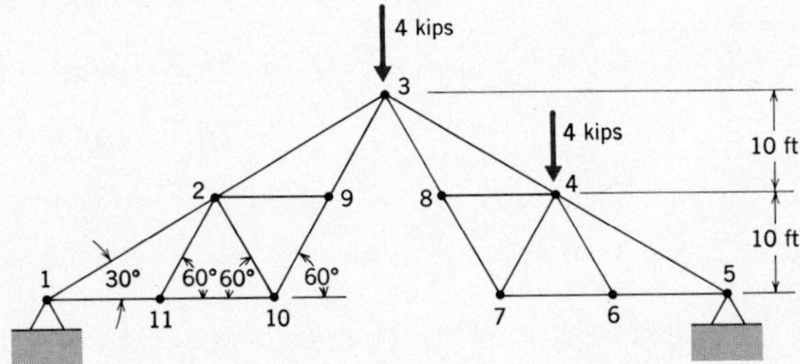

Solution

Perhaps the most notable feature of this truss is that it is supported by two pins, as shown by the reactions in the free body diagram.

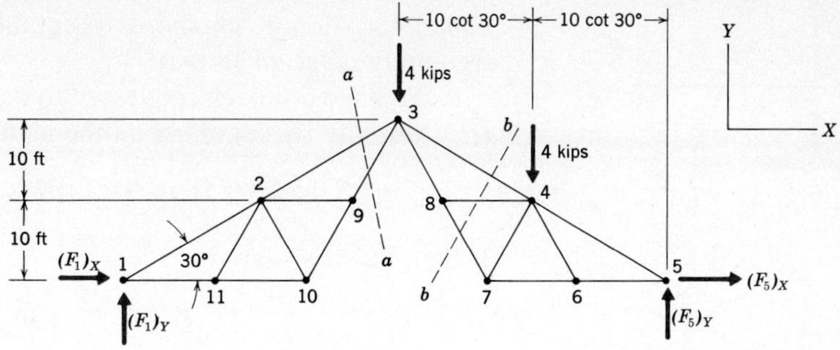

For equilibrium of the truss we have

$$\Sigma F_X = (F_1)_X + (F_5)_X = 0$$

$$\Sigma F_Y = (F_1)_Y + (F_5)_Y - 4 - 4 = 0$$

$$\Sigma M_{1Z} = (F_5)_Y [4(10 \cot 30°)] - 4[2(10 \cot 30°)] - 4[3(10 \cot 30°)] = 0$$

$$(F_5)_Y = 5 \qquad (F_1)_Y = 3 \qquad (F_5)_X = -(F_1)_X \text{ kips}$$

We cannot determine $(F_5)_X$ and $(F_1)_X$ from the foregoing equations. At first glance it might seem that these reactions are statically indeterminate. This is not so. Recall that an indeterminate structure has redundant supports that are not necessary for holding the system in static equilibrium. In the truss shown, changing one of the pin supports to a roller will result in a nonrigid structure, for the roller support will move, allowing joint 3 to fall down. On this basis we can intuitively conclude that the truss shown is statically determinate.

The reason for the apparent ambiguity is that the truss is really composed of two simple trusses, each of which is rigid by itself. The left truss is supported at pin 1 and joint 3, whereas the right truss is supported at pin 5 and joint 3. In general, a truss formed from several simple trusses is called a *compound truss*. The Fink truss that was shown in Figure 4 is another compound truss. It resembles the truss in this problem, but it is rigid and statically determinate with a pin and roller support because of the presence of the additional horizontal member connecting the two portions.

From these observations, we conclude that the determination of the values of $(F_1)_X$ and $(F_5)_X$ requires that we consider the equilibrium of the

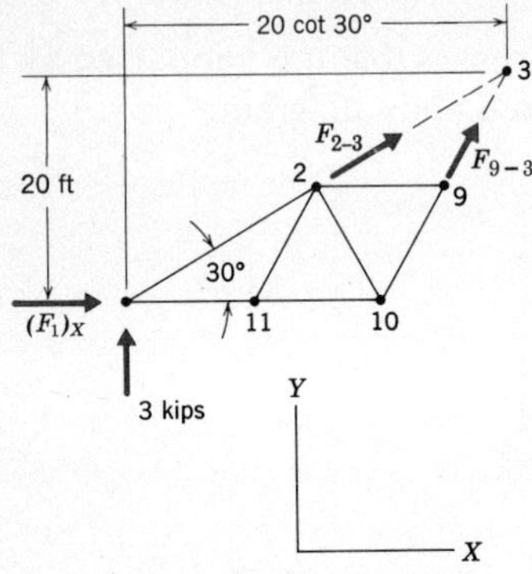

simple trusses from which the compound truss is formed, as well as of the entire structure. For this consideration note that cutting plane $a-a$ shown in the free body diagram only cuts through two members, although equilibrium of the sections formed by plane $a-a$ will yield three additional equations. Hence, choosing to equilibrate the left section, we have the free body diagram shown.

We have not been requested to determine F_{2-3} and F_{9-3}, so we obtain $(F_1)_X$ directly by summing moments about joint 3. Thus

$$\Sigma M_{3Z} = (F_1)_X(20) - 3(20 \cot 30°) = 0$$

It then follows that

$$(F_1)_X = 3 \cot 30° = 5.196 \text{ kips}$$

$$(F_5)_X = -(F_1)_X = -5.196 \text{ kips}$$

Now that we know the reactions, we can choose a cutting plane for determining the forces in members 4–8 and 7–8. We will use the cutting plane $b-b$ shown in the free body diagram of the entire truss because it cuts through both members of interest while cutting through a total of only three members. Consider the portion of the truss to the right of the cutting plane for equilibrium. The corresponding free body diagram is as shown.

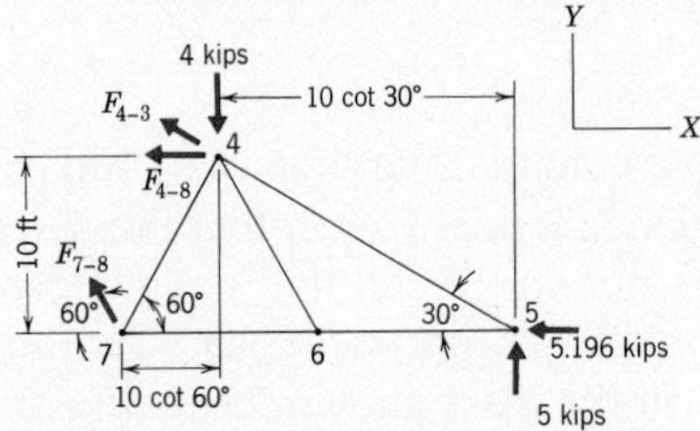

We choose joint 4 for the moment sum in order to eliminate F_{4-3} and F_{4-8}. The equilibrium equations are

$$\Sigma M_{4Z} = -F_{7-8} \sin 60° \, (10 \cot 60°) - F_{7-8} \cos 60°(10)$$
$$+ 5(10 \cot 30°) + (-5.196)(10) = 0$$

$$\Sigma F_X = -F_{7-8} \cos 60° - F_{4-8} - F_{4-3} \cos 30° - 5.196 = 0$$

$$\Sigma F_Y = F_{3-4} \sin 30° + F_{7-8} \sin 60° + 5 - 4 = 0$$

$$F_{7-8} = 3.464 \qquad F_{4-3} = -8.00 \qquad F_{4-8} = 0 \text{ kips}$$

The value of F_{4-3} was obtained in the process of solving the equations, even though it was not requested.

Finally, note that member 4–8 could have been readily shown to be a zero force member by the method of joints; a free body diagram shows that it is the only force acting on joint 8 in the direction perpendicular to the line from joint 3 to joint 7. However, determining the force in member 7–8 by the method of joints requires that other joints be solved first.

HOMEWORK PROBLEMS

Using the method of sections, determine the axial force in the members indicated.

IV.24 Member *BD* in the truss of Problem IV.5

IV.25 Members 3–4 and 3–6 in the truss of Problem IV.6

IV.26 Member *CE* in the truss of Problem IV.7

IV.27 Members *CD* and *DG* in the truss of Problem IV.8

IV.28 Members *BG* and *BF* in the truss of Problem IV.9

IV.29 Members 2–9 and 8–9 in the truss of Problem IV.10

IV.30 Members 2–3 and 2–7 in the truss of Problem IV.11

IV.31 Member *CF* in the truss of Problem IV.12

IV.32 Members *BC* and *GH* in the truss of Problem IV.13

IV.33 Members 2–7, 3–7, and 6–7 in the truss of Problem IV.14

IV.34 Determine the axial force in members *DE* and *DF* of the truss shown.

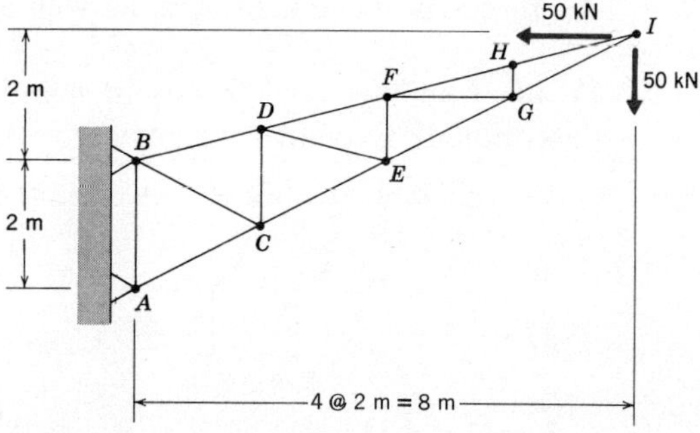

Prob. IV.34

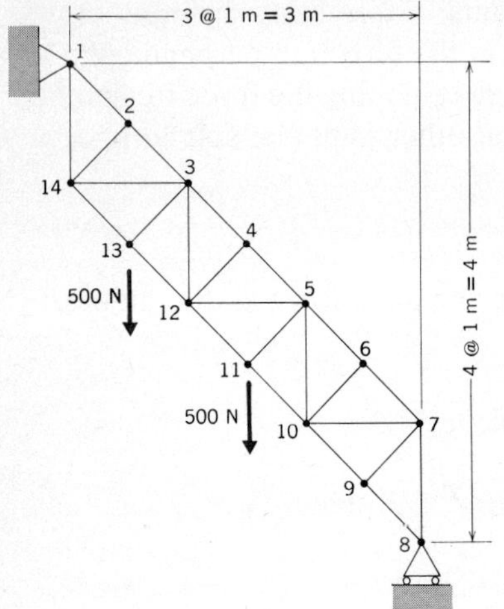

Prob. IV.35

IV.35 Determine the axial forces in members 4–12 and 5–12 of the truss shown.

IV.36 Determine the axial force in members *BH* and *BJ* of the truss shown. Note that the acute angles formed by the members are either 30°, or 60°.

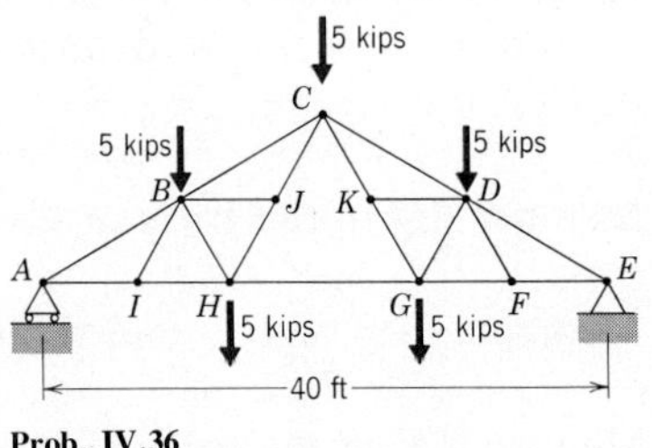

Prob. IV.36

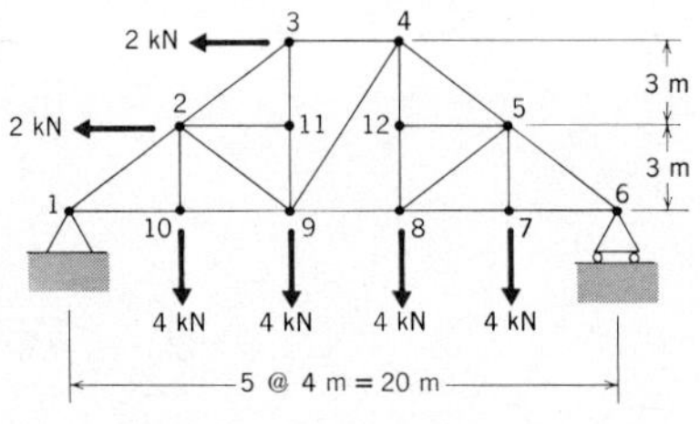

Prob. IV.37

IV.37 Determine the axial force in members 4–9 and 9–11 of the truss shown.

IV.38 Determine the axial forces in members *AB* and *AC* of the truss shown.

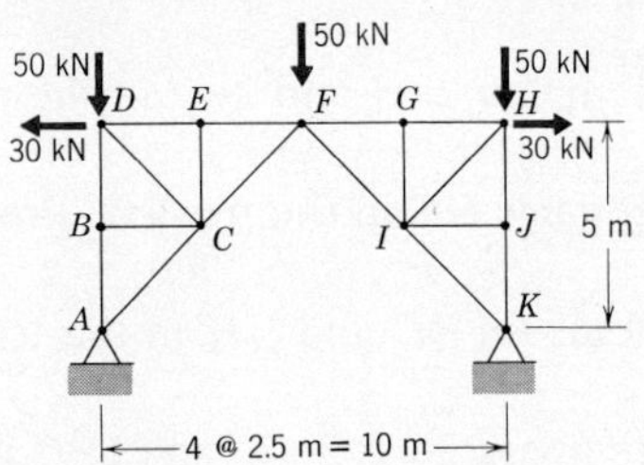

Prob. IV.38

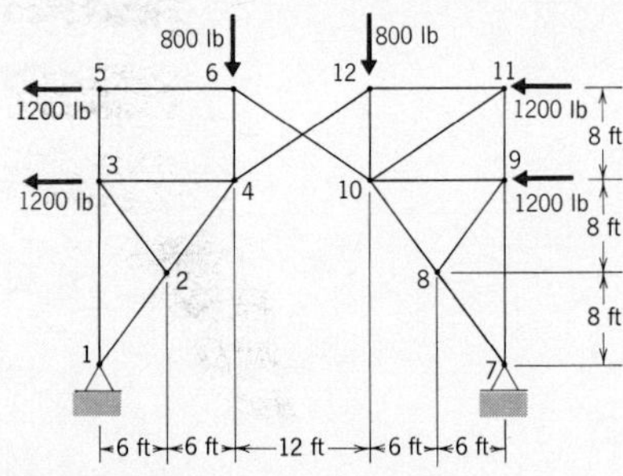

Prob. IV.39

IV.39 Members 4–12 and 6–10 cross, but are not connected. Determine the forces in these members, as well as in members 1–2 and 1–3.

IV.40 Can the axial forces in members *DE* and *GH* of the truss be determined? Explain your answer.

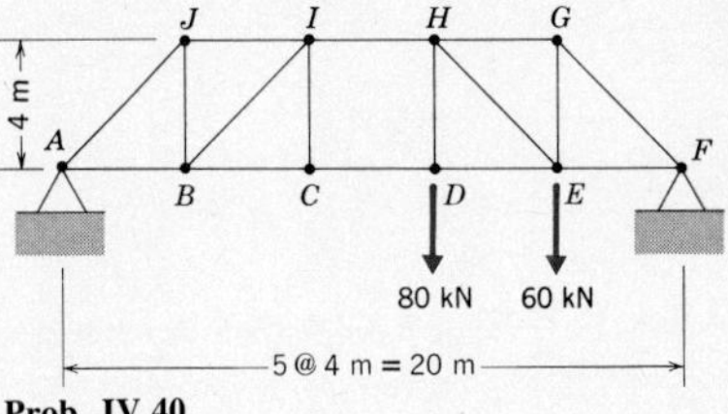

Prob. IV.40

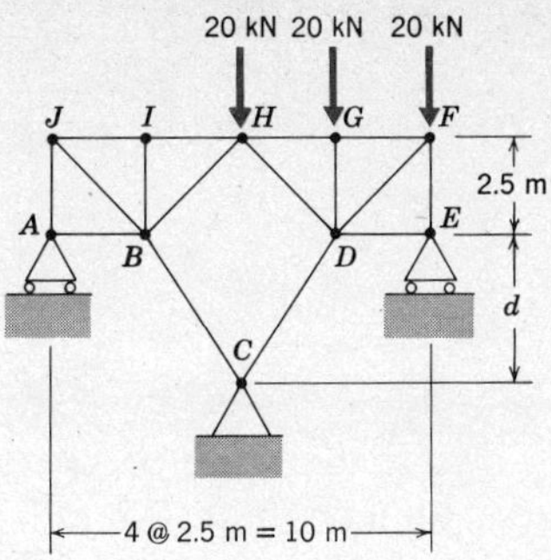

Prob. IV.41

IV.41 Determine the axial forces in members DC and FG of the truss shown when $d = 5$ m. Can these forces be found when $d = 2.5$ m? Explain your response.

IV.42 Determine the axial forces in members 5–8 and 6–8 of the truss shown.

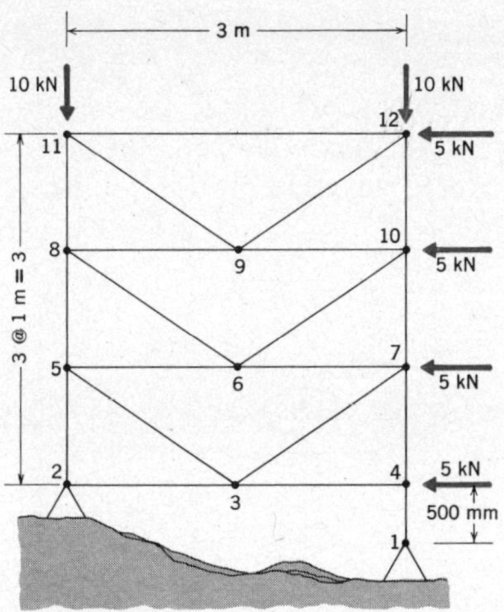

Prob. IV.42

IV.43 Determine the axial forces in members IJ and IO of the truss shown.

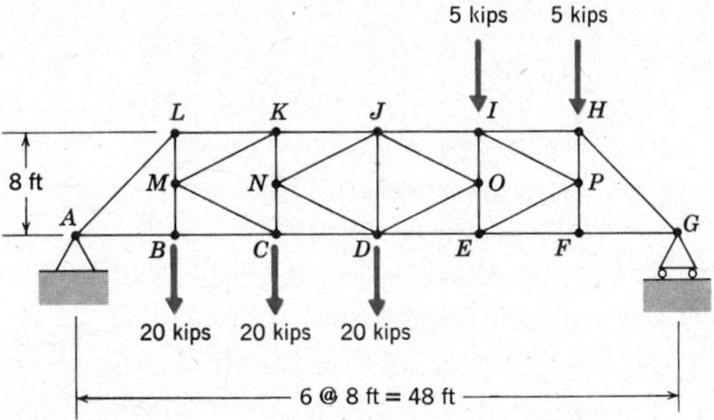

Prob. IV.43

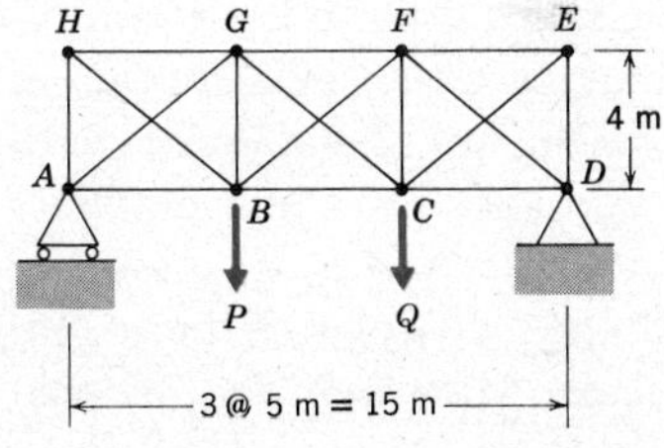

Prob. IV.44

IV.44 The diagonal members in the truss shown exert negligible forces when required to be in compression because of a phenomenon called "buckling." Hence, they may be considered to sustain tensile axial forces only. Such members are called *counters*. These counters cross, but are not connected to each other. Determine the axial forces in counters BF and CG when $P = 40$ kN and $Q = 0$.

IV.45 Solve Problem IV.44 for $P = 40$ kN and $Q = 80$ kN.

IV.46 Determine the forces in members *DE, DF,* and *DG* in the truss of Problem IV.22.

IV.47 Determine the force in members *DE, DF,* and *DG* in the truss of Problem IV.23.

IV.48 The tower of an oil drilling rig is formed in the shape of a pyramid. The supports *A, B,* and *C* are equally spaced on a circle of 1 m radius, as shown in the plan view, and the vertical line passing through the center of this circle intersects the top joint *D*. Determine the forces in members 1 and 2.

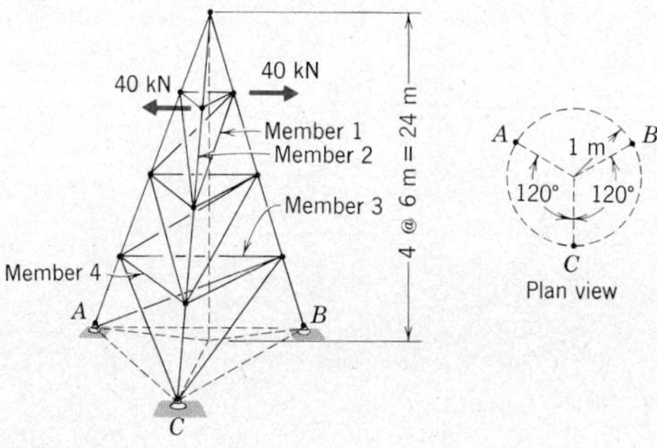

Prob. IV.48

IV.49 For the oil drilling rig of Problem IV.48, determine the forces in members 3 and 4.

B. FRAMES AND MACHINES

A truss is a special structure. It has only straight members which are connected concurrently at its joints, and the loads it transmits are applied only at these joints. We will now explore how to determine the forces transmitted by the members of more general structures for more general loading conditions.

In this section we shall examine two types of general structures. They differ only in their application. *Frames* are rigid structures intended to support a given loading, whereas *machines* are nonrigid structures that transmit and modify a given set of input loading forces into another set of output loadings. Our interest in machines in this text on statics is with the relationship of input and output forces necessary for static equilibrium. We leave the question of the possible movement of a machine for the companion text on dynamics. We shall not distinguish between the methods of solution for frames and machines. They both require that we

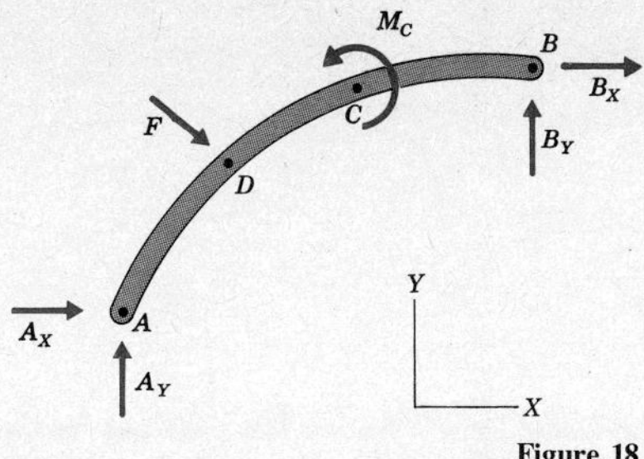

Figure 18

study the equilibrium of structures composed of many *multiforce members,* that is, members that carry forces that can be applied anywhere along its boundary.

In that a frame or machine is composed of multiforce members, we can no longer formulate the solution on the basis of all members carrying axial tension or compression forces. For example, consider the curved member AB in Figure 18, which is pinned at its ends to a larger structure. The nature of the reaction components at pins A and B cannot be determined in advance. Rather, it is necessary for us to write the equations of static equilibrium in order to relate these reactions to the given loading system, which in this case consists of a force at point D and a couple at point C. In Module VI, where we study the internal forces in structural members, we will learn how a multiforce member can sustain general internal forces and couples, as contrasted to just the axial tension or compression of a two force member.

It appears that the member in Figure 18 is statically indeterminate because there are only three equilibrium equations for the member, whereas there are four reaction components. However, recall that it was stipulated that this is a member in a structure. Hence, we will obtain additional equations when we consider the static equilibrium of the other members of the structure and of the structure as a whole. It is only after counting all such equations and the corresponding unknowns that we can judge the determinacy of the structure.

The foregoing is really a statement of the basic method of studying frames and machines. It is a process of simultaneously considering the static equilibrium (a) of the entire structure, and (b) of each of its members individually.

As always when dealing with bodies that are connected to each other, special care must be taken to satisfy Newton's third law of action and reaction when drawing free body diagrams. That is, when drawing free body diagrams, the forces exerted by one member on a second member must be equal, opposite, and collinear to the forces exerted by the second member on the first.

As an example, consider the frame in Figure 19, whose weight is negligible compared to the force $\bar{F}_D$. If we draw the free body diagram for the structure and the two members ACD and BC, we arrive at the results shown in Figure 20 on the following page, where the directions of the unknown forces are arbitrarily assumed using Newton's third law.

In these free body diagrams there are six unknown reaction components, for which we have nine equations of static equilibrium (three for each isolated body). As was true in our study of trusses, equilibrium of the entire truss is assured if the individual members are in equilibrium, so the excess equations may be regarded as checks for our calculations.

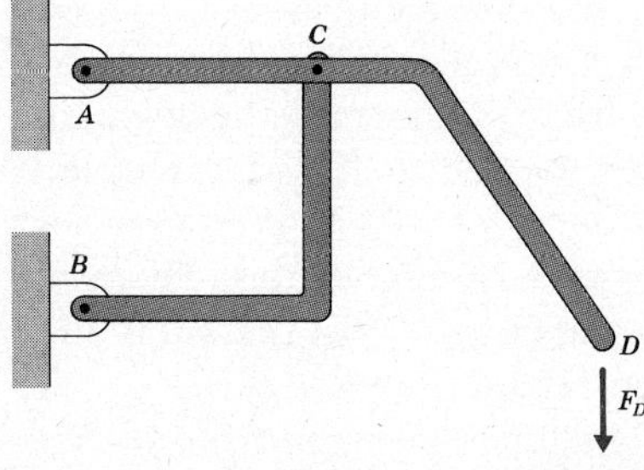

Figure 19

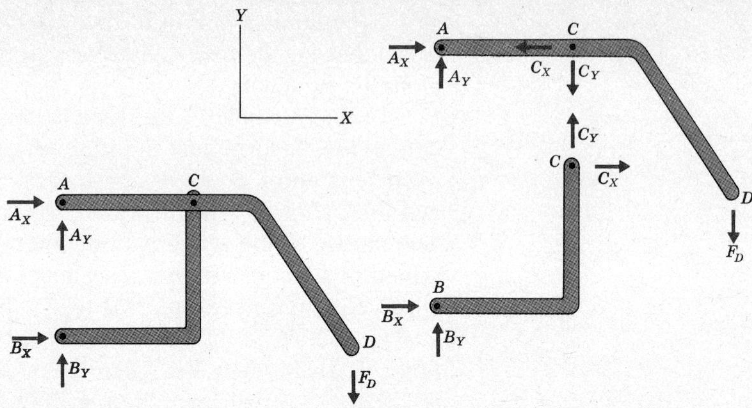

Figure 20

In drawing the free body diagrams in Figure 20 we guessed at the directions of the components of the force exerted by bar ACD on bar BC at joint C. We then drew the reaction forces exerted by bar BC on bar ACD *consistently* with this initial guess. As usual, we will know that a direction has been improperly assumed if the force component is found to be negative.

> A useful check on whether the free body diagrams for the individual members have been drawn consistent with Newton's third law is to superimpose the individual diagrams in order to form the original structure. If the reaction forces between the members cancel each other, with the result that the superposition exactly produces the free body diagram of the entire structure, then it can be concluded that the internal reaction forces are consistent.

We drew the free body diagrams in Figure 20 on the premise that we had not first carefully considered the nature of the members. However, with a little observation before we start sketching, we can save a good deal of computational effort. It will be noted that member BC is loaded only at its ends. Therefore, even though it is not a straight bar, it is a two-force member. The discussion of two-force members at the beginning of this module showed that the two forces applied to curved, as well as straight members must be equal in magnitude, opposite in direction, and collinear with the line connecting their points of application. Thus, the observation that bar BC is a two-force member results in the alternative set of free body diagrams shown in Figure 21 on the next page.

The computational savings previously alluded to result from the fact that member BC is a self-equilibrant member, as well as from the fact that we now have only one unknown, F_{BC}, in place of the four unknowns, B_X,

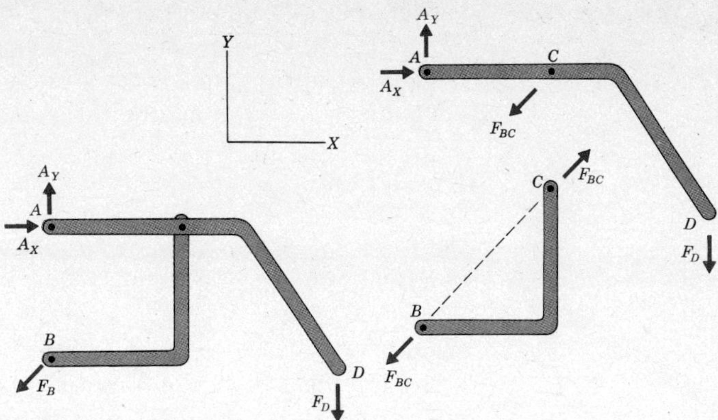

Figure 21

B_Y, C_X, and C_Y in Figure 19. Nevertheless, it should be emphasized that the observation that certain members are two-force members is useful, but not crucial, to the solution process.

The foregoing discussion may be generalized in the following series of steps for determining the forces exerted between the members of a frame or a machine.

1 Inspect the structure to determine if there are any two-force members.
2 Draw free body diagrams of the entire structure and of each member of the structure. In doing so, be certain to depict the forces acting on two force members consistently, that is, equal in magnitude, opposite in direction, and collinear. In drawing the free body diagrams, give special attention to satisfying Newton's third law for the forces exerted between the members.
3 Successively write and solve the equilibrium equations corresponding to the free body diagrams. The two-force members are automatically in equilibrium if their forces have been properly accounted for, so such members need not be considered. In following this step it is only necessary to formulate a sufficient number of equations to obtain the forces of interest. However, if all of the possible equilibrium equations are written, the excess equations may be used to check earlier computations.

These steps are equally valid for investigating planar and three-dimensional structures. For a given number of members, three-dimensional structures result in twice as many equilibrium equations as planar structures. The force and moment computations for three-dimensional structures are considerably more involved than those for planar structures. For these reasons, we shall restrict our attention to planar problems. In actuality, the solution of the equations for most space structures, as well as for many complicated planar structures, requires the aid of an electronic computer.

Finally, we note that it is possible for frames to be statically indeterminate. As always, this is manifested by an excess of unknown forces in comparison with the number of available equilibrium equations.

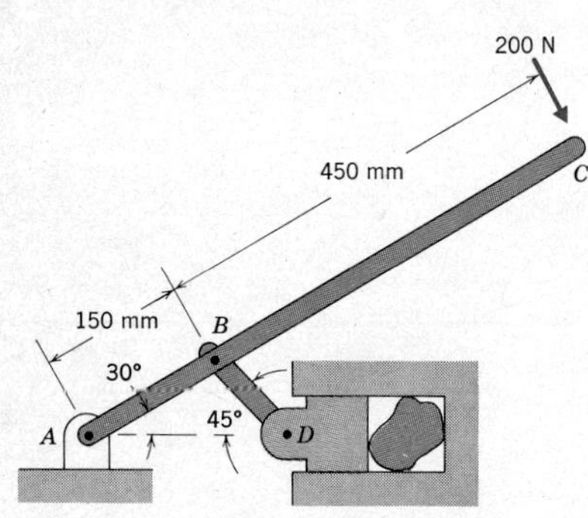

EXAMPLE 5

A 200-N force is applied to the crusher at point C, as shown. Determine the force the piston D exerts on the object within the cylinder and also determine the reaction at pin A.

Solution

There are three bodies of interest in this problem — the two bars and the piston. Of these, bar BD is a two-force member. In accounting for the reaction forces, we show a transverse force and a couple exerted by the walls of the cylinder on the piston, because the cylinder does not allow the piston to move transversely or to rotate. Thus, we have the following free body diagrams.

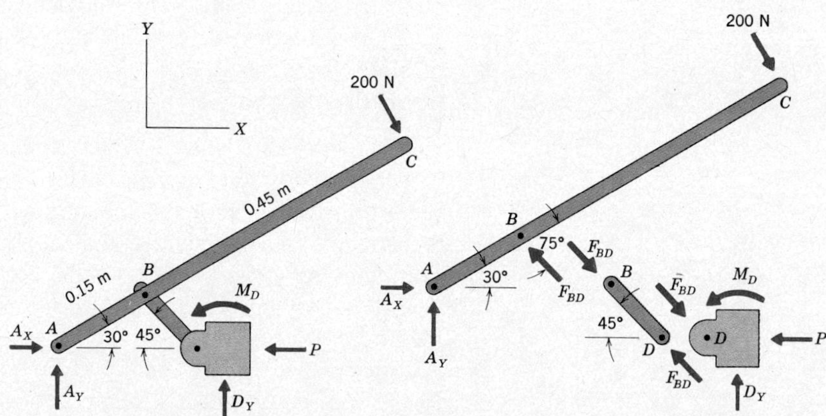

We have depicted $\bar{F}_{BD}$ as a compressive force, because it can be seen that the ends of member BD are being squeezed together. This is contrary to our general procedure when treating two-force members, in which we assume a state of tension. We ignored the practice here because there is no chance that we will obtain, and thereby misinterpret, a negative sign for the axial force.

The values of A_X, A_Y, and F_{BD} can be found from the equilibrium equations for bar ABC, which are

$$\Sigma M_{AZ} = F_{BD} \sin 75° \,(0.15) - 200(0.60) = 0$$

$$\Sigma F_X = A_X - F_{BD} \cos 45° + 200 \cos 60° = 0$$

$$\Sigma F_Y = A_Y + F_{BD} \sin 45° - 200 \sin 60° = 0$$

$$F_{BD} = 828.2 \qquad A_X = 485.6 \qquad A_Y = -412.4 \text{ N}$$

We may now relate F_{BD} to P by summing forces in the X direction for the piston. Thus

$$\Sigma F_X = -P + F_{BD} \cos 45° \equiv -P + 828.2(0.7071) = 0$$

$$P = 585.6 \text{ N}$$

Notice that the value of P could also have been obtained from a single equilibrium equation by summing forces in the X direction for the entire structure after determining the reaction $\bar{A}$. We leave it to you to write and use the checking equations for this structure.

EXAMPLE 6

In the compensating trailer hitch shown, the spring is stretched to a tension force $\bar{T}$ in order to better distribute the loading of the 1000-kg trailer on the automobile. Compare the reactions between all tires and the ground when $T = 3$ kN to that obtained when $T = 0$. The centers of mass of the automobile and the trailer are at points G_1 and G_2, respectively.

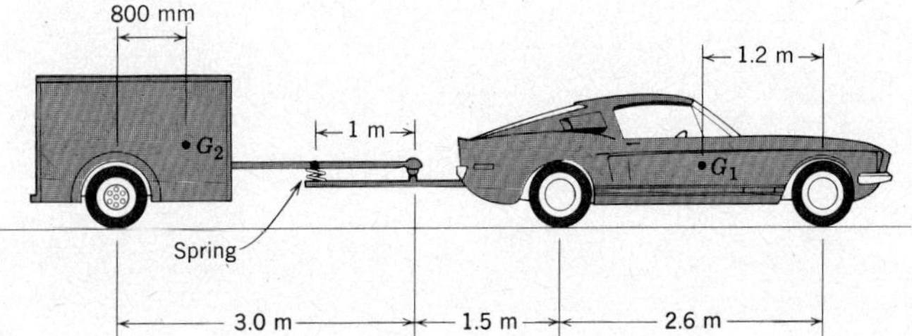

Solution

The connection between the trailer and the hitch is a ball joint, so we have the following free body diagrams. Notice that there are no horizontal forces acting on the tires, which is consistent with the entire system being at rest on a level surface.

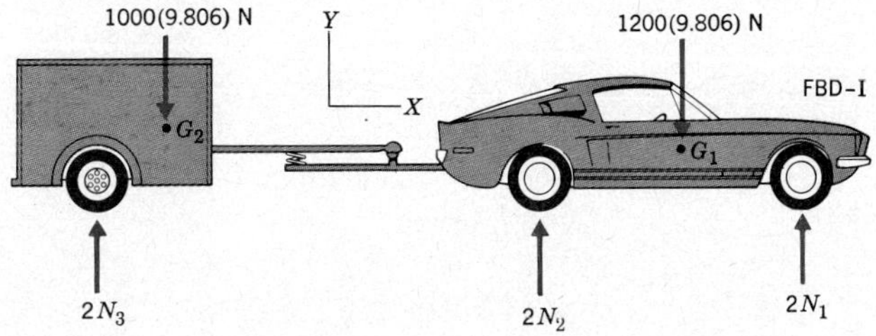

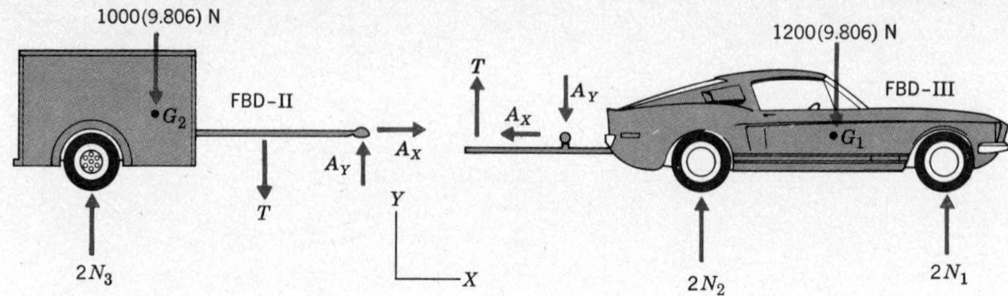

In this problem we are not interested in the reaction force components $\bar{A}_X$ and $\bar{A}_Y$ at the ball joint, so we write only the equilibrium equations that do not involve these forces. Thus, letting point C be the center of the tire of the trailer, we have

Car and trailer (FBD I):

$$\Sigma M_{CZ} = 2N_2(4.5) + 2N_1(7.1) - 1000(9.806)(0.8) - 1200(9.806)(5.9) = 0$$

$$\Sigma F_X \equiv 0$$

$$\Sigma F_Y = 2N_1 + 2N_2 + 2N_3 - 1200(9.806) - 1000(9.806) = 0$$

Car (FBD III):

$$\Sigma M_{AZ} = -T(1.0) + 2N_2(1.5) + 2N_1(4.1) - 1200(9.806)(2.9) = 0$$

Trailer (FBD II):

$$\Sigma M_{AZ} = T(1.0) + 1000(9.806)(2.2) - 2N_3(3.0) = 0$$

In the four algebraic equations above there are three unknowns: N_1, N_2, and N_3, because T is a known parameter. Thus, there is one check equation. Solution of the equation for the trailer yields

$$N_3 = \frac{T}{6} + 3596 \text{ newtons}$$

whereas the simultaneous solution of the two remaining moment equations is

$$N_1 = 0.2885T + 2414 \text{ newtons} \qquad N_2 = -0.4552T + 4777 \text{ newtons}$$

We now check these computations with the force sum in the Y direction, which gives

$$\Sigma F_Y = 2(0.2885T + 2414) + 2(-0.4552T + 4777) + 2\left(\frac{T}{6} + 3596\right) - 21573 \equiv 0$$

Hence, we conclude that the solutions are correct. Therefore, for T = 3000 newtons we find

$$N_1 = 3280 \qquad N_2 = 3411 \qquad N_3 = 4096 \text{ newtons}$$

whereas for $T = 0$ we find

$$N_1 = 2414 \qquad N_2 = 4777 \qquad N_3 = 3596 \text{ newtons}$$

These values show that an appreciable equalization of the forces on the front and rear wheels of the automobile is achieved through the action of the spring. Notice that it also increases the load on the wheels of the trailer.

EXAMPLE 7

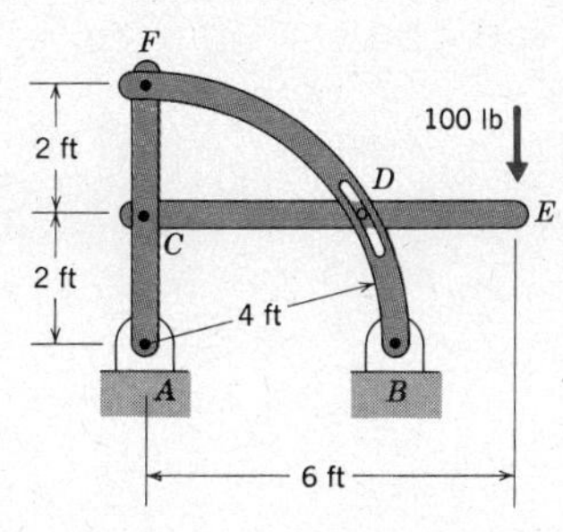

The frame shown supports a 100-lb load at point E. Determine the reactions at supports A and B.

Solution

None of the members are two-force members, so we show arbitrary force components at each pin joint. The pin at D rides in a groove, which means that the reaction is normal to the groove. This results in the following free body diagrams.

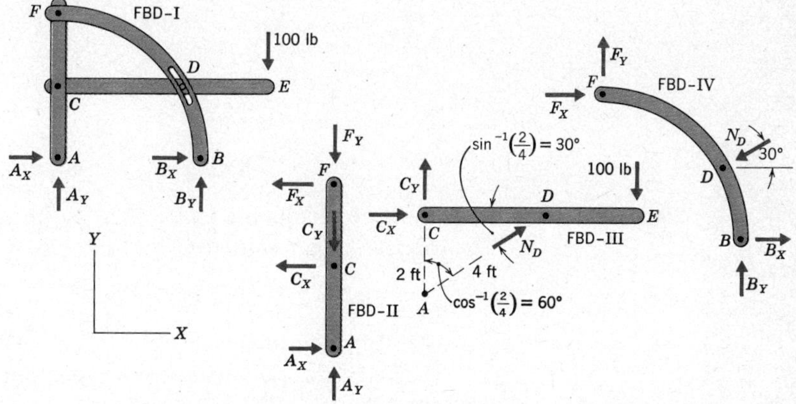

There are nine unknown force components in the free body diagrams. We begin by noting that the free body diagram for bar CD can be solved for all of the forces acting on that body, so we write:

Bar CD (FBD III):

$$\Sigma M_{CZ} = N_D \sin 30°(4 \sin 60°) - 100(6) = 0$$

$$\Sigma F_X = C_X + N_D \cos 30° = 0$$

$$\Sigma F_Y = C_Y + N_D \sin 30° - 100 = 0$$

The solutions are

$$N_D = 346.4 \qquad C_X = -300.0 \qquad C_Y = -73.2 \text{ lb}$$

The values of F_X and F_Y were not requested, so we next sum moments about joint F on bar AF and on bar BF. This yields:

Bar AF (FBD II):

$$\Sigma M_{FZ} = A_X(4) - C_X(2) \equiv 4A_X - (-300.0)(2) = 0 \qquad A_X = -150 \text{ lb}$$

Bar BF (FBD IV):

$$\Sigma M_{FZ} = -N_D \cos 30°(2) - N_D \sin 30°(4 \sin 60°) + B_Y(4) + B_X(4)$$

$$= -346.4(0.8660)(2) - 346.4(0.50)4(0.8660) + 4B_Y + 4B_X = 0$$

$$B_X + B_Y = 300.0 \text{ lb}$$

We now use the free body diagram of the entire structure to determine the values of A_Y, B_X, and B_Y. Thus:

Entire structure (FBD I):

$$\Sigma M_{AZ} = B_Y(4) - 100(6) = 0$$

$$\Sigma F_X = A_X + B_X = 0$$

$$\Sigma F_Y = A_Y + B_Y - 100 = 0$$

These equations yield

$$B_Y = 150 \qquad B_X = 150 \qquad A_X = -150 \qquad A_Y = -50 \text{ lb}$$

In these results the value of A_X is the same as that obtained by considering bar AF, thus validating (in part) the calculations.

HOMEWORK PROBLEMS

IV.50–IV.52 Determine the reactions at pins A and C in the frames shown.

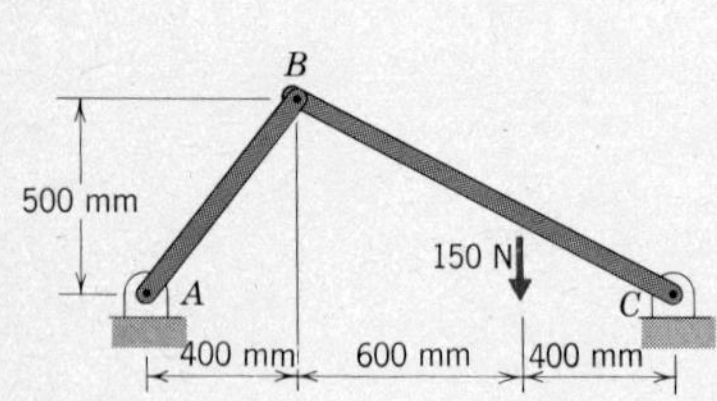

Prob. IV.50

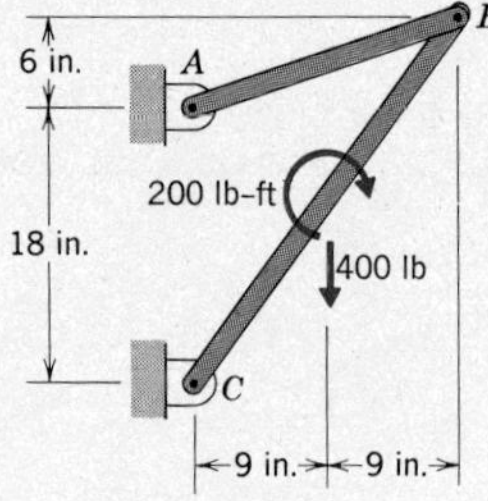

Prob. IV.51

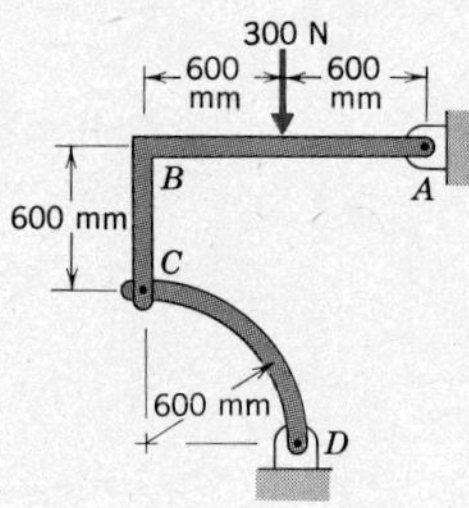

Prob. IV.52

IV.53–IV.54 Determine the clamping force on the bolt and the force on pin A for the pliers shown.

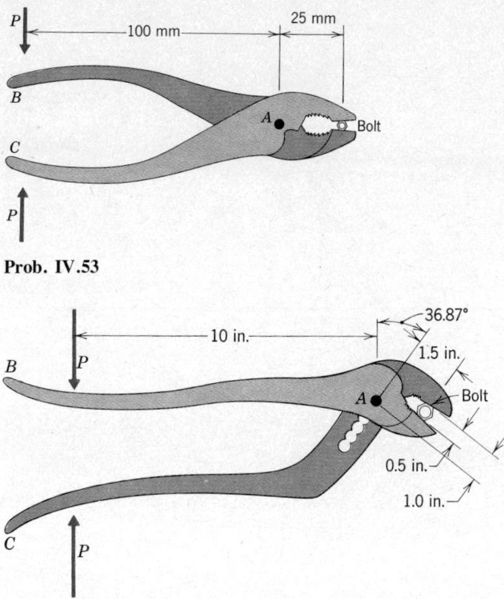

Prob. IV.53

Prob. IV.54

IV.55 A 200-N force is applied to the piston. For the position where $\theta = 90°$, determine the couple $\bar{M}$ that should be applied to the crankshaft to hold the system in static equilibrium.

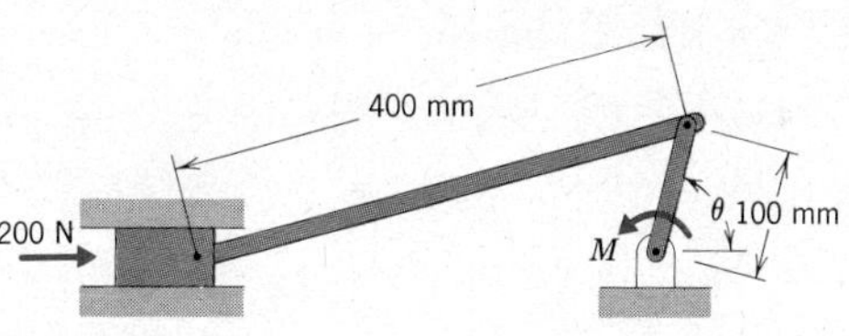

Prob. IV.55

IV.56 Solve Problem IV.55 for $\theta = 60°$.

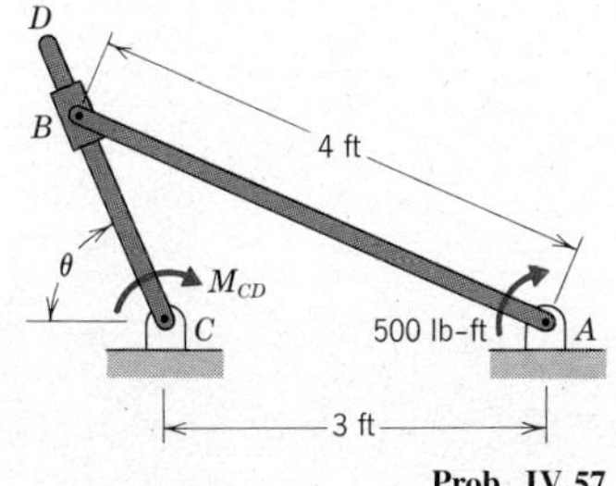

Prob. IV.57

IV.57 A 500 lb-ft couple is applied to bar AB, as shown. This bar is pinned to collar B, which slides on the smooth bar CD. Determine the couple $\bar{M}_{CD}$ that must be applied to bar CD to maintain static equilibrium at $\theta = 90°$.

IV.58 Solve Problem IV.57 for $\theta = 45°$.

IV.59 The mechanism shown is in equilibrium in the vertical plane. Collar C rides on the smooth vertical guide and the unstretched length of the

spring is 1 m. Knowing that the bars have a mass of 2 kg/m, determine the stiffness of the spring and the reaction at pin A.

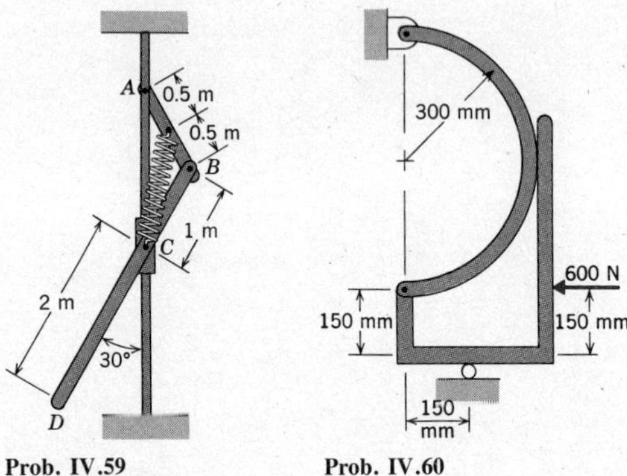

Prob. IV.59 Prob. IV.60

IV.60 Determine the forces acting on the smooth semicircular bar in the frame.

IV.61–IV.63 Determine the reactions at ends A and B for each system of beams shown.

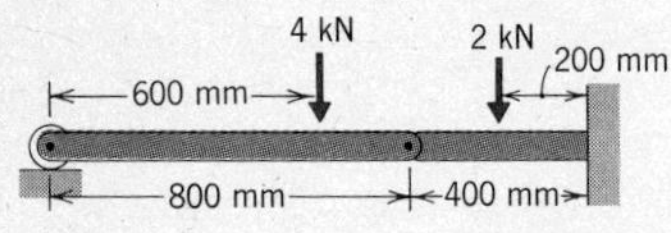

Prob. IV.61 Prob. IV.62

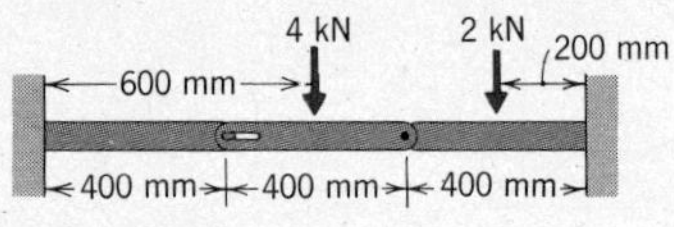

Prob. IV.63

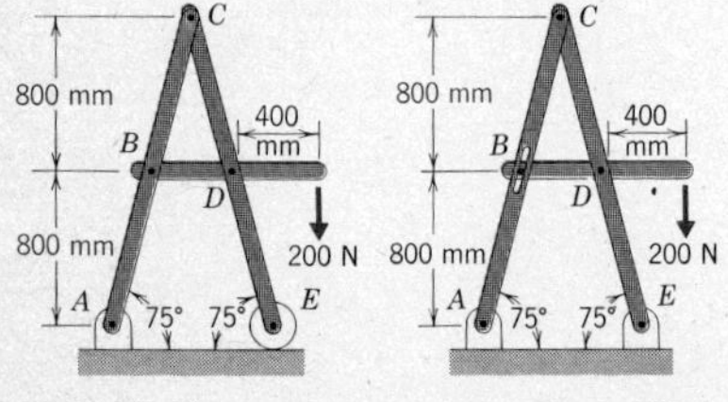

Prob. IV.64 Prob. IV.65

IV.64 and IV.65 Determine the forces acting on members AC and CE of the frame shown.

IV.66 Determine the forces acting on the members of the frame (a) if P = 200 lb and Q = 0, (b) if P = 0 and Q = 200 lb, (c) if P = Q = 200 lb.

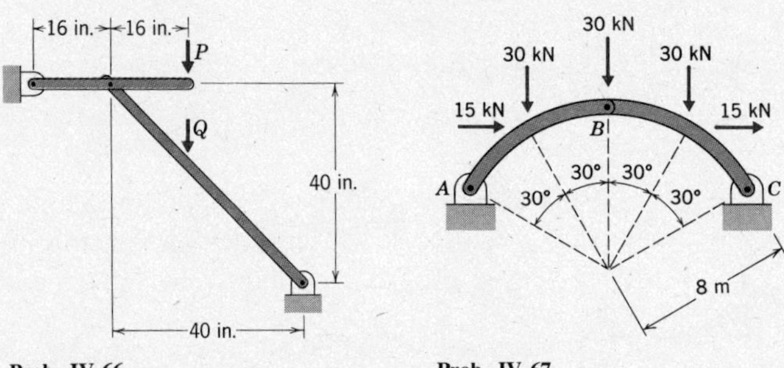

Prob. IV.66 Prob. IV.67

IV.67 The three-pin arch is subjected to the set of loads shown. Determine the reactions at pins A, B, and C.

IV.68 The spring has a stiffness of 50 lb/in. and an unstretched length of 20 in. Determine the reactions at pins A, B, and D.

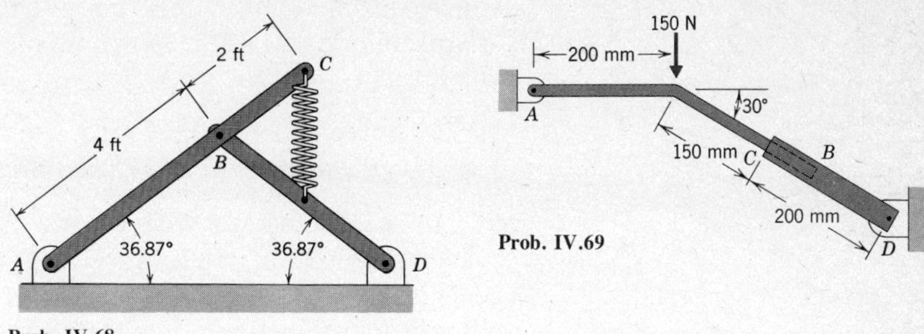

Prob. IV.68

Prob. IV.69

IV.69 Bar AB, having a circular cross section, fits closely in the smooth pipe CD. Determine the reactions at pins A and D.

IV.70 Determine the reactions at pins A, B, and C when the frame supports a 100-kg package.

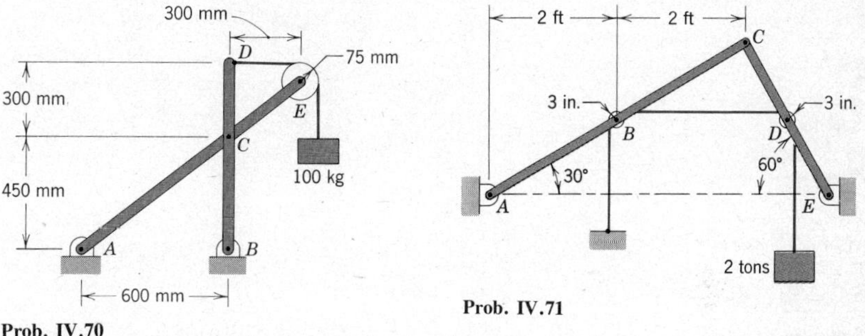

Prob. IV.70

Prob. IV.71

IV.71 The frame shown is used to support the 2-ton crate. Determine the reactions at pins A and E.

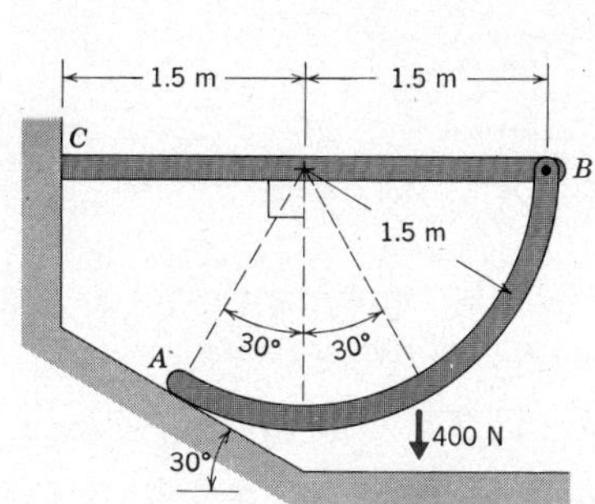

Prob. IV.72

IV.72 Curved bar AB is pinned at end B to the cantilevered bar BC. End A rests on the smooth incline. Determine the reactions at end C.

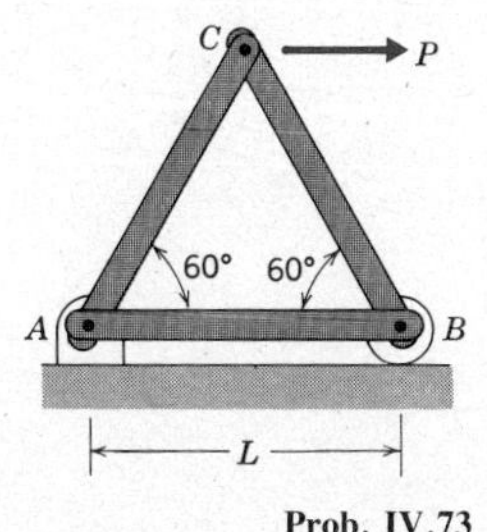

Prob. IV.73

IV.73 Each of the bars in the structure shown has weight W. Their centers of mass are at their midpoints. Consider the situation where $P = 0$. (a) Determine the reactions at the ends of each member. (b) Compare the results of part (a) to those obtained when the structure is considered to be a truss, where the weight *must* act through the joints. Thus, a fair approximation is that one-half the weight of each member is applied to the joints at the ends of that member. *Hint:* Use the method of joints for part (b).

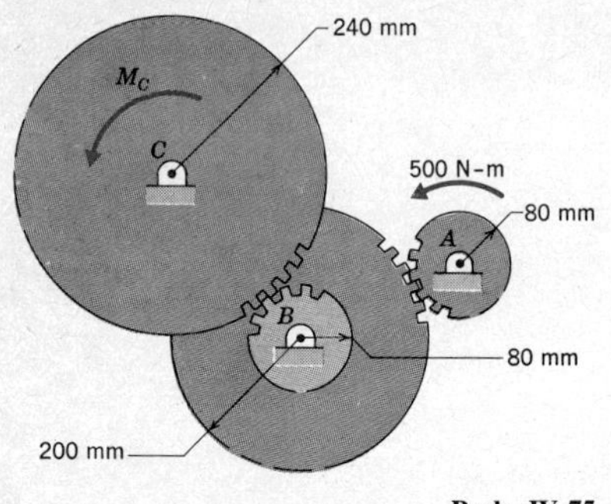

Prob. IV.75

IV.74 Solve Problem IV.73 for $P = 10W$.

IV.75 A 500 N-m couple is applied to gear A, as shown. Stepped gear B, which transmits this couple to gear C, rides freely on its shaft. Determine the magnitude and sense of the couple $\bar{M}_C$ acting on gear C necessary for equilibrium.

IV.76 Solve Problem IV.75 if a counterclockwise couple of 100 N-m is acting on the stepped gear B in addition to the 500 N-m couple on gears A.

IV.77 A counterclockwise couple $\bar{M}_1$ is applied to the planetary gear system shown. The radius of the central gear is r_1 and that of the planetary gears is r_2. Determine the couples $\bar{M}_2$ and $\bar{M}_3$ acting on the spider and the outer gear, respectively, for the system to be in static equilibrium.

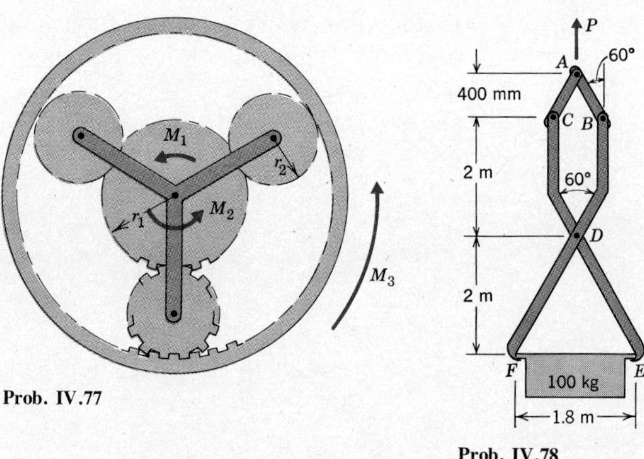

Prob. IV.77

Prob. IV.78

IV.78 A 100-kg radioactive container is held by applying the force $\bar{P}$ to the pair of tongs. Determine the forces acting on member BF.

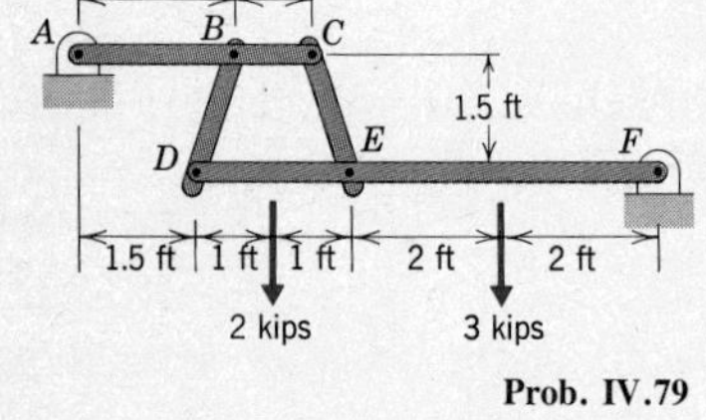

Prob. IV.79

IV.79 Determine the reactions at pins A and F for the loading shown.

IV.80 Determine all forces acting on member AC when $P = 8$ kN and $Q = 0$.

Prob. IV.80

IV.81 Solve Problem IV.80 for $P = 0$ and $Q = 8$ kN.

IV.82 Determine the force exerted on the bolt when the handle of the bolt cutter is gripped by the pair of 30-lb forces, as shown.

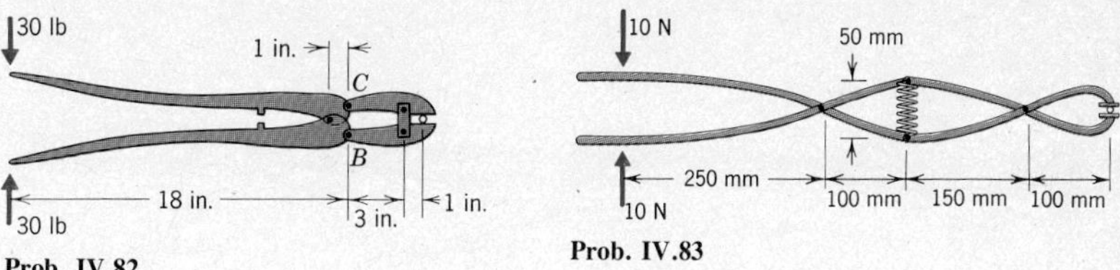

Prob. IV.82

Prob. IV.83

IV.83 The mechanism shown is a device for clamping onto inaccessible objects. The spring has a stiffness of 2 kN/m and an unstretched length of 40 mm. Determine the clamping force on the object if the handles are squeezed with a pair of 10-N forces.

IV.84 Determine the forces exerted on the bolt and on member CD when the locking pliers is gripped by the pair of 400-N forces, as shown.

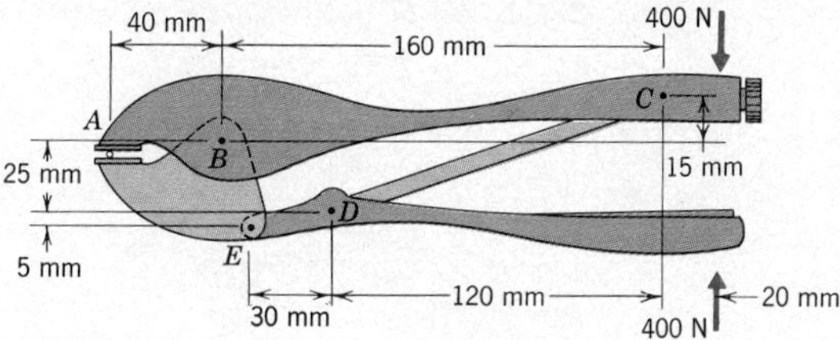

Prob. IV.84

IV.85 A 50-lb force is applied to the press punch, as shown. Determine the force exerted on the smooth plate F and the reaction forces at pins A and E.

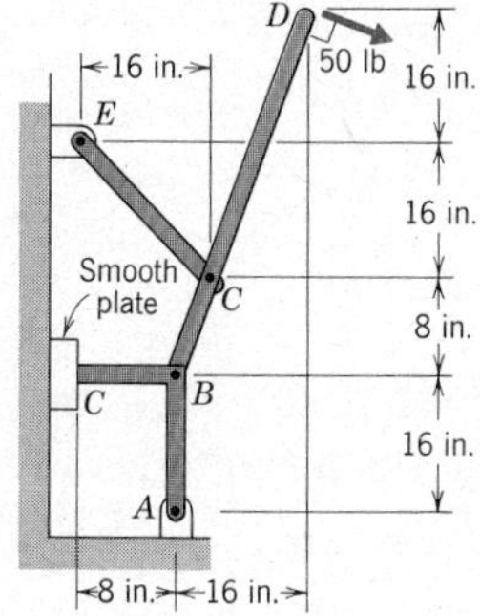

Prob. IV.85

IV.86 A 500-kg package rests on the platform, which is supported by the mechanism shown. Determine the force in the hydraulic cylinder and the reactions at pins A and F necessary for static equilibrium as a function of θ.

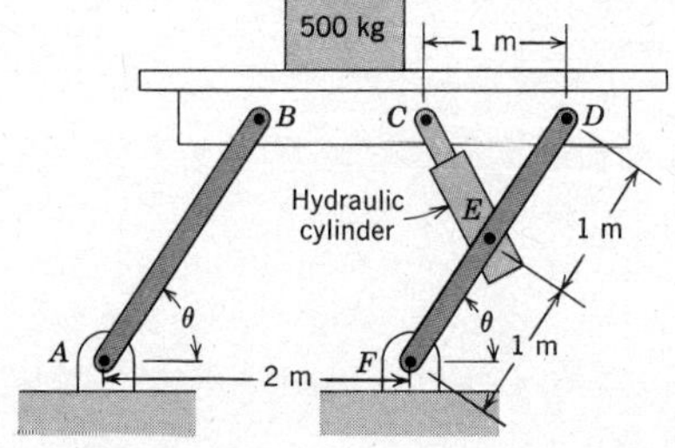

Prob. IV.86

IV.87 The position of the scoop on the tractor is controlled by two parallel mechanisms, only one of which appears in the side view. Determine the forces acting on the main arm ADF when the scoop is carrying an 8-kN load in the position shown.

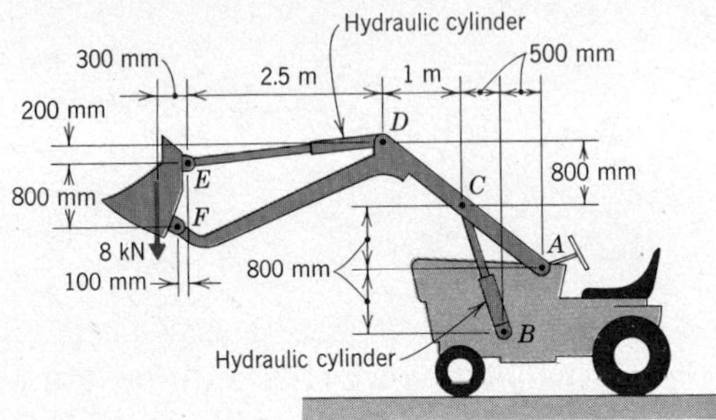

Prob. IV.87

IV.88 A mining trencher is subjected to the forces shown. The position of the bucket is controlled by the hydraulic cylinders BC, DE, and GH. Determine the forces acting on arms AF and EI.

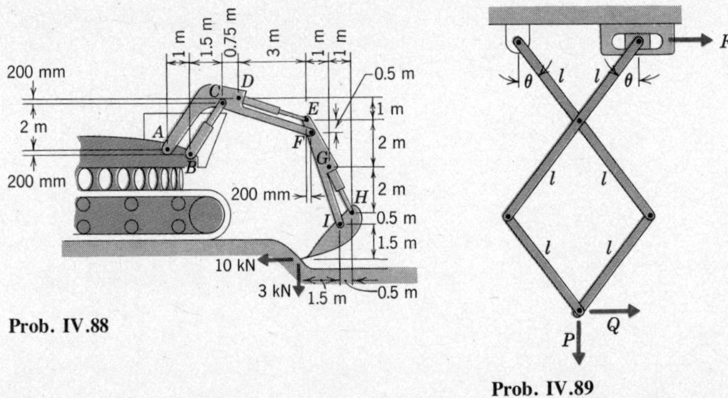

Prob. IV.88

Prob. IV.89

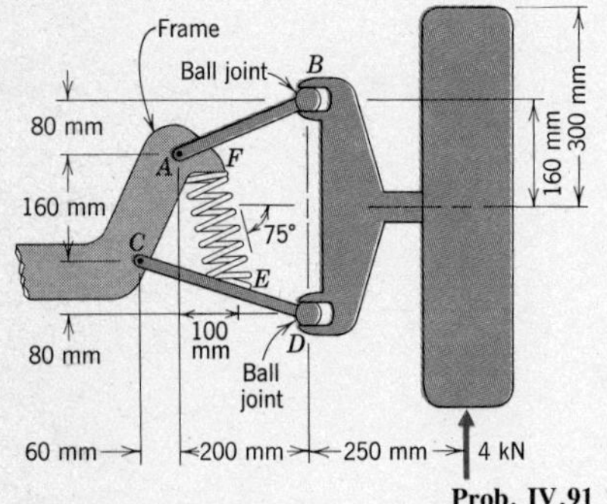

Prob. IV.91

IV.89 Determine the horizontal force $\bar{F}$ required to hold the mechanism in position as a function of the angle θ and the magnitude of the force $\bar{P}$ if $Q = 0$.

IV.90 Solve Problem IV.89 if $Q = 1.50P$.

IV.91 The suspension system shown transmits the 4-kN reaction between the tires and the ground to the frame of an automobile. Determine the force in the spring and the forces exerted by members AB and CD on the frame.

MODULE V
DISTRIBUTED FORCES

The focus of attention in this module will be situations where forces are spread out over a region of surface area or volume of a body. Such forces are called *distributed forces*. In the previous modules our concern was with the effects of forces acting at points, that is, *concentrated forces*. Such forces do not exist in reality, for any material will deform when a force is applied to it, resulting in the distribution of force over a small amount of contact area. Thus, when we consider a force to be concentrated, we are assuming that the contact area is sufficiently small to represent it as a point.

Our objective here will be to determine how to represent distributed forces by equivalent concentrated forces in situations where the region of application of the distributed force system is not negligible.

A. GRAVITATIONAL FORCE

A common distributed force system is the gravitational attraction exerted on a body by the planet earth. The *weight* of a body is the magnitude of the resultant of the gravity force on each particle of the body. The point of application of this resultant force is the *center of mass*, or equivalently, the *center of gravity*. Previously, we regarded the location of the center of mass as a given quantity. Now the task is to learn how to determine its location.

1 Centers of Mass and Geometric Centroids

We begin the investigation by recalling that the gravitational force acting on a particle of mass m near the surface of the earth is simply mg directed toward the center of mass of the earth. In the model of the earth that we employ in statics, the earth is a sphere and its center of mass is the center of the sphere.

We may consider a rigid body on the surface of the earth to be composed of many particles i. We will denote the mass of each particle as Δm_i. Because the center of the earth is at a large distance from these particles, the gravitational forces on all particles are essentially parallel, in the vertical direction. Without loss of generality, and also for consistency with the notion used in most references, let us denote a convenient coordinate system by the lower case letters xyz. Corresponding to this coordinate system are the unit vectors $\bar{i}, \bar{j}$, and $\bar{k}$. Thus, choosing the z axis to be vertically outward from the center of the earth, the gravitational forces acting on the body are as shown in Figure 1a on the following page.

The force system in Figure 1a consists of a set of parallel forces. From our studies in Module III we know that their effect can be represented by an equivalent resultant force $\bar{R}$, as shown in Figure 1b. The magnitude of this resultant force is the weight mg. To verify this we

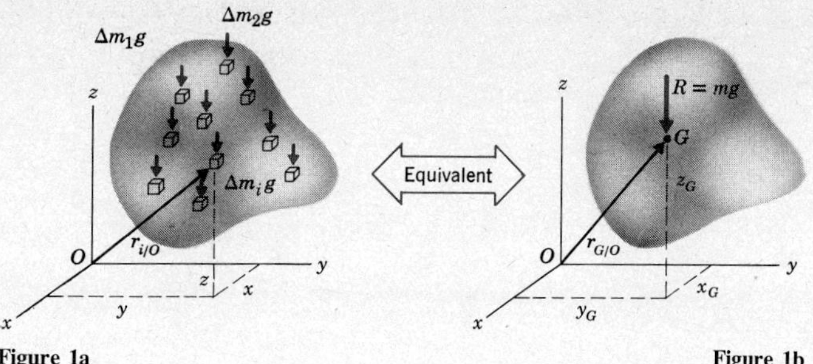

Figure 1a

Figure 1b

equate the sum of the forces in Figure 1a to the resultant $\bar{R}$.

$$\Sigma F_z = -\sum_{i=1}^{N} \Delta m_i g = -R$$

Because the mass m of the body is the sum of the mass of all of its particles, the force sum yields

$$R = \left(\sum_{i=1}^{N} \Delta m_i\right) g \equiv mg$$

Clearly, summing all of the mass elements Δm_i by summing the atoms of matter is not feasible. Instead, we consider the body to be a continuum of an infinite number of infinitesimal masses $\Delta m_i = dm$. The process of summation then becomes a process of integration over the space $\mathcal{V}$ occupied by the body. Thus

$$\boxed{\begin{aligned} R &= mg \\ m &= \int_{\mathcal{V}} dm \end{aligned}} \qquad (1)$$

The relationship between the space $\mathcal{V}$ and the elements dm will be addressed below.

The foregoing discussion defines the magnitude and direction of the resultant weight force. By definition, this force acts at the center of mass G shown in Figure 1b. To determine the coordinates of this point we use the requirement that the equivalent weight force in Figure 1b must exert the same moment about an arbitrary point as the distributed force system in Figure 1a. For this point let us choose the origin O of the coordinate system. The position of the center of mass is denoted $\bar{r}_{G/O}$. By equating the moments of the two-force systems we find

$$\Sigma \bar{M}_O = \bar{r}_{G/O} \times \bar{R} \equiv \bar{r}_{G/O} \times (-mg\bar{k})$$

$$= \sum_{i=1}^{N} \bar{r}_{i/O} \times (-\Delta m_i g \bar{k})$$

The rectangular Cartesian coordinates of mass element Δm_i are (x, y, z). Thus, as shown in Figure 1a, the components of $\bar{r}_{i/O}$ are these coordinates, that is,

$$\bar{r}_{i/O} = x\bar{i} + y\bar{j} + z\bar{k}$$

Similarly, the components of $\bar{r}_{G/O}$ are the coordinates of the center of mass G.

$$\bar{r}_{G/O} = x_G\,\bar{i} + y_G\,\bar{j} + z_G\,\bar{K}$$

Further, to account for the entire mass, we replace Δm_i by the infinitesimal mass element dm, thus transforming the summation in the moment equation to an integral. This gives

$$(x_G\,\bar{i} + y_G\,\bar{j} + z_G\,\bar{k}) \times (-mg\bar{k}) = \int_{\mathcal{V}} (x\bar{i} + y\bar{j} + z\bar{k})(-dm\; g\bar{k})$$

$$mgx_G\,\bar{j} \times mgy_G\bar{i} = g\int_{\mathcal{V}} (x\; dm\; \bar{j} - y\; dm\; \bar{i})$$

We can now equate the corresponding components in this equation, cancelling the factor g, with the following result:

$$mx_G = \int_{\mathcal{V}} x\; dm \tag{2a}$$

$$my_G = \int_{\mathcal{V}} y\; dm \tag{2b}$$

These integrals are the *first moments of mass* with respect to the x and y coordinates. They yield the coordinates of the mass center, x_G and y_G.

The value of z_G cannot be determined from the foregoing equations because the gravity forces in Figures 1 do not exert a moment about the z axis. This coordinate value may be determined by reorienting the gravity field of Figures 1, for instance, by orienting the force of gravity to be parallel to the x axis. Summing moments in the same manner as was done above then yields

$$mz_G = \int_{\mathcal{V}} z\; dm \tag{2c}$$

Equations (2) are the basic ones required to locate the center of mass of a body. Because we usually know the mass density ρ (mass per unit volume) of the material composing the body, we may perform the required integrations by writing

$$\boxed{dm = \rho \, d\mathcal{V}} \tag{3}$$

where $d\mathcal{V}$ is an infinitesimal element of volume.

For a body with arbitrary properties, equations (2) require the evaluation of three first moments of mass. However, a computational simplification can be achieved for bodies having a plane of symmetry. To investigate this matter, consider the body in Figure 2, where for convenience, we selected the x and y axes to coincide with the plane of symmetry.

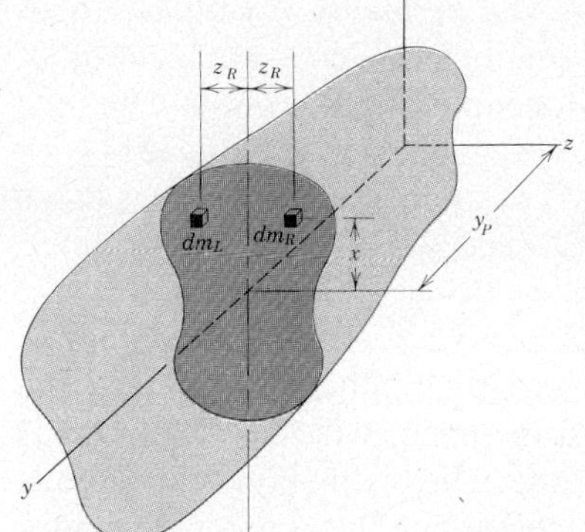

Figure 2

The property of symmetry with respect to the xy plane has the geometric significance that the portion of the body to the right of the plane is the mirror image of the portion to the left. Mathematically, this means that at every cross section perpendicular to the y axis at some value y_P, such as the one illustrated in Figure 2, each mass element dm_R having coordinates (x, y_P, z_R) to the right of the symmetry plane has a companion element dm_L to the left of this plane whose coordinates are $(x, y_P, -z_R)$. By virtue of the symmetry, we also know that both elements have the same density, $\rho_L = \rho_R$. These elements also have identical volumes, so it follows that $dm_L = dm_R$. Thus, in the integral for the first moment of mass with respect to the z coordinate, corresponding to each term $z_R \, dm_R$, there is a term $z_L \, dm_L \equiv - z_R \, dm_R$. Obviously, these terms cancel when added, so that the first moment of mass is zero. Equation (2c) then yields $z_G = 0$, which enables us to conclude that:

> The center of mass of a body having a plane of symmetry is situated somewhere on that plane.

Note that the foregoing statement applies only to those situations where the geometric shape *and* the mass distribution, as given by the density ρ, are symmetric with respect to a plane. In the common situation where a body is *homogeneous,* that is, where its material properties do not depend on the location within the body, the density ρ is constant. It is then only necessary to consider geometrical symmetry.

A further simplification resulting for a homogeneous body is that the density ρ, being a constant, may be removed from the integrals over the space $\mathcal{V}$. In this case equation (1) gives

$$\boxed{m = \int_{\mathcal{V}} dm = \rho \int_{\mathcal{V}} d\mathcal{V} = \rho \mathcal{V}} \tag{4}$$

Substituting equations (3) and (4) into equations (2), we find that the density ρ cancels, yielding the following results:

$$\mathcal{V} x_G = \int_{\mathcal{V}} x \, d\mathcal{V}$$
$$\mathcal{V} y_G = \int_{\mathcal{V}} y \, d\mathcal{V} \qquad (5)$$
$$\mathcal{V} z_G = \int_{\mathcal{V}} z \, d\mathcal{V}$$

The integrals in equations (5) are the *first moments of volume*. In all situations, homogeneous or otherwise, equations (5) define the location of the *geometric centroid of the volume* $\mathcal{V}$ occupied by the body. Thus we see that when the body is homogeneous, the center of mass coincides with the geometric centroid of the volume. It is not necessarily true that these points will coincide if the body is not homogeneous, *even* if it is a symmetric body.

It is important to understand the difference between the meanings of the center of mass and of the centroid. The former point is affected by the way the mass of the body is distributed in space, whereas the latter is affected only by the geometry of the volume. The distinction is analogous to the difference between the population center and the geographic center of a region, such as a state or nation.

Equations (2) and (5) are the basic ones we will need to locate the mass center. The most obvious way to employ them is to evaluate the integrals. To do this it will be necessary for us to describe the position of points in a body, as well as the body's element of volume $d\mathcal{V}$. The required techniques are the subject of the next section.

2 Integration Techniques

Let us recall two methods of analytic geometry for locating a point. One method is to write the rectangular Cartesian coordinates (x, y, z) of the point. The other method is to specify the location of the point using *cylindrical coordinates*. In the latter, first we choose a coordinate plane and locate the point with respect to the plane. For instance, arbitrarily choosing the xy plane, we have the situation depicted in Figure 3, where the distance z locates the point P above the xy plane. Then, we locate point P parallel to the xy plane by using the polar coordinates (R, ϕ). The resulting set of coordinates (R, ϕ, z) are the cylindrical coordinates. For the polar coordinates defined in Figure 3, we have the following coordinate transformation from the Cartesian to the polar variables:

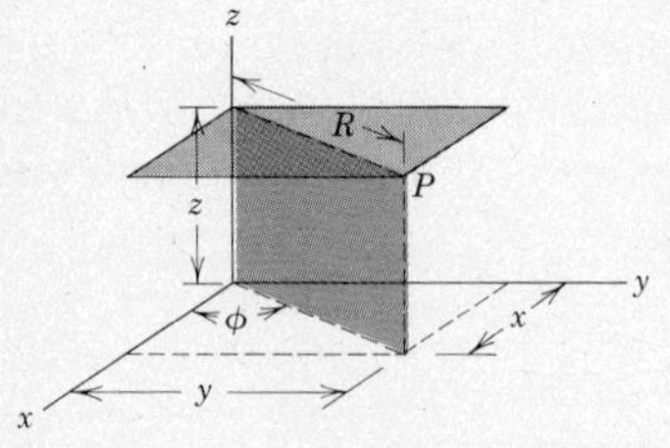

Figure 3

$$x = R \cos \phi \qquad y = R \sin \phi$$

Transformations such as these enable us to use either Cartesian or cylindrical coordinates in formulating the integrals in equations (2) and (5). However, the decision to use either of these sets of coordinates is not arbitrary. In general, the coordinates we shall select are the ones that most easily permit us to describe the boundaries of the body.

For an arbitrary body, we may form an element of mass dm from either Cartesian or cylindrical coordinates, as shown in Figure 4. The

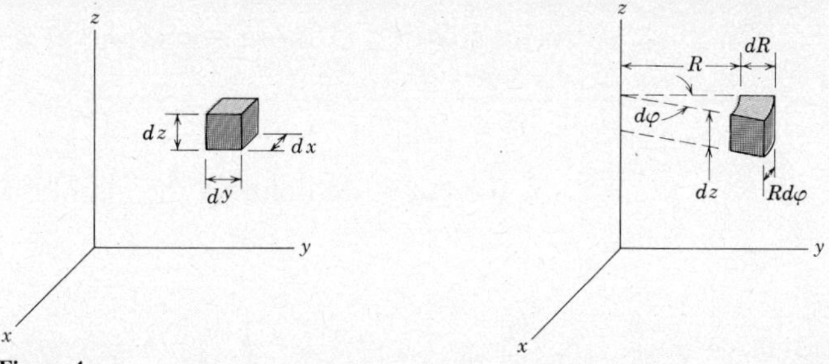

Figure 4

angle $d\phi$ is infinitesimal, so both elements dm in Figure 4 are parallelepipeds. Hence, using equation (3) we have

$$dm = \rho\, dx\, dy\, dz \quad \text{or} \quad dm = \rho R\, d\phi\, dR\, dz \tag{6}$$

We should note that larger elements of mass can be formed, but they seldom represent a significant advantage over the ones shown in Figure 4.

Another item that may require some review at this time is the formulation of the limits of an integral. This matter is treated in the following examples.

EXAMPLE 1

Determine by integration the (x, y, z) coordinates of the center of mass of the homogeneous orthogonal tetrahedron shown.

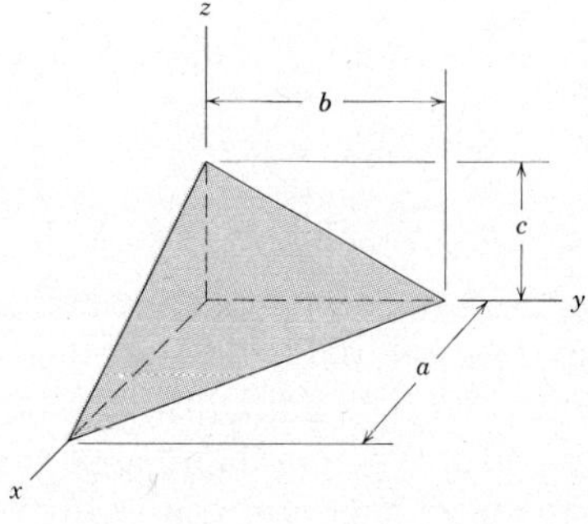

Solution

Because the body is homogeneous, the center of mass coincides with the geometric centroid, so we will use equations (5). The solution will be formulated in terms of the Cartesian coordinates (x, y, z) because three of the faces of the tetrahedron coincide with the coordinate planes and the fourth is a slanted plane. All lines lying in a flat plane are straight, so we know that the values of the (x, y, z) coordinates of points on this slanted plane are linearly related. Thus, the general equation of the plane is

$$z = k_1 x + k_2 y + k_3$$

where k_1, k_2, and k_3 are constants.

To determine these constants we note that three points on the plane are $(a, 0, 0)$, $(0, b, 0)$, and $(0, 0, c)$. Substituting these values into the general equation of the plane yields

$$0 = k_1 a \quad + k_2(0) + k_3$$

$$0 = k_1(0) + k_2 b \quad + k_3$$

$$c = k_1(0) + k_2(0) + k_3$$

The solution of these equations of the plane is

$$k_3 = c \qquad k_2 = -\frac{c}{b} \qquad k_1 = -\frac{c}{a}$$

so the equation defining the slanted face is

$$z = c\left(1 - \frac{x}{a} - \frac{y}{b}\right)$$

In Cartesian coordinates an element of volume is

$$d\mathcal{V} = dx\,dy\,dz$$

To integrate all of these elements we first consider a cross section corresponding to a fixed value of one of the coordinates. Choosing a cross section parallel to the yz plane at an arbitrary position x, we have the situation depicted in the following sketch.

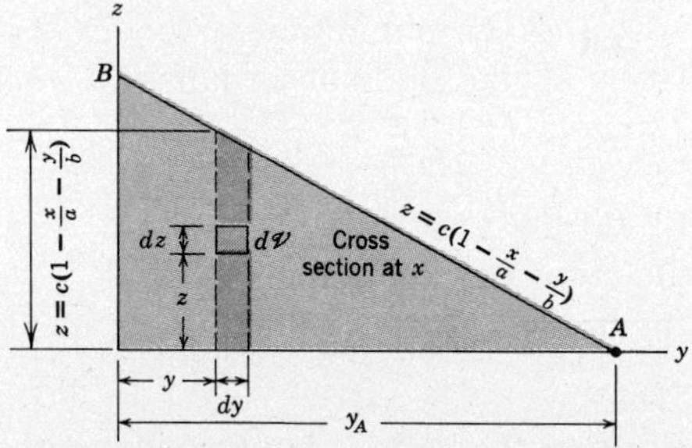

strip for any value of y on the cross section. We will account for all elements $d\mathcal{V}$ on the vertical strip by integrating first over z in the range 0 strip for any value of y on the cross section. We will acount for all elements $d\mathcal{V}$ on the vertical strip by integrating first over z in the range $0 \le z \le c(1 - x/a - y/b)$. Then we will account for all such vertical strips by integrating over all values of y on the cross section, which is the range $0 \le y \le y_A$. To determine the value of y_A, we note that at point A,

$$z = c\left(1 - \frac{x}{a} - \frac{y_A}{b}\right) = 0 \quad \text{so} \quad y_A = b\left(1 - \frac{x}{a}\right)$$

Finally, after having included all elements dm on the cross section, we account for all cross sections x by integrating over the range of values of x $(0 \leq x \leq a)$ for the tetrahedron.

To use equations (5) we will need the value for the volume of the tetrahedron. To this end we integrate each of the elements of volume (writing the integrals in the order in which they will be integrated) and find

$$\begin{aligned}
\mathcal{V} &= \int_0^a \int_0^{b(1-x/a)} \int_0^{c(1-x/a\ -\ y/b)} dz\ dy\ dx \\
&= \int_0^a \int_0^{b(1-x/a)} c\left(1 - \frac{x}{a} - \frac{y}{b}\right) dy\ dx \\
&= c \int_0^a \left[b\left(1 - \frac{x}{a}\right) - \frac{x}{a}(b)\left(1 - \frac{x}{a}\right) - \frac{1}{2b}(b)^2\left(1 - \frac{x}{a}\right)^2 \right] dx \\
&= \frac{1}{6} abc
\end{aligned}$$

Now we may calculate the first moments of volume. Beginning with the first of equations (5), we have

$$\begin{aligned}
\left(\frac{1}{6}abc\right) x_G &= \int_0^a \int_0^{b(1-x/a)} \int_0^{c(1-x/a\ -\ y/b)} x\ dz\ dy\ dx \\
&= \int_0^a \int_0^{b(1-x/a)} cx\left(1 - \frac{x}{a} - \frac{y}{b}\right) dy\ dx \\
&= c \int_0^a \left[bx\left(1 - \frac{x}{a}\right) - \frac{x^2}{a}(b)\left(1 - \frac{x}{a}\right) - \frac{x}{2b}(b)^2\left(1 - \frac{x}{a}\right)^2 \right] dx \\
&= \frac{1}{24} a^2 bc
\end{aligned}$$

The solution for x_G is

$$x_G = \frac{a}{4}$$

In other words, the centroid is at one-quarter of the distance from the base in the yz plane to the apex on the x axis.

At this point we could continue to evaluate the other first moments arising in equations (5). However, this is not necessary here, because the geometry with respect to all of the coordinate axes is identical. It follows that the results for the values of y_G and z_G can be obtained by permuting the symbols. This yields

$$y_G = \frac{b}{4} \qquad z_G = \frac{c}{4}$$

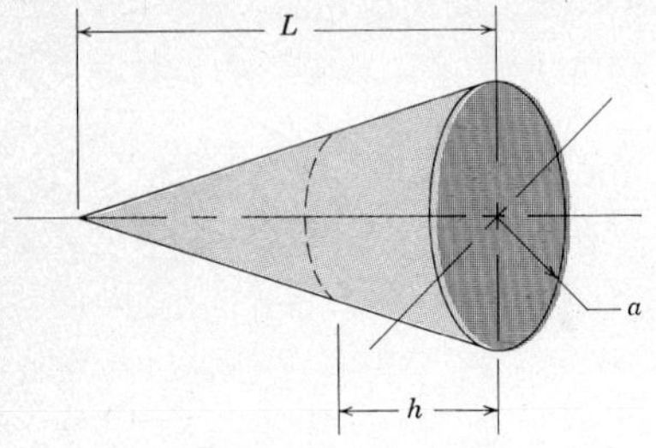

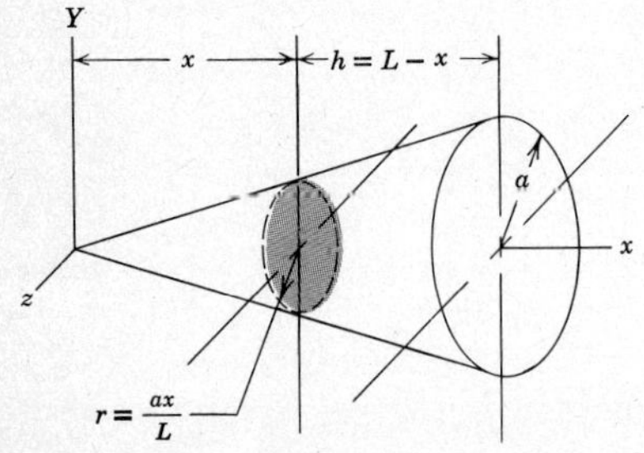

EXAMPLE 2

Determine the mass and the location of the center of mass of the right circular cone shown. Its density varies with distance h from the base according to $\rho = \rho_0(1 - h/2L)$.

Solution

Recall that when determining the center of mass, the concept of symmetry of the body about a plane requires that the density at matching points on either side of the geometric symmetry plane be identical. For the given cone the density at all points on a cross section perpendicular to the axis of the cone at any value of h is constant. Furthermore, this cross section is circular. Therefore, any plane containing the axis of the cone is a plane of symmetry, and it follows that the center of mass is situated on this axis.

In order to exploit the symmetry fully, we choose the xyz coordinate system such that one coordinate axis, say the x axis, coincides with the axis of the cone. For convenience we place the origin at the apex of the cone, as shown.

As was noted earlier, each cross section parallel to the yz plane, at any value x, is circular. The radius of this circle is shown in the sketch at the left to be $r = ax/L$. This suggests that we should use polar coordinates to locate the mass elements on a cross section. The polar coordinates to be employed are shown in the sketch below.

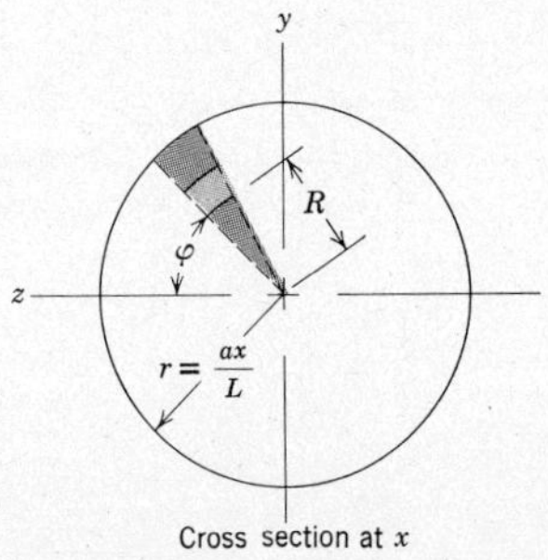

Cross section at x

For the polar coordinates selected, we see that

$$y = R \sin \phi \qquad z = R \cos \phi$$

Adapting equation (6) to the case where x measures the distance perpendicular to the plane of the polar coordinates, we have

$$dm = \rho R \, d\phi \, dR \, dx$$

The given expression for ρ is in terms of h. Noting that $h = L - x$, we have

$$\rho = \rho_0\left(1 - \frac{L - x}{2L}\right) = \frac{1}{2}\rho_0\left(1 + \frac{x}{L}\right)$$

To account for the elements of mass on the cross section we shall first integrate over the range of all possible radial distances, which is $0 \le R \le ax/L$. (This forms the shaded circular sector shown in the preceding sketch.) We then include all of these sectors by integrating over $0 \le \phi \le 2\pi$. (An alternative approach would be to form a circular ring by first integrating over $0 \le \phi \le 2\pi$ and then to include all such rings by integrating over $0 \le R \le ax/L$.) Finally, all cross sections are accounted for by integrating over $0 \le x \le L$.

The mass can be determined by using equation (1). Writing the integrals in the order in which they are to be evaluated, this yields

$$m = \int_0^L \int_0^{2\pi} \int_0^{ax/L} \left[\frac{1}{2}\rho_0\left(1 + \frac{x}{L}\right)\right] R\, dR\, d\phi\, dx$$

$$= \frac{1}{2}\rho_0 \int_0^L \int_0^{2\pi} \left(1 + \frac{x}{L}\right)\frac{1}{2}\left(\frac{ax}{L}\right)^2 d\phi\, dx$$

$$= \frac{1}{4}\rho_0 \int_0^L 2\pi\left(1 + \frac{x}{L}\right)\left(\frac{ax}{L}\right)^2 dx = \frac{7\pi}{24}\rho_0 a^2 L$$

Because the body is not homogeneous, its center of mass does not coincide with its geometric centroid. Hence, we must employ equations (2) to evaluate the first moments of mass. We know that the center of mass is situated on the x axis, so we need only the first moment of mass with respect to the x coordinate; specifically

$$\left(\frac{7\pi}{24}\rho_0\, a^2 L\right)x_G = \int_0^L \int_0^{2\pi} \int_0^{ax/L} x\left[\frac{1}{2}\rho_0\left(1 + \frac{x}{L}\right)\right] R\, dR\, d\phi\, dx$$

$$= \frac{1}{4}\rho_0 \int_0^L 2\pi x\left(1 + \frac{x}{L}\right)\left(\frac{ax}{L}\right)^2 dx = \frac{9\pi}{40}\rho_0 a^2 L^2$$

Solving for x_G we then have

$$x_G = 0.771L$$

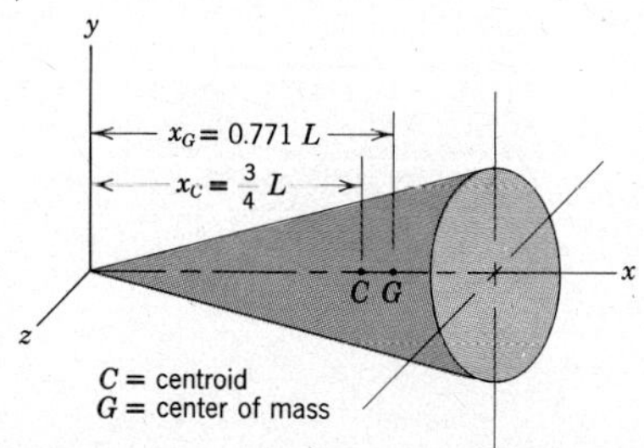

The location of point G is illustrated in the sketch. For contrast, the location of the centroid C of the volume of the cone is also shown. The center of mass is closer to the base of the cone because the density was prescribed to be greater near the base, resulting in a greater concentration of mass there.

HOMEWORK PROBLEMS

V.1–V.7 Set up and evaluate the integrals required for the determination of the location of the center of mass of the homogeneous body shown.

V.1 Triangular prism

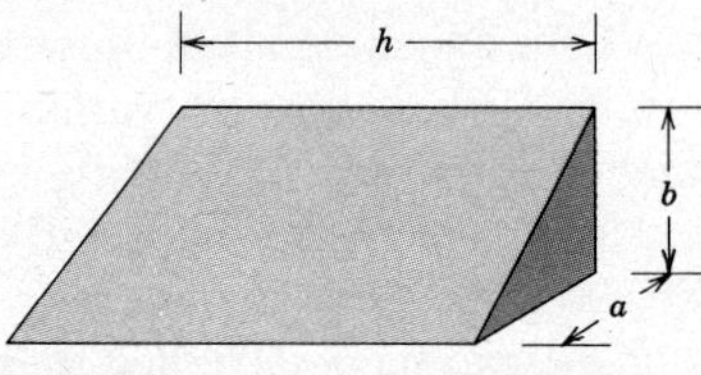

Prob. V.1

V.2 Hemisphere

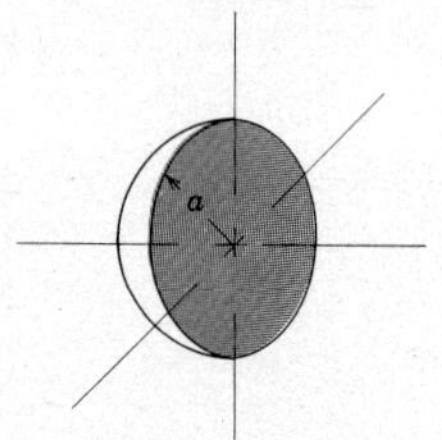

Prob. V.2

V.3 Paraboloid of revolution

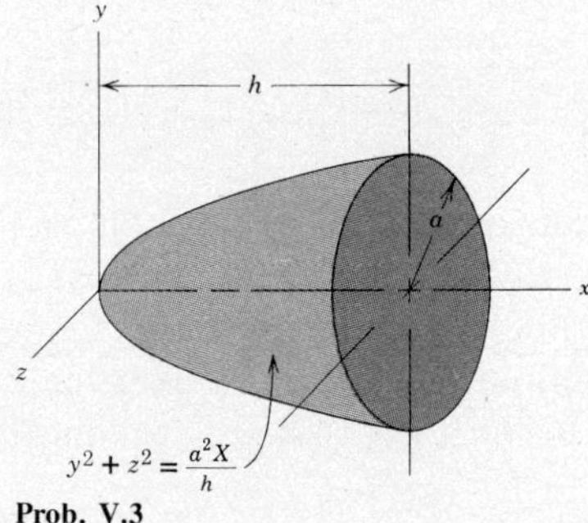

Prob. V.3

V.4 Spherical segment

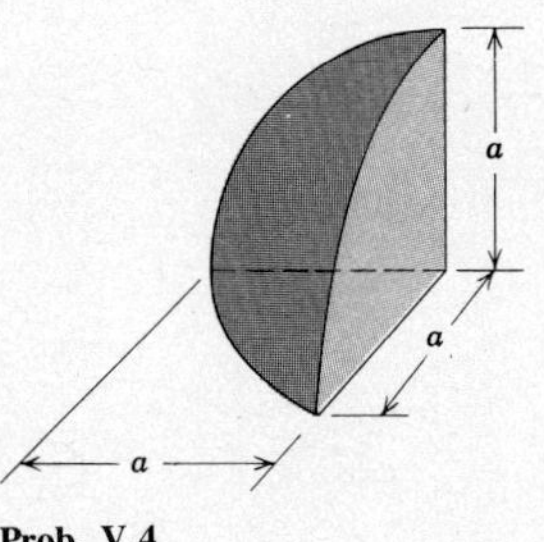

Prob. V.4

V.5 Semi-ellipsoid

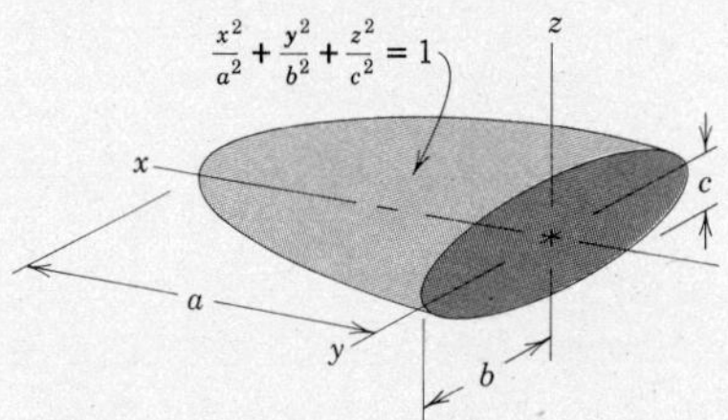

Prob. V.5

V.6 Circular cone segment

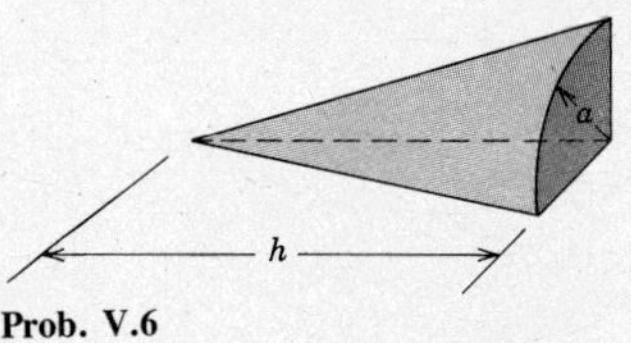

Prob. V.6

V.7 Sliced circular cylinder

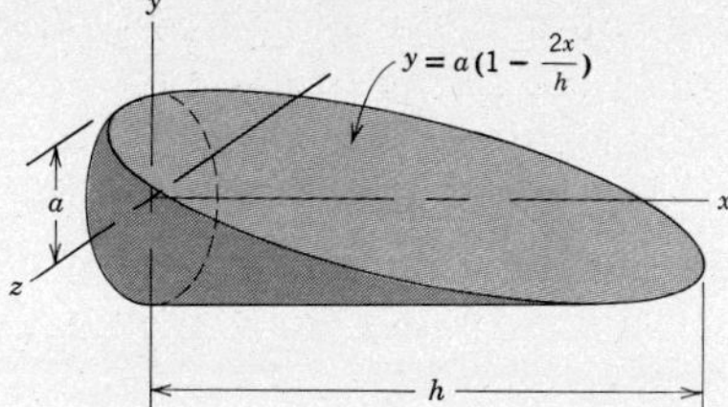

Prob. V.7

V.8 The density of the rectangular parallelepiped varies with the distance h from one face according to $\rho = \rho_0 \cos(\pi h/3a)$. Determine the mass and location of the center of mass of this body.

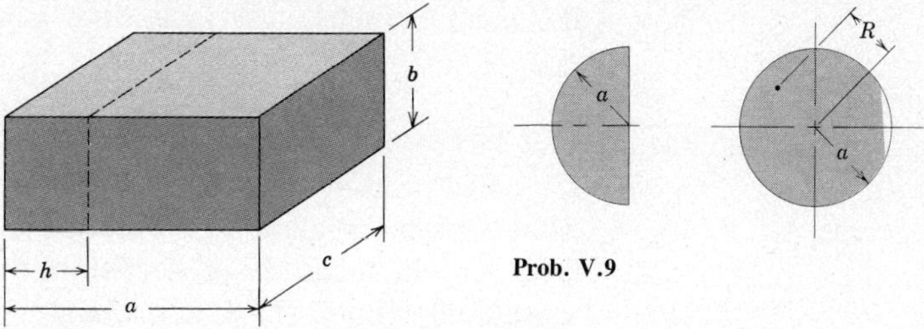

Prob. V.8

Prob. V.9

V.9 The density of a hemisphere of radius a varies with the radial distance R from the axis of symmetry, according to $\rho = \rho_0(1 + R/a)$. Determine the mass and the location of the center of mass of this body. Is the center of mass closer or farther from the flat base than the geometric centroid? Explain your answer.

V.10 The upper surface of the body shown is a ruled surface known as a hyperbolic paraboloid, which is defined by the equation $z = cxy/ab$. Knowing that the body is homogeneous, determine the mass and the location of the center of mass of this body.

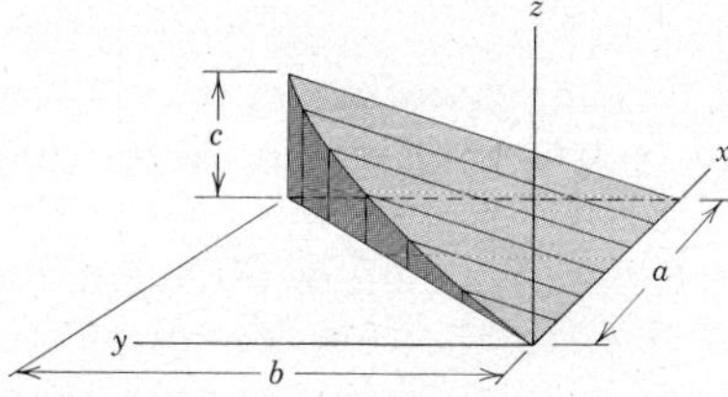

Prob. V.10

V.11 Solve Problem V.10 for the case where the mass density of the body varies according to $\rho = \rho_0(1 + z^2/c^2)$.

3 Centroids of Surfaces and Lines

The developments in the preceding sections enable us to determine the location of the center of mass for general three-dimensional bodies. Certain simplifications in the equations are possible in two cases, pertaining to thin bodies. The first type of thin body we shall consider is a thin *shell*. An egg shell is a familiar example of a thin shell. An arbitrary shell is shown in Figure 5, where t is the thickness and $\mathcal{A}$ is the surface area of the shell.

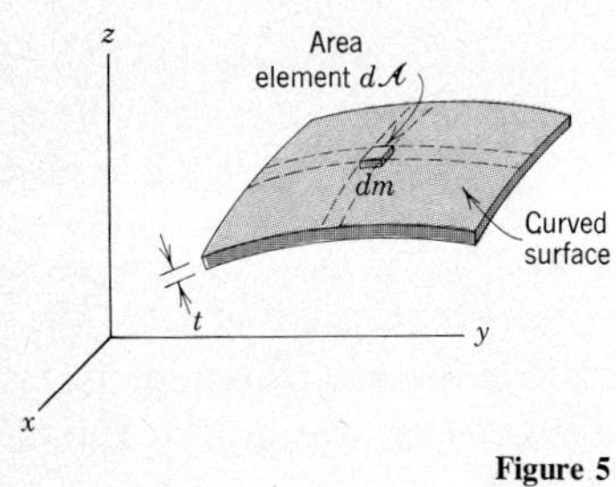

Figure 5

To determine the center of mass for a shell we consider the thickness to be so small that each element of mass dm is located at the curved surface. This means that the (x, y, z) coordinates locating the mass element are the same as those of the corresponding point on the surface. Furthermore, the infinitesimal volume of the element is then $t\,d\mathcal{A}$, where $d\mathcal{A}$ is an element of surface area. Equation (3) then gives

$$dm = \rho t\, d\mathcal{A}$$

The product ρt describes the mass per unit surface area. In most situations the shell will be homogeneous and it will have constant thickness, in which case ρt is a constant that may be factored out of integrals. Equations (1) and (2) then become

$$\boxed{\begin{aligned} m &= \rho t \int_{\mathcal{A}} d\mathcal{A} = \rho t \mathcal{A} \\ \mathcal{A} x_G &= \int_{\mathcal{A}} x\, d\mathcal{A} \\ \mathcal{A} y_G &= \int_{\mathcal{A}} y\, d\mathcal{A} \\ \mathcal{A} z_G &= \int_{\mathcal{A}} z\, d\mathcal{A} \end{aligned}} \tag{7}$$

Recall that the coordinates (x, y, z) in equations (7) locate points on the surface $\mathcal{A}$. The integrals in these equations are first moments of the surface area.

In general, equations (7) locate the *geometric centroid of the surface area* $\mathcal{A}$. Hence, the center of mass of a homogeneous shell with constant thickness coincides with the centroid of its surface area. In the case where the surface area is flat, the centroid of the area will be situated on the planar surface. It is then only necessary to utilize two of the first moments in equations (7). Conversely, if the surface is curved, the centroid need not, and generally will not, be situated on the surface.

As an aside, it is interesting to note that in addition to their application for thin shells, equations (7) may be employed to locate the center of mass of another type of homogeneous body. Suppose that a body has a constant thickness t (not necessarily small) in a direction perpendicular to two planar sides. Such bodies are called *cylinders* in analytic geometry. Figure 6 shows an arbitrary cylinder, for which the xy plane was chosen to be parallel to the planar sides whose area is $\mathcal{A}$. In view of the symmetry of this body, the center of mass G must be situated somewhere along the midplane that cuts the thickness in half. This is indicated by a long dashed curve in Figure 6. Hence, we need only determine the value of x_G and y_G to locate the center of mass. Because $t\, d\mathcal{A}$ is an element of volume for this body, just as it was for an arbitrary thin shell, we have $m = \rho t \mathcal{A}$. It then follows that the values of x_G and y_G are given by the corresponding equations (7). Thus, as indicated in Figure 6, we can locate the center of mass of a cylindrical body by merely locating the centroid of the area of one of its planar faces.

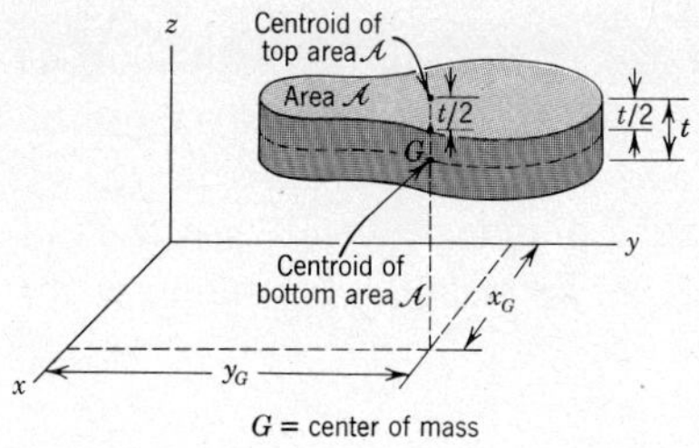

Figure 6

Let us now turn our attention to a second class of thin bodies. We have seen that a shell has one small dimension, its thickness, so its mass can be regarded to be concentrated at its surface. There are also bodies, such as the one depicted in Figure 7, whose mass is concentrated along a curve. This is the case of a *slender curved bar*. Such bodies have the characteristic that the dimensions defining the cross sectional area $\mathcal{A}_c$ perpendicular to its axis are very small compared to the arc length along the axis.

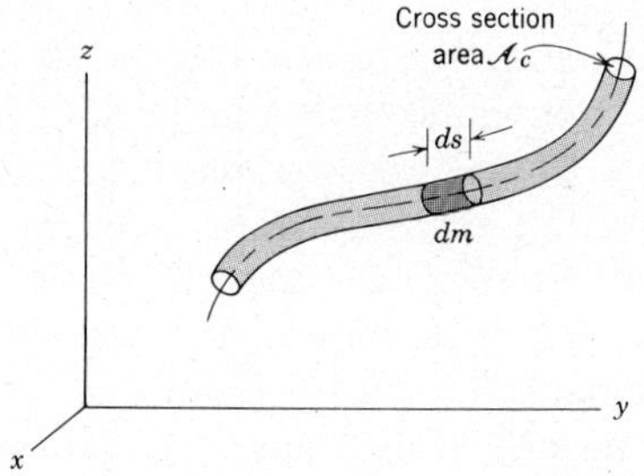

Figure 7

The element of mass for this body may be formed from a section of the bar of arc length ds, as indicated in Figure 7. The volume of this element is $\mathcal{A}_c\, ds$, so the corresponding expression for dm is

$$dm = \rho\, \mathcal{A}_c\, ds$$

The product $\rho\, \mathcal{A}_c$ is the mass per unit length of the bar. If the bar is homogeneous and has a constant cross-sectional area, we can factor $\rho\, \mathcal{A}_c$ out of the integrals in equations (1) and (2). Letting $\mathcal{S}$ denote the total arc length of the bar, these equations become

$$m = \rho \mathcal{A}_c \int_{\mathcal{S}} ds = \rho \mathcal{A}_c \mathcal{S}$$

$$\mathcal{S} x_G = \int_{\mathcal{S}} x\, ds$$

$$\mathcal{S} y_G = \int_{\mathcal{S}} y\, ds$$

$$\mathcal{S} z_G = \int_{\mathcal{S}} z\, ds \qquad (8)$$

Recall that we are considering the mass of the bar to be concentrated along the axis of the bar, so the coordinates (x, y, z) in equations (8) are those of points along this axis.

Equations (8) define the *geometric centroid of the curve* $\mathcal{S}$. It follows that the center of mass of a homogeneous bar of constant cross-sectional area coincides with the centroid of its axial curve. In general, this centroid probably will not be situated on the curve.

The primary difficulty in locating the centroids of surfaces and curves lies in relating the area $d\mathcal{A}$ for a surface, or arc length ds for a curve, to the x, y, and z coordinates. In the examples and homework problems we will restrict ourselves to geometries, particularly planar surfaces, surfaces of revolution, and planar curves, which can be readily formulated for solution.

EXAMPLE 3

Locate the center of mass of the parabolic steel plate of constant thickness t shown.

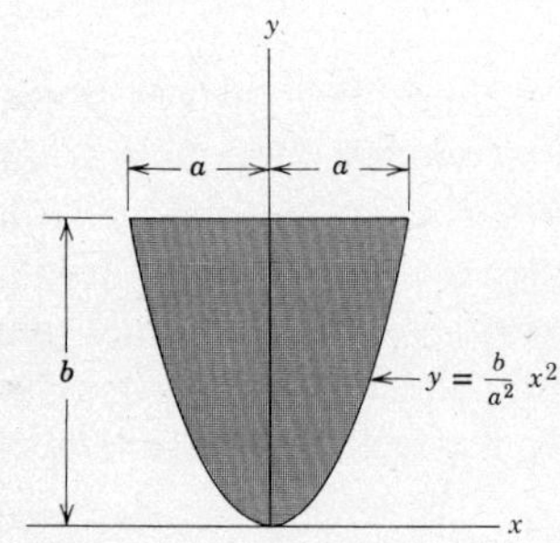

Solution

A flat plate is simply a planar shell. The given plate is composed of one material, steel, and it has a constant thickness. Therefore, the center of mass is situated midway in the thickness, having the same x and y coordinates as the centroid of the parabolic area. Further, noting that the z axis is outward from the plane of the diagram, we see that the yz plane is a plane of symmetry. This indicates that the center of mass must lie on the y axis. As a result, the problem reduces to the evaluation of the first moment of area with respect to the y coordinate.

The perimeter of the area is easily described in terms of equations relating the x and y coordinates. We shall formulate the solution in terms of Cartesian coordinates, for which an element of area is $d\mathcal{A} = \text{dx dy}$, as shown in the sketch.

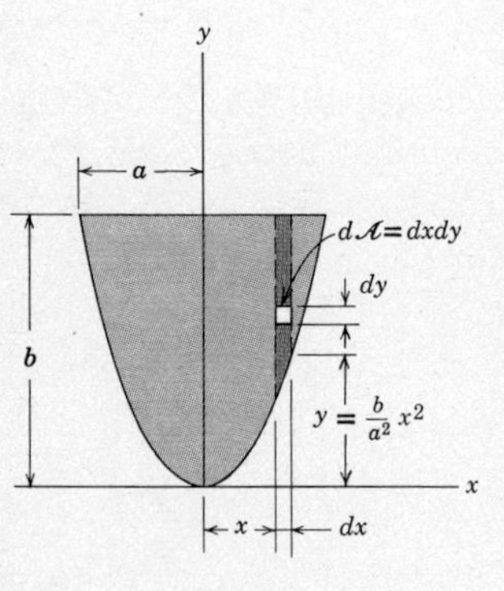

To account for all such elements, we could first integrate over all possible values of z for a particular value of y, and thus form a horizontal strip. However, as shown in the sketch, we choose to form a vertical strip by integrating over all possible values of y for a particular value of x. For this strip $bx^2/a^2 \leq y \leq b$. All vertical strips are then accounted for by integrating over $-a \leq x \leq a$.

The first step in applying equations (7) is to find the area. This is obtained by writing

$$\mathcal{A} = \int_{-a}^{a} \int_{bx^2/2}^{b} dy\, dx = \int_{-a}^{a} \left(b - \frac{bx^2}{a^2}\right) dx = 2ba - \frac{2ba^3}{3a^2} = \frac{4}{3}\,ba$$

Now, forming the first moment of area with respect to the y coordinate, equation (7) yields

$$\left(\frac{4}{3}\,ba\right)y_G = \int_{-a}^{a}\int_{bx^2/a^2}^{b} y\;dy\;dx = \frac{1}{2}\int_{-a}^{a}\left[b^2 - \left(\frac{bx^2}{a^2}\right)^2\right]dx = \frac{4}{5}\,b^2a$$

Solving for y_G, we find

$$y_G = \frac{3}{5}b$$

Thus the center of mass is midway in the plate thickness t with coordinates $x_G = 0$, $y_G = \frac{3}{5}b$.

EXAMPLE 4

An aluminum bar of constant mass per unit length σ is bent into a quarter-circle of radius r. Determine the (x, y) coordinates of the center of mass.

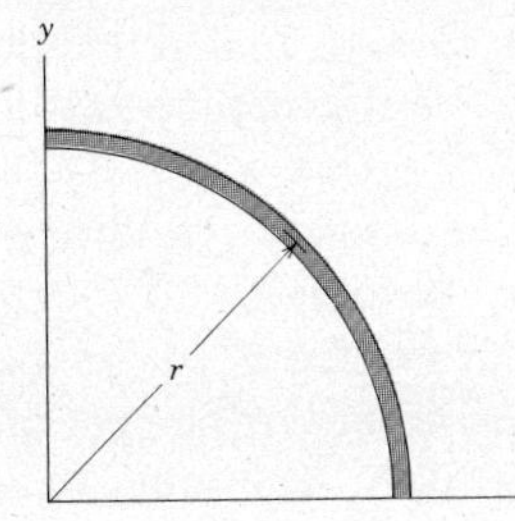

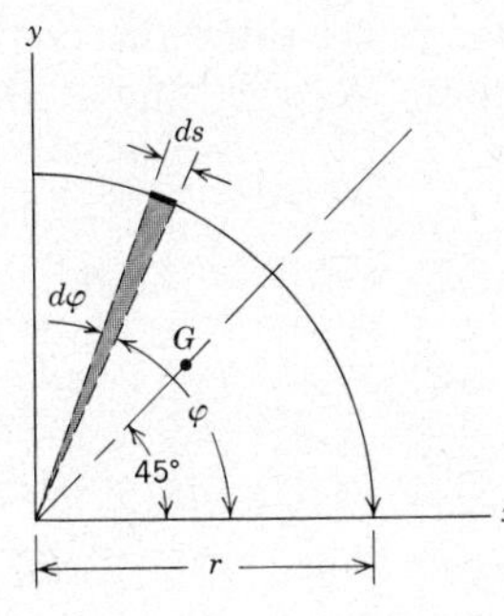

Solution

We begin by noting that the bar is symmetric about the plane formed by the z axis and the 45° line shown in the second sketch. This means that the centroid must lie along the 45° line, so $y_G = x_G$. Also, because the mass per unit length is constant, the center of mass coincides with the centroid of the circular arc.

We employ polar coordinates because the mass elements are all situated at a constant radial distance r from the origin. The coordinate transformations for the chosen coordinate systems are

$$x = r\cos\phi \qquad y = r\sin\phi$$

where the values of ϕ range from zero to $\pi/2$ radians. Noting that the element ds is the arc of the circle subtended by the angle $d\phi$, we have

$$ds = r\;d\phi$$

Similarly, we know that the total length of the circular arc is

$$\mathcal{S} = \frac{1}{4}(2\pi r) = \frac{\pi}{2}\,r$$

Thus, in this case it is not necessary to evaluate an integral to determine $\mathcal{S}$; we need only determine the value of x_G. From the given information we know that the mass per unit length is constant, so we can use the first moment with respect to the x coordinate in equations (8).

$$\left(\frac{\pi}{2}r\right)x_G = \int_0^{\pi/2} x\;ds = \int_0^{\pi/2}(r\cos\phi)(r\;d\phi) = r^2\sin\frac{\pi}{2} = r^2$$

$$x_G = \frac{2r}{\pi} \equiv y_G$$

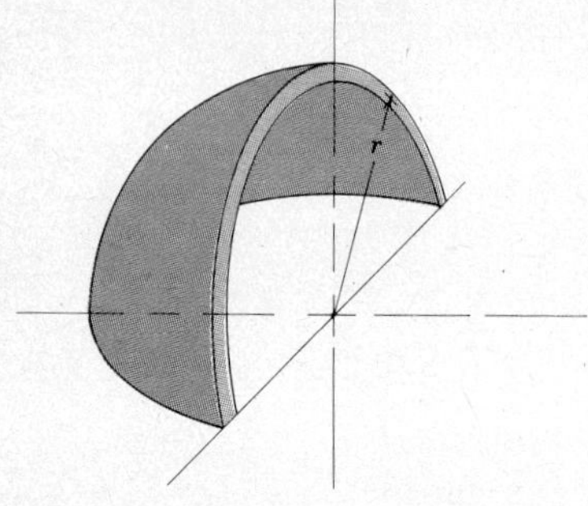

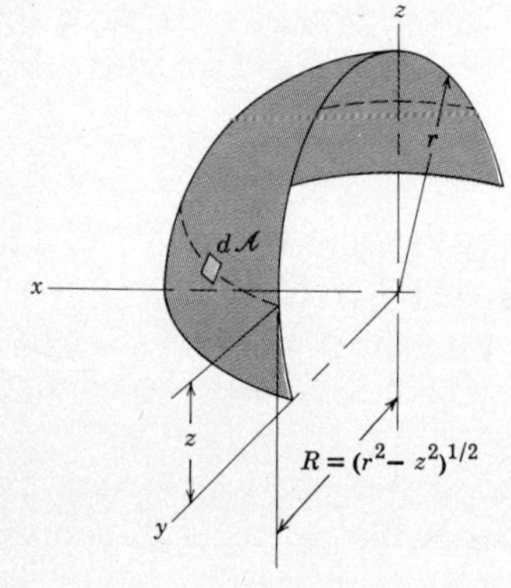

EXAMPLE 5

Locate the centroid of the aluminum shell of constant thickness formed by cutting a spherical shell of radius r into quarter-sections.

Solution

The shell is homogeneous and it has constant thickness. Therefore, its center of mass is coincident with the centroid of its spherical surface. We begin the solution by selecting an xyz coordinate system. In view of the shape of the body, we choose an origin at the center of the sphere and align the coordinate axes with the centerlines, as illustrated. Because the xz plane is a plane of symmetry, we know that $y_G = 0$. Furthermore, a plane containing the y axis at a 45° angle above the xy plane also cuts the surface in half, which means that $x_G = z_G$.

The spherical shape is conducive to the use of cylindrical coordinates. The polar coordinates R and ϕ may be defined to locate points parallel to any of the coordinate planes; our choice is the xy plane. As shown in the sketch to the left, cross sections of the surface parallel to the xy plane are semicircles of radius $R = (r^2 - z^2)^{1/2}$. The polar coordinates are defined with the aid of the sketch shown below of this cross section. Also, in order to depict the area element $d\mathcal{A}$, we draw the cross section formed by a cutting plane that contains the z axis and the element $d\mathcal{A}$.

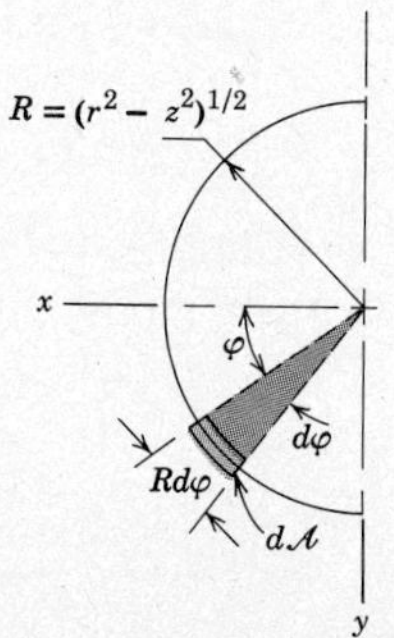

Cross section parallel to xy plane

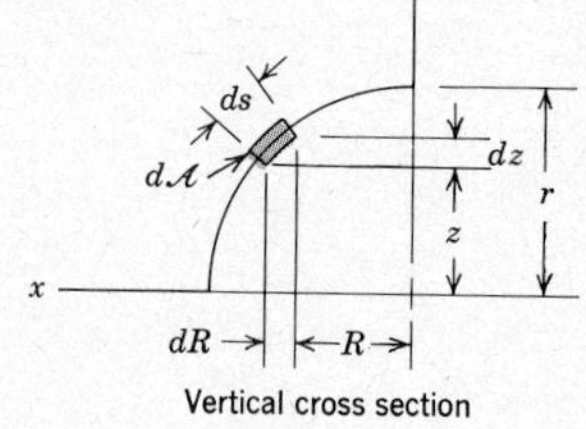

Vertical cross section

For the polar coordinates shown in the xy cross section, we may write

$$x = R \cos \phi = (r^2 - z^2)^{1/2} \cos \phi$$

$$y = R \sin \phi = (r^2 - z^2)^{1/2} \sin \phi$$

Deriving the expression for the area element $d\mathcal{A}$ is the only new feature of the formulation. The edge views of the element $d\mathcal{A}$ presented in the foregoing sketches of the cross sections show that

$$d\mathcal{A} = (R d\phi)\, ds = (r^2 - z^2)^{1/2}\, d\phi\, ds$$

The Pythagorean theorem is now used to relate the arc length ds to the increments of dz and dR illustrated in the sketch of the vertical cross section. This yields

$$ds = [(dz)^2 + (dR)^2]^{1/2}$$

Then, because $R = (r^2 - z^2)^{1/2}$, we find

$$dR \equiv \frac{dR}{dz}\, dz = \frac{1}{2}(r^2 - z^2)^{-1/2}(-2z)\, dz = \frac{-z\, dz}{(r^2 - z^2)^{1/2}}$$

$$ds = \left[(dz)^2 + \frac{(z\, dz)^2}{(r^2 - z^2)}\right]^{1/2} = \frac{r\, dx}{(r^2 - z^2)^{1/2}}$$

Hence

$$d\mathcal{A} = (r^2 - z^2)^{1/2}(d\phi)\left[\frac{r\, dz}{(r^2 - z^2)^{1/2}}\right] = r\, d\phi\, dz$$

As an aside, we note that this simple expression for $d\mathcal{A}$ is a result of the spherical geometry. For other surfaces of revolution, where the radial distance R is a different function of the axial distance z, this expression will not result.

Now that an expression for $d\mathcal{A}$ has been obtained, we may proceed to apply equations (7). All elements $d\mathcal{A}$ at an xy cross section are accounted for by integrating over $-\pi/2 < \phi < \pi/2$ radians. All cross sections are then accounted for by integrating over $0 \le z \le r$. Thus

$$\mathcal{A} = \int_0^r \int_{-\pi/2}^{\pi/2} d\mathcal{A} = \int_0^r \int_{-\pi/2}^{\pi/2} r\, d\phi\, dz = \pi r^2$$

As noted earlier, the symmetry of the shell means that we only have to calculate x_G (or z_G). Therefore, we form the first moment of surface area with respect to the x coordinate.

$$(\pi r^2)x_G = \int_0^r \int_{-\pi/2}^{\pi/2} x\, d\mathcal{A} = \int_0^r \int_{-\pi/2}^{\pi/2} [(r^2 - z^2)^{1/2} \cos\phi] r\, d\phi\, dz$$

$$= r \int_0^r (r^2 - z^2)^{1/2}\left(2 \sin\frac{\pi}{2}\right) dz = r\left[z(r^2 - z^2)^{1/2} + r^2 \sin^{-1}\frac{z}{r}\right]_0^r$$

$$= r^3 \sin^{-1}(1) = \frac{\pi}{2} r^3$$

A table of integrals was employed to evaluate the integral over the z variable. Solving for x_G, we find

$$x_G = \frac{r}{2} = z_G \qquad y_G = 0$$

HOMEWORK PROBLEMS

V.12–V.17 Determine the location of the center of mass of the homogeneous flat plate of constant thickness shown.

V.12 Right triangle

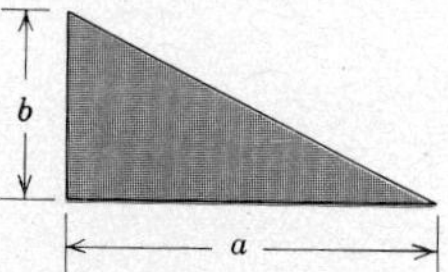

Prob. V.12

V.13 Elliptical quadrant

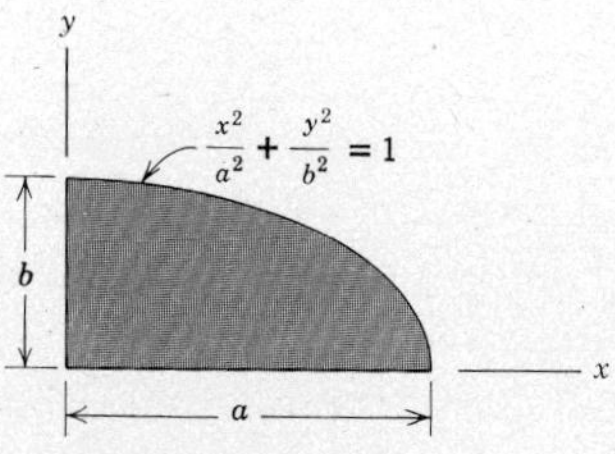

Prob. V.13

V.14 Circular sector

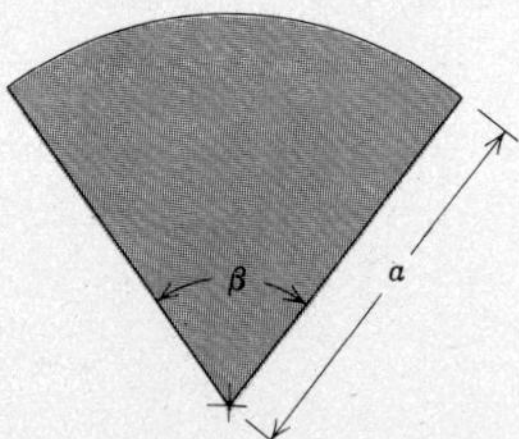

Prob. V.14

V.15 Cubic spandrel

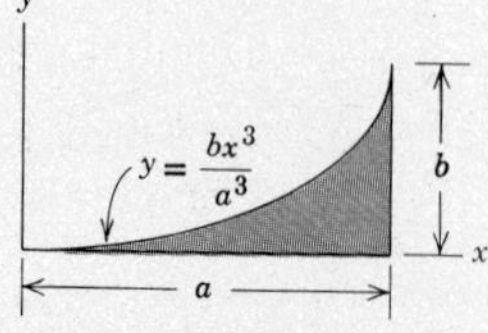

Prob. V.15

V.16 Spiral area

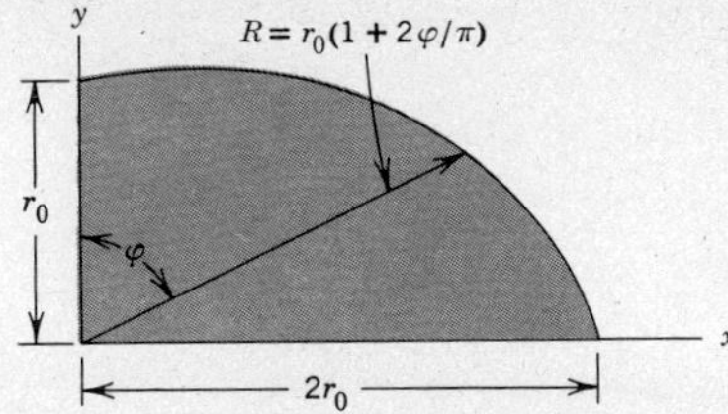

Prob. V.16

V.17 Hyperbolic quadrant

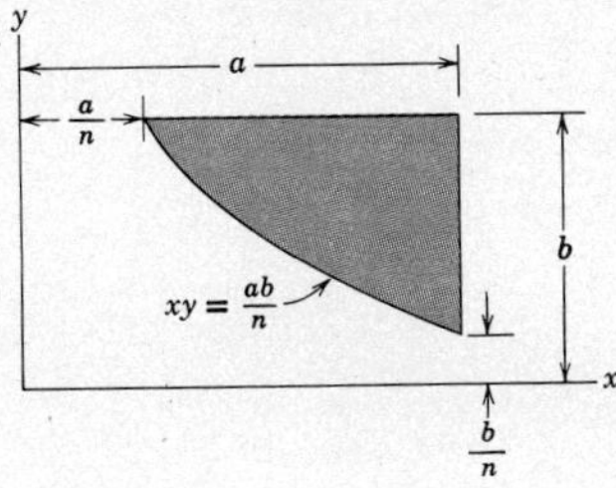

Prob. V.17

V.18–V.20 Determine the location of the center of mass of the bent wire shown. It has constant mass per unit length.

V.18 Circular arc

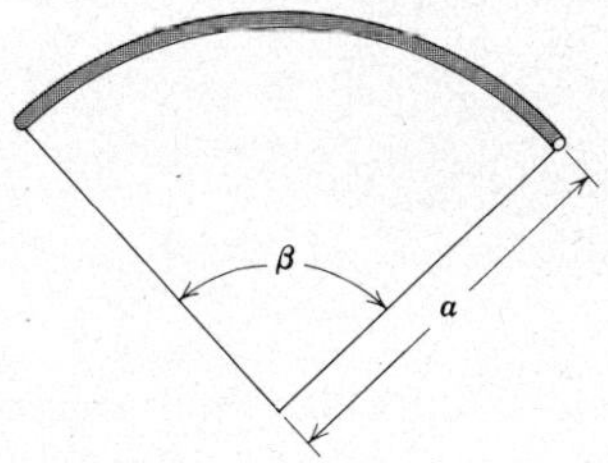

Prob. V.18

V.19 Parabolic arc

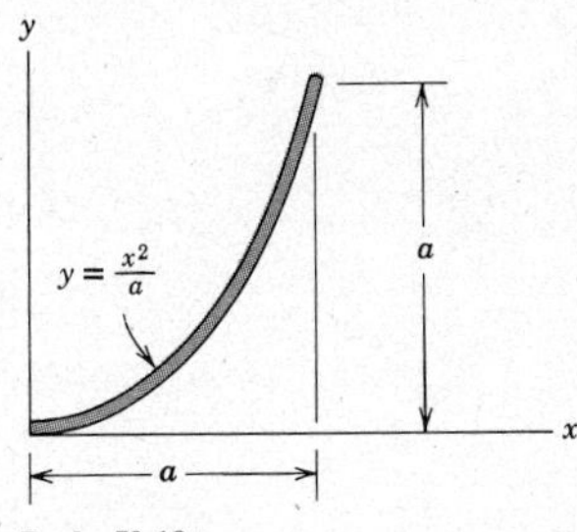

Prob. V.19

V.20 Spiral segment — *Hint:* In polar coordinates $ds = [(dR)^2 + (R\, d\phi)^2]^{1/2}$.

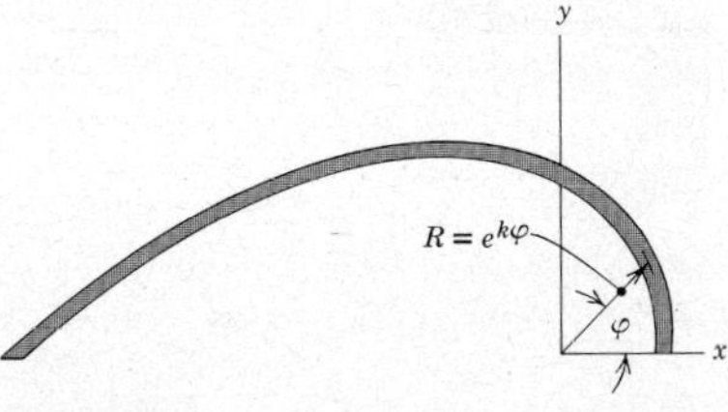

Prob. V.20

V.21–V.24 Determine the location of the center of mass of the thin shell shown having constant mass per unit length.

V.21 Conical shell

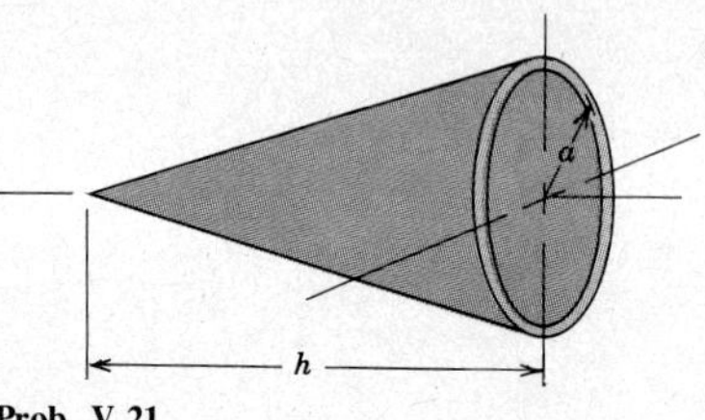

Prob. V.21

V.22 Conical shell segment

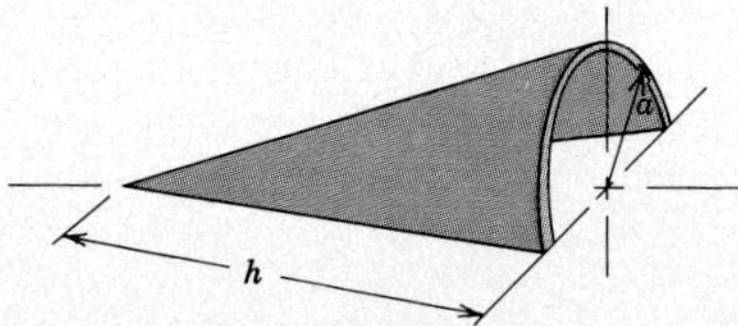

Prob. V.22

V.23 Parabolic shell

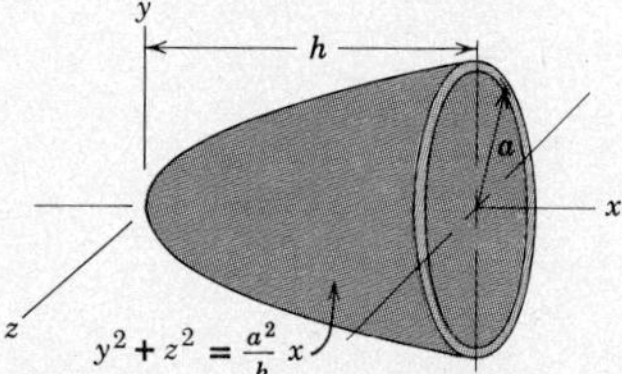

Prob. V.23

V.24 Circular cylindrical shell segment — *Hint:* Formulate the solution by locating points parallel to the xy plane in terms of polar coordinates and evaluate the integral over the z variable first.

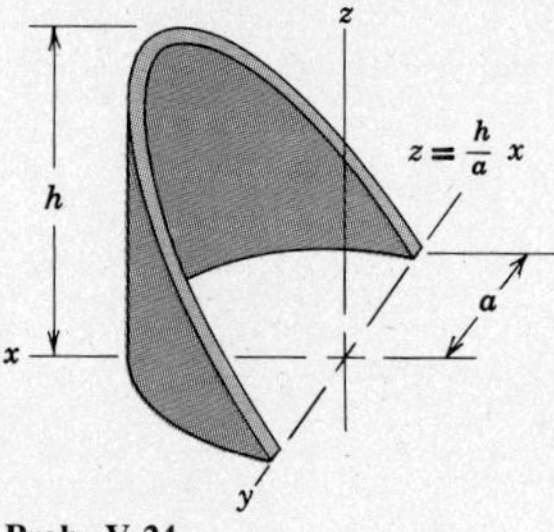

Prob. V.24

4 Composite Bodies

It would not make sense to evaluate integrals every time the location of the center of mass of a recognizable geometric shape is needed. To avoid doing so, it is common practice to tabulate the geometric and mass properties of a large number of basic (that is, recognizable) shapes. Typical tables are given at the end of this text; Appendix B contains information for planar bodies, and Appendix C contains information for three-dimensional shapes.

When a homogeneous body can be recognized as a tabulated shape, its center of mass may be located by substituting its dimensions into the appropriate tabulated formula. In other cases it may be possible to form the body of interest from two or more of the basic shapes. Such a body is called a *composite body*. An arbitrary composite body, formed from the basic bodies 1 and 2, is illustrated in Figure 8.

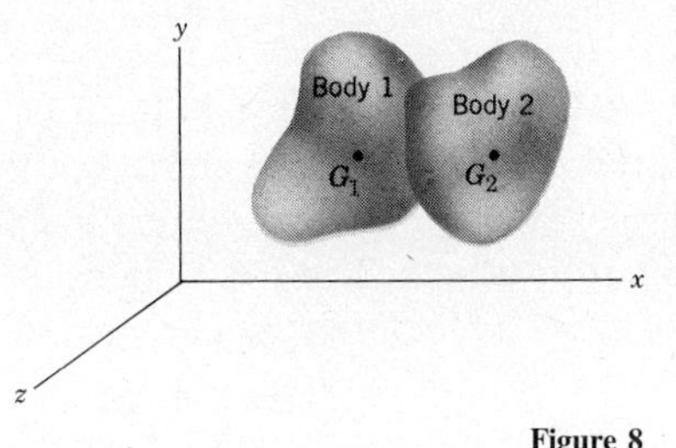

Figure 8

The space $\mathcal{V}$ occupied by the composite body is the combination of the spaces $\mathcal{V}_1$ and $\mathcal{V}_2$ of the basic recognizable shapes. From equation (1) we find that the total mass is

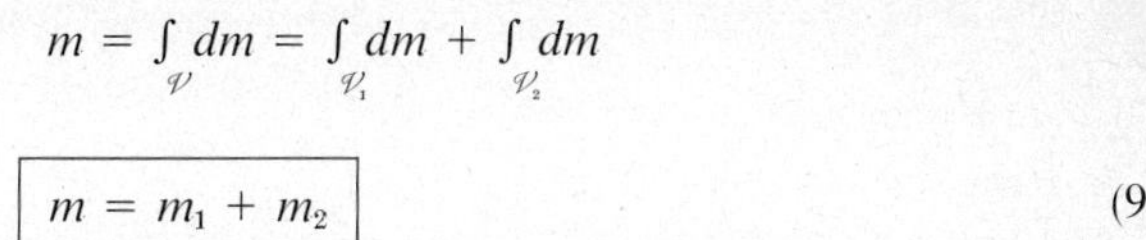

$$m = \int_{\mathcal{V}} dm = \int_{\mathcal{V}_1} dm + \int_{\mathcal{V}_2} dm$$

$$\boxed{m = m_1 + m_2} \qquad (9)$$

This result says that the total mass of the body is the sum of the mass of its individual components.

Similarly, first moments of mass may be decomposed into the contributions of the individual components. For instance, equation (2a) becomes

$$mx_G = \int_{\mathcal{V}} x\, dm = \int_{\mathcal{V}_1} x\, dm + \int_{\mathcal{V}_2} x\, dm$$

Here, the integrals are the first moments for components 1 and 2, individually. Applying equation (2a) again yields

$$\boxed{mx_G = m_1(x_G)_1 + m_2(x_G)_2} \qquad (10a)$$

Following similar steps for the other first moments, we get

$$\boxed{my_G = m_1(y_G)_1 + m_2(y_G)_2} \qquad (10b)$$

$$\boxed{mz_G = m_1(z_G)_1 + m_2(z_G)_2} \qquad (10c)$$

In equations (10), $((x_G)_i, (y_G)_i, (z_G)_i)$ are the coordinates of the center of mass of the ith component body. Knowing the mass and location of the center of mass of each component, it is a simple matter to solve equations (10) for the coordinates of the center of mass G of the entire composite body.

Equations (9) and (10) are easily extended to the case of composite bodies formed from more than two basic shapes by summing the contributions of all components. Also, in the case where the body contains a hole, equations (9) and (10) may be applied by considering the hole to contain a negative mass. Then for example, equation (10a) takes the form

$$mx_G = m_1(x_G)_1 - m_2(x_G)_2$$

where body 1 corresponds to the entire body considering the hole to be filled and body 2 corresponds to the material missing from the entire body because of the hole.

As was true in the case of a general body, when a composite body is homogeneous, we may locate its center of mass by locating the centroid of the volume occupied by the body. For such cases the equations for the first moments of volume are identical in form to equations (10). That is, noting that $m = \rho\mathcal{V}$, where ρ is a constant, the symbol m becomes the symbol $\mathcal{V}$ throughout the equations. Similarly, the equations for the center of mass of a shell of constant mass per unit surface area or a slender bent bar of constant mass per unit length are identical to equations (10), except that the symbol m becomes the symbol $\mathcal{A}$ or $\mathcal{S}$, whichever is applicable.

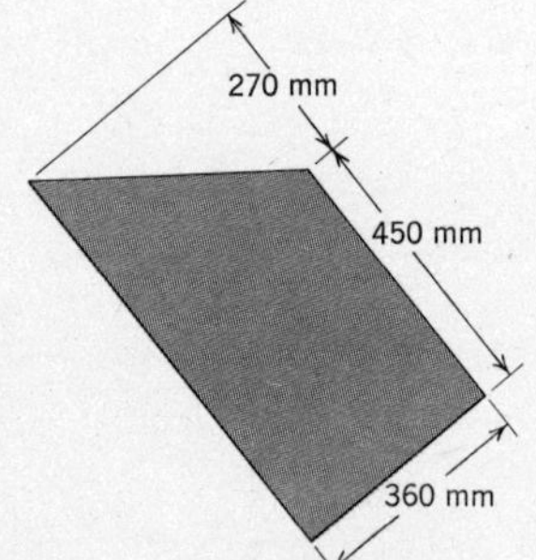

EXAMPLE 6

The trapezoidal plate shown is made of aluminum. It has a thickness of 40 mm. Determine the location of its center of mass.

Solution

The plate is homogeneous and it has a constant thickness. Hence, its center of mass is midway in its thickness, having the same location in the plane as the centroid of the trapezoid. Appendix B does not give the location of the centroid of a trapezoid, so we treat it as a composite shape. There are several possible decompositions that result in components having areas whose properties are tabulated. We will consider the trapezoid to be formed from a rectangle and triangle, as shown in the sketch.

The dimensions locating the center of mass of each component body are taken from Appendix B. The last step required before employing equations (10) is to select a coordinate system. We choose x and y axes

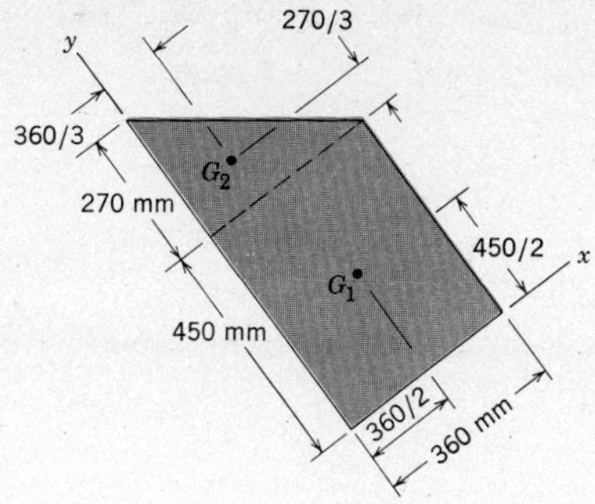

parallel to the known dimensions, as illustrated in the sketch.

The areas are found to be

$$\mathcal{A}_1 = 0.45(0.36) = 0.1620 \text{ m}^2$$

$$\mathcal{A}_2 = \tfrac{1}{2}(0.27)(0.36) = 0.0486 \text{ m}^2$$

$$\mathcal{A} = \mathcal{A}_1 + \mathcal{A}_2 = 0.2106 \text{ m}^2$$

Using the dimensions given in the sketch and the computed areas, the first moments of area are

$$0.2106x_G = \mathcal{A}_1(x_G)_1 + \mathcal{A}_2(x_G)_2$$

$$= 0.1620[\tfrac{1}{2}(0.36)] + 0.0486[\tfrac{1}{3}(0.36)]$$

$$x_G = 0.1662 \text{ m}$$

$$0.2106y_G = \mathcal{A}_1(y_G)_1 + \mathcal{A}_2(y_G)_2$$

$$= 0.1620[\tfrac{1}{2}(0.45)] + 0.0486[0.45 + \tfrac{1}{3}(0.27)]$$

$$y_G = 0.2977 \text{ m}$$

EXAMPLE 7

A 4-in.-diameter, 12-in.-long steel rod fits tightly into a hole in the rectangular aluminum bar, as shown. Determine the mass and the location of the center of mass of this assembly.

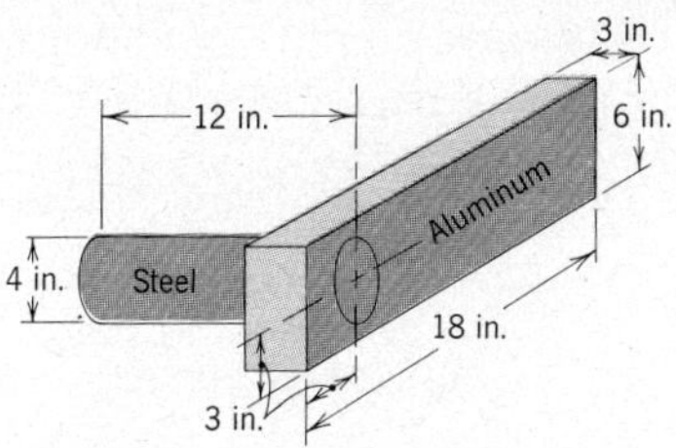

Solution

The assembly is composed of two different materials, so we must determine the center of mass using first moments of mass, not volume. We can break this composite body into three tabulated components. First, we have a solid rectangular parallelepiped, representing the aluminum bar. From this body we shall remove an aluminum cylinder 2 having the dimensions of the hole in the aluminum bar. The third body is, of course, the cylindrical steel rod.

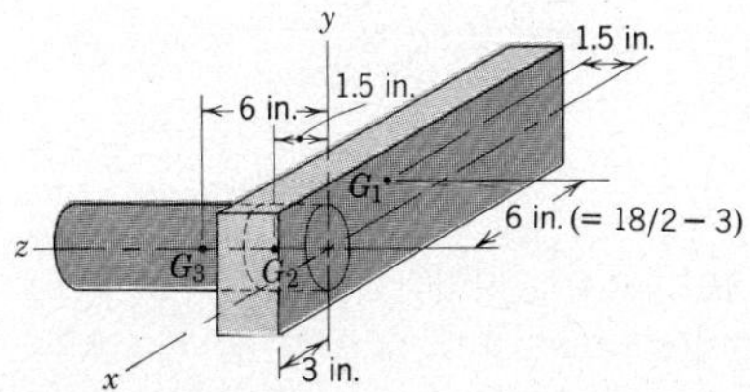

The origin of the xyz coordinate system may be located at any convenient point. Our choice for xyz is shown in the sketch to the left. In this sketch we also show the dimensions locating the center of mass G_i of each of the components. This sketch shows that the coordinates of these points are

$$\text{point } G_1: \quad (-6, 0, 1.5); \qquad \text{point } G_2: \quad (0, 0, 1.5); \qquad \text{point } G_3: \quad (0, 0, 6)$$

To avoid errors in writing these coordinates, particular care is required to place the points in the correct octants of the coordinate system.

We begin the calculation of the mass of the basic components by converting the mass densities of aluminum and steel given in Appendix D to units of inches.

$$\rho_{\text{al}} = \frac{168 \text{ lb/ft}^3}{g} = \frac{168 \text{ lb/ft}^3}{32.17 \text{ ft/s}^2}\left(\frac{1 \text{ ft}}{12 \text{ in.}}\right)^4 = 2.518(10^{-4}) \text{ lb-s}^2/\text{in.}^4$$

$$\rho_{\text{st}} = \frac{490 \text{ lb/ft}^3}{g} = 7.345(10^{-4}) \text{ lb-s}^2/\text{in.}^4$$

The mass of the hole in the aluminum bar is considered to be negative, because it is removed from the system. Thus, writing $m_i = \rho_i \mathcal{V}_i$, we have

$$m_1 = \rho_{\text{al}}(3)(6)(18) = 0.08158 \text{ lb-s}^2/\text{in.}$$

$$m_2 = -\rho_{\text{al}}\pi(2)^2(3) = -0.00949$$

$$m_3 = \rho_{\text{st}}\pi(2)^2(12) = 0.11076$$

$$m = m_1 + m_2 + m_3 = 0.18285$$

We may now employ equations (10), using the coordinates of the centers of mass determined earlier. This gives

$$0.18285 x_G = m_1(x_G)_1 + m_2(x_G)_2 + m_3(x_G)_3$$

$$= 0.08158(-6) + (-0.00949)(0) + 0.11076(0)$$

$$x_G = -2.68 \text{ in.}$$

$$0.18285 y_G = m_1(y_G)_1 + m_2(Y_G)_2 + m_3(Y_G)_3 = 0$$

$$y_G = 0$$

$$0.18285 z_G = 0.08158(1.5) + (-0.00949)(1.5) + 0.11076(6)$$

$$z_G = 4.23 \text{ in.}$$

Note that we could have foreseen that $y_G = 0$, because the xz plane is a plane of symmetry.

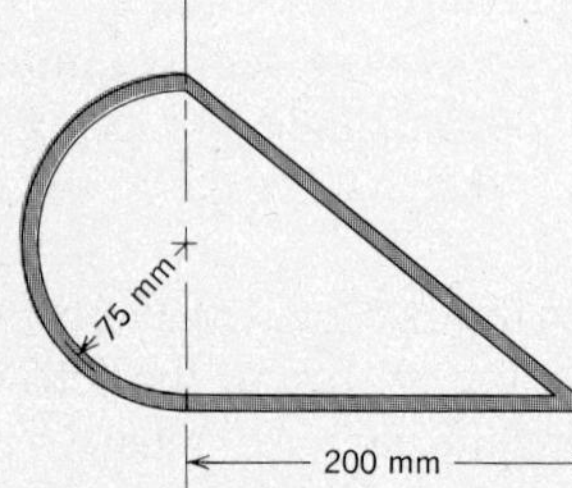

EXAMPLE 8

A copper wire having a circular cross section of 5 mm diameter is bent into the closed curve shown. Determine the mass and center of mass of the wire.

Solution

The density of copper is $8.91(10^3)$ kg/m^3, so the mass per unit length has the value

$$\rho \mathcal{A}_c = 8.91(10^3)\pi\left(\frac{0.005}{2}\right)^2 = 0.17495 \text{ kg/m}$$

Because this parameter is constant we may locate the center of mass by computing first moments of the arc length. The bent wire is a composite of a circular section 1 and two straight sections 2 and 3, as shown in the sketch. The origin of the chosen xyz coordinate system is placed at a convenient point. The sketch also contains the locations of each center of mass of the components of the composite. It is only necessary to refer to Appendix C for the location of point G_1, in that the locations of the other points are apparent from symmetry. This composite is a planar shape, so we need only the (x, y) coordinates of each component. The sketch shows that these are

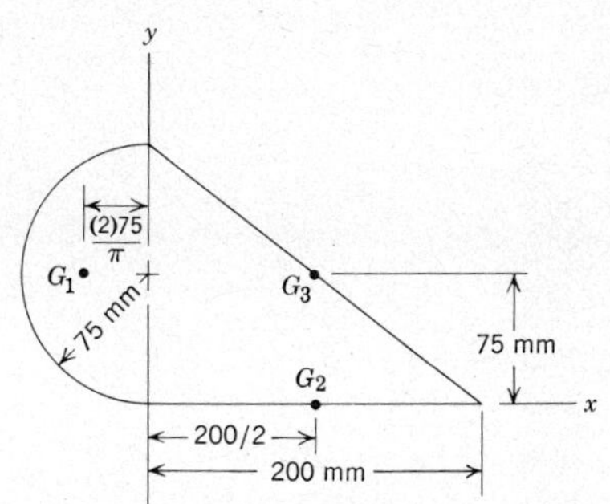

$$\text{point } G_1\text{: } \left(-\frac{0.150}{\pi}, 0.075\right); \qquad \text{point } G_2\text{: } (0.10, 0); \qquad \text{point } G_3\text{: } (0.10, 0.075)$$

The arc lengths of the basic components are readily calculated to be

$$\mathcal{S}_1 = \pi(0.075) = 0.2356 \text{ meters}$$

$$\mathcal{S}_2 = 0.20 \text{ meters}$$

$$\mathcal{S}_3 = \sqrt{(0.20)^2 + (0.15)^2} = 0.250 \text{ meters}$$

$$\mathcal{S} = \mathcal{S}_1 + \mathcal{S}_2 + \mathcal{S}_3 = 0.6856 \text{ meters}$$

The mass of the bent wire is its total length times its mass per unit length. Thus

$$m = \rho \mathcal{A}_c \mathcal{S} = (0.17495)(0.6856) = 0.1199 \text{ kg}$$

Adjusting the form of equations (10) to the calculation of first moments of arc length, we have

$$\mathcal{S} x_G = \mathcal{S}_1(x_G)_1 + \mathcal{S}_2(x_G)_2 + \mathcal{S}_3(x_G)_3$$

$$(0.6856)x_G = 0.2356\left(-\frac{0.150}{\pi}\right) + 0.20(0.10) + 0.25(0.10)$$

$$x_G = 0.0493 \text{ meters} = 49.3 \text{ mm}$$

$$\mathcal{S} y_G = \mathcal{S}_1(y_G)_1 + \mathcal{S}_2(y_G)_2 + \mathcal{S}_3(y_G)_3$$

$$(0.6856)y_G = 0.2356(0.075) + 0.20(0) + 0.25(0.075)$$

$$y_G = 0.0531 \text{ meters} = 53.1 \text{ mm}$$

HOMEWORK PROBLEMS

V.25–V.34 Determine the location of the center of mass of the homogeneous plates of constant thickness shown.

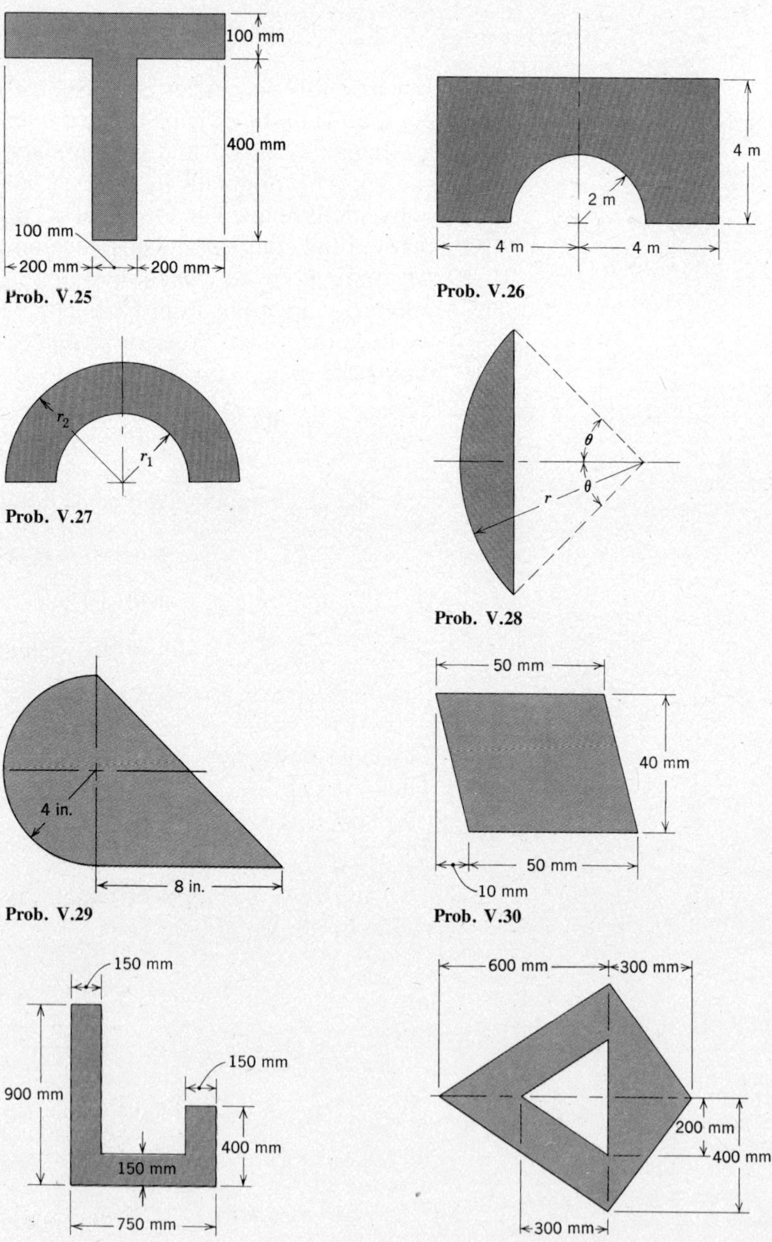

Prob. V.25

Prob. V.26

Prob. V.27

Prob. V.28

Prob. V.29

Prob. V.30

Prob. V.31

Prob. V.32

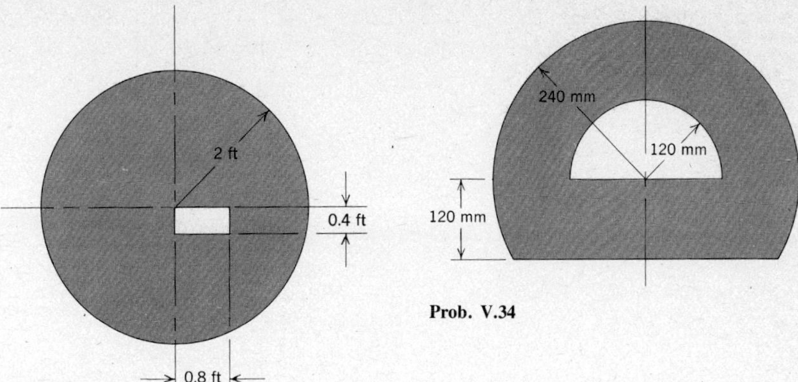

Prob. V.34

Prob. V.33

V.35–V.40 Determine the location of the center of mass of the bent wire shown having constant mass per unit length.

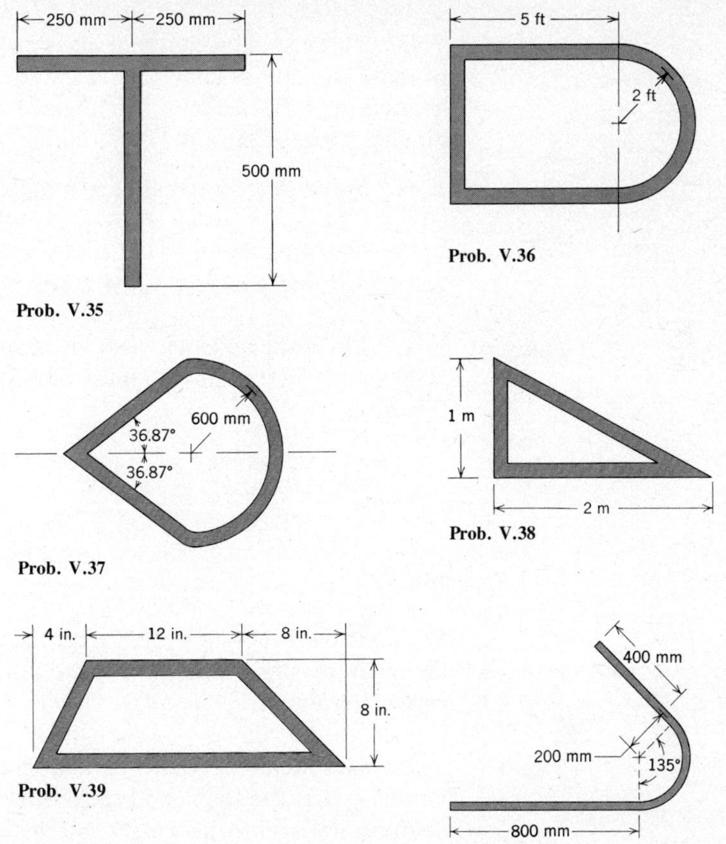

Prob. V.35

Prob. V.36

Prob. V.37

Prob. V.38

Prob. V.39

Prob. V.40

V.41 and V.42 The aluminum disk shown has a steel insert whose faces are flush with the faces of the disk. The thickness of the disk is as indicated. Determine the location of the center of mass of this body.

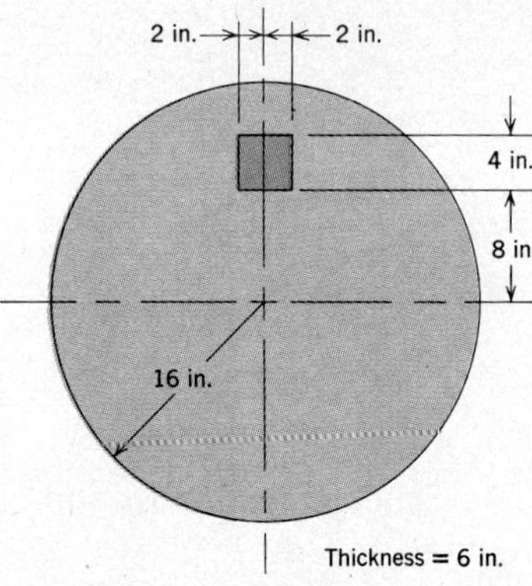

Prob. V.41

Prob. V.42

V.43 Three small weights of mass $m/50$ are attached to the circular disk of mass m. The assembly is suspended in the vertical plane from a pin located at A. Determine the angle θ between line OA and the vertical direction if $\alpha = 180°$, $\beta = 45°$, and $\gamma = 90°$.

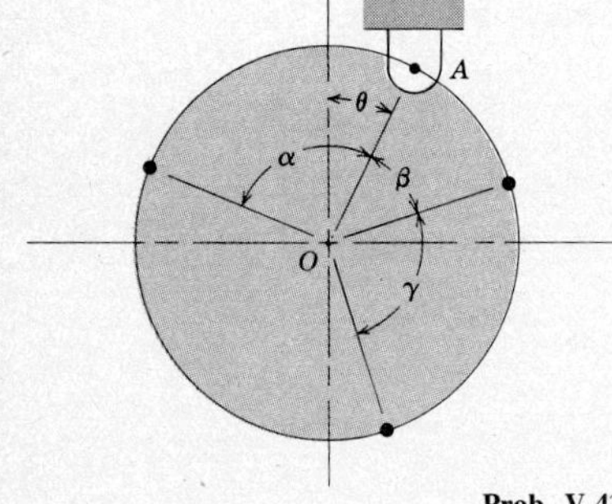

Prob. V.43

V.44 Solve Problem V.43 if $\alpha = 150°$, $\beta = 30°$, and $\gamma = 90°$.

V.45 Knowing that $\alpha = 130°$ and $\gamma = 30°$ in Problem V.43, determine the value of β that causes θ to be zero.

V.46 The body shown is a solid composed of plastic having a density of 2000 kg/m^3. Determine the mass and the location of the center of mass of this body.

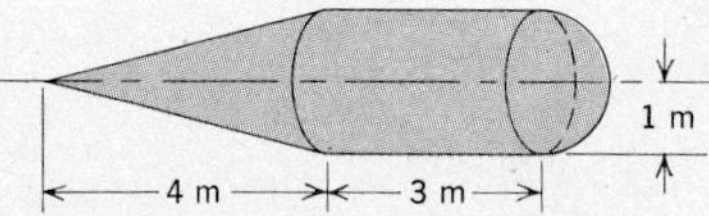

Prob. V.46

V.47 Solve Problem V.46 if only the cylindrical portion is composed of the given plastic, while the conical portion is steel and the hemispherical cap is aluminum.

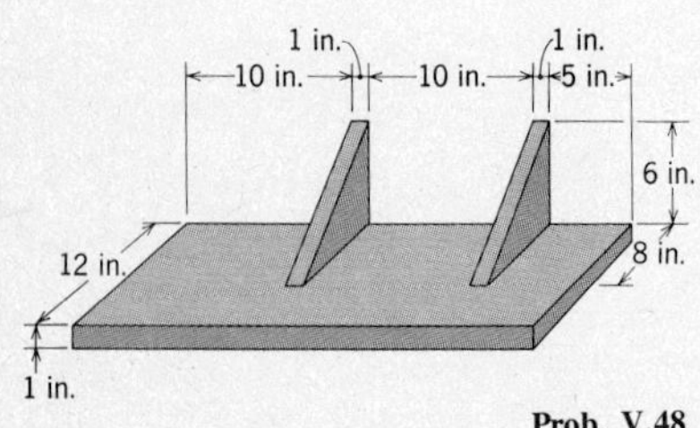

Prob. V.48

V.48 The identical shelf brackets are not spaced equally on the shelf. Knowing that the shelf and brackets are both composed of wood weighing 48 lb/ft^3, determine the weight and the location of the center of mass of the assembly.

V.49 The container shown is formed from 3-mm-thick aluminum. Determine the mass and the location of the center of mass of the container.

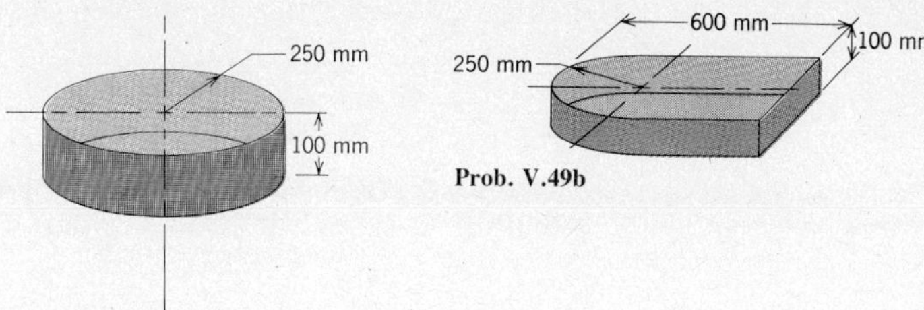

Prob. V.49b

Prob. V.49a

V.50 and V.51 The body shown consists of a solid pine interior that has been completely coated with lead to a uniform thickness of 10 mm. Determine the mass and the location of the center of mass of this body.

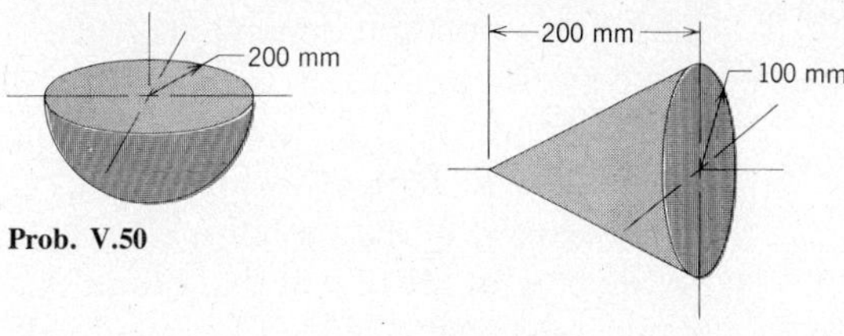

Prob. V.50

Prob. V.51

V.52 The crankpin shown is composed of steel. Determine the mass and the location of the center of mass of this body.

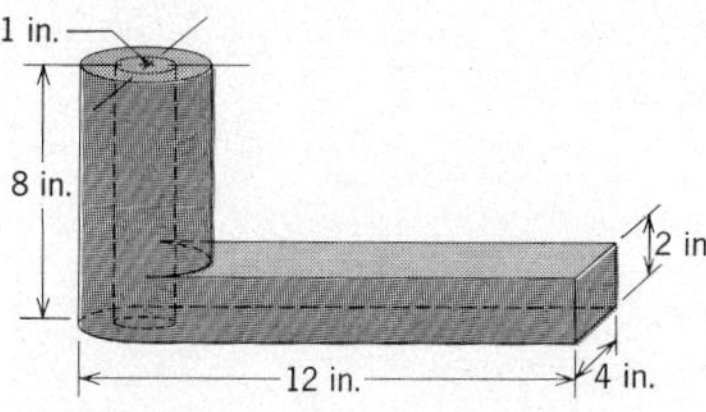

Prob. V.52

B. SURFACE LOADINGS

The gravitational forces we considered in Section A are examples of distributed force systems that are applied over the space occupied by a body; they are termed *body forces*. Another important type of force system is one that acts over the surface area of a body; this is the case of a *surface force*. Typically, distributed surface forces can result from the interaction of two bodies along an edge or contacting surfaces, or from pressure exerted by fluids, such as air and water.

Here we will study two types of surface forces. When a force is distributed over a line, we have a *line load*. The magnitude of a line load has the units of force per unit length. When a force is distributed over an area we have an *area load*. The magnitude of an area load has the units of force per unit area. In the following we will learn how to determine the resultant of each type of loading. For simplicity, we shall restrict our attention to cases where the loaded surface is flat.

1 Line Loads

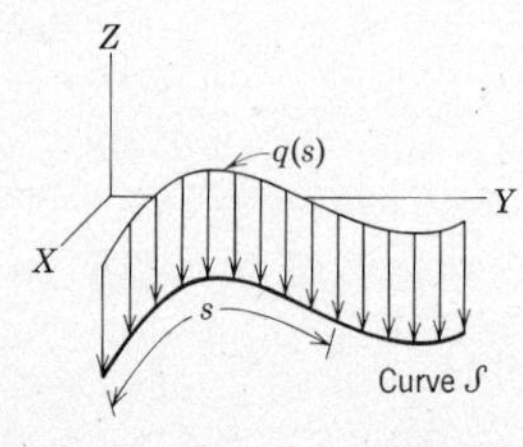

Figure 9

Frequently a structure having a flat surface, such as a plate or a beam, is required to support a normal force that is distributed along a line. We will denote this line load as the vector $\bar{q}$, and its magnitude as q, having units of force per unit length. A typical line load acting on a curve lying in the XY plane is shown in Figure 9.

The magnitude of q may vary with its location along the curve. To define this position dependency, we denote the arc length from an end of the curve to a point on the curve by the symbol s, as shown in the figure, and then give the function $q(s)$ defining the magnitude of q at each value of s. In order to depict this line load at a series of points along the curve, we draw arrows in the directions of the load whose heights are proportional to the magnitude of q at that point. As can be seen in Figure 9, the result is a curved surface, which we shall call the *loading surface*.

The forcing effect of the line load at some point on the surface may be described by multiplying the value of $\bar{q}$ at that point by the element of arc length ds situated at that point. The result is an infinite number of infinitesimal concentrated forces $\bar{q}\ ds$ acting along the curve. For clarity, only one of these concentrated forces is illustrated in Figure 10a.

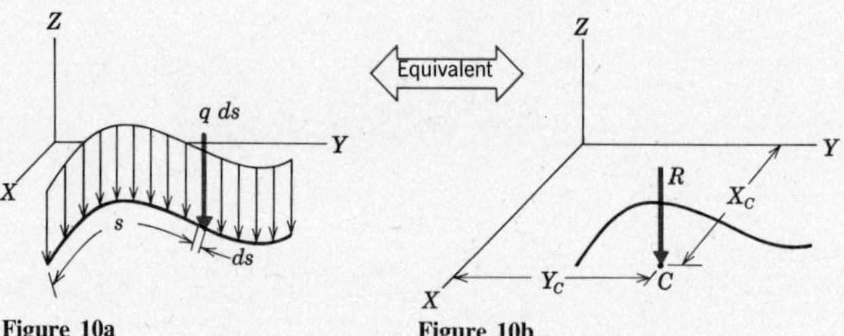

Figure 10a

Figure 10b

The system of concentrated forces formed in Figure 10a are all parallel to the Z axis. From our previous studies we know that such a system of forces can be represented by a single equivalent force $\bar{R}$, parallel to the Z axis. This force is depicted in Figure 10b. For equivalence, the magnitude of $\bar{R}$ must be identical to the sum of the magnitudes of the infinitesimal concentrated forces $\bar{q}\,ds$. The process of summing infinitesimal quantities requires integration. Specifically, we have

$$\boxed{R = \int_{\mathscr{S}} q\,ds} \tag{11}$$

where $\mathscr{S}$ denotes the curve carrying the load.

In view of the fact that the resultant is parallel to the Z axis, its line of action may be specified by determining the point of application C of this force in the XY plane. The coordinates of this point are determined by equating the moment of $\bar{R}$ to the sum of moments of all forces $\bar{q}\,ds$ about any convenient point. Computing moments about the origin of the coordinate system, we have

$$(X_C\bar{I} + Y_C\bar{J}) \times (-R\bar{K}) = \int_{\mathscr{S}} (X\bar{I} + Y\bar{J}) \times (-q\,ds\,\bar{K})$$

$$RX_C\bar{J} - RY_C\bar{I} = \int_{\mathscr{S}} (Xq\bar{J} - Yq\bar{I})\,ds$$

Equating corresponding components then yields

$$\boxed{\begin{aligned} RX_C &= \int_{\mathscr{S}} Xq\,ds \\ RY_C &= \int_{\mathscr{S}} Yq\,ds \end{aligned}} \tag{12}$$

Equations (12) apply to the situation where a line load perpendicular to the XY plane acts on a curve $\mathscr{S}$ in that plane. Formulas for line loads acting on the other coordinate planes may be obtained by permuting the symbols in equations (12).

The pictorial representation of a line load that we introduced in Figure 9 allows for a useful interpretation of equations (11) and (12). Recall how the loading surface was formed; at each point along the curve the height represents the magnitude of $\bar{q}$ at that point. Thus, a differential element of area for this surface can be formed from a rectangular strip whose height is q and whose base is ds. Several such strips are depicted in Figure 10a. It follows that an element of area for the loading surface is $q\,ds$. Hence, the integral in equation (11) gives the area of the loading surface and the integrals in equation (12) are first moments of area for this surface. In this interpretation, equations (12) have the same meaning as equations (7) for centroids of a surface. Hence, we have the following analogy.

> The resultant of a line load $\bar{q}$ perpendicular to a flat plane is the area of the loading surface, perpendicular to the flat plane in the sense of positive q. This resultant intersects the centroid of the loading surface.

The importance of this analogy is that it enables us to use the tabulated values in Appendices B and C to determine the location of the point of application of the resultant of a line load whenever the loading surface is a basic shape or a composite of basic shapes.

EXAMPLE 9

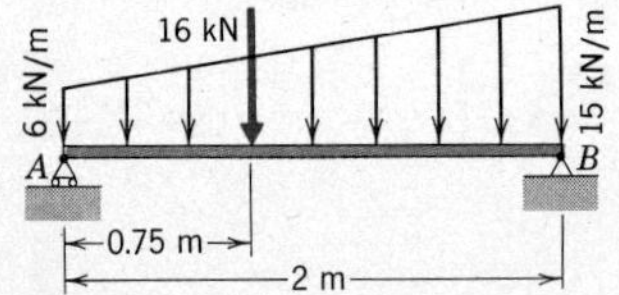

A beam supports the trapezoidal line load and the concentrated force shown. Determine (a) the resultant of the line load, (b) the reactions at supports A and B.

Solution

The given force diagram shows that the surface of the loading function is a planar trapezoid. To solve part (a) we sketch this loading surface and its resultant. Because the line load is downward, the resultant is also downward. Note that in order to distinguish between concentrated and distributed forces in a sketch, we draw the arrows for the former more boldly.

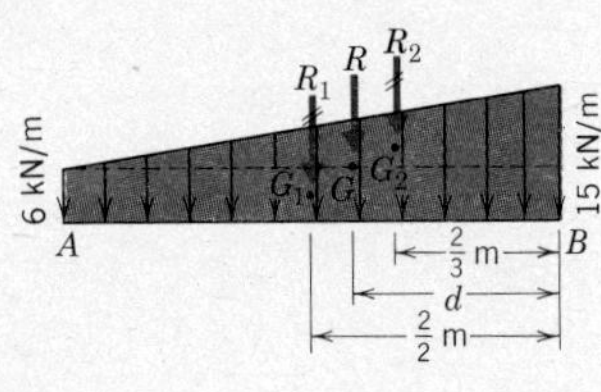

To employ the analogy between the loading surface and the resultant, we need the area of the trapezoid and the horizontal distance d to its centroid. As this information does not appear in Appendix B, we consider the trapezoid to be a composite consisting of a rectangle and a triangle, as shown. For each of these basic shapes, there is a resultant force $\bar{R}_i$ ($i = 1$, 2) that intersects the centroid of the corresponding area. These individual resultant forces are depicted in the sketch with hatchmarks in order to indicate that their combined effect is equivalent to the total resultant $\bar{R}$. The sketch also shows the location of the centroids of the triangular and rectangular areas.

The magnitude of each resultant is the area of the corresponding loading surface. Thus

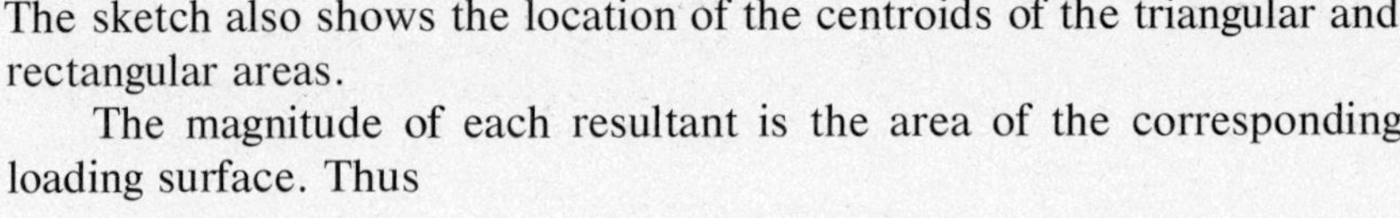

$$R_1 = (6 \text{ kN/m})\ (2 \text{ m}) = 12 \text{ kN}$$

$$R_2 = \tfrac{1}{2}[(15 - 6)\text{kN/m}](2 \text{ m}) = 9 \text{ kN}$$

It then follows that

$$R = R_1 + R_2 = 21 \text{ kN}$$

To determine the value of d, we equate the moment of $\bar{R}$ to that of the resultants of the basic areas. By summing moments about end B, we find

$$R(X_G) = R_1(1\text{ m}) + R_2(\tfrac{2}{3}\text{ m})$$

$$21(X_G) = 12(1) + 9(\tfrac{2}{3})$$

$$X_G = 0.8571\text{ m}$$

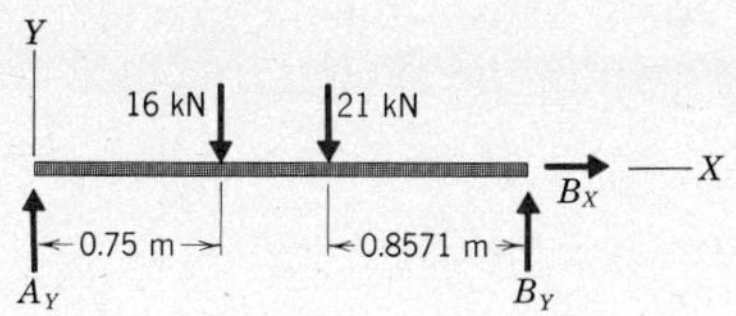

To solve part (b) we draw a free body diagram of the system, replacing the trapezoidal line load by its resultant. The XYZ coordinate system shown matches the directions in which dimensions are given. The equilibrium equations are

$$\Sigma M_{BZ} = -A_Y(2) + 21(0.8571) + 16(2-0.75) = 0$$

$$\Sigma F_X = B_X = 0$$

$$\Sigma F_Y = A_Y + B_Y - 16 - 21 = 0$$

Thus, we have

$$A_Y = 19\text{ kN} \qquad B_Y = 18\text{ kN} \qquad B_X = 0$$

Before we leave this problem, let us consider the situation where the problem statement requests that we solve part (b) only. In such a case, by combining the sketch for evaluating the line load with the free body diagram as illustrated in the sketch, we can eliminate the intermediate step of determining the total resultant $\bar{R}$. All that is necessary then is to determine the values of R_1 and R_2, as we did in the preceding solution, and apply the equilibrium equations to the system of concentrated forces appearing in the free body diagram.

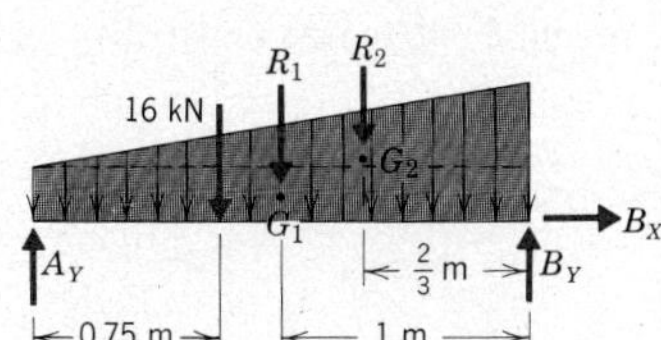

EXAMPLE 10

A bar, bent into a quarter-circular curve 18 in. in radius, is welded to a wall at end A. A line load whose intensity increases in proportion to the arc length along the bar from end A is applied transversely to the bar. The magnitude of the line load at end B is 4 lb/ft. Determine the reactions at end A.

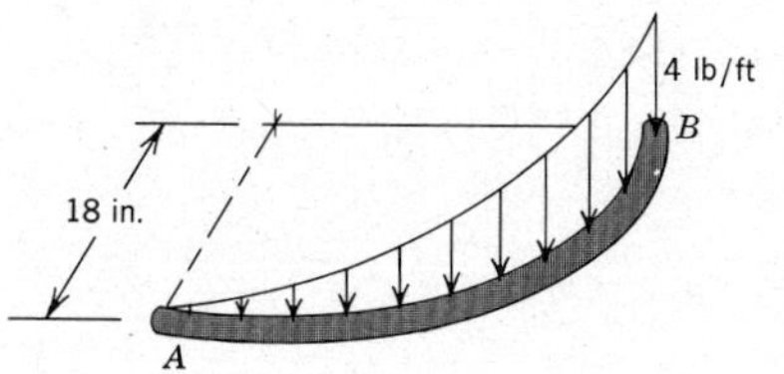

Solution

From the given information we deduce that

$$q(s) = \frac{4}{\mathcal{S}}s\text{ lb/ft}$$

where $\mathcal{S} = (\pi/2)(18\text{ in.})$ is the total length of the bar and s is the arc length from end A to a point on the bar. The resulting loading surface depicted in

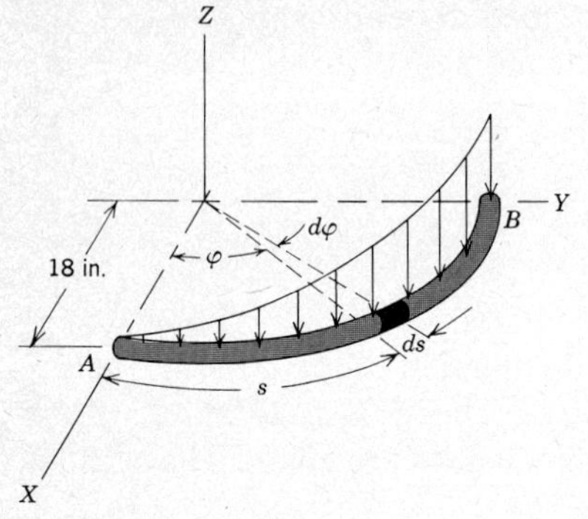

the given sketch does not appear in Appendix C, so the analogy between the resultant and the properties of the loading surface is of no use to us. We must therefore employ equations (11) and (12) to determine the resultant. For this we require a sketch describing the geometry of the system and a suitable XYZ coordinate system.

It will be necessary for us to relate the values of X and Y at some point on the bar to the value of s at that point. This is most conveniently done with the aid of the polar angle ϕ shown in the sketch, for then we have

$$X = 18 \cos \phi \qquad Y = 18 \sin \phi \text{ inches}$$

$$s = 18\ \phi \qquad ds = 18\ d\phi \text{ inches}$$

We see that points on the bent bar are contained in the range $0 \leq \phi \leq \pi/2$.

For consistency of units, we write q in units of pounds per inch.

$$q = \frac{4 \text{ lb/ft}}{\mathcal{S}} s = 4 \frac{\text{lb}}{\text{ft}}\left(\frac{1 \text{ ft}}{12 \text{ in.}}\right)\frac{s}{(\pi/2)18} = 1.1789(10^{-2})s \text{ lb/in.}$$

where s is measured in inches. Equation (11) then gives

$$R = \int q(s)\, ds = \int_0^{\pi/2} [1.1789(10^{-2})(18\phi)](18\ d\phi) = 4.712 \text{ lb}$$

This force is in the negative Z direction, as is $\bar{q}$. We next employ equations (12), which gives

$$\begin{aligned}
4.712X_C &= \int_{\mathcal{S}} Xq(s)\, ds \\
&= \int_0^{\pi/2} (18 \cos \phi)[1.1789(10^{-2})(18\phi)](18\ d\phi) \\
&= 1.1789(10^{-2})(18)^3\left[\cos \frac{\pi}{2} + \frac{\pi}{2} \sin \frac{\pi}{2} - \cos (0) - (0) \sin (0)\right] = 39.24
\end{aligned}$$

$$X_C = 8.328 \text{ in.}$$

$$\begin{aligned}
4.712Y_C &= \int_{\mathcal{S}} Yq(s)\, ds = \int_0^{\pi/2} (18 \sin \phi)[1.1789(10^{-2})(18\ \phi)](18\ d\phi) \\
&= 1.1789(10^{-2})(18)^3\left[\sin \frac{\pi}{2} - \frac{\pi}{2} \cos \frac{\pi}{2} - \sin (0) - (0) \cos (0)\right] = 68.75
\end{aligned}$$

$$Y_C = 14.590 \text{ in.}$$

Now that the resultant of the line load has been determined, we may proceed to the determination of the reactions at the support of the bar by drawing a free body diagram. The reactions consist of a force $\bar{F}_A$ and

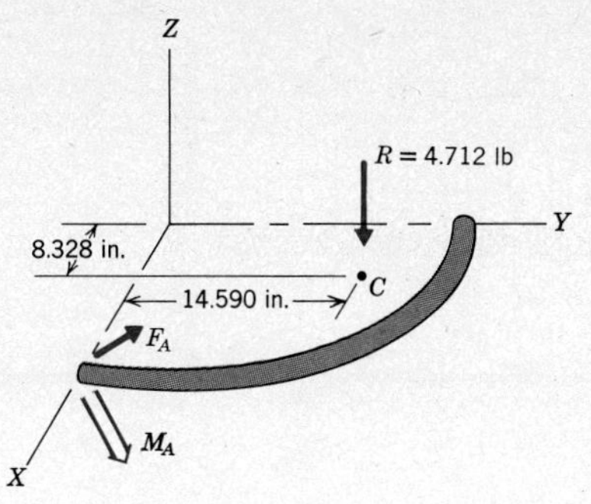

couple $\bar{M}_A$ at the welded end. The only other force is the resultant $\bar{R}$, so the free body diagram is as shown.

Summing the moments about point A, the equilibrium equations are

$$\Sigma\, \bar{M}_A = \bar{M}_A + [-(18 - 8.328)\bar{I} + 14.590\bar{J}] \times (-4.712\bar{K}) = 0 \text{ lb-in.}$$

$$\Sigma\, \bar{F} = \bar{F}_A - 4.712\bar{K} = 0 \text{ lb}$$

Thus

$$\bar{F}_A = 4.71\bar{K} \text{ lb}$$

$$\bar{M}_A = 68.7\bar{I} + 46.6\bar{J} \text{ lb-in.}$$

HOMEWORK PROBLEMS

V.53–V.57 Determine the resultant of the distributed loading acting on the beam shown. Also determine the support reactions.

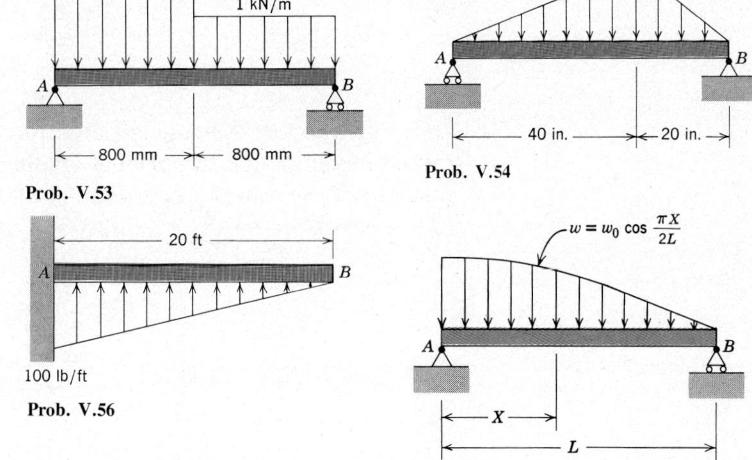

Prob. V.53

Prob. V.54

Prob. V.55

Prob. V.56

Prob. V.57

V.58–V.62 Determine the reactions at the supports of the beams shown.

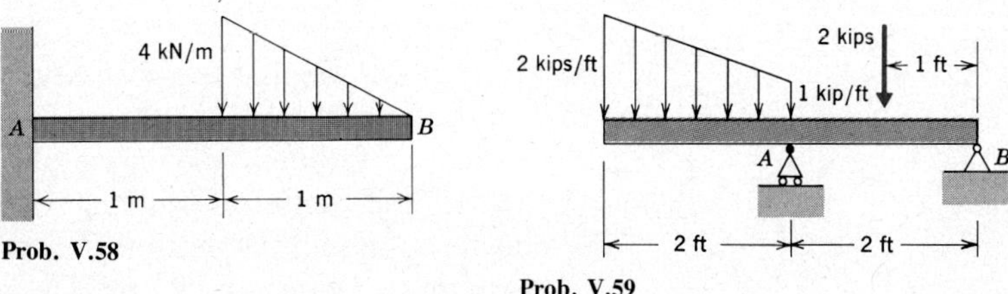

Prob. V.58

Prob. V.59

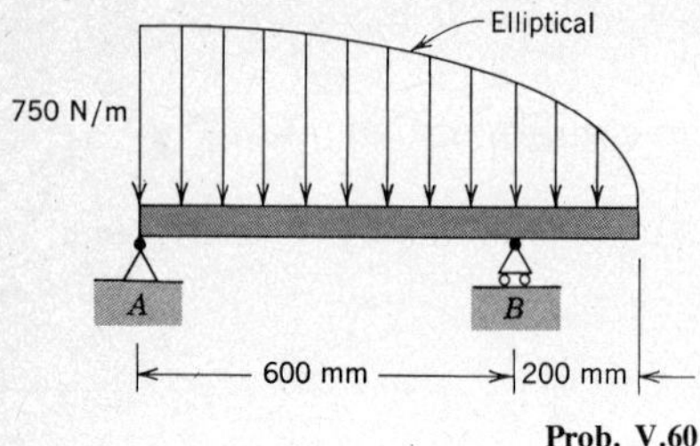

Prob. V.60

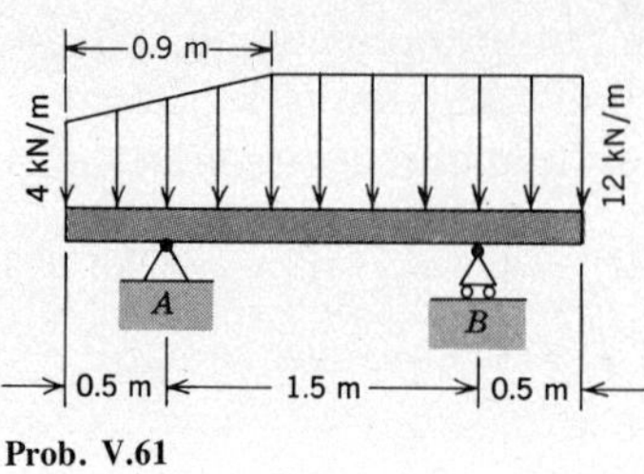

Prob. V.61

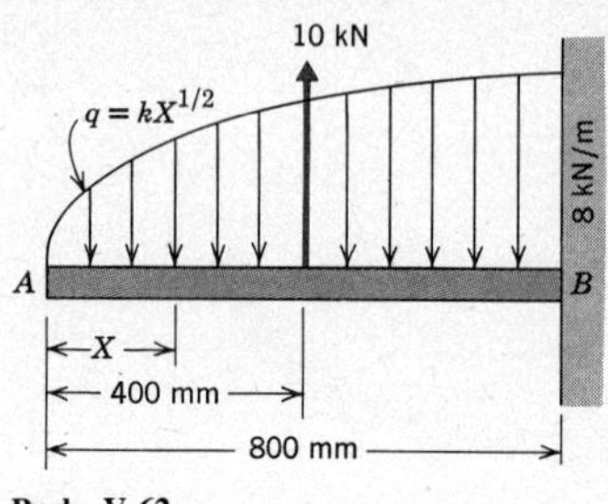

Prob. V.62

V.63 and V.64 For the type of loading shown, determine the values of q_1 and q_2 that result in the reactions at the fixed support A being (a) a force with no couple, (b) a couple with no force. In each case give the corresponding reaction.

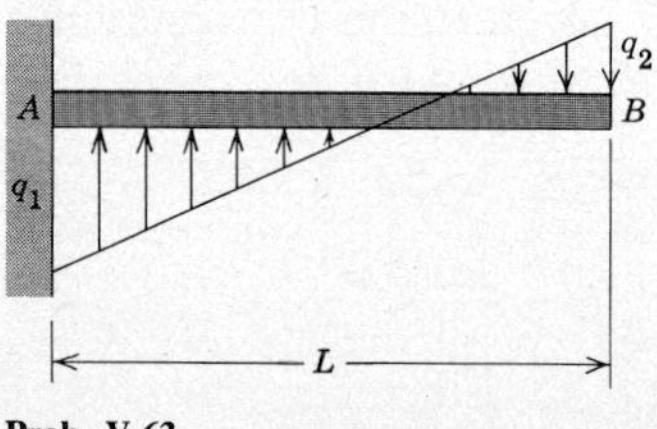

Prob. V.63

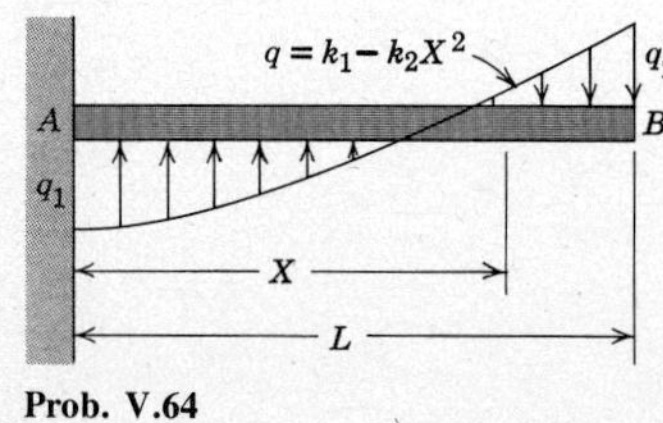

Prob. V.64

V.65 To support the concentrated force $\bar{R}$ acting on the foundation, the ground develops the trapezoidal loading shown. Determine the values of q_1 and q_2 in terms of the magnitude R of the force and the ratio of distances d/L. Answer: $q_1 = 2(2 - 3d/L)(R/L)$, $q_2 = 2(3d/L - 1)(R/L)$.

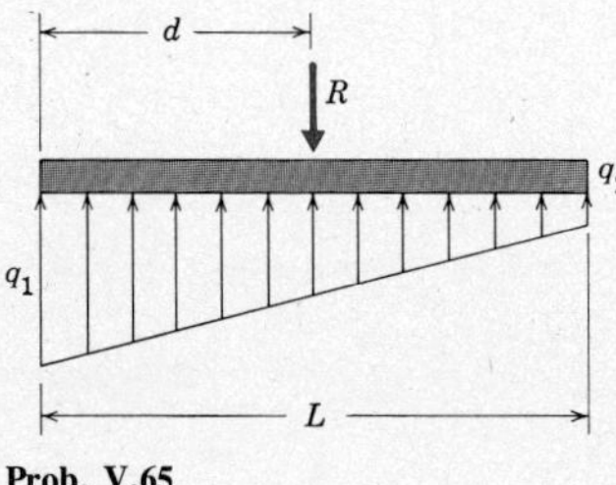

Prob. V.65

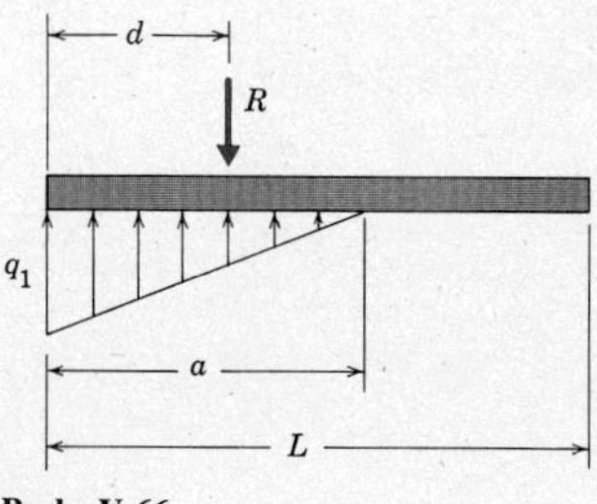

Prob. V.66

V.66 It is known that the ground can only push upward on the foundation in Problem V.65. (a) Determine the range of values of d/L for which the results of Problem V.65 are valid. (b) When the force $\bar{R}$ is applied to the left of the region of validity determined in part (a), the foundation loading is as shown. Determine the values of q_1 and the distance ratio a/L in terms of the values of R and d/L in this case.

V.67–V.70 Determine the reactions at the fixed end A of the curved beam shown.

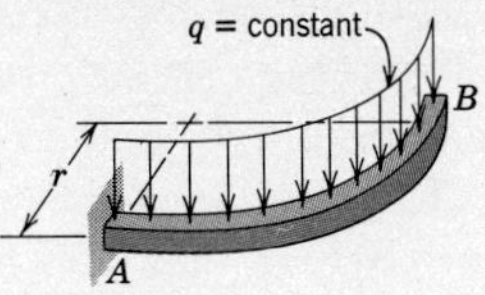

Prob. V.67

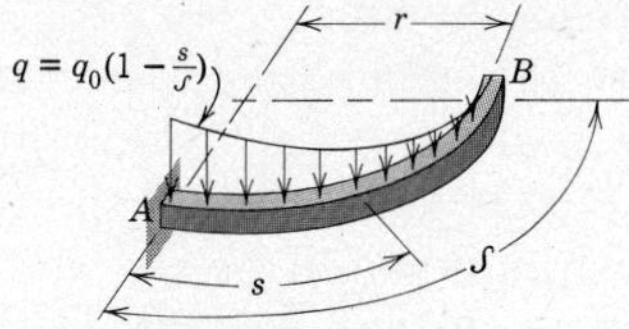

Prob. V.68

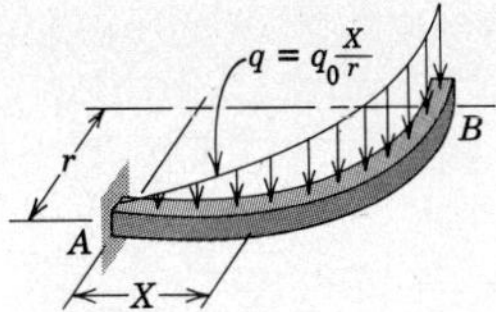

Prob. V.69

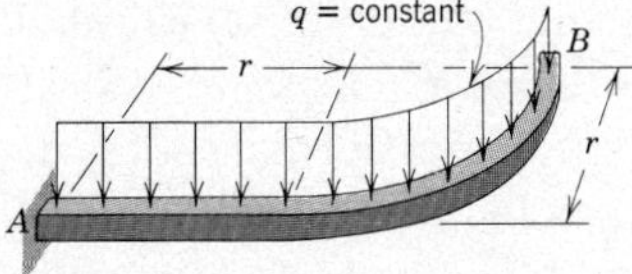

Prob. V.70

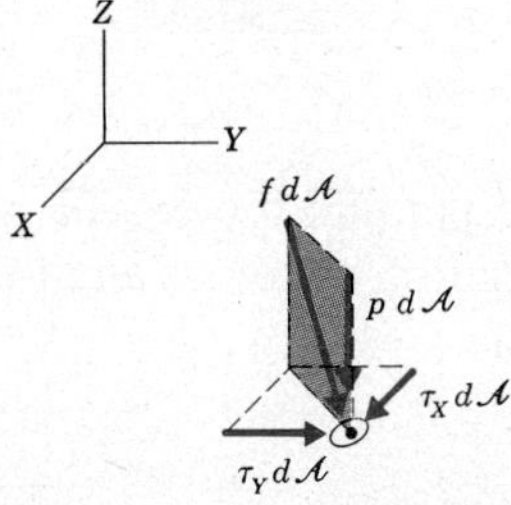

Figure 11

2 Surface Loads — Pressure Distributions

As is the case for all force systems, a surface load is a vector having magnitude and direction, which we shall denote as $\bar{f}$. In treating such a load, we can discuss its effect on each point of the surface by multiplying the vector $\bar{f}$ by the infinitesimal area $d\mathcal{A}$ on the surface. This is depicted in Figure 11, where the XY plane was chosen such that it coincides with the planar area $\mathcal{A}$ being loaded.

The surface load $\bar{f}$ may be replaced by three components, as depicted in Figure 11. The component $\bar{p}$, called the *pressure*, is normal to the surface, pushing on it. When $\bar{p}$ has a pulling (suction) effect, we will consider its magnitude to be negative.

In mechanics, quantities that have the units of force per unit area are termed *stresses*. Thus, the pressure can alternatively be termed a *normal stress*. The other two components of $\bar{f}$, $\bar{\tau}_X$ and $\bar{\tau}_Y$ tangent to the plane of the area $d\mathcal{A}$, are called *shear stresses*. In general, the magnitudes of the pressure and shear stresses can vary from point to point along the surface $\mathcal{A}$, and thus are functions of position. The function $p(X, Y)$ is the *pressure distribution* on the area, and $\tau_X(X, Y)$ and $\tau_Y(X, Y)$ are the *shear stress distributions* on the area.

Because the area load $\bar{f}$ is a vector quantity, we may consider the effect of each of these distributions separately, and then add the results vectorially. In the remainder of this section we will study the effects of pressure distributions on bodies and not consider the effects of shear stress distributions, although the results for the latter are straightforward extensions of those for pressure.

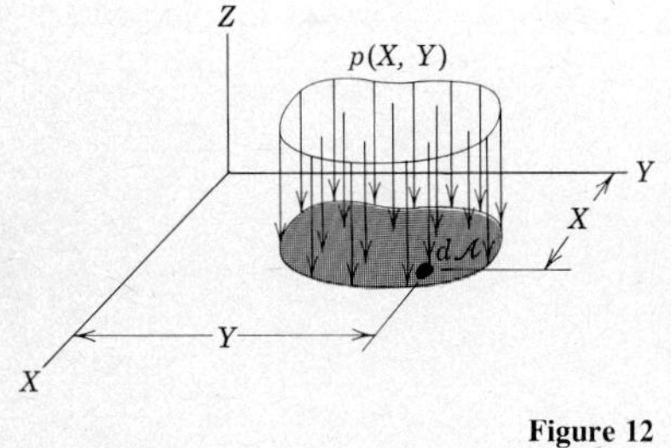

Figure 12

A typical pressure distribution $p(X, Y)$ acting on an area $\mathcal{A}$ in the XY plane is shown in Figure 12. In depicting $p(X, Y)$ we have used arrows whose length is proportional to the pressure p at the point where the arrow intersects the area $\mathcal{A}$. Thus the pressure distribution depicted in Figure 12 forms an imaginary body, which we shall call the *pressure space,* whose height above each point in the plane is the magnitude of the pressure function $p(X, Y)$ for that point.

Recall our earlier observation that the normal force acting on an infinitesimal area $d\mathcal{A}$ is $p d\mathcal{A}$. Therefore, there are an infinite number of parallel infinitesimal normal forces acting throughout the area $\mathcal{A}$. A set of parallel forces may be replaced by a single equivalent force. The pressure distribution of Figure 12 and its equivalent single resultant force $\bar{R}$ acting at point C are shown in Figure 13.

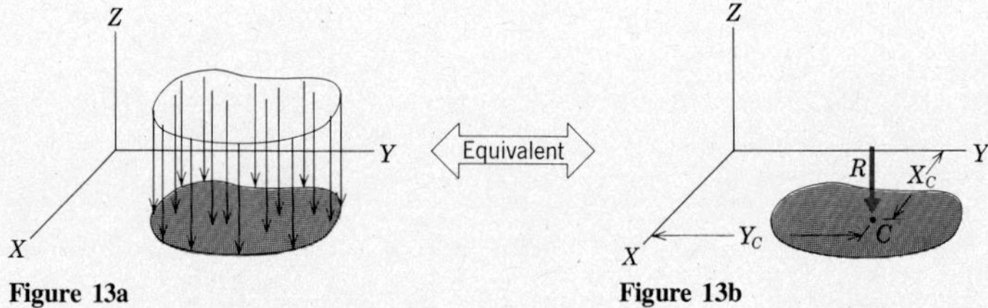

Figure 13a

Figure 13b

The force $\bar{R}$ is the sum of all of the parallel forces. To account for each of the infinitesimal forces, the summation becomes an integral. Thus

$$\boxed{R = \int_{\mathcal{A}} p \, d\mathcal{A}} \tag{13}$$

In view of the fact that $p(X, Y)$ is the height above the area $\mathcal{A}$ of the pressure space formed by the pictorial representation of the pressure distribution, equation (13) tells us that

> The resultant force $\bar{R}$ is the volume of the pressure space. It is normal to the planar surface being loaded, in the sense of the pushing effect for positive pressure.

A negative pressure should be regarded as a negative contribution to the volume of the pressure space.

The coordinates X_C and Y_C required to specify the line of action of the resultant are obtained by equating the moments of the two equivalent force systems in Figure 13. Computing the moment about the origin O, the position vector from point O to the point of application of the infinitesimal force $p d\mathcal{A}$ can be seen from the figure to be $X\bar{I} + Y\bar{J}$. Similarly, the corresponding position vector for the resultant $\bar{R}$ is $X_C\bar{I} + Y_C\bar{J}$. As before, accounting for all of the infinitesimal forces requires an integral formulation. Equating the moment sums yields

$$\Sigma \bar{M}_O = (X_C\bar{I} + Y_C\bar{J}) \times (-R\bar{K}) = \int_{\mathcal{A}} (X\bar{I} + Y\bar{J}) \times (-p \, d\mathcal{A} \, \bar{K})$$

$$R(X_C\bar{J} - Y_C\bar{I}) = \int_{\mathcal{A}} (X\bar{J} - Y\bar{I}) p \, d\mathcal{A}$$

and the result of matching corresponding components is

$$RX_C = \int_{\mathcal{A}} X p \, d\mathcal{A} \qquad RY_C = \int_{\mathcal{A}} Y p \, d\mathcal{A} \tag{14}$$

The position on the area $\mathcal{A}$ whose coordinates are (X_C, Y_C) is frequently called the *center of pressure*.

Equations (14) also have significance with respect to the properties of the pressure space. For this body $p \, d\mathcal{A}$ is an infinitesimal element of volume (height times area). Thus, the integrals in equation (14) are equivalent to the first moments of volume with respect to the X and Y coordinates. Recalling that R is the volume of the fictional body, by comparing equations (14) to equations (5), we deduce that X_C and Y_C are the X and Y coordinates of the centroid of this volume. This allows us to state that

> The line of action of the resultant force $\bar{R}$ intersects the geometric centroid of the pressure space.

The importance of the analogy between the resultant $\bar{R}$ and the pressure space is that it will frequently enable us to make use of tabulated geometric properties when determining the resultant force. This occurs whenever the pressure space can be represented in terms of the fundamental shapes given in Appendix C.

The results for a line load, which we obtained in the previous section, can be seen to be a special case of those for a pressure distribution, wherein the pressure space has the shape of a thin shell. Further, in the commonly occurring situation where we have a constant pressure distribution with respect to the width of a rectangular area, we can easily reduce the pressure space to an equivalent line load. The intensity of this line load is $q = pb$, where b is the width of the rectangle in the direction of constant pressure. This line load acts along the centerline of the rectangular area, as illustrated in Figure 14 for a pressure distribution which only depends on the coordinate X.

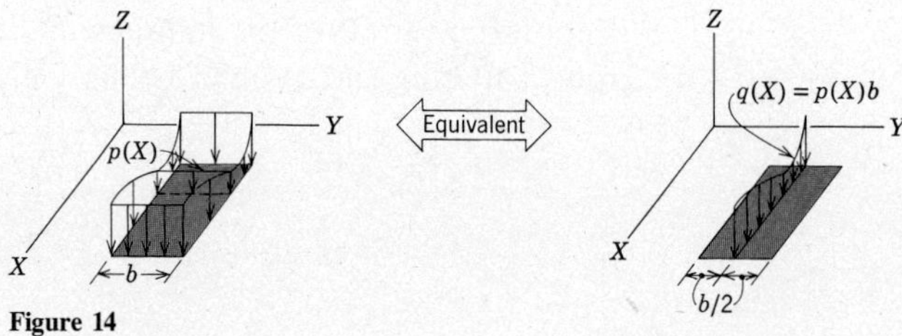

Figure 14

EXAMPLE 11

The pressure distribution on a rectangular plate is as shown, with a maximum value of $2(10^5)$ pascals. Determine the resultant of this pressure distribution and the corresponding center of pressure.

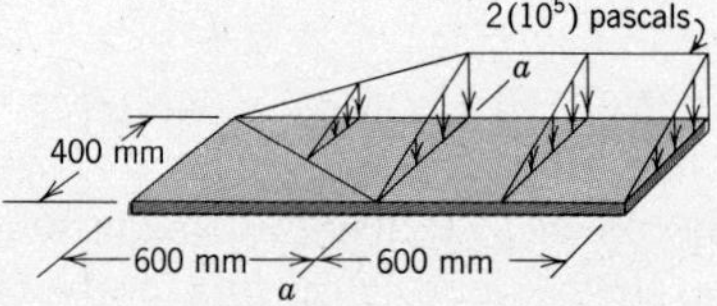

Solution

Although the pressure is applied to a rectangular area, it is not constant in the direction of either edge. Therefore, we do not seek an equivalent line load. Noting that the given diagram depicts the pressure space to be an orthogonal tetrahedron to the left of line a-a and a prism to the right of line a-a, we may employ the analogy between the resultant and the geometric properties of the pressure space.

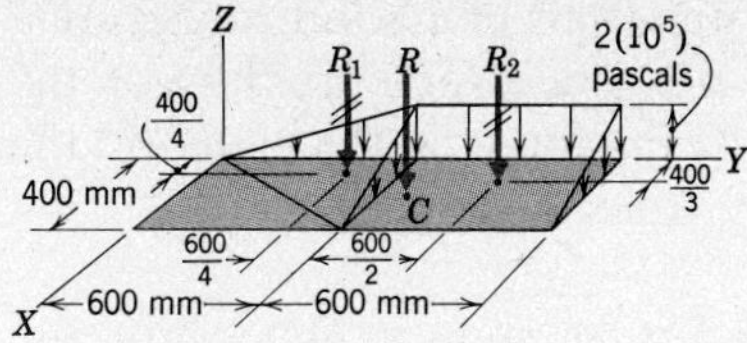

We begin with a sketch that shows the total resultant $\bar{R}$ and the resultants $\bar{R}_i$ ($i = 1, 2$) for the basic shapes forming the pressure space. The forces $\bar{R}_i$ intersect the centroids of their spaces, so the sketch also shows the location of the centroid of each basic shape in the XY plane, as found in Appendix C.

The values of R_1 and R_2 are calculated as the volume of the corresponding pressure spaces. By definition 1.0 pascal ≡ 1.0 N/m², so we have

$$R_1 = \tfrac{1}{6}(0.4)(0.6)2(10^5) \text{ N} = 8.00 \text{ kN}$$

$$R_2 = \tfrac{1}{2}(0.4)(0.6)2(10^5) \text{ N} = 24.00 \text{ kN}$$

$$R = R_1 + R_2 = 32.0 \text{ kN}$$

The center of pressure, being the point where the resultant $\bar{R}$ is applied to the plate, is determined by equating the moment of $\bar{R}$ to the total moment of $\bar{R}_1$ and $\bar{R}_2$. Referring to the sketch for the position vectors, we then have

$$(X_C\bar{I} + Y_C\bar{J}) \times \bar{R} = [0.10\bar{I} + (0.60 - 0.15)\bar{J}] \times \bar{R}_1 + [0.133\bar{I} + (0.60 + 0.30)\bar{J}] \times \bar{R}_2 \text{ N-m}$$

$$(X_C\bar{I} + Y_C\bar{J}) \times (-32.0\bar{K}) = (0.10\bar{I} + 0.45\bar{J}) \times (-8.0\bar{K}) + (0.133\bar{I} + 0.90\bar{J}) \times (-24.0\bar{K})$$

$$-32.0Y_C\bar{I} + 32.0X_C\bar{J} = -25.2\bar{I} + 3.992\bar{J}$$

Equating corresponding components then yields

$$X_C = 0.125 \text{ m} \qquad Y_C = 0.788 \text{ m}$$

HOMEWORK PROBLEMS

V.71–V.75 The traffic sign shown is subjected to a $5(10^3)$ N/m² uniform pressure caused by the wind. Determine the reactions at the fixed base of the pole resulting from this loading.

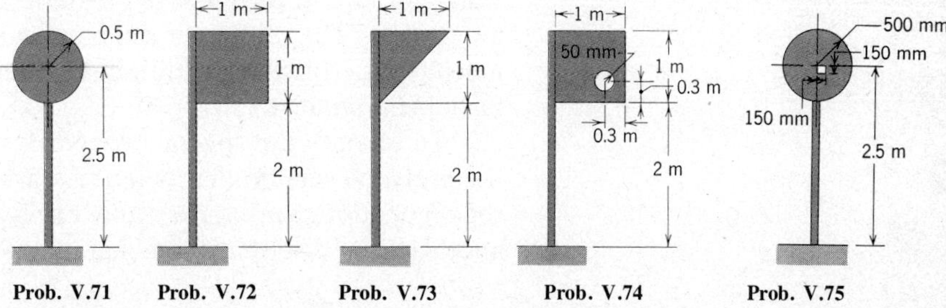

Prob. V.71 **Prob. V.72** **Prob. V.73** **Prob. V.74** **Prob. V.75**

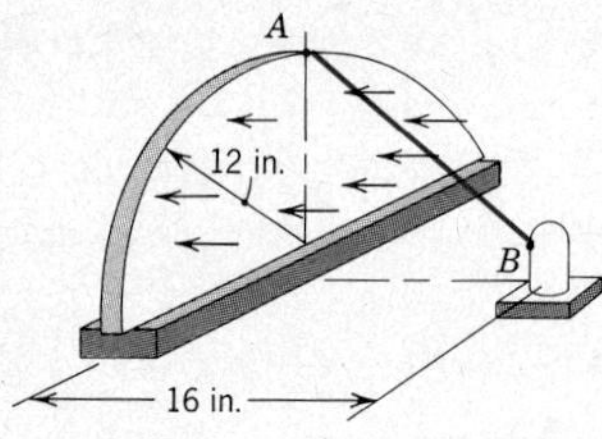

Prob. V.76

V.76 The semicircular plate shown is supported in a wind tunnel by a groove along its lower edge and by cable AB. The lateral pressure on the plate is 40 psi. Determine (a) the resultant wind force on the plate, (b) the tension in cable AB. The wind is blowing from the right.

V.77–V.79 The snow load on a flat roof is as shown. Determine the resultant force exerted by the snow on the roof. Specify the point of application of this force.

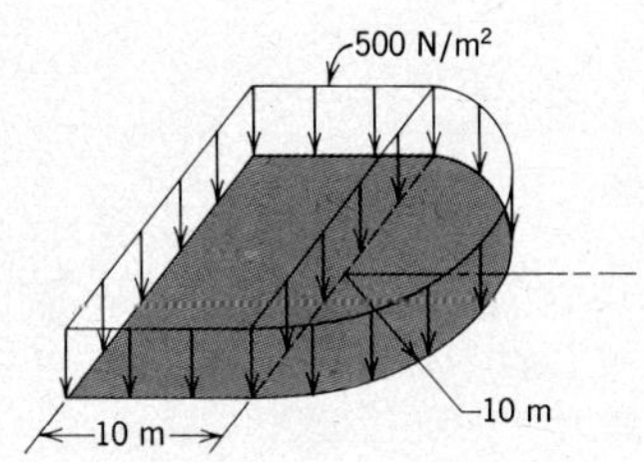

Prob. V.77

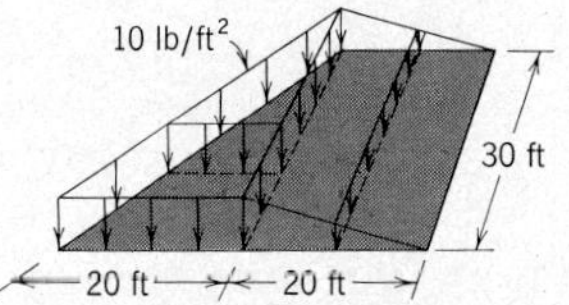

Prob. V.78

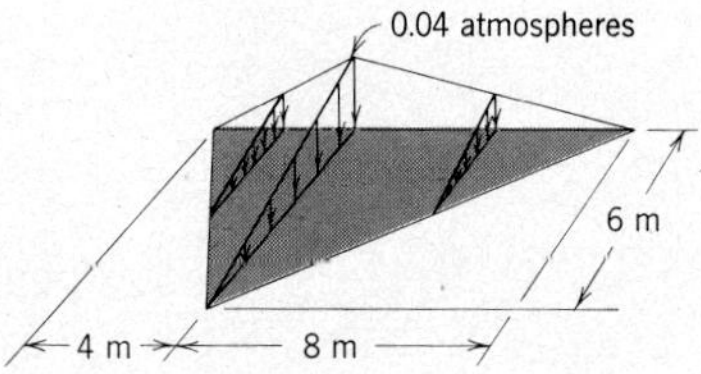

Prob. V.79

V.80–V.82 Determine the resultant of the pressure distribution acting on the flat plate shown. Also locate the corresponding center of pressure.

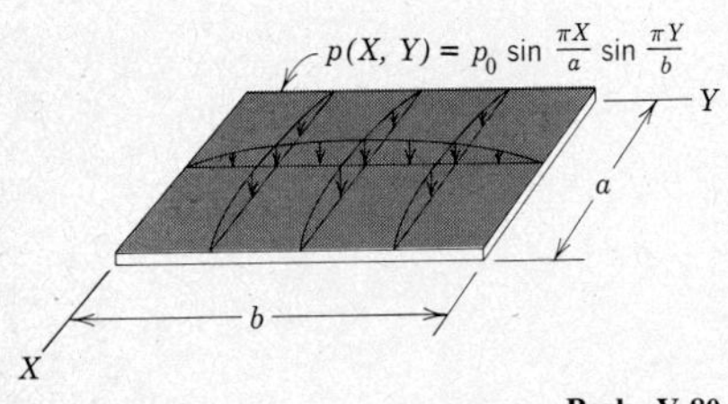

Prob. V.80

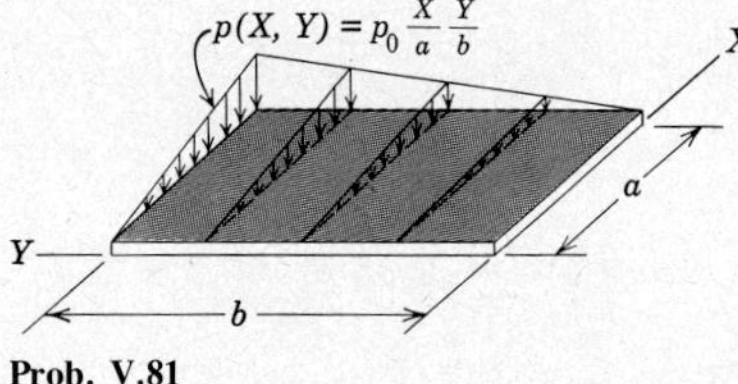

Prob. V.81

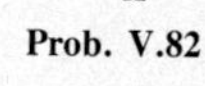

Prob. V.82

C. STATICS OF LIQUIDS

A common source of distributed loads on bodies is a fluid, such as oil, water, or air. In this course on statics we are concerned with fluids that are at rest. The matter of fluids in motion is treated in courses on fluid mechanics, which are traditionally offered after the elements of dynamics have been studied.

To identify the special properties of a fluid, recall our observation in the previous section that when two arbitrary bodies are in contact over a region of surface area, they may exert pressure and shear stresses on one another. The distinguishing characteristic of a *fluid* is that it is capable of applying shear stresses to a body only when it is in motion relative to the body. In other words, fluids at rest exert only pressure loads on stationary bodies.

The density of a *gas* changes radically with large changes in pressure, whereas the density of a *liquid* changes little with pressure, so it may be modeled as a fluid of constant density. In other words, a liquid is essentially *incompressible*. Here we shall only study liquids, for the study of gases requires the consideration of the temperature of the material.

When a liquid is placed in an open container it forms a horizontal surface, called the *free surface*. The pressure in a liquid is related directly to the vertical distance below the free surface. To develop this relationship, consider the rectangular parallelepiped element of liquid shown in Figure 15, which extends to a depth h below the free surface. Note that the XY plane is chosen to coincide with the free surface.

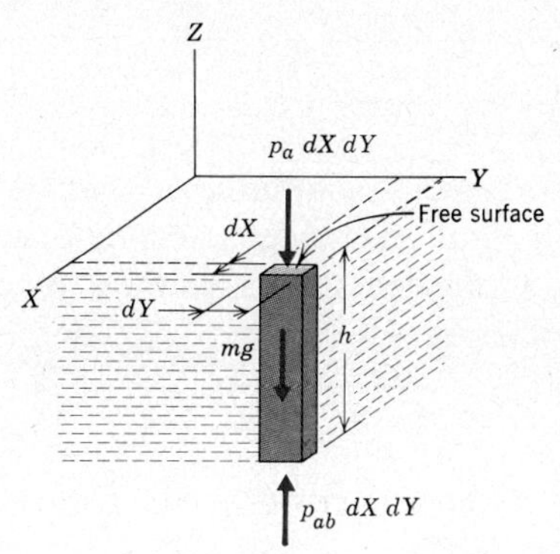

Figure 15

Only the vertical forces are depicted in Figure 15. The force $p_a\, dX\, dY$ is the result of the atmosphere pushing on the free surface (for standard conditions, $p_a = 1$ atm $= 1.0133(10^5)$ N/m^2 $= 14.7$ psi), whereas the force $p_{ab}\, dX\, dY$ is the result of the fluid below the element pushing on the lower horizontal face. Note that there are no vertical force components on the vertical faces, for these would require the existence of shear stresses. Summing forces in the vertical direction yields

$$\Sigma F_Z = p_{ab}\, dX\, dY - p_a\, dX\, dY - mg = 0$$

The mass m of the element of fluid may be calculated by multiplying the density ρ by the fluid volume $h\, dX\, dY$. Doing this and solving for p_{ab} gives

$$p_{ab} = p_a + \rho g h$$

The pressure ρ_{ab} defined by this equation is the *absolute pressure*. In most cases we will be concerned only with the increase of pressure above atmospheric, because for design purposes we usually regard bodies at atmospheric pressure to be unloaded. In order to describe this pressure increase we define a new pressure p, called the *gage pressure*, which is the difference between the absolute pressure and atmospheric pressure. Hence, the gage pressure is given by

$$\boxed{p = \rho g h} \tag{15}$$

Equation (15) defines a *hydrostatic pressure distribution*.

As it was derived, the pressure given by equation (15) is that acting on a horizontal surface. We will now prove that this pressure is exerted

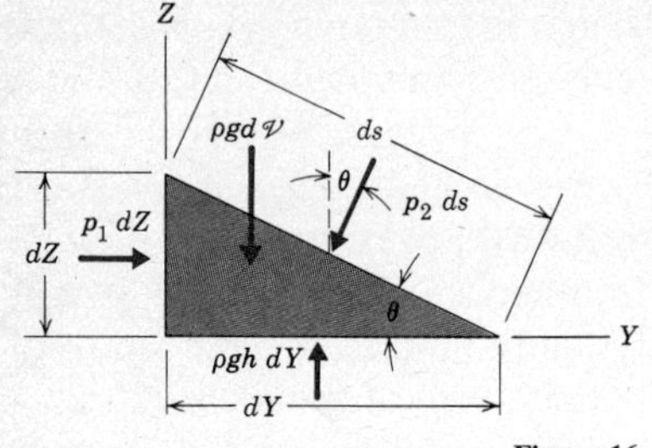

Figure 16

equally on surfaces having any orientation at the depth h below the free surface. To do this, consider the small prismatic element of fluid dm in Figure 16. Its width in the horizontal direction perpendicular to the plane of the diagram is unity.

We denote the pressures on the inclined and vertical faces of the element by the unknown values p_1 and p_2. The force on each face depicted in the figure is obtained by multiplying the pressure on that face by the area of the face (remembering that the width is unity).

The equations of equilibrium for the element of fluid are

$$\Sigma F_Y = p_1\,dZ - p_2\,ds \sin\theta = 0$$

$$\Sigma F_Z = \rho gh\,dY - p_2\,ds \cos\theta - \rho g\,d\mathcal{V} = 0$$

where ρg is the *specific weight* (units of force per unit volume) of the fluid. The weight of the element $\rho g\,d\mathcal{V} = \rho g\,dy\,dz/2$, being a second-order differential, is negligible when compared to the other terms in the Z force sum. The ΣF_Z equation then reduces to

$$\Sigma F_Z = \rho gh\,dY - p_2\,ds \cos\theta = 0$$

From Figure 16 we note that $ds \sin\theta = dZ$ and $ds \cos\theta = dY$. Thus the equilibrium equations yield

$$p_1 = p_2 = \rho gh$$

Noting that this result was obtained for an arbitrary value of θ, we may conclude that

> The gage pressure at a point in a fluid is proportional to the depth h from the free surface to the point ($p = \rho gh$). This pressure is exerted equally on all surfaces at that depth, regardless of their orientation.

The second sentence in this statement is known as Pascal's law. It should be noted that in the U.S.-British system of units it is common to prescribe the specific weight ρg of a material, whereas in the SI system the density ρ of a material is the preferred prescribed parameter.

Now that the relationship governing the pressure in a liquid has been established, we proceed to the study of the resultant force exerted by a fluid on a surface. Here we will consider three specific types of surfaces. They are typical of cases occurring in actual practice. The case of an arbitrary surface is left for texts on fluid mechanics.

1 Rectangular Surfaces

Consider the situation where one edge of a submerged rectangular surface is horizontal. Then any line in the rectangle parallel to this edge is situated

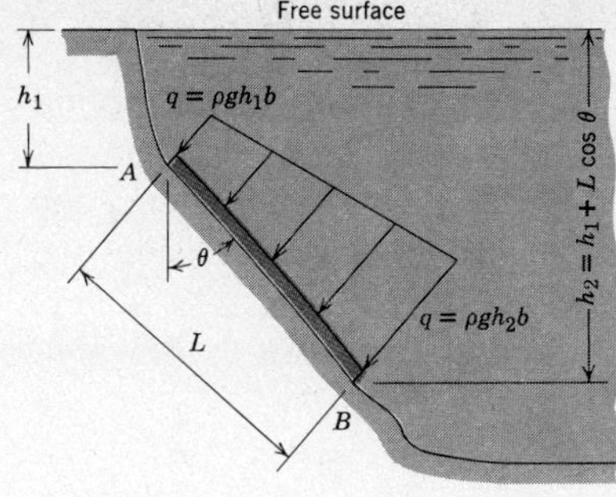

Figure 17

at a constant depth h. Therefore, the hydrostatic pressure is constant along this line. This permits us to convert the pressure distribution into a line load by multiplying the pressure by the horizontal width b of the surface. A typical case of the pressure on a rectangular plate of length L that is exposed to a liquid on only one side is shown in side view in Figure 17, where b is the width of the surface perpendicular to the plane of the figure.

At the upper edge A the depth is h_1. Here, the gage pressure is $\rho g h_1$, and the line load is $q = pb \equiv \rho g h_1 b$. At the lower edge B the depth is h_2, so the line load at this location is $q = \rho g h_2 b$. Noting that the distance along the line AB is linearly related to the depth, the line load is trapezoidal, as illustrated. Obvious exceptions occur when edge A is at the free surface, in which case $h_1 = 0$ and the loading is a triangular area; and when the plate is horizontal ($\theta = 90°$), for then $h_1 = h_2$ and the surface of the loading function is a rectangle.

EXAMPLE 12

The automatic valve shown in cross section consists of a 2- by 3-m rectangular plate AB that can rotate about a horizontal shaft passing through point C. Neglecting the weight of this plate, determine the depth of water d in the reservoir for which the valve will open.

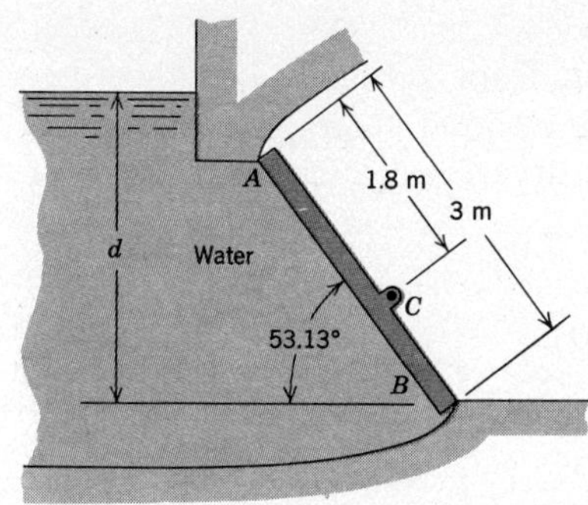

Solution

We are interested in the effect of the water on the valve plate. Hence, we wish to draw a free body diagram of the plate. In the condition where the valve is about to open, there will be no contact at edges A and B, so there will be no reactive force at either point. At edge B the depth is the unknown value d, whereas at edge A the depth is $d - 3 \sin 53.13°$. Recalling that the pressure is normal to the surface it acts upon, we obtain the free body diagram shown. In the values for the line load shown, the factor 2 is the width of the plate.

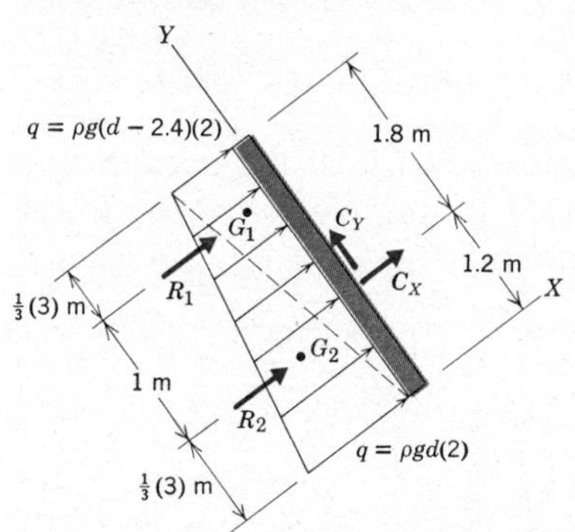

As indicated in the free body diagram, we consider the trapezoidal load to be a composite of two triangular loads. Computationally, this is somewhat more efficient than considering a composite of a triangular and rectangular load. Equating each force to the area of the corresponding triangle yields

$$R_1 = \tfrac{1}{2}[\rho g(d - 2.4)(2)](3) = \rho g(3d - 7.2) \text{ newtons}$$

$$R_2 = \tfrac{1}{2}[\rho g d(2)](3) = \rho g(3d) \text{ newtons}$$

We are not interested in the reactions at the shaft C, so the only equilibrium equation we need is the moment sum at that point.

$$\Sigma M_{CZ} = -R_1(1.8 - 1.0) + R_2(1.2 - 1.0) = 0 \text{ N-m}$$

Substituting the expressions for R_1 and R_2 found above, we find

$$-\rho g(3d - 7.2)(0.8) + \rho g(3d)(0.2) = 0 \qquad d = 3.20 \text{ m}$$

If d is less than this value, the portion of the trapezoidal loading below shaft C exerts a greater moment about this shaft than the portion of the loading above the shaft, thus keeping the valve shut. When d exceeds 3.20 m, the opposite is true and the valve swings open. As an aside, we note that if we had needed it, the density ρ would have been for water. However, because this parameter does not affect the value of d, the valve would open at a depth of 3.20 m for any liquid.

HOMEWORK PROBLEMS

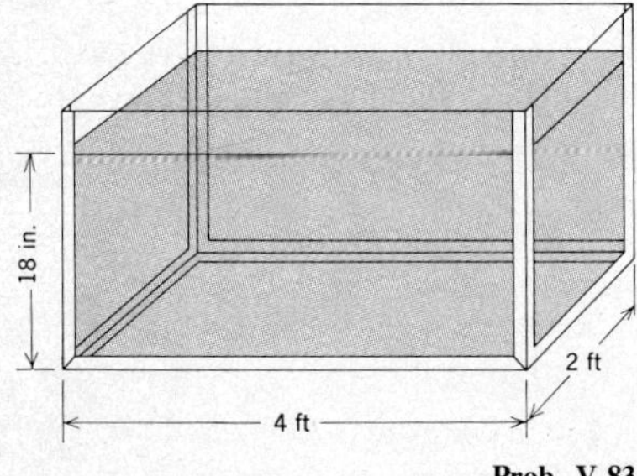

Prob. V.83

V.83 The fish tank shown holds water to a depth of 18 in. For each face, including the bottom, determine the resultant force exerted by the water. Also determine the point of application of each resultant force (the center of pressure).

V.84 and V.85 Determine the hydrostatic force acting on the 10-m section of seawall shown when the water is at flood stage, and locate the corresponding center of pressure. The density of sea water is 2.5% greater than that of fresh water.

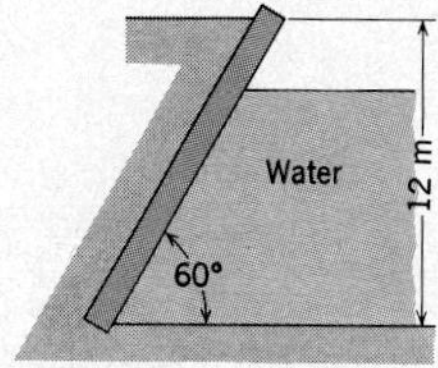

Prob. V.84

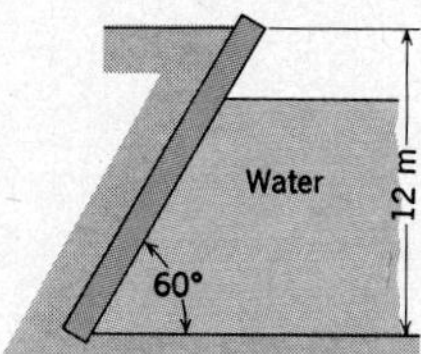

Prob. V.85

V.86 and V.87 The 3-m-wide sliding gate shown is at the bottom of a retaining wall. Determine the force exerted by the groove at the lower edge of the gate to resist the water pressure (a) when d = 3 m, (b) when d = 6 m.

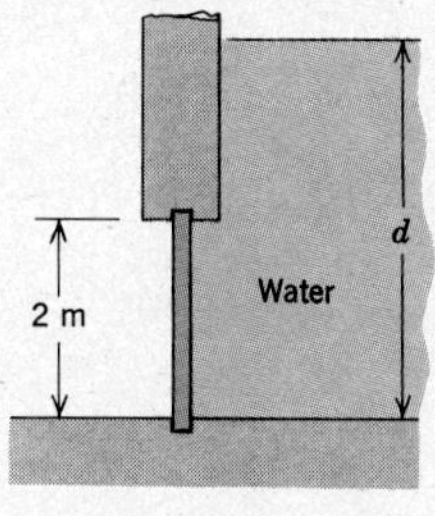

Prob. V.86

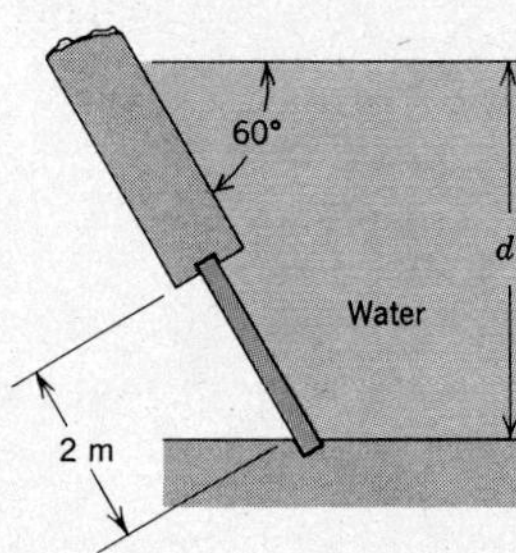

Prob. V.87

V.88 and V.89 The 1.5-m-wide gate shown, which may pivot about the horizontal shaft A, is held in the closed position by the preloaded spring. Determine the initial compressive force in the spring for which the gate will open when the depth of the water is (a) $d = 2$ m, (b) $d = 4$ m.

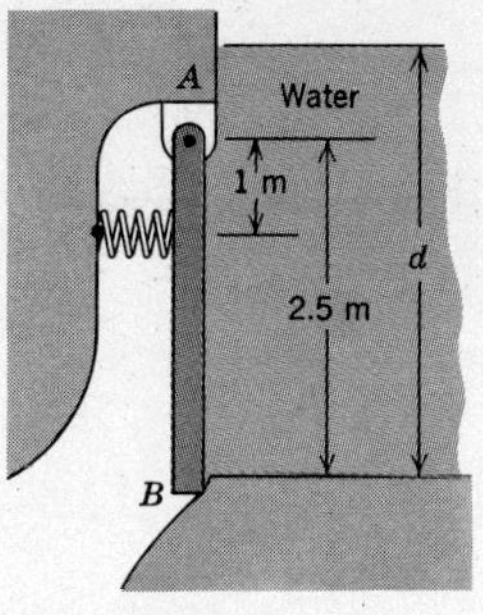

Prob. V.88

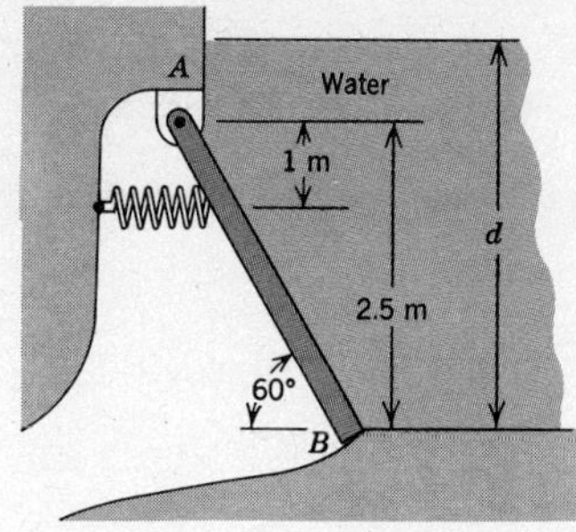

Prob. V.89

V.90 and V.91 The cross section of a 60-ft-long section of concrete formwork is shown. The left upright panel is attached to the anchored panel on the right by 40 equally spaced tie rods. Determine the tension in the tie rod when concrete, having a density of 150 lb/ft^3, is in its liquid state. Assume that the lower edge of the upright panel, edge A, is simply supported.

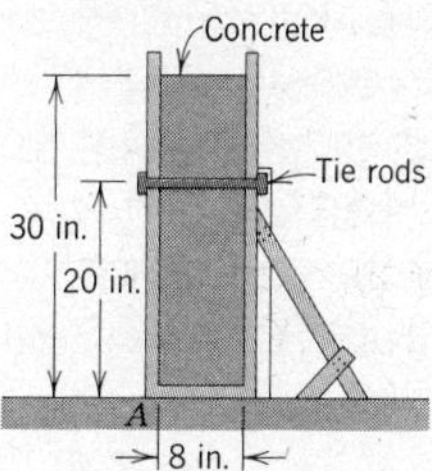

Prob. V.90

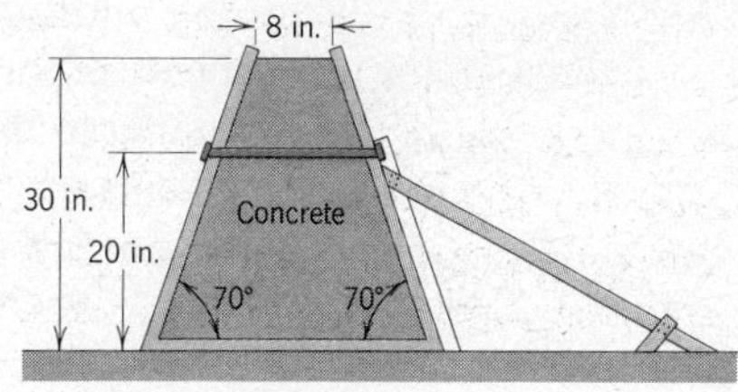

Prob. V.91

V.92 A 2-m-long tank, whose cross section is shown, is used to separate a pool of water from a pool of mercury. Four sets of opposing cables (only one set is shown in the cross-sectional view) support the metal panel separating the two fluids. The cables are slack when the tank is empty. Determine the force in the cables, specifying which side is tensioned, when the depth of mercury is (a) $d = 250$ mm, (b) $d = 500$ mm.

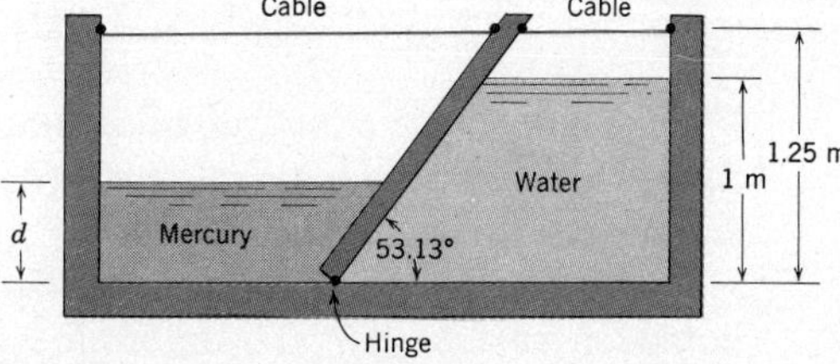

Prob. V.92

V.93 Water has accumulated in the trap of the oil storage tank, as shown. Determine the minimum horizontal force the hydraulic cylinder must exert to keep the 800-mm-wide gate shut.

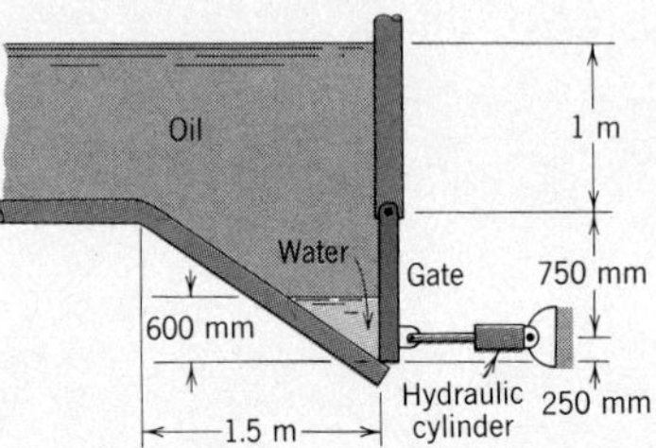

Prob. V.93

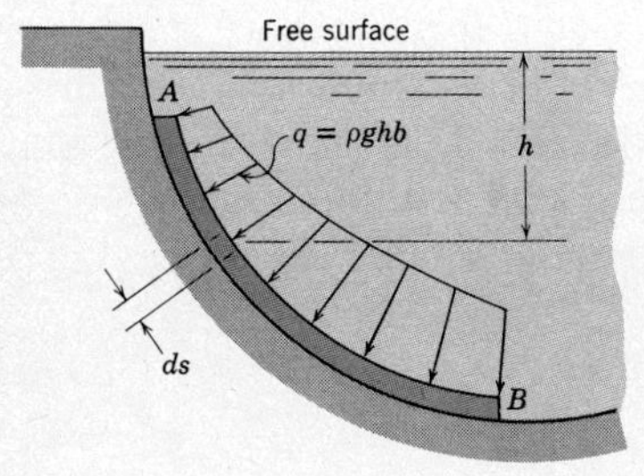

Figure 18

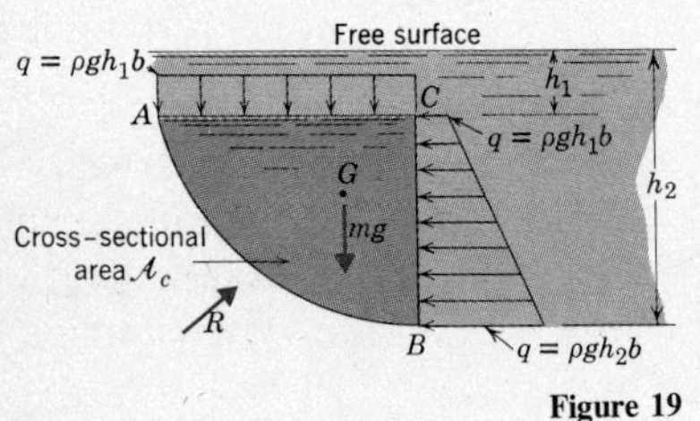

Figure 19

2 Cylindrical Surfaces

In this section we will consider a surface that can be described as a sector of a horizontal (arbitrary) cylinder. Such a surface, formed from the curve *AB*, is shown in the side view of Figure 18. Once again let the horizontal width of the surface be b.

For this case a line load may be formed by multiplying the pressure by the width b, as shown in Figure 18. The corresponding differential elements of force $\bar{q}\,ds$ change direction because the pressure at each point is parallel to the unit vector normal to the surface at each point. One way to account for this distributed force is to consider the horizontal and vertical components of $\bar{q}\,ds$ separately, expressing the slope of the curved surface at each location. However, a more convenient approach is available.

Consider the cylindrical body of liquid having cross-sectional area $\mathcal{A}_c$ depicted in Figure 19. For this body of liquid the line load is constant along the horizontal face *AC*, because the depth is constant for that face. The line load on the vertical face is trapezoidal. Another force acting on this body of liquid is its weight mg, which acts at center of mass G of the body. As viewed in Figure 19, the center of mass G coincides with the centroid of the cross-sectional area $\mathcal{A}_c$. Noting that this body has a constant width, we have

$$m = \rho \mathcal{V} = \rho \mathcal{A}_c b$$

Finally, to complete the free body diagram, we have the force $\bar{R}$, which is the reaction corresponding to the force exerted by the water on the curved surface *AB*.

The force $\bar{R}$ is what we seek when we wish to know the effect of the liquid on surface *AB*. To determine it we write the equations of static

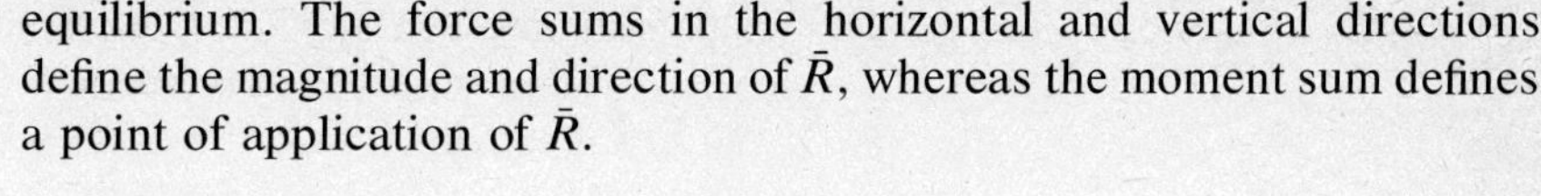

equilibrium. The force sums in the horizontal and vertical directions define the magnitude and direction of $\bar{R}$, whereas the moment sum defines a point of application of $\bar{R}$.

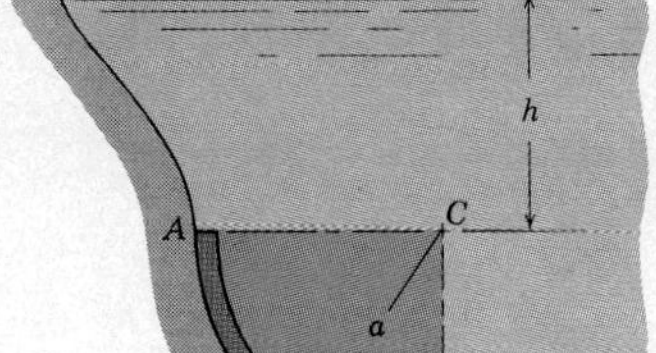

EXAMPLE 13

The cylindrical surface shown is formed from the circular arc AB. In this case, because the pressure at each point is normal to the local tangent plane, the distributed forces acting on the surface are concentric at the center of the arc, point C. It follows that the resultant must also pass through point C. Considering the equilibrium of the shaded element of fluid, prove this hypothesis and also determine the resultant force $\bar{R}$ acting on the cylindrical surface.

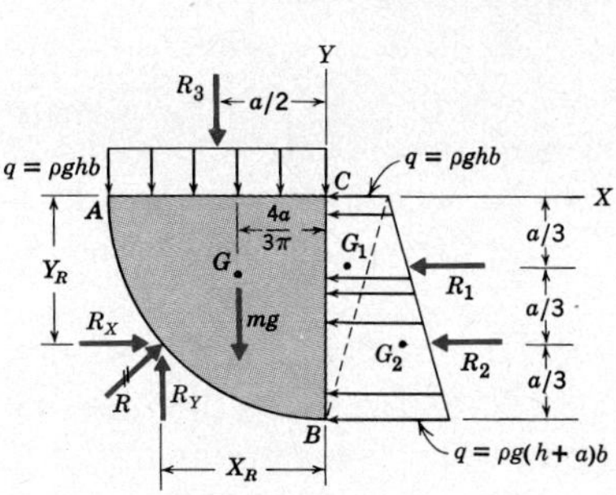

Solution

The free body diagram for the given body of liquid is shown to the left. For this diagram the location of the center of mass G is taken from Appendix B. Regarding the trapezoidal loading as the composite of the two triangular loadings shown, we have

$$R_1 = \tfrac{1}{2}(\rho ghb)a = \tfrac{1}{2}\rho ghab$$

$$R_2 = \tfrac{1}{2}[\rho g(h + a)b]a = \tfrac{1}{2}\rho g(ha + a^2)b$$

$$R_3 = (\rho ghb)a = \rho ghab$$

The cross-sectional area of the body of water is $\pi a^2/4$, so its mass is

$$m = \rho\, \mathcal{A}_c b = \frac{\pi}{4}\rho a^2 b$$

Choosing point C for a moment sum, the equilibrium equations are

$$\begin{aligned}
\Sigma\, M_{CZ} &= R_X(Y_R) - R_Y(X_R) - R_1\left(\frac{a}{3}\right) - R_2\left(\frac{2a}{3}\right) \\
&\quad + R_3\left(\frac{a}{2}\right) + mg\left(\frac{4a}{3\pi}\right) \\
&\equiv Y_R R_X - R_Y X_R - \tfrac{1}{6}\rho gha^2 b - \tfrac{1}{3}\rho g(ha + a^2)ab \\
&\quad + \tfrac{1}{2}\rho gha^2 b + \left(\frac{\pi}{4}\rho ga^2 b\right)\left(\frac{4a}{3\pi}\right) \\
&\equiv R_X Y_R - R_Y X_R = 0
\end{aligned}$$

$$\Sigma\, F_X = R_X - R_1 - R_2 \equiv R_X - \tfrac{1}{2}\rho g(2ha + a^2)b = 0$$

$$\Sigma\, F_Y = R_Y - mg - R_3 \equiv R_Y - \rho gab\left(\frac{\pi}{4}a + h\right) = 0$$

Because the moments of the other forces cancel each other, the moment equation states that the moment of $\bar{R}$ about point C is zero. This can be true only if the line of action of $\bar{R}$ intersects point C, thus proving the hypothesis.

The expression for R_X and R_Y are obtained by solving the force equilibrium equations. These give

$$R_X = \rho gab\left(h + \frac{1}{2}a\right)$$

$$R_Y = \rho gab\left(h + \frac{\pi}{4}a\right)$$

The force exerted by the water on the surface is the reaction corresponding to these values.

The problem did not request that we determine the value of X_R (or Y_R) defining the line of action of the resultant $\bar{R}$. If this information were desired, there are two ways in which we could proceed. The approach that is valid for all types of cylindrical surfaces, not just circular ones, is first to write the relationship between X_R and Y_R, which is this case is $Y_R = (a^2 - X_R{}^2)^{1/2}$. Substituting this relationship and the results for R_X and R_Y into the moment equation yields an equation for X_R.

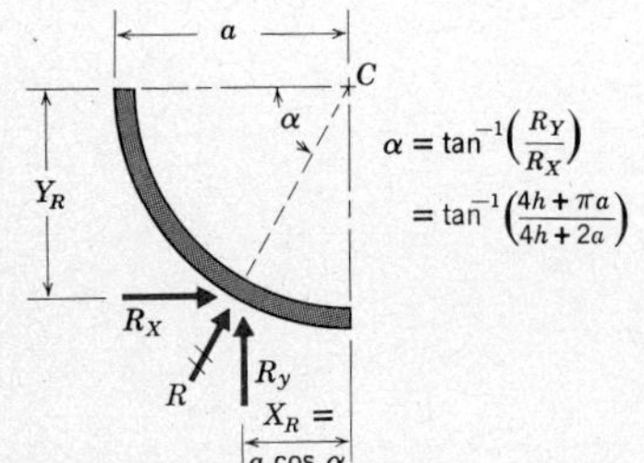

The alternative method is valid for circular cylindrical surfaces only. In this approach we perform a geometrical calculation using the fact that the resultant $\bar{R}$ intersects the center of the cylinder. This is illustrated in the sketch to the left.

EXAMPLE 14

The 3-ft-wide bent plate is supported by a horizontal shaft at edge B and bears against the bottom of the wall at edge A. Determine the reaction at edge A for a depth of water $d = 8$ ft.

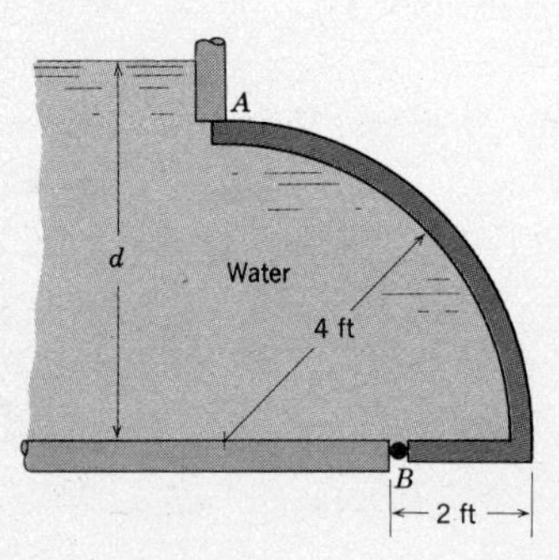

Solution

We could solve this problem in two parts. First, we could consider an isolated body of water to find the resultant, and then apply the resultant to the bent plate in order to evaluate the desired reaction. However, note that the problem is to determine the reaction force produced by the pressure of the water; it is not required that the resultant of the pressure force be found. This suggests that we can solve the problem in one step, by considering the equilibrium of the body of water and the plate *jointly*. In such an approach the resultant force $\bar{R}$, which is due to the interaction of the bent plate and the isolated body of water, becomes an internal force.

The isolated body of water we form is one having horizontal and vertical faces, with the remaining face being the curved surface of interest. The combination of this body of water and the bent plate is shown in the free body diagram on the following page.

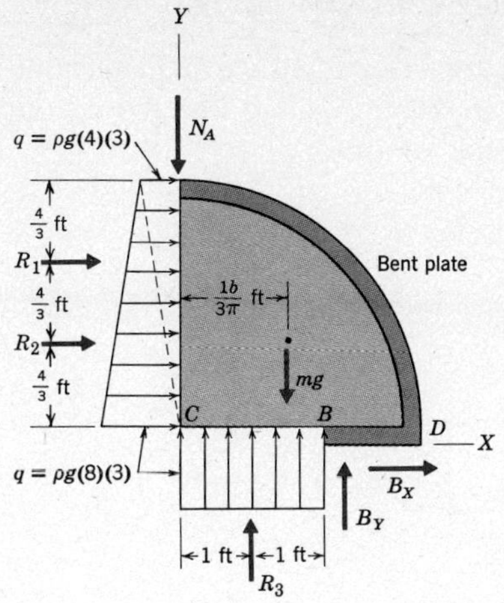

Note that on the horizontal face, the pressure acts only on the segment BC. For the segment BD the pressure is an internal force for the combination of bodies we have isolated.

Decomposing the trapezoidal loading into two triangular loadings, as indicated, we have

$$R_1 = \tfrac{1}{2}(12\rho g)(4) = 24\rho g \qquad R_2 = \tfrac{1}{2}(24\rho g)(4) = 48\rho g$$

$$R_3 = (24\rho g)(2) = 48\rho g$$

The weight of the body of water is

$$mg = \rho\left(\frac{\pi}{4}\right)(4^2)(3)g = 37.70\rho g$$

We only want the value of N_A, so we sum moments about shaft B. Thus

$$\Sigma M_{BZ} = -R_1\left(\frac{8}{3}\right) - R_2\left(\frac{4}{3}\right) - R_3(1) + mg\left(2 - \frac{16}{3\pi}\right) + N_A(2) = 0 \text{ lb-ft}$$

$$2N_A = 24\rho g\left(\frac{8}{3}\right) + 48\rho g\left(\frac{4}{3}\right) + 48\rho g(1) - 37.70\rho g(0.3023)$$

$$N_A = 82.30\rho g = 5.136 \text{ kips}$$

where the value of ρg in U.S.-British units is 62.4 lb/ft^3. The value of this reaction demonstrates the enormous forces required to contain large bodies of water.

HOMEWORK PROBLEMS

V.94 Cross sections of two alternative designs for the face of a dam are shown. Considering a 1-m-wide section, resolve the pressure of the water into an equivalent force-couple system at the base B when the water is at the level of the top of the dam.

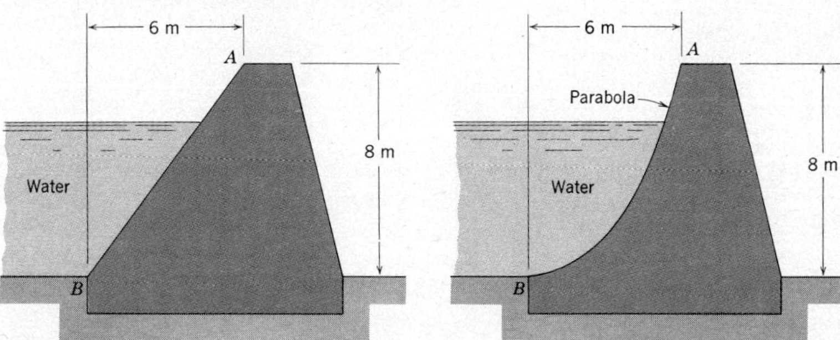

Prob. V.94

V.95 and V.96 A viewing window in an aquarium is in the form of a 3-m-long semicylinder, as shown in the cross-sectional view. Determine the magnitude, direction, and point of application along line AB of the resultant force exerted by the water on the window.

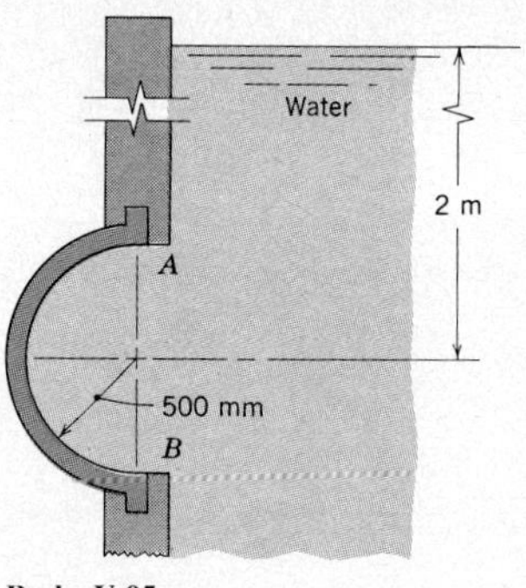

Prob. V.95

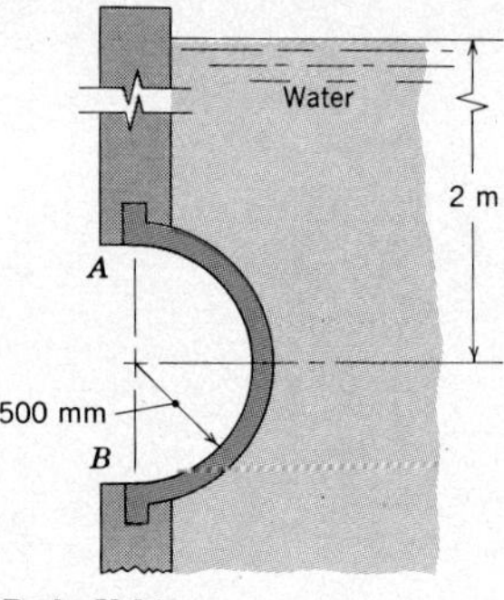

Prob. V.96

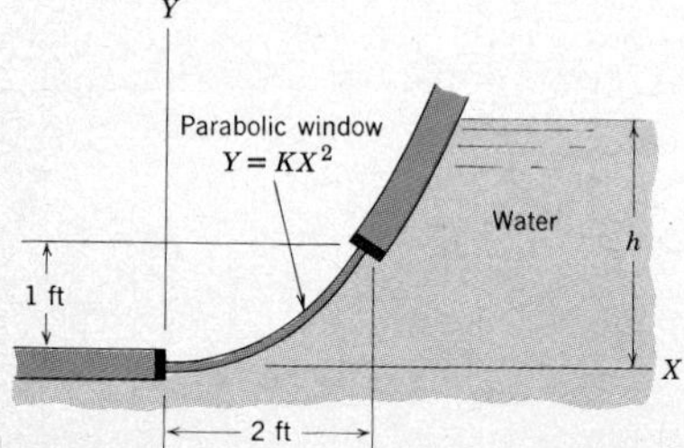

Prob. V.97

V.97 A window in the hull of a boat, through which fish can be observed, has the parabolic cross section shown. The window is 4 ft wide. Knowing that the depth $h = 2.5$ ft, determine the resultant force exerted on the window by the water. Specify the point of application on the window of this resultant.

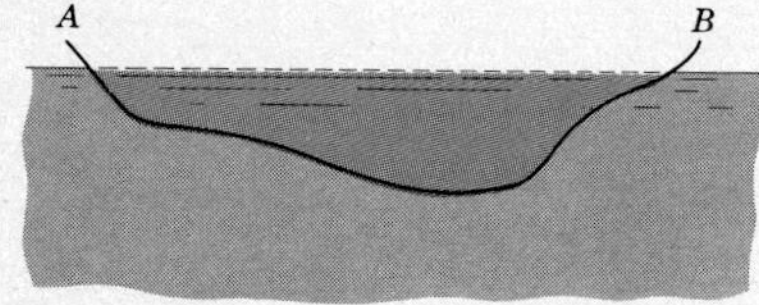

Prob. V. 98

V.98 According to Archimedes' principle of buoyancy, a liquid exerts an upward force on a body equal to the weight of the liquid displaced by the body. Demonstrate that this principle applies to the arbitrary cylindrical body shown, which is formed from the curve AB. Also determine the line of action of this buoyant force. The colored area defines the body of liquid that has been displaced.

V.99–V.101 A water conduit has the cross section shown. The conduit is supported by hinges along the lower edge A and by cables spaced at 400-mm intervals along edge B. Determine the tension in the cables when the conduit is filled to the top.

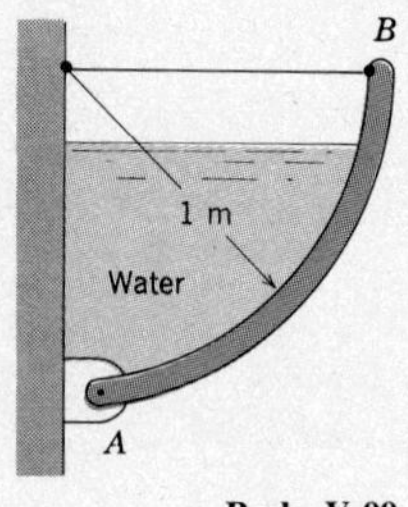

Prob. V.99

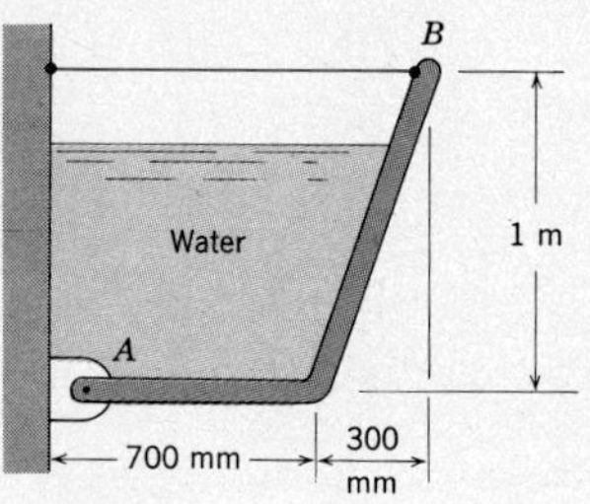

Prob. V.100

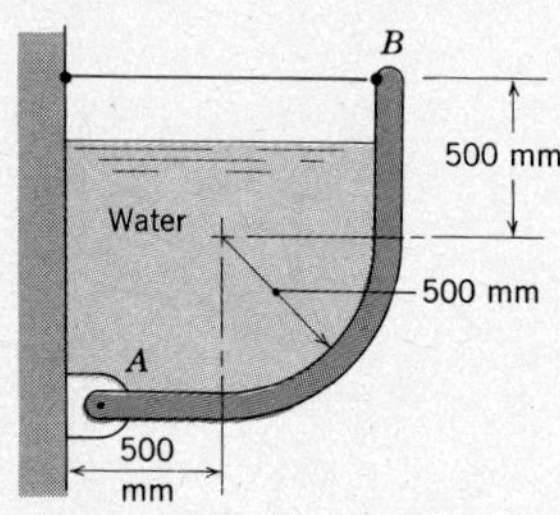

Prob. V.101

V.102 and V.103 A quarter-circular plate 6 ft long is used as a gate for a dam. Determine the minimum horizontal force exerted by the hydraulic cylinder needed to open the gate when the depth of the water is (a) $d = 3$ ft, (b) $d = 9$ ft.

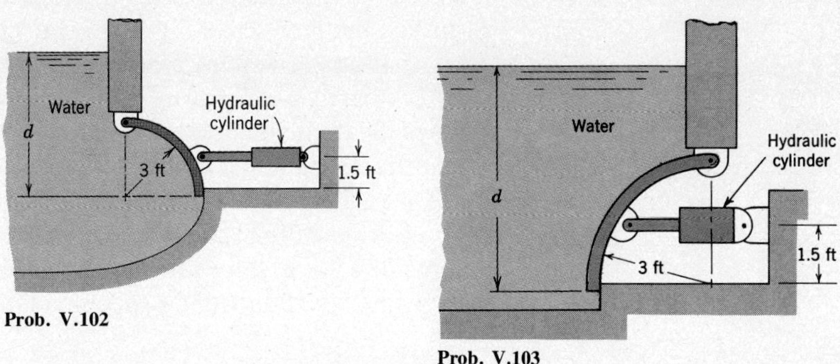

Prob. V.102

Prob. V.103

V.104 A 20-m-long section of a breakwater has the parabolic cross section shown. Determine the force-couple system at edge C representing the reaction of the *ground on the breakwater* when flooding is about to occur. The density of seawater is 1025 kg/m³ and that of concrete is 2400 kg/m³.

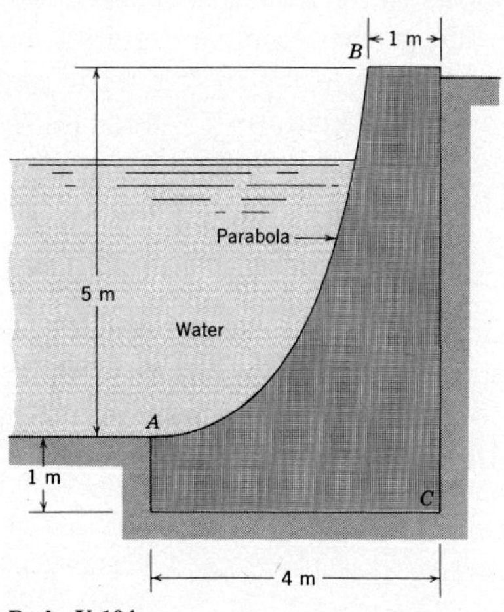

Prob. V.104

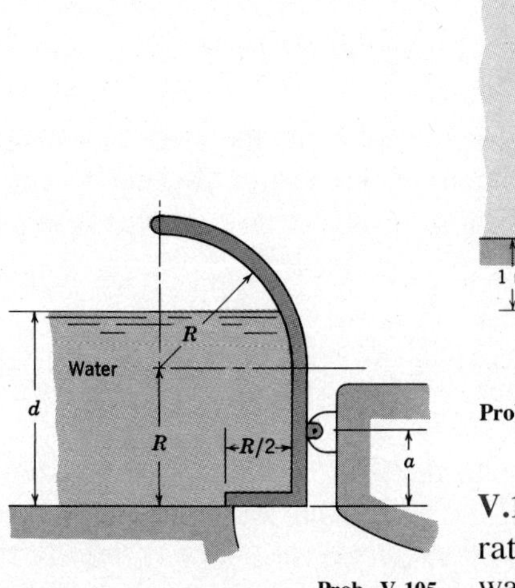

Prob. V.105

V.105 The cross section of a gate in a channel is shown. Determine the ratio of dimensions a/R for which the gate will open when the depth of water in the channel is $d = 2R$.

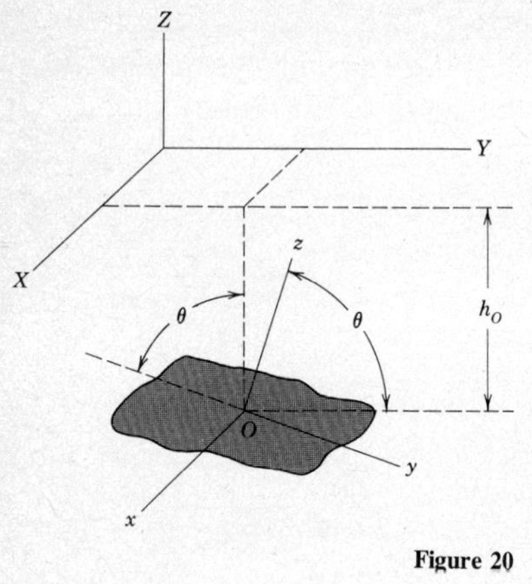

Figure 20

3 Arbitrary Planar Surfaces

We were able to treat the pressure distributions appearing the last two sections by converting them into line loads. In contrast, when the surface in contact with the liquid has an arbitrary width, we must deal with the actual pressure distribution. Consider the situation shown in Figure 20, where the XY plane locates the free surface and the colored area is a flat surface inclined at an angle θ from the vertical.

For convenience in locating points on this inclined surface, we define an auxiliary coordinate system xyz, for which the xy plane coincides with the plane of the inclined surface, with the x axis horizontal. It then follows that the yz plane is vertical and that the angle between the negative y axis and a vertical line is θ, as shown.

Because the x axis is horizontal, all points in the surface at a constant value of y are at the same depth. Letting h_O denote the depth of the origin O, whose location will be specified later, the depth of other points is given by

$$h = h_O + y \cos \theta$$

It follows that the pressure distribution is

$$p = \rho g h = \rho g (h_O + y \cos \theta) \tag{16}$$

To describe the resultant of this pressure distribution we could resort to the analogy with the geometric properties of the pressure space, as developed in Section B.2 of this module. However, except in rare situations, the pressure space resulting from equation (16) will not be easily represented in terms of the basic shapes in Appendix C. We therefore shall employ equations (13) and (14) found in the Section B.2. Letting $d\mathcal{A}$ be an element of area for the inclined surface, equation (13) gives

$$R = \int_{\mathcal{A}} p \, d\mathcal{A} = \rho g h_O \int_{\mathcal{A}} d\mathcal{A} + \rho g \cos \theta \int_{\mathcal{A}} y \, d\mathcal{A}$$

Clearly, the first integral on the right side gives the area $\mathcal{A}$, whereas the second integral is the first moment of area with respect to the y coordinate. In order to eliminate the latter term, we now *locate the origin* O *at the centroid of the surface area* $\mathcal{A}$. The result is

$$\boxed{R = \rho g h_O \mathcal{A}} \tag{17}$$

This resultant force is perpendicular to the plane in the negative z direction. An interpretation of equation (17) is to say that $\rho g h_O$, where h_O is the depth of the centroid of the area, is the *average pressure* acting on the area.

Now that the resultant force has been determined, we may proceed to locate the center of pressure. Note that $\bar{R}$ does *not* act at the centroid. Equations (14) give

$$Rx_R \equiv \rho g h_0 \mathcal{A} x_R = \int_{\mathcal{A}} xp \, d\mathcal{A}$$

$$= \rho g h_0 \int_{\mathcal{A}} x d\mathcal{A} + \rho g \cos\theta \int_{\mathcal{A}} xy \, d\mathcal{A}$$

$$Ry_R \equiv \rho g h_0 \mathcal{A} y_R = \int_{\mathcal{A}} yp \, d\mathcal{A}$$

$$= \rho g h_0 \int_{\mathcal{A}} y \, d\mathcal{A} + \rho g \cos\theta \int_{\mathcal{A}} y^2 \, d\mathcal{A}$$

As a result of the fact that the origin is located at the centroid, the first moments of area in the foregoing equations vanish. Let us define the following symbols:

$$I_x = \int_{\mathcal{A}} y^2 \, d\mathcal{A} \qquad I_{xy} = \int_{\mathcal{A}} xy \, d\mathcal{A} \tag{18}$$

The expressions for x_R and y_R may then be written as

$$x_R = \frac{I_{xy} \cos\theta}{h_0 \mathcal{A}} \qquad y_R = \frac{I_x \cos\theta}{h_0 \mathcal{A}} \tag{19}$$

The quantity I_x is called the *second moment of area about the x axis*, or alternatively, the *area moment of inertia about the x axis*. The reason for associating it with the x axis is that y is the distance of an element of area from the x axis. The value of y^2 can never be negative, so the integrand for I_x is never negative, and the value of I_x will always be a positive quantity. Hence, the center of pressure is never higher then the centroid of the surface area.

The parameter I_{xy} is the *area product of inertia*. This quantity is studied in detail in Module IX. Briefly, I_{xy} is a measure of the asymmetry of the area $\mathcal{A}$. If the area is symmetrical with respect to either the x or y axes, then $I_{xy} = 0$. In that case equations (19) give $x_R = 0$, which means that the center of pressure is situated on the y axis. In all of the problems we shall consider in this section the area will have this type of symmetry.

Values of the area moments of inertia for some basic geometric shapes about centroidal axes are given in Appendix B. Methods for calculating this property for geometric shapes not appearing in Appendix B are described in Module IX.

Before we apply equations (17) and (19), let us make some fundamental observations regarding them. The fact that the pressure at the centroid of the area is the average for the entire area is logical. What is surprising is that the angle of inclination θ does not affect the magnitude of the resultant. The expression for y_R shows that, for a given area $\mathcal{A}$, the center of pressure moves upward (y_R decreases) if the depth h_O or the angle of inclination θ is increased. These tendencies are a result of the fact that an increase of either of these parameters increases the ratio of the pressure of the highest point on the area to the pressure at the lowest point. It is also interesting to note that similar expressions arise in the study of mechanics of materials when the state of stress in beams carrying axial and transverse loads is studied.

EXAMPLE 15

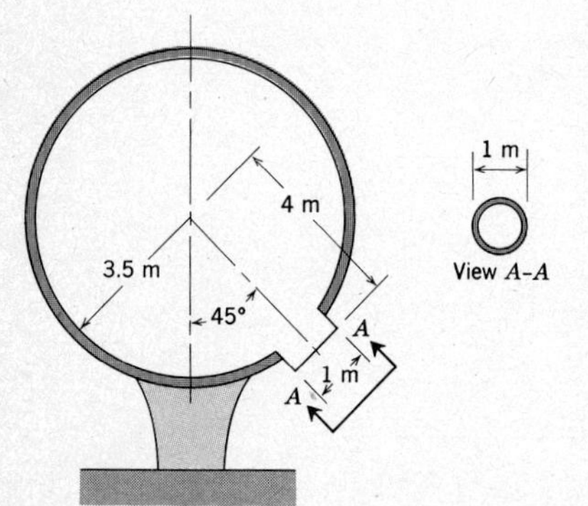

The 1-m diameter circular plate allows access to the interior of the spherical water tank when the tank is empty. Determine the resultant water force exerted on the plate and the location of the center of pressure when the water tank is full, but a free surface is still present.

Solution

We need a coordinate system xyz for the circular plate. In order to employ equations (17) and (19), we let x and y coincide with the plane of the plate with the x axis horizontal. Locating the origin at the centroid of the plate, which of course is the center of the circle, we obtain the coordinate system depicted in the next sketch.

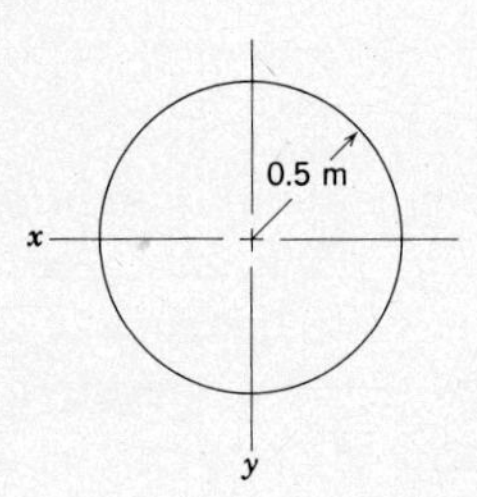

From the diagram accompanying the statement of the problem, we calculate that when the free surface is at the top of the tank, the depth of the center of the plate is

$$h_O = 3.5 + 4.0 \cos 45° = 6.328 \text{ m}$$

The angle of inclination of the plate is seen to be $\theta = 45°$. From Appendix B we have

$$\mathcal{A} = \pi(0.5)^2 = 0.7854 \text{ m}^2$$

$$I_x = \frac{\pi}{4}(0.5)^4 = 0.04909 \text{ m}^4$$

Also, the plate is symmetric about both the x and y axes, so we know that the center of pressure is situated on the y axis. (Symmetry about one axis is sufficient for this.)

Setting ρ = 1000 kg/m^3, equation (17) gives

$$R = 1000(9.806)(6.328)(0.7854) = 49.7 \text{ kN}$$

For the location of the center of pressure, equation (19) gives

$$y_R = \frac{0.04909(\cos 45°)}{6.328(0.7854)} = 6.98(10^{-3}) \text{ m}$$

Hence, the center of pressure is 6.98 mm below the center of the 500-mm-radius plate.

HOMEWORK PROBLEMS

V.106 A circular window in an aquarium is situated on a vertical wall. Determine the resultant force of the water on the window and the center of pressure if the depth is (a) d = 1 ft, (b) d = 10 ft.

Prob. V.106

V.107 Three possibilities for a throttle valve for the 1-m inside diameter circular pipe leading from the water tank are shown. In each design the plate forming the valve is pivoted about the horizontl axis BC. For a depth d = 1 m, determine the force exerted by the water on each plate and the corresponding center of pressure.

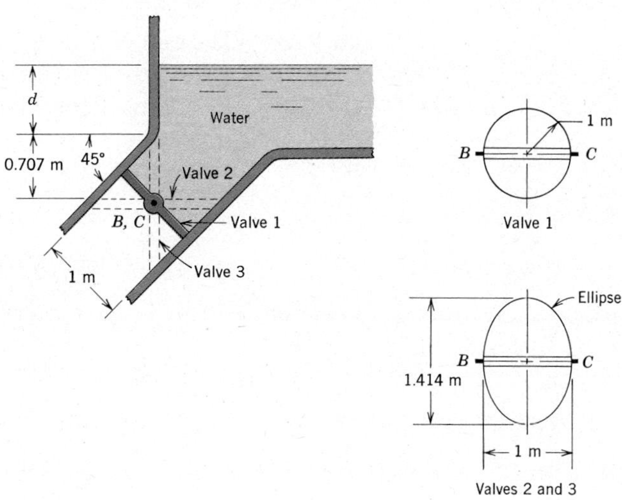

Prob. V.107

V.108 Solve Problem V.107 when d = 0.

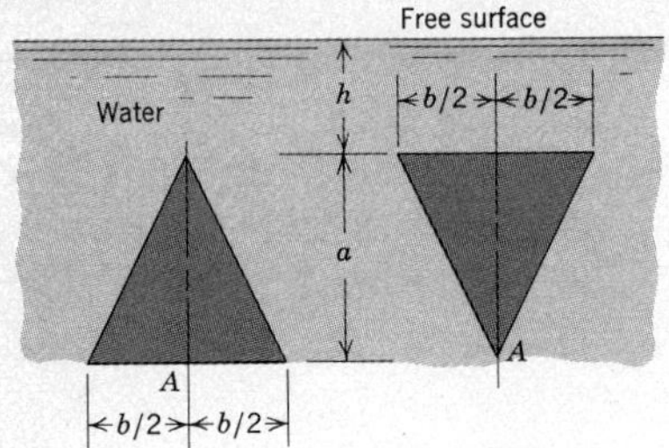

Prob. V.109

V.109 Two identical triangular plates are given opposite orientations on a vertical wall in the tank of water, as shown. (a) Determine the resultant force of the water on each plate and the corresponding center of pressure. (b) Replace the resultant forces determined in part (a) by equivalent force-couple systems acting at point A. Give a physical explanation for the difference in the results for the two plates.

V.110–V.112 The flat plate shown fits into a matching opening in the vertical wall of a tank of oil. The density of the oil is 800 kg/m³. The plate is hinged about the horizontal axis AB. Determine the force $\bar{P}$ applied at point C required to keep the plate in place.

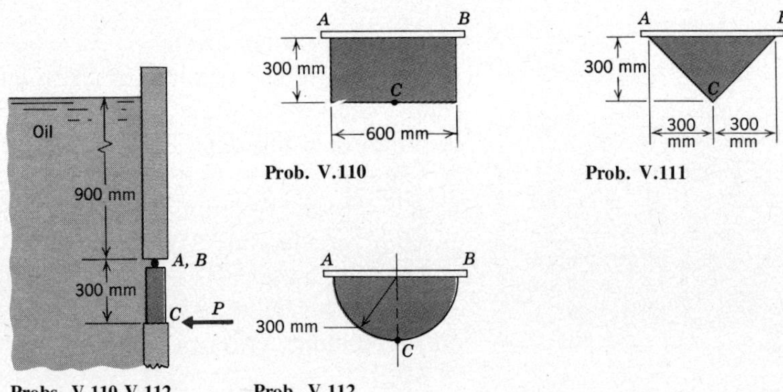

Probs. V.110-V.112 Prob. V.110 Prob. V.111 Prob. V.112

V.113 and V.114 The end of a water trough is inclined at 36.87° from the vertical. The end plate shown is attached to the walls of the trough by bolts at points A, B, and C. Determine the tension in the bolts when the trough is filled to the top.

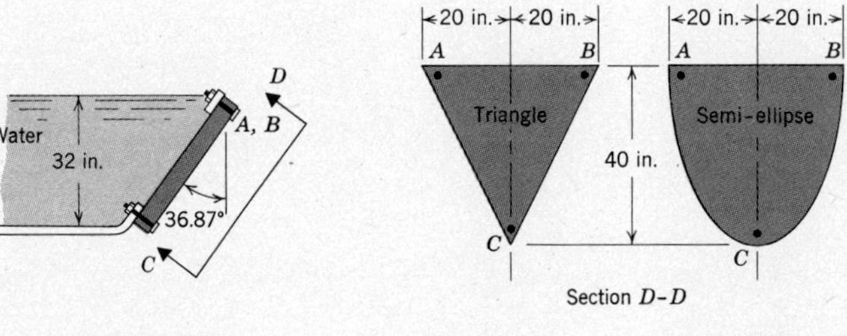

Probs. V.113 and V.114 Prob. V.113 Prob. V.114

MODULE VI
INTERNAL FORCE ANALYSIS

The focus of Module IV was structures and machines whose interconnected members support and modify the loads that are applied to them. In the process of studying the static equilibrium of such systems, we did not consider how the forces applied at one location on a member were transmitted internally to another location on the member. The determination of the internal forces is the subject of this module.

In earlier modules we implicitly considered the internal forces in some structural members. For example, a lightweight cable that is loaded only at its ends carries a constant tension force. We have also seen that a straight two-force member in a truss carries a constant axial tensile or compressive force. In fact, one of the methods we formulated for analyzing the forces in the members of a truss, the method of sections, is typical of the type of analysis we will perform here.

Recall that in the method of sections we passed an imaginary cutting plane through a truss in order to have the internal axial forces in certain members appear as external forces acting on the isolated section. The equilibrium equations for the isolated section then provided the required relations for determining the internal forces.

The methods we shall develop in this module will extend this type of analysis to treat more general types of structural members. We will utilize cutting planes to isolate a portion of the member of interest, in order to make the internal forces appear explicitly in a free body diagram. It is apparent that this approach is necessary, for the equations of static equilibrium only relate the external forces acting on a system.

Knowledge of the distribution and flow of internal forces in members of a structure is essential for their design. In the developments that follow we shall consider two of the basic elements of a structural system: (1) *beams* and (2) *cables* that serve as multiple force members.

A. BEAMS

1 Basic Definitions

Certain bodies occurring in previous modules were referred to as beams, without really considering the precise meaning of this term. A primary physical characteristic of a beam is that it is a slender bar. That is, the arc length of some line in the body is considerably larger than any dimension measured perpendicular to that line.

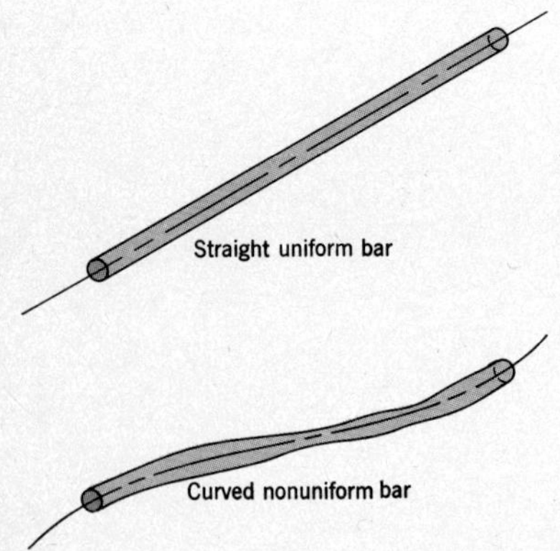

Figure 1

The line whose arc length we measure is the locus of the centroids of the cross sections. The bar is said to be uniform or nonuniform according to whether or not the cross sections have a constant shape. Also, the bar is said to be straight or curved in accordance with the shape of its centroidal axis, as depicted in Figure 1.

The property of slenderness is not sufficient to identify that a bar acts like a beam, although all beams must have that property. The second item of importance is the type of internal forces the bar transmits. To investigate this matter let us pass an imaginary cutting plane through an arbitrary bar and consider the force system exerted by one portion of the cut bar on the other.

In the most general situation the internal forces will be equivalent to a force-couple system. The point of reference that we choose for this force-couple system is the centroid of the cross-sectional area. The reason for choosing the centroid is that many formulas derived in mechanics of materials to describe the load-carrying ability of bars require such a resolution for the internal forces. Thus, considering the cut bar in Figure 2, the internal forces exerted on the cross-sectional surface to the right of the cutting plane by the portion of the bar to the left of the plane are equivalent to a resultant force $\bar{R}$ acting at the centroid C and a couple $\bar{M}_C$.

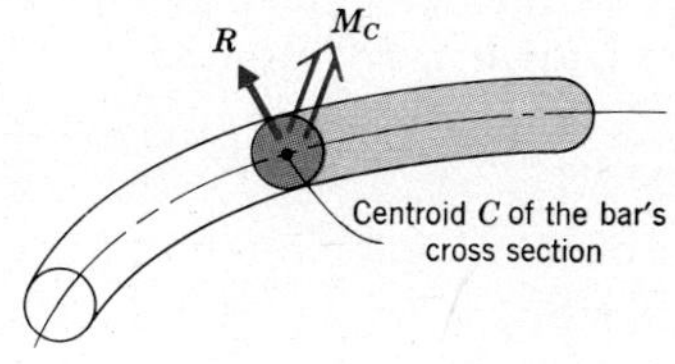

Figure 2

Although they are not depicted in Figure 2, it follows from Newton's third law that the internal forces exerted by the portion of the bar to the right of the cutting plane on the exposed cross section to the left of the cutting plane consist of a resultant force $-\bar{R}$, applied at the centroid, and a couple $-\bar{M}_C$.

To describe the internal forces further, let us choose an xyz coordinate system having its origin at the centroid C. The x axis is chosen to be outward from the exposed face, tangent to the centroidal axis of the bar. In other words, the x axis is the outward normal to the cross section. It follows that the yz plane is coincident with the cross section.

Lowercase letters are used to denote this coordinate system in order to emphasize that it may be different from the XYZ coordinate system used to evaluate the reactions acting on the beam. Also, aside from the case of a straight bar, where the cross sections all lie in parallel planes, the orientation of the xyz system will depend on which cross section of the bar is being considered.

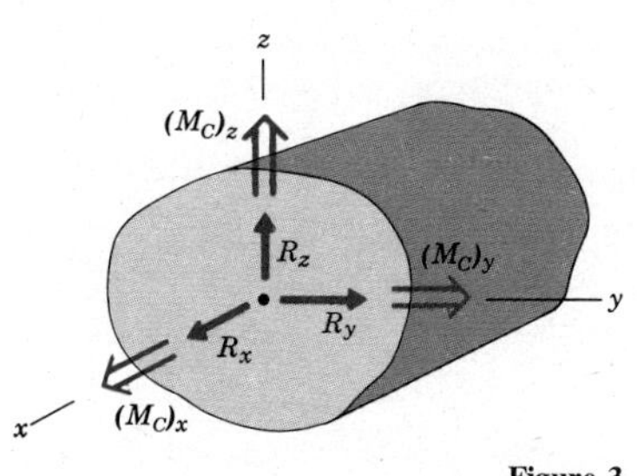

Figure 3

We shall use the xyz system to express the internal force $\bar{R}$ and couple $\bar{M}_C$ in terms of their components. Figure 3 depicts a blow-up of the cross section featured in Figure 2.

Special terms are used to describe the components of the internal force and moment. The force components $\bar{R}_y$ and $\bar{R}_z$, which are parallel to the cross section, are called the *shear forces*. As was true when we treated straight two-force members, the component $\bar{R}_x$, normal to the cross section, is called a *tensile* or *compressive* force, according to whether $\bar{R}_x$ is in the same or opposite sense of the x axis. The components of the internal moment parallel to the cross section, that is, $(\bar{M}_C)_y$ and $(\bar{M}_C)_z$, are called *bending moments* because they have the effect of tending to bend the bar. Finally, the component $(\bar{M}_C)_x$ normal to the cross section is called the *torsional moment* because it has the effect of tending to twist the bar about its centroidal axis.

The primary factor in deciding whether a bar acts like a beam is that

> In a beam the shear forces and bending moments are significant in comparison to the axial (tensile or compressive) force and the torsional moment.

Clearly, the foregoing definition excludes cables and the straight members of a truss, for both transmit only axial forces. It also excludes a straight bar, frequently called a torsion bar, whose only loads are torsional moments.

The majority of multiforce members in the frames and machines we analyzed in Module IV are beams. Beams of all descriptions may be found in the structural framework of buildings, automobiles, trucks and airplanes. They are crucial to many industrial machines, and parts of many hand tools, such as the handles of pliers and wrenches, which function like beams. A beam is one of the fundamental structural elements for a designer.

The basic method we will develop for evaluating the internal forces in a beam will be equally valid for curved and straight beams. We shall focus attention on straight beams, because of the simplicity of their analysis and because of the frequency with which they occur in practice. The added difficulty in analyzing curved beams is merely that of describing the geometry of its curvature.

It frequently proves convenient to classify straight beams according to the manner in which they are supported. Some typical types are displayed in Figure 4. They have been depicted in the horizontal position solely for convenience; in practice, beams may have any orientation in space.

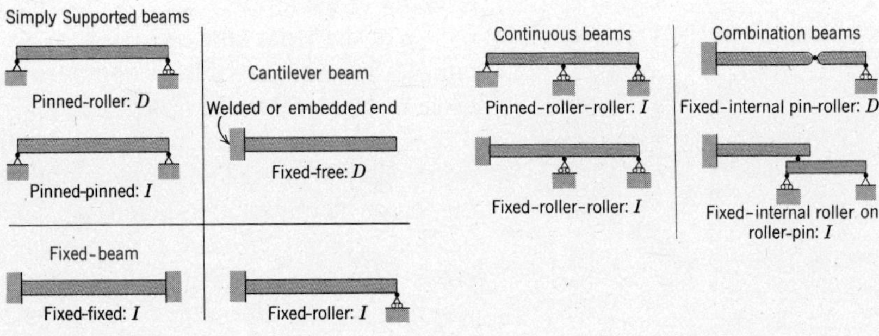

Figure 4

All except one of the beams exhibited in Figure 4 have two names. The more descriptive name explicitly states the type of supports, whereas the other name (heading each entry) is the one more commonly employed by engineers.

Another feature of the beams in Figure 4 is that in several cases, the supports exert more reactions than the number of available static equilibrium equations. These are statically indeterminate systems whose analysis requires consideration of the deformation of the beam caused by the beam loading. Note that each system in Figure 4 is labeled *D* or *I* to indicate that it is statically determinate or indeterminate, respectively. Statically indeterminate systems are treated in courses on mechanics of-materials and structures. We treat only statically determinate systems in this text.

2 Determination of the Internal Forces

A beam may be subjected to concentrated or distributed loads only, or to a combination of both. In view of the fact that we are restricting ourselves to the consideration of statically determinate beams, it follows that the reaction forces exerted by the supports of a beam to prevent the beam from moving can be fully determined by using the methods of Modules III, IV, and V to formulate the equations of equilibrium. After such a determination, we may treat the reactions as additional known loads.

At this juncture, we can discuss straight and curved beams with equal ease, so let us consider an arbitrary curved beam, for which the full set of loads, including the reactions, are known. Suppose that we wish to know the internal forces acting on a cross section at some specific location *C* along the centroidal axis of the beam. This determination requires that the beam be (fictitiously) separated into two portions by a cutting plane at location *C*, as depicted in Figure 5.

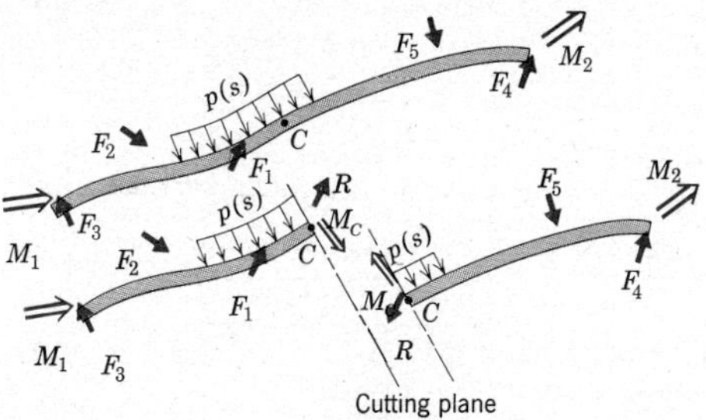

Figure 5

A significant feature of Figure 5 is that all forces acting on the portions of the beam to either side of the cutting plane are depicted. Thus, with the inclusion of the internal force $\bar{R}$ and the couple $\bar{M}_C$, Figure 5 shows a free body diagram for each portion of the cut beam.

From Figure 5, it is apparent that the internal force $\bar{R}$ and couple $\bar{M}_C$ are merely the force-couple system acting at the centroid *C* of the cross section required to hold each portion of the cut beam in equilibrium. We have stipulated that the entire beam is in static equilibrium under the system of external forces. It follows that the equilibrium equations for either cut portion of the beam give the same force $\bar{R}$ and couple $\bar{M}_C$, because the equilibrium of the entire beam and either portion guarantees the equilibrium of the other portion.

When we write the moment equilibrium equation, it is advantageous to sum moments about point C (the centroid of the cut cross section) in order to eliminate the unknown force $\bar{R}$ from the equation. Thus, the process of determining the internal force-couple system acting on a specific cross section of a beam can be summarized as follows.

> First, isolate in a free body diagram the portion of the beam and of the loading system to *either* side of the cutting plane for that cross section. Then, formulate the equilibrium equations for the chosen portion of the beam by summing moments about the centroid of the cross section and summing forces for the isolated portion.

It should be observed that two coordinate systems are utilized in our formulation: the XYZ that is used for determining the reactions and the xyz that is used for describing the internal forces. For simplicity, we shall make it standard practice to also employ the XYZ coordinate system when formulating the static equilibrium equations for the portion of the beam isolated by the cutting plane.

EXAMPLE 1

Determine the internal forces acting on the cross section of bar ABC at location E. Neglect the weight of the bar.

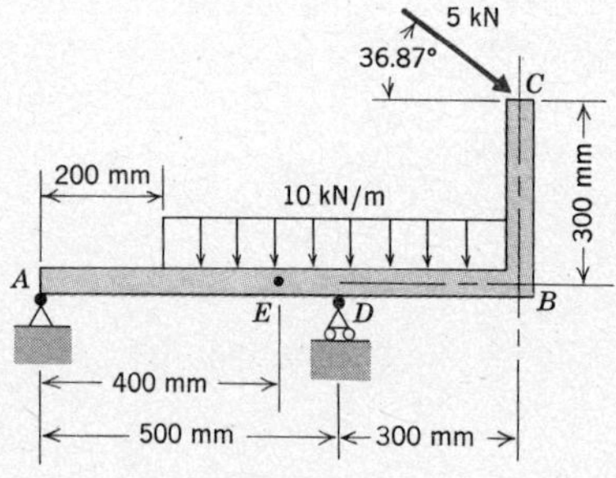

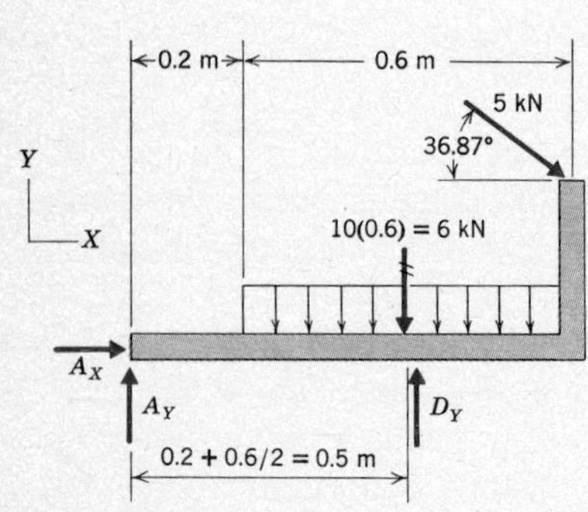

Solution

We first determine the reaction forces. For this, we draw a free body diagram replacing the 10-kN/m distributed load by its resultant acting at the centroid of the area of the loading function. The equilibrium equations for the entire beam are formulated in terms of any convenient coordinate system, such as the one shown. Choosing pin A for the moment sum, we have

$$\Sigma M_{AZ} = D_Y(0.5) - 6.0(0.5) - (5 \cos 36.87°)(0.3) - (5 \sin 36.87°)(0.8) = 0$$

$$\Sigma F_X = A_X + 5 \cos 36.87° = 0$$

$$\Sigma F_Y = A_Y + D_Y - 6.0 - 5 \sin 36.87° = 0$$

These equations give

$$D_Y = 13.2 \qquad A_X = -4 \qquad A_Y = -4.2 \text{ kN}$$

Now that the reactions have been found, we pass a cutting plane through point E, perpendicular to the centroidal axis of the bar. Either portion of the cut bar may be chosen for consideration; we will choose the left end because there are fewer forces on that side. In the free body diagram for the isolated portion chosen, the reaction forces at pin A are depicted in the direction in which they were calculated to act.

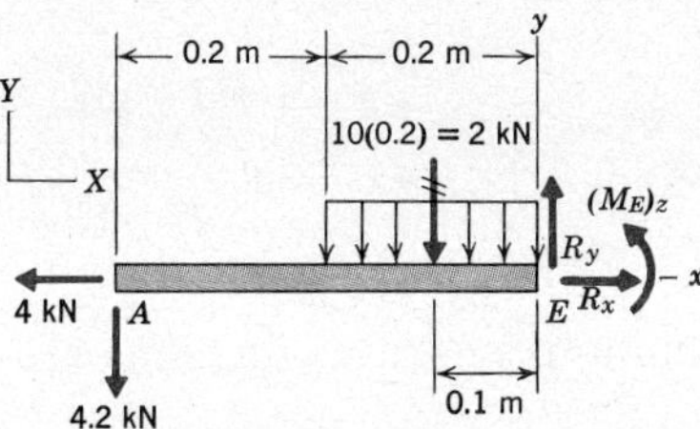

It should be noted that the xyz coordinate system illustrated in the sketch is chosen such that the x axis is the outward normal for the exposed face. We use this system solely for describing the internal force components. Because this is a planar problem, it can be seen that the only non-zero internal reactions are the axial force R_x, the shear force R_y, and the bending moment $(M_E)_z$. The XYZ coordinate system, which is also shown in the sketch, shall be employed for the formulation of the static equilibrium equations.

The free body diagram contains only that portion of the distributed load acting on the isolated section. The nontrivial static equilibrium equations are

$$\Sigma M_{EZ} = 2(0.1) + 4.2(0.4) + (M_E)_z = 0$$

$$\Sigma F_X = R_x - 4.0 = 0 \qquad \Sigma F_Y = -2.0 - 4.2 + R_y = 0$$

which give

$$(M_E)_z = -1.88 \text{ kN-m} \qquad R_x = 4.0 \text{ kN} \qquad R_y = 6.6 \text{ kN}$$

These results may be interpreted to mean that at point E the beam carries a tensile force (positive R_x) of 4.0 kN, a vertical shear force of 6.6 kN, and a bending moment of 1.88 kN-m about the horizontal axis parallel to the cross section.

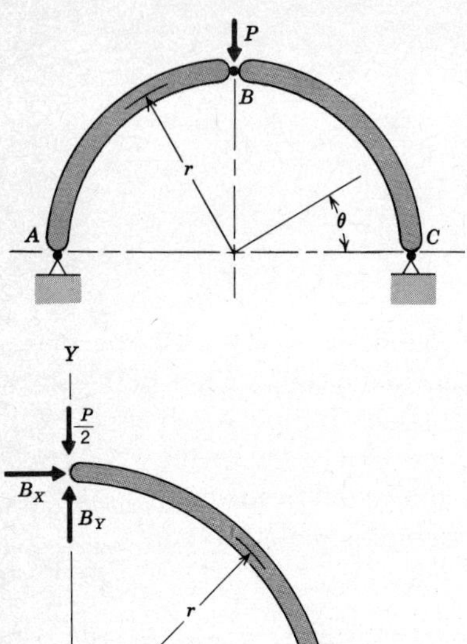

EXAMPLE 2

The three-pinned arch shown carries a vertical force $\bar{P}$ at pin B. Determine the internal forces acting on the cross section at $\theta = 30°$.

Solution

The arch consists of two curved bars AB and BC. The analysis of internal forces requires that we first find the reactions. By symmetry (or equivalently, from the equilibrium equations for the entire structure), we find that the vertical reactions at pins A and C are each $P/2$ upward. We also find that the horizontal reactions at pins A and C are equal in magnitude and opposite sense. To determine these reactions, as well as the force exerted between the two bars at pin B, we draw a free body diagram for one bar and write the corresponding equations for static equilibrium.

$$\Sigma M_{BZ} = C_X(r) + \frac{P}{2}(r) = 0$$

$$\Sigma F_X = C_X + B_X = 0$$

$$\Sigma F_Y = B_Y + \frac{P}{2} - \frac{P}{2} = 0$$

The solutions of these equations are

$$B_Y = 0 \qquad B_X = -C_X = \frac{P}{2}$$

Now that the reactions are known, we pass a cutting plane through the bar at the location where we wish to determine the internal forces. We have chosen to isolate the lower portion of the beam formed by the cutting plane, as shown in the free body diagram to the left. An xyz coordinate system with the x axis being the outward normal for the cross section is selected, as illustrated in the free body diagram. The internal forces are depicted as an axial force F in the sense of the x axis, a shear force V in the sense of the y axis, and a bending moment M about the z axis. There are no other components of internal force resulting from the planar system of forces.

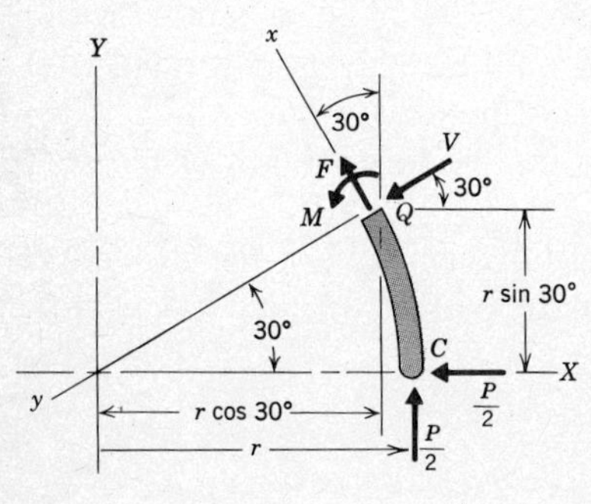

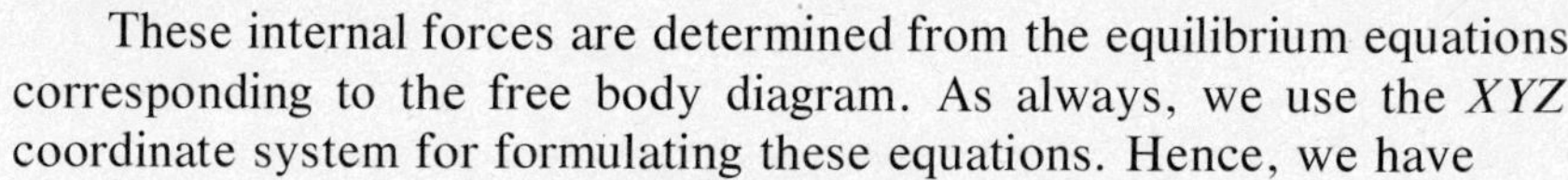
These internal forces are determined from the equilibrium equations corresponding to the free body diagram. As always, we use the XYZ coordinate system for formulating these equations. Hence, we have

$$\Sigma M_{QZ} = M + \frac{P}{2}(r - r\cos 30°) - \frac{P}{2}(r\sin 30°) = 0$$

$$\Sigma F_X = -V\cos 30° - F\sin 30° - \frac{P}{2} = 0$$

$$\Sigma F_Y = -V\sin 30° + F\cos 30° + \frac{P}{2} = 0$$

These equations yield

$$V = -0.1830P \qquad F = -0.6830P \qquad M = 0.1830Pr$$

The axial force is negative, so it is compressive. It should be noted that members AB and BC are two-force members, because loads are applied to them only at their ends. However, because the bars are curved there is internal shear and a bending moment. Such internal forces are not present in straight two-force members.

HOMEWORK PROBLEMS

VI.1–VI.14 Determine the axial and shear forces and the bending moment acting on the cross section at (a) location A, (b) location B for the bar shown.

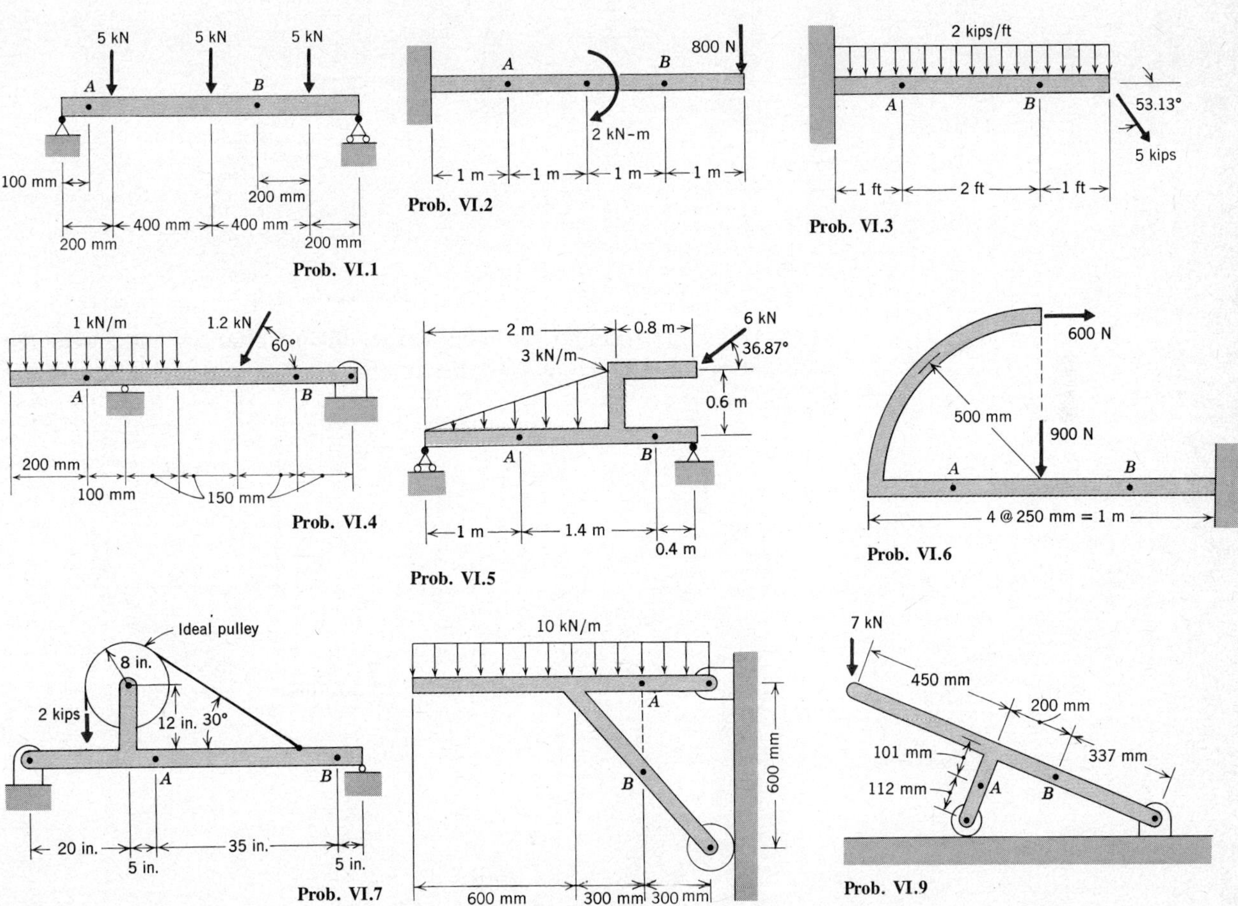

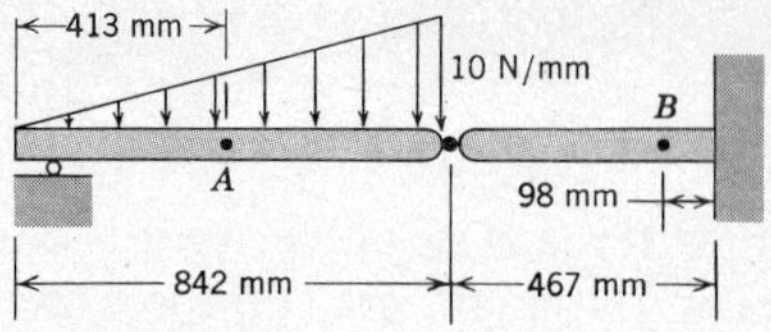

Prob. VI.10

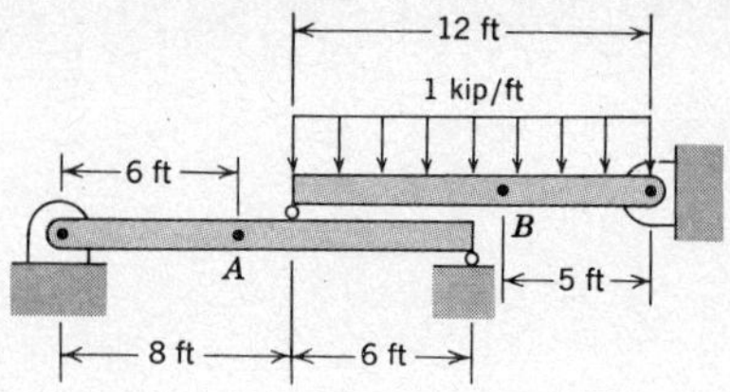

Prob. VI.11

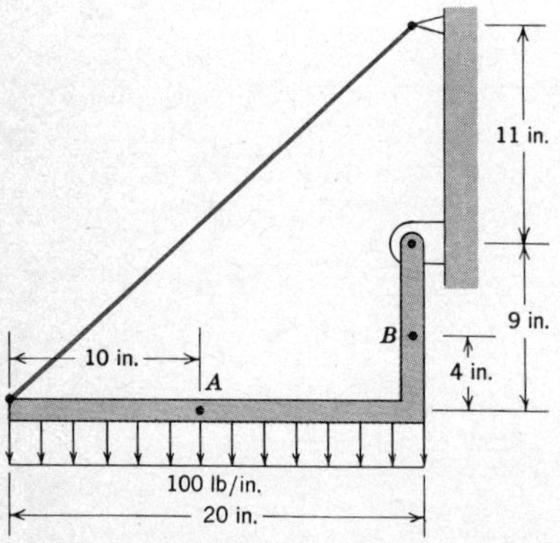

Prob. VI.12

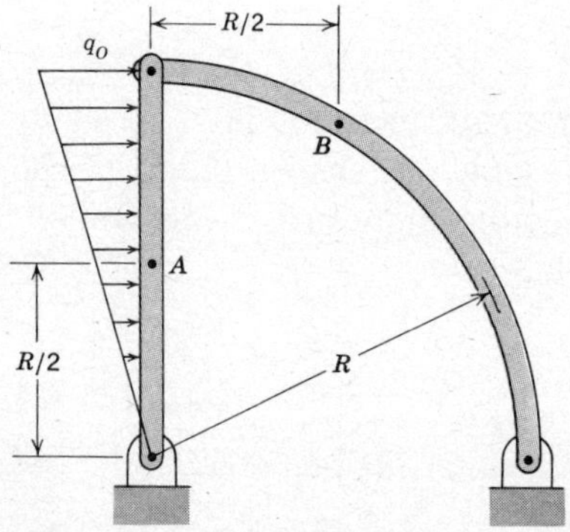

Prob. VI.13

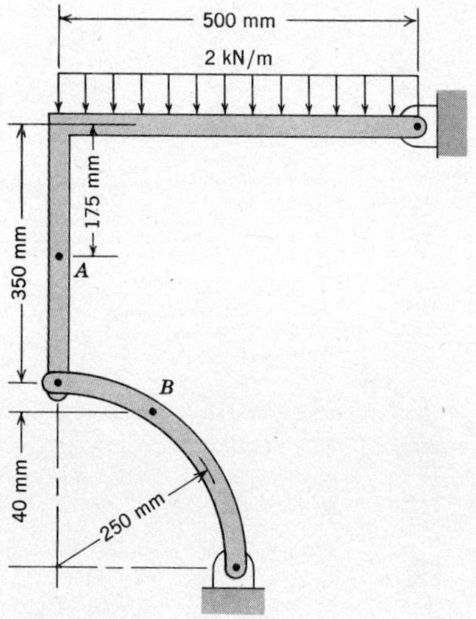

Prob. VI.14

VI.15–VI.17 Determine the axial force, the resultant shear force, the resultant bending moment, and the torsional moment at location *A* of the bar shown.

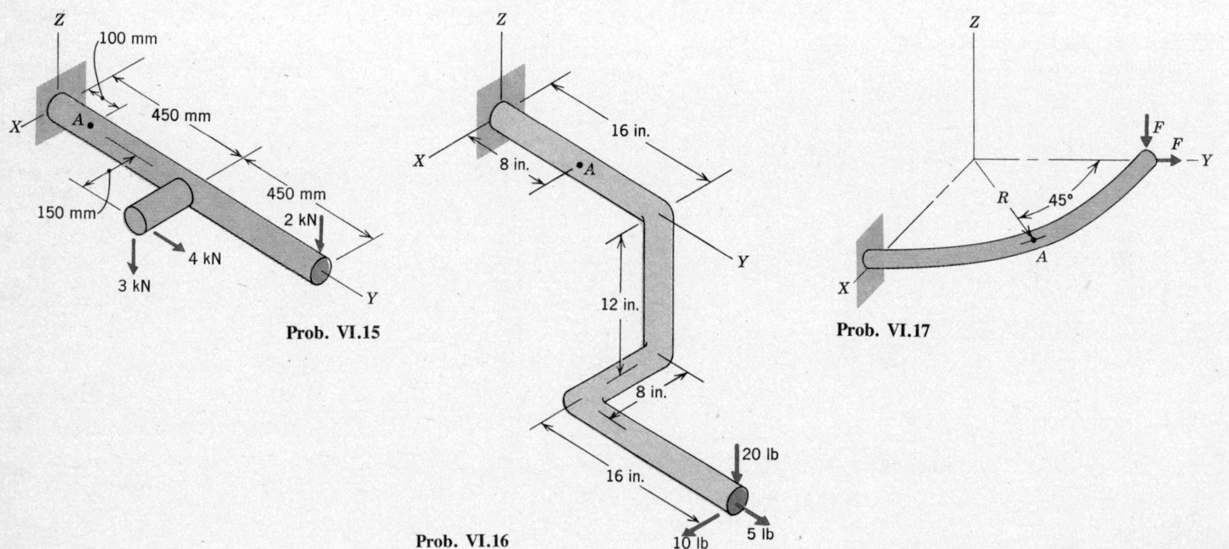

Prob. VI.15

Prob. VI.16

Prob. VI.17

3 Shear and Bending Moment Diagrams

The design of structural members requires knowledge of the internal forces at all locations within the member, not just at selected cross sections. This information can be obtained by systematically employing the method developed in the previous section for finding internal forces. The remainder of the material on beams in this module is devoted to such a determination.

We shall now abandon consideration of curved beams, and confine the development to the more common situation that arises in practice: straight beams that are loaded by a planar system of transverse forces and couples. In so doing we eliminate the need for an xyz coordinate system to describe the internal force component. This follows from the observation that the straight geometry means that the xyz and XYZ would always be mutually parallel.

A typical loading on a straight beam is depicted in Figure 6a. Now, the variable x is utilized to locate the cutting plane, consistent with standard usage.

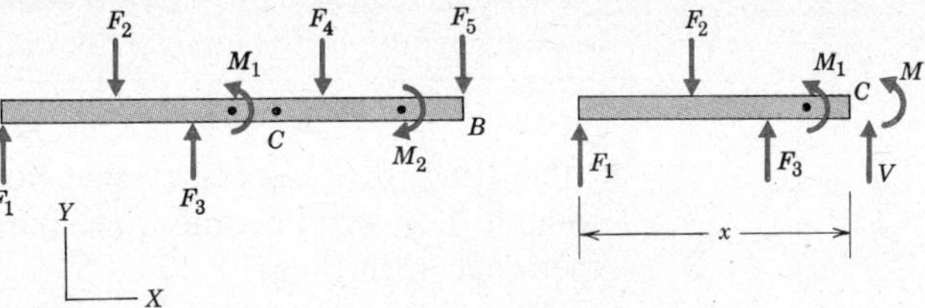

Figure 6a

Figure 6b

When we consider equilibrium of any section of the beam, such as the one of length x in Figure 6b, we see that the internal forces consist of a shear force V and a bending moment M. Therefore, an investigation of the internal forces on all cross sections of the beam reduces to the determination of the manner in which the shear force and bending moment depend on the distance x locating the cross section.

The functional dependence of the shear force and bending moments are usually exhibited by drawing *shear and bending moment diagrams*. Such diagrams require that we have nonambiguous definitions for positive shear and positive bending moment, both in order to avoid using inconsistent signs in the computations and in order to enable someone else to interpret correctly the diagrams we draw. These definitions are called *sign conventions*. (For example, we have already been using a sign convention by saying that an internal force normal to a cross section is positive if it is a tensile force.)

The sign convention we will use here is obtained by considering the type of deformation a section of the beam of length Δx would undergo if it carried only a shear force or only a bending moment. These situations are depicted on the following page in Figures 7a and b, respectively.

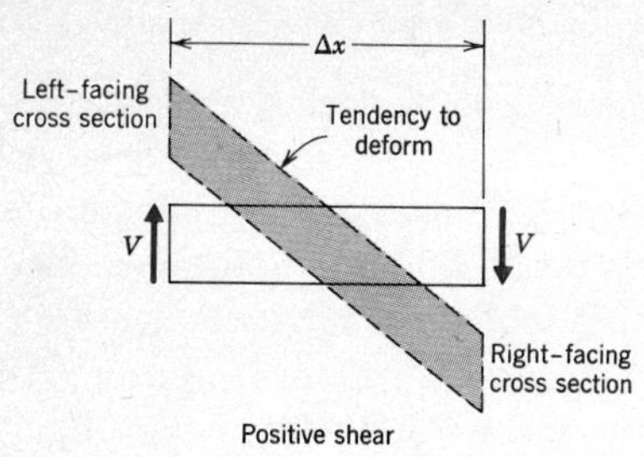

Figure 7a

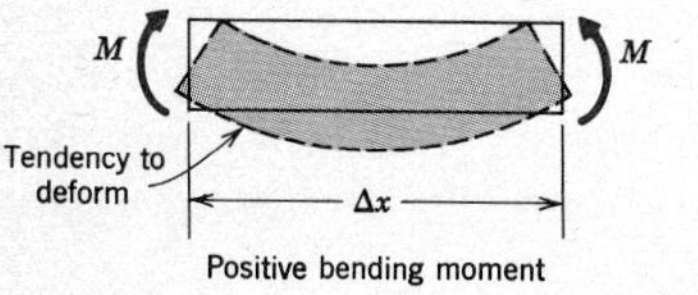

Figure 7b

For convenience, we will refer to a cross section whose outward normal faces left as a *left-facing cross section,* and a cross section whose outward normal faces right as a *right-facing cross section.* Figure 7a shows that in our sign convention,

> A shear force is positive if it pushes the left-facing cross section upward, or equivalently if it pushes the right-facing cross section downward.

Alternately, we can look upon this definition as: *Positive shear causes a clockwise rotation of the beam element being analyzed.*

Figure 7b shows that we will consider *positive bending moment* to be one that has the effect of bending the bar upward, that is, one that *causes the centroidal axis to assume a shape that is concave upward.* This means that in our sign convention,

> A bending moment is positive if it is clockwise on the left-facing cross section, or equivalently, if it is counterclockwise on the right-facing cross section.

At this time, it is necessary that you memorize these sign conventions, although they will become quite natural to use after you have had some experience with them.

Notice that we introduced deformation effects in Figure 7. Until now we considered a beam to be rigid. There is no contradiction here because, as is true for all systems that we model as rigid bodies, we are assuming that the beam undergoes a very small amount of deformation, so that the equations of static equilibrium may be formulated without considering deformations. (The validity of this assumption is shown in mechanics of materials to be quite good.)

Now that we have established the sign conventions for shear and bending moment, let us turn our attention to how we may go about constructing shear and bending moment diagrams. The most direct procedure is to construct a free body diagram of a section of beam of length x, for which the equations of static equilibrium yield the shear V and bending moment M in terms of x. Then the desired diagrams are plotted from the functions $V(x)$ and $M(x)$.

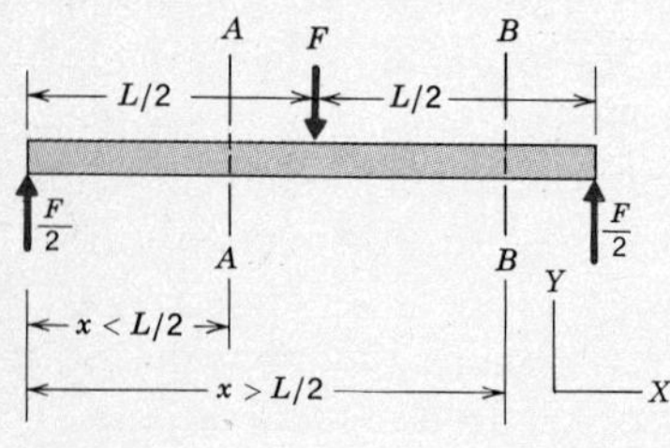

Figure 8

To develop this procedure, consider a lightweight simply supported beam loaded by a downward force $\bar{F}$ at its midpoint. After evaluating the reactions, the free body diagram of the entire beam is as shown in Figure 8. To determine V and M we must isolate a portion of the beam formed by a cutting plane. We may isolate the portion lying to either side of the cutting plane; let us use the left portion.

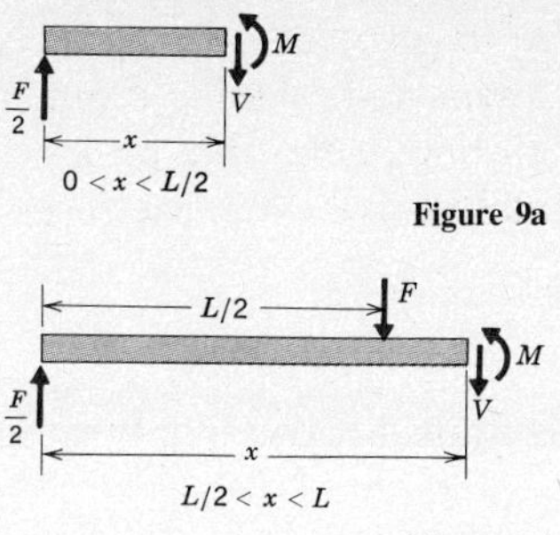

Figure 9a

Figure 9b

As depicted in Figure 8, we shall need to consider two cutting planes. This is because the concentrated force $\bar{F}$ represents a *singularity*. (The word "singularity" is used in the mathematical sense of an undefined situation.) The free body diagrams resulting from each cutting plane are shown in Figures 9a and b. It is important to note that in each diagram V and M are depicted in the positive direction as defined by our sign convention.

The static equilibrium equations for each free body diagram give

$$0 < x < L/2: \qquad V = \frac{F}{2} \qquad M = \frac{F}{2}x$$

$$\frac{L}{2} < x < L: \qquad V = -\frac{F}{2} \qquad M = \frac{F}{2}x - F\left(x - \frac{L}{2}\right) = \frac{F}{2}(L - x)$$

These functions are now plotted, to yield the shear and bending moment diagrams shown in Figure 10.

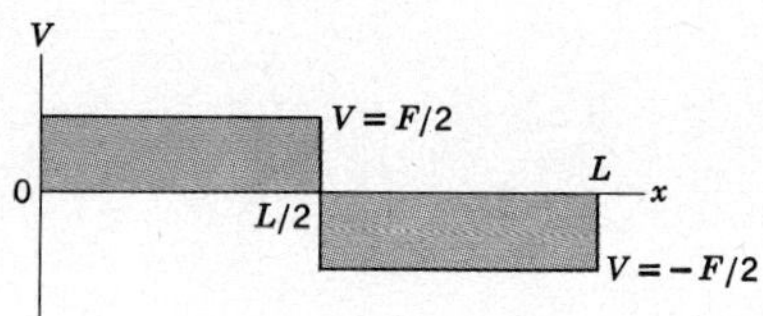

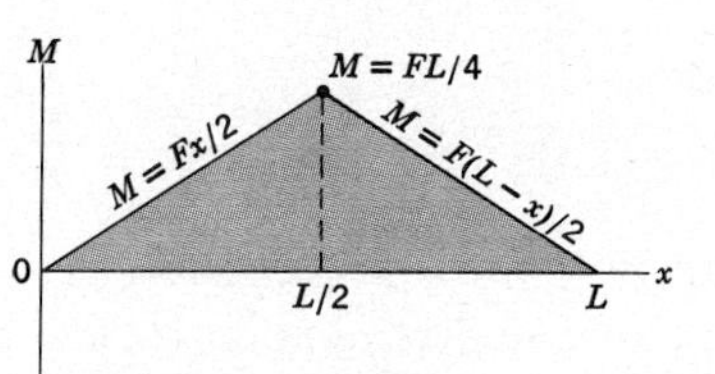

Figure 10

The correctness of these diagrams may be partially verified by checking that the values of the shear and bending moments at the ends match the force-couple systems acting at these locations. For instance, from Figure 8 we see that there is an upward force $F/2$ and no couple acting at the left end of the beam. According to our sign convention, a positive shear is upward on a left-facing cross section, which is exactly the situation for the left end. Therefore, the external force-couple system acting at the left end is equivalent to a positive shear force $F/2$ and no bending moment. These values are confirmed by the shear and bending moment diagrams in Figure 10.

The terminology we use to describe the foregoing check is to say that we have verified that the shear and bending moment diagrams *close* at the left end. The check for closure at both ends is very important, even though it may seem obvious. We leave it to you to verify closure at the right end of the diagrams in Figure 10.

A noteworthy feature of the diagrams in Figure 10 is that at the midpoint, where the force is applied, the value of the shear force and the derivative (slope) of the bending moment function are discontinuous. This is a result of the singularity introduced by the concentrated force $\bar{F}$.

Concentrated forces are one type of singularity resulting from the loads on the beam. Recall that this singularity resulted from a change in the nature of the force system acting on the isolated portion of the beam. Other types of singularities arise when a couple is applied to the beam at some location, and in the case of distributed loading, whenever the loading function changes.

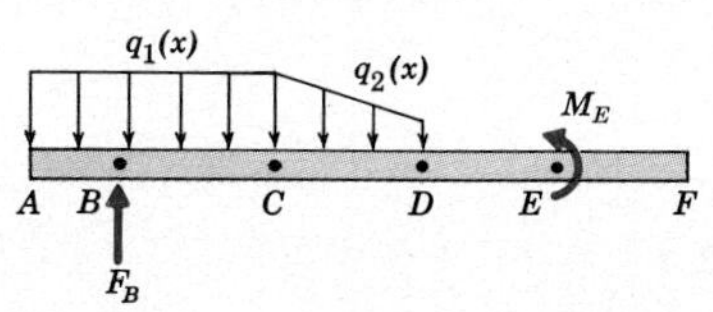

Figure 11

For example, consider the beam shown in Figure 11. Singularities occur at point B, where the concentrated force $\bar{F}_B$ is applied; at point C, where the distributed load function changes from q_1 to q_2; at point D, where the distributed load changes from q_2 to zero; and at point E, where the couple M_E is applied. To obtain the shear and bending moment diagrams in this case we would need to use cutting planes that pass between each pair of singular points, forming a total of five different cases to be considered.

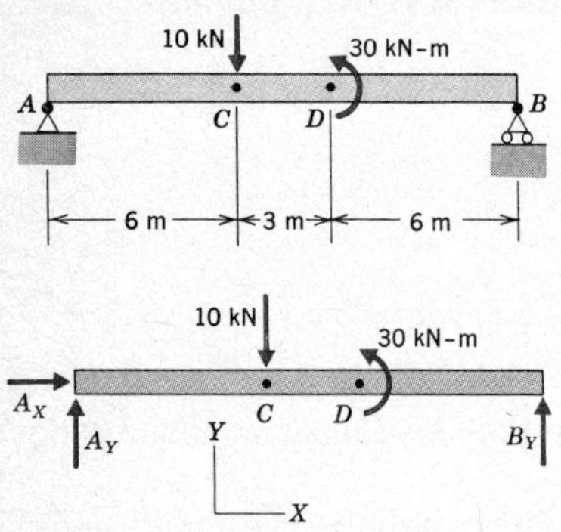

EXAMPLE 3

Sketch the shear and bending moment diagrams for the simply supported beam shown.

Solution

The first step in any analysis of internal forces is the determination of the reactions. Thus, we draw a free body diagram of the beam and write the equations of equilibrium.

$$\Sigma M_{AZ} = 30 - 10(6) + B_Y(15) = 0 \text{ kN-m}$$

$$\Sigma F_X = A_X = 0$$

$$\Sigma F_Y = A_Y + B_Y - 10 = 0 \text{ kN}$$

This gives

$$B_Y = 2 \qquad A_X = 0 \qquad A_Y = 8 \text{ kN}$$

We next ascertain where singularities occur in the loading. There is a concentrated force applied at point C and a couple applied at point D. This means that we will require three cutting planes, one between the left end A and point C, the second between points C and D, and the last between point D and the right end B.

It is best to perform the computation in an orderly fashion. Experience has shown that students are less prone to errors if they progress across the beam in one direction, usually the direction of increasing values of x. Measuring x from end A, we will isolate the section of the beam to the left of each of the three aforementioned cutting planes.

When we draw the free body diagrams, we depict the shear and bending moment in the positive direction according to the appropriate sign convention. On a right-facing cross section we consider downward shear and counterclockwise bending moment positive, so the free body diagrams and corresponding computations of V and M are as shown below.

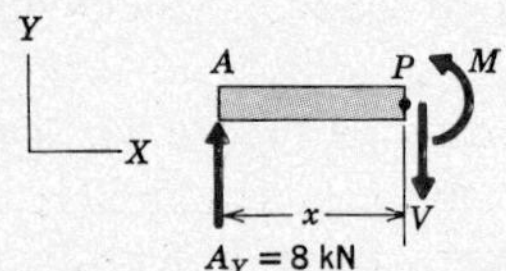

Region: $0 < x < 6$ m

$$\Sigma M_{PZ} = M - 8(x) = 0 \qquad \Sigma F_Y = -V + 8 = 0$$

$$V = 8 \text{ kN} \qquad M = 8x \text{ kN-m}$$

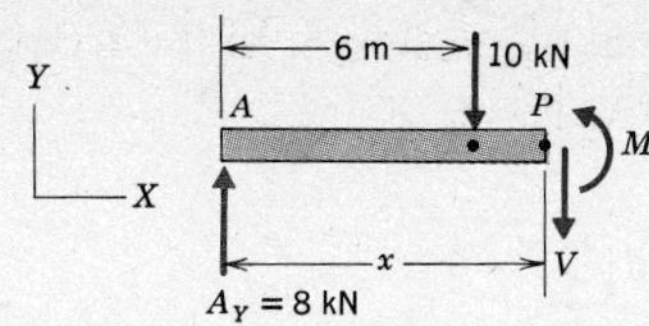

Region: $6 < x < 9$ m

$$\Sigma M_{PZ} = M + 10(x - 6) - 8(x) = 0 \qquad \Sigma F_Y = -V - 10 + 8 = 0$$

$$V = -2 \text{ kN} \qquad M = 30 - 2x \text{ kN-m}$$

Region: $9 < x < 15$ m

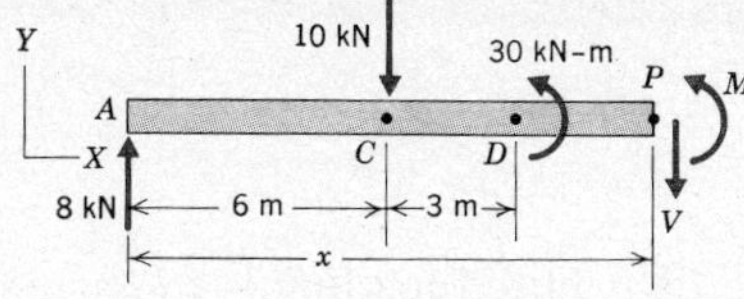

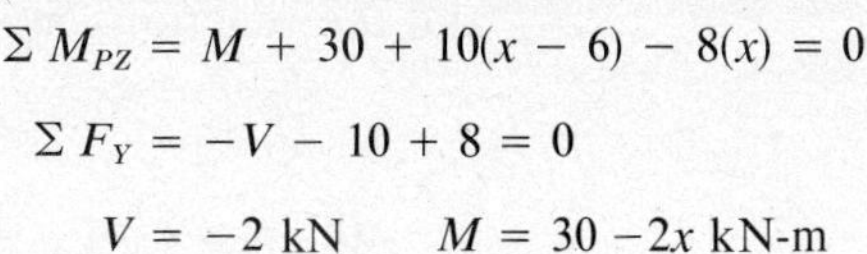

$$\Sigma M_{PZ} = M + 30 + 10(x - 6) - 8(x) = 0$$

$$\Sigma F_Y = -V - 10 + 8 = 0$$

$$V = -2 \text{ kN} \qquad M = 30 - 2x \text{ kN-m}$$

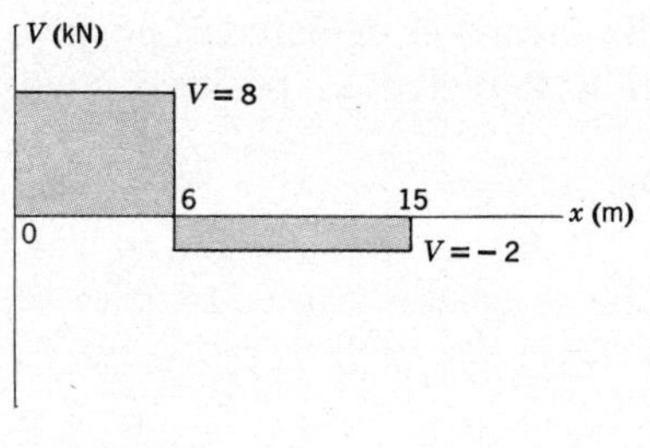

We now have sufficient information to plot the shear and bending moment diagrams. An accurate plot may be obtained by a point-by-point evaluation of the expressions. However, this is not really necessary, because we can evaluate the expressions corresponding to each region at their two limits of validity and then connect these points by the curve corresponding to the equation derived for the region. For example, in the range $0 < x < 6$ m, we calculate and plot the values of V and M at $x = 0$ and $x = 6$. The value of V is constant, so it is represented by a horizontal line. The value of M is linearly dependent on the value of x, so it is represented by a straight sloping line. The result of this procedure is the shear and bending moment diagrams shown. Notice that the singularity introduced by a couple loading is such that the value of the bending moment is discontinuous, but there is no discontinuity in the shear at that point. Also, notice that the singularity introduced by a concentrated force is such that the value of the shear is discontinuous, but there is no discontinuity in the bending moment at that point.

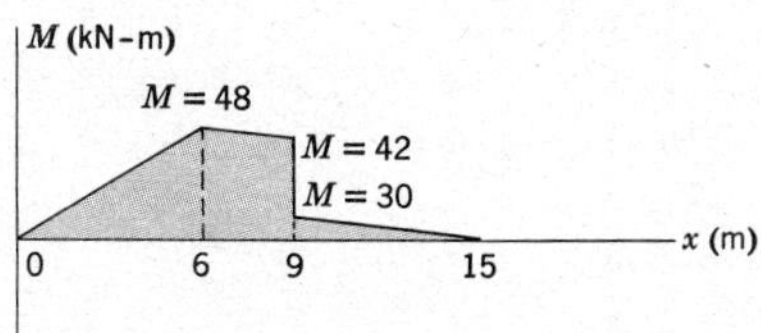

To verify that the shear and moment diagrams close, we refer back to the free body diagram of the entire beam. At end A we see that because $A_Y = 8$ kN upward and there is no couple, V should be positive 8 kN (end A is a left-facing cross section) and M should be zero. At end B we see that because $B_Y = 2$ kN upward and there is no couple, V should be negative 2 kN (right-facing cross section) and M should be zero. These values are all confirmed by, and consistent with, the diagrams we have drawn.

EXAMPLE 4

Sketch the shear and bending moment diagrams for the horizontal overhanging beam shown. The beam has constant weight per unit length w.

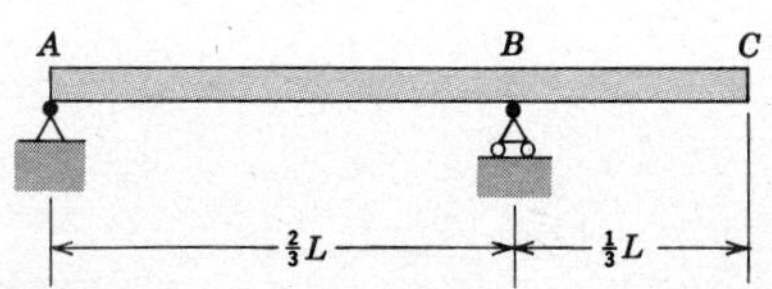

Solution

The first step in the solution is the determination of the reactions. The

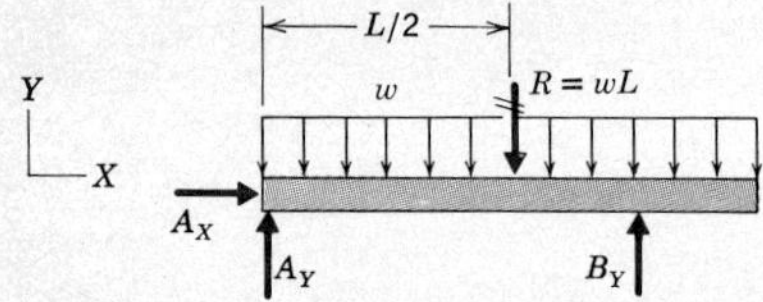

beam carries a constant distributed loading w, as shown in the free body diagram. The equilibrium equations for the beam are

$$\Sigma M_{AZ} = B_Y\left(\frac{2L}{3}\right) - wL\left(\frac{L}{2}\right) = 0$$

$$\Sigma F_X = A_x = 0 \qquad \Sigma F_Y = A_Y + B_Y - wL = 0$$

and the reactions are found to be

$$A_X = 0 \qquad A_Y = \tfrac{1}{4}wL \qquad B_Y = \tfrac{3}{4}wL$$

We next locate the singularities. The distributed loading is the same along the whole length of the beam, so the only change in the loading occurs when the cutting plane passes point B, where a reaction force is applied. Thus, we must analyze two regions, one to the left and one to the right of point B. Following the procedure outlined in the previous example, we measure x from the left end A and determine the internal forces by isolating the section of the beam to the left of the cutting plane. The required free body diagrams and equilibrium equations are given below. Notice that because we are concerned with the shear and bending moment acting on the right-facing cross section, the shear V is depicted as positive downward and the bending moment M is depicted as positive counterclockwise.

Region: $0 < x < 2L/3$

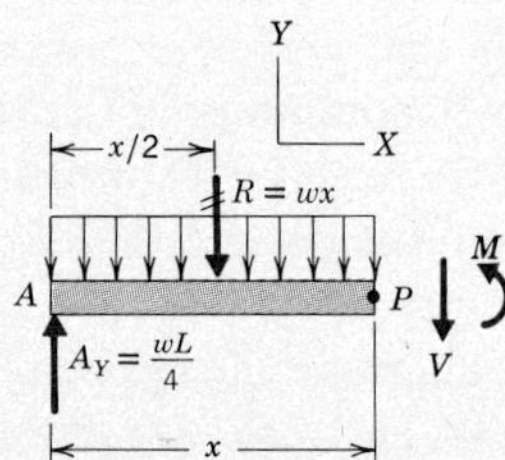

$$\Sigma M_{PZ} = M + wx\left(\frac{x}{2}\right) - \frac{wL}{4}(x) = 0$$

$$\Sigma F_Y = -V - wx + \frac{wL}{4} = 0$$

$$V = w\left(\frac{L}{4} - x\right) \qquad M = -\frac{w}{4}(2x^2 - Lx)$$

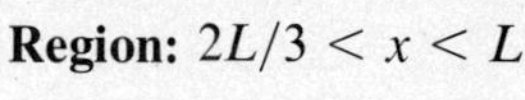

Region: $2L/3 < x < L$

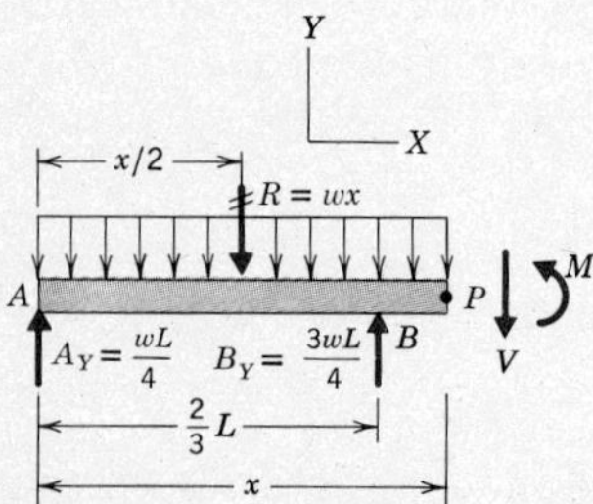

$$\Sigma M_{PZ} = M - \frac{3wL}{4}\left(x - \frac{2L}{3}\right) + wx\left(\frac{x}{2}\right) - \frac{wL}{4}(x) = 0$$

$$\Sigma F_Y = -V + \frac{3wL}{4} + \frac{wL}{4} - wx = 0$$

$$V = w(L - x) \qquad M = -w\left(\frac{x^2}{2} - Lx + \frac{1}{2}L^2\right)$$

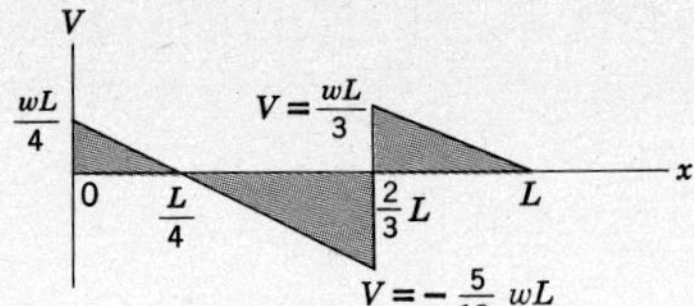

To sketch the shear diagram, we note that both expressions for V are linear in x, so the curves for V in both ranges are straight lines. Evaluating each function for V at the limits of its region of validity and plotting the corresponding points, we obtain the shear diagram shown.

To sketch the bending moment, we note that both expressions for M are quadratic polynomials in which the coefficient of the x^2 term is negative. This means that the curves representing M in either range are parabolas that are concave downward (in the mathematical sense of curvature). We may determine the location of the horizontal tangent of each parabola by setting $dM/dx = 0$. This yields

Region: $0 < x < 2L/3$

$$\frac{dM}{dx} = -\frac{w}{4}(4x - L) = 0 \quad \text{so} \quad x = L/4$$

Region: $2L/3 < x < L$

$$\frac{dM}{dx} = -w(x - L) = 0 \quad \text{so} \quad x = L$$

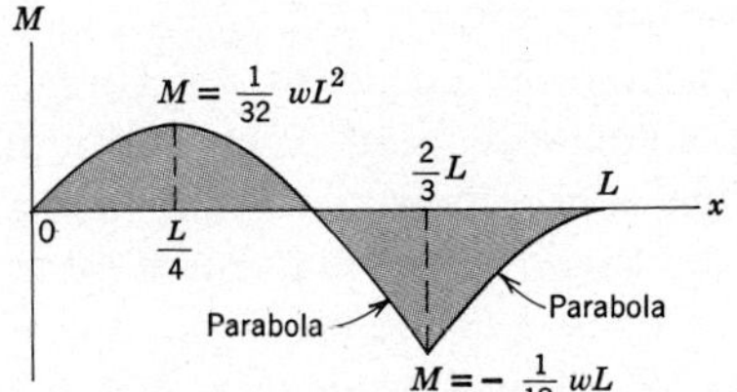

The foregoing knowledge of the shape of the bending moment curve, together with the plotted points representing the values of each bending moment function at the limits of its range of validity, allow us to draw the bending moment diagram. Can you verify that the shear and bending moment diagrams close?

It is interesting to note that the bending moment has its extreme values ($dM/dx = 0$) at the locations $x = L/4$ and $x = L$ where the shear is also zero. This is not a coincidence, as will be proven in the next section.

EXAMPLE 5

Sketch the shear and bending moment diagrams for the cantilever beam carrying the triangular distributed load shown.

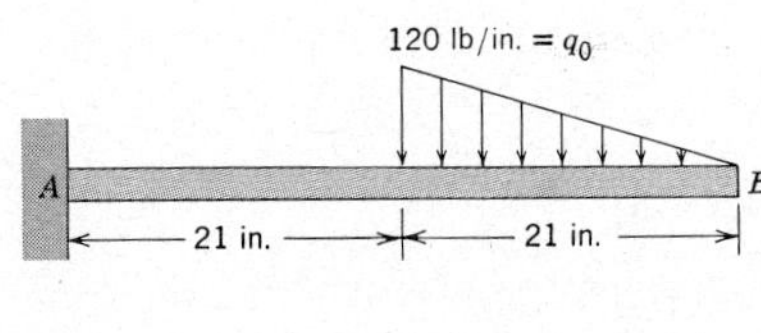

Solution

We begin with a free body diagram of the entire beam, which we use to determine the reactions at the fixed end A. In the free body diagram the force $\bar{R}$ is the resultant of the distributed load.

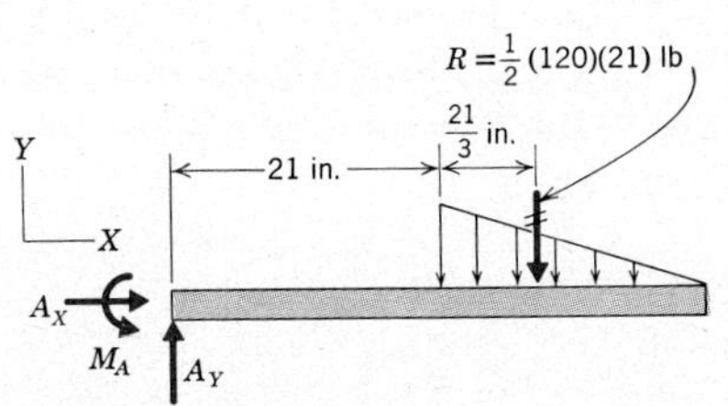

$$\Sigma M_A = M_A - 1260(21 + 7) = 0$$

$$\Sigma F_X = A_X = 0 \qquad \Sigma F_Y = A_Y - 1260 = 0$$

$$A_X = 0 \qquad A_Y = 1260 \text{ lb} \qquad M_A = 35{,}280 \text{ lb-in.}$$

We next identify the singularities. For the loading shown there is only one singular point, specifically the midpoint where the distributed load begins. Measuring x from the left end, we first consider a cutting plane to the left of the distributed load. In accordance with the procedure we have developed, we draw a free body diagram and write the equilibrium equations for the section of beam of length x formed by the plane. This is described below.

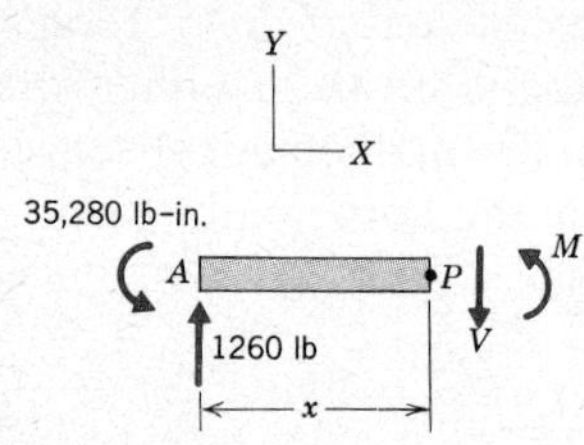

Region: $0 < x < 21$ in.

$$\Sigma M_{PZ} = M + 35{,}280 - 1{,}260(x) = 0 \qquad \Sigma F_Y = -V - 1{,}260 = 0$$

$$V = 1{,}260 \text{ lb} \qquad M = 1{,}260x - 35{,}280 \text{ lb-in.}$$

We now consider the situation arising from a cutting plane to the right of the midpoint of the beam. If we were to follow the approach developed in the two previous examples, we would isolate the portion of the beam to the left of the cutting plane. It is not difficult to see that this section carries a trapezoidal distributed loading. (Recall that we consider only the portion of the loads that are actually applied to the isolated section of the beam.) Solution of the equilibrium equations would then require that we determine the resultant of a trapezoidal loading.

Alternatively, should we choose to isolate the portion of the beam to the right of the cutting plane, we find that the resulting section carries a triangular loading, whose resultant is considerably easier to determine. Thus, the free body diagram and corresponding equations of equilibrium are as shown below. Note that in our sign convention, for a left-facing cross section positive shear is upward and positive bending moment is clockwise, as depicted. Also note that the position of the cross section of the beam is still measured from the left end. By doing this we minimize the chance of error when plotting the shear and moment equations obtained for each region of the beam.

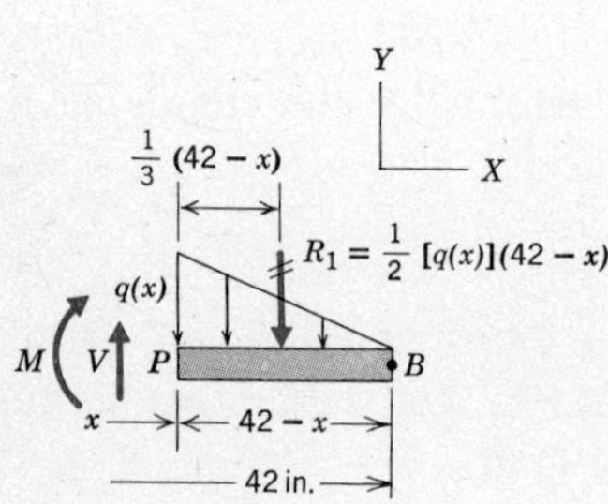

Region: $21 < x < 42$ in.

$$\Sigma M_{PZ} = -M - R_1[\tfrac{1}{3}(42 - x)] = 0 \qquad \Sigma F_Y = V - R_1 = 0$$

$$V = R_1 = \tfrac{1}{2}q(x)(42 - x)$$

$$M = -\tfrac{1}{3}R_1(42 - x) = -\tfrac{1}{6}q(x)(42 - x)^2$$

The determination of V and M in terms of x now requires that we determine the expression for the loading function $q(x)$. This is obtained by observing that the load decreases linearly with increasing values of x, so $q(x) = k_1 - k_2x$. The values of k_1 and k_2 are found by equating $q(x)$ to 120 lb/in. at $x = 21$ in. and to zero at $x = 42$ in. Thus

$$q(21) = k_1 - 21k_2 = 120 \qquad q(42) = k_1 - 42k_2 = 0$$

so

$$k_1 = 240 \qquad k_2 = \frac{120}{21} \quad \text{and} \quad q(x) = 120\left(2 - \frac{x}{21}\right)$$

The equations for V and M then become

Region: $21 < x < 42$ in.

$$V = \frac{1}{2}\left[120\left(2 - \frac{x}{21}\right)\right](42 - x) = 1260\left(2 - \frac{x}{21}\right)^2 \text{ lb}$$

$$M = -\frac{1}{6}\left[120\left(2 - \frac{x}{21}\right)\right](42 - x)^2 = -8820\left(2 - \frac{x}{21}\right)^3 \text{ lb-in.}$$

We may now sketch the shear and bending moment diagrams. To do this we note that in the region $0 < x < 21$ in., V is constant and M increases linearly with x, so both functions plot as straight lines. In the region $21 < x < 42$ in., V is a quadratic polynomial and M is a cubic polynomial. The curve for V is readily identified as an upward-curving parabola (in the mathematical sense of curvature) having its lowest point at $x = 42$ in. where $V = 0$. Similarly, we see that the curve for M is a downward-curving cubic whose maximum value is zero at $x = 42$ in. Knowing this, we evaluate the expressions for M and V in the region $0 < x < 21$ at $x = 0$ and $x = 21$, and the expressions for M and V in the region $21 < x < 42$ at $x = 21$. Then, by plotting these points and connecting them by the appropriate curves, we obtain the accompanying diagrams.

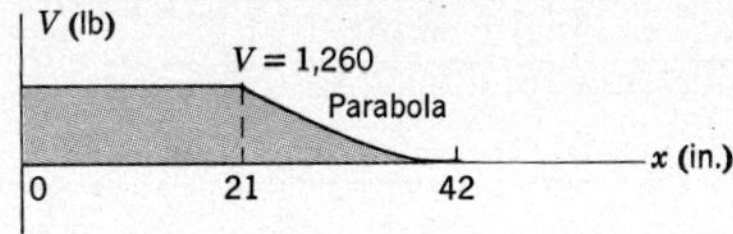

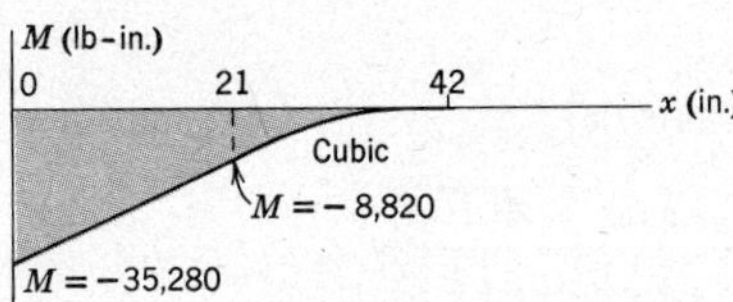

These diagrams close at both ends, because there is no force-couple system acting at the right end B, and the reactions at end A, which is a left-facing cross section, are equivalent to a positive (upward) shear and a negative (counterclockwise) bending moment.

HOMEWORK PROBLEMS

VI.18–VI.35 Draw shear and bending moment diagrams for the beam shown. Give all critical values.

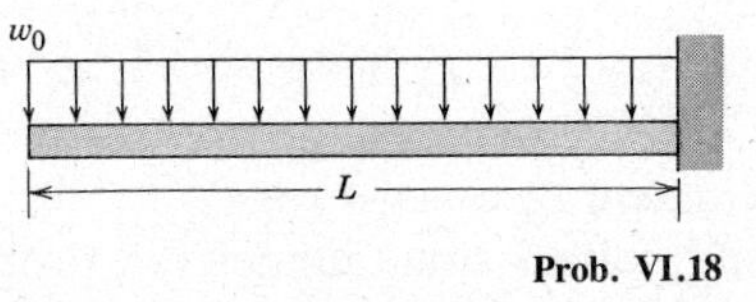

Prob. VI.18

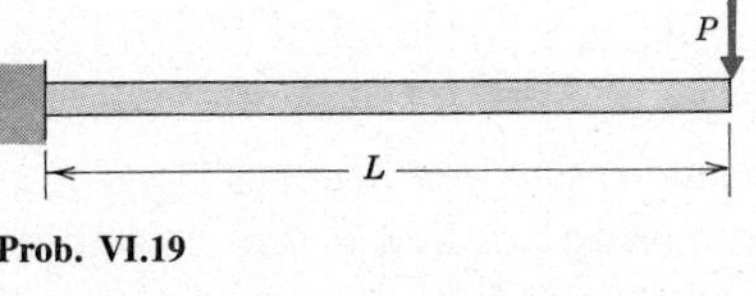

Prob. VI.19

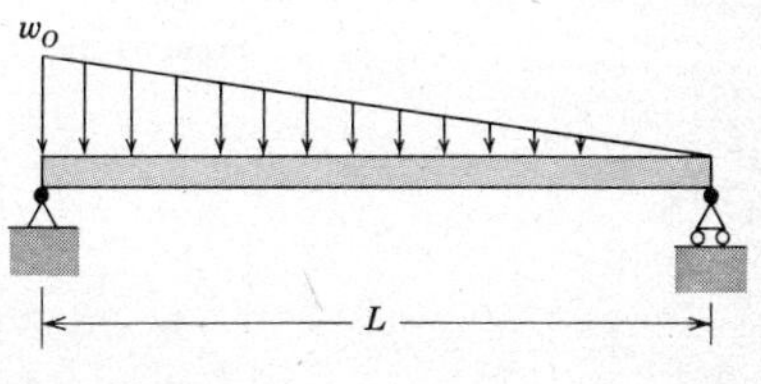

Prob. VI.20

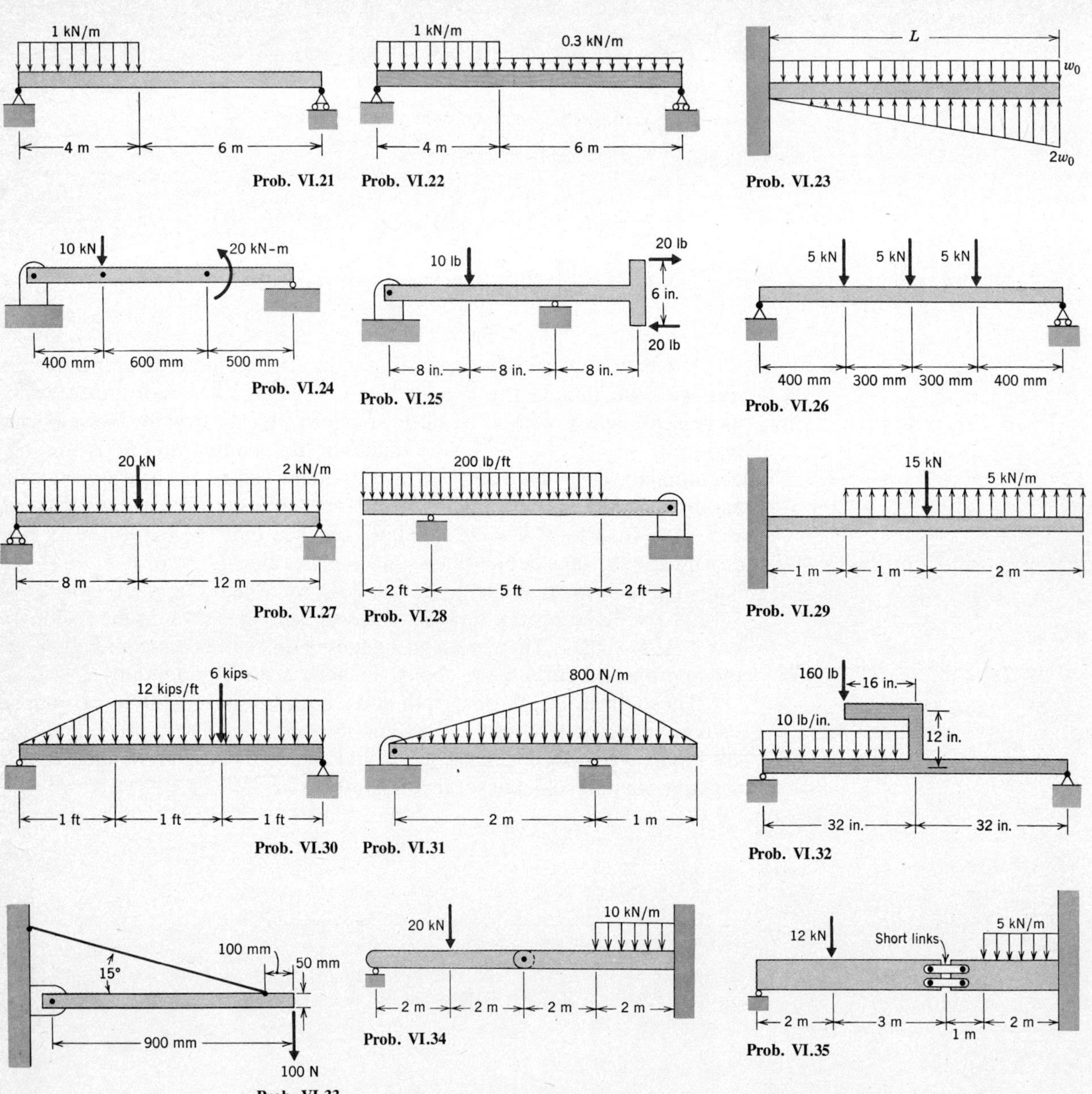

4 Relations Between Bending Moment, Shear, and Distributed Load

After having studied a few examples and solved some homework problems, it should be apparent to you that although the theory for the determination of the shear and bending moment diagrams is conceptually straightforward, in practice the computations can get quite involved. To reduce the amount of calculations, in this section we develop a method for constructing shear and bending moment diagrams without explicit consideration of the equilibrium of isolated sections of a beam.

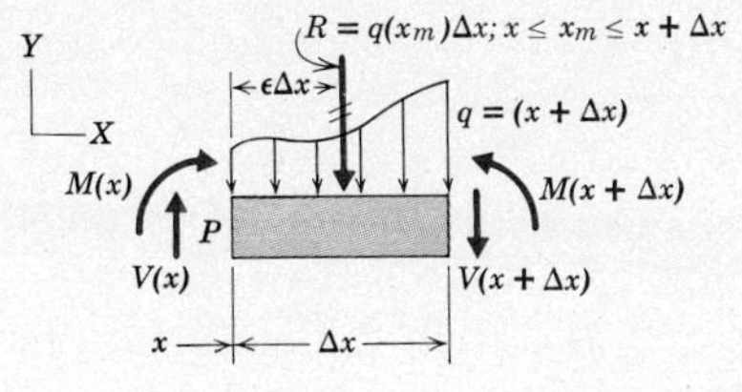

Figure 12

Let us begin by determining the general relations between the internal forces and the loading on a beam. In Figure 12 we isolate a small segment of a beam, of length Δx, carrying a distributed load $\bar{q}$, but no concentrated loads or couples.

Acting on the isolated segment are the internal forces on the left-facing cross section at position x and on the right-facing cross section at position $x + \Delta x$. The shear and bending moment are shown in Figure 12 in their positive sense on each cross section. As indicated there, these quantities need not be the same on both faces. The free body diagram for the element is completed by replacing the distributed load by its equivalent concentrated force. As shown, the magnitude of this force is obtained by multiplying the mean value of q, corresponding to some point x_m within the region $x \leq x_m \leq x + \Delta x$, by the length Δx. Further, the force is applied at a distance $\epsilon\, \Delta x$ from the left end, where ϵ is a finite number.

Let us write the equilibrium equations for this beam element. Summing moments about point P, we have

$$\Sigma M_{PZ} = M(x + \Delta x) - M(x) - [V(x + \Delta x)]\Delta x - q(x_m)\, \Delta x(\epsilon\, \Delta x) = 0$$

$$\Sigma F_X \equiv 0 \qquad \Sigma F_Y = V(x) - V(x + \Delta x) - q(x_m)\, \Delta x = 0$$

After we rearrange terms, these relations become

$$\frac{V(x + \Delta x) - V(x)}{\Delta x} = -q(x_m) \tag{1}$$

$$\frac{M(x + \Delta x) - M(x)}{\Delta x} = V(x + \Delta x) - \epsilon q(x_m)\, \Delta x \tag{2}$$

Consider the situation where the length of the element becomes smaller and smaller, that is, $\Delta x \to 0$. In the limit the left sides of equations (1) and (2) are the definitions of dV/dx and dM/dx, respectively. Further, as $\Delta x \to 0$, $x_m \to x$, because the value of x_m is intermediate between x and $x + \Delta x$. Therefore, in the limit, equations (1) and (2) yield

$$\boxed{\frac{dV}{dx} = -q(x)} \tag{3a}$$

$$\boxed{\frac{dM}{dx} = V(x)} \tag{3b}$$

Equations (3) are called *differential equations* because they equate the derivative of one variable to another function. We will find these equations to be useful when plotting the shear and bending moment diagrams. For example, equation (3b) tells us that the bending moment has a horizontal tangent at the location where the shear is zero. (We noted this fact in Example 4.)

Another use for equations (3) is associated with the determination of the shear and bending moment resulting from a given loading. This calls for an integral representation of the relations. Suppose that we know the shear and bending moment at some value x_A and wish to determine the value of these forces at some other value x_B. Equations (3) may be used for this determination by multiplying each equation by dx and integrating between the limits corresponding to the two points. This gives

$$\int_{V_A}^{V_B} dV \equiv V_B - V_A = -\int_{x_A}^{x_B} q(x)\,dx$$

$$\int_{M_A}^{M_B} dM \equiv M_B - M_A = \int_{x_A}^{x_B} V(x)\,dx$$

Rearranging terms, we obtain the desired result

$$V_B = V_A - \int_{x_A}^{x_B} q(x)\,dx \tag{4a}$$

$$M_B = M_A + \int_{x_A}^{x_B} V(x)\,dx \tag{4b}$$

From the fact that an integral of a function is the area under the curve representing that function, equation (4a) can be seen to be equivalent to the statement that

> The shear force at x_B is the shear force at x_A *minus* the area under the distributed load curve in the region between x_A and x_B.

Similarly, equation (4b) states that

> The bending moment at x_B is the bending moment at x_A *plus* the area under the shear curve in the region between x_A and x_B.

Our interest in this text is the quantitative determination of the internal forces at a few key locations along a beam, for example, where the bending moment and shear assume extreme values, as well as the determi-

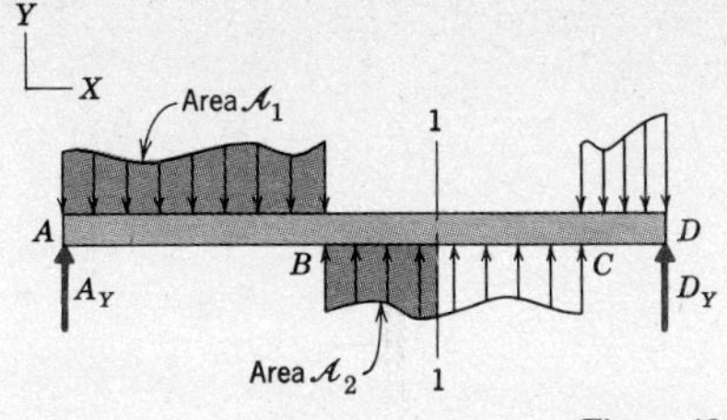

Figure 13

nation of a pictorial representation of the way in which the internal forces vary with location. To see how we can use the foregoing statements toward this goal, consider the situation depicted in Figure 13, where a beam is carrying a distributed load and concentrated forces (reactions) at its ends.

According to our sign conventions, the vertical force $\bar{A}_Y$ at end A is equivalent to a positive shear force. Using equation (4a), we see that the shear force at the location of the cutting plane 1-1 appearing in the figure is given by

$$V_1 = V_A - (\mathcal{A}_1 - \mathcal{A}_2) = A_Y - \mathcal{A}_1 + \mathcal{A}_2$$

where $\mathcal{A}_1$ and $\mathcal{A}_2$ are areas formed by the loading function, as indicated in Figure 13. In regard to the area $\mathcal{A}_2$, it is important to realize that between points B and C, the load, and thus the area, is negative because we considered downward loads to be positive in deriving equations (3).

Let us determine the shear diagram for the beam in Figure 13. We begin by recalling that discontinuous changes in the value of shear occur only at locations where there are concentrated transverse forces. Thus, the curves representing the shear distribution in the regions bounded by the singular points B and C are continuous. We may use equation (4a) to obtain the value of shear at any critical point, particularly the ends of the beam and the singular points B and C. Then, the character of the shear curves connecting the plotted points may be explored by using equation (3a), $dV/dx = -q$, to ascertain how dV/dx (the slope of the shear curve) and V depend on x.

The foregoing procedure for constructing the shear diagram, and a similar procedure for bending moment diagrams, will be demonstrated in the next two examples. Before proceeding, however, we will consider the cases of loadings by concentrated forces and couples in order to gain further insight into the discontinuities they cause in the shear and bending moment diagrams.

To investigate concentrated forces and couples, consider an isolated beam element of length Δx. We apply either a concentrated downward force $\bar{F}_C$ or a couple $\bar{M}_C$ at the midpoint C of the element, having position x_C, as illustrated in Figures 14a and b, respectively.

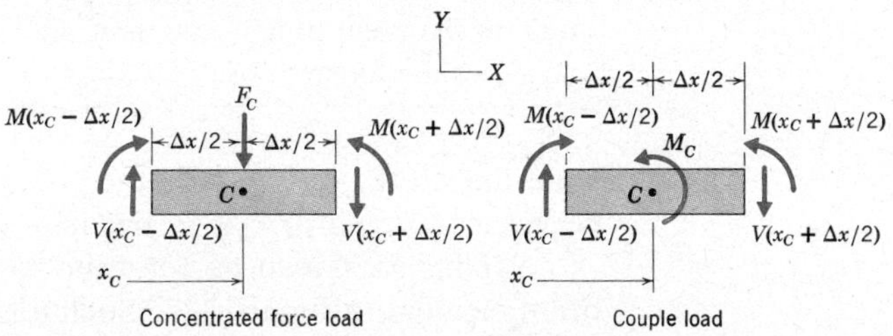

Figure 14a

Figure 14b

Equilibrium of the element in Figure 14a requires that

$$\Sigma M_{CZ} = M\left(x_C + \frac{\Delta x}{2}\right) - M\left(x_C - \frac{\Delta x}{2}\right) - \left[V\left(x_C - \frac{\Delta x}{2}\right) + V\left(x_C + \frac{\Delta x}{2}\right)\right]\left(\frac{\Delta x}{2}\right) = 0 \tag{5a}$$

$$\Sigma F_Y = V\left(x_C - \frac{\Delta x}{2}\right) - V\left(x_C + \frac{\Delta x}{2}\right) - F_C = 0 \tag{5b}$$

Because of the presence of the finite term F_C in the force equation, we cannot simply divide each term in equations (5) by Δx and take the limits as $\Delta x \to 0$, as we did in the case of a distributed load. This would give rise to the improper division of a finite number by zero.

Instead consider what happens to equations (5) as Δx becomes infinitesimal, but nonzero. As Δx decreases, the left and right cross sections both approach point C. Denoting the position of the cross section to the left of point C by x_C^-, and the position of the cross section to the right of point C as x_C^+, and letting Δx become the infinitesimal quantity dx, equations (5) become

$$\Sigma M_{CZ} = M(x_C^+) - M(x_C^-) - [V(x_C^-) + V(x_C^+)]\left(\frac{dx}{2}\right) = 0$$

$$\Sigma F_Y = V(x_C^-) - V(x_C^+) - F_C = 0$$

Neglecting terms of order dx, these equations reduce to

$$V(x_C^+) = V(x_C^-) - F_C \tag{6a}$$

$$M(x_C^+) = M(x_C^-) \tag{6b}$$

In words, equations (6) state that

> The value of the bending moment is continuous at the singular point of a downward concentrated force, but the shear is discontinuous, being less to the right of the point of application of the force by an amount equal to the magnitude of the force.

Note that in the case where the concentrated force is upward, it is only necessary to regard F_C as a negative quantity in equation (6a).

Similar steps can be followed to investigate the equilibrium of the beam element in Figure 14b. Such an investigation shows that

$$V(x_C^+) = V(x_C^-) \tag{7a}$$

$$M(x_C^+) = M(x_C^-) - M_C \tag{7b}$$

In words, equations (7) state that

> The value of the shear is continuous at the singular point of a clockwise couple, but the bending moment is discontinuous, being less to the right of the point of application of the couple by an amount equal to the magnitude of the couple.

Note that a clockwise couple is treated by regarding M_C as a negative quantity.

Referring back to any of the shear and bending moment diagrams previously determined, particularly in Examples 3 and 4, you will observe that equations (6) and (7) validate the results given there.

Placing equations (6) and (7) in the context of our general goal of determining shear and moment diagrams, we see that equations (4) may be used to relate shears or bending moments within the regions lying between points where concentrated forces and couples are applied, whereas equations (6) and (7) may be used to go from one side of these singular points to the other. In this manner, the values of shear and bending moment at the singular points and other critical points may be determined. After these values are plotted, the characteristics of the curves connecting the plotted values in each region bounded by the singular points can be investigated with the aid of equations (3).

The remarkable feature of this method for constructing the shear and bending moment diagrams is that once the reactions have been obtained, it is no longer necessary to write the equations of static equilibrium, as is demonstrated in the following examples.

EXAMPLE 6

The simply supported beam carries the loading shown. Use the derived relations between load, shear, and bending moment to sketch the shear and bending moment diagrams. Give all critical values.

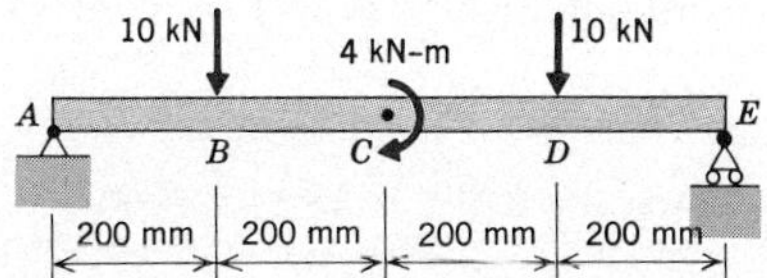

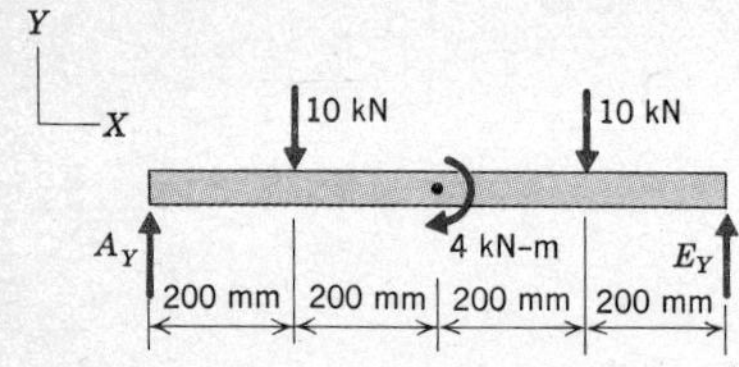

Solution

We begin by determining the reactions. The necessary free body diagram and computations are given below.

$$\Sigma M_{AZ} = -10(0.20) - 4 - 10(0.60) + E_Y(0.80) = 0$$

$$\Sigma F_X \equiv 0 \qquad \Sigma F_Y = A_Y + E_Y - 20 = 0$$

$$E_Y = 15 \text{ kN} \qquad A_Y = 5 \text{ kN}$$

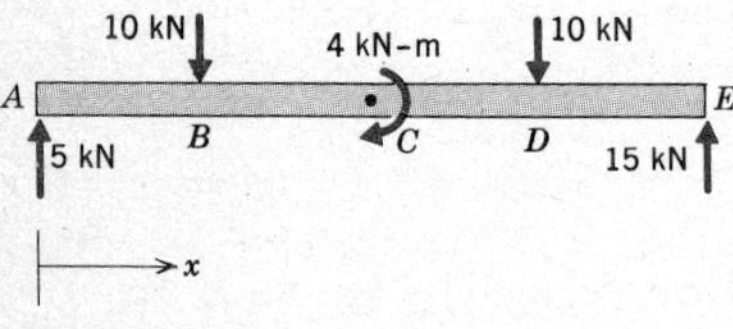

We now draw the loading diagram by simply repeating the free body diagram of the beam, with the unknown reactions replaced by their calculated values. Singularities are readily identified to exist at points B, C, and D.

From the load diagram we may construct the shear diagram. At point A, the force is equivalent to a positive shear (upward on a left-facing cross section), $V = 5$ kN. From equation (6a) we observe that the concentrated *downward* forces result in discontinuous *decreases* of the shear by 10 kN each as points B and D are passed from left to right when sketching the shear diagram. Further, with respect to the couple, we observe from equation (7a) that the shear is continuous at point C. Finally, noting that there is no loading between the singular points, we may conclude that the shear is constant in these regions. These observations lead to the shear diagram shown. As a check, we note that the shear diagram closes at end E, because the upward 15-kN force at this end is equivalent to a negative shear.

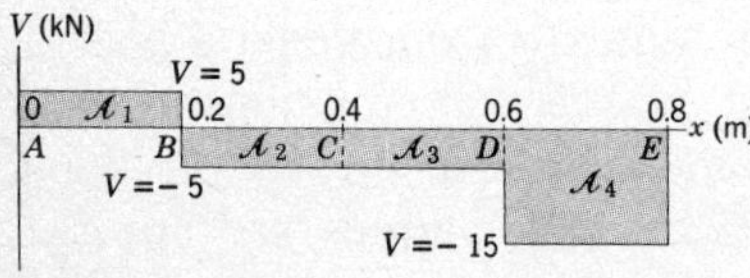

We now use the shear diagram to construct the moment diagram. From the load diagram, we see that there is no moment at end A. From equation (6b) we note that the bending moment is continuous at points B and D, where there are concentrated forces, whereas from equation (7b) we observe that the *clockwise* 4 kN-m couple at point C results in a discontinuous *increase* in bending moment in that amount as point C is passed in the diagram from left to right.

The values of bending moment at the singularities may then be calculated by using equation (4b) to proceed from one singularity to the next. In terms of the areas indicated in the shear diagram, this equation gives

$$M(x_B = 0.20) = M(x_A = 0) + \mathcal{A}_1 = 0 + 5(0.20) = 1.0 \text{ kN-m}$$

$$M(x_C = 0.40^-) = M(x_B = 0.20) - \mathcal{A}_2 = 1.0 - 5(0.20) = 0$$

$$M(x_C = 0.40^+) = M(x_C = 0.40^-) + 4 = 4.0 \text{ kN-m}$$

$$M(x_D = 0.60) = M(x_C = 0.40^+) - \mathcal{A}_3 = 4.0 - 5(0.20) = 3.0 \text{ kN-m}$$

$$M(x_E = 0.80) = M(x_D = 0.60) - \mathcal{A}_4 = 3.0 - 15(0.2) = 0$$

Note that the bending moment closes at end E.

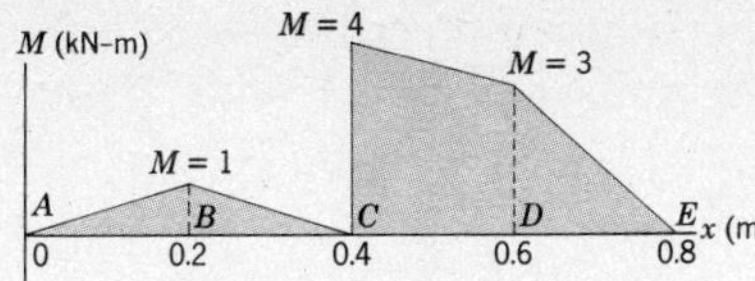

We next plot these values. To determine the type of curves connecting the plotted points, we recall equation (3b), $dM/dx = V$. Because V is constant between each pair of singular points, we see that each curve has constant slope. In other words, the curves are straight lines. Thus, we obtain the bending moment diagram shown to the left.

EXAMPLE 7

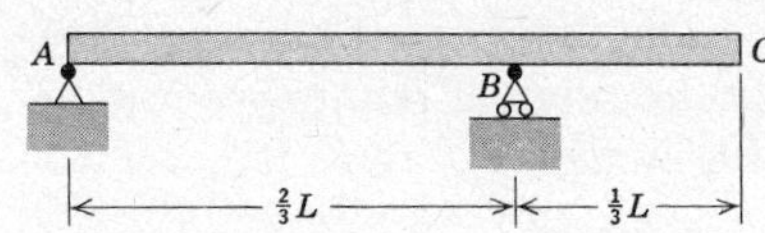

The overhanging beam of weight per unit length w studied in Example 4 is depicted again in the sketch. Use the relations between distributed load, shear, and bending moment to sketch the shear and bending moment diagrams. Give all critical values.

Solution

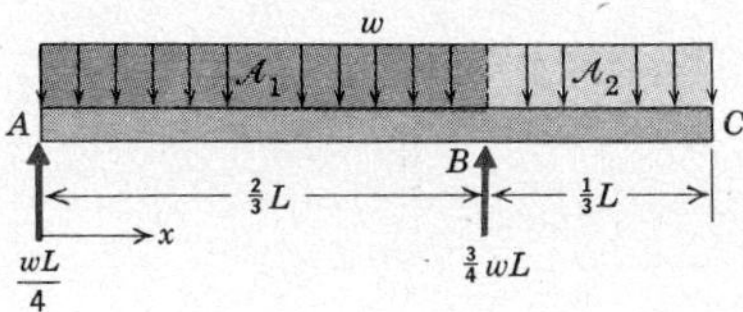

The necessary preliminary step of determining the reactions was performed in Example 4, so it is not repeated here. After the reactions are determined the loading diagram is drawn by showing all loads, including the reactions, in terms of their known magnitude and sense. This diagram indicates that the only singularity occurs at point B, where there is a concentrated upward force.

The shear diagram is obtained from the loading diagram. The force $wL/4$ at end A is equivalent to a positive shear. From equation (6a) we find that the concentrated *upward* force at point B results in a discontinuous *increase* in shear by $3wL/4$ as this point is passed in the diagram from left to right. Then, using equation (4a) and the areas indicated in the loading diagram, we calculate the shear at the singularities. This yields

$$V\left(x_B = \frac{2L^-}{3}\right) = V(x_A = 0) - \mathcal{A}_1 = \frac{1}{4}wL - w\left(\frac{2L}{3}\right) = -\frac{5}{12}wL$$

$$V\left(x_B = \frac{2L^+}{3}\right) = V\left(x_B = \frac{2L^-}{3}\right) - F_B = -\frac{5}{12}wL - \left(-\frac{3}{4}wL\right) = \frac{1}{3}wL$$

$$V(x_C = L) = V\left(x_B = \frac{2L^+}{3}\right) - \mathcal{A}_2 = \frac{1}{3}wL - w\left(\frac{L}{3}\right) = 0$$

As a check, we note that the shear closes at end C, because no force is applied at this end.

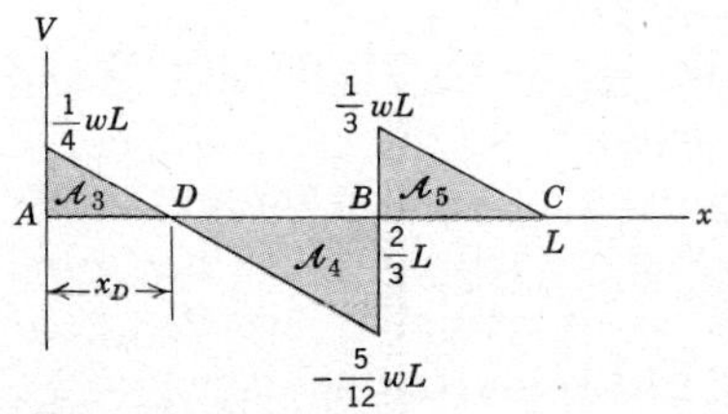

These points are now plotted and equation (3a), $dV/dx = -q$, is used to ascertain the types of curves connecting the plotted points. Because q is constant, it follows that V is linearly dependent on x (integration of a constant). In other words, the shear curves are straight lines. Further, because q is positive, these straight lines are sloping downward. The shear diagram resulting from these considerations is as shown.

The determination of the bending moment proceeds in a similar manner, using the areas indicated in the shear diagram to evaluate the integrals arising from equation (4b). The determination of these areas requires that we first obtain the distance x_D locating point D where $V = 0$. Using similar triangles, we have

$$\frac{x_D}{(\frac{1}{4}wL)} = \frac{(2L/3)}{(\frac{1}{4}wL + \frac{5}{12}wL)} \quad \text{so} \quad x_D = \frac{L}{4}$$

The area $\mathcal{A}_4$ is negative because the shear is negative. Also, the beam does not carry any couple loads, so the moment diagram is continuous. Thus, the computations of the bending moment at the singular points proceed as follows, starting from $x_A = 0$, where we know that $M = 0$.

$$M\left(x_D = \frac{L}{4}\right) = M(x_A = 0) + \mathcal{A}_3 = 0 + \frac{1}{2}\left(\frac{1}{4}wL\right)\left(\frac{L}{4}\right) = \frac{1}{32}wL^2$$

$$M\left(x_B = \frac{2L}{3}\right) = M\left(x_D = \frac{L}{4}\right) - \mathcal{A}_4$$

$$= \frac{1}{32}wL^2 - \frac{1}{2}\left(\frac{5}{12}wL\right)\left(\frac{2L}{3} - \frac{L}{4}\right) = -\frac{1}{18}wL^2$$

$$M(x_C = L) = M\left(x_B = \frac{2L}{3}\right) + \mathcal{A}_5$$

$$= -\frac{1}{18}wL^2 + \frac{1}{2}\left(\frac{1}{3}wL\right)\left(\frac{L}{3}\right) = 0$$

Notice that the bending moment closes at end C because there is no couple at that location.

The bending moment diagram is obtained by first plotting the values of bending moment that were determined. To draw the curves connecting the plotted points, we recall equation (3b), $dM/dx = V$. The shear curves are straight lines between the singularities, so V is linearly dependent on x. It then follows that M is dependent on x^2 (integrating x), so the moment curves are parabolas.

The nature of these parabolas is investigated by using $dM/dx = V$ to study their slope. Proceeding first from point A to point D, we see that the slope of the parabola decreases from a positive value at $x_A = 0$ to zero at $x_D = L/4$ and then becomes large negatively for increasing values of x beyond x_D. This can be true only if the parabola between points A and B curves downward with a horizontal tangent at x_D.

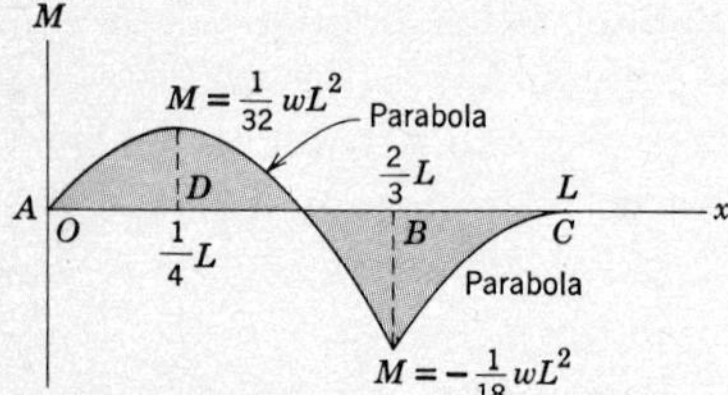

Similar reasoning may be applied to the region between points B and C. There is a discontinuity in the shear diagram, so the parabolic curve depicting the bending moment in this region is different from that in the region between points A and B. Noting that V, and therefore the slope dM/dx, decreases from a positive value at x_B to zero at x_C, we may conclude that the parabola curves downward with a horizontal tangent at x_C. Thus, we obtain the following bending moment diagram.

Clearly, the diagrams we obtained here are identical to those obtained in Example 4.

The advantages of the procedure we have just presented become evident when we note that the majority of the effort in the solution of the examples was devoted to explanations, not calculations. By evaluating the areas under a few curves, we can quickly and accurately obtain the required diagrams. It is no longer necessary to study the free body diagrams corresponding to different cutting planes. Needless to say, the formulas for the areas of some elementary shapes given in Appendix C may be valuable for computations by this method.

HOMEWORK PROBLEMS

VI.36–VI.47 Using the relationships between the transverse load, the shear, and the bending moment, sketch shear and bending moment diagrams for the beams indicated below. Give all critical values and show all computations.

Homework Problem	VI.36	VI.37	VI.38	VI.39	VI.40
Beam in Problem	VI.19	VI.24	VI.26	VI.25	VI.18

Homework Problem	VI.41	VI.42	VI.43	VI.44	VI.45
Beam in Problem	VI.20	VI.22	VI.27	VI.29	VI.30

Homework Problem	VI.46	VI.47
Beam in Problem	VI.31	VI.34

VI.48 and VI.49 The shear diagram for a simply supported beam is partially known, as depicted in the accompanying figure. It is also known that no couples are applied to the beam. (a) Determine the distributed and concentrated forces acting on the beam, (b) the length of the beam and all distances required to specify where the loads are applied, (c) the magnitude and location of the maximum bending moment in the beam.

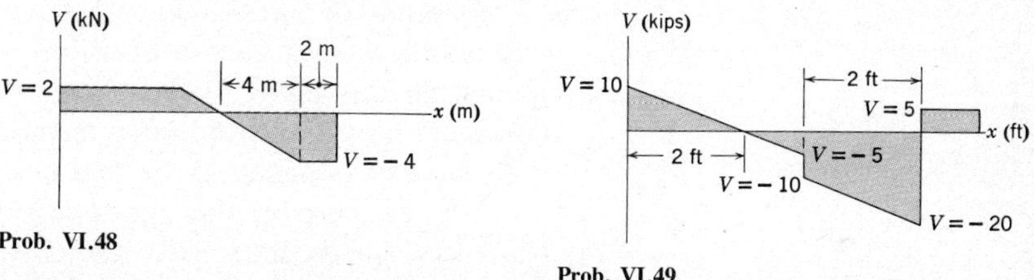

Prob. VI.48

Prob. VI.49

VI.50 The shear diagram for a simply supported beam carrying no couples is known to the extent shown. The maximum bending moment in

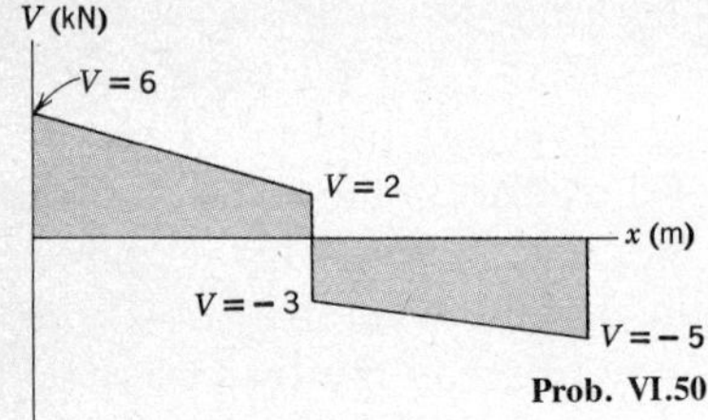

Prob. VI.50

this beam is $M = 20$ kN-m. Make a sketch of the beam, showing all loads, reaction forces, and lengths along the beam.

VI.51 The overhanging beam is loaded by a sinusoidal distributed force, given in units of newtons per meter by $q(x) = 400 \sin \pi x/20$, where x is the distance in meters from the left end. Sketch the shear and moment diagrams for the beam, specifying all critical values.

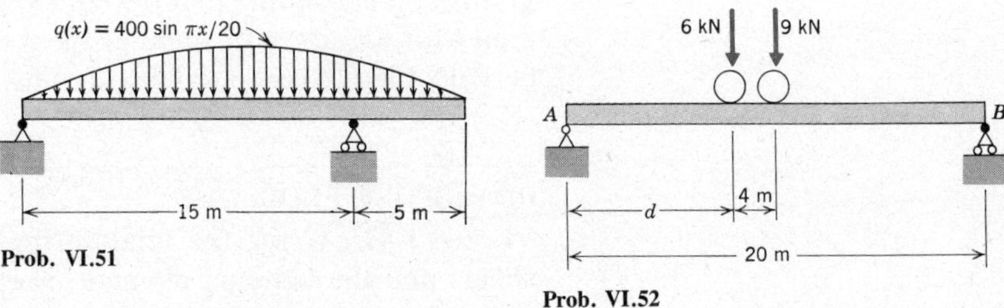

Prob. VI.51

Prob. VI.52

VI.52 The load exerted by the chassis of a movable crane on one set of wheels is shown. Determine the distance d, specifying the location of the chassis that results in the maximum possible bending moment in the supporting beam AB. What is the corresponding value of this moment?

B. FLEXIBLE CABLES

Like a beam, a *flexible cable* is a slender body. The primary difference between these two types of structural components is that the ratio of the length to the largest cross-sectional dimension for a cable (of the order of 1000 or more) is much larger than that for a beam. Because of this, a cable offers negligible resistance to bending and it bows out if one tries to apply a compressive force to it. Thus, the only internal force of any significance that it may withstand is an axial tension force.

Flexible cables are used in a variety of ways in structures and machines. They are a main element in suspension bridges, conveyor belts, and alpine cable cars. Their behavior is important to the design of electric and telephone transmission lines.

One use of a cable that we have already treated is to tie it between two points having known positions as a means of interconnecting these points. In this usage the tensile force is usually so large that we may consider the cable to be taut, meaning that the tensile force is collinear with the line connecting the end points.

Our concern in this module is with situations where cables carry transverse loads that are sufficiently large to make it unacceptable to consider the cable to be stretched in a straight line between its ends. We then say that the cable *sags*.

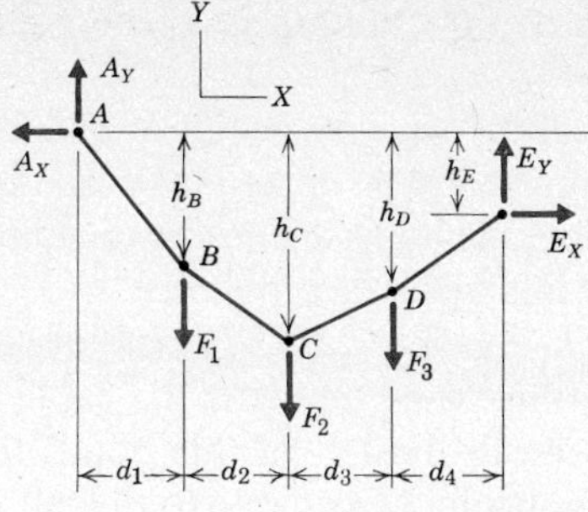

Figure 15

1 Discrete Transverse Loads

Here we will consider the case where a cable supports a set of transverse loads at discrete points along its length. Such a situation is exemplified by Figure 15, which shows the free body diagram of a cable that is anchored to pins at ends A and E as it supports a series of transverse loads $\bar{F}_i$. The lines of action of the vertical loads are specified by the dimensions d_1, d_2, d_3, and d_4, which we will regard as known quantities.

If the loads are much larger than the weight of the cable, we may regard the segments of the cable between the points of application of the loads to be taut. Each segment then carries a constant tensile force, although the tension may vary from segment to segment. Thus, the analysis of the forces within the cable in Figure 15 reduces to the determination of the tensile forces T_{AB}, T_{BC}, T_{CD}, and T_{DE}.

Let us assume that the location of ends A and E of the cable are known, so that h_E is a known dimension. Then, additional information we seek is the vertical distances h_B, h_C, and h_D. These distances represent the *sag* of the cable. Including the four unknown reaction components, the analysis of this cable requires that we determine 11 unknowns (4 tensions, 4 reactions, and 3 sag distances).

Considering that the cable is functioning as a series of straight two-force members, it might seem that the problem can be solved by the same methods as those used for trusses. This is true, but there is one new feature here that complicates the solution. Specifically, in a truss the location of all joints is known, whereas the locations of joints B, C, and D for the cable are part of the information we seek.

Let us investigate the number of available equations. There are five joints (points A through E). Because the forces at each joint are collinear, there are two equilibrium equations for each joint, making a total of 10 available equilibrium equations. In practice, we usually consider three of the equilibrium equations for the joints as checks. These equations are replaced by the three equilibrium equations corresponding to the free body diagram of the entire structures, Figure 15. Nevertheless, there are only 10 independent equations of static equilibrium, whereas there are 11 unknowns; we need one more equation.

The cable is *not* statically indeterminate. Rather, the difficulty lies in the fact that we have not fully specified the geometry of the system.

One class of problems we can pursue pertains to situations where we know the length of the cable. In that case we would use the Pythagorean theorem to express the length of each segment of the cable in terms of the sags h_B, h_C, and h_D. The required additional equation is then obtained by equating the sum of the individual lengths to the total length.

A complication arises in a problem of this type, because the Pythagorean theorem gives rise to terms in the equation for the length that are the square roots of the unknown sags. Thus, although the formulation of the required equations is fairly straightforward, the solution of these equations is laborious.

A simpler situation occurs when one of the sags is specified. This, of course, reduces by one the number of unknowns to be determined, so that the static equilibrium equations are sufficient. There is a simple procedure available for formulating the solution for this class of problems.

For example, suppose that we know h_C in Figure 15. We employ the method of sections for trusses, using a cutting plane that cuts the cable to one side of pin C. Choosing a cutting plane to the left of this point, the corresponding free body diagram for the portion of the cable to the left of the plane is shown in Figure 16.

Figure 16

When we sum moments about point C, we obtain an equation relating A_Y and A_X. In conjunction with the three equilibrium equations for the entire system, we can determine all of the reactions. Further, the force equilibrium equations for the section in Figure 16 give the tension T_{BC} and the angle of elevation θ_{BC} of cable BC.

Once the reactions are known, the method of joints may be used to determine the tension and angle of elevation of each cable segment. Finally, the sag of each point can be determined from the laws of trigonometry.

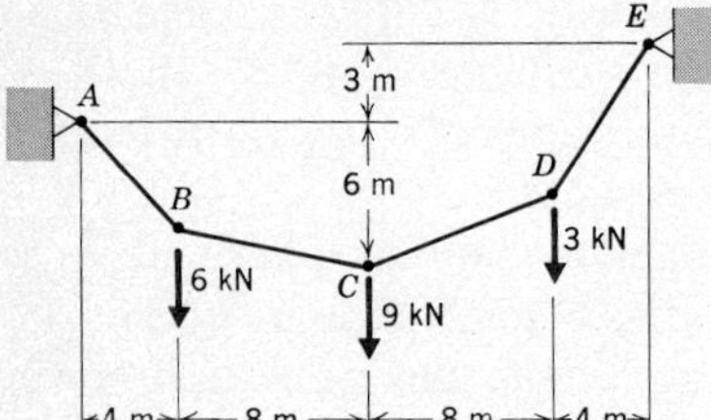

EXAMPLE 8

The cable carries three transverse loads, as shown. Determine (a) the elevations of joints B and D, (b) the maximum tension, (c) the maximum angle of elevation.

Solution

The free body diagram and equilibrium equations for the entire system are as given below.

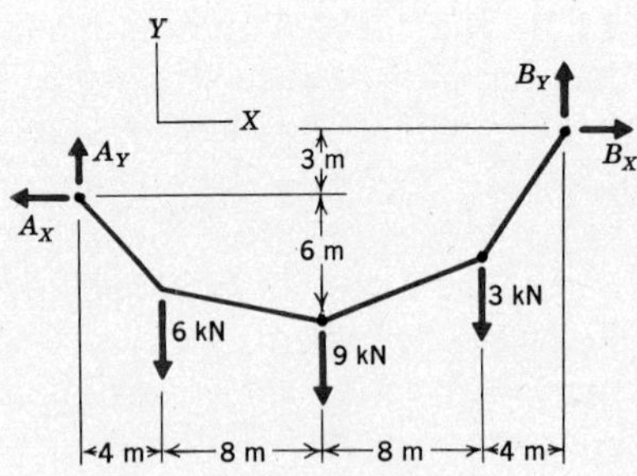

$$\Sigma M_{AZ} = 24B_Y - 3B_X - 6(4) - 9(12) - 3(20) = 0 \text{ kN-m}$$

$$\Sigma F_X = -A_X + B_X = 0$$

$$\Sigma F_Y = A_Y + B_Y - 18 = 0 \text{ kN}$$

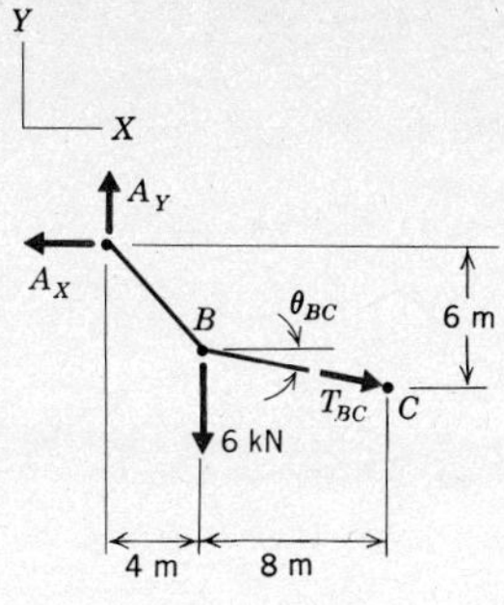

These equations are not sufficient to solve for the reactions. Additional equations are obtained by passing a cutting plane through the segment of cable on either side of point C, whose elevation is known, and isolating one of the cable sections formed by the plane. Our choice is shown in the free body diagram. The corresponding equilibrium equations are

$$\Sigma M_{CZ} = A_X(6) - A_Y(12) + 6(8) = 0$$

$$\Sigma F_X = T_{BC} \cos \theta_{BC} - A_X = 0$$

$$\Sigma F_Y = A_Y - T_{BC} \sin \theta_{BC} - 6 = 0$$

The simultaneous solutions of the moment equation for the section and the equations for the entire system are

$$A_X = B_X = 9.60 \qquad A_Y = 8.80 \qquad B_Y = 9.20 \text{ kN}$$

The force equations for the section of cable then become

$$T_{BC} \cos \theta_{BC} = A_X = 9.60$$

$$T_{BC} \sin \theta_{BC} = A_Y - 6 = 2.80$$

The solutions of these equations are

$$\theta_{BC} = \tan^{-1}\left(\frac{2.80}{9.60}\right) = 16.26°$$

$$T_{BC} = \frac{9.60}{\cos 16.26°} = 10.00 \text{ kN}$$

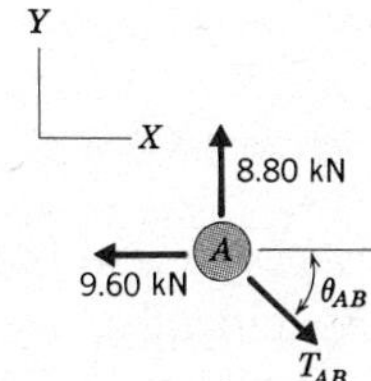

We now proceed to investigate the joints, starting with either end. Considering pin A first, the free body diagram and equilibrium equations are as follows.

$$\Sigma F_X = T_{AB} \cos \theta_{AB} - 9.60 = 0$$

$$\Sigma F_Y = 8.80 - T_{AB} \sin \theta_{AB} = 0$$

The foregoing equations yield

$$\theta_{AB} = 42.51° \qquad T_{AB} = 13.023 \text{ kN}$$

Because we now know the tension and angle of elevation of segments AB and BC, the equilibrium equations for joint B only give checks on the computations. Skipping this step, we proceed to joint C. The free body diagram and equilibrium equations for this joint are described on the next page, assuming that point D is higher than point C.

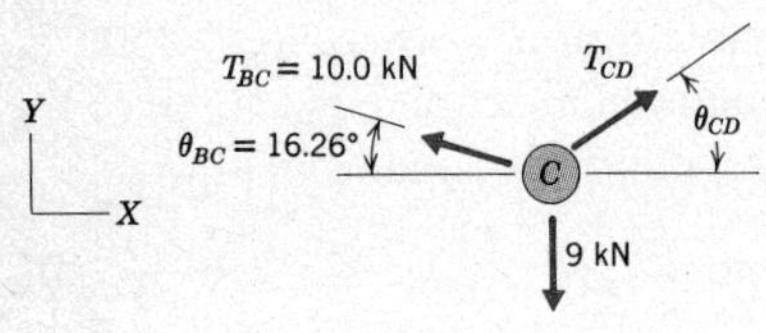

$$\Sigma F_X = T_{CD} \cos \theta_{CD} - 10.00 \cos 16.26° = 0$$

$$\Sigma F_Y = T_{CD} \sin \theta_{CD} + 10.00 \sin 16.26° - 9 = 0$$

$$\theta_{CD} = 32.86° \qquad T_{CD} = 11.428 \text{ kN}$$

Finally, for joint D we have

$$\Sigma F_X = T_{DE} \cos \theta_{DE} - 11.428 \cos 32.86° = 0$$

$$\Sigma F_Y = T_{DE} \sin \theta_{DE} - 11.428 \sin 32.86° - 3 = 0$$

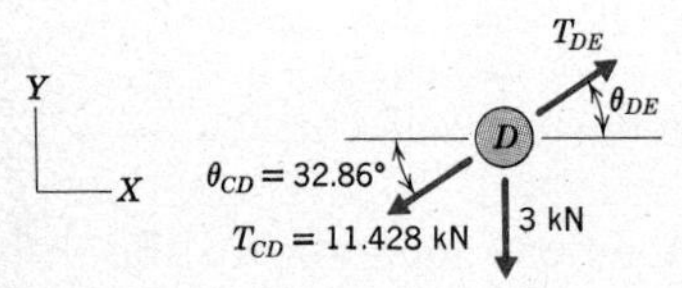

Thus

$$\theta_{DE} = 43.78° \qquad T_{DE} = 13.297 \text{ kN}$$

We have found all of the tensions and all of the angles of elevation, so we omit consideration of the equilibrium of joint E, which would only serve as a check on the calculations. From these results we may answer parts (b) and (c), for we see that the maximum tension and angle of elevation are those for segment DE, being

$$\theta_{\max} = \theta_{DE} = 43.8° \qquad T_{max} = T_{DE} = 13.30 \text{ kN}$$

We should not be surprised that the tension is maximum in the cable having the steepest slope, for equilibrium requires that the horizontal component T_i and θ_i of the tension in the ith segment be a constant equal to A_X $(= -E_X)$.

To determine the sags from the angle of elevation we draw a sketch of the geometry. From this sketch we see that

$$h_B = 4 \tan \theta_{AB} = 3.667 \text{ m}$$

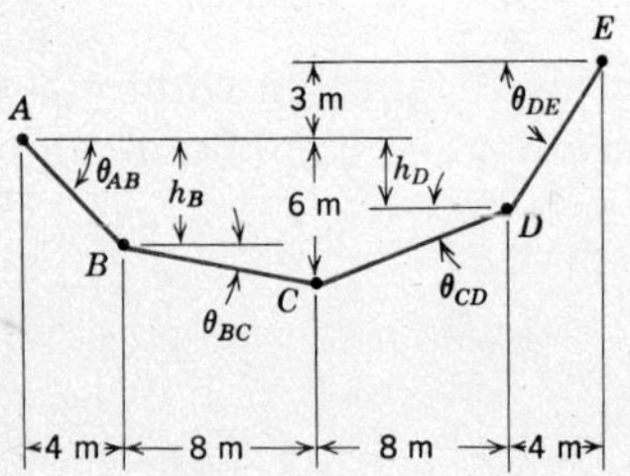

and

$$h_D + 3 = 4 \tan \theta_{DE} \quad \text{so} \quad h_D = 0.833 \text{ m}$$

The sketch also suggests another check, in addition to the excess equilibrium equations, on the computations. Specifically, we may ascertain whether, starting from either pin A or E, the sag h_C of point C actually is found to be 6 m. Starting from pin A, we have

$$h_C = h_B + 8 \tan \theta_{BC} = 6.000 \text{ m}$$

whereas starting from pin E, we find

$$h_C = h_D + 8 \tan \theta_{CD} = 6.001 \text{ m}$$

Because both values of h_C are in agreement with the given value, the check validates the results.

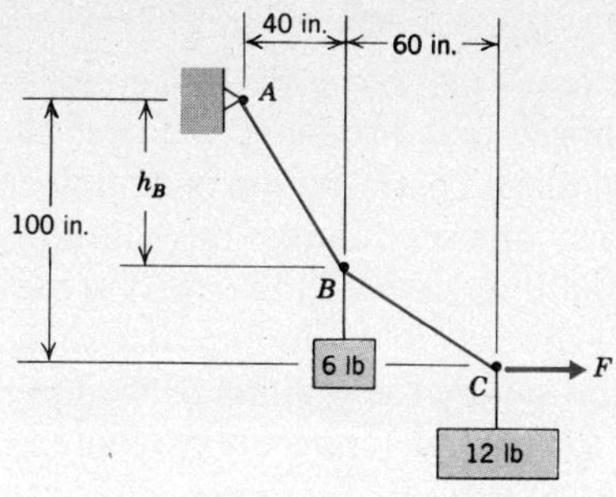

Prob. VI.53

HOMEWORK PROBLEMS

VI.53 Two blocks are supported by the lightweight cable ABC, as shown. Determine (a) the magnitude of the horizontal force $\bar{F}$, (b) the corresponding sag h_B.

VI.54 A lightweight cable carries the two concentrated forces shown. Knowing that $h_B = 8$ m, determine (a) the sag h_C, (b) the maximum tension, (c) the maximum angle of elevation.

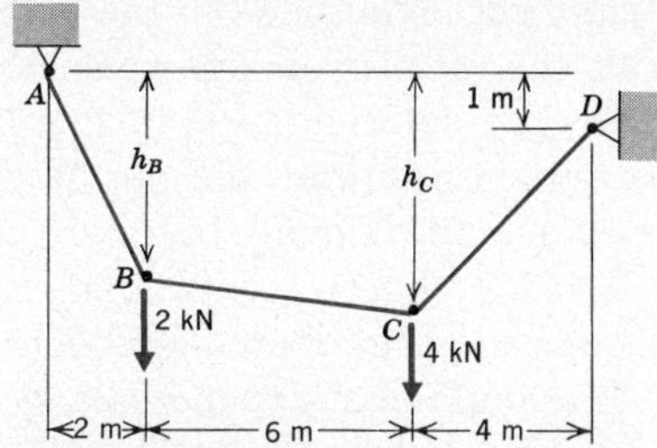

Prob. VI.54 and VI.55

VI.55 A lightweight cable supports the two concentrated forces shown. Knowing that $h_C = 8$ m, determine (a) the sag h_B, (b) the maximum tension, (c) the maximum angle of elevation.

VI.56 A cable supports three forces, as shown. Determine (a) the sags h_B and h_D, (b) the maximum tension, (c) the maximum angle of elevation.

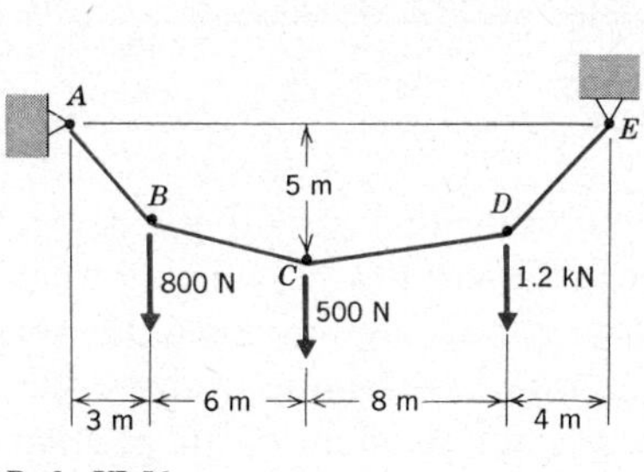

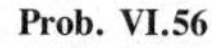
Prob. VI.56

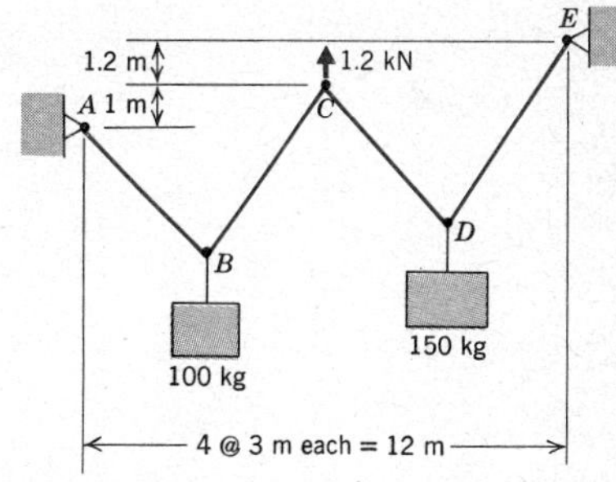

Prob. VI.57

VI.57 A 1.2-kN vertical force is applied to the cable at point C in order to hold the blocks in the position shown. Determine (a) the maximum tension in the cable, (b) the length of the cable.

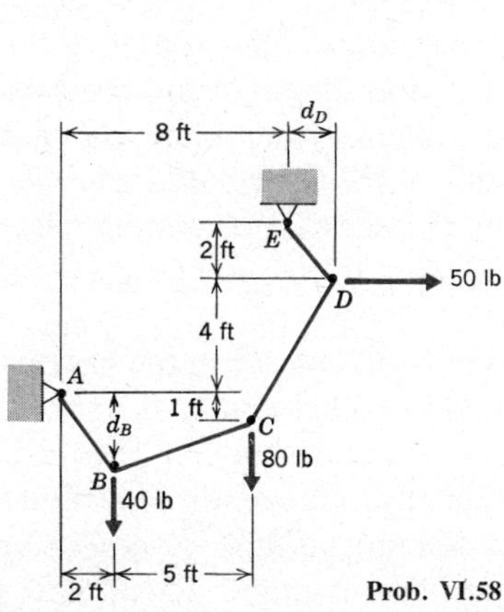

Prob. VI.58

VI.58 A set of horizontal and vertical forces are applied to cable $ABCDE$. Determine (a) the distances d_B and d_D, (b) the maximum tension, (c) the maximum angle of elevation.

2 Distributed Loads — Parabolic Sag

Consider a cable supporting many discrete loads that are applied at equally spaced intervals a distance h apart in the horizontal direction. We wish to determine how the cable sags under the loading. This situation is depicted in Figure 17. It was demonstrated in the previous section that the solution of this situation requires additional information, such as the sag at some point along the cable or the total length of the cable. Assuming that such information is given, we may determine the sag by the tedious procedure of considering the equilibrium equations for each discrete point where a load is applied, as outlined in the preceding section.

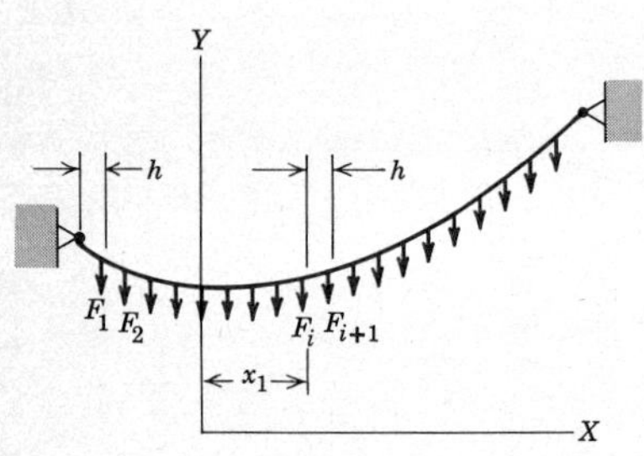

Figure 17

With the advent of high-speed electronic computers, such an approach for the determination of the sag is not unreasonable. However, engineers in earlier times did not have such computational aids, so they developed an approximate, but more convenient, method for determining the sag. This method was based on replacing the discrete loads in Figure 17 by a continous distributed load $q(x)$, where q has units of force per unit length. The functional dependence of $q(x)$ is defined such that at point i along the cable, having the horizontal coordinate x_i as shown in the figure, $q(x_i) = F_i/h$. In other words, $q(x)$ is the average line load.

The case where $q(x)$ is a constant value, q_0, is the one that is most commonly encountered in practice. This is the situation that arises in the design of a suspension bridge, such as the one depicted in Figure 18, where a number of vertical cables are suspended from the spanning cables

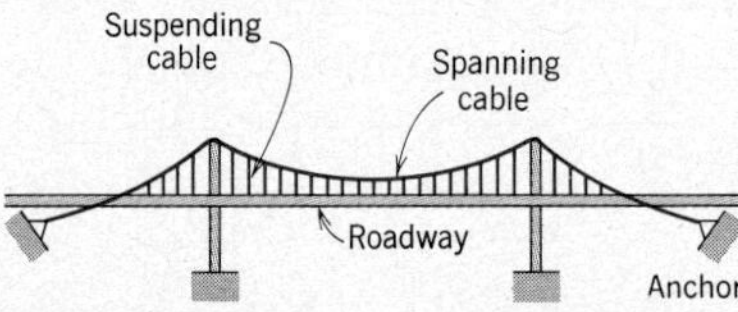

Figure 18

to support the roadway. If we neglect the weight of the cables and the stiffness of the roadway, we may treat the spanning cables as though they were loaded only by the constant weight per unit length of the roadway.

Equations pertaining to the sag of the cable may be obtained in general terms, independent of the precise details of the distributed load $q(x)$. For the convenience it affords, we select an xyz coordinate system whose origin is placed at the point where the cable has a horizontal tangent, that is, at the lowest point of the cable.

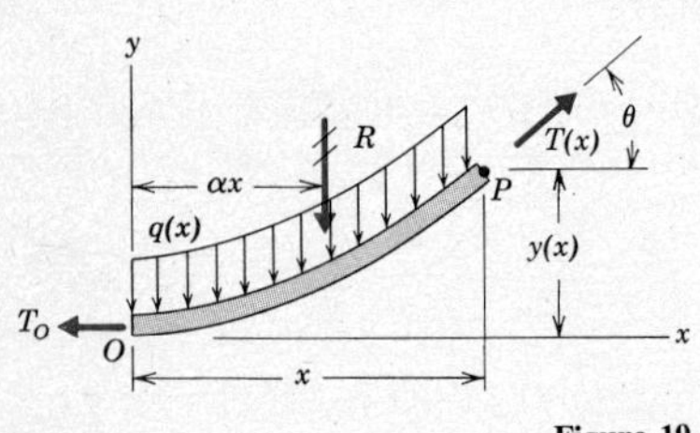

Figure 19

Let us consider a finite section of the cable of length x in the horizontal direction, having one end at the origin and the other end at an arbitrary point P. The free body diagram for such an element is shown in Figure 19.

The force $\bar{R}$ shown in the figure is the resultant of the distributed load. Its magnitude and the distance αx defining its line of action are known if the loading function $q(x)$ is known. For example, for a uniform load q_0, $R = q_0 x$ and $\alpha = \frac{1}{2}$. The other forces acting on the element are the tensile forces tangent to the cable at each end. The term T_O denotes the tension at the horizontal tangent, point O.

Figure 19 shows that the sag of the cable is defined by the function $y(x)$. To determine this function, as well as the tension within the cable, we formulate the equations of equilibrium for the element. Choosing point P on the right for the moment sum, we obtain

$$\Sigma M_{PZ} = R(x - \alpha x) - T_O(y) = 0$$

$$\Sigma F_x = T \cos \theta - T_O = 0 \qquad \Sigma F_y = T \sin \theta - R = 0$$

The solution of these equations is

$$T = (T_O^2 + R^2)^{1/2} \tag{8a}$$

$$\tan \theta = \frac{R}{T_O} \tag{8b}$$

$$y = \frac{(1 - \alpha)xR}{T_O} \tag{8c}$$

Two features of these results are of particular importance. First, from the equation for ΣF_X we see that the cable sags such that the tension within it always has a constant horizontal component T_O. This quantity will be a key parameter for the solution of problems. Second, because θ defines the tangent to the cable, it follows that

$$\tan \theta = \frac{dy}{dx} \tag{9}$$

In other words, equations (8b) and (8c) are not independent. After substitution of the expression for R into equation (8c), the resulting expression for y as a function of x must give the same result for $\tan \theta$ in equation (9) as that obtained from equation (8b). This serves to verify our assumption that the cable carries only a tensile force.

To proceed further, we must define the load function. As noted earlier, our primary interest is with the case of a uniform load. Setting $q(x) = q_0$, $R = q_0 x$, and $\alpha = \frac{1}{2}$ in equations (8a and c), we obtain

$$\boxed{y = \frac{q_0}{2T_O} x^2} \tag{10a}$$

$$\boxed{T = (T_O^2 + q_0^2 x^2)^{1/2}} \tag{10b}$$

The equation for y shows that a cable carrying a *uniform horizontal load* sags into a *parabolic* shape. It is readily verified that the equation for $\tan \theta$ obtained from equations (9) and (10a) is identical to that given by equation (8b), confirming the correctness of the analysis.

As was true in the case of cables carrying concentrated loads, there are two standard types of problems we can analyze, assuming that the locations of the ends of the cables are known. In one, we know the sag of some point on the cable. Equations (10) will prove sufficient for such cases.

The second type of problem is the one where we know the length of the cable. To solve problems of this type we need an expression relating the length of the cable to the other parameters for the system. This expression is obtained by evaluating the arc length of the parabola formed by the cable.

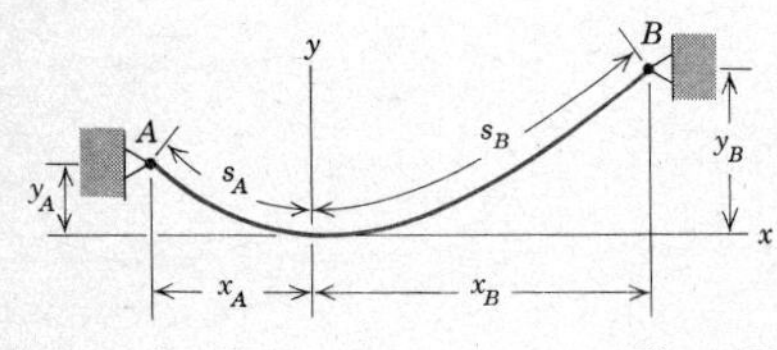

Figure 20

Consider the situation depicted in Figure 20. The total length of the cable is the sum of the arc lengths s_A and s_B measured from the origin to each end. Both quantities are arc lengths along the parabola measured from the origin, so we need only consider how an expression for one arc length is obtained. In courses on calculus, it is shown that the arc length s_B may be obtained by evaluating the following integral.

$$s_B = \int_0^{x_B} \left[1 + \left(\frac{dy}{dx}\right)^2\right]^{1/2} dx$$

Upon substitution for y from equation (10), the integral becomes

$$s_B = \int_0^{x_B} \left[1 + \frac{q_0^2}{T_O^2} x^2\right]^{1/2} dx \tag{11}$$

An exact solution for this integral may be found using a table of integrals; however, the resulting expression is cumbersome for calculations. A more convenient expression may be obtained by expanding the integrand in equation (11) in an infinite series using the binomial theorem, which is

$$(1 + Z)^n = 1 + nZ + \frac{n(n-1)}{2!} Z^2 + \frac{n(n-1)(n-2)}{3!} Z^3 + \ldots$$

With this expression, equation (11) becomes

$$\begin{aligned} s_B &= \int_0^{x_B} \left(1 + \frac{q_0^2 x^2}{2T_O^2} - \frac{q_0^4 x^4}{8T_O^4} + \ldots\right) dx \\ &= x_B\left(1 + \frac{q_0^2 x_B^2}{6T_O^2} - \frac{q_0^4 x_B^4}{40T_O^4} + \ldots\right) \end{aligned}$$

Finally, using equation (10) to eliminate q_0/T_O, we find

$$\boxed{s_B = x_B\left[1 + \frac{2}{3}\left(\frac{y_B}{x_B}\right)^2 - \frac{2}{5}\left(\frac{y_B}{x_B}\right)^4 + \ldots\right]} \tag{12}$$

The infinite series in equation (12) is convergent when $y_B/x_B < 0.5$. If y_B/x_B is substantially lower than this value, as it is in many engineering applications, then the convergence is very rapid and the listed terms provide sufficient accuracy. On the other hand, if y_B/x_B exceeds 0.5, the exact integral of equation (11) is required.

EXAMPLE 9

A cable supporting a pipeline weighing 1 kip/ft is to be suspended between pins A and B, whose positions are as shown. The design calls for the pipeline to be 8 ft below the elevation of pin A and for the low point on the cable to be at the pipeline. Determine (a) the maximum tension in the cable, (b) the required length of the cable.

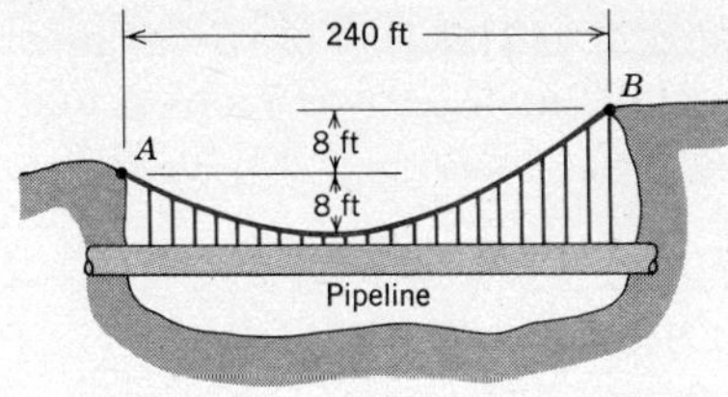

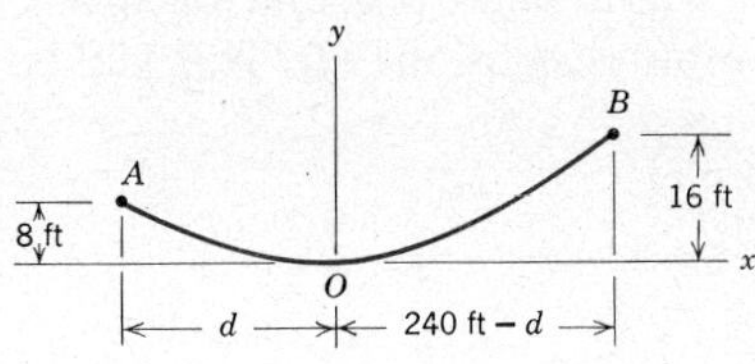

Solution

In this problem the location of the horizontal tangent (the low point) is not completely given. This point is the origin from which the position of all points on the cable are measured, so its location must be found. This evaluation is achieved with the aid of a sketch of the geometry and the coordinate system.

Letting d be the horizontal distance of the origin O from pin A, we see from the sketch that the coordinates of point A are ($x_A = -d$, $y_A = 8$), whereas those for point B are ($x_B = 240 - d$, $y_B = 16$). Both of these points lie on the parabola defined by equation (10). Therefore, we have

$$y_A = 8 = \frac{q_0}{2T_O}(x_A)^2 = \frac{q_0}{2T_O}d^2$$

$$y_B = 16 = \frac{q_0}{2T_O}(x_B)^2 = \frac{q_0}{2T_O}(240 - d)^2$$

The distributed load is the weight of the pipeline, $q_0 = 1000$ lb/ft. Therefore, the foregoing represents two equations whose simultaneous solution gives the values of d and T_O. We first solve for q_0/T_O from the first equation.

$$\frac{q_0}{T_O} = \frac{8}{d^2}$$

The second equation then becomes a quadratic in d. Specifically, it gives

$$16 = \left(\frac{8}{d^2}\right)(240 - d)^2$$

$$d^2 + 480d - 5.76(10^4) = 0$$

$$d = 99.41, \; -579.4 \text{ ft}$$

The negative value is an extraneous root for this problem, so $d = 99.41$ ft. Substituting the value of q_0 and d, we then find T_O.

$$T_O = q_0\left(\frac{d^2}{8}\right) = 1000\frac{(99.41)^2}{8} = 1.2353(10^6)\text{ lb}$$

We now have the parameters required to obtain the information requested. From equation (10b) we see that the tension is a maximum at the point that is farthest from the horizontal tangency. This is at pin B because the distance d is less than half the horizontal distance between pins A and B. Thus, from equation (10b) we have, with $x_B = 240 - d = 140.59$ ft,

$$\begin{aligned} T_{\max} &= (T_O{}^2 + q_0{}^2x_B{}^2) = [(1.2353)^2(10^{12}) + (10)^6(140.59)^2] \\ &= 1.2433(10^6)\text{ lb} \end{aligned}$$

To determine the length of the cable we note that y_A/x_A and y_B/x_B are both extremely small. We therefore may employ equation (12). This yields

$$\begin{aligned} s_A &= x_A\left[1 + \frac{2}{3}\left(\frac{y_A}{x_A}\right)^2 - \frac{2}{5}\left(\frac{y_A}{x_A}\right)^4\right] \\ &= 99.41\left[1 + \frac{2}{3}\left(\frac{8}{99.41}\right)^2 - \frac{2}{5}\left(\frac{8}{99.41}\right)^4\right] \\ &= 99.41 + 0.4275 \end{aligned}$$

$$\begin{aligned} s_B &= 140.59\left[1 + \frac{2}{3}\left(\frac{16}{140.59}\right)^2 - \frac{2}{5}\left(\frac{16}{140.59}\right)^4\right] \\ &= 140.59 + 1.2045 \end{aligned}$$

$$s = s_A + s_B = 240 + 1.632 = 241.63\text{ ft}$$

These results show that because of the relatively small amount by which the cable sags, the tension within the cable is much larger than the total weight of the pipeline, (1000 lb/ft)(240 ft), and the required length of the cable is only 1.63 ft longer than the horizontal distance spanned.

HOMEWORK PROBLEMS

VI.59 The span of a suspension bridge between two towers of equal height is 1000 m and the sag of its cables at midspan is 100 m. The roadway, having a mass of $20(10^3)$ kg/m, is supported equally by two suspension cables. Determine (a) the maximum and minimum tensions in the cable, (b) the length of the cable between the towers.

VI.60 Cable AB supports a 200-kN total load which is distributed uniformly along the horizontal. The slope of the cable is zero at the lower

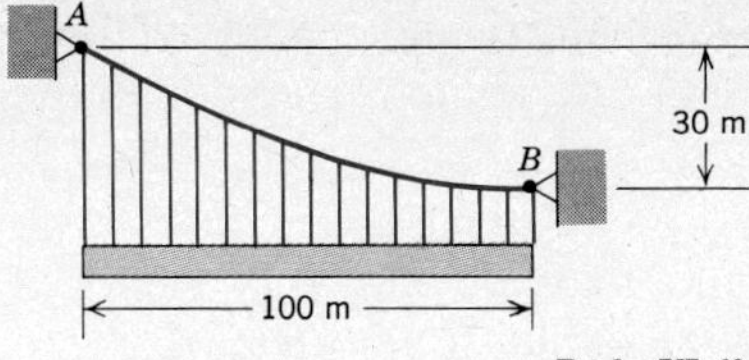

Prob. VI.60

end. Determine (a) the maximum and minimum tensions in the cable, (b) the location of the point on the cable where the tension is the average of the values determined in part (a).

VI.61 A cable carrying a constant load q_0 along the horizontal direction is anchored at ends A and C. The cable passes over a roller at the intermediate support D, as shown. Derive an expression for L_2/h_z in terms of L_1, h_1, and L_2.

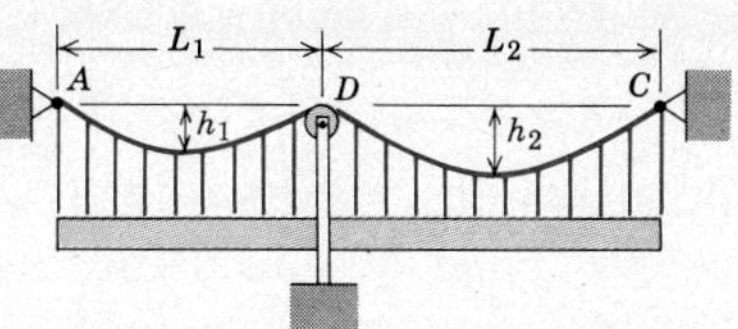

Prob. VI.61

VI.62 A cable spanning a distance of 300 ft between pins located at the same elevation supports a uniform load of 3 lb/ft. The maximum tension in the cable is 2000 lb. Determine (a) the maximum sag, (b) the slope of the cable at its supports, (c) the length of the cable.

VI.63 Cable AB, carrying a uniform load of 500 lb/ft, has a span of 300 ft. Knowing that the tangent to the cable at an end is at a 45° angle from the vertical, determine the maximum sag and the maximum tension in the cable.

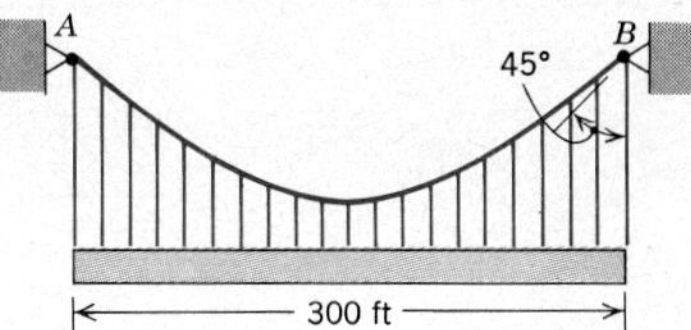

Prob. VI.63

VI.64 Cables AB and BC supporting a uniform horizontal load of 4 kN/m are both pinned to tower B, as shown. Determine the resultant force exerted by both cables on the tower.

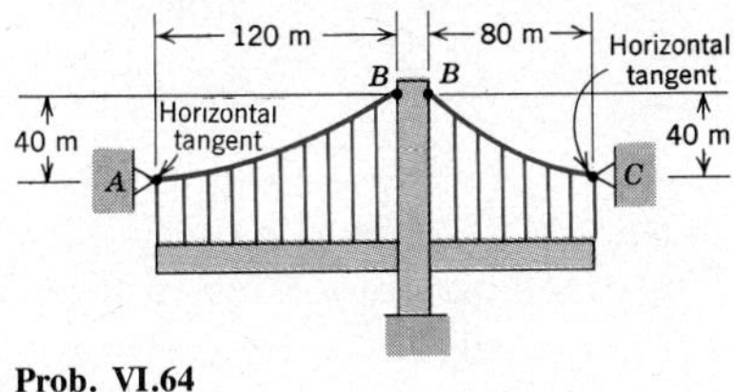

Prob. VI.64

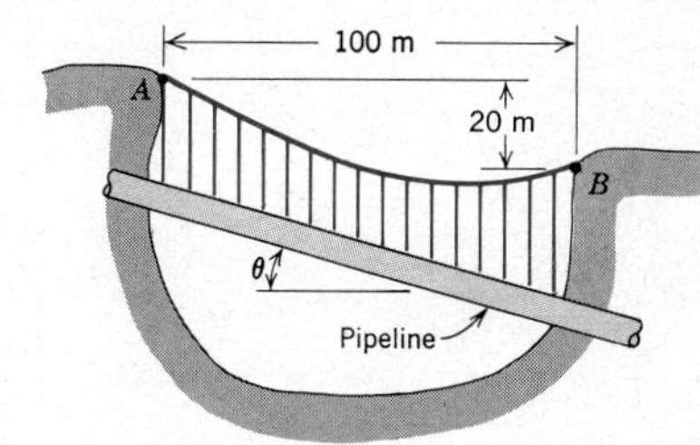

Prob. VI.65

VI.65 The pipeline, whose mass per unit length is 200 kg/m, is oriented horizontally ($\theta = 0$) as it passes over the gorge. The ends of the cable are located as shown. It is known that at its lowest point the cable is 5 m below pin B. Determine (a) the maximum tension within the cable, (b) the length of the cable.

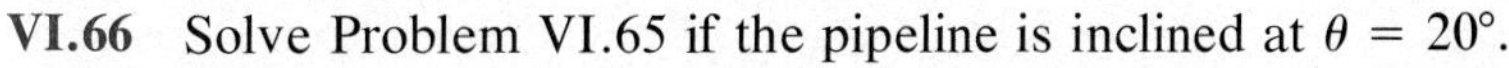

VI.66 Solve Problem VI.65 if the pipeline is inclined at $\theta = 20°$.

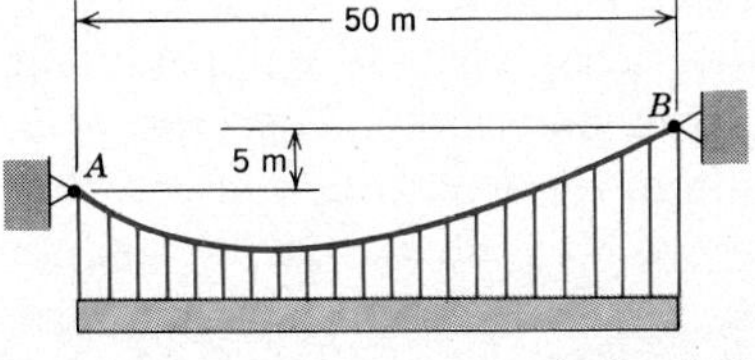

Prob. VI.67

VI.67 A cable spanning the distance between pins A and B supports a uniform load of 200 N/m, distributed along the horizontal. Knowing that the lowest point on the cable is 10 m horizontally from pin A, determine the lowest elevation of the cable below pin A and the length of the cable.

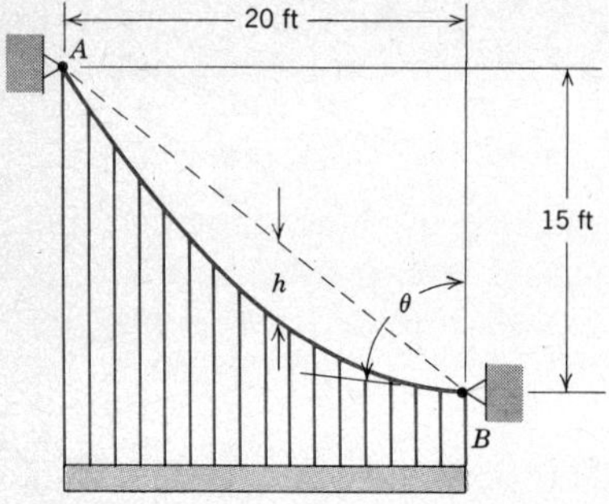

Prob. VI.69

VI.68 Solve Problem VI.67 if, instead of knowing the horizontal location of the low point of the cable, it is known that the minimum tension in the cable is 20 kN.

VI.69 A cable carrying a uniform horizontal load of 4 kips/ft is anchored at pins A and B. It is known that at end B the cable slopes upward at $\theta = 75°$ from the vertical. Determine (a) the maximum tension, (b) the length of the cable, (c) the maximum value, h_{max}, by which the cable sags below line AB.

VI.70 and VI.71 The homogeneous body shown has a weight W. It is known that the cable has a horizontal tangency at end A. Considering the span L and the constant horizontal component T_O of the tension force to be known parameters, derive expressions for (a) the equation of the curve formed by the sagging cable, (b) the distance h by which pin B is higher than pin A, and (c) the maximum tension in the cable.

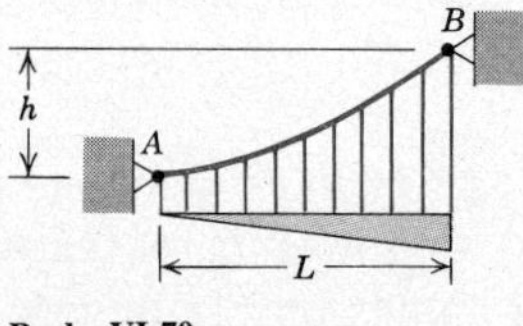

Prob. VI.70

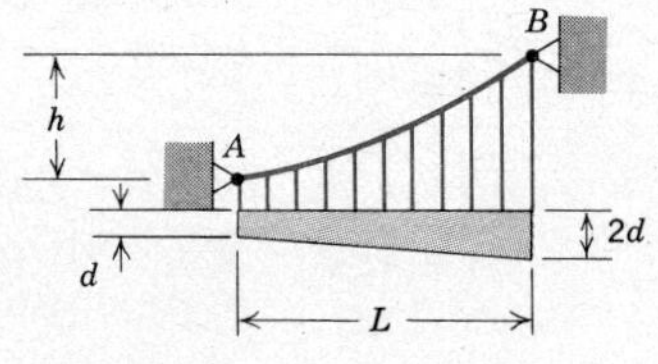

Prob. VI.71

3 Catenary

Frequently, we are interested in the sag of a cable under its own weight. Restricting ourselves to the case of a uniform cable having a weight per unit length w_0, it follows that the gravitational force acting on an element of cable of infinitesimal arc length ds is $w_0\, ds$, as depicted in Figure 21.

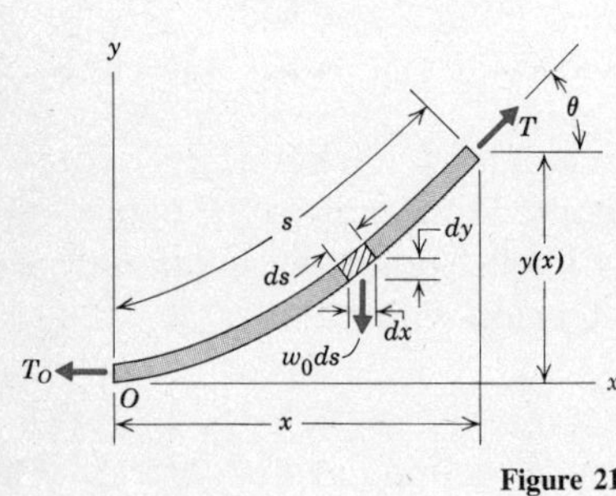

Figure 21

The situation in Figure 21 is a special case of a cable carrying an arbitrary distributed load, described by Figure 19 and equations (8). Let us therefore consider the resultant of the distributed gravity force acting on the section of cable. Because w_0 is constant, it follows that $w_0 s$ is the resultant gravitational force, or equivalently, the weight of the section of cable. Hence, $R = w_0 s$. Further, by definition, the weight force is applied at the center of mass. This creates a problem, because the location of the center of mass depends on the shape of the curve formed by the cable in Figure 21, and this shape is what we wish to determine. Therefore, we cannot locate the line of action of the resultant force $\bar{R}$, which means that the parameter α in equation (8c) is unknown. It follows that we cannot determine the function $y(x)$ describing the sag of the cable solely by solving the equilibrium equations for the element.

To resolve this difficulty, we recall that equations (8b) and (8c) are not independent because equation (9), $\tan\theta = dy/dx$, must be satisfied. Equation (8b) does not depend on the parameter α, so we will use that equation in place of equation (8c). Substituting $R = w_0 s$ into equations (8a) and (8b), we get

$$T = [T_O^2 + w_0^2 s^2]^{1/2} \tag{13a}$$

$$\tan\theta = \frac{dy}{dx} = \frac{w_0}{T_O}s \tag{13b}$$

Equation (13b) is a *differential equation*. It relates the differentials dy and dx of the infinitesimal element shown in Figure 21. This figure also shows that, by the Pythagorean theorem,

$$(ds)^2 = (dx)^2 + (dy)^2$$

Noting that s appears in the right side of equation (13b), we use the foregoing relationship to eliminate dy by writing

$$\frac{dy}{dx} = \frac{[(ds)^2 - (dx)^2]^{1/2}}{dx} = \frac{w_0}{T_O}s$$

$$\left[\left(\frac{ds}{dx}\right)^2 - 1\right]^{1/2} = \frac{w_0}{T_O}s$$

$$\frac{ds}{dx} = 1 + \left(\frac{w_0}{T_O}\right)^2 s^2$$

Solving for dx, we find

$$dx = \frac{ds}{1 + (w_0/T_O)^2 s^2} \tag{14}$$

This last equation may be integrated with the aid of a table of integrals. As shown in Figure 21, s is measured from the origin, so $s = 0$ when $x = 0$. Thus, a definite integral of equation (14) gives

$$\int_0^x dx = \int_0^s \frac{ds}{[1 + (w_0/T_O)^2 s^2]^{1/2}}$$

$$x = \frac{T_O}{w_0}\left[\sinh^{-1}\left(\frac{w_0}{T_O}s\right) - \sinh^{-1}(0)\right] \tag{15}$$

The function $v = \sinh u$ is the hyperbolic sine of u, and the function $u = \sinh^{-1} v$ is its inverse, called the arc hyperbolic sine. There is some similarity between the hyperbolic and trigonometric functions, resulting from their definitions in terms of complex numbers. For our purposes here, we will need the following properties of the hyperbolic sine and cosine.

$$\sinh u \equiv \tfrac{1}{2}(e^{u} - e^{-u}) \qquad \cos u \equiv \tfrac{1}{2}(e^{u} + c^{-u}) \tag{16a}$$

$$\frac{d}{du}\sinh u = \cosh u \qquad \frac{d}{du}\cosh u = \sinh u \tag{16b}$$

$$\cosh^2 u - \sinh^2 u = 1 \tag{16c}$$

From the definitions, equations (16a), we see that

$$\sinh(-u) = -\sinh u \qquad \cosh(-u) = \cosh u \tag{16d}$$

$$\sinh(0) = 0 \qquad \cosh(0) = 1 \tag{16e}$$

Values of the hyperbolic functions may be found in standard numerical tables, as well as by evaluating them directly. In addition, they are keyboard functions on some electronic calculators.

Returning to equation (15), we apply equations (16e) and solve for s, with the result that

$$\boxed{s = \frac{T_O}{w_0}\sinh\left(\frac{w_0 x}{T_O}\right)} \tag{17}$$

This result is important because it provides a relationship between the horizontal distance and the length of the cable, analogous to equation (12) for a cable supporting a uniform horizontal load. Equation (17) also enables us to determine the sag $y(x)$.

Substitution of the expression for s into equation (13) yields

$$\frac{dy}{dx} = \sinh\left(\frac{w_0 x}{T_O}\right)$$

It then follows from the second of equations (16b) that

$$y = \frac{T_O}{w_0}\cosh\frac{w_0 x}{T_O} + C$$

where C is a constant of integration. Because the cable passes through the origin, it must be that $y = 0$ when $x = 0$, which means that

$$0 = \frac{T_O}{w_0}\cosh(0) + C \quad \text{so} \quad C = -\frac{T_O}{w_0}$$

Thus, the expression for y becomes

$$y = \frac{T_O}{w_0}\left[\cosh\left(\frac{w_0 x}{T_O}\right) - 1\right] \tag{18}$$

The curve formed by plotting $y(x)$ is called a *catenary*.

In addition to the curve formed by the sagging cable, we also need to know the tension within the cable. The required expression is found by substituting equation (17) for s into equation (13a) and then using the identity given by equation (16c). This yields

$$T = \left[T_O^2 + w_0^2\left(\frac{T_O}{w_0}\right)^2 \sinh^2\left(\frac{w_0 x}{T_O}\right)\right]$$

$$T = T_O \cosh\left(\frac{w_0 x}{T_O}\right) \tag{19}$$

Equations (17), (18), and (19) are sufficient for solving problems. Assuming that we know the location of the pins holding the ends of the cable, a key parameter to be determined is the constant T_O, which is the constant horizontal component of the tension. Let us consider what is involved in such a determination for a cable whose pins are at the same elevation, as illustrated in Figure 22.

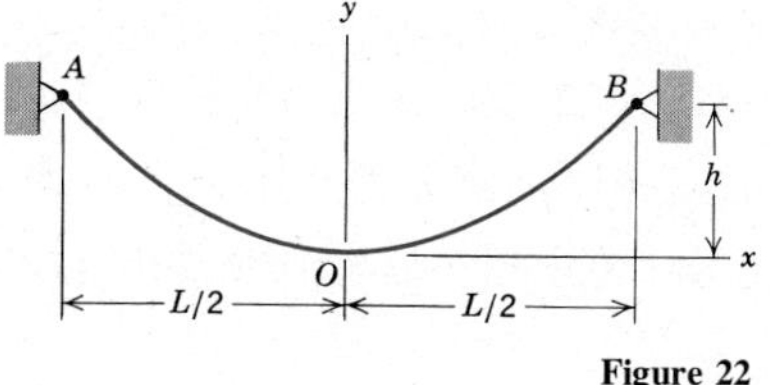

Figure 22

By symmetry, the origin O, which is located at the horizontal tangent, must be located midway in the span, as shown. Letting h be the maximum sag, we see that the end points, having coordinates ($x_A = -L/2$, $y_A = h$) and ($x_B = L/2$, $y_B = h$), are points on the catenary. Substituting these coordinates into equation (18), we find that

$$h = \frac{T_O}{w_0}\left[\cosh\left(\frac{w_0 L}{2T_O}\right) - 1\right]$$

Two *nondimensional* combinations of the parameters may be identified in the foregoing, specifically, h/L and $w_0 L/T_O$. In terms of these quantities, we then find

$$\frac{h}{L}\left(\frac{w_0 L}{T_O}\right) = \cosh\left(\frac{w_0 L}{2T_O}\right) - 1 \tag{20}$$

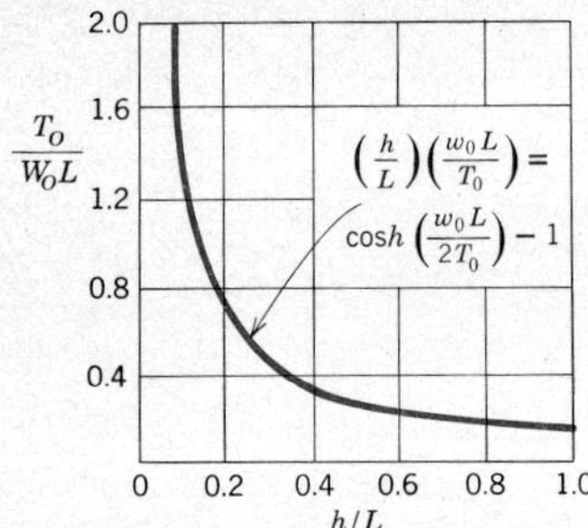

Figure 23

If we know the sag h and span L, the solution of equation (20) will give the value of T_O. Unfortunately, a closed-form solution is not possible. Nevertheless, it is possible to solve equation (20) by a trial-and-error procedure. The graph in Figure 23 shows the relationship between the nondimensional parameters (h/L) and (wL/T_O) resulting from such a solution. This graph may be used by itself to solve problems, or alternatively, as the starting point of your own trial-and-error computation if more accurate results are required.

In closing, it should be noted that there is no resemblance between the formulas obtained here and those obtained in the previous section for a cable carrying a uniform horizontal load. This is attributable to the fact that the weight of a cable is distributed uniformly along its length, not along the horizontal. However, in the case where the tension T_O is sufficiently large to make the sag ratio very small ($h/L < 0.10$), the cable is almost oriented horizontally. In such a case we may obtain an approximate solution by considering the weight to be a uniform horizontal load, and thus set $q_0 = w_0$ in the formulas for a uniform load in the previous section.

EXAMPLE 10

A uniform cable whose mass per unit length is 10 kg/m hangs as shown. Determine the maximum tension in the cable and its length.

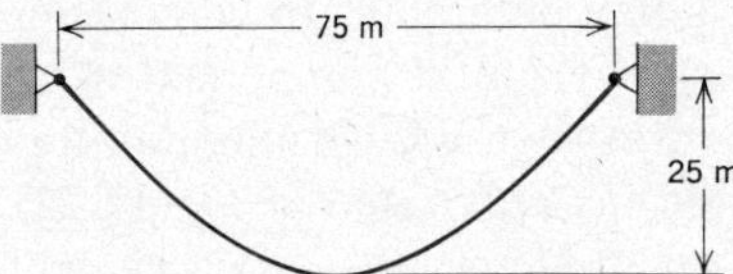

Solution

From the given information we calculate the weight per unit length and the ratio of maximum sag to span

$$w_0 = (10 \text{ kg/m})(9.806 \text{ m/sec}^2) = 98.06 \text{ N/m}$$

$$\frac{h}{L} = \frac{25}{75} = \frac{1}{3}$$

The other important parameter for an anlysis is $T_O/w_0 L$. The value of $T_O/w_0 L$ that we read from the graph given in Figure 23 corresponding to $h/L = 1/3$ is

$$\frac{T_O}{w_0 L} = 0.42$$

It may be that this is all the accuracy that is desired, in which case substitution of the values of w_0 and L will yield T_O. However, let us assume that we want more accuracy. We may follow a trial-and-error procedure by using the estimate of T_O/w_0L to evaluate the difference between the left- and right-hand sides of equation (20). The estimate is then improved by modifying the value of T_O/w_0L to reduce this difference. This computation may be performed in tabular form, as indicated below, where

$$\text{LHS} = \left(\frac{h}{L}\right)\left(\frac{w_0L}{T_O}\right) = \left(\frac{1}{3}\right)\left(\frac{w_0L}{T_O}\right)$$

$$\text{RHS} = \cosh\left[\frac{1}{2}\left(\frac{w_0L}{T_O}\right)\right] - 1$$

T_O/w_0L	LHS	RHS	LHS – RHS
0.42	0.7937	0.7964	−0.0027
0.419	0.7955	0.8006	−0.0051
0.421	0.7918	0.7921	−0.0003
0.4211	0.7916	0.7917	−0.0001
0.4212	0.7914	0.7913	0.0001

The second trial value of T_O/w_0L in the tabulation was based on a guess that the initial value was too high. This guess was shown to be wrong by the fact that there was a larger difference between the values of LHS and RHS corresponding to the second estimate than there was for the initial estimate. The succeeding values of T_O/w_0L were chosen in accord with the trend indicated by the first two trial values. The last two trials indicate that the value T_O/w_0L is somewhere between 0.4211 and 0.4212. Using the former value, we calculate that

$$T_{\rightarrow} = 0.4211 w_0 L = 0.4211(98.06)(75) = 3097 \text{ N}$$

The tension in the cable is given by equation (19). This equation shows that T has its maximum value at the location where x is a maximum, which is one of the ends. Setting $x = L/2 = 37.5$ m, we find

$$T_{\max} = T\left(x = \frac{L}{2}\right) = 3097 \cosh\left[\frac{(98.06)(37.5)}{3097}\right] = 5549 \text{ N}$$

The total length of the cable is twice the arc length from the lowest point on the cable, which is where the origin is located. Therefore, from equation (17) we find

$$s_{\text{total}} = 2s\big|_{x=37.5} = 2\left(\frac{3097}{98.06}\right)\sinh\left[\frac{(98.06)(37.5)}{3097}\right] = 93.9 \text{ m}$$

HOMEWORK PROBLEMS

VI.72 A uniform 1-kg/m cable is suspended as shown. (a) Determine the maximum and minimum values of the tension in the cable, and the cable length. (b) Compare the values found in part (a) to the approximate values obtained by considering the weight of the cable to be 1 kg/m along the horizontal.

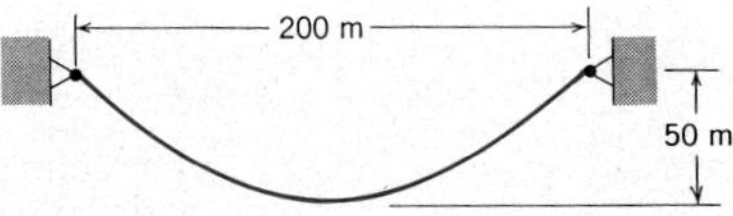

Prob. VI.72

VI.73 The maximum tension that an electrical transmission cable weighing 10 lb/ft can withstand without failure is 4 kips. The cable is suspended from two towers at equal elevations 200 ft apart. Determine the minimum length of cable that can be used and the corresponding sag. *Hint:* There are two values of T_0 that give a maximum tension of 4 kips.

VI.74 A 0.5-kg/m cable is suspended between two buildings with its ends at equal elevations. The length of the cable is 24 m and the maximum sag is 10 m. Determine (a) the horizontal distance spanned by the cable, (b) the minimum and maximum tensions. *Hint:* Formulate expressions for the length and the maximum sag of the cable and then use identity (16c).

VI.75 and VI.76 The cable shown, whose weight per unit length is w_0, is pinned at end B and attached to collar A which may slide along the smooth bar. Knowing that $h = d$, determine the magnitude of the force $\bar{P}$ required to hold the collar in position in terms of d and w_0.

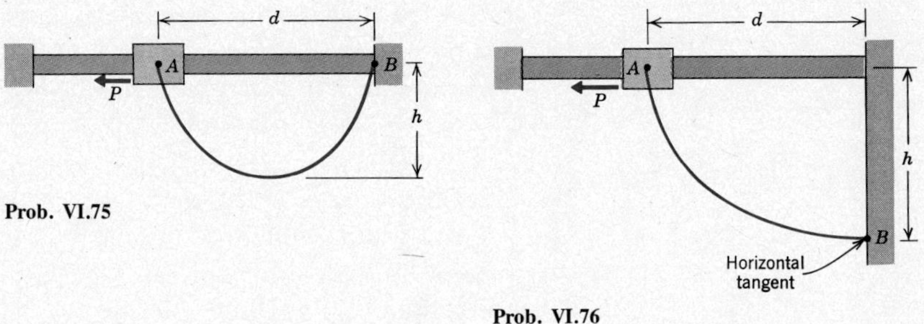

Prob. VI.75

Prob. VI.76

VI.77 Solve Problem VI.75 in the case where (a) $h = 2d$, (b) $h = 3d$.

VI.78 An electrical transmission cable having a mass per unit length of 10 kg/m is suspended by a series of towers, as shown. Determine the resultant force exerted by the cables on tower B.

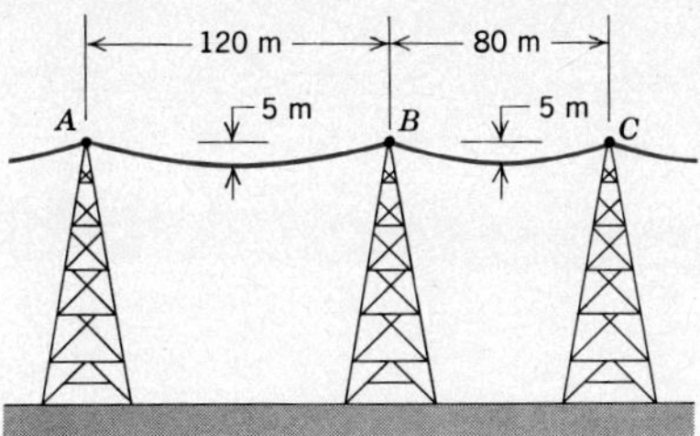

Prob. VI.78

VI.79 A 50-ft-long cable weighing 2 lb/ft spans a distance L between two pins at equal elevation. Knowing that the maximum sag $h = 0.2L$, determine (a) the distance L, (b) the horizontal and vertical components of the reaction force at one of the pin supports.

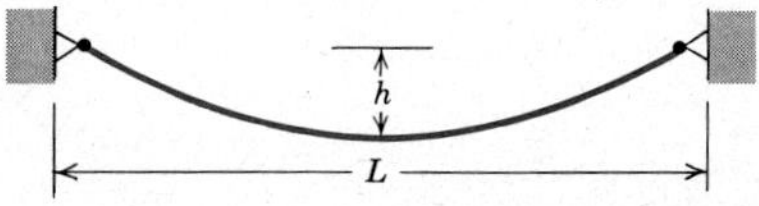

Probs. VI.80 and VI.81

VI.80 Determine the sag to span ratio h/L of a cable whose total weight equals the minimum tension in the cable.

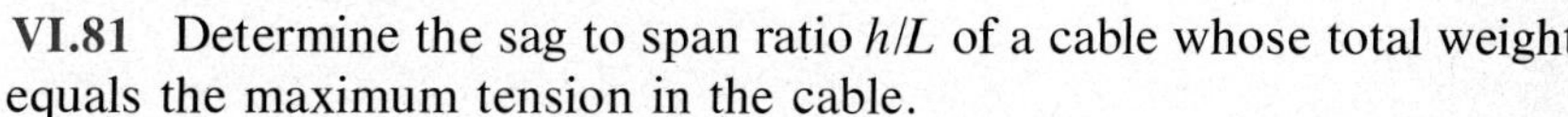

VI.81 Determine the sag to span ratio h/L of a cable whose total weight equals the maximum tension in the cable.

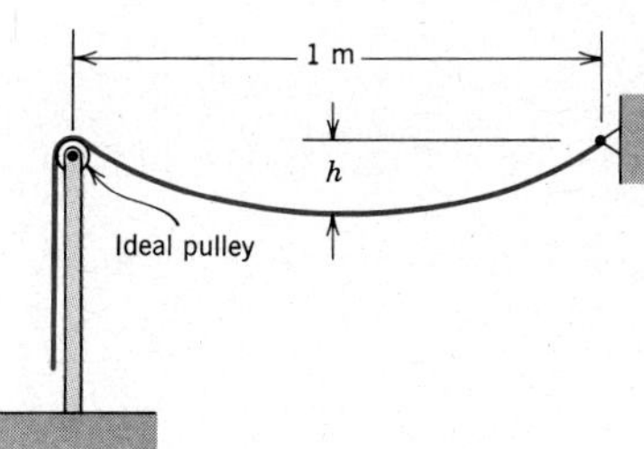

Prob. VI.82

VI.82 A uniform 2-m-long cable having a total mass of 1 kg is suspended as shown. Determine the *two* values of h for which the cable is in equilibrium.

VI.83 The mass per unit length of the cable is 0.4 kg/m. Determine the *two* values of the sag h for which the maximum tension in the cable is 300 N.

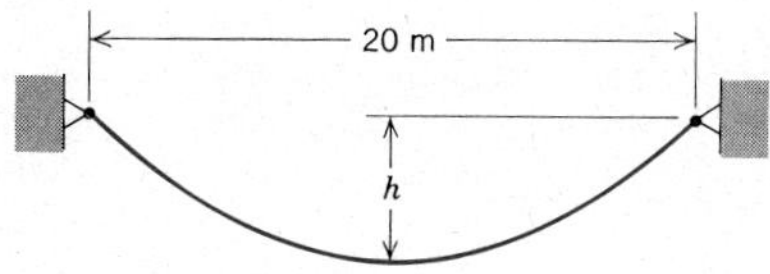

Prob. VI.83

MODULE VII
FRICTION

Reactive forces that frequently appeared in earlier modules were the normal forces exerted between two bodies having surfaces in contact. It was usually stipulated in those situations that the surfaces were smooth. As a result, the normal forces (which have the effect of preventing the contact surfaces from penetrating each other) were the only possible reactions.

In this module we shall consider the situation where the contact surfaces are rough. This results in the possibility of forces tangent to the plane of contact between the two bodies, whose effect is to oppose the movement of one surface relative to the other. Such forces are called *friction forces*.

Obviously, there is no surface that is absolutely smooth, but the concept of a system without friction is frequently a useful model to study. We refer to the condition of a smooth surface as an *ideal* model. The friction that we account for in a *real* model may originate from several sources. In general, friction forces can be classified on the basis of whether the contacting surfaces are dry, or whether a fluid medium is involved.

Fluid friction forces arise partially, but not entirely, from the viscosity of fluids. Such forces are important phenomena in the design of conduits for conveying liquids, in the design of transportation systems such as airplanes, automobiles, and ships, and in the study of lubricants. The study of fluid friction requires knowledge of the mechanics of fluids, and will not be treated here.

Dry friction is itself a complicated phenomenon that is still not fully understood. Current thinking is that it results from the microscopic irregularities present in all surfaces (hence the phrase: rough surface) and also from molecular attraction. Fortunately, elementary laws, first reported by Coulomb in 1781, are available for forming a simple model to describe dry friction. Coulomb's laws, which are the subject of the next section, should not be regarded as exact, because they possess certain anomalies. Nevertheless, they do provide accurate information in a broad range of applications that we will study in the following sections.

A negative connotation is usually associated with the word friction, because it gives rise to additional energy expenditures, and because it results in wear of surfaces in contact. This represents a limited view, because the existence of friction is very often useful, if not essential. For instance, how could we walk without friction?

A. DRY FRICTION

1 Coulomb's Laws

A primary feature of the friction forces exerted between surfaces in contact is that they oppose the direction in which the surfaces move, or would *tend* to move, relative to each other. To illustrate this, consider the block

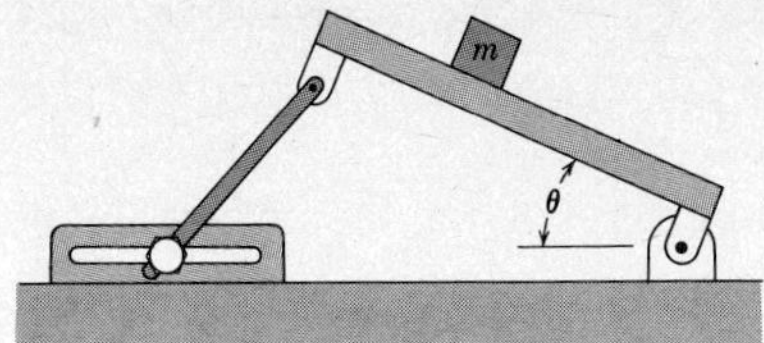

Figure 1a

on the inclined plane AB in Figure 1, whose angle of elevation θ can be set at any desired value.

In the absence of friction, the block would certainly slide down the plane. Thus, the free body diagram, Figure 1b, shows the friction force $\bar{f}$ acting on the block to be in the negative X direction, opposite to the direction in which the block is tending to move relative to the incline.

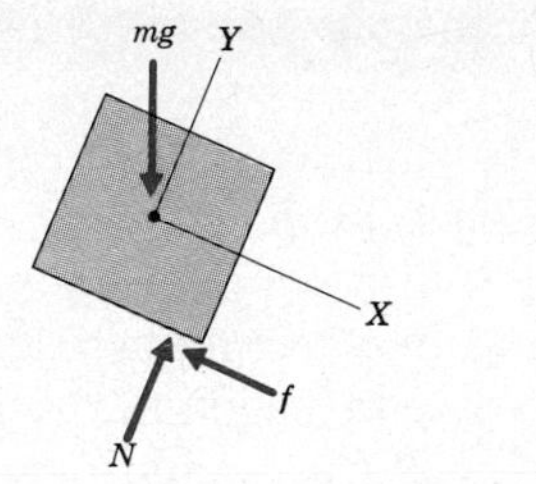

Figure 1b

Note that we have not considered a free body diagram of the incline because, in effect, it is the ground. However, if we did draw a free body diagram of the incline, the friction force $-\bar{f}$, which is the reaction exerted on the incline by the block, would be in the positive X direction. This is consistent with the foregoing general characteristic of a friction force, because if the block were to slide down the plane the inclined surface would seem to move in the negative X direction when viewed from the block. Remember that the friction force is opposite to the *relative* movement.

Let us now consider a simple experiment with the apparatus in Figure 1a, in which the value of θ is set at successively larger values, starting with $\theta = 0$, after which the block is placed at rest on the incline. There is a range of values of the angle θ for which the block will not slide. In this situation, the total reaction force exerted on the block $\bar{R} = -f\bar{I} + N\bar{J}$ is known from the equations of equilibrium to be as shown in Figure 2.

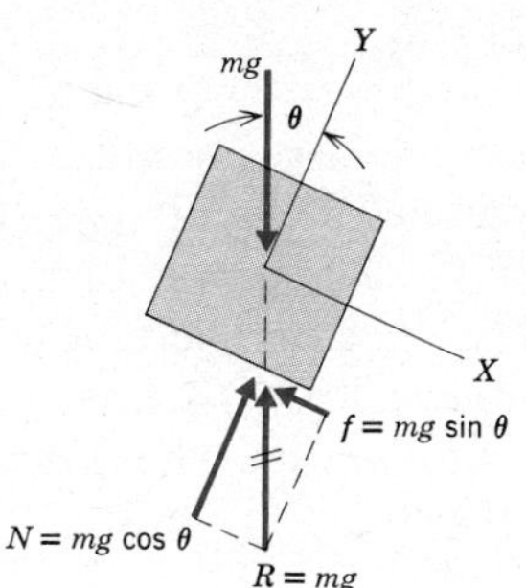

Figure 2

In the figure, $mg \sin\theta$ is the component of the external (gravity) force acting on the block in the direction in which the block would tend to slide, whereas f is the component of the reaction force resisting this tendency to move. For static equilibrium,

$$
\begin{aligned}
f &= mg \sin\theta \\
N &= mg \cos\theta
\end{aligned}
\tag{1}
$$

Now consider the situation for larger values of the angle θ. At some angle, call it the *critical angle* θ_{cr}, the block will slide, rather than remaining at rest. The friction force during this sliding motion may be determined by means of Newton's second law.

Let us plot a graph showing along the ordinate the magnitude of the friction force f in this series of trials for different values of θ. Along the abscissa of this graph we plot the component of the gravitational load parallel to the direction of movement, $mg \sin\theta$. The result is shown in Figure 3.

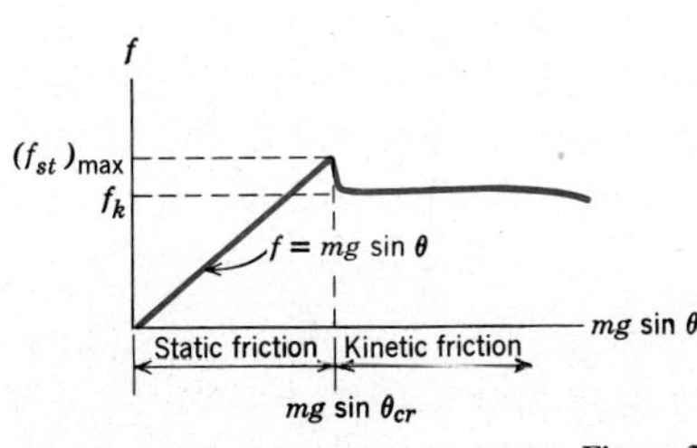

Figure 3

The graph illustrates that when there is static equilibrium, $f = mg \sin\theta$. The maximum friction force, $(f_{st})_{\max}$ acting on the block occurs at an angle θ infinitesimally smaller than θ_{cr} for which sliding occurs, so

$$(f_{st})_{\max} = mg \sin\theta_{\mathrm{cr}}$$

The graph also shows that once sliding begins, the friction force decreases to the value f_k and varies only slightly as the force component in the direction of motion increases. In the region where we have static equilibrium, we say that $\bar{f}$ is a *static friction force*, $\bar{f}_{st}$, whereas $\bar{f}$ is said to be a *kinetic friction force*, $\bar{f}_k$, when movement occurs.

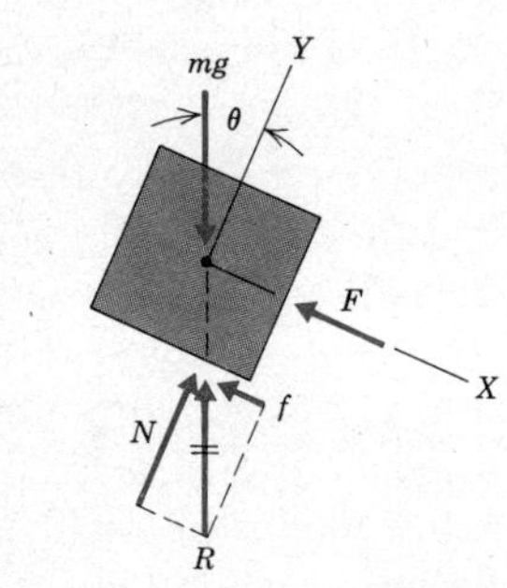

Figure 4

Let us now consider a second experiment involving the apparatus of Figure 1, where we apply a force $\bar{F}$ tangent to the incline, in the uphill sense. Let the angle of elevation be set at a constant value θ smaller than θ_{cr}. The resulting free body diagram of the block is shown in Figure 4. The component of the external load acting on the block in the X direction is now $mg \sin \theta - F$.

Let us plot f versus $mg \sin \theta - F$ for this experiment, as F is increased from zero. From Figure 4 it is apparent that $f = mg \sin \theta - F$ when there is static equilibrium. The resulting graph appears in Figure 5.

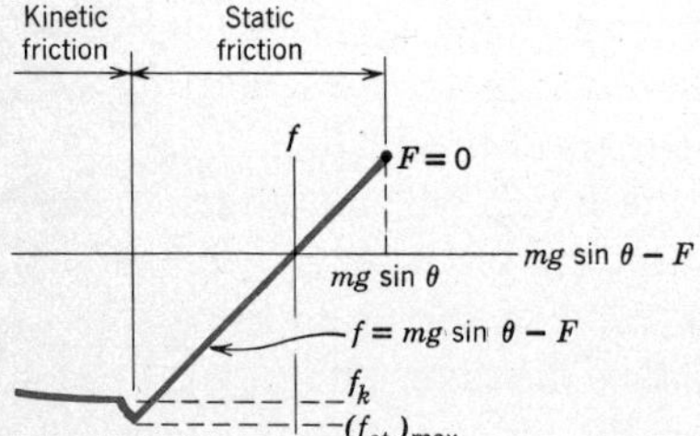

Figure 5

From this graph we see that there is again a maximum magnitude for the static friction force, followed by a decrease in magnitude to the kinetic friction force. Note that the plotted values of $(f_{st})_{max}$ and f_k are not the same as those plotted in Figure 3. Also, the negative values for f represent a reversal in the direction of the friction force, corresponding to $F \geq mg \sin \theta$. This is because the friction force opposes the relative movement that would occur if the surfaces were smooth.

Suppose that we perform a series of these experiments for many values of θ. We now that $N = mg \cos \theta$ in all cases, because even when the block slides, the sliding motion is parallel to the X axis so there is always a force balance in the Y direction. When we form the ratios $|(\bar{f}_{st})_{max}|/N$ and $|\bar{f}_k|/N$ for each value of θ, it is found that these ratios are constants. The first is called the *coefficient of static friction*, μ_s, and the second is called the *coefficient of kinetic friction*, μ_k. These observations may be generalized as follows.

a *Case of no relative movement*. For any pair of contact surfaces that are not moving relative to each other, the magnitudes of the friction and normal forces satisfy

$$\boxed{|\bar{f}| \leq \mu_s |\bar{N}|} \tag{2}$$

where the equality sign corresponds to the situation where sliding is about to occur, that is, the case of *impending motion*. It should be emphasized that the question of whether or not movement will occur depends on the magnitude, but not the sense, of the friction force. The sense in which the friction force actually acts on one of the surfaces is opposite to the sense in which that surface would move relative to the other surface if there were no friction.

b *Case of relative movement*. For any pair of surfaces that are sliding over each other, the magnitudes of the friction and normal forces satisfy

$$\boxed{|\bar{f}| = \mu_k|\bar{N}|} \tag{3}$$

where μ_k is a lower value than μ_s. In this situation the direction of $\bar{f}$ is opposite the direction of the relative movement.

Equations (2) and (3) are Coulomb's laws of friction. As originally stated, the values of μ_s and μ_k are regarded as dependent only on the types of materials in contact (for example, steel in contact with aluminum) and the condition of each surface (for example, highly polished or corroded). This is an approximation, for it gives rise to certain subtle differences with observed phenomena, such as the different tractive capabilities of two automobile tires of the same construction, but different dimensions. Further, as is exhibited by Figures 3 and 5, the value of μ_k is not truly a constant. A more refined experiment would show that, in reality, μ_k is slightly dependent on the relative velocity of the sliding surfaces.

Equations (2) and (3) are quite useful, even though they have the limitations discussed above. Typical values of μ_s and μ_k are presented in Table 1. Note that there is no reason why the coefficients of friction should be less than unity, as exemplified by the value of μ_s given for contact between two cast iron surfaces. In actuality a broad range of values exists for each case given. The values shown in Table 1 merely give an idea of the general phenomena. For specific cases these values may be incorrect by more than 100%. Thus the tabulation should be employed only if values resulting from experiments on the actual system are unavailable.

Table 1

MATERIALS	μ_s	μ_k
Mild steel on mild steel	0.74	0.57
Aluminum on mild steel	0.61	0.47
Copper on mild steel	0.53	0.36
Cast iron on cast iron	1.10	0.15
Brake material on cast iron	0.40	0.30
Oak on cast iron	0.60	0.32
Leather on cast iron	0.60	0.56
Stone on cast iron	0.45	0.22
Rubber on metal	0.40	0.30
Rubber on wood	0.40	0.30
Rubber on pavement	0.90	0.80
Leather on wood	0.40	0.30
Glass on nickel	0.78	0.56

The coefficient of static friction is sometimes represented in terms of the angle θ_{cr} at which the block in Figure 1 begins to slide down the incline. Corresponding to θ_{cr} (impending motion), we have $f = \mu_s N$. Equations (1) then give

$$N = mg \cos \theta_{cr}$$

$$f = mg \sin \theta_{cr} = \mu_s N = \mu_s\, mg \cos \theta_{cr}$$

from which we find

$$\mu_s = \tan \theta_{cr} \tag{4}$$

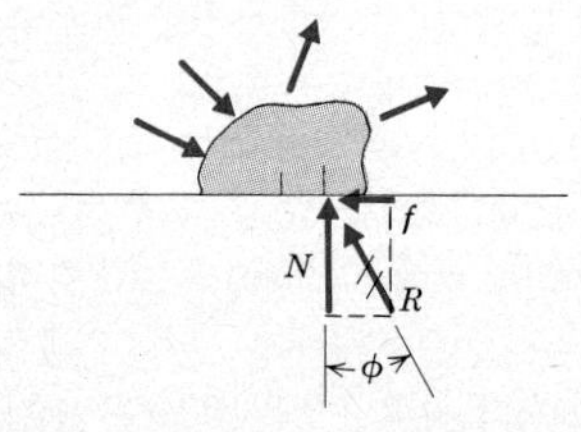

Figure 6

Equation (4) is useful for a pictorial representation of the static friction force. For example, let us consider a body in contact with a rough plane and subjected to an arbitrary set of loads, as shown in Figure 6. The friction force $\bar{f}$ and normal force $\bar{N}$ are equivalent to the resultant reaction $\bar{R}$ at an angle ϕ from the normal to the plane of contact. Clearly,

$$\tan \phi = \frac{f}{N}$$

Substituting for f from equations (2) and (4) results in

$$\tan \phi \leq \left(\frac{\mu_s N}{N}\right) = \tan \theta_{cr}$$

$$\phi \leq \theta_{cr}$$

In words, if there is no sliding, the angle ϕ for the total reaction $\bar{R}$ at the contact surface will be less than θ_{cr}, whereas the case of impending motion corresponds to

$$\phi_s = \theta_{cr} = \tan^{-1} \mu_s \tag{5}$$

The angle ϕ_s is called the *angle of static friction.*

For the situation where slipping occurs, equation (3) tells us that $f = \mu_k N$, so we may replace μ_s with μ_k in equation (5). Thus

$$\phi_k = \tan^{-1} \mu_k \tag{6}$$

where ϕ_k is the *angle of kinetic friction*. From the fact that $\mu_k < \mu_s$, it follows that $\phi_k < \phi_s$. It can be shown that ϕ_k is the angle of elevation of the apparatus in Figure 1 for which the block will continue to slide down the incline at a constant speed once it is set in motion.

Equations (5) and (6) represent alternative ways of specifying the coefficient of friction. However, our approach to statics involves treating reaction forces in terms of their components rather than their magnitude and direction, so we shall have little use for the concept of the angle of friction when we solve problems.

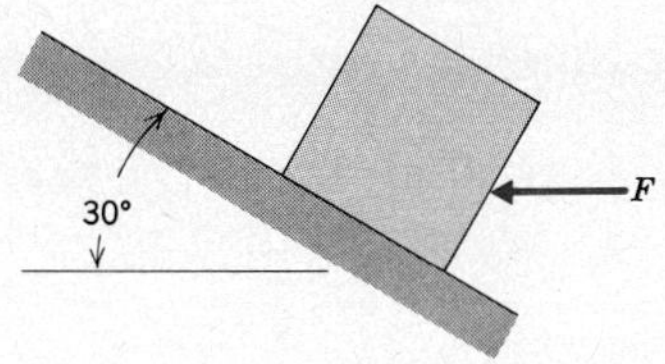

EXAMPLE 1

The coefficients of friction between the inclined plane and the 2.5-kg block are $\mu_s = 0.25$ and $\mu_k = 0.20$. The block is initially at rest when the horizontal force $\bar{F}$ is applied to it. Determine (a) the friction force when $F = 20$ N, (b) the friction force when $F = 5$ N, (c) the minimum magnitude of $\bar{F}$ for which the block will slide uphill.

Solution

In parts (a) and (b) we cannot tell by inspection whether there is sliding or not, for this depends on the comparison of the friction and normal reactions. We therefore will assume that there is no sliding, and compare the values of the friction force $\bar{f}$ and normal force $\bar{N}$ obtained from the equations of static equilibrium. On the other hand, in part (c) we are interested in the condition where uphill motion is impending, so for that case, we set $|\bar{f}| = \mu_s|\bar{N}|$. Note that in parts (a) and (b) we can assume a direction for f, because a negative sign in the solution for the magnitude of $\bar{f}$ will indicate that the direction was assumed incorrectly. However, in part (c) the friction force must be opposite the direction of the impending movement of the block relative to the hill, so $\bar{f}$ *must* be depicted as being downhill.

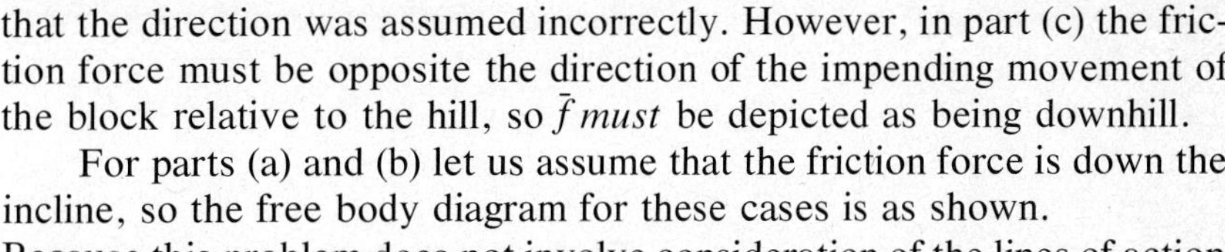

For parts (a) and (b) let us assume that the friction force is down the incline, so the free body diagram for these cases is as shown. Because this problem does not involve consideration of the lines of action of the forces, we need only consider the equations for force equilibrium. These are

$$\Sigma F_X = -F\cos 30° + f + 2.5(9.806)\sin 30° = 0 \text{ newtons}$$

$$\Sigma F_Y = N - F\sin 30° - 2.5(9.806)\cos 30° = 0 \text{ newtons}$$

Thus

$$f = 0.866F - 12.26$$

$$N = 0.50F + 21.23$$

Substituting the given values of the applied force F in parts (a) and (b) yields

$$\text{part (a):} \quad f_a = 5.06 \qquad N_a = 31.73 \text{ newtons}$$

$$\text{part (b):} \quad f_b = -11.83 \qquad N_b = 23.73 \text{ newtons}$$

The negative value of f_b tells us that the friction force is up the incline, which means that the tendency of the block is to move downward.

We cannot accept the preceding values of f until they have been checked against the maximum static friction force possible in each case, $|\bar{f}| = \mu_s N$. Thus

$$\text{part (a):} \quad \mu_s N_a = 0.25(31.73) = 7.93 \text{ newtons} > |\bar{f}_a|$$

$$\text{part (b):} \quad \mu_s N_b = 0.25(23.73) = 5.93 \text{ newtons} < |\bar{f}_b|$$

We observe that equation (2) is satisfied in part (a), but not in part (b). Thus, in part (a) there is no sliding, so

$$\bar{f} = f_a \bar{I} = 5.06\bar{I} \text{ newtons}$$

In part (b) the contact surfaces cannot sustain a friction force sufficient to prevent sliding. The direction in which the sliding will occur can be deduced to be down the plane, because that is the tendency indicated by the result for f_b. Therefore

$$\bar{f} = -\mu_k N_b \bar{I} = -4.75\bar{I} \text{ newtons}$$

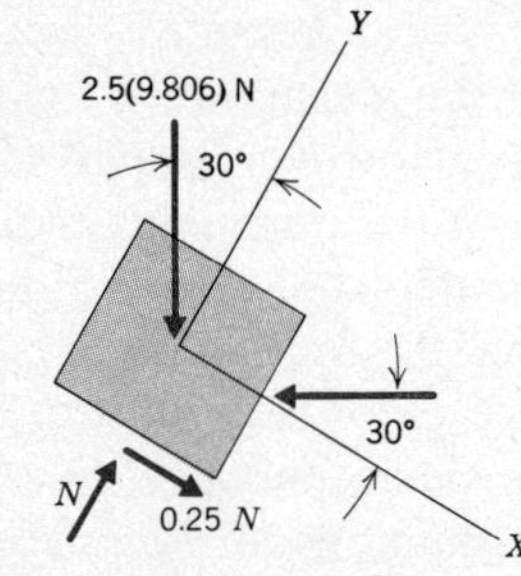

To solve part (c) we draw a new free body diagram, because in this situation of impending motion, $\bar{f}$ is $\mu_s N$ down the plane. The equilibrium equations are now

$$\Sigma F_X = -F \cos 30° + 0.25N + 2.5(9.806) \sin 30° = 0 \text{ newtons}$$

$$\Sigma F_Y = N - F \sin 30° - 2.5(9.806) \cos 30° = 0 \text{ newtons}$$

Solving these equations simultaneously yields

$$F = 23.7 \text{ newtons}$$

HOMEWORK PROBLEMS

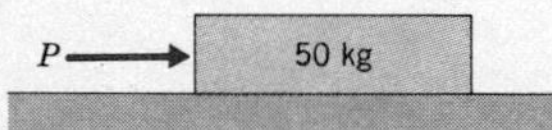

Prob. VII.1

VII.1 Determine the horizontal force $\bar{P}$ that must be exerted on the 50-kg block to cause the block to move. The coefficient of static friction is 0.30.

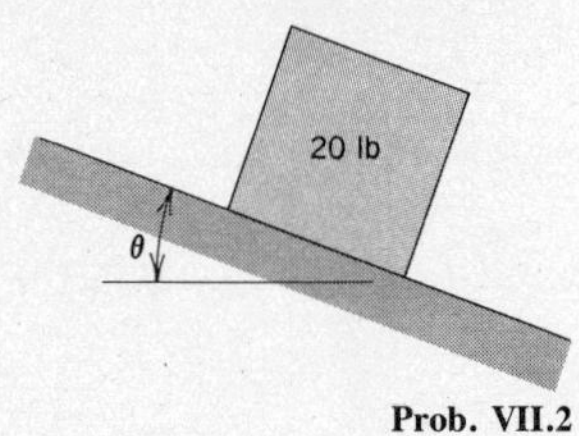

Prob. VII.2

VII.2 The coefficients of friction between the 20-lb block and the inclined plane are $\mu_s = 0.20$ and $\mu_k = 0.15$. Determine the friction force acting on the block if (a) $\theta = 5°$, (b) $\theta = 20°$.

VII.3 For the system in Example 1, determine the minimum value of the force $\bar{F}$ for which the block will not slide down the incline.

VII.4 The coefficients of friction between the 5-kg block and the inclined

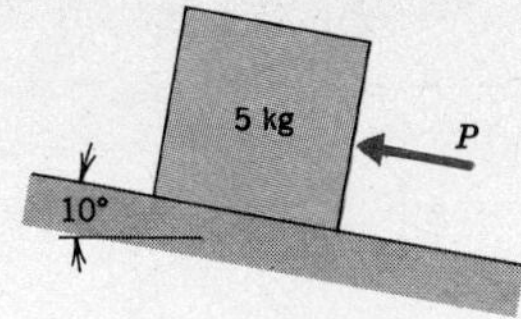

Prob. VII.4

plane are $\mu_s = 0.40$, $\mu_k = 0.32$. Draw a graph showing how the magnitude of the friction force acting on the block depends on the magnitude of the force $\bar{P}$, which is always oriented up the plane.

VII.5 The coefficient of static friction between the block and the inclined plane is 0.60. A 100-N force is applied to the block at an angle α with respect to the normal to the incline, as shown. Knowing that this force moves the block up the plane if $\alpha > 45°$, determine the mass of the block.

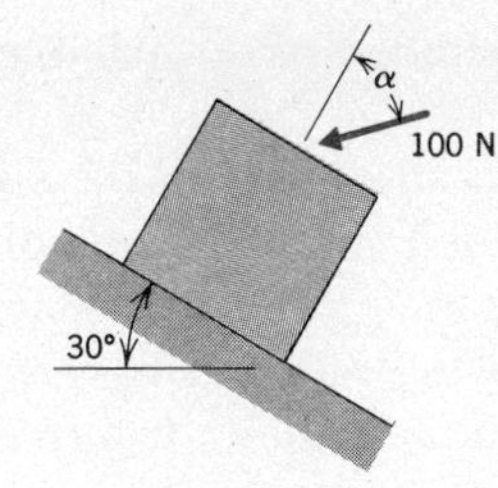

Probs. VII.5 and VII.6

VII.6 The mass of the block is 6 kg, and the 100-N force is applied at an angle $\alpha = 15°$. Knowing that the block is held in position by the force, determine (a) the friction force acting on the block, (b) the corresponding minimum allowable value of the coefficient of static friction.

2 Systems with Friction

The example and homework problems in the last section illustrate the fact that there are basically two classes of static friction problems. The first class of problems involves situations where the external loads are given and we must evaluate the friction force. To do this, we start by considering the case where sliding does not occur. The friction force is depicted in the free body diagram as being tangent to the contact plane in any sense we choose. Then, the equations of equilibrium yield the unknown friction force, including its actual sense. If it is found that the magnitude f of the friction force required for static equilibrium exceeds the maximum possible static friction force $\mu_s N$, it can be concluded that sliding will occur. This means that the friction force is $\mu_k N$. Furthermore, the kinetic friction force will be in the same direction as the friction force required for static equilibrium, because the static force is opposite the tendency of the motion.

The second class of static friction problems involves situations where motion is impending and we must evaluate some characteristic of the system or of the set of loads acting on the system. The friction force is then $\mu_s N$ (the maximum value for which sliding will not occur), acting opposite the direction of the impending relative movement. The equations of static equilibrium then yield the desired information.

The preceding discussion focuses on only one aspect of a system, specifically, the reaction force generated between two rough surfaces in contact. The system itself may be a single particle, a system of particles, a single rigid body, or a system of rigid bodies. The equations of static equilibrium to be employed are those that are appropriate to the system. Thus, the basic methods for studying static equilibrium developed in earlier modules will be required for a successful solution of a static friction problem.

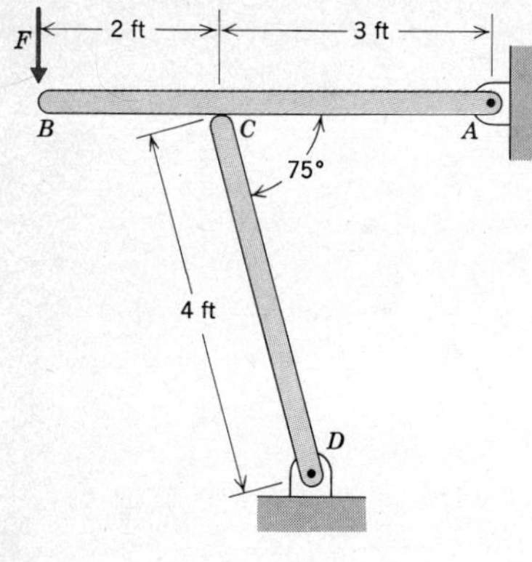

EXAMPLE 2

The system of bars supports a vertical force $\bar{F}$ at end B. Each bar weighs 2 lb/ft. Determine (a) the friction and normal forces acting on bar CD at end C, (b) the minimum value of the coefficient of static friction between the bars required for the system to be in static equilibrium.

Solution

The system consists of two multiforce members. We wish to examine the internal interaction force at point C. Following the method developed in Module IV for frames, we draw free body diagrams for the entire system and for each member, as shown in the left margin. The contact forces exerted at point C are depicted by noting that the contact plane is parallel to bar AB, so the normal component of the reaction is vertical and the friction force is horizontal. The weight forces are known from the given information.

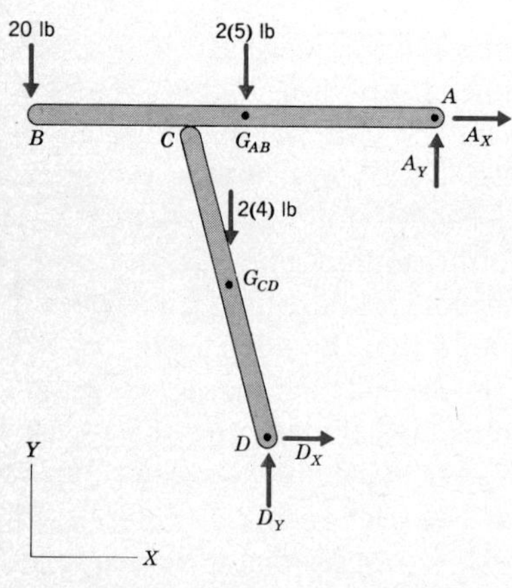

In drawing the diagrams, we assumed a sense for the friction force acting on bar AB, and then depicted the friction force acting on bar CD consistent with Newton's third law. There is no need to concern ourselves with depicting the friction forces in their proper sense. If we are wrong in our guess, the equilibrium equations will indicate it by giving a negative value for f_C. (In this system it is possible to determine in advance the sense of the friction forces by considering the direction in which the bars would move relative to each other if there were no friction between them. We have not pursued this consideration here because it is difficult to apply for systems where the tendency to move is not obvious.)

Returning to the solution of the problem, we write only the equations for moment equilibrium of each bar about its fixed pin, because we are not interested in the reactions at these pins. Therefore, we have

$$\text{bar } AB: \quad \Sigma M_{AZ} = -N_C(3) + 10(2.5) + 20(5) = 0 \text{ lb-ft}$$

$$\text{bar } CD: \quad \Sigma M_{DZ} = 8(2\cos 75^\circ) + N_C(4\cos 75^\circ) + f_C(4\sin 75^\circ) = 0 \text{ lb-ft}$$

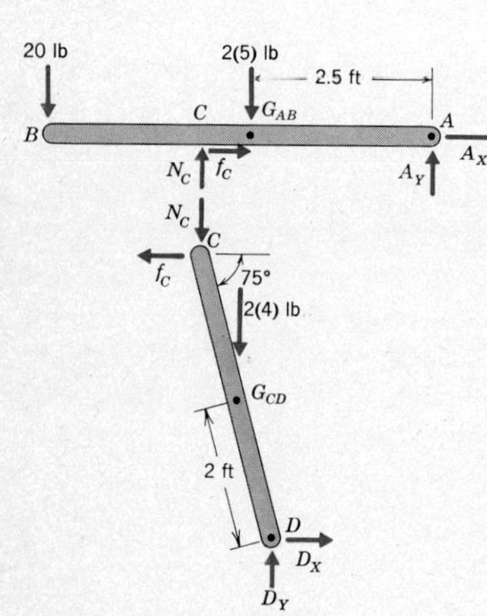

Solving the first equation for N_C and substituting the result into the second equation enables us to solve for f_C, with the result that

$$N_C = 41.67 \qquad f_C = -12.24 \text{ lb}$$

This is the solution to part (a). To solve part (b) we use the fact that the minimum allowable coefficient of static friction corresponds to the condition of impending motion: $|\bar{f}_C| = \mu_s N_C$. Thus, we have

$$(\mu_s)_{\min} = \frac{|\bar{f}_C|}{N_C} = \frac{12.24}{41.67} = 0.293$$

EXAMPLE 3

The coefficients of static friction between collar A and the vertical rod, and between bar AB and the semicylindrical surface, are 0.25 and 0.20, respectively. Determine if the bar is in static equilibrium in the position shown.

Solution

One approach to this problem is to try to determine the friction forces acting on the bar, assuming that the system is in static equilibrium, and to compare the ratio of the friction and normal force at each point to the corresponding given coefficient of static friction. The difficulty with this approach is that there are four unknown forces acting on the bar (the friction and normal forces between the bar and the semicylindrical surface, and between the vertical rod and collar A). Because there are only three equilibrium equations for the bar, the problem is statically indeterminate when formulated in this manner.

We require a different approach. Note that in the condition of impending sliding, the friction forces are known in terms of the normal forces. Let us therefore apply a force $\bar{P}$ of unknown magnitude that can bring the system to a condition of impending motion. We may determine the range of values of the magnitude of $\bar{P}$ for which the system remains in static equilibrium. Then, if $|\bar{P}| = 0$ is within this range, we know that the given system (without $\bar{P}$) is in equilibrium.

The only requirement in choosing the type of force $\bar{P}$ to apply is that it must have the possibility of causing the system to move. Thus, it cannot be normal to a contact plane, acting at the point of contact for that plane. The force $\bar{P}$ we will apply is a vertical force acting at collar A. If such a force is sufficiently large, collar A will certainly slide.

We can draw a general free body diagram by denoting the friction forces as f_A and f_C, and then explore the two types of impending motion by substituting either $f = \mu_s N$ or $f = -\mu_s N$, according to the direction in which the contact surfaces tend to move. This diagram is shown below, along with a separate diagram for geometric calculations.

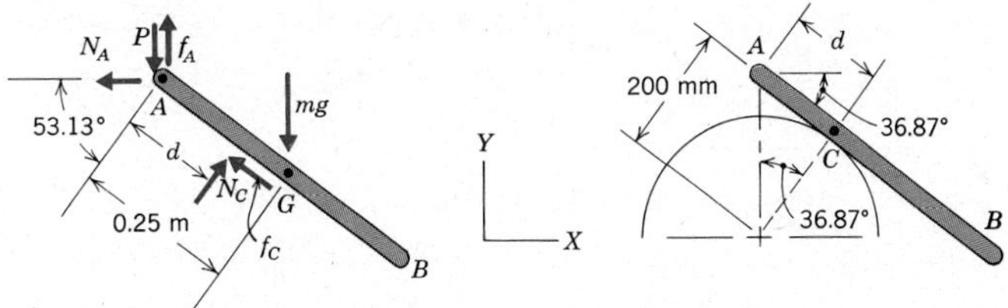

The distance d is found from the geometric sketch to be

$$d = 0.20 \tan 36.87^\circ = 0.15 \text{ m}$$

Choosing point A for the moment sum, the equilibrium equations are

$$\Sigma M_{AZ} = N_C(0.15) - mg(0.25 \cos 36.87°) = 0$$

$$\Sigma F_X = N_C \sin 36.87° - f_C \cos 36.87° - N_A = 0$$

$$\Sigma F_Y = f_A - P - mg + N_C \cos 36.87° + f_C \sin 36.87° = 0$$

We shall first consider impending downward motion of collar A. In this case, point C on bar AB has a tendency to slide downward, tangent to the semicylinder. Considering the friction forces to oppose these tendencies, and using the given values of the coefficients of friction, we have

$$f_A = 0.25N_A \qquad f_C = 0.20N_C$$

Thus, the equilibrium equations become

$$\Sigma M_{AZ} = 0.15N_C - 0.20mg = 0$$

$$\Sigma F_X = 0.60N_C - 0.80(0.20N_C) - N_A = 0$$

$$\Sigma F_Y = (0.25N_A) - P - mg + 0.80N_C + 0.60(0.20N_C) = 0$$

The solutions of these equations are

$$N_C = 1.3333mg \qquad N_A = 0.5867mg \qquad P = 0.3733mg$$

Next we consider impending upward motion of collar A. In this case the friction forces are reversed from their sense in the preceding case of impending motion. Therefore,

$$f_A = -0.25N_A \qquad f_C = -0.20N_C$$

The general equilibrium equations then become

$$\Sigma M_{AZ} = 0.15N_C - 0.20mg = 0$$

$$\Sigma F_X = 0.60N_C - 0.80(-0.20N_C) - N_A = 0$$

$$\Sigma F_Y = (-0.25N_A) - P - mg + 0.80N_C + 0.60(-0.20N_C) = 0$$

The solutions are

$$N_C = 1.3333mg \qquad N_A = 1.0133mg \qquad P = -0.3467mg$$

From these results it can be deduced that sliding will not occur in the range $-0.3467mg < P < 0.3733mg$. The system of interest corresponds to $P = 0$, so we can conclude that the system is in static equilibrium.

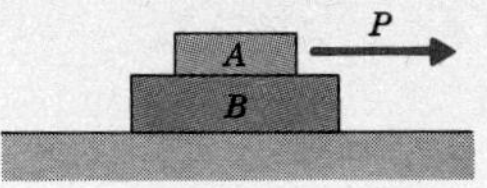

Prob. VII.7

HOMEWORK PROBLEMS

VII.7 Determine the force $\bar{P}$ required to cause motion in the system shown. At which surfaces will sliding occur? The coefficient of static friction between the 2-kg block A and the 8-kg block B is $\mu_s = 0.2$ and the coefficient of static friction between block B and the horizontal plane is $\mu_s = 0.1$.

VII.8 The coefficient of static friction between the person and the floor is 0.50, whereas the coefficient of static friction between the crate and the floor is 0.25. Determine the largest mass for the crate that the person can move by pulling on the cable, if the person's mass is 70 kg.

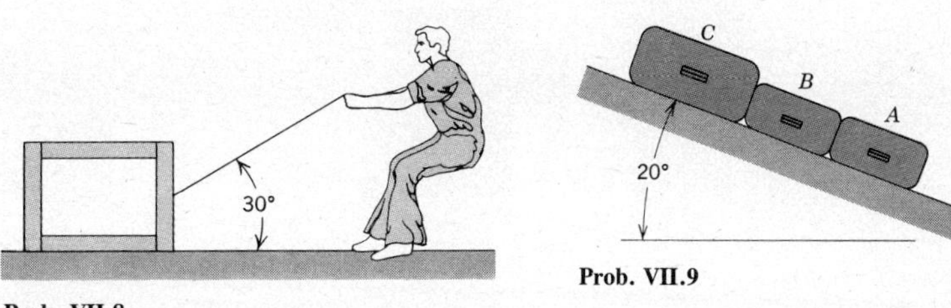

Prob. VII.8

Prob. VII.9

VII.9 Three suitcases A, B, and C are placed on a chute. Their weights are $w_A = w_B = 30$ lb, $w_C = 45$ lb. The coefficients of static friction between the suitcases and the chute are $\mu_A = \mu_B = 0.40$, $\mu_C = 0.20$. Determine which, if any, of the suitcases will slide down the chute. *Hint:* Successively consider the possibility of impending motion of suitcase A alone, then of both suitcases A and B, and finally of all three suitcases.

VII.10 Solve Problem VII.9 if the suitcases are arranged such that A is the lowest and B is the highest.

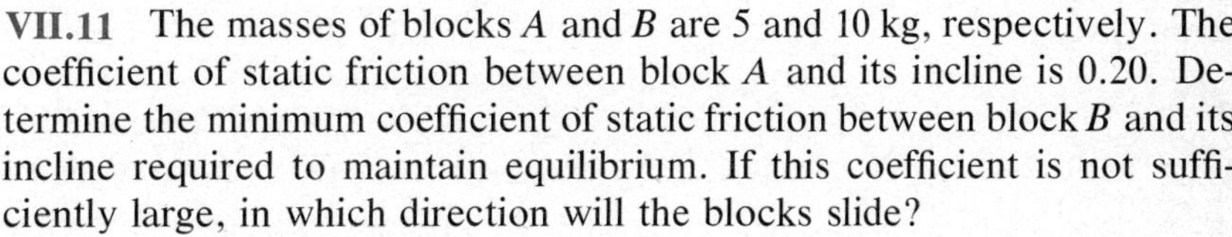

VII.11 The masses of blocks A and B are 5 and 10 kg, respectively. The coefficient of static friction between block A and its incline is 0.20. Determine the minimum coefficient of static friction between block B and its incline required to maintain equilibrium. If this coefficient is not sufficiently large, in which direction will the blocks slide?

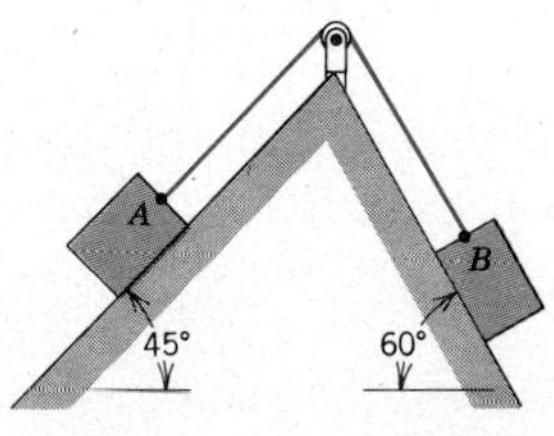

Prob. VII.11

VII.12 Solve Problem VII.11 if block A has a mass of 10 kg and block B has a mass of 5 kg.

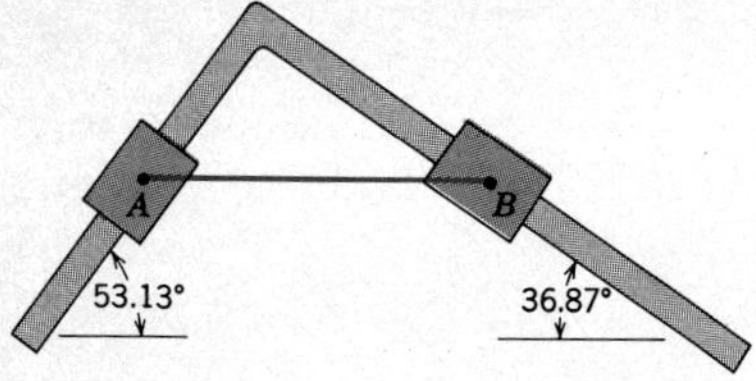

Prob. VII.13

VII.13 Two 3-kg collars are connected by a cord and are in equilibrium in the position shown. The coefficient of static friction between collar A and its inclined guide is 0.30. Determine the minimum allowable value of the coefficient of static friction between collar B and its guide.

VII.14 The masses of blocks A and C are 10 and 20 kg, respectively. These blocks are interconnected by cord ABC. Knowing that the system is in equilibrium, determine (a) the friction force acting on each block, (b) the minimum allowable coefficient of static friction between each block and its surface for static equilibrium.

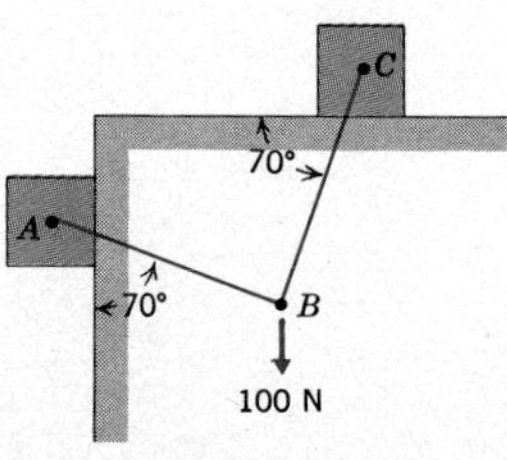

Prob. VII.14

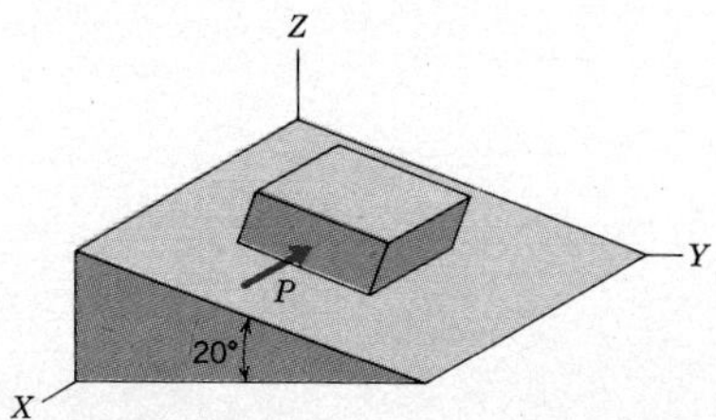

Prob. VII.15

VII.15 A 10-lb block resting on a plane inclined at 20° above the horizontal is subjected to a horizontal force $\bar{P}$ parallel to the plane, as shown. It is observed that the block will start to move if the magnitude of $\bar{P}$ exceeds 3 lb. Determine (a) the coefficient of static friction between the block and the inclined plane, (b) the direction in which the block starts to slide if the magnitude of $\bar{P}$ exceeds 3 lb.

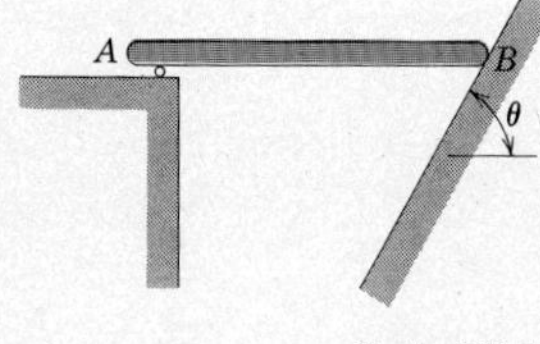

Prob. VII.16

VII.16 Bar AB, whose mass is 100 kg, is supported by a roller resting on the horizontal surface at end A and by the rough inclined plane at end B. Knowing that the angle $\theta = 60°$ and that the bar is in static equilibrium, determine (a) the friction force acting on the bar at end B, (b) the minimum value of the coefficient of static friction between the bar and the incline for which static equilibrium is possible.

VII.17 For the bar in Problem VII.16, the coefficient of static friction between the inclined plane and end B is 0.15. Determine the maximum value of the angle θ for which equilibrium is possible.

VII.18 The uniform 60-kg beam rests against the smooth vertical wall and the rough horizontal wall ($\mu_s = 0.25$, $\mu_k = 0.21$). Knowing that the

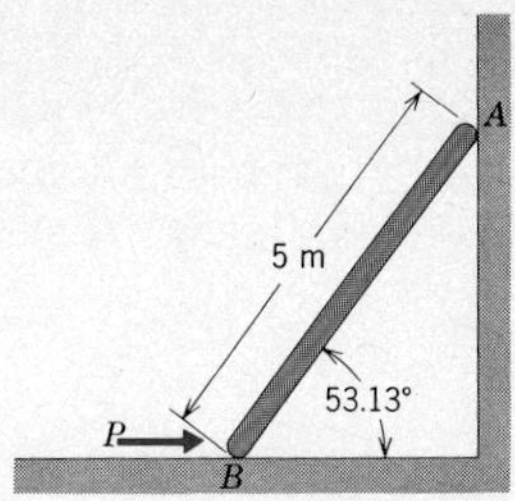

Probs. VII.18 and VII.19

horizontal force $\bar{P}$ is 100 N, acting to the right, determine the friction force acting on the lower end of the beam.

VII.19 The uniform 60-kg beam AB rests against the smooth vertical wall and the rough horizontal floor ($\mu_s = 0.25$). A horizontal force $\bar{P}$ is applied to end B. Determine the range of values for the magnitude of $\bar{P}$ such that the beam will not move from the position shown.

VII.20 Solve Problem VII.19 if the wall is also rough ($\mu_s = 0.15$).

VII.21 The 3-ft-diameter cylinder weighs 500 lb. The coefficient of friction between the cylinder and both surfaces is $\mu_s = 0.32$. Determine the maximum magnitude of the counterclockwise couple $\bar{M}$ for which sliding will not occur.

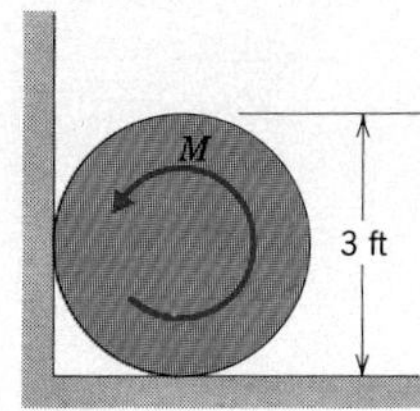

Prob. VII.21

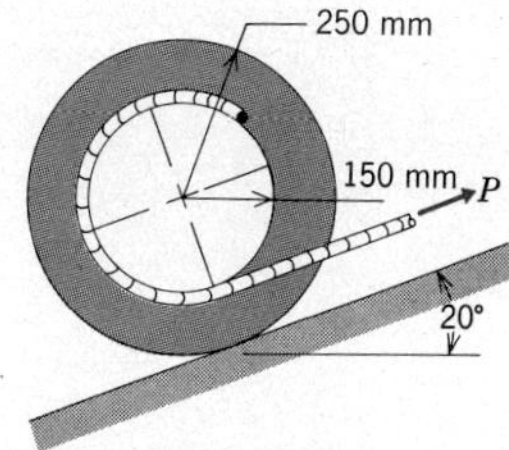

Prob. VII.22

VII.22 A cable is wrapped around the 50-kg stepped drum and a tensile force $\bar{P}$ is applied to the free end of the cable, as shown. The drum is in static equilibrium. Determine (a) the magnitude of $\bar{P}$ and of the friction force acting on the drum, (b) the direction in which the point on the drum in contact with the inclined surface is tending to move, (c) the minimum value of the coefficient of static friction for which static equilibrium is possible.

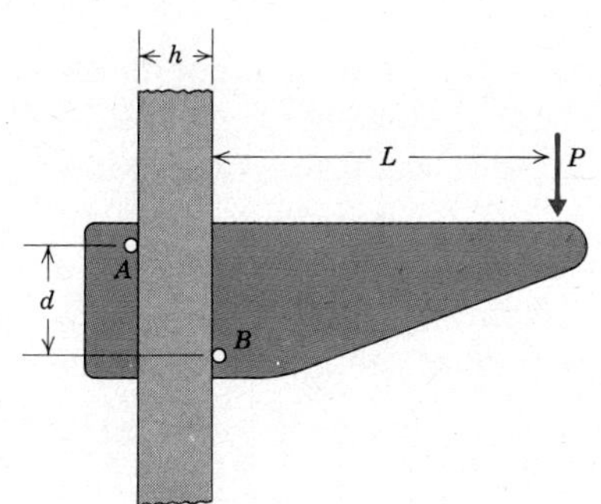

Prob. VII.23

VII.23 The movable bracket transfers the downward load $\bar{P}$ to the vertical support arm. The coefficient of static friction between pins A and B and the vertical arm are both μ. Derive an expression for the minimum distance L for which the bracket will not slip regardless of the magnitude of $\bar{P}$.

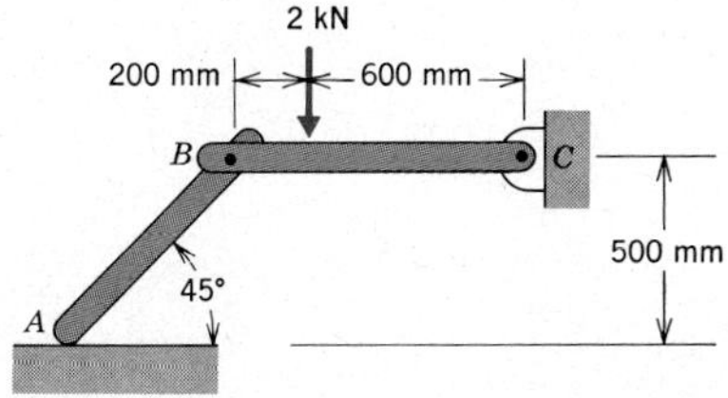

Prob. VII.24

VII.24 A 2-kN load is applied to the linkage, as shown. Knowing that the system is in equilibrium, determine (a) the friction force acting on bar AB at end A, (b) the minimum value of the coefficient of static friction for which static equilibrium is possible. The mass of the bars is negligible.

VII.25 Solve Problem VII.24 for the case where the uniform bars have masses AB = 100 kg and BC = 150 kg.

VII.26 Determine the smallest horizontal distance d between the center of mass G of the 150-lb worker and the wall for which the 30-lb ladder will not slip. The coefficients of static friction between the ladder and the house, and the ladder and the ground are 0.25 and 0.50, respectively.

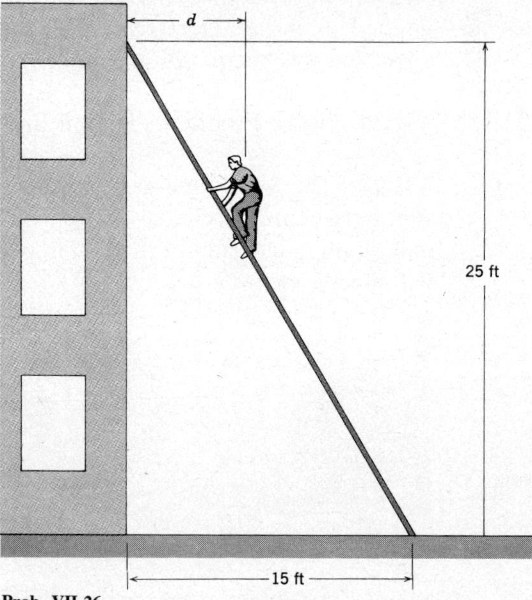

Prob. VII.26

VII.27 The semicylinder of mass m and radius R is in a condition of impending motion. The coefficients of static friction between the semicylinder and the inclined planes are the identical values μ. Determine μ.

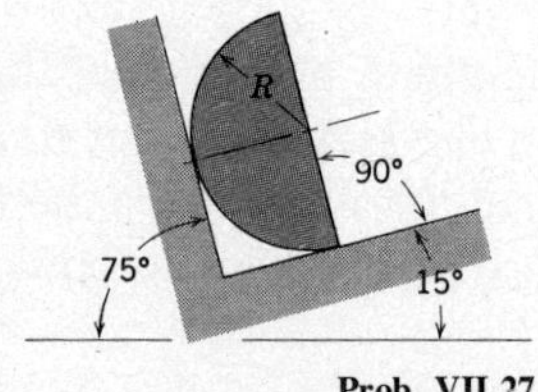

Prob. VII.27

VII.28 The coefficient of static friction between the 20-kg block C and the incline is 0.40. Knowing that the mass of bars AB and BC is negligible, determine the range of values of the magnitude of the horizontal force $\bar{P}$ for which the block will not move.

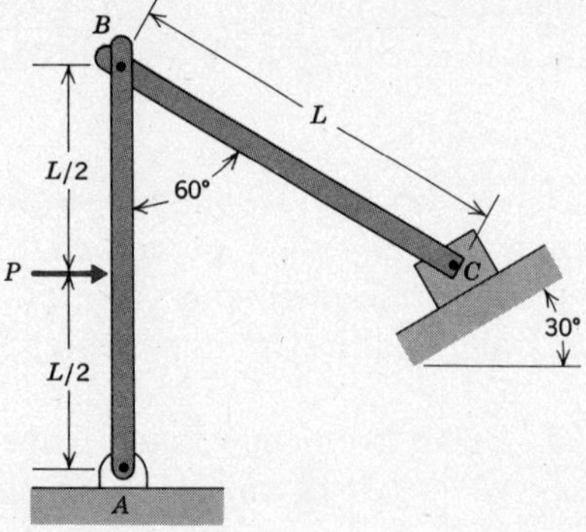

Prob. VII.28

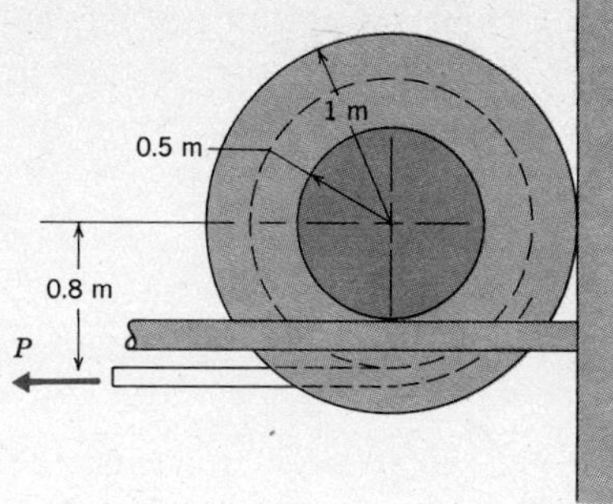

Prob. VII.29

VII.29 A spool of wire having a mass of 400 kg is supported by two horizontal rails (only one is shown in the side view) and rests against the vertical wall. The coefficient of static friction between all surfaces of contact is 0.40. Determine the smallest horizontal force $\bar{P}$ applied to the end of the wire which will cause the spool to rotate.

VII.30 The coefficient of static friction between the oil drum and all surfaces it contacts is 0.50. The drum has a mass of 500 kg. Determine the minimum force the hydraulic cylinder must exert to move the oil drum up the incline. At which surface is slip impending for this force? *Hint:* There is no contact between the drum and the horizontal surface when the drum begins to move up the hill. Assume slipping at one contact surface and check if the solution satisfies this assumption.

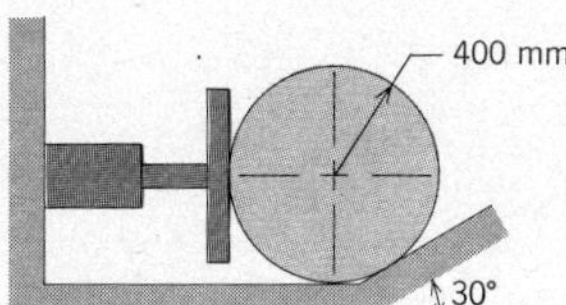

Prob. VII.30

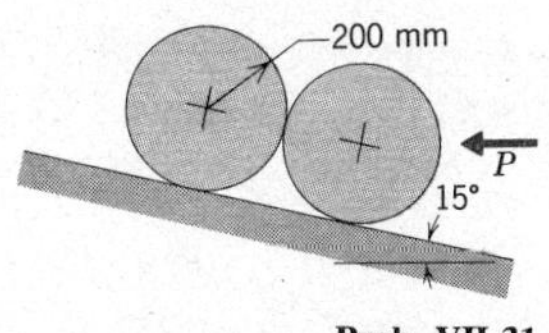

Prob. VII.31

VII.31 Two identical 30-kg cylinders are held on the ramp by the horizontal force $\bar{P}$. The coefficient of friction between all contacting surfaces is 0.25. Determine the maximum force $\bar{P}$ that can be applied without causing the cylinders to move. *Hint:* Assume that the cylinders tend to roll over the inclined plane without slipping, and then verify this assumption.

VII.32 The coefficient of friction between the uniform 30-kg bar and both surfaces is $\mu_s = 0.20$. Is the system in static equilibrium?

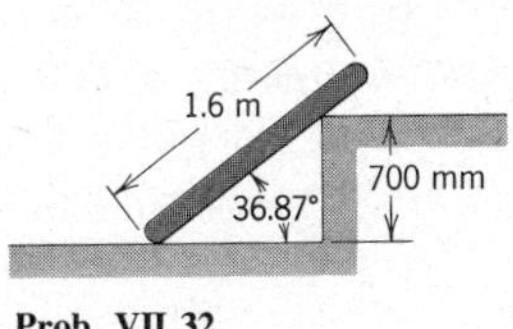

Prob. VII.32

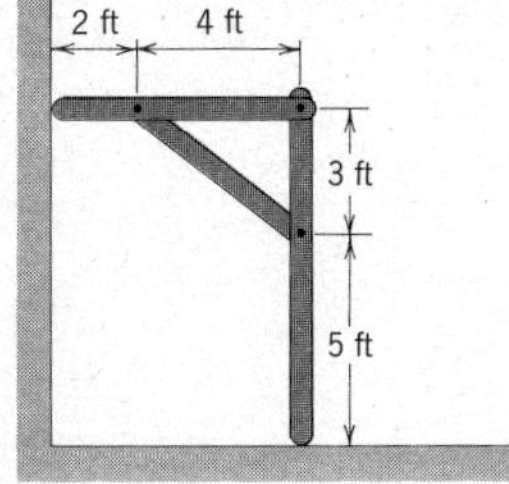

Prob. VII.33

VII.33 The coefficients of static friction between the frame and the wall, and the frame and the floor are 0.20 and 0.05, respectively. Each member of the frame weighs 10 lb/ft. Is the frame in static equilibrium?

VII.34 A hydraulic cylinder applies the vertical force $\bar{P}$ to the lifting frame that is supporting a 1000-kg homogeneous crate. The crate is

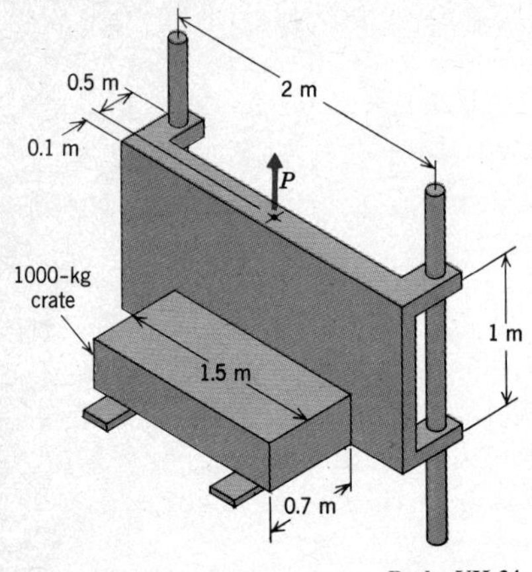

Prob. VII.34

situated symmetrically on the frame and the weight of the frame is negligible. Knowing that the coefficient of static friction at each contact surface between the frame and the vertical guide rails is 0.10, determine the minimum force $\bar{P}$ required to move the frame upward from rest.

VII.35 In Problem VII.34, determine the magnitude and sense of the minimum force $\bar{P}$ required to move the frame downward from rest.

VII.36 The 200 lb-in. couple is applied to rod AB in order to resist the 70-lb force applied to piston C. Knowing that the linkage is in static equilibrium at the position where $\theta = 90°$, determine (a) the friction force acting on the piston, (b) the minimum coefficient of static friction between the piston and its cylinder for which this situation is possible.

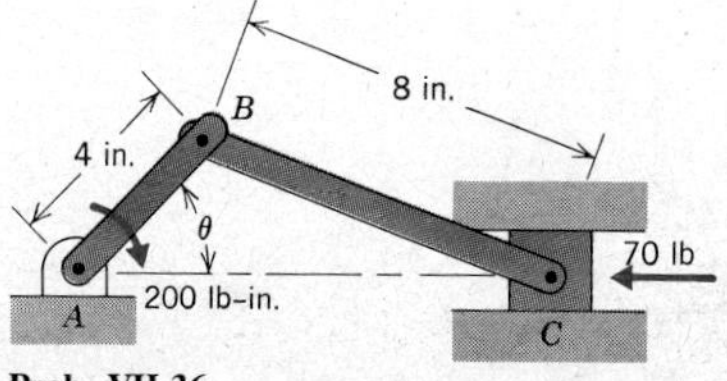

Prob. VII.36

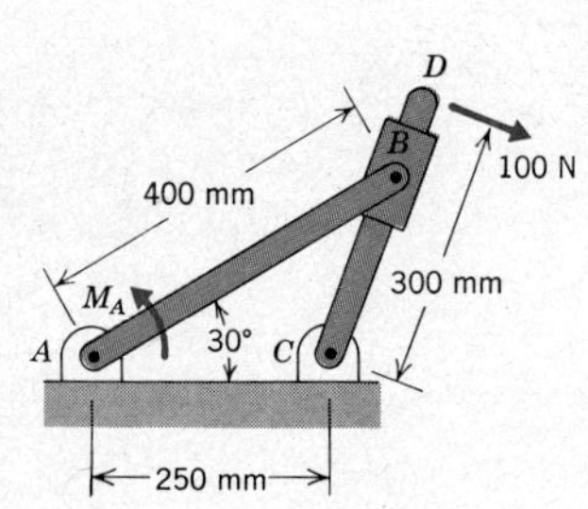

Prob. VII.38

VII.37 Solve Problem VII.36 for the position where $\theta = 135°$.

VII.38 Collar B may slide over rod CD. The coefficient of static friction between these bodies is 0.16. Determine the range of values of the couple $\bar{M}_A$ for which the linkage is in static equilibrium under the 100-N load at end D.

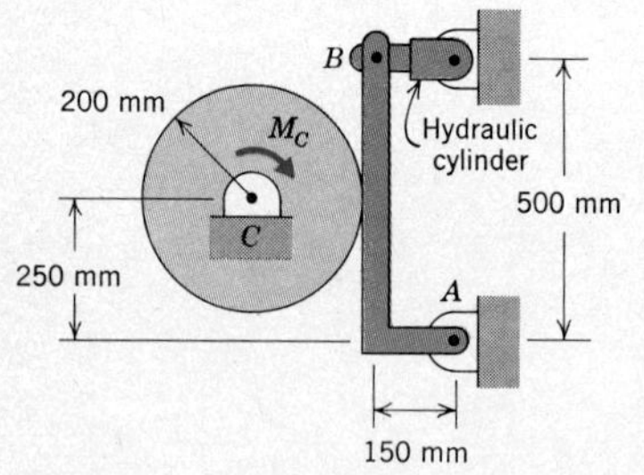

Probs. VII.39 and VII.40

VII.39 A couple $\bar{M}_C$ is applied to the brake drum C. Determine the smallest force exerted by the hydraulic cylinder on brake arm AB for which the brake drum will rotate if $\bar{M}_C$ is 300 N-m (a) clockwise, (b) counterclockwise. The coefficient of static friction between the drum and the arm is 0.75.

VII.40 The coefficients of static and kinetic friction between the brake drum C and the brake arm AB are 0.50 and 0.40, respectively. Knowing that the hydraulic cylinder pushes end B to the left with a force of 1.0 kN, determine the friction force acting on the drum if the couple $\bar{M}_C$ is 300 N-m (a) clockwise, (b) counterclockwise.

VII.41 Rod CD, which weighs 4 lb, is used to prop up board AB, which weighs 16 lb. Determine the minimum values of the coefficients of static friction between each pair of surfaces in contact for which equilibrium can be maintained.

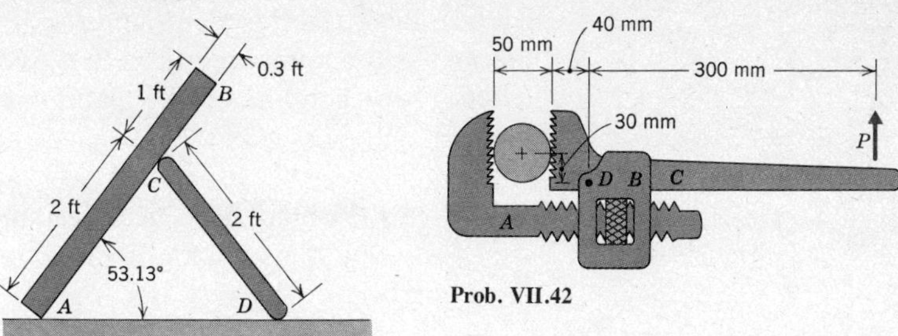

Prob. VII.41

Prob. VII.42

VII.42 A 50-mm-diameter rod is being gripped by the Stilson wrench shown. Members A and B may be regarded as a single rigid body connected to member C only by pin D. It is observed that, regardless of the magnitude of the force $\bar{P}$, the wrench does not slip over the rod (the wrench is said to be *self-locking*). Determine the minimum coefficients of static friction at both points of contact between the wrench and the pipe.

VII.43 The thin hemispherical shell of mass m and radius R is acted upon by a horizontal force $\bar{P}$ applied at its rim. Determine the angle ϕ for impending motion if the coefficient of static friction between the shell and the surface is 0.18.

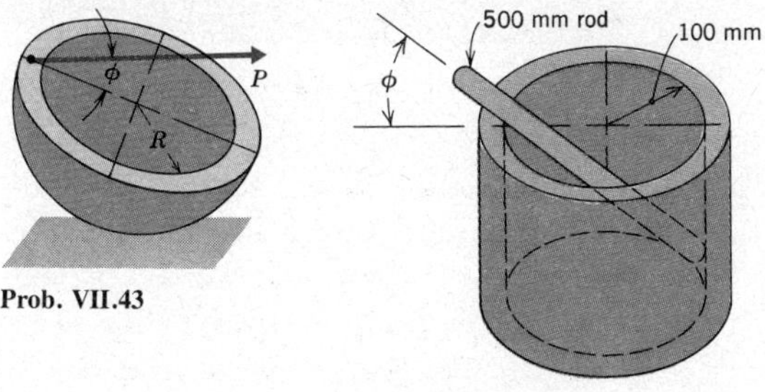

Prob. VII.43

Prob. VII.44

VII.44 A uniform rod 500 mm long is placed against the rim of a cylinder as shown. Determine the range of values of the angle of elevation ϕ of the rod for which the rod is in static equilibrium. The coefficient of static friction between the cylinder and the rod is 0.25.

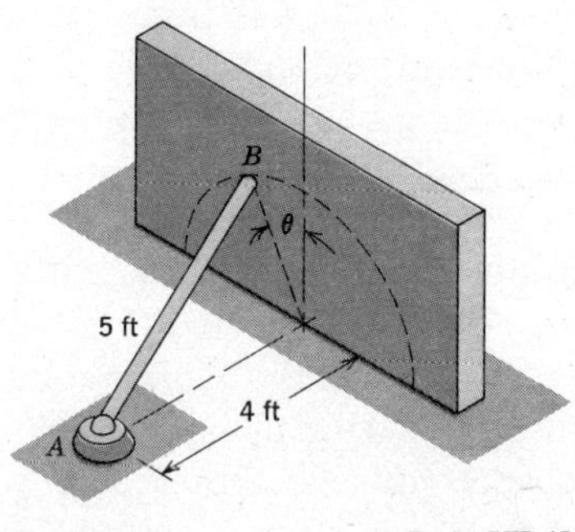

Prob. VII.45

VII.45 A 5-ft-long bar is connected to the floor by a ball-and-socket joint that is 4 ft from the vertical wall supporting the other end of the bar. The bar weighs 20 lb and the coefficient of static friction between the bar and the wall is 0.35. Determine the maximum value of the angle θ for which static equilibrium is possible. *Hint:* Because of the support at end A, end B tends to move tangent to the dashed circle.

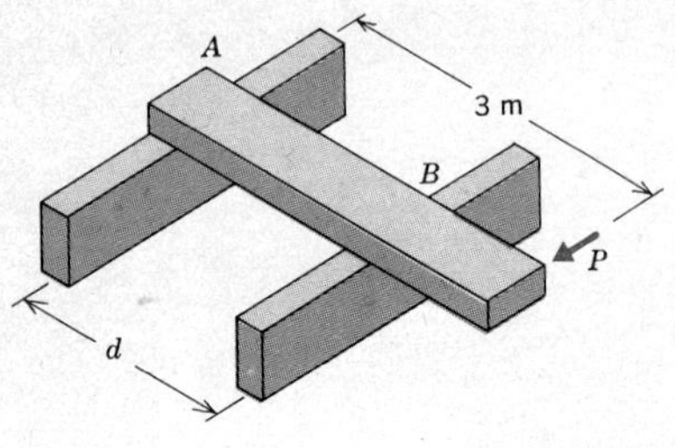

Prob. VII.46

VII.46 A 30-kg plank is resting on two horizontal joists, perpendicular to the axis of the joists in the horizontal plane. The coefficient of static friction between the plank and the joists is 0.40. Determine the force $\bar{P}$, parallel to the joists, required to move the plank when $d = 2$ m. *Hint:* The friction forces acting on the plank may be considered to be parallel to the joists.

VII.47 In Problem VII.46, determine the range of values of the dimension d for which a sufficiently large force $\bar{P}$ would cause the plank to slide (a) at joist A, (b) at joist B.

3 Sliding or Tipping

The accidental overturning of a glass of water that was resting on a table is a classic illustration of the fact that trying to overcome the force of friction can have surprising consequences. A comparable problem arises when a filing cabinet is pushed across a floor, as depicted in Figure 7a.

Figure 7a Figure 7b Figure 7c

In Figure 7b we see the cabinet slide along the floor, as we would expect it to, whereas in Figure 7c we see the unexpected result in which the cabinet tips over. To determine which situation will occur, let us draw a free body diagram of the filing cabinet, as shown in Figure 8, and evaluate the forces. Note that $\bar{P}$ is the force being applied to move the cabinet.

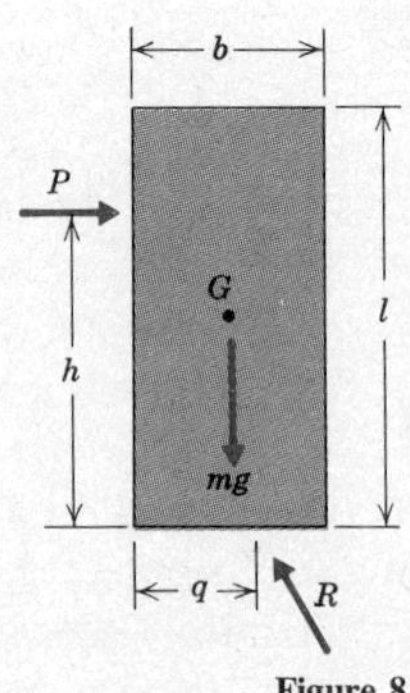

Figure 8

The force $\bar{R}$ is the resultant of the normal and frictional forces exerted by the floor, and the distance q locates the point of application of this force. Physically, because the cabinet is standing on edge in the tipping condition of Figure 7c, it must be that the reaction $\bar{R}$ acts at $q = b$ for tipping.

Note that because we are limited to the equations of static equilibrium, we cannot determine whether or not the cabinet will actually fall over on its side. This is so because, as soon as the bottom of the cabinet rotates away from the floor, the cabinet is in motion, requiring the application of the laws of dynamics. We can, however, determine if the force $\bar{P}$ will cause tipping to begin.

To determine if the cabinet will tip, we use the equations of static equilibrium to find the distance q at which the resultant force must act for static equilibrium. If $q < b$, we can conclude that the bottom of the cabinet remains in contact with the floor. In contrast, if $q > b$, we have a geometrically impossible situation, for the resultant force required for equilibrium cannot be applied to the cabinet. We then would conclude that tipping will occur. Clearly, $q = b$ is the transitional condition for *impending tipping*.

It is best to be consistent in the approach to problems involving the possibility of both slipping and tipping. In situations where we are uncertain about which, if either, of these conditions will occur, we shall begin by assuming that slipping will occur. Then, if the normal force $\bar{N}$ required for equilibrium is physically possible (for example, if $q < b$ for the filing cabinet in Figure 8) we know that the impending motion is one in which there is no tipping. On the other hand, if we determine that the normal force $\bar{N}$ corresponds to an impossible situation, then we can conclude that tipping will occur. This is the case when $q > b$ for the filing cabinet. The tipping condition may then be investigated by placing the normal force at the point of contact when tipping occurs, for instance, the lower right-hand corner of the crate in Figure 8. Note that in the tipping condition, the point of contact will generally not be in a condition where slip is also impending. Therefore, the friction force will be an unknown; that is, in general, $|\bar{f}| < \mu_s|\bar{N}|$.

These observations are illustrated in the following example.

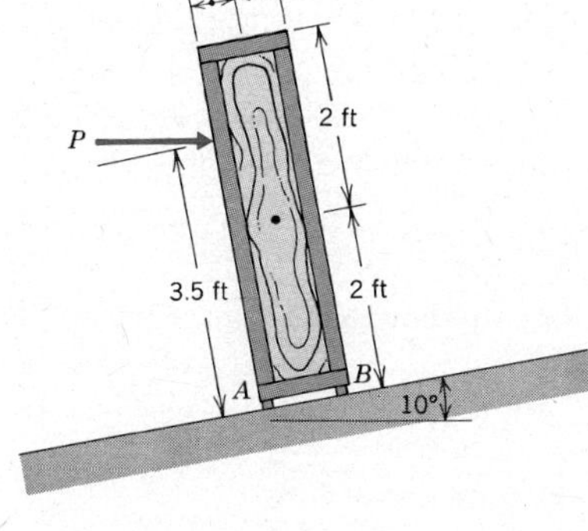

EXAMPLE 4

The 200-kg crate, whose center of mass is point G, is at rest on the hill. The crate is supported by small skids at points A and B. The coefficient of static friction between the incline and the skids is 0.25. Determine the minimum horizontal force $\bar{P}$ that will move the crate.

Solution

We will start by assuming that if the magnitude of $\bar{P}$ is sufficiently large, the crate will slide uphill without tipping. Thus, the friction force at each skid will be considered to be μ_s times the normal force at that skid, acting downhill, as shown in the free body diagram below.

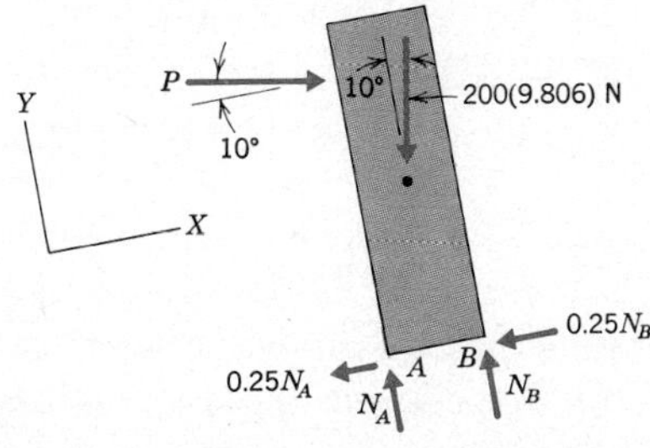

Because we are concerned with the points of application of forces, we write the equations for force and moment equilibrium of the crate. Choosing point A for the moment sum, we have

$$\Sigma M_{AZ} = N_B(1.0) - (P \cos 10^\circ)(3.5) - 200(9.806) \cos 10^\circ (0.5)$$
$$+ 200(9.806) \sin 10^\circ (2.0) = 0 \text{ N-m}$$
$$\Sigma F_X = P \cos 10^\circ - 200(9.806) \sin 10^\circ - 0.25N_A$$
$$- 0.25N_B = 0 \text{ newtons}$$
$$\Sigma F_Y = N_A + N_B - P \sin 10^\circ - 200(9.806) \cos 10^\circ = 0 \text{ newtons}$$

We now have three equations for the three unknowns N_A, N_B, and P. Their solutions are

$$P = 875 \qquad N_B = 3299 \qquad N_A = -1216 \text{ newtons}$$

These are the magnitudes that the unknown forces must have if the crate is to be in a condition of impending slipping without tipping. Notice that the value of N_A is negative. Clearly, this is physically impossible, for it requires that the incline pull downward on corner A. We may therefore conclude that applying the force $\bar{P}$ in the indicated manner will cause the crate to tip before it slides.

To determine the force $\bar{P}$ required to produce the condition of impending tipping, we set N_A to zero and draw a free body diagram of the crate in its given equilibrium position. Recall that in this condition, the friction force at the point of contact is unknown. Thus, the free body diagram for this portion of the investigation is as shown.

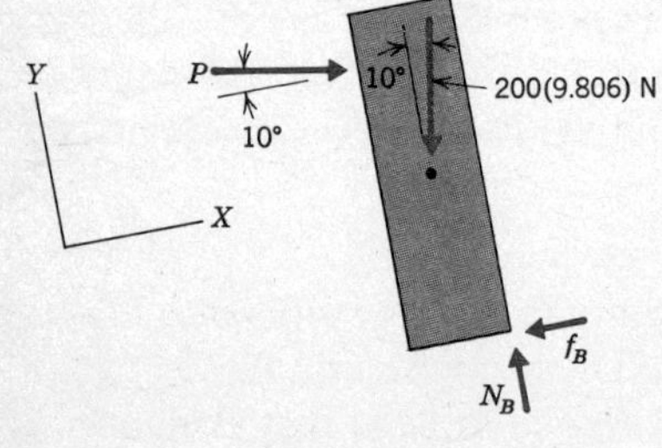

Choosing point B for the moment sum, the equilibrium equations for this case are

$$\Sigma M_{BZ} = -(P \cos 10^\circ)(3.5) + (P \sin 10^\circ)(1.0) + 200(9.806) \cos 10^\circ (0.5)$$
$$+ 200(9.806) \sin 10^\circ (2.0) = 0 \text{ N-m}$$
$$\Sigma F_X = P \cos 10^\circ - 200(9.806) \sin 10^\circ - f_B = 0 \text{ newtons}$$
$$\Sigma F_Y = -P \sin 10^\circ - 200(9.806) \cos 10^\circ + N_B = 0 \text{ newtons}$$

These equations yield

$$P = 503 \text{ newtons}$$
$$f_B = 155 \qquad N_B = 2019 \text{ newtons}$$

Notice that the value of P we have obtained is smaller than that found earlier by considering the slipping condition. This confirms our earlier conclusion that the crate would tip before it slides. Also, note that the friction force f_B given above should be smaller than $\mu_s N_B$, for if it were not, we would have a contradiction with the results of the study of the condition of impending slip. It is readily verified that this requirement is fulfilled.

An interesting aside to this problem is its similarity to the problem posed in part (c) of Example 1. It can be seen that concern about the possibility of impending tipping, which always requires that we consider moment equilibrium, complicates the solution of problems involving impending motion.

HOMEWORK PROBLEMS

VII.48 The coefficient of static friction between the 1-ton granite block and the floor is 0.35. Knowing that $h = 2.5$ ft, determine the smallest horizontal force $\bar{P}$ that will cause the block to move.

Prob. VII.48

VII.49 Determine the range of values of h in Problem VII.48 for which slipping will occur before tipping.

VII.50 In Example 4, let h denote the distance above the bottom of the crate at which the horizontal force $\bar{P}$ is applied. Determine the maximum value of h for which the crate can be pushed up the hill without tipping.

VII.51 A 60-kg refrigerator, having center of mass G, is mounted on four casters. The coefficient of static friction between a locked caster and the floor is 0.50, and the frictional resistance of a rolling caster is negligible. Determine the magnitude of the horizontal force $\bar{P}$ required to move the refrigerator when $h = 1.0$ m, (a) if all casters are locked, (b) if only casters B are locked, (c) only casters A are locked.

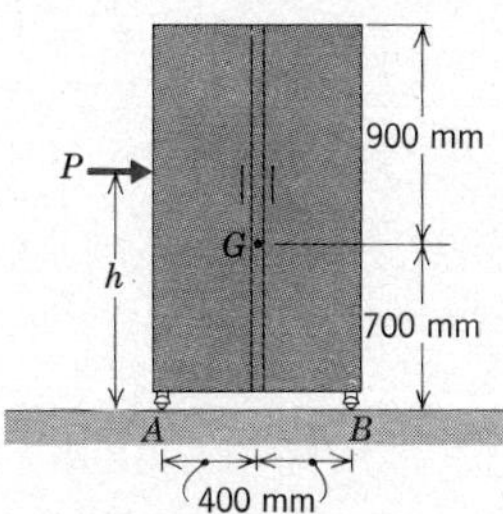

Prob. VII.51

VII.52 In Problem VII.51, determine the range of values of h for which the refrigerator will not tip (a) if all casters are locked, (b) if only casters B are locked. (c) Explain why it is impossible to tip the refrigerator over with the force $\bar{P}$ when only casters A are locked.

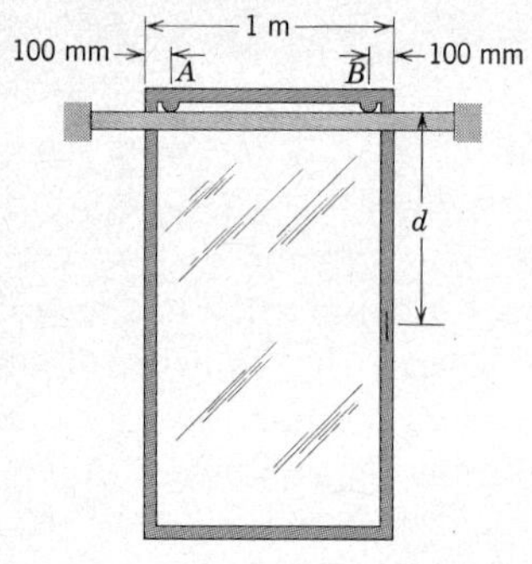

Prob. VII.53

VII.53 The 25-kg door is mounted on the horizontal rail by means of runners A and B. The coefficients of friction for these runners are $\mu_A = \mu_B = 0.15$. The door handle is pulled to the right to open the door. Determine (a) the maximum distance d for the door handle for which the door will not tip when it is opened, (b) the force required to open the door if d equals the value found in part (a), (c) the force required to open the door if d equals one-half the value found in part (a).

VII.54 Solve Problem VII.53 if $\mu_A = 0.20$, $\mu_B = 0.10$.

VII.55 Solve Problem VII.53 for the case where the door is being closed (moved to the left) if $\mu_A = 0.20$, $\mu_B = 0.10$.

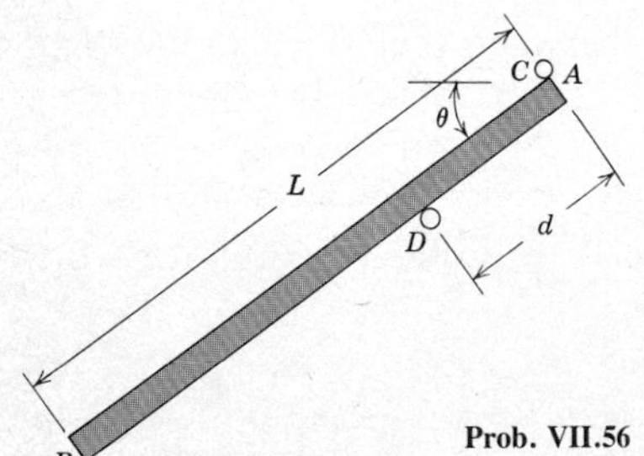

Prob. VII.56

VII.56 Bar AB of mass m and length L is supported in the vertical plane by pins C and D, which are separated by the distance d. The coefficient of static friction between the pins and the bar is μ_s. Derive an expression for the range of values of L/d for which static equilibrium is possible.

VII.57 The angle β of the incline is very gradually increased until the block of mass m moves. Develop formulas that indicate whether the block slides or tips in terms of the coefficient of friction μ, the ratio b/h, and the angle β at which movement occurs.

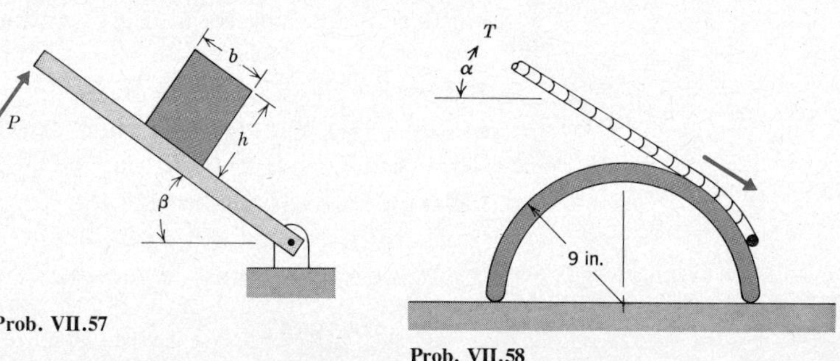

Prob. VII.57

Prob. VII.58

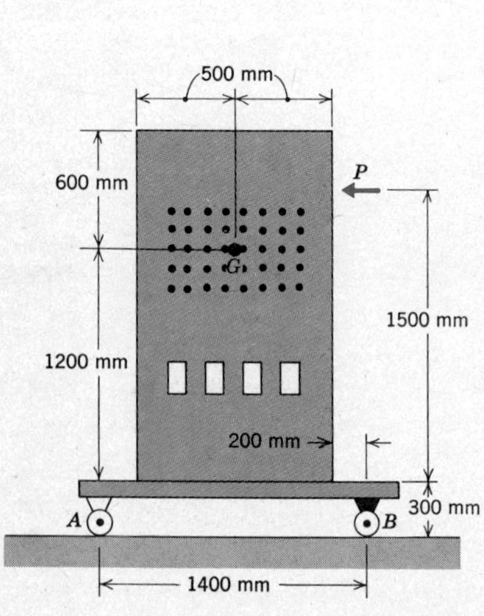

Prob. VII.59

VII.58 The 8-lb thin semicylindrical shell is to be towed to the left. The coefficient of static friction between the shell and surface is 0.35. Determine the largest angle α at which the shell may be towed without causing it to tip. What is the corresponding cable tension?

VII.59 A 500-kg electronic computer, having center of mass G, is placed on the 20-kg dolly. The casters A of the dolly are locked. The coefficients of friction between the computer and the dolly, and between the casters and the floor, are 0.50 and 0.30, respectively. Determine the maximum force $\bar{P}$ that can be applied without causing the computer to move.

B. BASIC MACHINES HAVING FRICTION

There are a multitude of simple mechanical devices, which we refer to as basic machines, that can be used to either move or prevent the motion of physical objects. These devices can magnify or convert the input forces applied to them into a different set of output forces that are applied to other systems. For example, a wheel fixed to an axle has the capability of transforming a torque exerted by its axle into a forward force that can propel a vehicle.

In earlier modules, particularly Module IV, we developed the basic techniques for investigating the equilibrium of machines. The common feature of the machines we shall consider here is the important, and sometimes useful, effect of friction.

1 Wedges

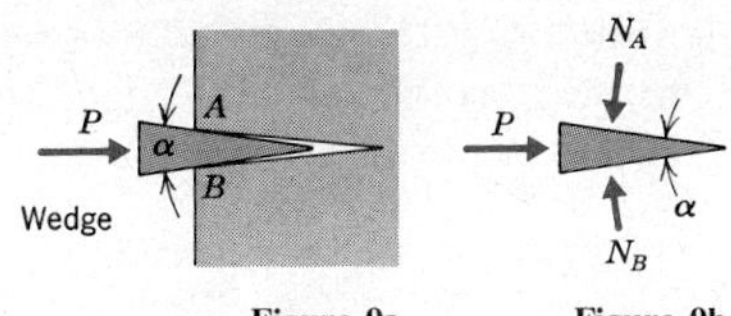

Figure 9a Figure 9b

A wedge is merely a block having two flat surfaces that form a small angle relative to each other. Consider the triangular wedge in Figure 9a, which is being pushed into a crack in a large body by the force $\bar{P}$.

In the situaton where the surfaces of contact between the wedge and the large body are smooth, the free body diagram of the wedge is as shown in Figure 9b. It is not difficult to see that, if the angle α is sufficiently small, the normal forces $\bar{N}_A$ and $\bar{N}_B$ will be much larger than the applied force $\bar{P}$. Thus, this wedge could be used to enlarge the crack. An example of this type of action is the head of an axe.

However, Figure 9b is not a correct representation of the action of the wedge, because friction cannot be neglected. To study the effect of friction we must distinguish between the situation where the tendency is for the wedge to be forced further into the crack, as shown in the free body diagram in Figure 10a, and the case where the tendency is for the wedge to be squeezed out of the crack, as shown in the free body diagram in Figure 10b.

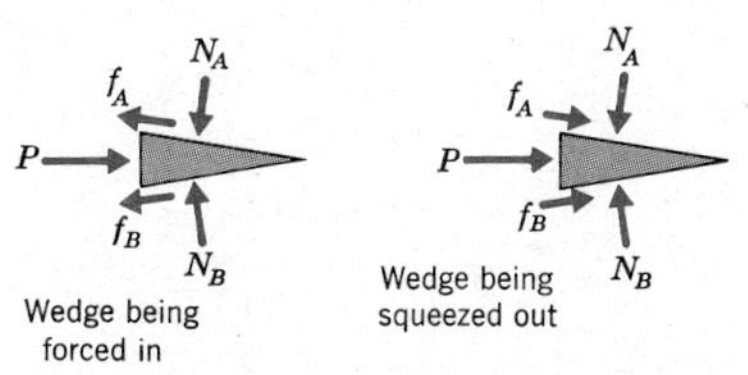

Figure 10a Figure 10b

Considering $\bar{P}$ to be known, we see that there are four unknown forces acting on the wedge in each case illustrated in Figure 10. Because we are not concerned here with the points of application of the forces acting on the wedge, there are only two scalar equations of force equilibrium. We can conclude then that the free body diagrams in Figure 10 represent statically indeterminate systems having two redundant forces.

Fortunately, our interest in wedges lies in the study of impending motion. One problem we wish to investigate is how the wedge can be forced further into the crack. This is the case of impending inward motion. It may be studied by using the free body diagram of Figure 10a with $f = \mu_s N$ for each friction force. For such a problem the friction forces are no longer independent unknowns, so the system becomes a statically determinate one.

Another problem we wish to investigate is how the wedge can be pulled out of the crack. In this case we have impending outward motion, so we use the free body diagram of Figure 10b, setting $f = \mu_s N$ for each friction force. Once again, the friction forces are no longer independent unknowns, so the system is a statically determinate one.

The preceding discussion enables us to observe that the existence of friction has both good and bad effects in wedges. The bad effect arises because the friction force for impending inward motion, Figure 10a, opposes the input force $\bar{P}$. Therefore, the existence of friction means that a larger force $\bar{P}$ is needed to drive the wedge in. The beneficial effect of friction is associated with removing the wedge from the crack. If the coefficients of friction are very small, impending outward motion of the wedge, Figure 10b, still corresponds to an inward force $\bar{P}$. In other words, it will be necessary to hold the wedge in position. In contrast, if the coefficients of static friction are sufficiently large, for impending outward motion a force $\bar{P}$ in the opposite sense of the one depicted in the free body diagram (that is, a pulling force) will be required to remove the wedge. In such cases the wedge is said to be *self-locking*.

EXAMPLE 5

Wedges A and B are used to raise block C against the 5-kN load that is acting on it. The coefficient of friction between wedge B and all surfaces in contact with it is 0.10. Cleat D holds wedge A in position. (a) Derive an expression for the minimum force $\bar{P}$ required to raise block C. (b) Determine whether the wedge is self-locking.

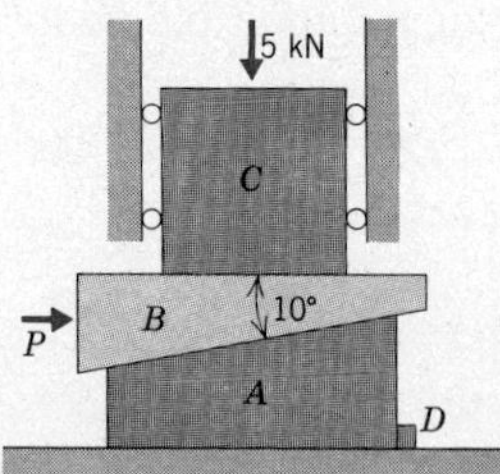

Solution

The first consideration here is which free body diagrams to draw. In view of the presence of cleat D, the wedge A is in effect part of the ground, so there is no reason to consider this wedge. On the other hand, wedge B and block C are individual bodies whose equilibrium must be assured. Thus, we will draw free body diagrams of these bodies. In part (a) we are interested in impending motion of wedge B to the right, whereas, to investigate self-locking we want to consider impending motion of wedge B to the left. In order to have one set of free body diagrams for both cases, we denote

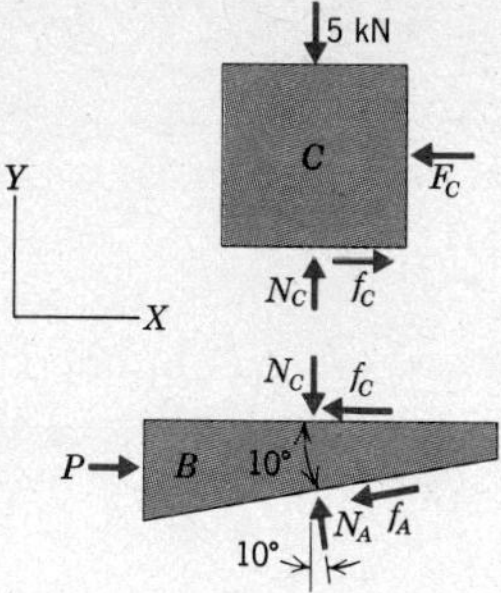

the friction forces on this wedge as $\bar{f}_C$ and $\bar{f}_A$ and then set $f = \pm\mu_s N$, choosing the sign to fit the type of impending motion. For the purposes of drawing free body diagrams, $\bar{f}_C$ and $\bar{f}_A$ have been depicted in the diagrams as though block B where tending to move to the right. Note that the forces exerted by the wedge on block C must satisfy Newton's third law.

The equilibrium equations for these bodies are

$$\text{wedge } B: \quad \Sigma F_X = P - N_A \sin 10° - f_C - f_A \cos 10° = 0$$

$$\Sigma F_Y = N_A \cos 10° - f_A \sin 10° - N_C = 0$$

$$\text{block } C: \quad \Sigma F_Y = N_C - 5 = 0 \text{ kN}$$

We are not interested in $\bar{F}_C$, so we have disregarded ΣF_X for block C. First, to solve part (a) we set

$$f_C = +0.10N_C \qquad f_A = +0.10N_A$$

Substituting these expressions, as well as $N_C = 5$ kN found from the last equilibrium equation, into the first two equilibrium equations, we obtain

$$P - 0.1736N_A - (0.50) - 0.9848(0.10N_A) = 0$$

$$0.9848N_A - 0.1736(0.10N_A) - 5 = 0$$

Solution of these equations yields

$$N_A = 5.168 \qquad P = 1.961 \text{ kN}$$

We now turn our attention to part (b). In order to evaluate whether the wedge is self-locking, we will determine if the force $\bar{P}$ should be to the right or left when motion is impending to the left. In this case, we have

$$f_A = -\mu_s N_A \qquad f_C = -\mu_s N_C$$

The equilibrium equation for block C still gives $N_C = 5$ kN, so the equilibrium equations for the wedge now become

$$P - 0.1736N_A - (-0.50) - 0.9848(-0.10N_A) = 0$$

$$0.9848N_A - 0.1736(-0.10N_A) - 5 = 0$$

The corresponding solutions are

$$N_A = 4.989 \qquad P = -0.125 \text{ kN}$$

The fact that P is negative means that the force $\bar{P}$ must pull on the wedge to the left to move the wedge in that direction. Thus, the wedge is self-locking.

HOMEWORK PROBLEMS

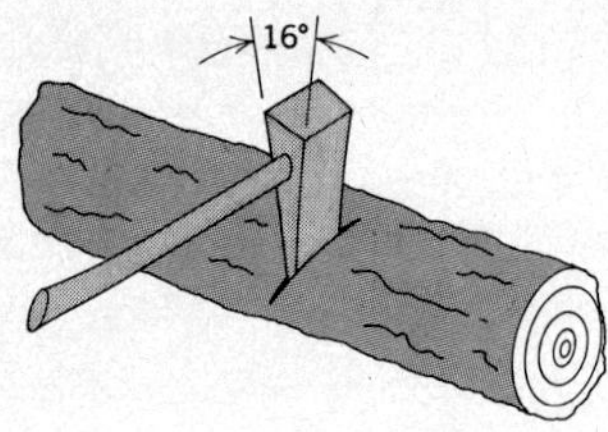

Prob. VII.60

VII.60 The faces of the blade of the axe form a 16° angle. Determine the minimum coefficient of static friction between the blade and the log for which the axe is self-locking.

VII.61 and VII.62 Determine the downward force $\bar{P}$, applied to the 20-kg block, required to move the 500-kg block. The coefficient of static frction between all surfaces is 0.15.

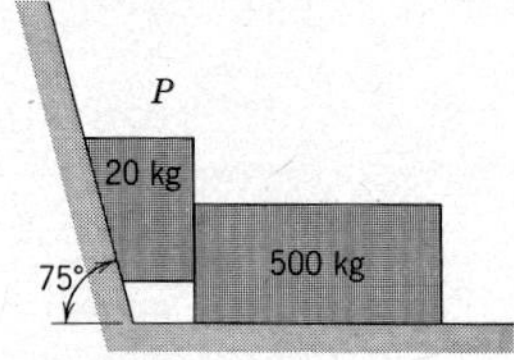

Prob. VII.61

Prob. VII.62

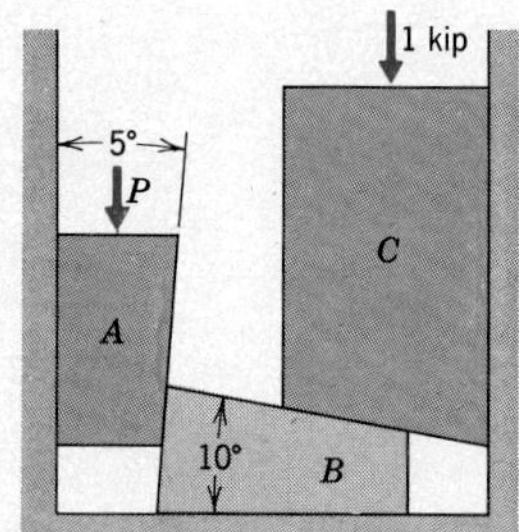

Prob. VII.63

VII.63 Determine the downward force $\bar{P}$, applied to wedge A, required to raise block C against the 1-kip load. The coefficient of static friction between all surfaces is 0.45.

VII.64 Two 100-lb flat plates resting on horizontal surfaces support a lightweight wedge, as shown. The coefficients of static friction between the plates and the wedge, and between the plates and the horizontal surfaces, are 0.40 and 0.30, respectively. Determine (a) the maximum overhang d for which the wedge will cause the plates to slide without tipping, (b) the vertical force $\bar{F}$ that should be applied to the wedge in order to move the plates when d is smaller than the value obtained in part (a).

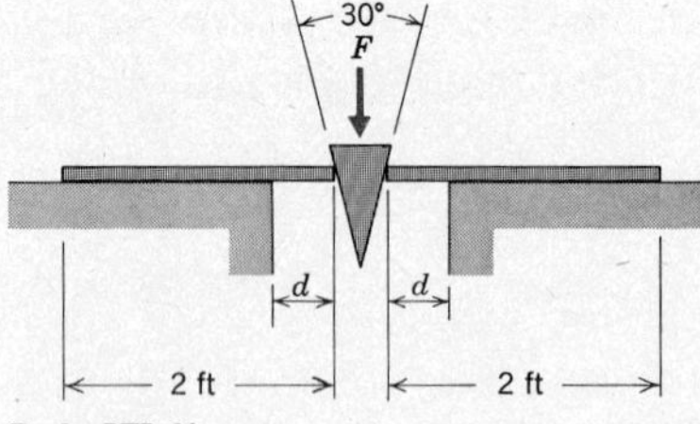

Prob. VII.64

VII.65 A 2000-kg lathe having center of mass G is to be leveled by means of a wedge at corner B. The coefficient of static friction between the wedge and the surfaces it contacts is 0.10. Determine (a) the minimum

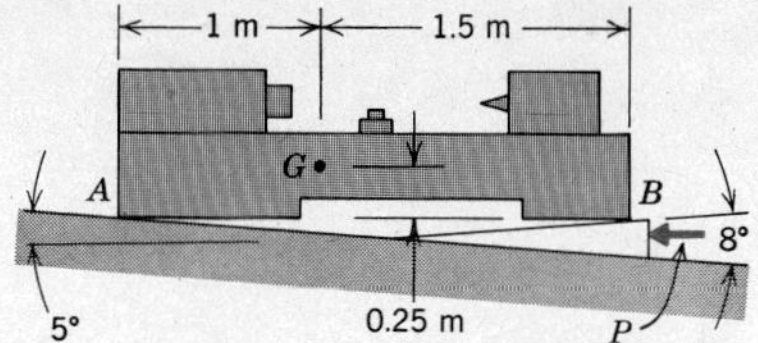

Prob. VII.65

horizontal force $\bar{P}$ required to raise corner B of the lathe, (b) the minimum coefficient of static friction between corner A of the lathe and the floor for which the lathe will not slide when the force $\bar{P}$ in part (a) is applied to the wedge. (c) Is the system self-locking? The lathe may be considered to be in the horizontal position.

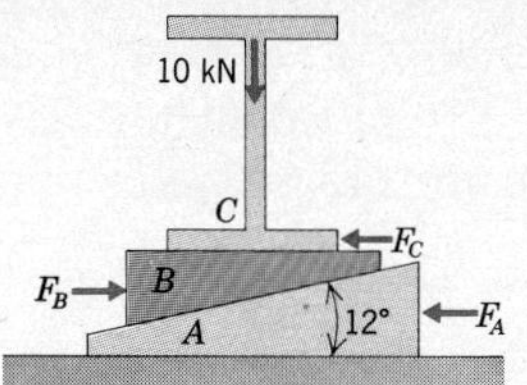

Prob. VII.66

VII.66 The H-beam is to be adjusted into its final position by applying the force $\bar{F}_B$ to wedge B. The reaction on the end of the beam is 10 kN. The coefficient of static friction between all surfaces is 0.20. (a) What is the minimum force $\bar{F}_B$ required to raise the beam? (b) What are the corresponding forces $\bar{F}_C$ and $\bar{F}_A$ required to prevent the beam and lower wedge from shifting?

VII.67 Solve Problem VII.66 in the case where the sense of $\bar{F}_B$ is reversed in order to lower the beam.

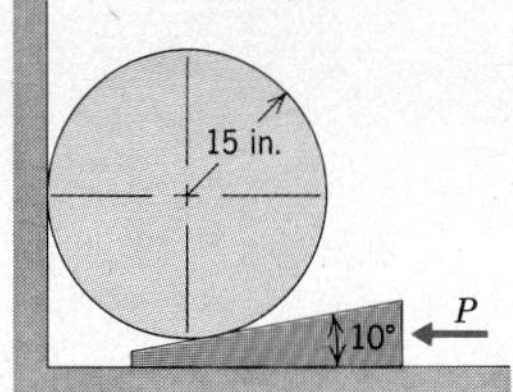

Prob. VII.68

VII.68 Determine the minimum force $\bar{P}$ required to lift the 500-lb drum shown. The mass of the wedge is negligible and the coefficient of friction between all surfaces is 0.25. Which surfaces are on the verge of sliding for this force $\bar{P}$?

VII.69 A vertical force $\bar{P}$ is to be applied to wedge E in order to push end C of bar CD to the right. The coefficient of static friction between all surfaces is 0.15. Determine the minimum force $\bar{P}$ that will achieve this goal.

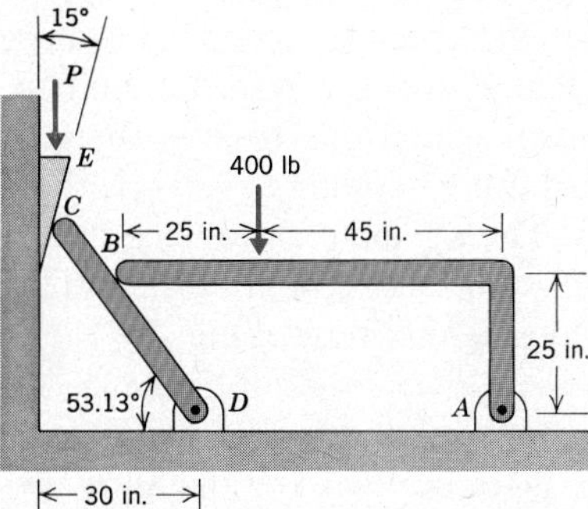

Prob. VII.69

VII.70 Determine whether the system in Problem VII.69 is self-locking. That is, will the wedge remain in place if the force $\bar{P}$ is no longer applied to the wedge?

2 Square Threaded Screws

A screw is a basic machine for transforming a couple applied about an axis into an axial force that can be used to move objects or to fasten objects together. Familiar examples of screw-type machines are an automobile bumper jack (for moving objects) and a bolt and nut set (for fastening objects).

There are several common shapes for the threads of a screw. We will limit our efforts here to the analysis of square threads, because the analysis of such threads is quite similar to that of wedges. A blow-up of a square-threaded screw carrying a couple $\bar{M}$ that would advance the screw in opposition to a load $\bar{F}$ is shown in Figure 11.

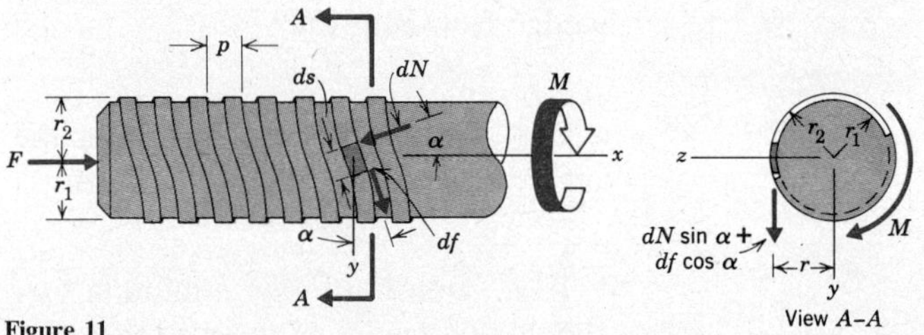

Figure 11

In order for the screw to sustain the axial force $\bar{F}$ and the couple $\bar{M}$, there are reaction forces distributed along the screw threads resulting from their contact with the body to which the screw is fastened. Let us consider an infinitesimal portion of a screw thread of length ds, such as the one shown in Figure 11.

A screw thread is helical, so it can be thought of as being developed by wrapping an inclined plane around a cylinder. The inclined plane analogy indicates that we can consider the forces acting on the infinitesimal element of thread ds to be a normal force $d\bar{N}$ and a frictional force $d\bar{f}$. Note that if the screw is tending to advance (move in the negative x direction) against the force $\bar{F}$, then $d\bar{f}$ is in the sense shown in Figure 11, whereas if it is tending to retract, then $d\bar{f}$ has the opposite sense from that shown. The advancing movement corresponds to tightening a screw, and the retracting movement corresponds to loosening it.

The angle α describing the orientation of the normal and frictional forces is the *lead angle* of the helix, and is a constant. As we will soon see, this angle is related to the *pitch* p of the screw thread, which is shown in Figure 11 to be the axial distance between matching points on two adjacent screw threads.

Considering the screw threads to be rigid and to mesh perfectly with the threads of the body to which the screw is fastened, it follows that dN and df are the same for all elements ds.

Let us now write the equilibrium equations for the screw. We require the equations for the force sum in the x direction and the moment sum about the x axis; the polar symmetry of the many elements ds about the x axis automatically satisfies the other equilibrium equations. From the side view in Figure 11 we see that the forces acting on the element ds have a resultant axial component dF_x, given by

$$dF_x = df \sin \alpha - dN \cos \alpha \tag{7}$$

Further, the sectional view in Figure 11 shows that the y component of these forces, which is $df \cos \alpha + dN \sin \alpha$, exerts a positive moment dM_z about the z axis, given by

$$dM_z = (df \cos \alpha + dN \sin \alpha)r \tag{8}$$

where r is the average of the inner and outer radii for the screw thread, that is,

$$r = \tfrac{1}{2}(r_1 + r_2) \tag{9}$$

The sum of an infinite number of infinitesimal quantities is an integral, so the basic equilibrium equations for the screw are

$$\Sigma F_x = F + \int dF_x = 0 \tag{10}$$

$$\Sigma M_z = -M + \int dM_z = 0 \tag{11}$$

Substituting the expressions for dF_x and dM_z from equations (7) and (8) into equations (10) and (11) and noting that the angle α is constant, the equilibrium equations become

$$\begin{aligned} \Sigma F_X &= F + \int(df \sin \alpha - dN \cos \alpha) \\ &\equiv F + f \sin \alpha - N \cos \alpha = 0 \end{aligned} \tag{12}$$

$$\begin{aligned} \Sigma M_Z &= -M + r \int(df \cos \alpha + dN \sin \alpha) \\ &\equiv -M + r(f \cos \alpha + N \sin \alpha) = 0 \end{aligned} \tag{13}$$

When using a screw we are usually interested in the relationship between the force $\bar{F}$ and the couple $\bar{M}$. Considering F as a known quantity, equations (12) and (13) are only two equations for the three unknowns, $N, f,$ and M. Thus, as it stands, the system is statically indeterminate.

Our interest, however, lies solely in cases of impending motion. Basically, we would like to know the couple required to advance (tighten) a screw and the couple required to retract (loosen) a screw. Noting the sense in which the friction force was assumed to act in Figure 11, we have

$$f = +\mu_s N: \quad \text{advancing} \tag{14a}$$

$$f = -\mu_s N: \quad \text{retracting} \tag{14b}$$

Either of equations (14) provides the additional relationship that will enable us to determine M as a function of F.

Before we present the solution, let us see how we can determine the lead angle α for a particular type of screw. Clearly, this is a key parameter in the equations. To do this, recall that a helix is formed by wrapping a triangle around a cylinder. If we unwrap a portion of the helix equivalent to one circumference ($2\pi r$), we obtain the triangle shown in Figure 12.

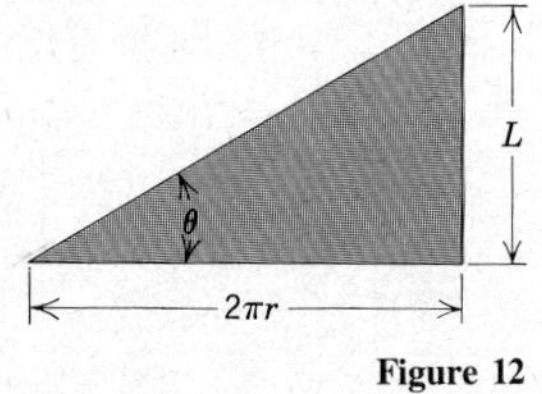

Figure 12

The distance L is called the *lead*. It is the distance the screw will advance (or retract) when given one full turn. In most common screws there is only one thread, which is continuously wrapped around the axis of the screw. Such a screw is said to be *single threaded,* and the lead L equals the pitch p. Other screws have two or three parallel sets of threads. Such cases are termed *double-* or *triple-threaded screws,* respectively. By induction we can state the for an *n-threaded screw,* we have

$$L = np$$

It is then apparent from Figure 12 that

$$\sin \alpha = \frac{np}{[(np)^2 + (2nr)^2]^{1/2}} \qquad \cos \alpha = \frac{2\pi r}{[(np)^2 + (2\pi r)^2]^{1/2}} \tag{15}$$

When we substitute the preceding expressions and either of the friction relations of equations (14) into the equilibrium equations, we can eliminate N to find

$$\text{advance:} \qquad M = Fr\,\frac{np + 2\pi\mu_s r}{2\pi r - \mu_s np} \tag{16a}$$

$$\text{retract:} \qquad M = Fr\,\frac{np - 2\pi\mu_s r}{2\pi r + \mu_s np} \tag{16b}$$

Equations (16) describe the situation where the screw is at rest and we wish to advance or retract it further. In the situation where the screw is rotating slowly, the threads are sliding over each other. The force analysis in this case is essentially unchanged (assuming that inertial effects from the rotation are negligible). Thus, equations (16) may be modified to describe a rotating screw by replacing μ_s by μ_k, the coefficient of kinetic friction.

This analysis demonstrates that a square screw thread functions like a wedge wrapped around a cylinder. Characteristic of the similarities between a screw and a wedge is the fact that both systems are statically indeterminate when motion is not impending. A further similarity is the possibility that a screw, like a wedge, may be self-locking. A self-locking screw will retract only if the couple $\bar{M}$ is in the sense of the rotation required to retract the screw. This property is exhibited by the value of M in equation (16b) being negative. Thus,

$$\mu_s > \frac{np}{2\pi r} \equiv \tan \alpha \text{ for self-locking} \tag{17}$$

This result has a simple interpretation. Consider once again the element of thread ds in Figure 11. Now we wish to consider the case where the screw is about to retract. This situation is depicted in Figure 13. The

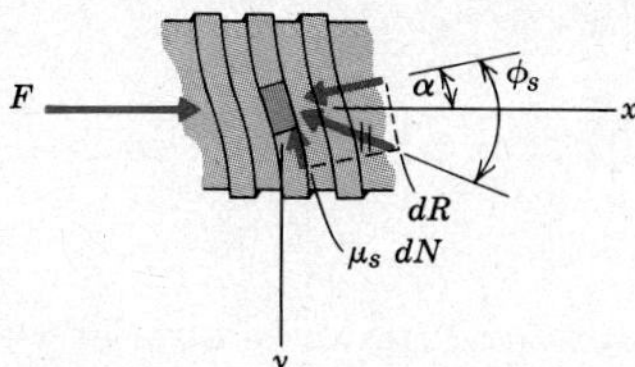

Figure 13

friction and normal forces on the thread element are equivalent to the resultant $d\bar{R}$, which for the condition depicted, is a force at an angle ϕ_s measured from the normal to the thread. Angle ϕ_s is the angle of static friction; that is, $\phi_s = \tan^{-1} \mu_s$. If ϕ_s is larger than α, we see that the y component of $d\bar{R}$ is negative. This means that $d\bar{R}$ will oppose a rotation that would tend to retract the screw. Now, noting the relationship between ϕ_s and μ_s, it can be deduced that equation (17) is equivalent to the requirement $\phi_s > \alpha$.

EXAMPLE 6

A square-threaded single-pitch bolt whose thread has a mean diameter of 12 mm and a pitch of 4 mm is screwed into a fixed metal block. The coefficients of static and kinetic friction between the block and the bolt are 0.35 and 0.27, respectively. Friction between the head of the bolt and the block is negligible. Determine the axial force developed in the bolt if it is tightened with a socket wrench that applies a 10 N-m-couple in order to (a) turn the bolt slowly, (b) start the bolt rotating from rest.

Solution

The primary difference between the two given cases is that in case (a) the bolt is actually turning as it advances, so there is kinetic friction. This requires that we replace μ_s by $\mu_k = 0.27$ in equation (16a). In contrast, the advancing movement is impending in case (b), so we will apply equation (16a) directly, with $\mu_s = 0.35$. For both cases, the given information tells us that

$$r = \tfrac{1}{2}(12 \text{ mm}) = 0.006 \text{ m} \qquad p = 0.004 \text{ m} \qquad n = 1$$

Solving part (a) first, for a 10 N-m couple equation (16a) yields

$$10 = F(0.006)\frac{(1)(0.004) + 2\pi(0.27)(0.006)}{2\pi(0.006) - (0.27)(1)(0.004)}$$

$$F = 4304 \text{ N}$$

Similarly, for part (b) we have

$$10 = F(0.006)\frac{(1)(0.004) + 2\pi(0.35)(0.006)}{2\pi(0.006) - (0.35)(1)(0.004)}$$

$$F = 3518 \text{ N}$$

The difference between these results has an important, subtle consequence in the general problem of tightening screws to a predetermined clamping force by means of a torque wrench. (A torque wrench has a dial that indicates the moment being applied to the screw.) Clearly, it is easy to see that a screw is rotating, and almost impossible to sense by hand that the screw is on the verge of rotating. Therefore, torque specifications corresponding to a desired axial force in a screw usually require that the screw be turned with the wrench until the wrench indicates the specified torque. After achieving this adjustment it would require a considerably higher torque to start the screw rotating again to tighten it further.

HOMEWORK PROBLEMS

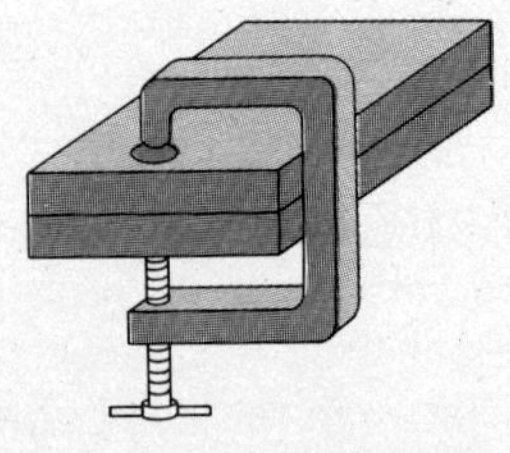

Prob. VII.71

VII.71 The C clamp holds two pieces of wood together. The clamp has a single square thread with a pitch of 3 mm and a mean diameter of 18 mm. The coefficient of static friction is 0.20. Determine the torque required to tighten the clamp further when it is at rest, applying a clamping force of 800 N.

VII.72 In Problem VII.71, determine the torque required to tighten the clamp further if it is triple threaded.

VII.73 In Problem VII.71, determine the torque required to loosen the clamp.

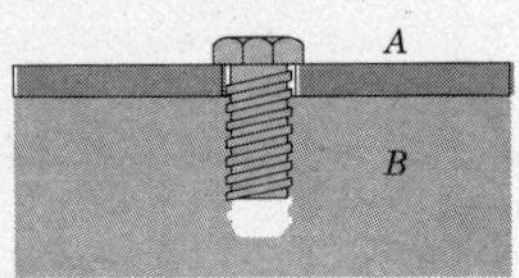

Prob. VII.74

VII.74 The metal plate A shown in the diagram is fastened to block B by the double-threaded screw, whose square threads have a pitch of $\frac{1}{16}$ in. The mean radius of the threads is $\frac{1}{4}$ in. The coefficients of static and kinetic friction for the threads are 0.30 and 0.25, respectively, and friction between plate A and the head of the screw is negligible. The screw is tightened by means of a torque wrench. Determine the clamping force on plate A if the torque wrench reads 40 ft-lb (a) when the screw is about to turn, (b) when the screw is turning.

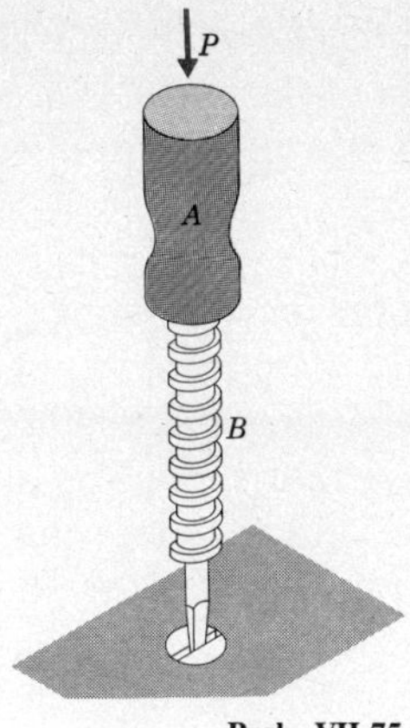

Prob. VII.75

VII.75 The automatic screwdriver shown operates by pushing the handle A downward with a force $\bar{P}$ to turn the threaded shaft B, which holds the bit. The square threads have a pitch of 90 mm, and the mean radius of the thread is 15 mm. The coefficients of static and kinetic friction for the threads are 0.15 and 0.12, respectively. If shaft B is single threaded, determine the force $\bar{P}$ required to tighten the wood screw to 2 N-m in one slow continuous motion from the position where the screw is first started in the hole.

VII.76 Solve Problem VII.75 if the shaft of the screwdriver is double threaded and all other properties are unchanged.

VII.77 The turnbuckle is used to keep the cable taut. Its eyebolts have square single threads whose mean radius and pitch are 0.25 in. and 0.10 in., respectively. The coefficient of static friction between the threaded sleeve and the eyebolts is 0.40. Determine the torque that must be applied to the sleeve to tighten the turnbuckle further if it has already been adjusted to produce a tensile force of 1000 lb in the cable. Eyebolt A has a right-hand thread and eyebolt B has a left-hand thread.

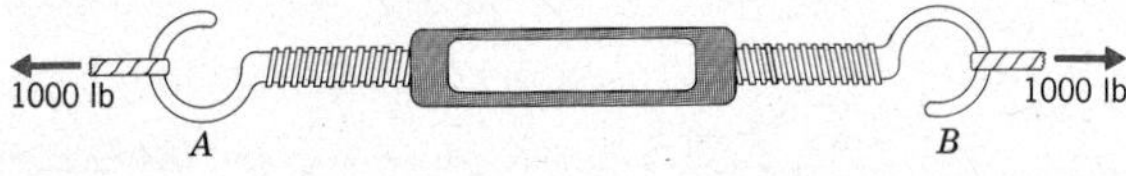

Prob. VII.77

VII.78 Solve Problem VII.77 if the eyebolts are double threaded and all other information is as given.

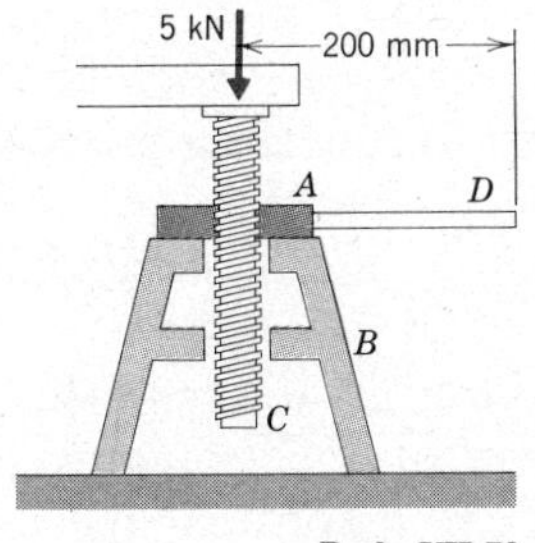

Prob. VII.79

VII.79 The jack shown in cross section consists of the threaded collar A, which bears on the frame B as it rides on screw C. The collar may be rotated by turning bar AD. The screw itself is prevented from rotating by the object being supported. The single square thread has inner and outer radii of 20 mm and 25 mm, and its pitch is 10 mm. The coefficient of friction for the screw threads is 0.40. The friction between collar A and the frame is negligible. Determine the smallest force applied perpendicular to the diagram at end D of bar AD that will raise the object carrying a 5-kN reaction.

VII.80 For the jack in Problem VII.79, determine the smallest force applied perpendicular to the diagram at end D of bar AD that will lower the object carrying a 5-kN reaction.

VII.81 Worm gear CD having a single square thread of 50 mm mean radius and 10 mm pitch meshes with gear A. Gear A is rigidly attached to

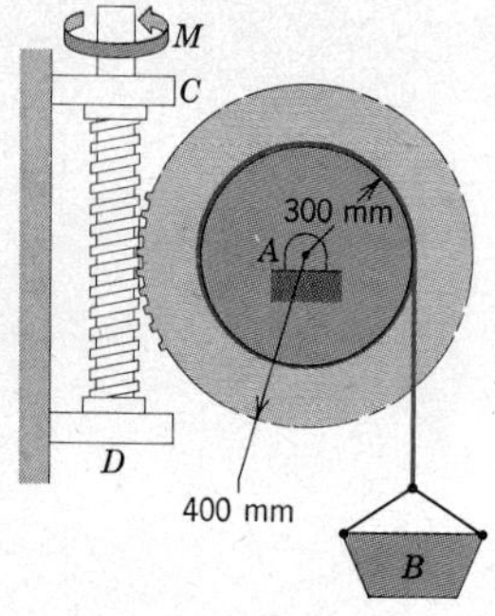

Prob. VII.81

the drum that holds the cable suspending the 100-kg bucket B. The coefficient of static friction between the threads of the worm gear and the teeth of gear A is 0.20. Determine the torque $\bar{M}$ that must be applied to the shaft CD to lift the bucket slowly.

VII.82 The scissors jack shown supports an automobile exerting the 800-lb load. The screw has a double square thread whose pitch is $\frac{1}{8}$ in. The mean radius of the thread is $\frac{3}{4}$ in. Knowing that the coefficient of static friction is 0.33, determine the couple $\bar{M}_{AB}$ that should be applied to the screw (a) to raise the automobile further, (b) to lower the automobile. The screw fits loosely in collar A. Friction between the shoulder in the screw and this collar is negligible.

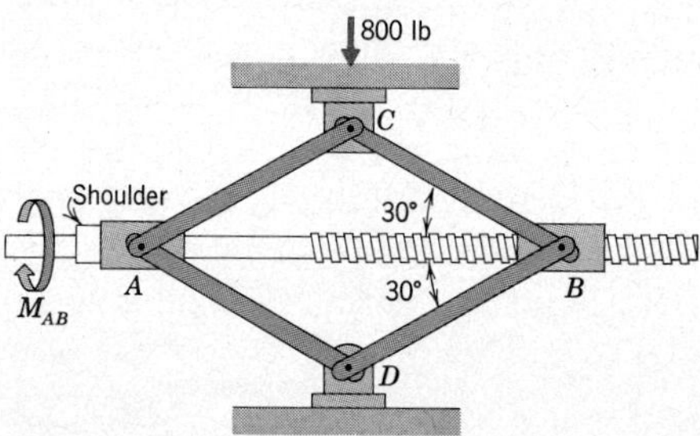

Prob. VII.82

3 Belt Friction

Cables that are guided over pulleys are effective devices for transferring forces between bodies. If it were not for friction, a cable would slide freely over a pulley and the tension in the cable on either side of the pulley would be identical. In some, but not all, cases this would be a desirable result. At times, friction may be required to provide a propelling force. For example, friction is needed for the functioning of the belt drive used in gasoline engines to drive electric generators and other auxiliary equipment.

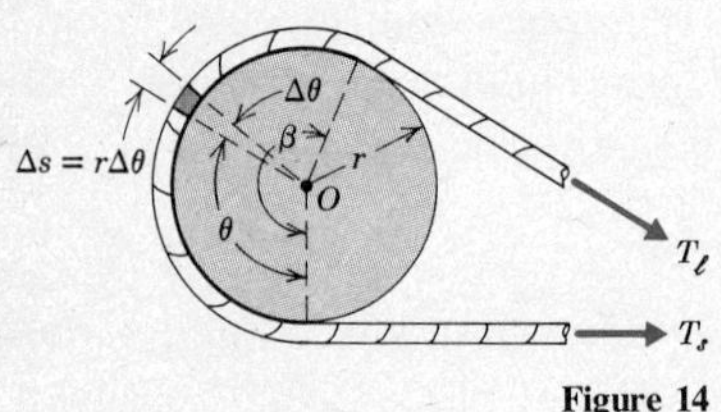

Figure 14

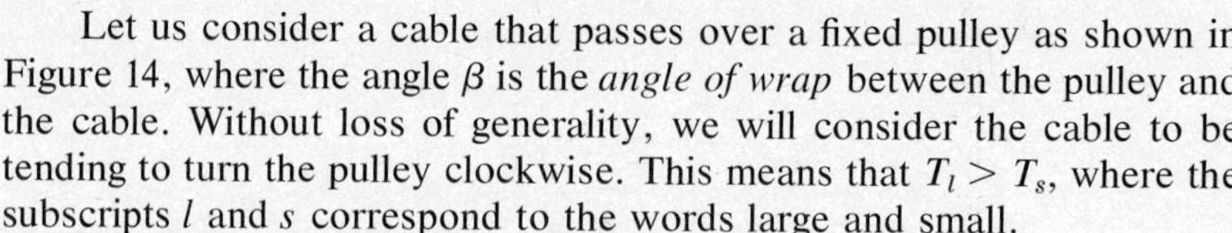

Let us consider a cable that passes over a fixed pulley as shown in Figure 14, where the angle β is the *angle of wrap* between the pulley and the cable. Without loss of generality, we will consider the cable to be tending to turn the pulley clockwise. This means that $T_l > T_s$, where the subscripts l and s correspond to the words large and small.

It then follows that the cable exerts a moment $\bar{M}_O$ about the center of the pulley whose magnitude is given by

$$M_O = (T_l - T_s)r \tag{18}$$

and whose sense is the same as that of the moment of $\bar{T}_l$ about center O.

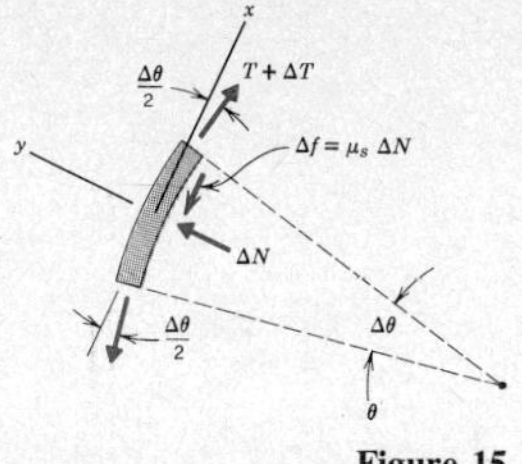

Figure 15

We wish to explore the relationship between T_l and T_s in the case where sliding between the cable and the pulley is impending. To do this, let us isolate a small element of cable Δs that is in contact with the pulley in the situation where the tendency of the cable is to rotate the pulley clockwise. Because of the contact between the cable and the pulley, a frictional force Δf and normal force ΔN are developed. The element Δs is tending to move tangentially to the pulley, in the positive x direction shown in Figure 15. Therefore, when sliding is impending, the friction force acting on the element is $\Delta f = \mu_s \Delta N$ in the negative x direction. To complete the free body diagram we have the tension force T at the position θ on one side of the element and the tension force $T + \Delta T$ at the position $\theta + \Delta\theta$ on the other side of the element. The x axis is chosen to be tangent to the pulley at the midpoint of the element, so it follows that the angle between either tensile force and the x axis is $\Delta\theta/2$.

The equations for force equilibrium of the element are then found to be

$$\Sigma F_x = -T\cos\frac{\Delta\theta}{2} + (T + \Delta T)\cos\frac{\Delta\theta}{2} - \mu_s \Delta N = 0$$

$$\Sigma F_y = -T\sin\frac{\Delta\theta}{2} - (T + \Delta T)\sin\frac{\Delta\theta}{2}\Delta N = 0$$

These reduce to

$$\Delta T\cos\frac{\Delta\theta}{2} = \mu_s \Delta N \tag{19a}$$

$$(2T + \Delta T)\sin\frac{\Delta\theta}{2} = \Delta N \tag{19b}$$

Let us now consider what happens when we isolate smaller and smaller elements, which means that the value of $\Delta\theta$ is reduced. In the limit of an infinitesimal element, $\Delta\theta$ approaches $d\theta$. The trigonometric terms in equations (19) then become

$$\cos\frac{d\theta}{2} = 1 \qquad \sin\frac{d\theta}{2} = \frac{d\theta}{2}$$

Therefore, retaining only first-order differentials, equations (19) reduce to

$$dT = \mu_s\, dN \quad \text{and} \quad T\, d\theta = dN \tag{20}$$

Our interest is with the tension force, not the normal force, so we eliminate dN and obtain

$$\frac{dT}{T} = \mu_s\, d\theta \tag{21}$$

Equation (21) may be integrated. Referring to Figure 14, the tension at $\theta = 0$ is T_s, while that at $\theta = \beta$ is T_l. Thus we have

$$\int_{T_s}^{T_l} \frac{dT}{T} = \int_0^{\beta} \mu_s \, d\theta$$

$$\ln T_l - \ln T_s \equiv \ln \frac{T_l}{T_s} = \mu_s \beta$$

A more useful form for this result is

$$\boxed{\frac{T_l}{T_s} = e^{\mu_s \beta}} \tag{22}$$

Let us review the meaning of each term in equation (22). The angle β, measured in radians, is the angle of wrap of the cable around the pulley. It is possible for β to be larger than 2π. For instance, if there are n full turns of the cable around the pulley, then $\beta = 2\pi n$. The symbols T_l and T_s represent the larger and the smaller of the tension forces on either side of the pulley; the tendency of the cable is to slide toward the side where the tension is T_l.

Note that equation (22) describes the situaton where a cable is about to slide over *any fixed cylindrical object,* such as a binding post. It may also be employed to describe a cable that is wrapped around a rotating cylindrical object when there is no slipping of the cable, provided that inertial effects are negligible. Further, equation (22) may be modified to describe the situation where the cable is actually sliding over the cylindrical object. In this case, the symbol μ_s should be replaced by the symbol μ_k, the coefficient of kinetic friction.

Finally, note that equation (22) was derived for the case of sliding or impending motion. It is not valid for the general, statically indeterminate case. A further restriction on equation (22) is that it is valid only if contact between the cable and the fixed object is along a cylindrical surface. It must be modified if we wish to analyze a V-belt and its pulley. This modification is the subject of Homework Problem VII.95.

EXAMPLE 7

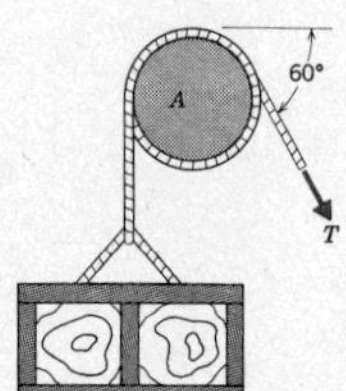

The tensile force $\bar{T}$ is applied to the free end of a cable that is wrapped around the horizontal binding post A for one complete turn and then attached to the 60-lb crate. The coefficient of static friction between the post and the cable is 0.30. Determine (a) the minimum tension T for which the crate will not descend, (b) the maximum tension T for which the crate will not rise.

Solution

The tensile force in the vertical portion of the cable is obviously the weight of the crate, 60 lb. When the crate is to be lowered, the cable is tending to move in the direction that the crate is pulling it, so we have $T_l =$ 60 lb. The value T we seek then corresponds to T_s when the cable is on the verge of slipping. Conversely, when the crate is to be raised, the cable is about to move in the direction in which its free end is being pulled. In that case T corresponds to T_l and $T_s = 60$ lb.

The angle of wrap is determined from the diagram and the given information. Adding the 150° of contact depicted in the diagram to the one additional turn specified in the problem, we have

$$\beta = 150°\left(\frac{\pi}{180°}\right) + 2\pi = 8.901 \text{ rad}$$

Equation (22) for part (a) then yields

$$\frac{60}{T} = e^{(0.30)(8.901)} = 14.444$$

$$T = 4.15 \text{ lb}$$

Similarly, for part (b) we have

$$\frac{T}{60} = e^{(0.30)(8.901)}$$

$$T = 866 \text{ lb}$$

We can see that it is easy for a person to lower the crate and quite difficult for a person to raise it. This is a consequence of the fact that the exponential function increases rapidly with an increase in its argument.

EXAMPLE 8

A motor applies a torque $M_A = 20$ N-m to the small pulley A. The coefficient of static friction between the flat belt and the pulley is 0.40. Knowing that the given torque is the maximum for which the belt will not slip over pulley A, determine (a) the tension in the belt on either side of pulley A, and (b) the torque M_B on pulley B for equilibrium. (c) Prove that the belt is not slipping over pulley B.

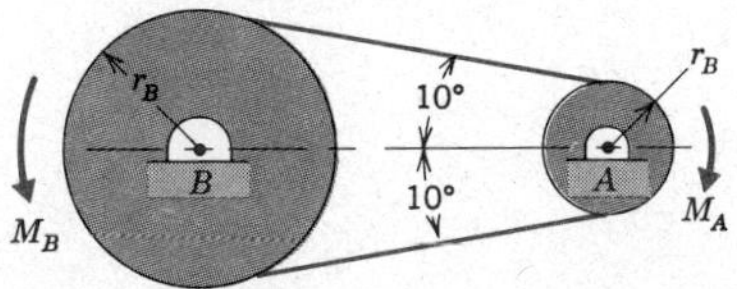

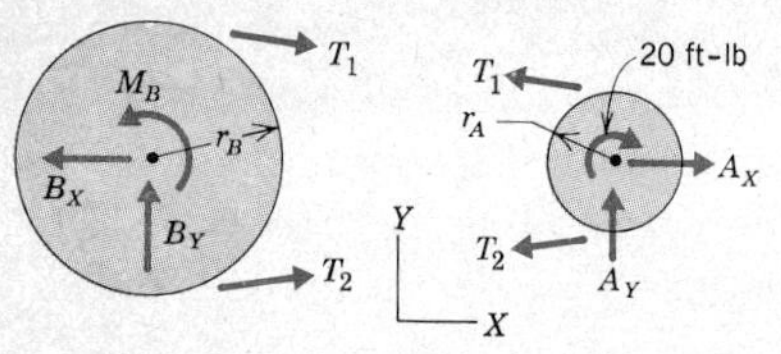

Solution

There are two rigid bodies, specifically, the pulleys, whose equilibrium must be assured. Therefore, let us cut the cable and draw the free body diagrams of each pulley. In doing this, we must not forget to include the forces exerted by the shaft of each pulley to withstand the pulling effect of the cables.

We are not interested in the reaction forces exerted by the shaft, so we write only the equations for moment equilibrium of each pulley about its center. It is given that $M_A = 20$ N-m in the sense shown, so we have

$$\text{pulley } A: \quad \Sigma M_{AZ} = -20 + (T_1 - T_2)r_A$$

$$\text{pulley } B: \quad \Sigma M_{BZ} = M_B + (T_2 - T_1)r_B$$

These equations allow us to solve part (b) immediately, for they show that

$$T_1 - T_2 = \frac{20}{r_A} = \frac{M_B}{r_B}$$

$$M_B = 20\frac{r_B}{r_A} \text{ N-m}$$

The ratio of the radii of the pulleys is the *mechanical advantage* of the system.

To solve part (a) we require another relationship for the belt tensions, specifically, equation (22). The information given in the problem states that slip is impending at pulley A when $M_A = 20$ N-m. Therefore, this is the pulley we shall describe. From the given diagram of the system, we determine that the belt is wrapped around pulley A by

$$\beta = 160°\left(\frac{\pi \text{ rad}}{180°}\right) = 2.793 \text{ rad}$$

Further, the equation for moment equilibrium of pulley A tells us that T_1 is larger than T_2, so T_1 corresponds to T_l. Then, for $\mu_s = 0.40$, equation (22) gives

$$\frac{T_1}{T_2} = e^{(0.40)(2.793)} \quad \text{so} \quad T_2 = 0.3272T_1$$

Substituting this expression into the moment equilibrium equation results in

$$T_1 - T_2 \equiv T_1 - 0.3272T_1 = \frac{20}{r_A}$$

$$T_1 = \frac{29.73}{r_A} \qquad T_2 = \frac{9.73}{r_A} \text{ N-m}$$

We now must show that the belt is not slipping over pulley B. To do this, we note that the angle of wrap for pulley B is larger than that for pulley A. Therefore, because the value of μ_s is the same for both pulleys, the ratio $T_1/T_2 \equiv T_l/T_s$ required for impending slip over pulley B is larger than that required for pulley A. It follows that the value of T_1/T_2 in the system, which is just sufficient to cause impending slip over pulley A, is insufficient to cause slip over pulley B.

HOMEWORK PROBLEMS

VII.83 A 160-lb person stands on a 20-lb scaffold attached to one end of a rope. The rope passes over a fixed cylinder and the free end of the cable is grasped by the person. The coefficient of static friction between the cable and the cylinder is 0.50. Determine the force the person must apply to (a) lower the scaffold, (b) raise the scaffold.

Prob. VII.83

VII.84 A ship is secured by wrapping a rope around the capstan. A dock worker can apply a 200-N force to counteract a 7-kN force by the ship. Determine the number of complete turns of the rope about the capstan required to keep the rope from slipping if the coefficient of static friction is 0.25.

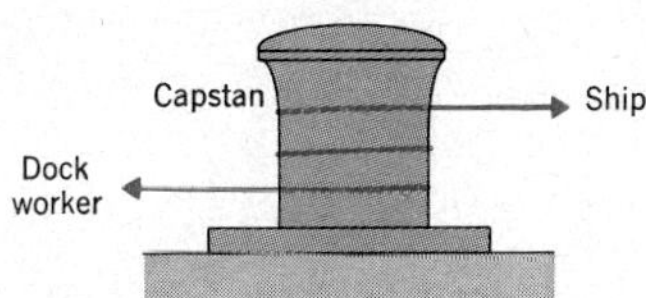

Probs. VII.84 and VII.85

VII.85 A dock worker wraps a rope around the capstan four times and then applies a tensile force of 200 N to one end. Knowing that the coefficient of static friction is 0.25, determine the range of values of the tensile force applied by the ship for which the cable will not slip over the capstan.

VII.86 A cable is wrapped around a fixed cylinder through an angle β. If the tension T on one end of the cable is constant, show that $T = (T_1 T_2)^{1/2}$, where T_1 and T_2 are the maximum and minimum values of the tension on the other end of the rope for which the cable will not slip.

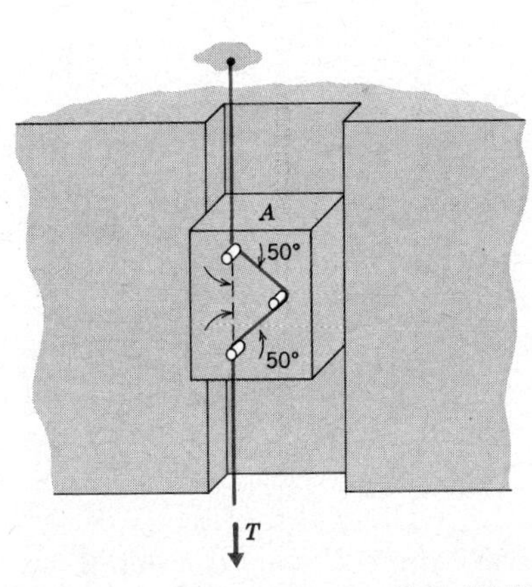

Prob. VII.87

VII.87 Three rough pegs protrude from the 50-kg block A, which slides in smooth vertical guides. Determine the tension T at the lower end of the cable for which block A is in static equilibrium in a condition of impending motion. The coefficient of static friction between the cable and the pegs is 0.60.

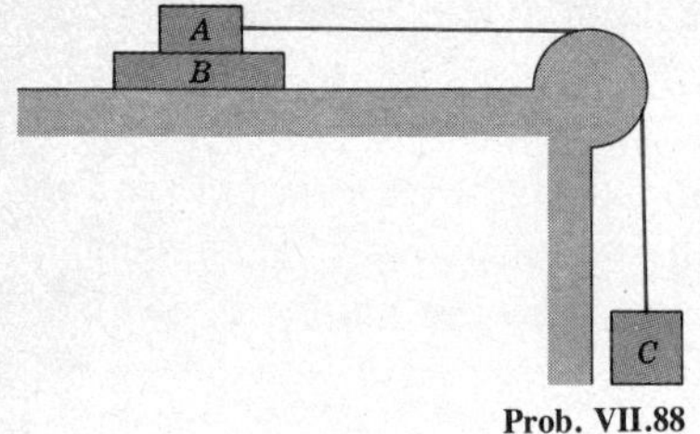

Prob. VII.88

VII.88 Block B weighs 10 lb and block C weighs 12 lb. The coefficients of static friction are 0.30 between the cable and the fixed circular guide, 0.50 between blocks A and B, and 0.40 between block B and the horizontal surface. Determine the minimum weight of block A for which the system is in static equilibrium.

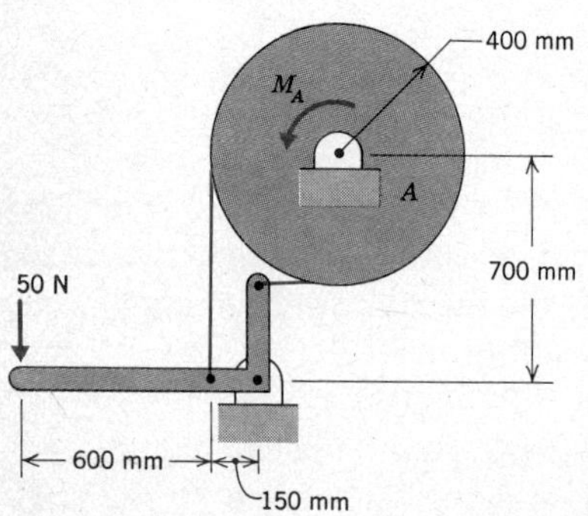

Prob. VII.89

VII.89 A 50-N force is applied to a band brake which restrains flywheel A against the counterclockwise couple $\bar{M}_A$. The coefficients of static and kinetic friction between the belt of the brake and the flywheel are 0.40 and 0.30, respectively. Determine the couple $\bar{M}_A$ required to (a) hold the flywheel at rest in a condition of impending slip, (b) rotate the flywheel in the direction of the couple at a constant rate of rotation.

VII.90 Solve Problem VII.89 for the case where the couple $\bar{M}_A$ is clockwise.

VII.91 and VII.92 Determine the maximum couple $\bar{M}_A$ that the motor may apply to pulley A without exceeding the maximum allowable belt tension of 2500 N. Also determine the corresponding couple $\bar{M}_B$ exerted on pulley B by its drive shaft. Where will slipping first occur? The coefficient of static friction between the belt and the pulleys is 0.35 and the pulleys are rotating at a constant rate.

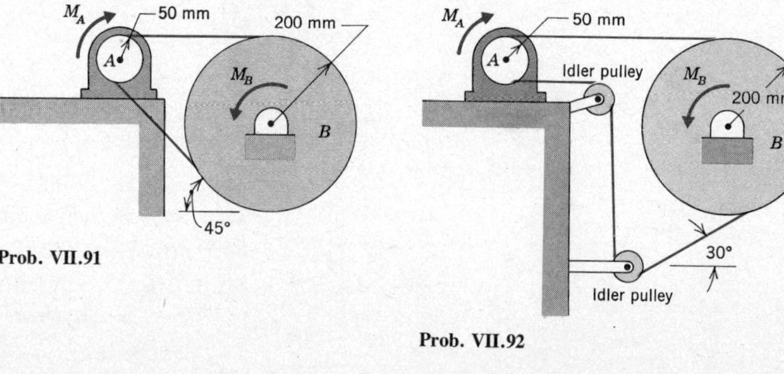

Prob. VII.91

Prob. VII.92

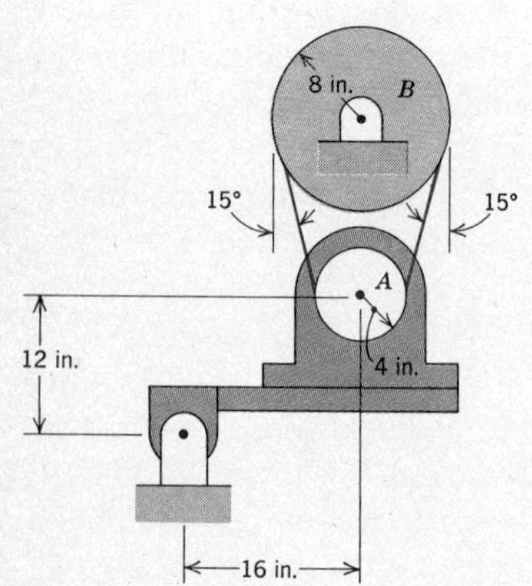

Prob. VII.93

VII.93 The belt drive shown is called a Rockwood mount. In it the 40-lb motor, whose center of mass coincides with its shaft, is mounted on a pivoted platform. Determine the maximum clockwise couple $\bar{M}_A$ the motor may apply to the drive pulley A, and also determine the corresponding couple $\bar{M}_B$ applied to pulley B by its shaft. The coefficient of friction between the belt and the pulley is 0.50. Neglect the mass of the platform.

VII.94 Solve Problem VII.93 in the case where the motor applies a counterclockwise couple to pulley A.

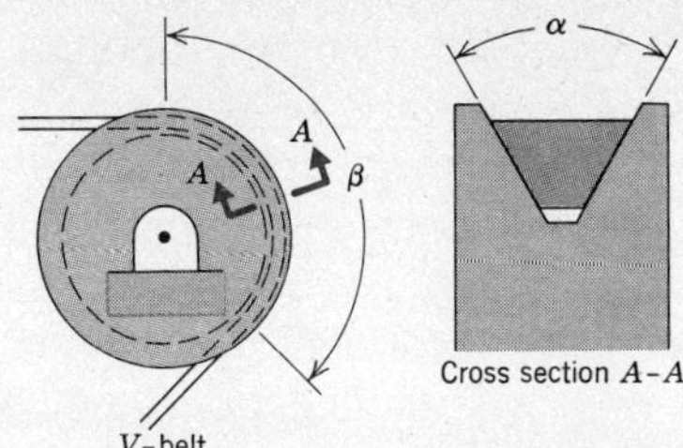

Prob. VII.95

VII.95 A V-belt and its pulley are in contact along the nonparallel sides of the belt, as shown in the cross sectional view A-A. Contact is not made along a cylindrical surface, so equation (22) is not valid. Isolate an element ds of the V-belt and follow similar steps to those employed in deriving equation (22) to show that, for impending slip,

$$\frac{T_l}{T_s} = e^{\mu_s \beta / \sin(\alpha/2)}$$

Hint: Determine the radial component of the resultant of the normal forces exerted by the pulley on the sides of an element ds of the V-belt.

4 Rolling Friction

Perhaps the simplest and most important of the basic machines is the wheel. However, it is not without reason that we have made consideration of the resistance to rolling of a wheel the last item in this module, for as you will see it is a very complicated phenomenon to analyze.

Consider a homogeneous disk of mass m that is rolling to the right along a flat horizontal surface at a constant speed v, being propelled by a horizontal force $\bar{P}$ applied to the center O. Assuming that the disk and the surface are rigid, the only reaction forces possible are friction and normal forces exerted by the surface on the disk at the point of contact C. These are shown in the free body diagram, Figure 16. The disk is rolling along at a constant speed, so we may apply the equations of static equilibrium. By summing moments about center O, we find that $\bar{f} = 0$, which means that $\bar{P}$ must also equal zero. This leaves us with a paradox, for its says that there can be no resistance, so that the wheel will roll at a constant velocity without being pushed. However, we know that there must be resistance, for all freely rolling bodies slow down.

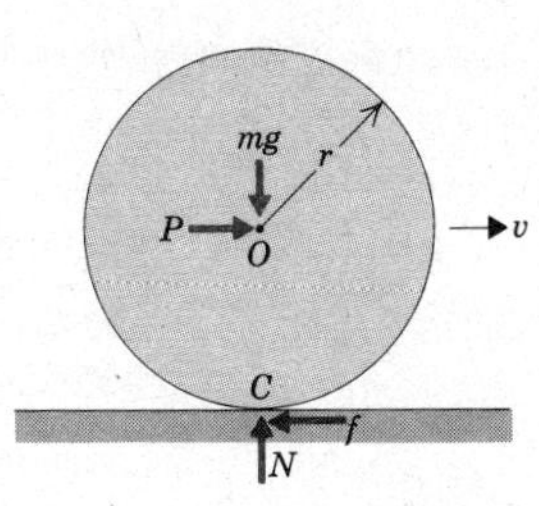

Figure 16

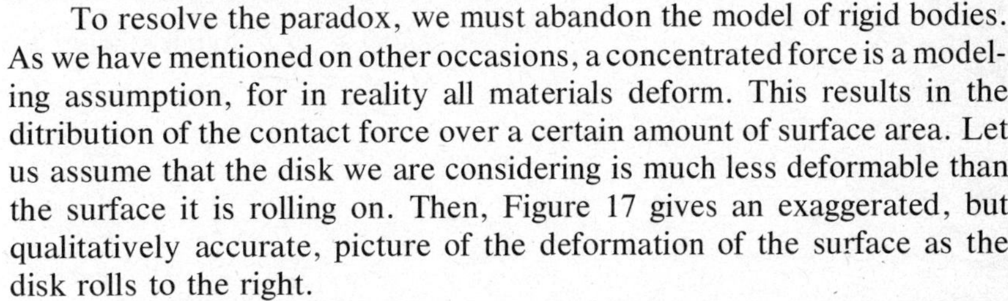

To resolve the paradox, we must abandon the model of rigid bodies. As we have mentioned on other occasions, a concentrated force is a modeling assumption, for in reality all materials deform. This results in the ditribution of the contact force over a certain amount of surface area. Let us assume that the disk we are considering is much less deformable than the surface it is rolling on. Then, Figure 17 gives an exaggerated, but qualitatively accurate, picture of the deformation of the surface as the disk rolls to the right.

The deformation pattern of the surface moves ahead with the disk, in a fashion similar to the bow wave preceding a boat in water. Thus, the phenomenon of rolling resistance bears little resemblance to the concept of dry friction described by Coulomb's laws, for which the model of rigid bodies are sufficient.

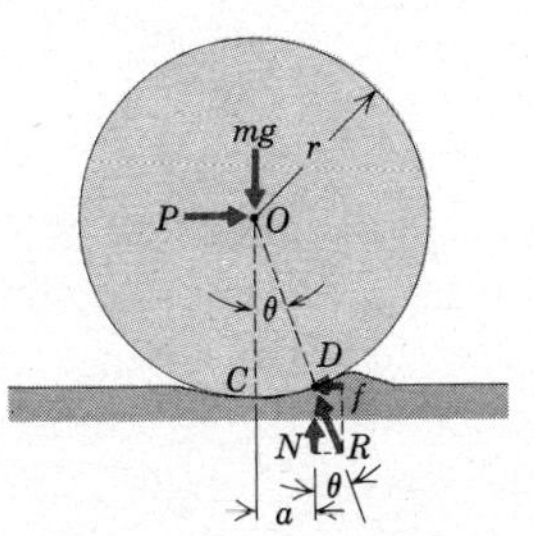

Figure 17

As shown in Figure 17, the creation of an area of contact between the disk and the surface means that the resultant contact force $\bar{R}$ exerted between these bodies need no longer be applied at the lowest point C. Instead, it may be applied at point D, which is an angle θ ahead of point C. Now, the equilibrium equation obtained by summing moments about the center of the disk requires that the line of action of the resultant force $\bar{R}$ be the line OD.

As shown in Figure 17, the horizontal component of $\bar{R}$ may be interpreted as the friction force $\bar{f}$, whereas the vertical component of $\bar{R}$ may be interpreted as the reaction force $\bar{N}$ normal to the undeformed surface of the floor. Thus, we have

$$
\begin{aligned}
f &= R \sin \theta \\
N &= R \cos \theta
\end{aligned} \tag{23}
$$

It is apparent from the figure that the horizontal force $\bar{P}$ required to move the wheel forward at a constant velocity is no longer zero. We have resolved the paradox caused by assuming that the floor is rigid.

We noted earlier that Figure 17 gives an exaggerated picture of the deformation pattern. In most instances, the area of contact, and therefore the angle θ, is very small. It proves more convenient in such cases to replace the angle θ by the horizontal distance a locating the point D where the resultant force acts. It can be seen from Figure 17 that

$$
\sin \theta = \frac{a}{r} \qquad \cos \theta = \left(1 - \frac{a^2}{r^2}\right)^{1/2}
$$

Substituting these expressions into equations (23) and eliminating the magnitude R of the resultant force yields

$$
f = \frac{\sin \theta}{\cos \theta} N = \frac{a/r}{(1 - a^2/r^2)^{1/2}} N
$$

Then, because a/r is much smaller than unity, this reduces to

$$
\boxed{f = \frac{a}{r} N} \tag{24}
$$

The distance a is called the *coefficient of rolling resistance,* and it has units of length. Early investigaors, such as Coulomb, regarded it as a parameter that is independent of the magnitude of the normal force $\bar{N}$, and also independent of the radius r. If this were true, a/r would be a factor of proportionality between the friction and normal forces, similar to the coefficients of dry friction. A further implication of equation (24) in that case is that the rolling resistance on a wheel is inversely proportional to its radius. Typical experimental values for the magnitude of the coefficient of rolling resistance a that have been reported are: mild steel on mild steel, $a = 0.18$ to 0.38 mm; hardened steel on hardened steel, $a = 0.005$ to 0.012 mm.

The foregoing analysis is contradicted by recent experimental and theoretical work. Although differing among each other, the investigators found that a is slightly dependent on the magnitude of $\bar{N}$ and strongly dependent on the radius r. In view of this disagreement, equation (24) should be regarded as useful for discussion and preliminary analysis purposes only. We will therefore not pursue the application of this formulation to solve problems.

MODULE VIII
VIRTUAL WORK AND ENERGY METHODS

The methods developed thus far for studying the static equilibrium of systems require that we isolate each of the rigid bodies forming the system. The equations for force and moment equilibrium then provide the basis for determining all unknown forces, provided that the system is statically determinate. In such analyses the interaction forces exerted between bodies in a system must be accounted for, even when knowledge of these forces is not of interest.

Unlike these methods, the *principle of virtual work* and its corollaries considers a system of rigid bodies in its entirety. This *systems viewpoint* frequently allows us to ascertain unknown forces without considering the internal interaction forces within the system. The principle of virtual work has many specialized applications, particularly in the field of structural mechanics.

We will see that the usefulness of the principle of virtual work decreases as the complexity of the geometry of the system increases, whereas in treatments employing the equilibrium equations the geometry of the system has little effect on the procedures followed. Thus, the principle of virtual work is merely a supplement to, but not a replacement for, the basic equations of static equilibrium.

The starting point for a treatment of this topic is the concept of the *work done* by a force. For simplicity, we shall limit ourselves to planar problems, although the resulting principles are equally applicable to three-dimensional problems.

A. WORK DONE BY A FORCE

When we think of doing work to perform a physical task, we intuitively think of the force required to move an object by a specific distance. Loosely speaking, that is the basis for our analyses.

Let us consider the work done by a force $\bar{F}$ applied to a point P when that point follows a fixed path S in space. A typical situation is shown in Figure 1.

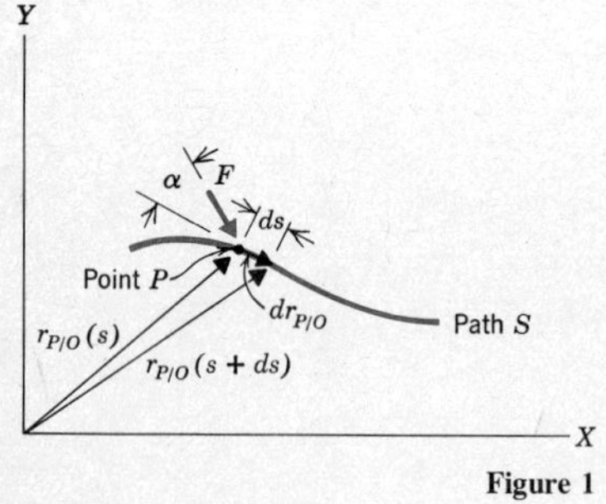

Figure 1

When point P moves an infinitesimal arc length ds, the position vector $\bar{r}_{P/O}$ of that point changes by a small amount $d\bar{r}_{P/O}$. Because the arc is infinitesimal, the magnitude of $d\bar{r}_{P/O}$ is ds and the direction of $d\bar{r}_{P/O}$ is tangent to the path, as illustrated in the figure.

The result of this differential displacement is that the force does a differential amount of work dU. Mathematically, this can be expressed as

$$dU \equiv \bar{F} \cdot d\bar{r}_{P/O} \equiv |\bar{F}|\, ds \cos \alpha \tag{1}$$

where, as shown in Figure 1, α is the angle between $\bar{F}$ and the local tangent.

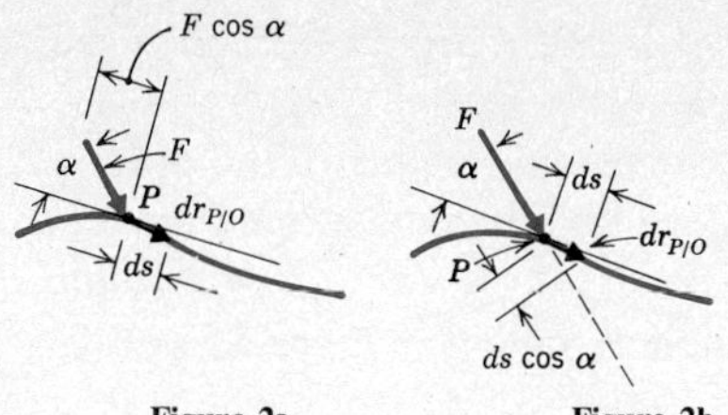

Figure 2a Figure 2b

Recalling the definition of the dot product of two vectors, the work dU given by equation (1) may be pictured in either of two ways. We can say that dU is the component of force tangent to the path S multiplied by the distance ds traveled along the path, as depicted in Figure 2a. Alternatively, we can say that dU is the component of the displacement vector $d\bar{r}_{P/O}$ parallel to the force $\bar{F}$ multiplied by the magnitude of this force, as shown in Figure 2b.

Some special features of the concept of work can be found by studying equation (1). The most obvious is that work is a scalar quantity having units of force times length. The value of the work dU may be positive or negative, depending on whether α is less than, or greater than 90°, respectively. The case where dU is negative means that the force opposes the movement, because in that case the component of force parallel to the path is opposite the direction of movement along the path. Also, note that when the force is perpendicular to the displacement, that is, when $\alpha =$ 90°, the work done by the force is zero. Finally, note that we can state that the work done is the force times the distance traveled only when $\alpha = 0$, that is, only when the force is parallel to the displacement.

So far, we have only considered the work done by a single force. Because work is a scalar quantity, the total work done by a system of forces acting on a body may be obtained by algebraically adding the work done by each force. Hence, letting dU_i denote the work done by the ith force in a set of N forces, we have

$$\boxed{dU = \sum_{i=1}^{N} dU_i} \tag{2}$$

Equation (2) expresses the means by which we may determine the infinitesimal work done by a system of forces. In the special case where two forces form a couple, it may be specialized even further. A couple $\bar{F}$ and $-\bar{F}$ acting on an arbitrary rigid body is depicted in Figure 3. The moment of this couple is $M = F\,l$ counterclockwise, or vectorially, $\bar{M} = F\,l\,\bar{K}$. We now wish to determine the work done by this couple.

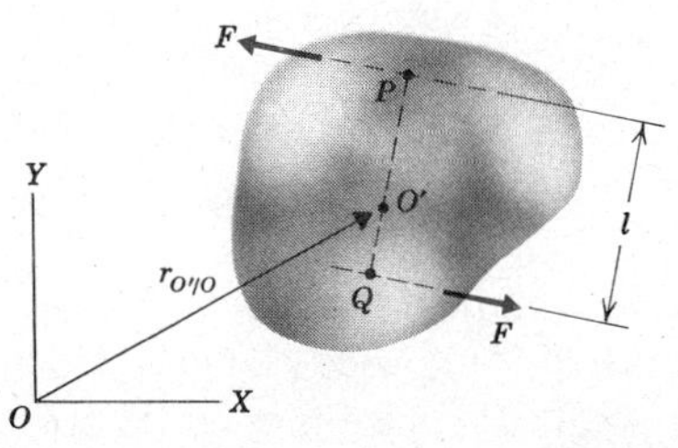

Figure 3

The most general infinitesimal change in the position of a rigid body can be shown to consist of a superposition of an infinitesimal displacement of an arbitrarily chosen point O', such as the point shown in Figure 3, and an infinitesimal rotation about point O'. (Proofs of this statement are contained in Modules IV and VI of the companion text on dynamics.) With this in mind, to simplify the development that follows we use the principle of transmissibility to transfer the forces $\bar{F}$ and $-\bar{F}$ to points P and Q on their respective lines of action, as illustrated in Figure 3. Note that line PQ intersects point O' and is perpendicular to the forces.

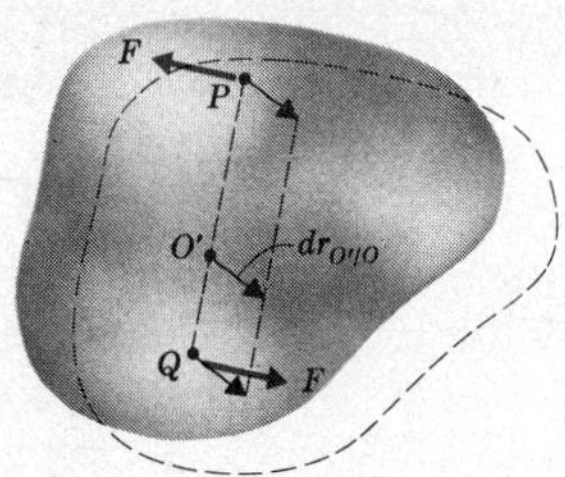

Figure 4a

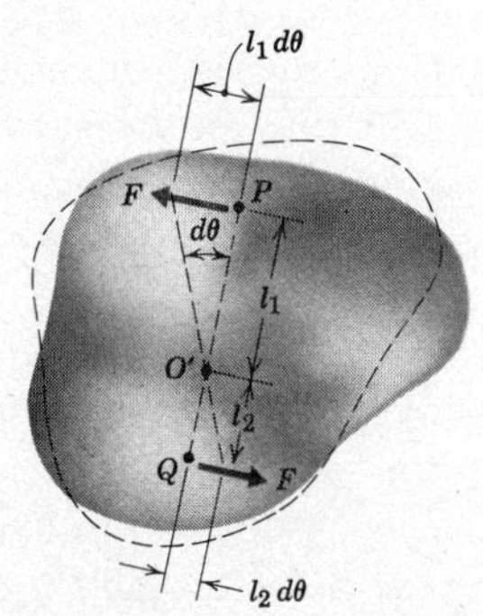

Figure 4b

Let us consider the infinitesimal displacement $d\bar{r}_{O'/O}$ of point O separately from the infinitesimal rotation $d\theta$ about point O'. The first type of movement is illustrated in Figure 4a and the second is shown in Figure 4b. The total work done by the couple will then be the sum of the work done in each type of movement.

For the situation in Figure 4a, which is called a translational displacement, points O', P, and Q all undergo the same displacement $d\bar{r}_{O'/O}$. Hence, the work done by the couple is

$$dU_{\text{trans}} = \bar{F} \cdot d\bar{r}_{O'/O} + (-\bar{F}) \cdot d\bar{r}_{O'/O} \equiv 0$$

thus demonstrating that a couple does no work when a body translates. This makes sense, because we have seen that a couple does not exert a pushing effect on a rigid body.

Considering the rotational movement in Figure 4b, we see that point P moves in the direction of $\bar{F}$ and point Q moves in the direction of $-\bar{F}$, so each force does positive work. Also, because the angle of rotation is infinitesimal, each displacement is parallel to the corresponding force, and its magnitude equals the arclength. Adding the work done by each force, we have

$$dU_{\text{rot}} = F(l_1\,d\theta) + F(l_2\,d\theta) = F(l_1 + l_2)\,d\theta$$

where l_1 and l_2 are the distances from point O' to points P and Q, respectively.

It can be seen from Figure 4b that the lever arm l for the couple is $l = l_1 + l_2$. It then follows that the total work done by the couple in the infinitesimal translation and rotation is

$$dU = dU_{\text{rot}} + dU_{\text{trans}} = Fl\,d\theta$$

$$\boxed{dU = |\bar{M}|\,d\theta} \tag{3}$$

Note that the rotation $d\theta$ was considered to be in the same sense as the moment of the couple, that is, counterclockwise, so the work done is positive. In situations where the rotation is in the opposite sense from the moment, the work done is negative, indicating that the couple is opposing the rotation. In general, the best way to avoid a sign error is to choose the signs of moments and rotations according to the right-hand rule with respect to the Z axis normal to the plane.

Equations (1) and (3) bring to mind the important matter of physical units. Dimensionally, equation (1) has units of force times length. This is consistent with equation (3), because the angle of rotation must be measured in radians. In order to avoid confusing moment and work, we generally describe a quantity of work by giving its length unit first and then its force unit, which is the opposite order from that used to describe a moment. Also, in the SI system work has its own derived unit, the joule (J), 1 J being defined as 1 m-N.

B. VIRTUAL MOVEMENT

Because static systems are at rest, and we now know that a force does no work unless its point of application moves, the question that arises is how the concept of work can be used to solve statics problems. To answer this question we resort to the concept of a *virtual movement*. The word *virtual* is a classic term; what is really meant is *fictitious* or *imaginary*. In a virtual movement we imagine that the system moves away from its static equilibrium position. In the virtual movement all forces acting on the system are considered to maintain their magnitude and direction. Also, the points at which the forces are applied to the bodies in the system are unchanged. For example, in the case of a fully constrained structure, in which a virtual movement can be obtained only by imagining that a physical support is removed, the force corresponding to this support is still considered to act on the system.

In order to indicate that we are not considering a real displacement of the system, we use the Greek symbol δ instead of d to denote a differential virtual quantity. Thus, the *virtual displacement* of a point P resulting from the virtual movement we give the system is denoted as $\delta\bar{r}_{P/O}$, and the *virtual work* δU of a force $\bar{F}$ applied at this point is

$$\boxed{\delta U = \bar{F} \cdot \delta\bar{r}_{P/O}} \tag{4}$$

A simple system that is useful for illustrating a virtual movement is the inclined bar shown in Figure 5, which is carrying a load $\bar{F}$ perpendicu-

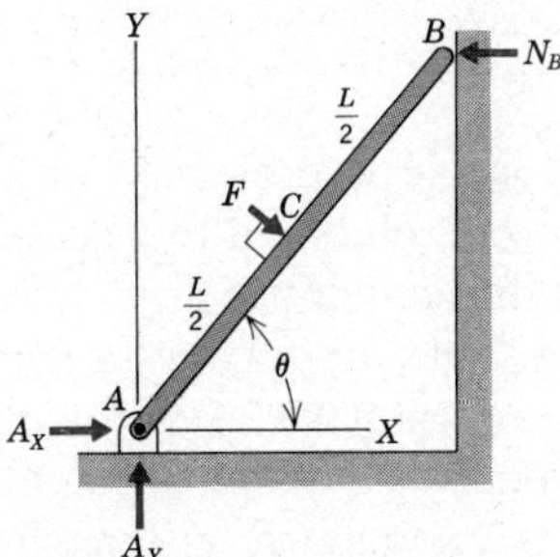

Figure 5

lar to the bar. A virtual movement of this bar may be produced by imagining what would happen if the wall at end B were removed, thus allowing the angle θ to change. Because the virtual displacement is entirely a figment of our imagination, we can consider the rotation of the bar in this virtual movement to be either clockwise or counterclockwise. In order to minimize the opportunity for making an error is sign, the virtual movement depicted in Figure 6 shows a counterclockwise rotation corresponding to an increase of the angle θ by $\delta\theta$.

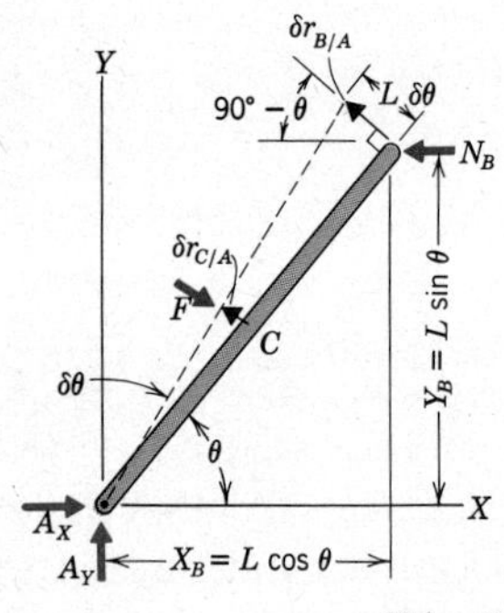

Figure 6

As a result of the virtual rotation, the forces acting on the bar do work. The computation of this virtual work requires that the virtual displacement of the point of application of each force be related to the value of $\delta\theta$. There are basically two ways of achieving this. One, called the geometric approach, involves applying the laws of geometry and trigonometry to the diagram of the virtual displacement, which is Figure 6 in this case.

The virtual movement depicted in Figure 6 is particularly suitable for the geometric approach. The movement depicted there is a pure planar rotation. That is, considering the points of the body lying in the plane, one point remains fixed as the other points follow circular arcs centered at the fixed point. We are dealing with an infinitesimal angle of rotation $\delta\theta$, so a point at a radial distance r from this fixed point will displace a distance $r\,\delta\theta$. Furthermore, because of the smallness of $\delta\theta$, the direction of the virtual displacement of a point will be perpendicular to the radial line from the fixed point, in the sense of the rotation.

These general observations enable us to construct directly the required expressions for the virtual displacements. For instance, point B is at a distance L from the fixed point A. Thus, the virtual displacement $\delta\bar{r}_{B/O}$ is $L\,\delta\theta$ perpendicular to the radial line AB. From the fact that the virtual rotation is counterclockwise, it follows that point B displaces upward and to the left. By referring to Figure 6 for the appropriate direction cosines, we find that

$$\begin{aligned}\delta\bar{r}_{B/O} &= L\,\delta\theta[-\cos(90° - \theta)\,\bar{I} + \sin(90° - \theta)\,\bar{J}] \\ &= L\,\delta\theta[-\sin\theta\,\bar{I} + \cos\theta\,\bar{J}] \qquad (5)\end{aligned}$$

Clearly, comparable expressons for the virtual displacements of other points may also be obtained by this method.

Although this geometric approach is particularly expedient for situations where a body undergoes a pure planar rotation, it has serious shortcomings. As we have seen, the most general movement that may be imparted to a body is a combination of a translation and a rotation. Most students find the geometric approach difficult and prone to error when treating such a movement.

Let us now develop an alternative method, called the analytical method, for expressing the virtual displacements. Returning to the system and virtual movement depicted in Figure 6, we need an XYZ coordinate system whose origin is located at a point that does not move during the virtual movement of the system. The coordinate system appearing in Figure 6 fits this requirement. The position vector of the point of application of each force with respect to the origin of the coordinate system must be expressed in terms of an *arbitrary* value of the geometric parameter that is being incremented in the virtual movement. From Figure 6 we find that, for an arbitrary value of θ,

$$\bar{r}_{B/A} = L \cos\theta\, \bar{I} + L \sin\theta\, \bar{J}$$

$$\bar{r}_{C/A} = \tfrac{1}{2}\bar{r}_{B/A} \qquad \bar{r}_{A/A} = \bar{0} \tag{6}$$

The virtual displacement of each point is then found by evaluating the differential change in each position vector resulting from the infinitesimal virtual change in the geometric parameter θ. The unit vectors $\bar{I}$ and $\bar{J}$ are constants. Therefore, the differential changes of the position vectors in equations (6) are

$$\delta\bar{r}_{B/A} = \frac{d\bar{r}_{B/A}}{d\theta}\,\delta\theta = \left[\frac{d}{d\theta}(L\cos\theta)\,\bar{I} + \frac{d}{d\theta}(L\sin\theta)\,\bar{J}\right]\delta\theta$$

$$= -L\,\delta\theta \sin\theta\, \bar{I} + L\,\delta\theta\cos\theta\,\bar{J}$$

$$\delta\bar{r}_{C/A} = \frac{1}{2}\,\delta\bar{r}_{B/A} = -\frac{L}{2}\,\delta\theta\sin\theta\,\bar{I} + \frac{L}{2}\,\delta\theta\cos\theta\,\bar{J}$$

$$\delta\bar{r}_{A/A} = 0 \tag{7}$$

Note that the expression for $\delta\bar{r}_{B/A}$ is the same result as that obtained in equation (5) by the geometric approach. Also, note that if θ had a specific numerical value, this value could be substituted into equations (7) *after* the derivatives have been evaluated.

Once the virtual displacements are expressed, the virtual work of the forces is readily obtained from equation (4) by expressing each force in terms of its components. Using Figure 6 to determine the force components, the virtual work resulting from the virtual rotation $\delta\theta$ is

$$\delta U = (-N_B\bar{I}) \cdot \delta\bar{r}_{B/A} + (F\sin\theta\,\bar{I} - F\cos\theta\,\bar{J}) \cdot \delta\bar{r}_{C/A}$$

$$+ (A_X\bar{I} + A_Y\bar{J}) \cdot \delta\bar{r}_{A/A}$$

$$= (-N_B)(-L\,\delta\theta\sin\theta) + (F\sin\theta)\left(-\frac{L}{2}\,\delta\theta\sin\theta\right)$$

$$+ (-F\cos\theta)\left(\frac{L}{2}\,\delta\theta\cos\theta\right) + 0$$

$$= [N_B L\sin\theta - \frac{FL}{2}(\sin^2\theta + \cos^2\theta)]\,\delta\theta$$

$$= \left[N_B L \sin\theta - \frac{FL}{2}\right]\delta\theta \tag{8}$$

It could have been foreseen by inspection that the reactions $\bar{A}_X$ and $\bar{A}_Y$ do no work, because point A is fixed during the virtual movement.

In the foregoing we discussed one type of virtual movement of the bar in Figure 5, specifically, the virtual rotation produced when we imagine that the support at end B is removed. Another type of virtual movement occurs if we pretend that the pin support at point A is changed to a roller, thus allowing for a virtual displacement of point A horizontally. As shown in Figure 7, this results in a vertical virtual displacement of end B.

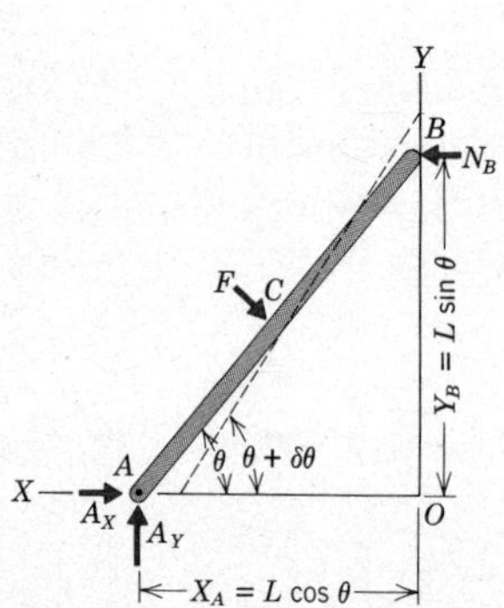

Figure 7

The coordinate system in Figure 7 is different from that used in Figure 6, because we need a set of axes having an origin that is fixed during the virtual movement. However, the steps in computing the virtual displacement analytically follow the prescribed order. First we express the position vectors of the points where the forces are applied. Using θ as the independent geometric parameter, this gives

$$\begin{aligned}\bar{r}_{B/O} &= Y_B\bar{J} = L\sin\theta\,\bar{J}\\ \bar{r}_{C/O} &= \tfrac{1}{2}X_B\bar{I} + \tfrac{1}{2}Y_B\bar{J} = \tfrac{1}{2}L\cos\theta\, I + \tfrac{1}{2}L\sin\theta\,\bar{J}\\ \bar{r}_{A/O} &= X_A\bar{I} = L\cos\theta\,\bar{I}\end{aligned} \tag{9}$$

The virtual displacements are then found by computing the differentials. Thus,

$$\begin{aligned}\delta\bar{r}_{B/O} &= \frac{d\bar{r}_{B/O}}{d\theta}\,\delta\theta = L\,\delta\theta\cos\theta\,\bar{J}\\ \delta\bar{r}_{C/O} &= \frac{d\bar{r}_{C/O}}{d\theta}\,\delta\theta = \tfrac{1}{2}L(-\sin\theta\,\bar{I} + \cos\theta\,\bar{J})\,\delta\theta\\ \delta\bar{r}_{A/O} &= \frac{d\bar{r}_{A/O}}{d\theta}\,\delta\theta = -L\,\delta\theta\sin\theta\,\bar{I}\end{aligned} \tag{10}$$

The corresponding virtual work is then found from equation (4) to be

$$\begin{aligned}\delta U &= (N_B\bar{I})\cdot\delta\bar{r}_{B/O} + (-F\sin\theta\,\bar{I} - F\cos\theta\,\bar{J})\cdot\delta\bar{r}_{C/O}\\ &\quad + (-A_X\bar{I} + A_Y\bar{J})\cdot\delta\bar{r}_{A/O}\\ &= 0 + \tfrac{1}{2}FL\,\delta\theta(\sin^2\theta - \cos^2\theta) + A_XL\,\delta\theta\sin\theta\\ &= (-\tfrac{1}{2}FL\cos 2\theta + A_XL\sin\theta)\,\delta\theta\end{aligned} \tag{11}$$

In this case, the fact that $\bar{N}_B$ does no work could have been anticipated without calculation, because $\bar{N}_B$ is perpendicular to the virtual displacement given to point B.

A feature of the concept of virtual movement that sometimes proves troublesome for students is that the choice of the geometric parameter to represent the movement is arbitrary. For instance, for the virtual movement of Figure 7, we could regard X_A as the geometric parameter describing the virtual movement because X_A is known if θ is known; specifically, $X_A = L\cos\theta$. The Pythagorean theorem then gives $Y_B = (L^2 - X_A{}^2)^{1/2}$, so the position vectors are

$$\begin{aligned} \bar{r}_{B/O} &= Y_B\bar{J} = (L^2 - X_A{}^2)^{1/2}\bar{J} \\ \bar{r}_{C/O} &= \tfrac{1}{2}X_A\bar{I} + \tfrac{1}{2}Y_B\bar{J} = \tfrac{1}{2}X_A\bar{I} + \tfrac{1}{2}(L^2 - X_A{}^2)^{1/2}\bar{J} \\ \bar{r}_{A/O} &= X_A\bar{I} \end{aligned} \tag{12}$$

The virtual displacements are then

$$\begin{aligned} \delta\bar{r}_{B/O} &= \frac{d\bar{r}_{B/O}}{dX_A}\,\delta X_A = -\frac{X_A\,\delta X_A}{(L^2 - X_A{}^2)^{1/2}}\bar{J} \\ \delta\bar{r}_{C/O} &= \frac{d\bar{r}_{C/O}}{dX_A}\delta X_A = \tfrac{1}{2}\,\delta X_A\,\bar{I} - \tfrac{1}{2}\frac{X_A\,\delta X_A}{(L^2 - X_A{}^2)^{1/2}}\bar{J} \\ \delta\bar{r}_{A/O} &= \delta X_A\,\bar{I} \end{aligned} \tag{13}$$

The corresponding virtual work may then be found from dot products, as before. This demonstrates that any convenient parameter may be chosen to describe the virtual movement.

Perhaps the most essential observation regarding the virtual work in the two types of virtual movements we considered is that when we imagine that a support is removed, the reaction force associated with this support does virtual work. This result occurs because the geometric constraint condition imposed by the support is violated in the virtual movement. In contrast, the reaction forces corresponding to supports that are unmodified during a virtual movement do no work.

For the two types of virtual movement discussed above, in each case one support was removed, so the virtual displacement could be described in terms of the infinitesimal change of one geometrical parameter. This parameter is called a *generalized coordinate*. After removal of the support, the system is said to have one *degree of freedom;* the number of degrees of freedom is defined as the number of generalized coordinates the system has. It follows that if we consider removing more than one support at a time, more than one generalized coordinate would be required. This added complication is unnecessary in statics, so it will not be pursued further.

The bar in Figure 5 is a fully constrained system, because its supports fully prevent movement. A primary goal of a statics analysis of such systems is the determination of the reaction forces. There are systems that are only partially constrained, that is, that move unless some unknown force (or forces) has the necessary value for static equilibrium. For our purposes, such systems may be analyzed from the same viewpoint as that used for fully constrained systems. The number of generalized coordinates is then the number of geometrical quantities required to describe the virtual movement of the system consistent with the existing supports. The unknown external forces required to prevent the system from moving, in effect, are then additional constraint forces.

We have seen how the concept of a virtual movement leads to the virtual work of the forces acting on the system. In the following sections we shall develop the principles that enable us to use this concept to solve statics problems. However, because the determination of virtual displacements is the key to evaluating the virtual work, the following examples and homework problems are provided to gain practice in this technique.

EXAMPLE 1

Bar AB rests on the 300-mm radius semicylinder and its lower end A rests on the floor, as shown. Determine the virtual work done by the force $\bar{F}$ at end B in a virtual movement in which the bar remains tangent to the semicylinder as end A moves horizontally.

Solution

The starting point of the solution is a sketch depicting the virtual movement.

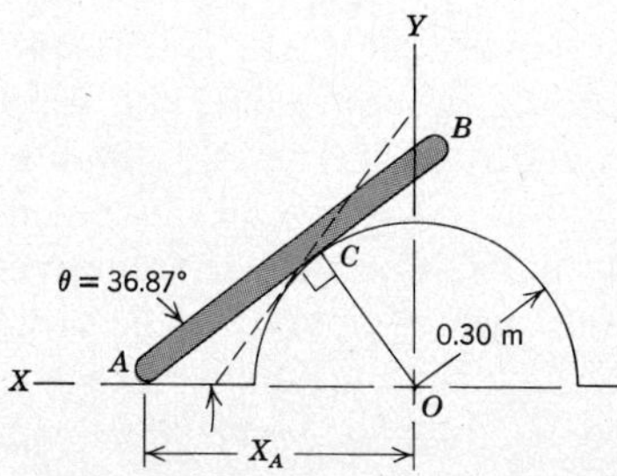

In general, we require a coordinate system whose origin is fixed in the virtual movement. The XYZ coordinate system shown in the sketch fits this requirement.

Next, we select the generalized coordinate for the virtual movement. One possible choice is the angle θ between the bar and the horizontal, which is shown in the sketch to change from its initial value of 36.87° as a result of the virtual movement. Another possible choice is the distance X_A. A way in which to choose between alternative choices for the generalized coordinate is to consider the construction of the position vectors for the points of application of the forces. For the problem at hand, the vector $\bar{r}_{B/O}$ may be obtained by replacing the 0.70-m length of bar AB

by its horizontal and vertical projections. This gives

$$\bar{r}_{B/O} = (X_A - 0.70 \cos \theta)\, \bar{I} + 0.70 \sin \theta\, \bar{J}$$

If a parameter is to serve as a generalized coordinate, it must be possible to express all other geometrical parameters that change in the virtual movement in terms of the chosen parameter. Thus, regardless of whether we choose X_A or θ, the complete description of $\bar{r}_{B/O}$ requires that we determine the relationship between these parameters. This relationship is readily determined, once it is noted that in the sketch of the system X_A forms the hypotenuse of the right traingle ACO. Thus

$$X_A = \frac{0.30}{\sin \theta} \text{ m}$$

It now appears that choosing the angle θ for the generalized coordinate gives the simpler expression for $\bar{r}_{B/O}$. Hence, for an arbitrary value of θ, we have

$$\bar{r}_{B/O} = \left(\frac{0.30}{\sin \theta} - 0.70 \cos \theta\right) \bar{I} + 0.70 \sin \theta\, \bar{J}$$

The virtual displacement of point B is obtained as a differential. This gives

$$\delta \bar{r}_{B/O} = \frac{d\bar{r}_{B/O}}{d\theta}\, \delta\theta = \left[\left(-\frac{0.30}{\sin^2 \theta} \cos \theta + 0.70 \sin \theta\right) \bar{I} + 0.70 \cos \theta\, \bar{J}\right] \delta\theta$$

In the position of interest, $\theta = 36.87°$. Substituting this value into the foregoing expression for $\delta\bar{r}_{B/O}$ yields

$$\delta\bar{r}_{B/O} = \left\{\left[-\frac{0.30}{(0.60)^2}(0.80) + 0.70(0.60)\right] \bar{I} + 0.70(0.80)\bar{J}\right\} \delta\theta$$

$$= (-0.2467\bar{I} + 0.56\bar{J})\, \delta\theta \text{ m}$$

In the position where $\theta = 36.87°$, the force $\bar{F}$, which is perpendicular to bar AB, is given by

$$\bar{F} = F(-\sin 36.87°\, \bar{I} - \cos 36.87°\, \bar{J})$$

$$= -F(0.6\bar{I} + 0.8\bar{J}) \text{ m}$$

Hence, the virtual work done by this force is

$$\delta U = \bar{F} \cdot \delta\bar{r}_{B/O} = -F(0.6\bar{I} + 0.8\bar{J}) \cdot (-0.2467\bar{I} + 0.56\bar{J})\, \delta\theta$$

$$= -0.300\, F\delta\theta \text{ m-N} \equiv -0.300\, F\delta\theta \text{ J}$$

A particular fact to bear in mind is that one choice for a generalized coordinate is better than another only in the sense that the resulting expressions for the position vectors of the points of application of the forces are simpler. However, regardless of what your choice for the generalized coordinate is, it must be used *consistently* throughout the solution.

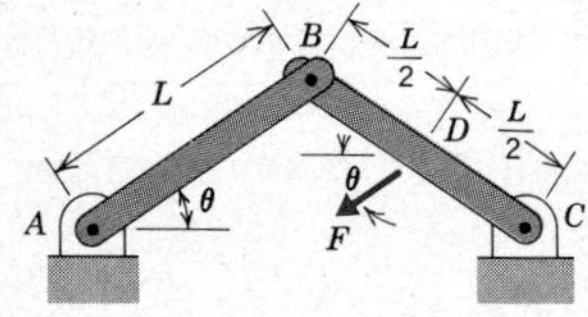

EXAMPLE 2

A force $\bar{F}$ is applied to the two-bar frame, as shown. Determine the virtual work done by this force in a virtual movement in which pin C is given a horizontal displacement, while the connections at pins A and B are maintained.

Solution

The following sketch illustrates the prescribed virtual movement.

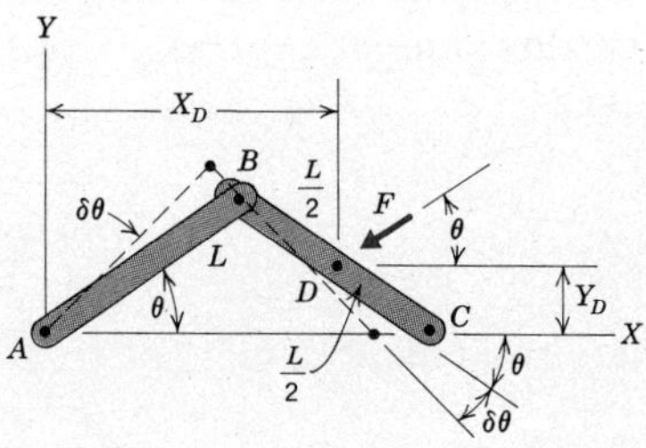

Because pin A remains fixed in the virtual movement, we select this point as the origin for the coordinate system.

As shown in the sketch, the position of the system can be established from the fact that triangle ABC is isosceles with side AC horizontal, provided that the angle θ is known. This suggests the use of θ as the generalized coordinate. Then, considering the aforementioned isosceles triangle, we see that the angle between bar BC and the X axis is also θ. Therefore, the general position vector for point D is

$$\bar{r}_{D/A} = X_D\,\bar{I} + Y_D\,\bar{J} = \frac{3L}{2}\cos\theta\,\bar{I} + \frac{L}{2}\sin\theta\,\bar{J}$$

The virtual displacement of point D is then found as a differential. This yields

$$\delta\bar{r}_{D/A} = \frac{d\bar{r}_{D/A}}{d\theta}\,\delta\theta = \left(-\frac{3L}{2}\sin\theta\,\bar{I} + \frac{L}{2}\cos\theta\,\bar{J}\right)\delta\theta$$

The angle between the force $\bar{F}$ and the horizontal is also θ, so the virtual work done by this force is

$$\delta U_F = \bar{F} \cdot \delta \bar{r}_{D/A} = F(-\cos\theta\,\bar{I} - \sin\theta\,\bar{J}) \cdot \left(-\frac{3L}{2}\sin\theta\,\bar{I} + \frac{L}{2}\cos\theta\,\bar{J}\right)\delta\theta$$

$$= FL(\sin\theta\cos\theta)\,\delta\theta = \tfrac{1}{2}FL\sin 2\theta\,\delta\theta$$

HOMEWORK PROBLEMS

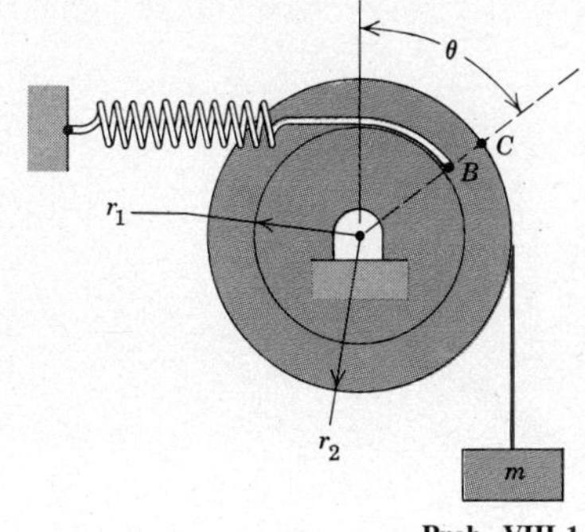

Prob. VIII.1

VIII.1 The mass m is supended by a cable that is attached at point C to the outer drum of the stepped pulley. A spring of stiffness k is attached at point B to the inner drum. There is no force in the spring when $\theta = 0$. Using the geometrical method, compute the virtual work done by the weight of the mass and by the spring force in a virtual rotation of the stepped pulley for an arbitrary value of θ.

VIII.2 The balance scale holds the small masses m_1 and m_2 in the position shown. Determine the work done by the weight forces acting on these masses in a virtual rotation about the pivot O. Compare this result to the moment of the weight forces about pivot O.

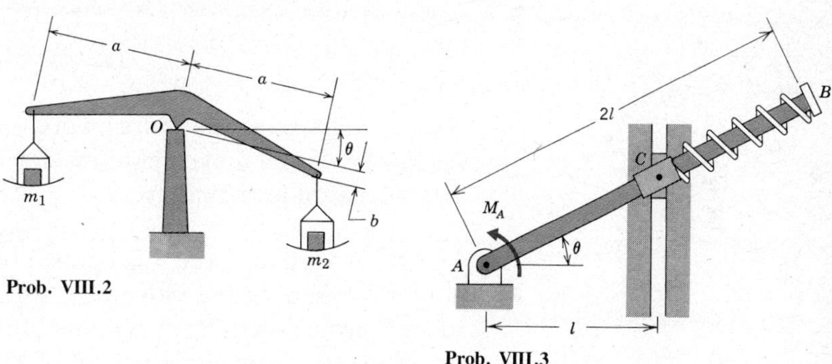

Prob. VIII.2

Prob. VIII.3

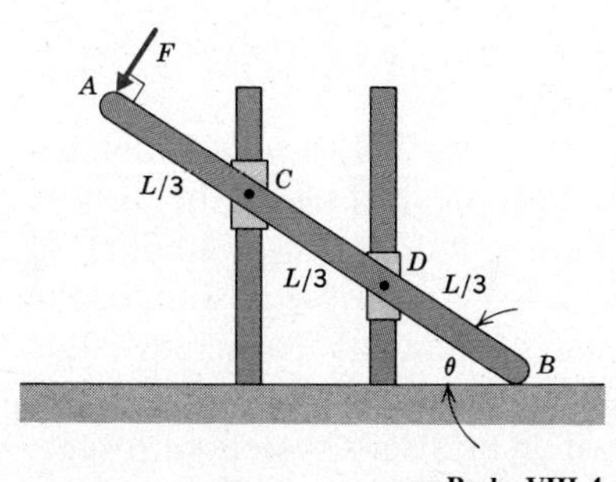

Prob. VIII.4

VIII.3 Collar C, which rides on bar AB, is pinned to a block that is guided by the straight groove. The spring, whose stiffness is k, is unstressed when $\theta = 0$. Determine the virtual work done by the spring in a virtual rotation of bar AB about pin A.

VIII.4 Bar AB is pinned to collars C and D, which ride on vertical rods. The lower end B is supported by the smooth ground. Determine the work done by the force $\bar{F}$ in a virtual movement that violates (a) only the constraint of the ground against the vertical movement of end B, (b) only the constraint of the left bar against horizontal movement of collar C.

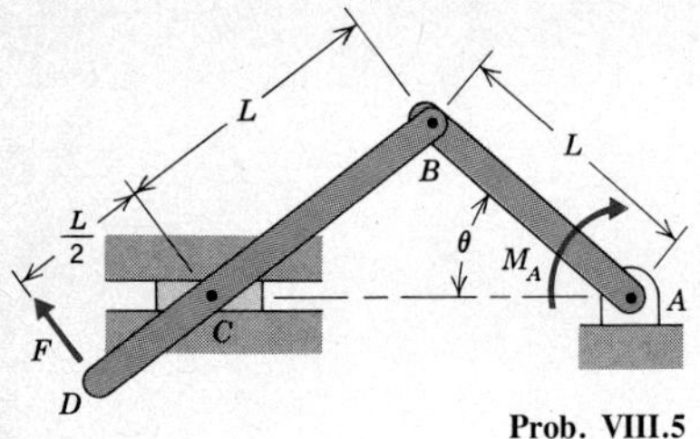

Prob. VIII.5

VIII.5 Block C, which rides in the groove, is pinned to bar BD. At an arbitrary angle θ the system is given a virtual movement in which all pin connections remain intact as block C moves parallel to the groove. Determine the virtual work done by the force $\bar{F}$ and by the couple $\bar{M}_A$.

VIII.6 A spring-cushioned platform supports a package of mass m. The spring, whose stiffness is k, is unstressed when $h = 800$ mm. At an arbitrary value of h the system is given a virtual movement in which all pin connections remain intact as roller A displaces horizontally. Determine the virtual work done by the spring force.

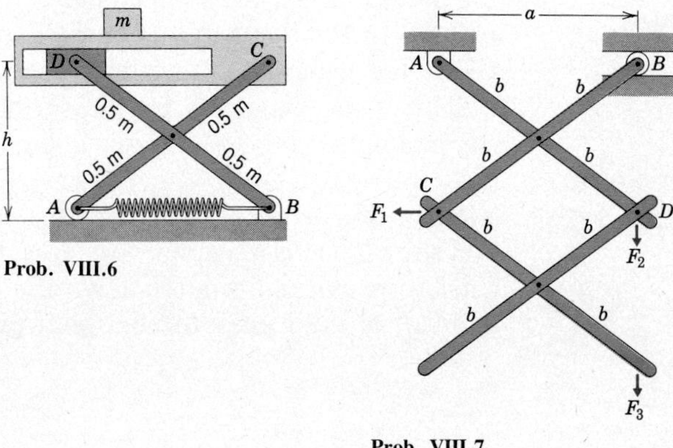

Prob. VIII.6

Prob. VIII.7

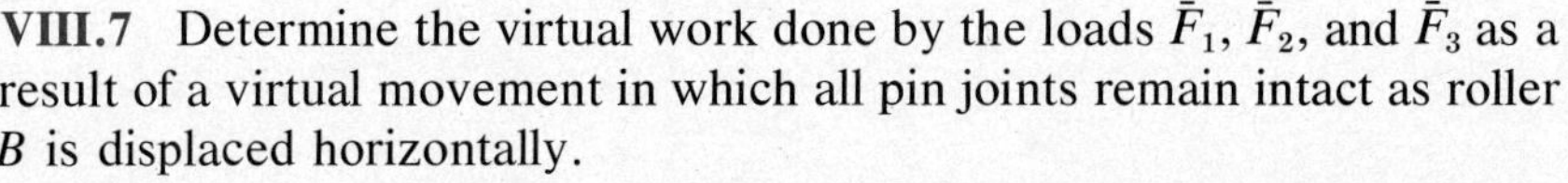

VIII.7 Determine the virtual work done by the loads $\bar{F}_1$, $\bar{F}_2$, and $\bar{F}_3$ as a result of a virtual movement in which all pin joints remain intact as roller B is displaced horizontally.

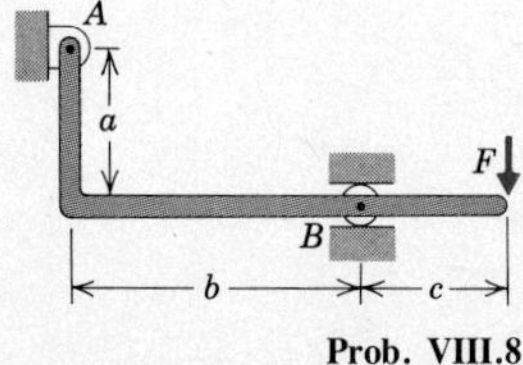

Prob. VIII.8

VIII.8 A vertical force $\bar{F}$ is applied to the bent bar. Determine the work done by $\bar{F}$ in a virtual movement that violates (a) only the constraint of roller B against vertical movement of that point, (b) only the constraint of pin A against vertical movement of that point.

C. THE PRINCIPLE OF VIRTUAL WORK

From the preceding section we know that it is possible to select a virtual movement such that only selected unknown forces acting on the system do work. Therefore, if we could determine what the total amount of virtual work should equal, we would be able to form a relationship for the unknown forces doing work in the virtual movement. To achieve this goal, consider the body shown in Figure 8a on the following page, which is in equilibrium under the loading of an arbitrary system of forces.

We learned in Module III that the system of forces in Figure 8a is

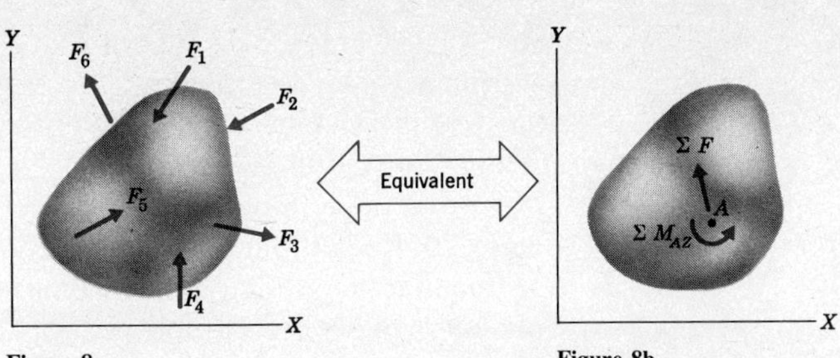

Figure 8a

Figure 8b

equivalent to the force-couple system at an arbitrary point A shown in Figure 8b. In this figure $\Sigma\, \bar{F}$ denotes the sum of all forces acting on the body, and $\Sigma\, M_{AZ}$ denotes the sum of the moments of all forces about point A.

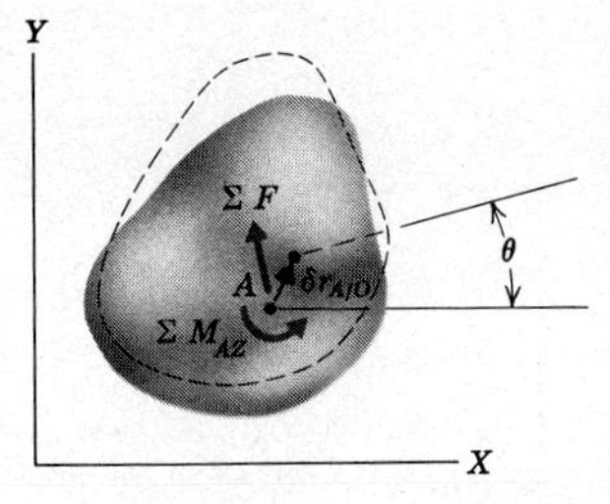

Figure 9

Recall that the most general infinitesimal movement of a body is the superposition of a translation corresponding to the infinitesimal displacement of a point plus an infinitesimal rotation about that point. Choosing point A for this point, a general virtual movement is as shown in Figure 9. In the translation the couple does no work, and in the rotation the force does no work. Hence, the total virtual work done by the forces acting on the body is

$$\delta U = \Sigma\, \bar{F} \cdot \delta\bar{r}_{A/O} + \Sigma\, M_{AZ}\, \delta\theta \tag{14}$$

This is a general expression for the virtual work done by an arbitrary set of forces. However, we have specified that the body is in equilibrium. The equilibrium conditions require that $\Sigma\, \bar{F} = 0$ and $\Sigma\, M_{AZ} = 0$. It then follows from equation (14) that

$$\boxed{\delta U = 0} \tag{15}$$

This simple equation is the *principle of virtual work*. As derived, it applies only to an isolated rigid body. However, the principle is also valid for systems of interconnected rigid bodies, because the total virtual work done by the forces acting on the system is the sum of the virtual work done in moving each body in the system, which is zero. Hence, the principle of virtual work may be stated as

> The virtual work δU done by all forces acting on any system in static equilibrium is zero.

The key feature of this principle is related to the fact that the type of virtual movement given to the system is at our discretion. A movement that violates only one constraint condition results in the appearance of only the associated constraint force (out of the full set of unknown reactions), in the virtual work δU. Setting $\delta U = 0$ then yields a single scalar equation for the unknown force.

For example, consider the bar of Figure 5, which is shown again in Figure 10. Recall that $\bar{F}$ is a given loading. When we wish to find $\bar{N}_B$ we imagine that the vertical wall is removed, leaving the pin at point A intact. This leads to a virtual rotation about point A, in which the reactions $\bar{A}_X$ and $\bar{A}_Y$ do no work. The resulting virtual work was found in equation (8). Setting that expression for $\delta U = 0$ yields

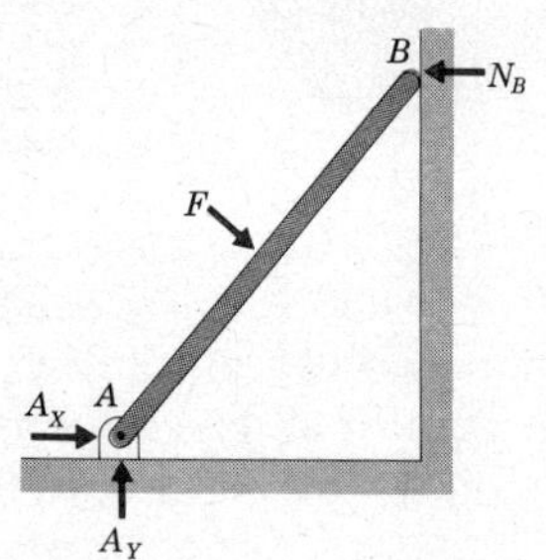

Figure 10

$$\delta U = \left(N_B L \sin\theta - \frac{FL}{2}\right)\delta\theta = 0$$

Because $\delta\theta$ is an arbitrary nonzero quantity, it must be true that

$$N_B = \frac{F}{2\sin\theta}$$

Should we wish to determine the reaction $\bar{A}_X$, we would imagine that the constraint of pin A against horizontal movement is removed. We would then maintain the constraint of pin A against vertical movement, as well as the constraint against horizontal motion of end B imposed by the wall. In the resulting virtual movement forces $\bar{A}_Y$ and $\bar{N}_B$ do no work. The virtual work resulting from this type of virtual movement was found in equation (11). Setting that expression for $\delta U = 0$ yields

$$\delta U = (-\tfrac{1}{2}FL\cos 2\theta + A_X L\sin\theta)\,\delta\theta = 0$$

$$A_X = \frac{F\cos 2\theta}{2\sin\theta}$$

Clearly, the preceding requires a large effort to solve some simple problems. The values of N_B and A_X for the bar can be determined from the equations of static equilibrium. In fact, summing moments about pin A would yield the value of N_B in a single equation, whereas a moment sum about the point of intersection of the horizontal line through end B and the vertical line through pin A would yield the value of A_X in a single equation. Nevertheless, even for this simple system, the principle of virtual work gives us a new viewpoint of the influence of the forces acting on the system.

The significant area of application of the principle of virtual work is for systems consisting of interconnected rigid bodies. To demonstrate this, consider the situation depicted on the next page in Figure 11a, wherein two arbitrary bodies are connected by a frictionless pin.

According to Newton's third law, the interaction forces exerted be-

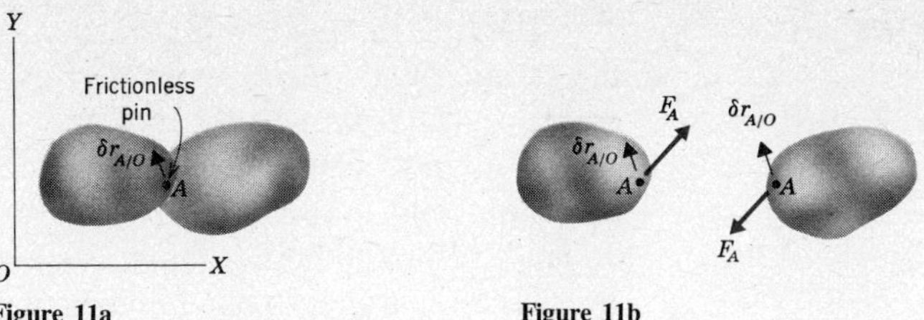

Figure 11a **Figure 11b**

tween the two bodies are $\bar{F}_A$ and $-\bar{F}_A$, as shown in Figure 11b. Thus, in a virtual movement of the two bodies in which the bodies remain connected, so that the displacement of their common point of connection is the vector $\delta\bar{r}_{A/O}$, the virtual work of $\bar{F}_A$ and $-\bar{F}_A$ is

$$\delta U_{\text{pin}} = \bar{F}_A \cdot \delta\bar{r}_{A/O} + (-\bar{F}_A) \cdot \delta\bar{r}_{A/O} \equiv 0$$

Thus, frictionless pins do no work in a virtual movement in which the connected bodies remain joined.

In general, when we consider friction to have a negligible effect in the connections between rigid bodies, we are forming an *ideal* model of the system. In effect, the internal forces in ideal systems are constraint forces. For example, the pin forces $\bar{F}_A$ and $-\bar{F}_A$ in Figure 11 are the constraint forces required to ensure that the points on each body that are joined by pin A have the same virtual displacement $\delta\bar{r}_{A/O}$. As we have seen, if a virtual movement is consistent with a constraint, then the corresponding constraint force does no work.

To gain further insight into this result, and also to learn how the presence of friction affects the problem, consider the two bars in Figure 12a which are connected by a collar that may slide on bar 1. As

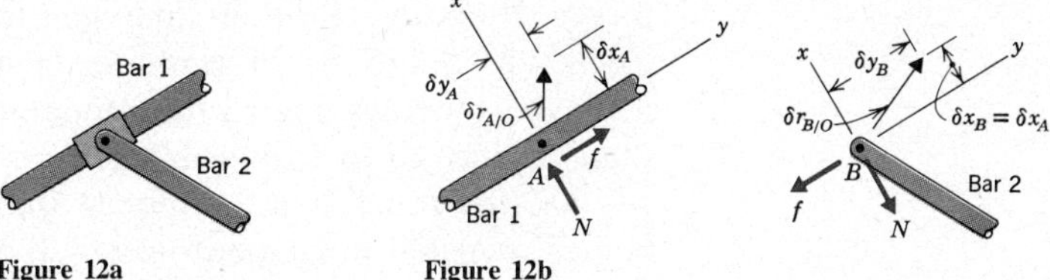

Figure 12a **Figure 12b**

shown in Figure 12b, the interaction forces exerted between the two bars consist of a pair of forces $\bar{N}$ and $-\bar{N}$ normal to bar 1, and a pair of frictional forces $\bar{f}$ and $-\bar{f}$ tangent to bar 1. Also, as illustrated, there are two points we must consider, because the virtual displacement $\delta\bar{r}_{A/O}$ of point A on bar 1 will generally not be the same as $\delta\bar{r}_{B/O}$ of point B on bar 2. Note that for convenience in discussing the various vectors, an xyz coordinate system is defined, for which x is normal and y is tangent to bar 1.

The first thing we will consider about this connection is the nature of the geometrical constraint it imposes. If we were to position ourselves on bar 1, a virtual movement that is consistent with this type of connection would seem to cause the collar to displace tangentially, that is, in the y direction. This is the only possible movement of the collar relative to bar 1. Thus, although points A and B may displace by entirely different values δy_A and δy_B parallel to bar 1 in a virtual movement, the collar has the effect of making them displace by the same amount δx_A perpendicular to bar 1. This is the constraint condition. The reactions $\bar{N}$ and $-\bar{N}$ are the corresponding constraint forces, because they restrict the motion of the bars in the x direction.

Let us now compute the virtual work. Writing the virtual displacements of points A and B in terms of their x and y components, we have

$$\begin{aligned} \delta U_{\text{collar}} &= (N\bar{\imath} + f\bar{\jmath}) \cdot \delta\bar{r}_{A/O} + (-N\bar{\imath} - f\bar{\jmath}) \cdot \delta\bar{r}_{B/O} \\ &= (N\bar{\imath} + f\bar{\jmath}) \cdot (\delta x_A\,\bar{\imath} + \delta y_A\,\bar{\jmath}) \\ &\quad + (-N\bar{\imath} - f\bar{\jmath}) \cdot (\delta x_A\,\bar{\imath} + \delta y_B\,\bar{\jmath}) \\ &= f(\delta y_A - \delta y_B) \end{aligned} \tag{16}$$

In this result the quantity $(\delta y_A - \delta y_B)$ represents the displacement of the collar relative to bar 1. We now observe that if bar 1 is smooth (the ideal model), then f is zero and the internal forces do no work. On the other hand, in a model that includes friction, the internal friction forces do work because they have the effect of resisting the movement of the collar relative to bar 1.

In general, if we apply the principle of virtual work to systems of rigid bodies where there is friction in the connections, these unknown frictional forces do work in addition to the work done by the unknown external forces. Equation (15) then no longer leads to a single equation for a single unknown. Solutions of problems pertaining to such systems are best achieved by returning to the fundamental methods of static equilibrium that we derived in earlier modules.

On the other hand, suppose that we give an ideal system of rigid bodies a virtual movement in which all of the bodies remain connected, thus satisfying the constraint conditions associated with the connections. The internal forces do not hinder such a movement, and therefore do no virtual work. They may therefore be ignored. This is one of the beneficial features of the method of virtual work in comparison to consideration of the equilibrium equations.

One qualifying statement applies to the preceding result. Specifically, in situations where a spring connects two bodies, so that the pair of forces it applies to the bodies are internal to the system, the spring forces will generally do work in a virtual movement. (The reason for this will be discussed in the section on potential energy.) Nevertheless, this does not represent an obstacle to the application of the principle of virtual work, for the force of a spring is known whenever its elongation is known.

D. PROBLEM SOLVING

Before we proceed to utilize the principle of virtual work, it is best that we synthesize the foregoing development into a series of steps for solving problems.

1 Draw a free body diagram of the entire system in the position of interest as an aid in accounting for all forces. General reaction forces, such as those exerted by pin supports, may be broken up into any convenient set of components, for instance horizontal and vertical.

2 Draw another sketch showing the position of the system before any virtual movement, and also showing the supports of the system.

3 Determine the type of virtual movement desired. For fully constrained systems, this is done by violating only the constraint corresponding to the reaction of interest. For partially constrained systems the virtual movement should be the type the system would undergo if the external loads were not those required for static equilibrium. Show the position of the system after the virtual movement in the sketch of step 2.

4 Choose an XYZ coordinate system whose origin is located at a point that does not move during the virtual displacement. As always, the coordinate axes should have orientations that suit the geometry of the system. Show this coordinate system both in the free body diagram of step 1 and the sketch of step 2.

5 Choose any convenient geometrical parameter whose value changes during the virtual movement as the generalized coordinate.

6 Evaluate the virtual displacement of the points of application of the forces. In general, follow the analytical approach for this determination by first expressing the position vectors of the points in terms of an arbitrary value of the generalized coordinate. The required virtual displacements are then the differential changes in the position vectors resulting from an infinitesimal increase of the generalized coordinate. If the generalized coordinate has a specific value in the equilibrium position, substitute that value *after* evaluating the derivatives.

Optionally, in situations where a body is given a pure planar rotation in a virtual movement, the geometrical method may prove easy to apply. The virtual displacement of a point is then the product of the angle of the virtual rotation and the radial distance from the point to the fixed point, perpendicular to the radial line in the sense of the rotation.

7 Express the forces doing work in terms of their components, using the free body diagram as an aid. Compute the total virtual work δU as the sum

of the dot product of each force and the virtual displacement of its point of application. Equate δU to zero.

8 Solve the equation $\delta U = 0$ for the unknown force.

Clearly, the key step among those outlined above is step 6. In the illustrative problems that follow we will focus on the analytical viewpoint for constructing virtual displacements because it is useful for all types of virtual movement. Equally important, because this method is more mathematical, it requires less insight into the nature of the virtual movement of each body and therefore reduces the chance of making sign errors.

ILLUSTRATIVE PROBLEM 1

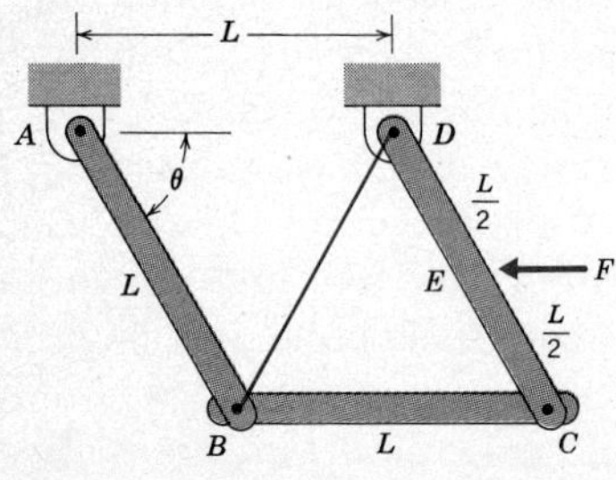

The frame shown supports the applied load $\bar{F}$. Determine the tension in cable BD using the principle of virtual work.

Solution

Step 1 Letting T denote the desired tension, the free body diagram of the system of rigid bars is as shown.

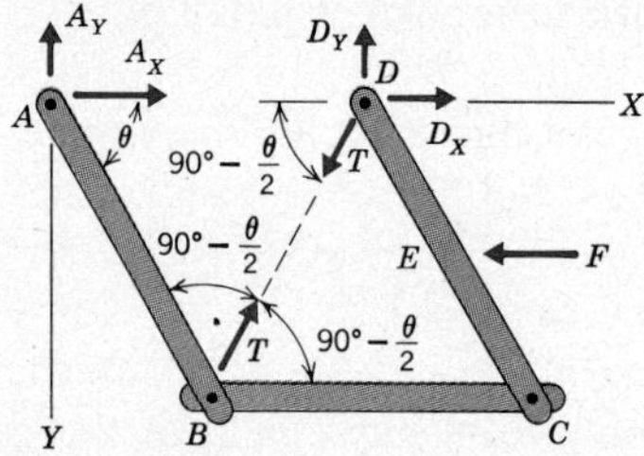

Step 2 The physical picture of the system before any virtual displacement is shown by the heavy solid lines.

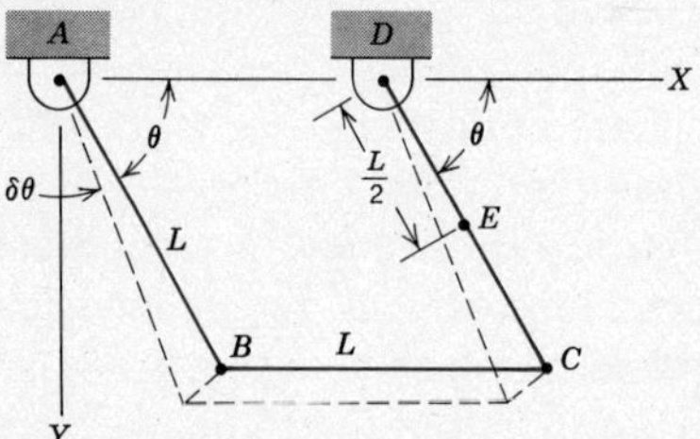

Step 3 Without the string for support, the system is only partially constrained, because the angle θ would then be a variable parameter. Thus, the virtual movement we give the system is one in which the angle θ is increased by $\delta\theta$ with all pins remaining intact, as shown by the dashed line in the sketch of step 2.

Step 4 Points A and D are both fixed during the virtual movement, so either may be chosen to be the origin. We will use the former. For convenience in representing position vectors, we choose the Y axis downward, as shown in the sketches of steps 1 and 2.

Step 5 We will use the angle θ as the generalized coordinate because we have already seen that the position of the system subsequent to the virtual movement can be described in terms of the increase in this angle.

Step 6 The only forces doing work are the tension force $\bar{T}$ at pin B and the horizontal force $\bar{F}$ at point E; the other forces appearing in the free body diagram are applied at fixed points.

We first write the position vectors for points B and E by referring to the sketch in step 2 for the dimensions. This gives

$$\bar{r}_{B/A} = L(\cos\theta\,\bar{I} + \sin\theta\,\bar{J}) \quad \bar{r}_{E/A} = \left(L + \frac{L}{2}\cos\theta\right)\bar{I} + \frac{L}{2}\sin\theta\,\bar{J}$$

The virtual displacements are then

$$\delta\bar{r}_{B/A} = \frac{d\bar{r}_{B/A}}{d\theta}\,\delta\theta = L(-\sin\theta\,\bar{I} + \cos\theta\,\bar{J})\,\delta\theta$$

$$\delta\bar{r}_{E/A} = \frac{d\bar{r}_{E/A}}{d\theta}\,\delta\theta = \frac{L}{2}(-\sin\theta\,\bar{I} + \cos\theta\,\bar{J})\,\delta\theta$$

Step 7 The force components may be obtained from the free body diagram of step 1, so the virtual work is

$$\begin{aligned}\delta U &= \bar{T}\cdot\delta\bar{r}_{B/A} + \bar{F}\cdot\delta\bar{r}_{C/A} \\ &= T\left[\cos\left(90^\circ - \frac{\theta}{2}\right)\bar{I} - \sin\left(90^\circ - \frac{\theta}{2}\right)\bar{J}\right]\cdot(-L\sin\theta\,\bar{I} + L\cos\theta\,\bar{J})\,\delta\theta \\ &\quad + (-F\bar{I})\cdot\left(-\frac{L}{2}\sin\theta\,\bar{I} + \frac{L}{2}\cos\theta\,\bar{J}\right)\delta\theta \\ &= \left[-TL\left(\sin\frac{\theta}{2}\sin\theta + \cos\frac{\theta}{2}\cos\theta\right) + \frac{FL}{2}\sin\theta\right]\delta\theta = 0\end{aligned}$$

Step 8 Cancelling all common factors in the equation for δU, we find that

$$TL\left(\sin\frac{\theta}{2}\sin\theta + \cos\frac{\theta}{2}\cos\theta\right) = \frac{FL}{2}\sin\theta$$

Using standard trigonometric identities this becomes

$$TL\cos\frac{\theta}{2} = \frac{FL}{2}\left(2\sin\frac{\theta}{2}\cos\frac{\theta}{2}\right)$$

Hence

$$T = F\sin\frac{\theta}{2}$$ ◺

We note in closing that step 6 could have been formulated by the geometric method. You should try it, and verify that the results are identical to those obtained above.

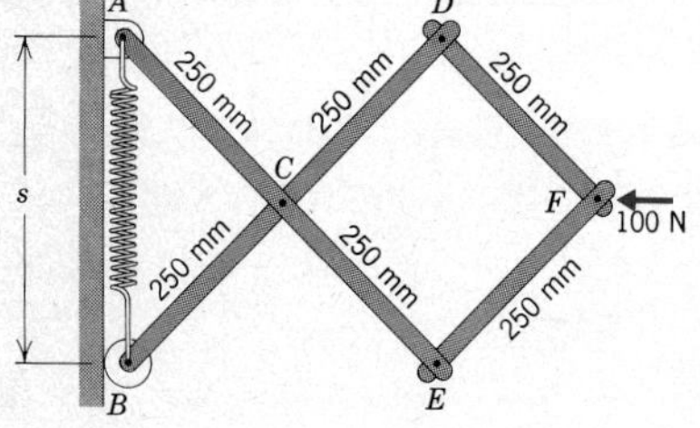

ILLUSTRATIVE PROBLEM 2

The parallelogram frame is loaded by a horizontal 100-N force. The unstretched length of the spring is 350 mm. Determine the required stiffness k of the spring if $s = 400$ mm in the static equilibrium position.

Solution

Step 1 The spring is extended by $s - 0.35$ m, so the tension in the spring is $k(s - 0.35)$. (For the sake of increased generality, we will substitute the given value of s later.) Thus, the free body diagram of the entire frame is as shown.

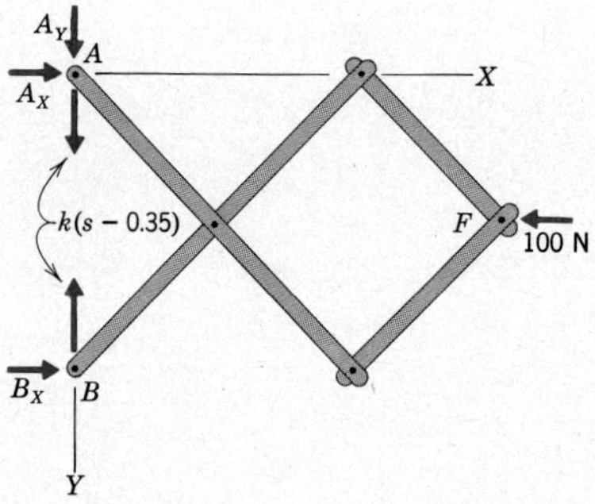

Step 2 The heavy solid lines in the sketch depict the frame before any virtual movement.

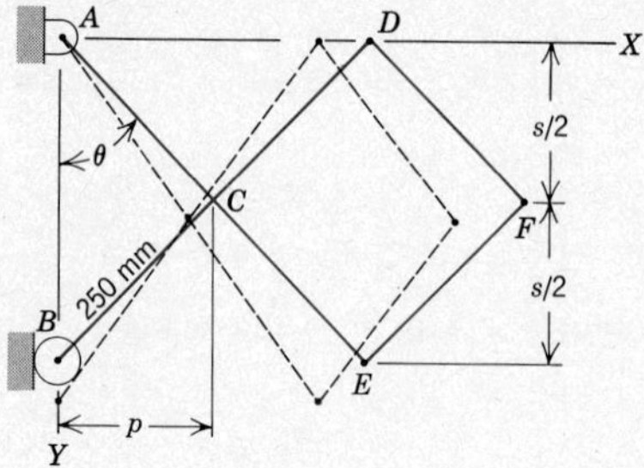

Step 3 The frame is a partially constrained system, because the distance s depends on the loads being carried by the frame. Thus the virtual displacement we give to the system is one in which the value of s is changed, with the constraints of the pins and the roller maintained. As depicted by the dashed lines in the sketch of step 2, the result of this virtual movement is that roller B moves vertically.

Step 4 The sketch of the virtual displacement shows that pin A is fixed, so it is the point chosen as the origin. The Y axis shown in the free body diagram and the sketch of the geometry is chosen downward for convenience.

Step 5 Suitable choices for the generalized coordinate are the distance s or the angle θ shown in the sketch of step 2, because the locations of points in the frame are easily described in terms of either of these variable parameters. We choose s solely because the given information gives the value of s in the equilibrium position.

Step 6 Point A is fixed, so the reaction forces there do no work. Hence, we need only construct the general position vectors for point B and F. In terms of the dimension p shown in the sketch in step 2, these vectors are

$$\bar{r}_{B/A} = s\bar{J} \qquad \bar{r}_{F/A} = 3p\bar{I} + \frac{s}{2}\bar{J}$$

Because s is the generalized coordinate, we express p in terms of s using the Pythagorean theorem. This gives

$$\bar{r}_{B/A} = s\bar{J}$$

$$\bar{r}_{F/A} = 3\left[(0.25)^2 - \left(\frac{s}{2}\right)^2\right]^{1/2}\bar{I} + \frac{s}{2}\bar{J}$$

The virtual displacements are the differential increase in the position vectors resulting from an infinitesimal increase δs. Therefore

$$\delta\bar{r}_{B/A} = \frac{d\bar{r}_{B/A}}{ds}\,\delta s = (\bar{J})\,\delta s$$

$$\delta\bar{r}_{F/A} = \frac{d\bar{r}_{F/A}}{ds}\,\delta s = \left\{3\left(\frac{1}{2}\right)\left[(0.25)^2 - \left(\frac{s}{2}\right)^2\right]^{-1/2}(-s)\bar{I} + \frac{1}{2}\bar{J}\right\}\delta s$$

Step 7 The reaction force $\bar{B}_X$ does no work because it is horizontal and point B displaces vertically (or from another viewpoint, because the constraint imposed by the roller has not been violated.) Thus, the principle of virtual work gives

$$\delta U = [-(s - 0.35)k\,\bar{J}] \cdot \delta\bar{r}_{B/A} + (-100\bar{I}) \cdot \delta\bar{r}_{F/A}$$

$$= -(s - 0.35)k\,\delta s + (-100)\left(-\frac{3}{2}s\right)\left(0.0625 - \frac{s^2}{4}\right)^{-1/2} \delta s$$

$$= \left[-(s - 0.35)k + \frac{150s}{(0.0625 - s^2/4)^{1/2}}\right] \delta s = 0 \text{ J}$$

Step 8 Cancelling the nonzero factor δs, the solution of the preceding expression is

$$k = \frac{300s}{(s - 0.35)(0.250 - s^2)^{1/2}} \text{ N/m}$$

The value of k for the desired equilibrium position is then found by substituting $s = 0.400$ m, which yields

$$k = 8000 \text{ N/m}$$

An interesting feature of the principle of virtual work demonstrated by our analysis of this problem is that it is usually just as easy to determine the results in terms of arbitrary parameters, for instance k in terms of s, rather than for specific values.

ILLUSTRATIVE PROBLEM 3

A 2-kip load is applied to beam AB which rests on beam CD. Determine the reaction force at roller C in terms of the distance r locating the load.

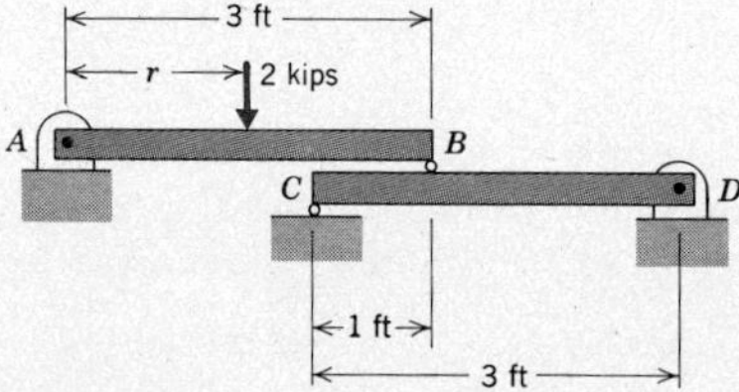

Solution

Step 1 The reaction at roller B is an internal force exerted between the beams, so the free body diagram of the system of beams is as shown.

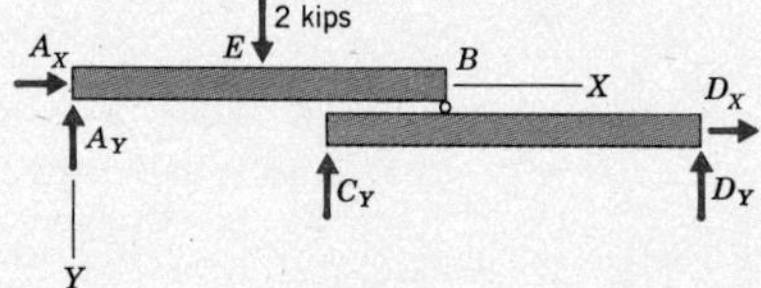

Step 2 The configuration of the system before any virtual displacement is shown as solid lines in the following sketch.

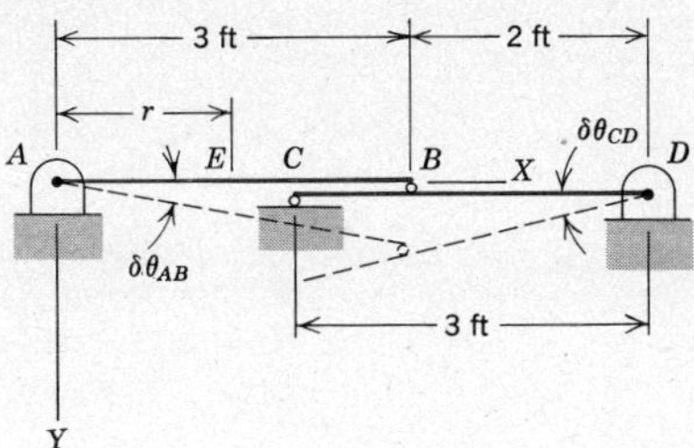

Step 3 The system of beams is fully constrained by its supports. Because it is our desire to determine the reaction at roller C, the virtual movement we give the system is one that violates only the constraint of this roller against vertical displacement. The reaction $\bar{C}_Y$ will then be the only reaction doing virtual work.

Step 4 As shown in the sketch, the result of the virtual movement we have chosen is that each bar executes a virtual rotation about its pin support. Either of the fixed pins A or D is suitable for the origin. The coordinate system we will use has an origin at point A, as shown in the sketches.

Step 5 From the sketch in step 2 we see that the position of the system after the virtual movement may be established in terms of the virtual rotation $\delta\theta_{AB}$. If this angle is known, we can locate end B. This in turn allows us to determine $\delta\theta_{CD}$ by drawing a straight line through pin D tangent to roller B. Hence, we choose θ_{AB} as the generalized coordinate.

Step 6 The virtual movement depicted in the sketch is one in which both beams execute pure rotations about their fixed ends. Because the radial line from each point on a beam to the fixed end of that beam is horizontal, all virtual displacements are vertical. For the rotations depicted in the sketch in step 2, these displacements are downward. It is therefore a straightforward matter to employ the geometrical method. We need the virtual displacements of points C and E, which are the points of application of forces doing work. Point C is on beam CD, 3 ft from the fixed end D, whereas point E is on beam AB, at a distance r from the fixed end A. Thus

$$\delta\bar{r}_{C/A} = 3\ \delta\theta_{CD}\ \bar{J}$$

$$\delta\bar{r}_{E/A} = r\ \delta\theta_{AB}\ \bar{J}$$

We now relate the angle $\delta\theta_{CD}$ to the change in the generalized coordinate θ_{AB} by noting that the roller support at point B, and hence its constraint, was not altered in producing the virtual movement. Therefore, the virtual displacement of point B at the end of beam AB must be the same as that of the point where roller B rests on beam CD. This means that

$$\delta\bar{r}_{B/A} = 3\ \delta\theta_{AB}\ \bar{J} = 2\ \delta\theta_{CD}\ \bar{J}$$

$$\delta\theta_{CD} = \tfrac{3}{2}\ \delta\theta_{AB}$$

Therefore,

$$\delta\bar{r}_{C/A} = 3(\tfrac{3}{2}\ \delta\theta_{AB})\ \bar{J} = \tfrac{9}{2}\ \delta\theta_{AB}\ \bar{J}$$

$$\delta\bar{r}_{E/A} = r\ \delta\theta_{AB}\ \bar{J}$$

Step 7 The forces at points E and C are shown in the free body diagram, so the virtual work is

$$\delta U = 2\bar{J}\cdot\delta\bar{r}_{E/A} + (-C_Y\bar{J})\cdot\delta\bar{r}_{C/A}$$

$$= 2r\ \delta\theta_{AB} - \tfrac{9}{2}C_Y\ \delta\theta_{AB} = 0 \text{ ft-kips}$$

Step 8 Cancelling the common factor $\delta\theta_{AB}$, the value of C_Y is found to be

$$C_Y = \tfrac{2}{9}(2r) = \tfrac{4}{9}r \text{ kips}$$

As an aside to this solution, we should note that the analytical method for evaluating the virtual displacements requires that the system of beams first be considered for arbitrary values of the angles θ_{AB} and θ_{CD}. This method is much more cumbersome than the geometrical method for this problem because the virtual displacements here were obvious from the sketch.

ILLUSTRATIVE PROBLEM 4

A linkage is formed by pinning collar C to bar BD. This collar may ride on the smooth horizontal guide EG. Determine the couple $\bar{M}_A$ that should be applied to bar AB to hold the linkage in the position shown when a vertical 8-kN force is applied at end D.

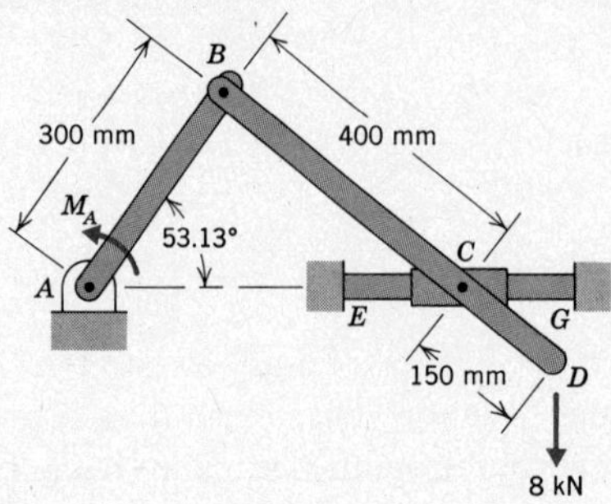

Solution

Step 1 The free body diagram of the system of bars is as shown.

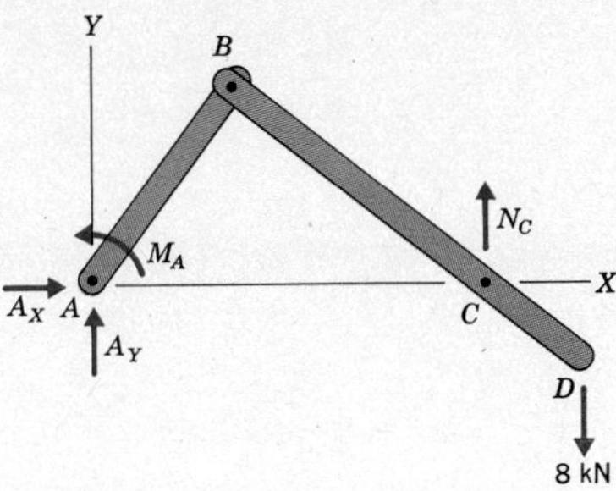

Step 2 The position of the system prior to a virtual movement is shown by solid lines in the following sketch.

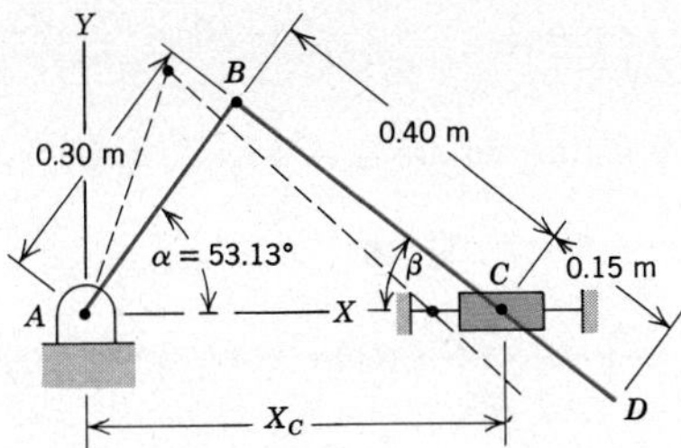

Step 3 The system is partially constrained, because the angles between the bars will change according to the loads being applied. Furthermore, we are requested to relate the values of F and M_D. Hence, the virtual movement we shall consider is one in which the constraints of the supports are unmodified, thereby preventing their reactions from doing work. As shown by the dashed lines in the sketch of step 2, the result is that bar AB executes an infinitesimal rotation about pin A, causing collar C to slide to the left.

Step 4 Pin A is the only fixed point in the system, so it is chosen as the origin for the coordinate system shown in the sketches.

Step 5 Knowledge of the angle α, the angle β, or the distance X_C would allow us to determine the other dimensions in an arbitrary position of the linkage. In view of the fact that we will need to know the virtual rotation of bar AB in order to determine the virtual work done by the couple $\bar{M}_A$, we select α to be the generalized coordinate.

Step 6 With the aid of the sketch in step 2, the position vector for point D, where the 8-kN load is applied, is

$$\bar{r}_{D/A} = (0.30 \cos \alpha + 0.55 \cos \beta)\, \bar{I} - 0.15 \sin \beta\, \bar{J}$$

It is now necessary to relate the variable angle β to the generalized coordinate α. This is done by applying the law of sines to triangle ABC, which yields

$$\frac{\sin \alpha}{0.40} = \frac{\sin \beta}{0.30} \qquad \sin \beta = 0.75 \sin \alpha$$

Replacing β by α in the expression for $\bar{r}_{D/A}$ will involve inverse trigonometric functions. Therefore, we proceed to the determination of $\delta\bar{r}_{D/A}$, bearing in mind that β is a function of α as given above. Thus

$$\delta\bar{r}_{D/A} = \frac{d\bar{r}_{D/A}}{d\alpha}\,\delta\alpha = \left[\left(-0.30 \sin \alpha - 0.55\frac{d\beta}{d\alpha}\sin \beta\right)\bar{I} - 0.15\frac{d\beta}{d\alpha}\cos \alpha\ \bar{J}\right]\delta\alpha$$

The simplest way to determine $d\beta/d\alpha$ is to differentiate the relationship between β and α implicitly. This yields

$$\frac{d}{d\alpha}(\sin \beta) \equiv \frac{d\beta}{d\alpha}\cos \beta = 0.75 \cos \alpha$$

In the position where $\alpha = 53.13°$, we have

$$\sin \beta = 0.75 \sin 53.13° = 0.6 \qquad \beta = 36.87°$$

$$\frac{d\beta}{d\alpha}\cos 36.87° = 0.75 \cos 53.13° \qquad \frac{d\beta}{d\alpha} = 0.5625$$

Thus, in the position of interest

$$\begin{aligned}\delta\bar{r}_{D/A} &= \{[-0.30 \sin 53.13° - 0.55(0.5625)\sin 36.87°]\,\bar{I} \\ &\quad - 0.15(0.5625)\cos 36.87°]\,\bar{J}\}\,\delta\alpha \\ &= (-0.4256\bar{I} - 0.0675\bar{J})\,\delta\alpha \text{ m}\end{aligned}$$

Step 7 Equation (3) gives the work done by a couple. Bar AB rotates counterclockwise with an increase in the angle α, so the counterclockwise couple $\bar{M}_A$ does positive work. Thus

$$\begin{aligned}\delta U &= M_A\,\delta\alpha + (-8\bar{J}) \cdot (-0.4256\bar{I} - 0.0675\bar{J})\,\delta\alpha \\ &= (M_A + 0.540)\,\delta\alpha \text{ kN-m}\end{aligned}$$

Step 8 When the common factor $\delta\alpha$ is cancelled, the value of M_A is found to be

$$M_A = -0.540 \text{ kN-m}$$

The negative value for M_A merely indicates that the couple applied to bar AB should be clockwise, opposite the direction in which it was assumed.

An overview of the solution we have presented shows that the complications introduced by the lack of *simple* geometric relations, such as the need to evaluate derivatives implicitly, decreases the attractiveness of using the principle of virtual work for this problem. A straightforward force and moment analysis would give the same result, without such complications.

HOMEWORK PROBLEMS

VIII.9 and VIII.10 Bar AB, having a mass m, is pinned to the collar, which rides on the vertical rod. Friction is negligible. Determine the value of θ for static equilibrium.

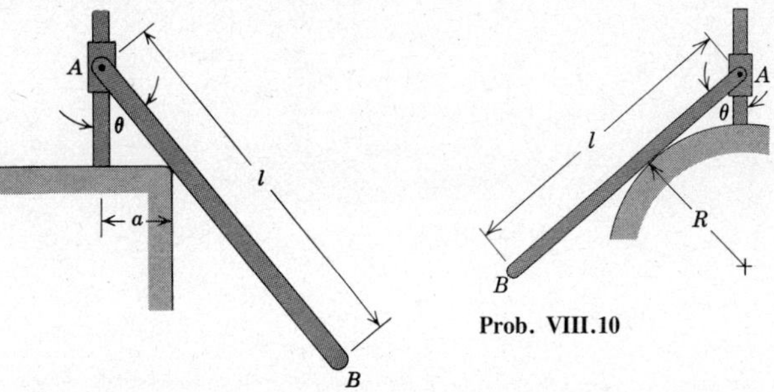

Prob. VIII.9

Prob. VIII.10

VIII.11 Bar AB is pinned to collars C and D, which ride on the vertical rods. The bar weighs 10 lb and the ground is smooth. Determine the reaction exerted by the vertical rod on collar C.

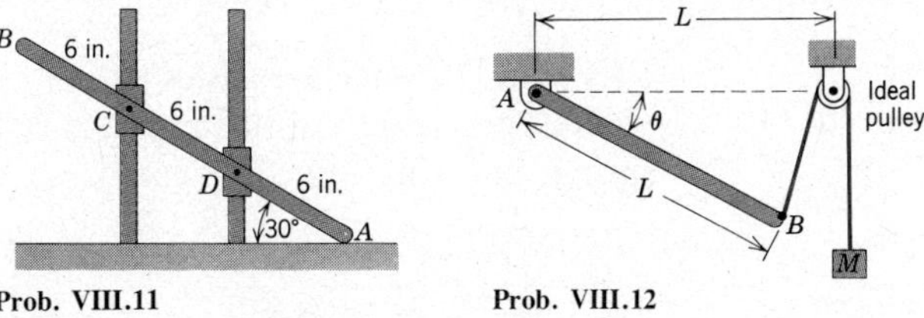

Prob. VIII.11

Prob. VIII.12

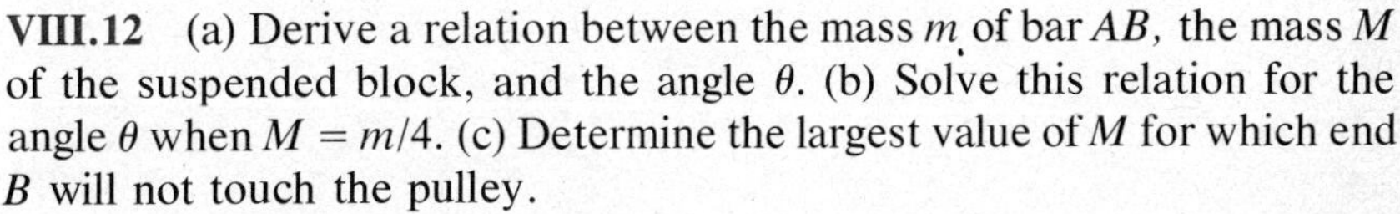

VIII.12 (a) Derive a relation between the mass m of bar AB, the mass M of the suspended block, and the angle θ. (b) Solve this relation for the angle θ when $M = m/4$. (c) Determine the largest value of M for which end B will not touch the pulley.

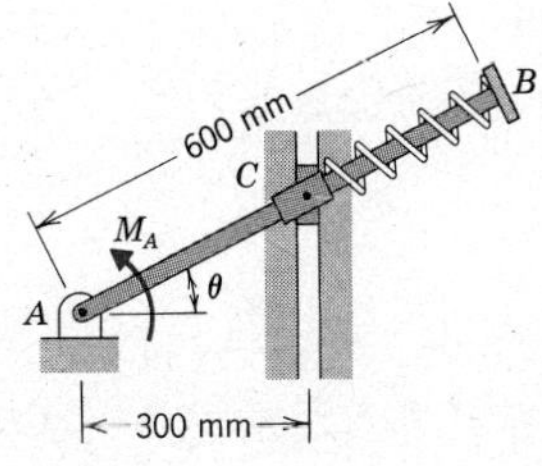

Prob. VIII.13

VIII.13 Collar C, which rides on the smooth bar AB, is pinned to a block that is guided by the smooth straight groove. The spring is unstressed when $\theta = 0$ and its stiffness is 5 kN/m. Determine the magnitude of the couple $\bar{M}_A$ required to maintain equilibrium in terms of θ.

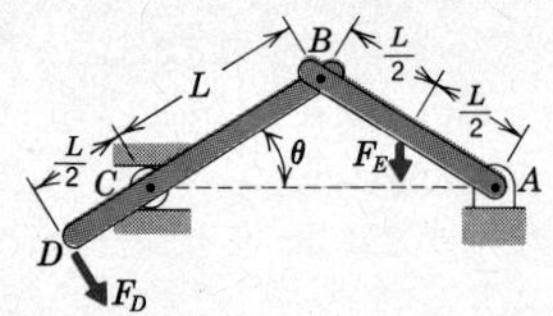

Prob. VIII.14

VIII.14 Roller C, which rides in the groove, is pinned to bar BD. Derive an expression for the force $\bar{F}_E$ required to hold the system in equilibrium against the force $\bar{F}_D$.

VIII.15 A 40-lb force is applied to arm ABC of the press punch shown in cross section. Determine the force the punch applies to the piece of sheet metal in the position shown.

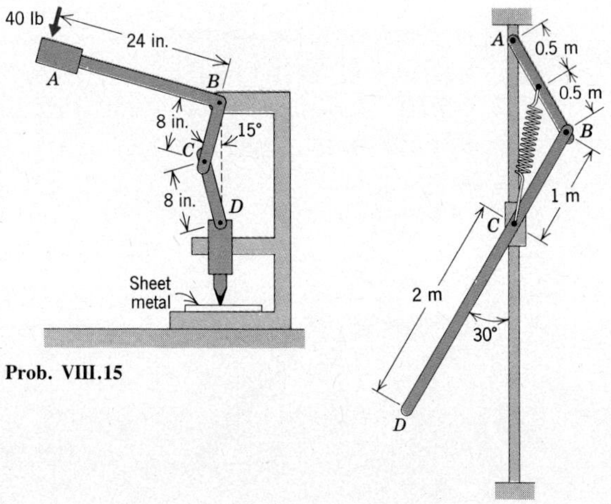

Prob. VIII.15

Prob. VIII.16

VIII.16 The mechanism shown lies in a vertical plane and is in equilibrium. Collar C rides on the smooth vertical guide, and the unstretched length of the spring is 1 m. Knowing that each of the bars has a mass of 2 kg/m, determine the stiffness of the spring.

VIII.17 The spring-cushioned platform supports a 200-kg package. The unstretched length of the spring is 700 mm. Determine the stiffness of the spring, knowing that the system is in equilibrium in the position shown.

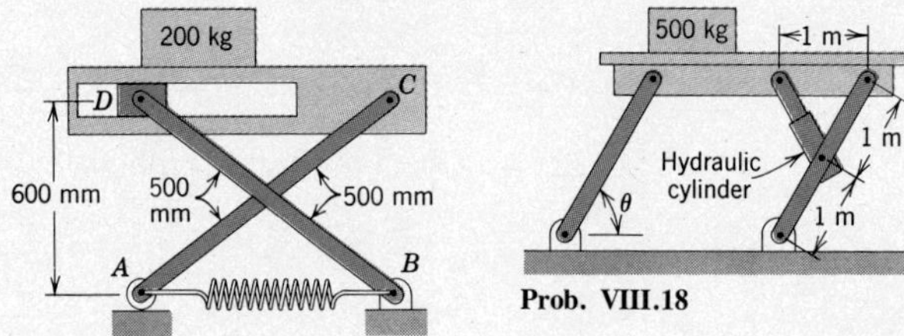

Prob. VIII.17

Prob. VIII.18

VIII.18 A 500-kg crate rests on the platform, which is supported by the mechanism shown. Determine the force in the hydraulic cylinder as a function of θ.

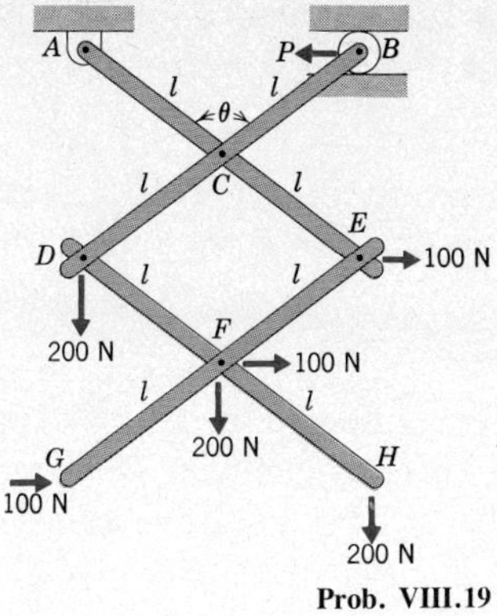

Prob. VIII.19

VIII.19 The parallelogram frame carries the loads shown. In terms of the angle θ, determine the horizontal force $\bar{P}$ that is required to maintain equilibrium.

VIII.20 The load $\bar{P}$ is always perpendicular to bar DF. The spring has stiffness k and it is unstressed when $\theta = 60°$. Derive an expression for the magnitude of $\bar{P}$ required for equilibrium at an arbitrary angle θ.

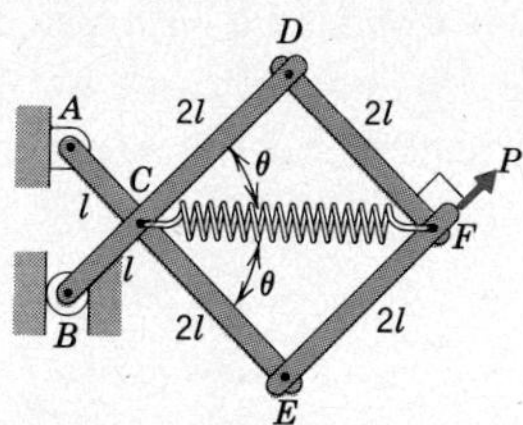

Prob. VIII.20

VIII.21 and VIII.22 Determine the reaction moment at support B of the compound beam shown.

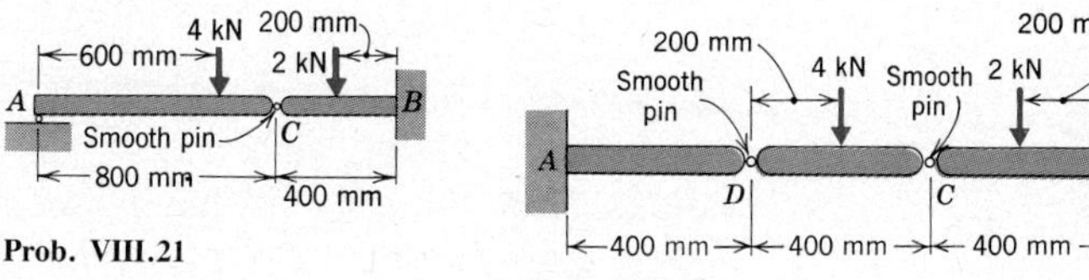

Prob. VIII.21

Prob. VIII.22

VIII.23 Determine the reaction at roller C resulting from the set of loads shown.

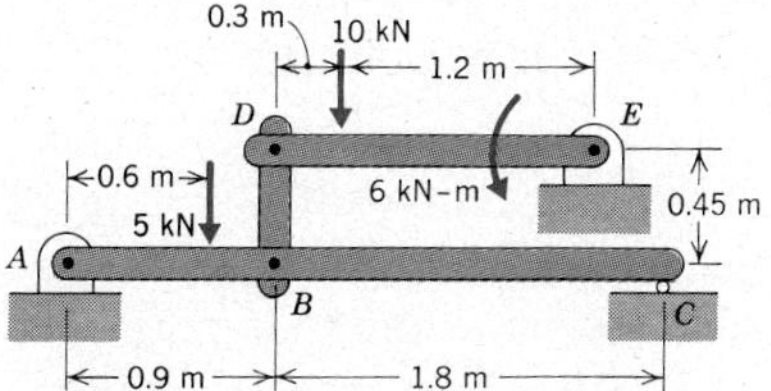

Prob. VIII.23

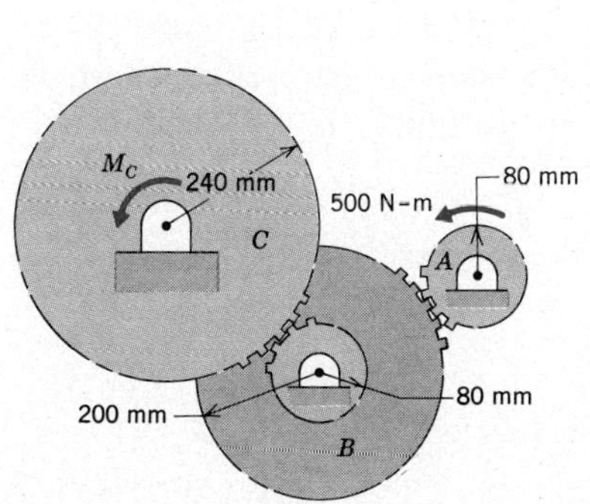

Prob. VIII.24

VIII.24 A 500 N-m couple is applied to gear A, as shown. Gear B, which transmits this couple to gear C, rides freely on its shaft. Determine the magnitude and sense of the couple $\bar{M}_C$ acting on gear C necessary for equilibrium. *Hint:* The constraint of the gear teeth requires that, in a virtual rotation of the gears, the teeth in contact on a pair of gears move the same distance circumferentially.

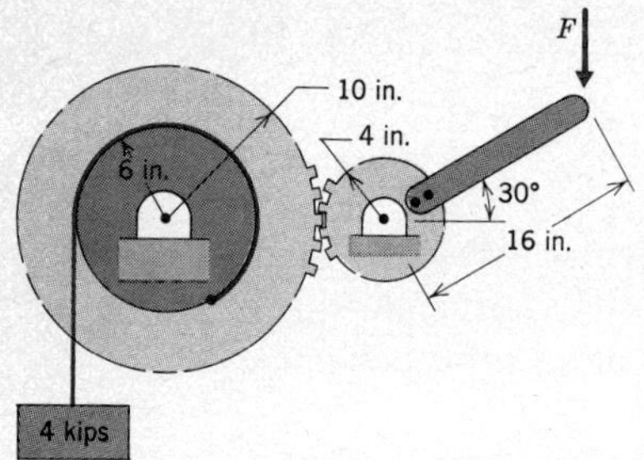

Prob. VIII.26

VIII.25 Solve Problem VIII.24 if a counterclockwise couple of 100 N-m is acting on the stepped gear B in addition to the couples on gears A and C.

VIII.26 The winch shown is used to suspend a 4-kip crate. Determine the force $\bar{F}$ required for equilibrium in the position shown. (See the hint for Problem VIII.24.)

VIII.27 The two blocks, which are interconnected by a cable passing over a pulley, are in equilibrium on the smooth inclines. Determine an expression for the mass ratio m_B/m_A. *Hint:* The length of the cable is constant in a virtual movement.

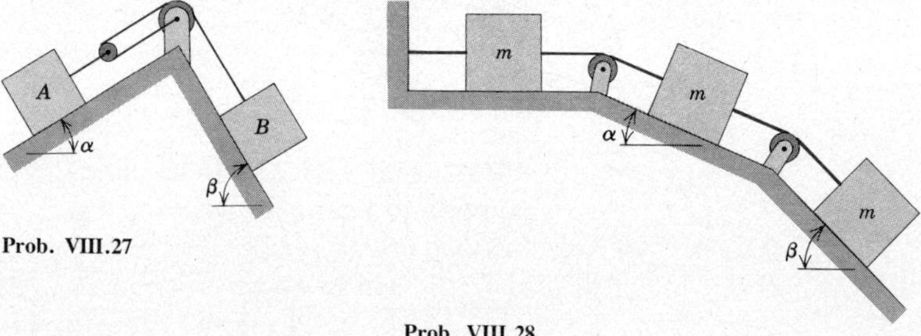

Prob. VIII.27

Prob. VIII.28

VIII.28 The three blocks having identical masses m are tied together by cables. Knowing that all contact surfaces are smooth, determine the tension in the cable that is fastened to the wall. (See the hint for Problem VIII.27.)

VIII.29 The 100-kg radioactive container is held by applying a force $\bar{P}$ to the pair of tongs. Determine the horizontal component of the force acting on the container at point F.

VIII.30 The mechanism shown is a device for clamping onto inaccessible objects. The spring has a stiffness of 2 kN and an unstretched length of 40 mm. Determine the clamping force on the object if the handles are squeezed with a pair of 10-N forces.

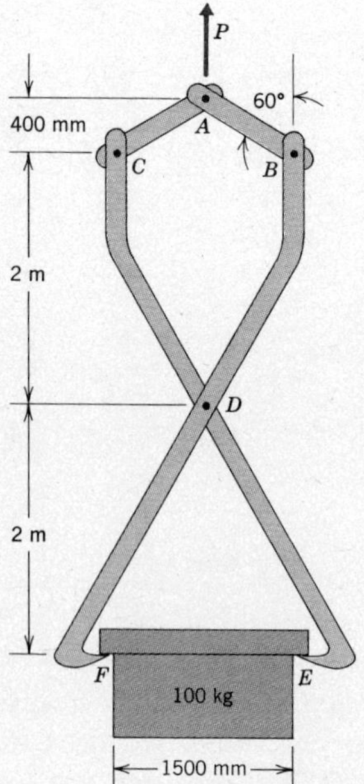

Prob. VIII.29

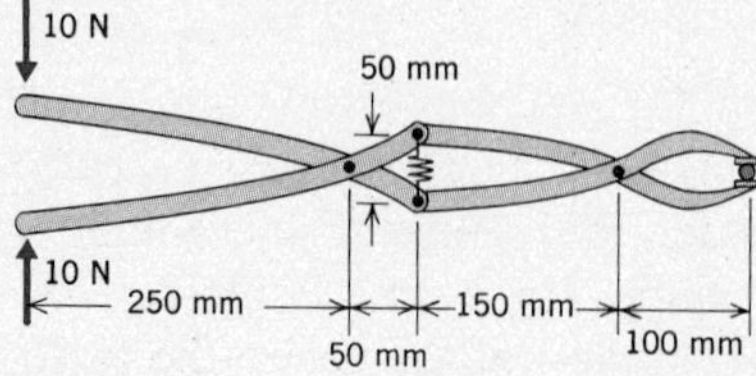

Prob. VIII.30

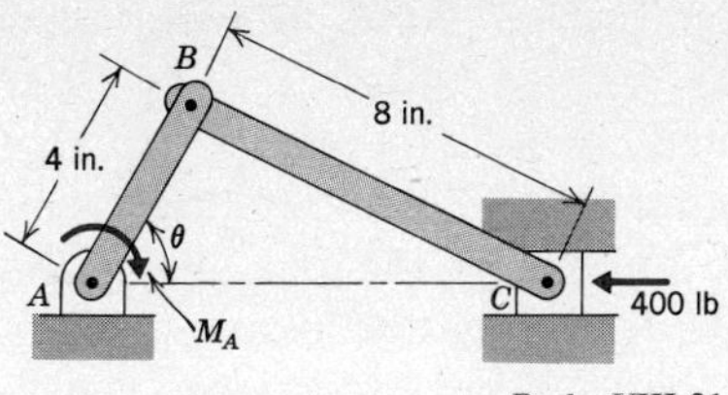

Prob. VIII.31

VIII.31 A 400-lb force is applied to the piston. Determine the couple $\bar{M}_A$ required for equilibrium when (a) $\theta = 30°$, (b) $\theta = 150°$.

VIII.32 Determine the force $\bar{F}$ required to hold the system shown in equilibrium if the couple $\bar{M}_A$ is 2 kN-m counterclockwise when (a) $\theta = 90°$, (b) $\theta = 0$, (c) $\theta = 45°$.

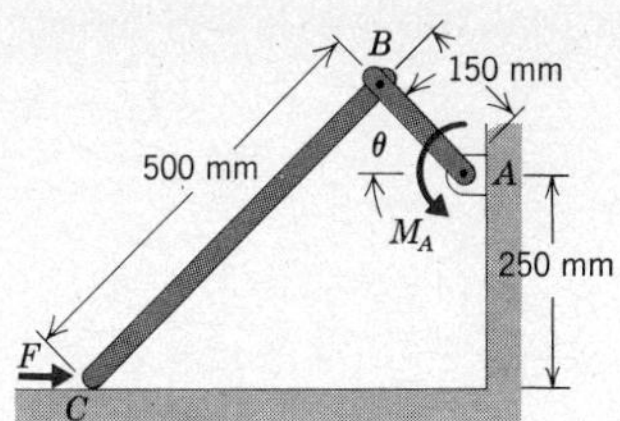

Prob. VIII.32

E. THE PRINCIPLE OF VIRTUAL WORK AND POTENTIAL ENERGY

Previous investigations were limited to the determination of the work done by a force in an infinitesimal movement of a system. This is the only consideration required in statics for the principle of virtual work. In contrast, for a system that is actually in motion, the point of application of a force will have a finite displacement. The determination of the work done in a finite movement of a system plays an important role in the study of dynamics. One feature of such a study is also relevant to statics, for it will greatly simplify the determination of the virtual work done by certain types of forces.

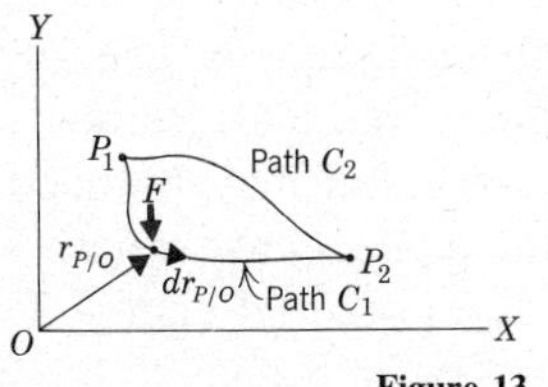

Figure 13

To investigate this matter, consider Figure 13, where we wish to investigate the work done by the force $\bar{F}$ in moving the point P from position P_1 to position P_2 in the case where the movement can take place over two different arbitrary paths C_1 and C_2.

The infinitesimal work of the force is defined as $dU = \bar{F} \cdot d\bar{r}_{P/O}$, so the total work will be the sum of all infinitesimal terms. Therefore, the work done when point P moves from position P_1 to position P_2 is

$$U_{1\to 2} = \int_{r_1}^{r_2} \bar{F} \cdot d\bar{r}_{P/O} \tag{17}$$

where r_1 and r_2 denote the coordinates of points P_1 and P_2, respectively.

In general, both the magnitude and direction of $\bar{F}$ will depend on the location of point P. Furthermore, the angle between $\bar{F}$ and the infinitesimal displacement vector $d\bar{r}_{P/O}$ will depend on the orientation of the tangent to the path. We can therefore expect that the value of the work obtained from equation (17) will depend on whether path C_1 or path C_2 has been followed in the finite displacement.

Our interest in statics, however, rests in the exceptional situation where the work done does not depend on the path followed. In that case it must be true that the work $U_{1\to 2}$ is a function only of the locations of the initial point P_1 and the final point P_2. Considering that the locations of the initial and final points appears only in the limits of the integral for $U_{1\to 2}$, it then follows that the work will have the form

$$\boxed{U_{1\to 2} = V_1 - V_2} \tag{18}$$

where V represents a scalar function whose value depends on the position of point P, so that V_1 is the value of this function at position P_1 and V_2 is the value of the function at position P_2.

The function V is called the *potential energy* of the force $\bar{F}$ and the force is said to be *conservative*. To see why, suppose that point P were to follow a closed path by moving from position P_1 to position P_2 along path C_1, and then returning to position P_1 along path C_2. According to equation (18), the work done in the second portion of the movement is

$$U_{2\to 1} = V_2 - V_1 = -U_{1\to 2} \tag{19}$$

This means that the total work done in following the closed path is zero, because

$$U_{1\to 2\to 1} = U_{1\to 2} + U_{2\to 1} = 0 \tag{20}$$

Equations (19) and (20) show that the work done by the force $\bar{F}$ when point P moves from one position to another is not lost. Rather, it is stored (thus, the word conservative) in the system and may be used to return point P back to its original position. In the special case where $V_2 = 0$, we have $U_{1\to 2} = V_1$. Thus, we may regard the potential energy V to be the stored ability of the force to do work when point P goes to the position of zero potential energy.

Two types of conservative forces that especially interest us are the weight force acting on a body at the surface of the earth and the force of a spring. Let us consider the weight force first. Figure 14 shows a body whose center of mass G moves from position P_1 to position P_2. Note that for the purpose of computing the potential energy of the weight force shown, we need not consider what causes the movement.

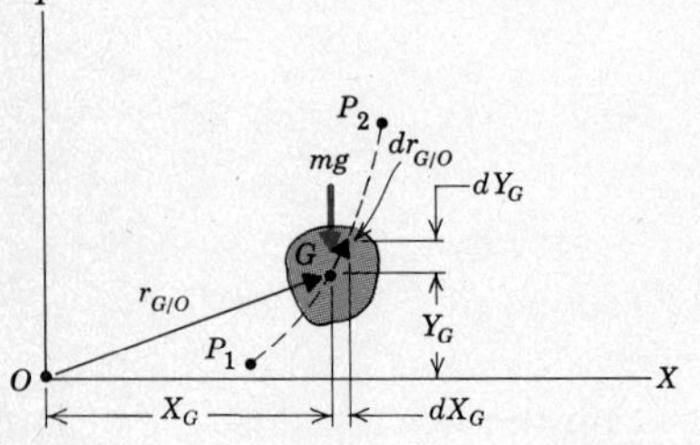

Figure 14

To describe the vectors shown in Figure 14, we shall utilize an XYZ coordinate system whose Y axis is vertical and whose origin O is located at an arbitrary position, as illustrated. Denoting the coordinates of point G as (X_G, Y_G), an infinitesimal displacement $d\bar{r}_{G/O}$ can be seen from Figure 14 to be equivalent to a displacement dX_G to the right and dY_G upward. That is,

$$d\bar{r}_{G/O} = dX_G\,\bar{I} + dY_G\,\bar{J}$$

The work done by the weight force as a result of this displacement is

$$dU = (-mg\bar{J}) \cdot d\bar{r}_{G/O} = -mg\,dY_G$$

Thus, the total work done by the weight force in the movement from position P_1 to position P_2 is

$$U_{1\to2} = \int_{r_{P_1/O}}^{r_{P_2/O}} (-mg\,dY_G) = -mg\,Y_G\Big|_{r_{P_1/O}}^{r_{P_2/O}}$$
$$= -mgY_2 + mgY_1 \tag{21}$$

where Y_1 and Y_2 denote the Y coordinates of points P_1 and P_2, respectively.

Equation (21) was obtained without knowing any details about the type of path followed by the center of mass, so we can conclude that gravity is a conservative force. By comparing the specific result of equation (21) to the general result for a conservative force, equation (18), we can see that

$$\boxed{V_{\text{grav}} = mgY_G} \tag{22}$$

where Y_G represents the vertical height of the center of mass G above the origin.

An alternative way of regarding the distance Y_G is to think of the X axis in Figure 14 as defining a reference elevation, called the *datum*. Then Y_G is the elevation of the center of mass above the datum. When point G is below the datum, the value of Y_G is negative, so V_{grav} is also negative. This simply means that the weight force will do negative work if the center of mass is moved back to the datum.

It is important to realize that the preceding derivation applies equally well to other forces having constant magnitude and direction, not just the weight force acting on a body near the surface of the earth. Specifically, if a force $\bar{F}$ has a constant magnitude and direction, regardless of where its point of application moves to, we can form an analogy with the gravity force by defining a line perpendicular to $\bar{F}$ to be the datum for this force. Letting h be the height of the point of application *above* the datum, that is, in the direction opposite that of $\bar{F}$, the potential energy of the force is $V_F =$

Fh. Hence, all forces having constant magnitude and direction are conservative. This fact will be useful for our study in the next section of the stability of the equilibrium position.

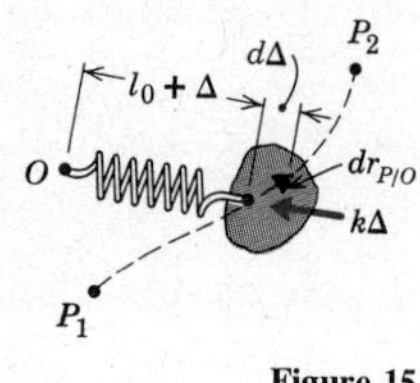

Figure 15

Let us now consider the force exerted by a spring. In Figure 15 there is a spring anchored at the fixed point O and attached at its other end to point P in a moving body. Again, we consider the situation where point P follows an arbitrary path as it moves from the initial position P_1 to the final position P_2.

In the situation depicted in the figure, the spring has an extension Δ beyond the unstretched length l_0. Thus, in a differential displacement $d\bar{r}_{P/O}$ in which point P moves outward from point O, the spring is extended by an additional amount $d\Delta$. Noting that the spring force applied to the body is $F = k\Delta$ directed from point P to point O, the component of the displacement $d\bar{r}_{P/O}$ perpendicular to line OP is also perpendicular to the spring force, so no work is done in this type of movement. On the other hand, the component of displacement $d\Delta$ outward along line OP is opposite from the spring force, so it results in negative work in the amount $-k\Delta\, d\Delta$. Therefore, the total work in moving from position P_1 to position P_2 is

$$\begin{aligned} U_{1\to 2} &= \int_{r_{P_1/O}}^{r_{P_2/O}} (-k\Delta\, d\Delta) = -\tfrac{1}{2}k\Delta^2 \Big|_{r_{P_1/O}}^{r_{P_2/O}} \\ &= -\tfrac{1}{2}k{\Delta_2}^2 + \tfrac{1}{2}k{\Delta_1}^2 \end{aligned} \tag{23}$$

where Δ_1 and Δ_2 denote the extension of the spring at positions P_1 and P_2, respectively.

Equation (23) was obtained without knowledge of the path followed by point P; it follows that the spring force is conservative. Comparison of equation (23) with the general result of equation (18) then shows that

$$\boxed{V_{\text{spr}} = \tfrac{1}{2}k\Delta^2} \tag{24}$$

Although this result was obtained by considering a spring that is extended, it is equally valid for springs that can sustain a compressive force $k\Delta$ due to a decrease Δ in its length. We therefore see that regardless of whether a spring is elongated or shortened, its potential energy is positive. This means that it will do positive work when the system returns to the position where the spring is undeformed.

We could go on to consider other types of conservative forces, but equations (22) and (24) will suffice for the systems we wish to consider. An important aspect of the potential energy of a force is that it is a scalar, so the potential energy of each of several conservative forces is additive as a scalar.

Let us now return to the reason why we consider the concept of potential energy in a course on statics. Suppose that we have chosen a generalized coordinate, which we will label as q. By definition, the position of all points in the system may be described in terms of the value of q. It follows that the total potential energy V of all conservative forces is a function of q; that is, $V = V(q)$. Then, with the aid of equation (18), we see that the virtual work done by the conservative forces in an infinitesimal increase δq of the generalized coordinate is

$$\begin{aligned}\delta U^{(\text{cons})} &= V(q) - V(q + \delta q) \\ &= -\delta V(q) \equiv -\frac{dV}{dq}\,\delta q \end{aligned} \tag{25}$$

We now separate the total virtual work δU into the portion $\delta U^{(\text{cons})}$, attributable to the set of conservative forces acting on the system, and the portion $\delta U^{(\text{nc})}$, attributable to the nonconservative forces acting on the system. Then, upon application of equation (25), the principle of virtual work becomes

$$\delta U = \delta U^{(\text{nc})} + \delta U^{(\text{cons})} = \delta U^{(\text{nc})} - \frac{dV}{dq}\,\delta q = 0$$

$$\boxed{\delta U^{(\text{nc})} = \frac{dV}{dq}\,\delta q} \tag{26}$$

Equation (26) is the principle of virtual work and energy. In words, it states that

> In the static equilibrium position the virtual work done by the nonconservative forces equals the virtual change in the potential energy of the conservative forces.

The usefulness of this principle is that we no longer have to evaluate the virtual displacements of the points of application of conservative forces. All that is necessary to account for the conservative forces is that we express their potential energy V in terms of the generalized coordinate q and then evaluate the derivative dV/dq. Thus, equation (26) is merely the result of a refinement in the method of computing the work done by a certain class of forces. Of course, all forces of uncertain nature should be regarded as nonconservative.

In the special situation where all of the forces doing work are conservative, in which case we say that we have a *conservative system,* equation (26) reduces to

$$\boxed{\frac{dV}{dq} = 0} \tag{27}$$

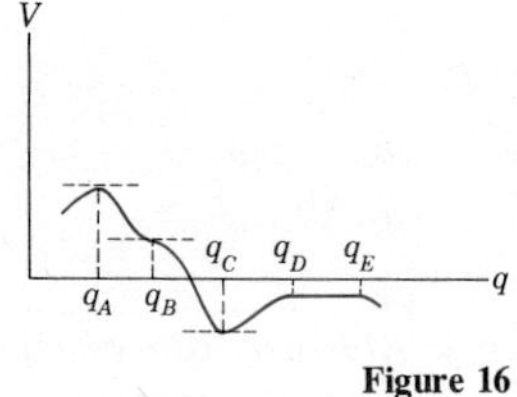

Figure 16

This is known as the *principle of stationary potential energy,* because it states that in the static equilibrium position of a conservative system, the potential energy is either a maximum or minimum value, or possibly a horizontal inflection point. A typical potential energy function for a conservative system is illustrated in Figure 16. The values q_A, q_B, q_C, and the region between q_D and q_E all correspond to values of the generalized coordinate for which the system is in static equilibrium.

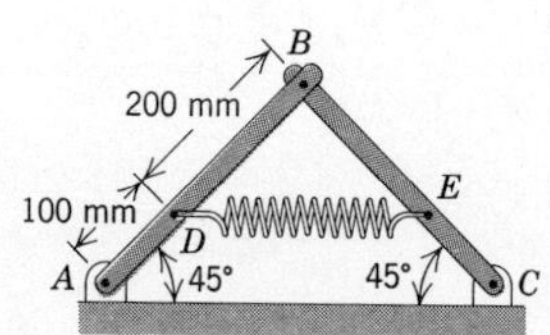

EXAMPLE 3

Bars AB and BC in the system shown each have a mass of 5 kg. The spring has an unstretched length of 200 mm. Determine the value of the stiffness k for which the horizontal reaction at pin C is reduced to one-half the value that would be obtained if $k = 0$.

Solution

Aside from the fact that it is no longer necessary to compute the virtual displacement of the points of application of conservative forces, the basic method for solving problems that was developed in the last section is essentially unmodified by the utilization of the concept of potential energy. We begin with a free body diagram and a sketch of the geometry of the system.

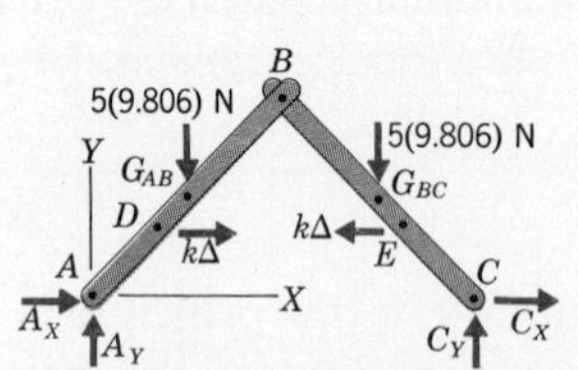

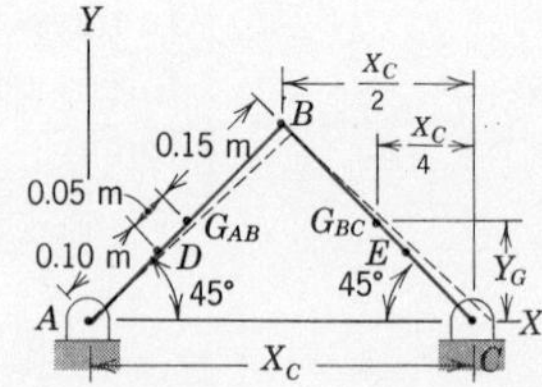

We have been requested to determine the reaction $\bar{C}_X$. Therefore, the virtual movement we give the system is one that violates the constraint of pin C against horizontal movement. The resulting virtual movement is pictured in the geometric sketch. Note that the other reaction forces do no work because the reactions $\bar{A}_X$ and $\bar{A}_Y$ are applied at a fixed point and $\bar{C}_Y$ is perpendicular to the horizontal virtual displacement of pin C. Point A is selected as the origin for the coordinate system shown because it is the only point in the system that is fixed in the virtual movement.

Suitable choices for the generalized coordinate are either the angle θ describing the angle of elevation of the bars, or else the horizontal distance X_C. For the sake of variety from previous problems, let us use the latter. The only nonconservative force doing work in the virtual displacement is the reaction $\bar{C}_X$. Thus, the only virtual displacement we need to formulate is that of point C. The position vector of this point in terms of X_C is simply

$$\bar{r}_{C/A} = X_C\bar{I}$$

so

$$\delta\bar{r}_{C/A} = \frac{d\bar{r}_{C/A}}{dX_C}\,\delta X_C = (\bar{I})\,\delta X_C$$

It then follows that the work $\delta U^{(nc)}$ is

$$\delta U^{(nc)} = C_X\bar{I}\cdot\delta\bar{r}_{C/A} = C_X\,\delta X_C$$

We now formulate the potential energy of the conservative weight and spring forces. The X axis is a convenient datum for the potential energy of gravity, as indicated in the free body diagram. Adding the potential energy of each of the conservative forces, we have

$$\begin{aligned} V &= V_{\text{grav}} + V_{\text{spr}} \\ &= 2(5)(9.806)Y_G + \tfrac{1}{2}k\Delta^2 \end{aligned}$$

where Y_G is the elevation of the centers of mass and Δ is the elongation of the spring.

The application of equation (26) requires that we express V in terms of the generalized coordinate $q \equiv X_C$. From the sketch of the geometry we see that the horizontal distance from point C to the center of mass G_{BC} is $X_C/4$, so the Pythagorean theorem gives

$$Y_G = \left[(0.15)^2 - \left(\frac{X_C}{4}\right)^2\right]^{1/2}\ \text{m}$$

To evaluate the spring extension Δ, we subtract the unstretched length of 0.20 m from the stretched length l_{DE}. Considering the similar triangles BDE and BAC shown in the sketch, we find

$$\frac{l_{DE}}{0.20\ \text{m}} = \frac{X_C}{0.30\ \text{m}}$$

$$l_{DE} = \tfrac{2}{3}X_C$$

Thus

$$\Delta = \tfrac{2}{3}X_C - 0.20 \text{ m}$$

Upon substitution of these expressions for X_C and Δ, the potential energy becomes

$$V = 98.06\left(0.0225 - \frac{X_C^2}{16}\right)^{1/2} + \tfrac{1}{2}k(\tfrac{2}{3}X_C - 0.20)^2$$

We now employ the work-energy principle, equation (26). Thus

$$\delta U^{(\text{nc})} = \frac{dV}{dX_C}\,\delta X_C$$

$$C_X\,\delta X_C = \left[\frac{1}{2}(98.06)\left(0.0225 - \frac{X_C^2}{16}\right)^{-1/2}\left(-\frac{X_C}{8}\right) + k\left(\frac{2}{3}X_C - 0.200\right)\left(\frac{2}{3}\right)\right]\delta X_C \text{ J}$$

We know that $X_C = 0.30\sqrt{2}$ m in the static equilibrium position, because triangle ABC in that case is a 45° right triangle. Therefore, after the factor δX_C is cancelled, the equilibrium equation is

$$C_X = (49.03)\left[0.0225 - \frac{0.090(2)}{16}\right]^{-1/2}\left(-\frac{0.30\sqrt{2}}{8}\right) + k\left[\left(\frac{2}{3}\right)0.30\sqrt{2} - 0.20\right]\left(\frac{2}{3}\right)$$

$$= -12.258 + 0.05523k \text{ N}$$

When $k = 0$ this equation gives $C_X = -12.258$ N. Thus, setting C_X equal to one-half this value, we find that

$$-\tfrac{1}{2}(12.258) = -12.258 + 0.05523k$$

$$k = 111.0 \text{ N/m}$$

EXAMPLE 4

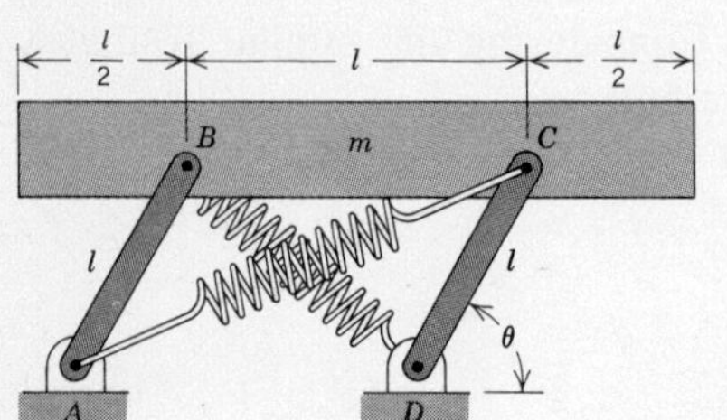

The table top of mass m is supported by the lightweight bars AB and CD, which are braced by two identical springs of stiffness k and unstretched length $1.20l$. Determine an expression for the value of k for which the system is in equilibrium at $\theta = 60°$.

Solution

As always, we start with a free body diagram of the system and a separate sketch describing the geometry of the system. In the latter, h is an unspecified dimension required to locate the center of mass G.

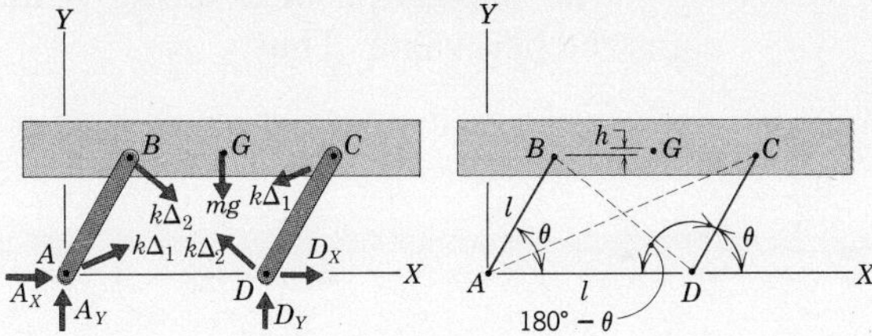

Noting that we are interested in the value of k for equilibrium, and not the reaction forces at pins A and D, the virtual movement of interest is one in which the angle θ is changed while the pin connections are maintained. Clearly, in such a movement the only forces doing work will be those of the springs and gravity, both of which are conservative. Therefore, we may employ the principle of stationary potential energy, equation (27), and it will not be necessary to evaluate virtual displacements of points.

For a generalized coordinate we choose the angle θ. The choice of the datum for the potential energy of gravity is arbitrary. The origin of the XYZ coordinate system defined in the sketch of the geometry is located at a convenient fixed point, so we shall choose the X axis for the datum. From the second sketch we see that the height of point G above the datum is $l \sin\theta + h$, so the potential energy of the gravity force is

$$V_{\text{grav}} = mg(l \sin\theta + th)$$

The potential energy of a spring is given by equation (24), so the total potential energy of both springs is

$$V_{\text{spr}} = \tfrac{1}{2}k\Delta_{AC}^2 + \tfrac{1}{2}k\Delta_{BD}^2$$

In order to employ equation (27), we will need to know the relationship between the deformation of each spring and the generalized coordinate θ. This relationship is obtained with the aid of the sketch of the geometry. Applying the law of cosines to triangle ABD, we find that the length of the spring when stretched between pins B and D is defined by

$$l_{BD}^2 = l^2 + l^2 - 2(l)(l)\cos\theta = 2l^2(1 - \cos\theta)$$

$$l_{BD} = \sqrt{2}\, l(1 - \cos\theta)^{1/2} \equiv 2l \sin\frac{\theta}{2}$$

A similar analysis of triangle ACD shows that the length of the spring joining pins A and C is

$$l_{AC} = 2l \cos\frac{\theta}{2}$$

Note that these expressions were obtained with the aid of the half-angle identities for the sine and cosine functions.

The elongation of a spring is the increase in its length beyond its unstretched value. Thus

$$\Delta_{AC} = l_{AC} - 1.20l = l\left(2 \cos \frac{\theta}{2} - 1.20\right)$$

$$\Delta_{BD} = l_{BD} - 1.20l = l\left(2 \sin \frac{\theta}{2} - 1.20\right)$$

The total potential energy of the system as a function of θ is then found to be

$$V = V_{\text{grav}} + V_{\text{spr}} = mg(h + l \sin \theta) + \frac{1}{2} kl^2\left(2 \cos \frac{\theta}{2} - 1.20\right)^2 + \frac{1}{2} kl^2\left(2 \sin\frac{\theta}{2} - 1.20\right)^2$$

Equation (27) now requires that in the position of static equilibrium

$$\frac{dV}{d\theta} = mgl \cos \theta + kl^2\left(2 \cos \frac{\theta}{2} - 1.20\right)\left(- \sin\frac{\theta}{2}\right) + kl^2\left(2 \sin \frac{\theta}{2} - 1.20\right)\left(\cos \frac{\theta}{2}\right)$$

$$= mgl \cos \theta + 1.20kl^2\left(\sin\frac{\theta}{2} - \cos \frac{\theta}{2}\right) = 0$$

One possible solution of this equation is $\theta = 90°$, corresponding to the position where bars AB and CD are vertical. However, in this problem we are interested in the existence of another equilibrium position, corresponding to $\theta = 60°$. Substituting this value into the equation for $dV/d\theta$ yields

$$0.50mgl + 1.20kl^2(0.50 - 0.8660) = 0$$

$$k = 1.138\frac{mg}{l}$$

HOMEWORK PROBLEMS

VIII.33 Two identical springs, which are anchored at points A and B and attached together, have a tension T_0 when they are not supporting the block. Derive an expression for the mass m of the block that will give equilibrium at a specific value of the angle θ. Show that, for a fixed value of θ, this value of m increases if either the stiffness k or the initial tension

T_0 is increased. Explain this trend physically by considering the amount of energy stored in a spring for a given increase in its elongation.

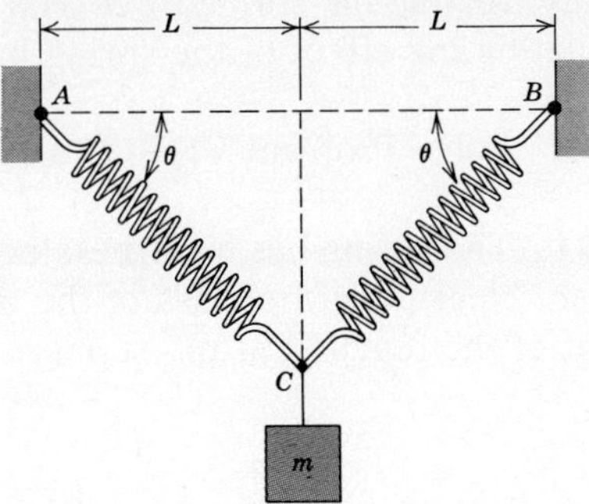

Prob. VIII.33

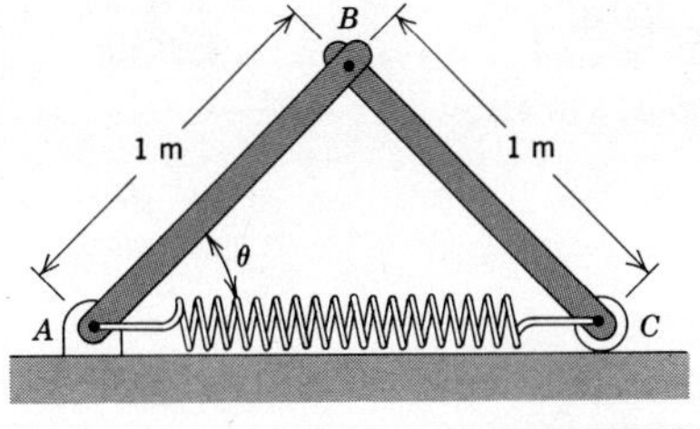

Prob. VIII.36

VIII.34 Solve Problem VIII.9 using the concept of potential energy.

VIII.35 Solve Problem VIII.13 using the concept of potential energy.

VIII.36 Identical bars AB and BC, each having a mass of 10 kg, form a linkage in the vertical plane. The spring has a stiffness of 400 N/m. Knowing that the equilibrium position corresponds to $\theta = 45°$, determine the undeformed length of the spring.

VIII.37 and VIII.38 Determine the relationship between the mass m of the suspended body and the angle θ for equilibrium. The identical springs have stiffness k and are unstressed when $\theta = 30°$.

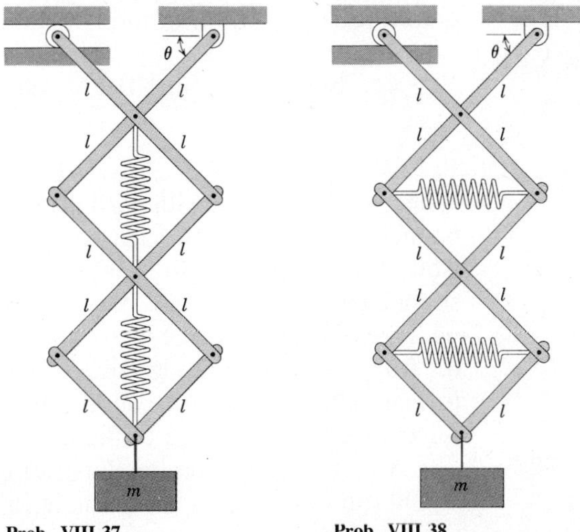

Prob. VIII.37 Prob. VIII.38

VIII.39 Consider the system studied in Problem VIII.16. Explain why the virtual work of the spring in the position shown is more easily formulated by treating the spring force as a nonconservative force, rather than formulating the effect of the spring in terms of potential energy.

VIII.40 Solve Problem VIII.17 using the concept of potential energy.

VIII.41 The spring has a stiffness of 50 lb/in. and an unstretched length of 20 in. Neglecting the mass of the bars, determine the horizontal component of the reaction at pin support D.

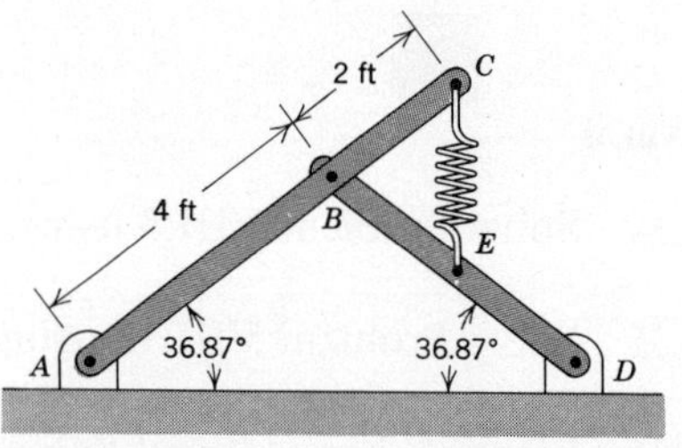

Prob. VIII.41

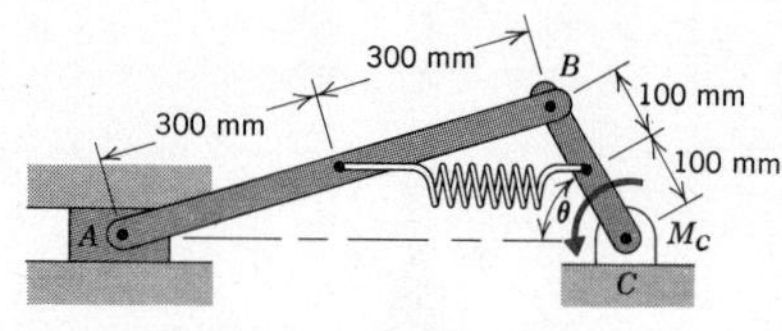

Prob. VIII.42

VIII.42 The spring, whose stiffness is 0.80 kN/m, is unstretched in the position where $\theta = 90°$. Determine the couple $\bar{M}_C$ required for equilibrium in terms of the angle θ.

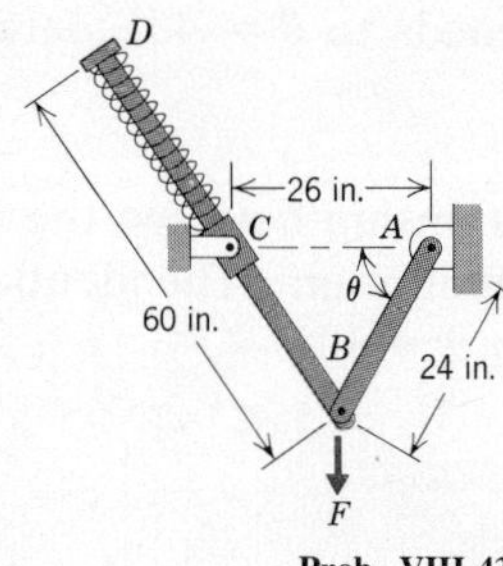

Prob. VIII.43

VIII.43 Collar C, which rides on the smooth bar BD, is pinned to the ground. The spring, whose stiffness is 30 lb/in., is unstressed when $\theta = 22.62°$. Determine an expression for the vertical force $\bar{F}$ at point B required for equilibrium in terms of the angle θ.

VIII.44 Solve Problem VIII.20 using the concept of potential energy.

F. STABILITY OF THE EQUILIBRIUM POSITION

When we are dealing with partially constrained systems, it will frequently happen that several different equilibrium positions are theoretically possible. For example, in the case of the table top of Example 4, which was supported by two parallel bars, we found that equilibrium was possible with the bars in the vertical position, as well as when they were inclined at 60° from the horizontal. In such situations it may not actually be possible to maintain some of the equilibrium positions determined theoretically, for the reason that some of them may be unstable.

A very simple situation illustrating the question of the stability of the equilibrium position is the particle in Figure 17, which is placed on the smooth surface whose vertical profile is shown. At positions A, B, and C

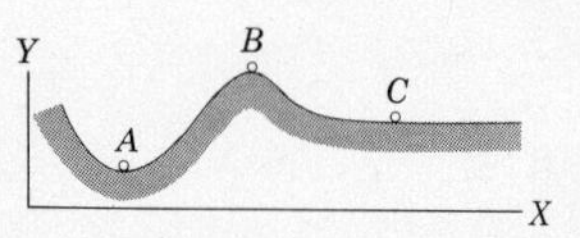

Figure 17

the tangent to the hill is horizontal. It then follows that each of these locations is a theoretical static equilibrium position for the block. Let us consider each of these.

Position A is the lowest point in a valley. After being moved away from this position by a slight disturbance, such as a gust of wind, the particle will return to the low point. This is said to be a *stable* equilibrium position. On the other hand, point B is the top of a hill. There, a small disturbance away from the top causes the particle to fall down the hill. This is an *unstable* equilibrium position. At location C the particle will remain at any point along the level portion of the hill, because there is neither a tendency to move away nor return to location C after a disturbance. This is the condition of *neutral* equilibrium.

Two observations regarding the situation in Figure 17 are significant. First, notice that in each case the only constraint acting on the particle is that of the hill, which exerts a normal force on the particle in order to prevent movement of the particle in the normal direction. In the movement satisfying this constraint, that is, in a displacement of the particle tangent to the hill, the only force doing work is gravity. Thus, the particle on the hill forms a partially constrained system that is conservative with respect to virtual movements satisfying the existing constraints. The second observation regarding the system of Figure 17 is that when the X axis is chosen as the datum for the potential energy, then the potential energy of the system is proportional to the coordinate Y; that is, $V = mgY$. If we now choose the horizontal coordinate X as the generalized coordinate, a graph of the potential energy as a function of the generalized coordinate has the same shape as the vertical profile of the hill shown in Figure 17.

The preceding discussion suggests that the type of stability of a partially constrained conservative system is determined by the properties of its potential energy function. This result can be proven, and the results are analogous to those for the particle on the hill.

Let q denote the generalized coordinate that describes the movement of a partially constrained system satisfying whatever constraints there are. Suppose that Figure 16, which is repeated here as Figure 18, represents

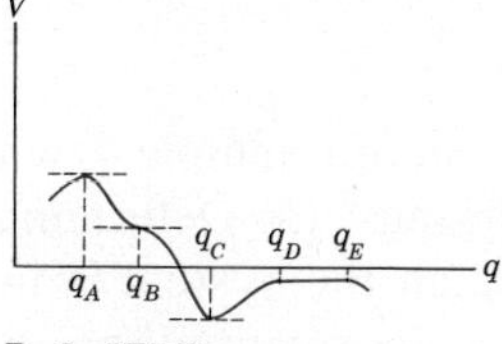

Prob. VII.18

the potential energy of such a system. When $q = q_A$, where the potential energy is a maximum (analogous to a particle at the peak of a hill), there is unstable equilibrium. If the system is given a slight disturbance from this location, the conservative forces will act to move the system farther away from q_A. At location C, where the potential energy is minimum (a particle at the bottom of a valley), the conservative forces act to return the system to this location after a disturbance. This is a stable equilibrium position. Thus we have the following criteria for conservative systems.

$$\left.\frac{dV}{dq}\right|_{q=q_0} = 0 \qquad \left.\frac{d^2V}{dq^2}\right|_{q=q} > 0 \qquad q_0 \text{ represents a stable equilibrium position}$$

$$\left.\frac{dV}{dq}\right|_{q=q_0} = 0 \qquad \left.\frac{d^2V}{dq^2}\right|_{q=q_0} < 0 \qquad q_0 \text{ represents an unstable equilibrium position} \tag{28}$$

In the special case where the second derivative of the potential energy is zero, one must distinguish between the horizontal inflection point at location B in Figure 18 and the level plateau lying between locations D and E. At the former, after a disturbance that increases the value of q slightly from q_B, the conservative forces will cause the system to continue to move away from this location. Even though the system will return to this location after a slight decrease of q from q_B, a horizontal inflection point must be regarded as representing an unstable equilibrium position. In contrast, all values of q lying in the range between q_D and q_E, where the potential energy is constant, represent possible equilibrium positions ($dV/dq \equiv 0$). Therefore, this is a region of neutral stability.

It should be noted that the stability theorems developed here apply only to conservative systems. The stability of partially constrained systems in which nonconservative forces do work is a more complicated question, being concerned with the dynamics of the system. Thus, before applying equations (28) it must be ascertained that all forces doing work are conservative.

EXAMPLE 5

Consider the system of Example 4, where a table top is supported by bars and braced by springs. (a) Determine the range of values of the spring stiffness k for which the system is in stable equilibrium at $\theta = 90°$. (b) Letting k be the value required for equilibrium at $\theta = 60°$, determine the stability of that equilibrium position.

Solution

As we saw in the solution of Example 4, the system is conservative. The derivative $dV/d\theta$ was found to be

$$\frac{dV}{d\theta} = mgl \cos \theta + 1.20kl^2\left(\sin \frac{\theta}{2} - \cos \frac{\theta}{2}\right)$$

This derivative is zero when $\theta = 90°$. It also was made to be zero at $\theta = 60°$ by setting $k = 1.138\ mg/l$, which was the solution to Example 4.

To evaluate the stability we calculate the second derivative of the potential energy. This is

$$\frac{d^2V}{d\theta^2} = -mgl \sin \theta + \left(\frac{1}{2}\right)(1.20)kl^2\left(\cos \frac{\theta}{2} + \sin \frac{\theta}{2}\right)$$

Part (a) is then solved by setting $\theta = 90°$, with the result that

$$\left.\frac{d^2V}{d\theta^2}\right|_{\theta=90°} = -mgl + 0.8485kl^2$$

If k is too small, then $d^2V/d\theta^2$ is negative and the equilibrium is unstable. If the springs are sufficiently stiff (large k), then $d^2V/d\theta^2$ is positive, and the system is stable. The allowable range of values of k for stability is therefore found by setting

$$\left.\frac{d^2V}{d\theta^2}\right|_{\theta=90°} = -mgl + 0.8485kl^2 > 0$$

$$k > 1.179\frac{mg}{l}$$

To solve part (b) we set $\theta = 60°$ and $k = 1.138\ mg/l$ in the general expression for $d^2V/d\theta^2$. This gives

$$\frac{d^2V}{d\theta^2} = [-1 + 0.8485(1.138)]mgl < 0$$

The second derivative is negative, so $\theta = 60°$ is an unstable equilibrium position for the given value of k.

EXAMPLE 6

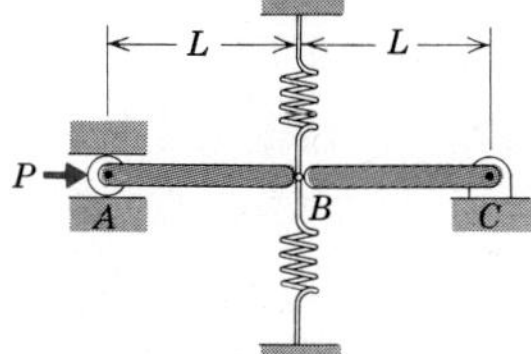

The system of lightweight rigid bars are loaded by a compressive force $\bar{P}$. In the horizontal position shown, the springs of stiffness k have identical tensile forces T. Determine the maximum value of P for which this horizontal equilibrium position is stable.

Solution

The system of bars is only partially constrained, because joint B may move up or down, causing roller A to move to the right. To assist us in

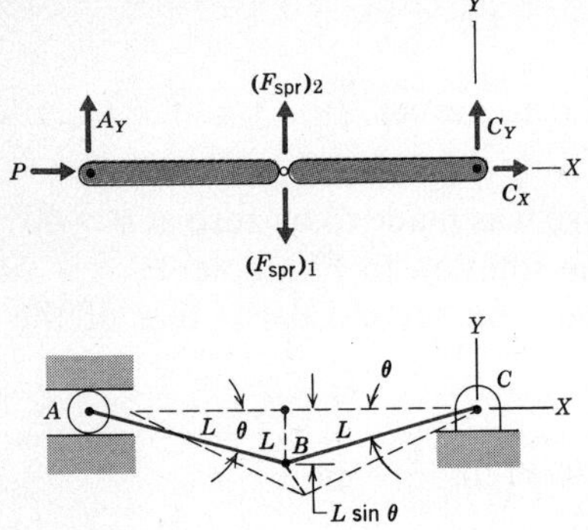

discussing the forces, we draw a free body diagram, and as an aid in locating points in general, we draw a sketch of the system in an arbitrary (in this case, nonhorizontal) position. These are shown to the left.

The angle θ can be seen from the geometric sketch to serve for describing the movement of the system consistent with the constraint conditions imposed by the supports. The equilibrium position of interest corresponds to $\theta = 0$.

In the virtual movement appearing in the geometric sketch, the only forces doing work are those of the springs and $\bar{P}$. The application of the criteria of equations (28) requires that all forces doing work are conservative ones. Assuming the force $\bar{P}$ to be constant in magnitude and direction, we may then regard it as a horizontal "weight" force. Choosing the Y axis (a line perpendicular to $\bar{P}$) as the datum for this force, the point of application A of this force is at a distance $2L \cos\theta$ above the datum, so

$$V_P = P(2L \cos\theta)$$

In order to formulate the potential energy of the springs in the form required for application of the energy principles, we need to relate their extension to the generalized coordinate θ. If we were interested in the equilibrium of the system for large values of θ, these expressions would be somewhat complicated to obtain because it would then be necessary to consider the horizontal and vertical distances locating pin B in the displaced position. However, because our interest here is in the equilibrium position at $\theta = 0$, we only need an expression for potential energy that is accurate for small values of θ.

When θ is a very small angle, pin B in the displaced position may be regarded as being a distance $L \sin\theta$ vertically below its position when $\theta = 0$. Further, for small values of θ, we may approximate $\sin\theta \approx \theta$, $\cos\theta = 1 - \theta^2/2$. Thus, the upper spring is elongated by an additional amount $L \sin\theta \approx L\theta$ in the displaced position, whereas the lower spring is shortened by the same amount. From the given information, we know that in the position $\theta = 0$ each spring has a tension T and thus an elongation T/k. Combining this elongation with the deformation introduced by a nonzero value of θ, we find that

$$V_{spr} = \frac{1}{2}k\left(\frac{T}{k} + L\theta\right)^2 + \frac{1}{2}k\left(\frac{T}{k} - L\theta\right)^2$$
$$= \frac{T^2}{k} + kL^2\theta$$

As an aside, notice that the only effect of the pretensioning T given the springs is to change the reference level of the potential energy.

We now add V_{spr} and V_P and compute the required derivatives of the total potential energy V. This gives

$$V = 2PL\left(1 - \frac{\theta^2}{2}\right) + \frac{T^2}{k} + kL^2\theta^2$$

$$\frac{dV}{d\theta} = -2PL\theta + 2kL^2\theta$$

$$\frac{d^2V}{d\theta^2} = -2PL + 2kL^2$$

The expression for $dV/d\theta$ is identically zero at $\theta = 0$, confirming that this is an equilibrium position. The requirement for stability then reveals that

$$\frac{d^2V}{d\theta^2}\bigg|_{\theta=0} \equiv -2PL + kL^2 > 0$$

$$P < \tfrac{1}{2}\, kL$$

Any value of P exceeding this limit will cause the system to seek an equilibrium position where θ is nonzero. We cannot determine this position on the basis of the foregoing analysis, because we have limited the analysis to very small values of θ.

It is interesting to note that many aspects of the analysis are similar to those of the much more complicated problem of the buckling of structures due to compressive loads. Buckling is treated in courses on strength of materials and structures.

HOMEWORK PROBLEMS

VIII.45 Blocks A and B of mass m, resting against the smooth wall and floor, are connected by a bar of mass $2m$. The spring attached to block A has a stiffness k and is unstressed when the bar is vertical. (a) Determine the equilibrium positions of the bar. (b) Determine the range of values of k for which the equilibrium positions found in part (a) are stable.

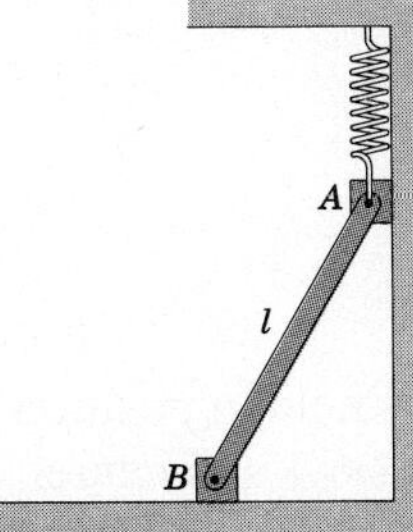

Prob. VIII.45

VIII.46 Determine the type of stability of the equilibrium position $\theta = 45°$ for the linkage of Problem VIII.36.

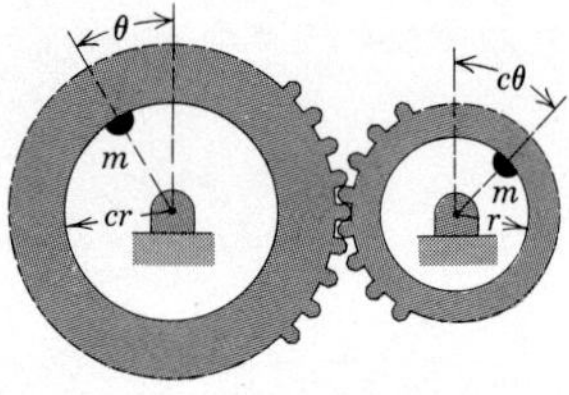

Prob. VIII.47

VIII.47 The two gears lying in the vertical plane carry identical eccentric masses m at radial distances that are in the same ratio c as the ratio of the radii of the gears. (a) Determine an expression for the angle θ at the equilibrium positions. (b) For the case $c = 1$, solve the equation obtained in part (a) and evaluate the stability of each possible equilibrium position. (c) Repeat part (b) for $c = 2$.

VIII.48 In each case illustrated, a bar of mass m is supported by two identical bars of negligible mass. For each system determine the type of stability of the equilibrium position shown.

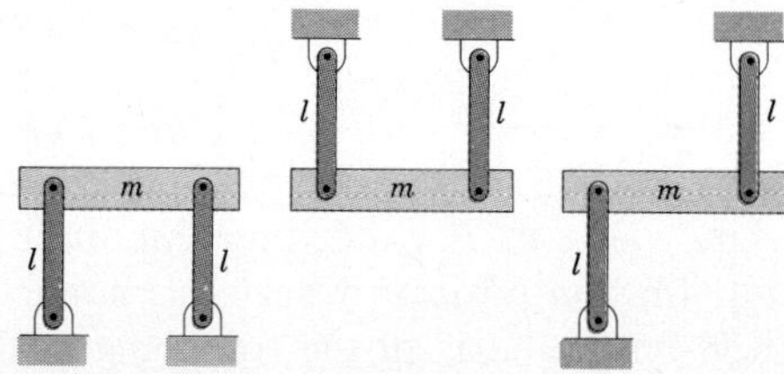

Prob. VIII.48

VIII.49 Each of the bars in the mechanism shown has a mass per unit length m. Bar BD is pinned to collar C which is supported by the spring of stiffness k and unstretched length $3l$. Determine the values of the angle θ for equilibrium and evaluate the stability of each position.

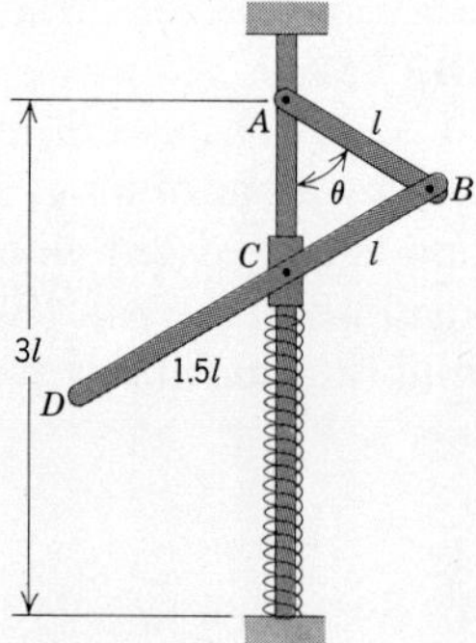

Prob. VIII.49

VIII.50 The lamp is held in position by the mechanism shown. The bulb and reflector has mass m and center of mass G, whereas the mass of the bars is negligible. The spring is unstressed in the condition corresponding to $\theta = 180°$. Determine the stiffness k required for equilibrium at $\theta = 120°$ and evaluate the stability of this equilibrium position.

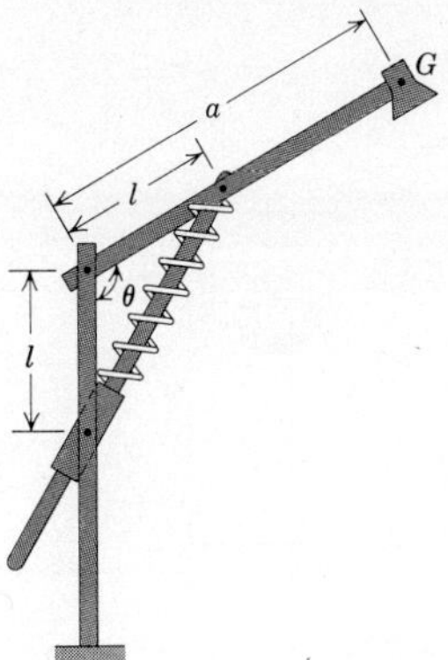

Prob. VIII.50

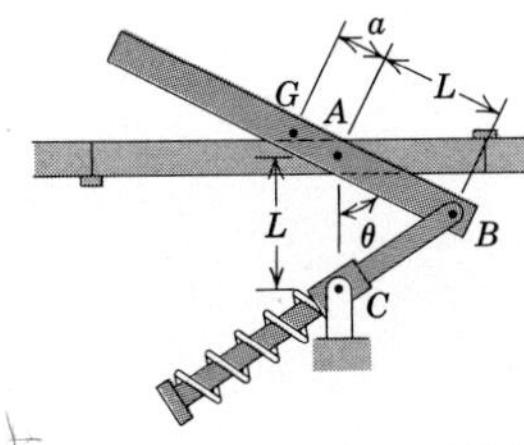

Prob. VIII.51

VIII.51 A ventilating door in a roof, having mass m and center of mass G, is hinged at point A. The angle of inclination θ, measured from the vertical, is controlled by the mechanism shown. The spring is unstressed in the condition corrresponding to $\theta = 0°$. Determine the stiffness k of the spring required for equilibrium at a specific angle θ and determine the stability of the corresponding equilibrium position.

VIII.52 Evaluate the stability of the equilibrium position of the mechanism of Problem VIII.43.

VIII.53 The mechanism shown is a component of a seismometer. When the block is directly over the pivot point B, the springs of stiffness k have equal compression forces T. Considering the mass of all elements except the block to be negligible, determine the range of values of k for which the vertical position is stable.

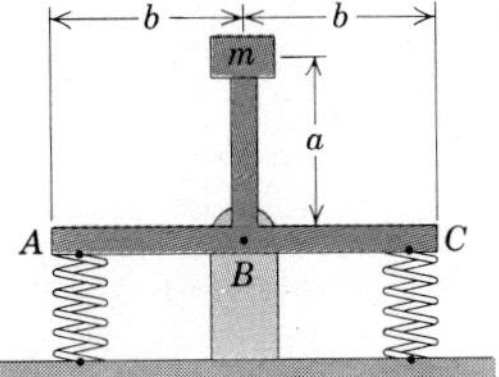

Prob. VIII.53

VIII.54 and VIII.55 A horizontal force $\bar{P}$ is applied at end A of the mechanism shown. Derive an expression for the maximum force $\bar{P}$ for which the mechanism is stable in the position shown. The springs have a stiffness k and carry a compressive force T in the position shown.

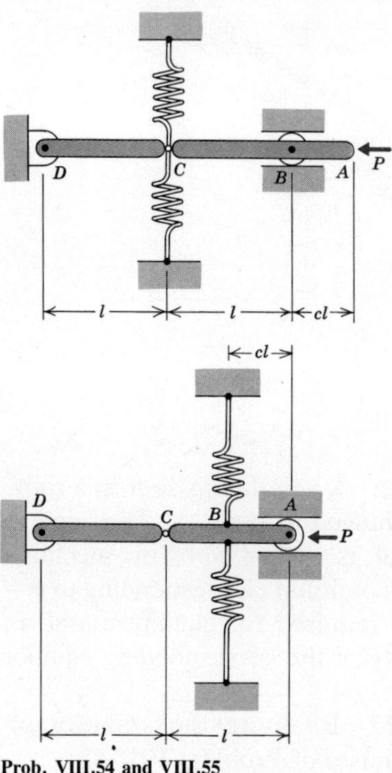

Prob. VIII.54 and VIII.55

MODULE IX
AREA MOMENTS AND PRODUCTS OF INERTIA

We studied two important properties of a planar figure in Module V, specifically, the area and the location of the centroid. However, neither of these properties conveys much information about the geometry of the shape. In this module we will study numerical parameters, alternatively called *second moments of area* or *area moments and products of inertia*, which are related to the shape of a planar figure.

There are many applications for the inertia parameters in the field of engineering. It was shown in Module V that the effect of a hydrostatic pressure distribution on a flat surface can be described in terms of a moment and a product of inertia. Another subject where these parameters occur is in mechanics of materials, where they are needed to describe the stress distribution in beams resulting from transverse and torsional loads. Also, they have certain similarities to parameters which occur in dynamics, called mass moments of inertia.

A. BASIC DEFINITIONS

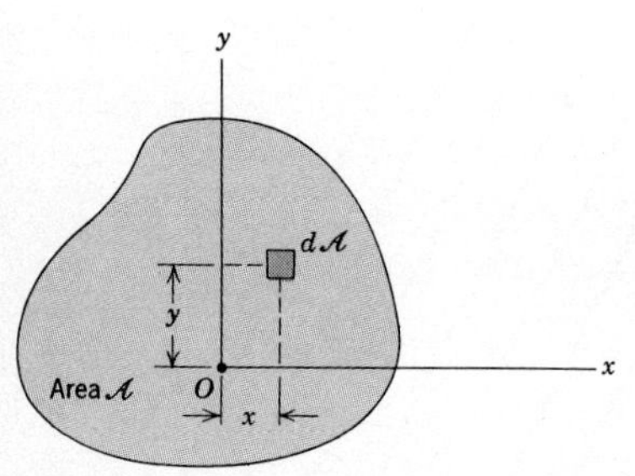

Figure 1

Consider the area $\mathcal{A}$ shown in Figure 1. To locate points on this area we shall employ an xyz coordinate system whose xy plane coincides with the plane of the area. The origin O of this coordinate system is located at an arbitrary point. The area $\mathcal{A}$ is composed of infinitesimal elements $d\mathcal{A}$ having coordinates $(x, y, 0)$. Suppose that we want to define a parameter that conveys information about the "average" distance of these elements from the x axis. One possible parameter is $y\,d\mathcal{A}$, which is the integrand in the first moment of area with respect to the y coordinate. However, recall that the first moment of area provides information about the location of the centroid. Therefore, $y\,d\mathcal{A}$ will not provide the information we seek here.

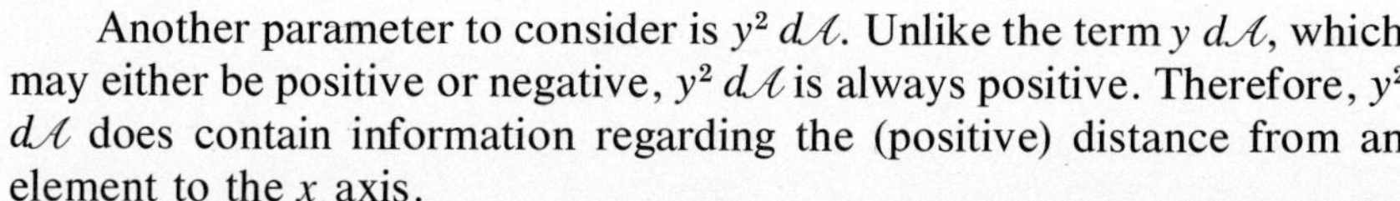

Another parameter to consider is $y^2\,d\mathcal{A}$. Unlike the term $y\,d\mathcal{A}$, which may either be positive or negative, $y^2\,d\mathcal{A}$ is always positive. Therefore, $y^2\,d\mathcal{A}$ does contain information regarding the (positive) distance from an element to the x axis.

On the basis of these observations we may define the *area moment of inertia* (or alternatively, the *second moment of area*) about the x axis which we shall denote by the symbol I_x. Clearly, similar reasoning when discussing the distribution of area about the y axis will lead us to define the area moment of inertia about the y axis, I_y. The sum of an infinite number of infinitesimal terms with respect to $d\mathcal{A}$ is an integral, so the definitions are

$$\boxed{\begin{aligned} I_x &= \int_{\mathcal{A}} y^2\,d\mathcal{A} \\ I_y &= \int_{\mathcal{A}} x^2\,d\mathcal{A} \end{aligned}} \tag{1}$$

Observe that equations (1) are functions of the coordinate values squared, hence the phrase, *second moments*. The phrase *moment of inertia* about the x axis is a result of the similarity between the properties of an area as defined above and the properties of the mass distribution of a rigid body. Thus, although the word *area* is frequently omitted when referring to the moments of inertia, it is acceptable to do so only when there is no possibility of confusion with the mass properties of a rigid body.

The values of I_x and I_y are always positive, because their integrands are always positive or zero. It follows, then, that if two areas have the same shape (for instance, circular, or rectangular in a fixed proportion), the larger area will have the larger values for the moments of inertia. Also, for two different shapes having the same area, the one whose area is distributed farther from an axis will have the larger value for the moment of inertia about that axis.

The parameters we have just defined convey information about the distribution of area about the x and y axes. A third moment of inertia is concerned with the distribution of area about the z axis. Referring to Figure 1, we see that the z axis is perpendicular to the plane of the area, intersecting the origin O. Hence, the distance from this axis to an area element $d\mathcal{A}$ is $(x^2 + y^2)^{1/2}$. Squaring this distance and multiplying it by $d\mathcal{A}$, we form the integral for the (area) *polar moment of inertia* J_z.

$$\boxed{J_z = \int_{\mathcal{A}} (x^2 + y^2)\, d\mathcal{A}} \tag{2}$$

The reason for using the symbol J_z rather than I_z for the polar moment of inertia is that it is conventional to do so.

The polar moment of inertia is not an independent quantity. To see this, it is only necessary to express equation (2) as a sum of integrals rather than an integral of a sum. Employing equations (1) then leads to the identity

$$\boxed{J_z = \int_{\mathcal{A}} x^2\, d\mathcal{A} + \int_{\mathcal{A}} y^2\, d\mathcal{A} = I_y + I_x} \tag{3}$$

A common way to describe the moments of inertia of a shape about an axis is to give the *radius of gyration* k about that axis. This quantity is simply the square root of the ratio of the corresponding moment of inertia to the area $\mathcal{A}$. That is,

$$\boxed{k_x = \sqrt{\frac{I_x}{\mathcal{A}}} \qquad k_y = \sqrt{\frac{I_y}{\mathcal{A}}} \qquad k_z = \sqrt{\frac{I_z}{\mathcal{A}}}} \tag{4}$$

A radius of gyration has units of length, because the area moments of inertia have units of the fourth power of length. The radii of gyration have a simple interpretation. Suppose that we have a thin rectangular strip whose area $\mathcal{A}$ is the same as that for the general shape of interest. Let us orient this strip parallel to the x axis at a distance equal to the value of k_x, as shown in Figure 2a. Because all of the elements of area are then

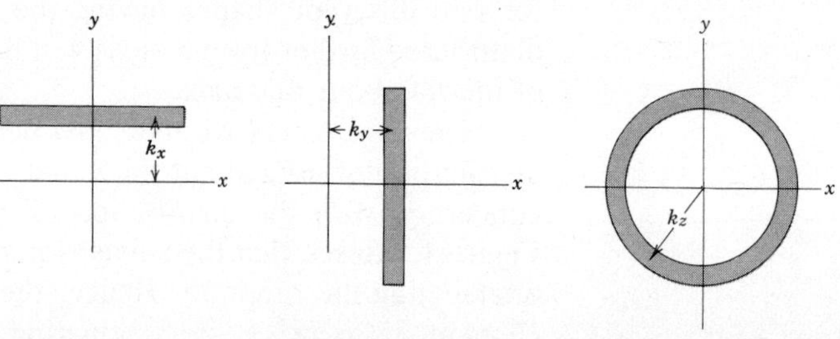

Figure 2a **Figure 2b** **Figure 2c**

essentially at the distance k_x from the x axis, the moment of inertia of this strip about the x axis is $k_x^2\mathcal{A} \equiv I_x$. Similarly, the moment of inertia about the y axis of the strip in Figure 2b is identical to I_y for the general shape, and the polar moment of inertia about the z axis for the thin ring in Figure 2c is identical to J_z for the general shape. In view of equation (3), the radii of gyration are related by

$$k_z^2 = k_x^2 + k_y^2 \tag{5}$$

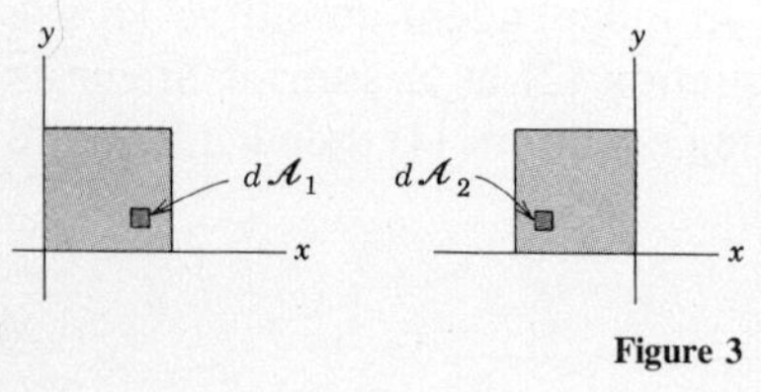

Figure 3

Having seen how the moments of inertia characterize the distribution of area about the axes, it may seem that these are the only parameters we need. This is not so, as exhibited by the two squares shown in Figure 3. For every element of area $d\mathcal{A}_1$ in the square on the left, there is a companion element $d\mathcal{A}_2$ in the right square. Both of these elements are situated at the same distances from each of the coordinate axes, so both squares have identical moments of inertia. Yet clearly, these squares differ, in the sense that they are located in different coordinate quadrants.

To describe how the area of a planar figure is situated in the coordinate quadrants, we define the area *product of inertia* as follows:

$$\boxed{I_{xy} \equiv I_{yx} = \int_{\mathcal{A}} xy \, d\mathcal{A}} \tag{6}$$

Because xy is positive if an element of area is located in the first or third quadrant, and negative in the second or fourth quadrant, we conclude that I_{xy} being positive means that area predominates in the first and/or third quadrants. Obviously, when I_{xy} is negative, the area predominates in the second and/or fourth quadrants.

In the important case where a shape is symmetric about one of the coordinate axes, the value of xy for an element to the left of the axis of symmetry is cancelled by the value of xy for the mirror-image element to the right. This is demonstrated in Figure 4 for a shape that is symmetric about the x axis. Thus

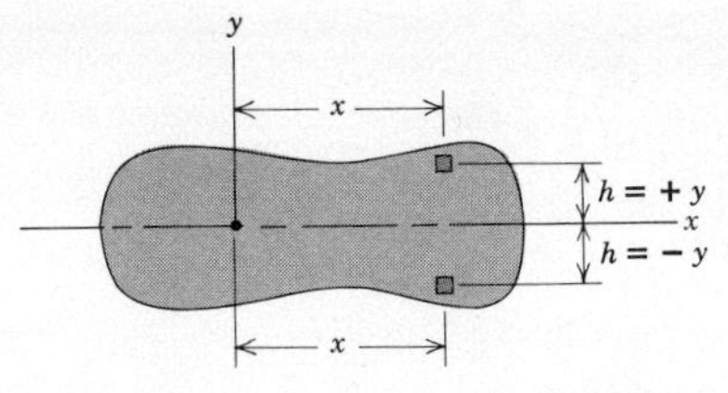

Figure 4

> Whenever a planar shape has an axis of symmetry that is either the x or y axis, then $I_{xy} \equiv 0$.

In general, when the x and y coordinate axes give a zero product of inertia, we say that they are *principal axes*. All shapes, not just symmetric ones, have principal axes. Later in this module we will see how to locate the principal axes for an arbitrary figure.

For the remainder of this module, we will consider various techniques for evaluating the area moments and products of inertia. We could go on to define higher moments of area. However, these would only be of theoretical interest, as they do not find application in the field of mechanics.

B. INTEGRATION TECHNIQUES

In view of the fact that moments and product of inertia are defined in terms of integrals, the obvious way to evaluate these properties is to evaluate the integrals. Doing this requires that we describe the position of points contained within the closed curve forming the area $\mathcal{A}$.

Recall that in Module V, when we located centroids and centers of mass by integration, we formulated the necessary integrals in terms of either Cartesian or cylindrical coordinates, whichever best suited the description of the boundaries of the body. The same is true here. However, in describing a planar area, we are concerned only with the Cartesian coordinates (x, y) or the polar coordinates (R, ϕ) locating points in the plane. A possible choice for the polar coordinates is shown in Figure 5.

For the coordinates defined in the figure, we have

$$x = R \cos \phi \qquad y = R \sin \phi$$

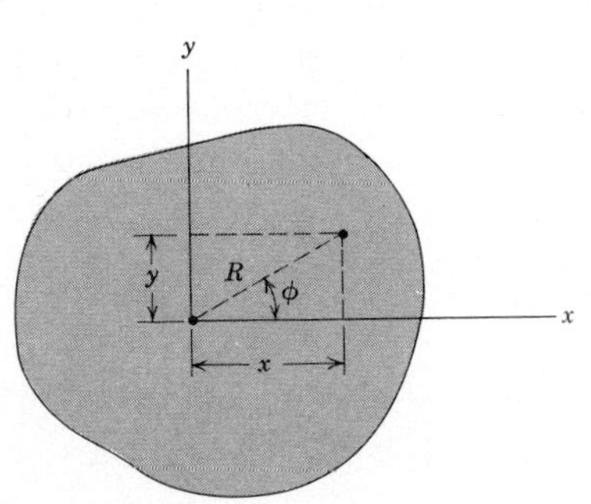

Figure 5

It should be noted that other choices for the polar coordinates are possible, for instance, by measuring the angle ϕ from a different axis.

After having made a decision to use Cartesian or polar coordinates on the basis of the shape of the curve forming the boundary of the area, we may describe the area $d\mathcal{A}$. The expressions for $d\mathcal{A}$ were derived in Module V, but for completeness we have illustrated the two basic types of elements again in Figure 6 below.

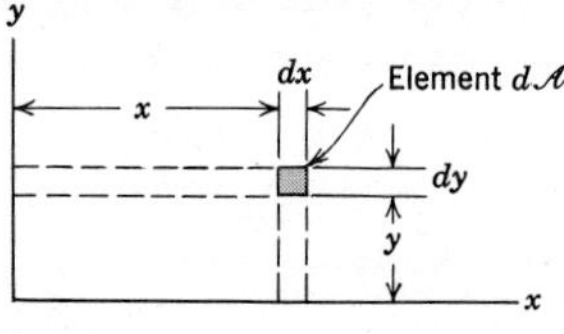

Figure 6

From Figure 6 we see that

$$d\mathcal{A} = dx\,dy \qquad \text{or} \qquad d\mathcal{A} = R\,d\phi\,dR \tag{7}$$

The following examples will demonstrate that the methods for formulating the limits of an integral extending over all of the area are the same as those we employed in locating centers of mass by integration.

EXAMPLE 1

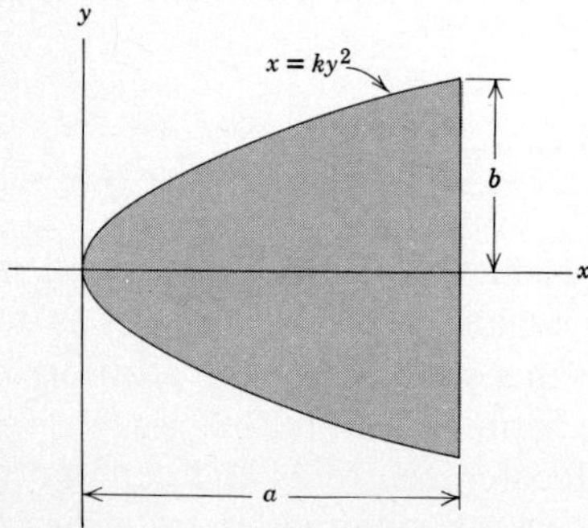

Calculate the moments and product of inertia of the parabolic sector shown about the axes of the given coordinate system.

Solution

The parabolic curve is defined in terms of the x and y coordinates, so we will formulate the solution in terms of these coordinates. The constant k is related to the given dimensions because the points (a, b) and $(a, -b)$ are points on the parabola. Thus, we have $a = kb^2$, so $k = a/b^2$. For the formulation of the various integrals we will use the element of area shown in the diagram below.

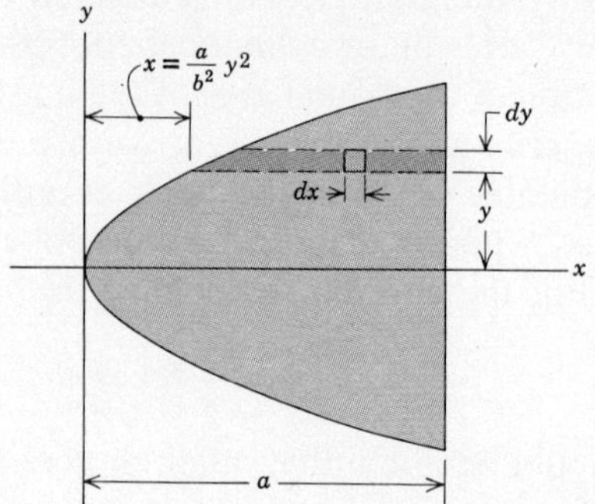

An integration over the entire area may be achieved by integrating first over the range of all x variables or all y variables, thus forming horizontal or vertical strips, respectively. As indicated in the preceding sketch, the equation for the parabola gives the value of x on the parabola for each value of y. Therefore, we will form horizontal strips by integrating x over the range $ay^2/b^2 \leq x \leq a$. All horizontal strips are then accounted for by integrating y over the range $-b \leq y \leq b$.

The parabolic sector is symmetrical about the x axis, so we have

$$I_{xy} = 0$$

Also, because the polar moment of inertia J_z may be obtained from equation (3), it is only necessary for us to evaluate the integrals in equation (1). These are

$$I_x = \int_{-b}^{b} \int_{ay^2/b^2}^{a} y^2\, dx\, dy = \int_{-b}^{b} y^2\left(a - \frac{ay^2}{b^2}\right) dy$$

$$= \frac{4}{15} ab^3$$

$$I_y = \int_{-b}^{b} \int_{ay^2/b^2}^{a} x^2\, dx\, dy = \tfrac{1}{3}\int_{-b}^{b} \left(a^3 - \frac{a^3y^6}{b^6}\right) dy$$

$$= \frac{4}{7} a^3 b$$

From equation (3) we then find

$$J_z = I_x + I_y = \frac{4}{105} ab(7b^2 + 15a^2)$$

These results demonstrate a typical property of moments of inertia. Specifically, each value increases in proportion to the cube of the dimension perpendicular to the corresponding axis.

EXAMPLE 2

Calculate the moments and product of inertia of the circular sector shown with respect to the xyz coordinate system. Also, determine the corresponding radii of gyration.

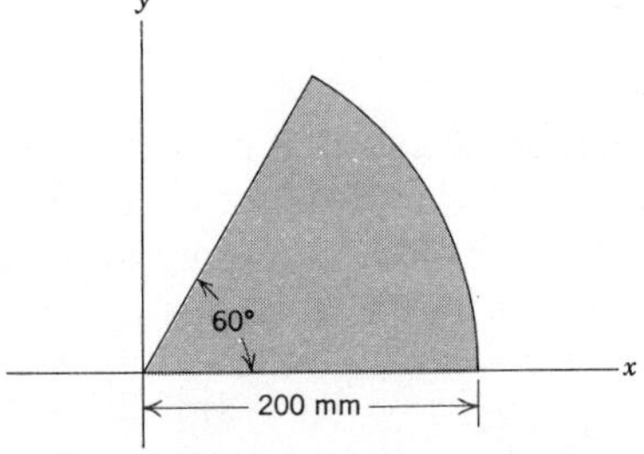

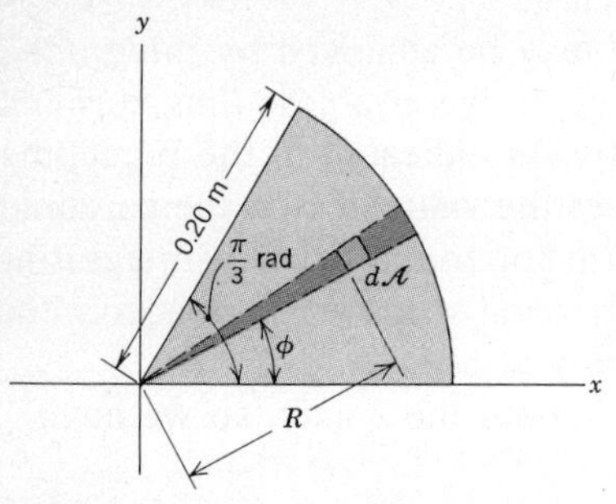

Solution

We could formulate the solution using (x, y) coordinates by writing the equation for the circular arc in terms of x and y. However, it is easier to employ polar coordinates in this problem, because points on the arc are at a constant radial distance from the origin, and equally important, points on the straight sides are at constant polar angles ϕ from the coordinate axes. Our choice for the polar coordinates is shown in the sketch.

For the coordinates defined in the sketch we have

$$x = R\cos\phi \qquad y = R\sin\phi$$

and the element of area is $d\mathcal{A} = R\,d\phi\,dR$. As indicated in the sketch, the sequence of integrations we choose will form a differential sector by integrating over $0 \le R \le 0.20$ m. All of these sectors will then be accounted for by integrating over $0 \le \phi \le \pi/3$ radians.

We rearrange the differentials in the expression for $d\mathcal{A}$ to match the order of integration, so the moments of inertia are

$$I_x = \int_0^{\pi/3}\int_0^{0.2} y^2 R\,dR\,d\phi = \int_0^{\pi/3}\int_0^{0.2} R^3 \sin^2\phi\,dR\,d\phi$$

$$= \tfrac{1}{4}(0.2)^4 \int_0^{\pi/3} \sin^2\phi\,d\phi = 1.228(10^{-4})\text{ m}^4$$

$$I_y = \int_0^{\pi/3}\int_0^{0.2} x^2 R\,dR\,d\phi = \int_0^{\pi/3}\int_0^{0.2} R^3 \cos^2\phi\,dR\,d\phi$$

$$= \tfrac{1}{4}(0.2)^4 \int_0^{\pi/3} \cos^2\phi\,d\phi = 2.960(10^{-4})\text{ m}^4$$

$$I_{xy} = \int_0^{\pi/3}\int_0^{0.2} xy\,R\,dR\,d\phi = \int_0^{\pi/3}\int_0^{0.2} R^3 \cos\phi\sin\phi\,dR\,d\phi$$

$$= \tfrac{1}{4}(0.2)^4 \int_0^{\pi/3} \cos\phi\sin\phi\,d\phi = 1.50(10^{-4})\text{ m}^4$$

Then, from equation (3), we find

$$J_z = I_x + I_y = 4.188(10^{-4})\text{ m}^4$$

The radii of gyration may now be determined using their definitions, equations (4). For this we need the area $\mathcal{A}$. In this problem it is not necessary to evaluate an integral to determine $\mathcal{A}$, because the sector is one-sixth of a full circle. Thus

$$\mathcal{A} = \tfrac{1}{6}\pi(0.2)^2 = 2.094(10^{-2})\text{ m}^2$$

Hence, for the radius of gyration about the x axis we have

$$k_x = \sqrt{\frac{1.228(10^{-4})}{2.094(10^{-2})}} = 0.0766 \text{ m}$$

Similarly,

$$k_y = \sqrt{\frac{2.960(10^{-4})}{2.094(10^{-2})}} = 0.1189 \text{ m}$$

$$k_z = \sqrt{\frac{4.188(10^{-4})}{2.094(10^{-2})}} = 0.1414 \text{ m}$$

HOMEWORK PROBLEMS

IX.1–IX.6 By integration, determine the area moments of inertia I_x and I_y for the shaded area shown.

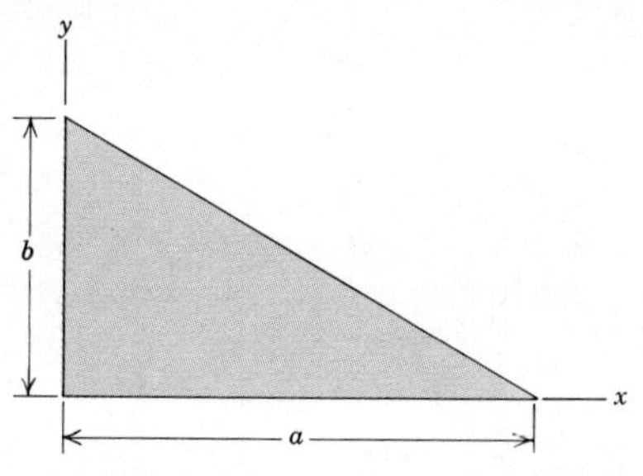

Probs. IX.1 and IX.7

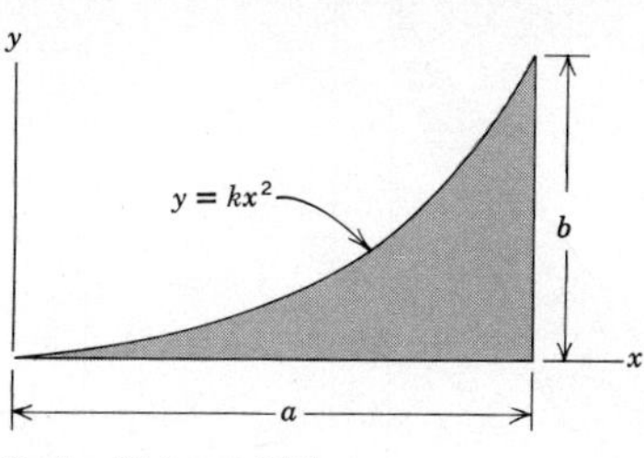

Probs. IX.2 and IX.8

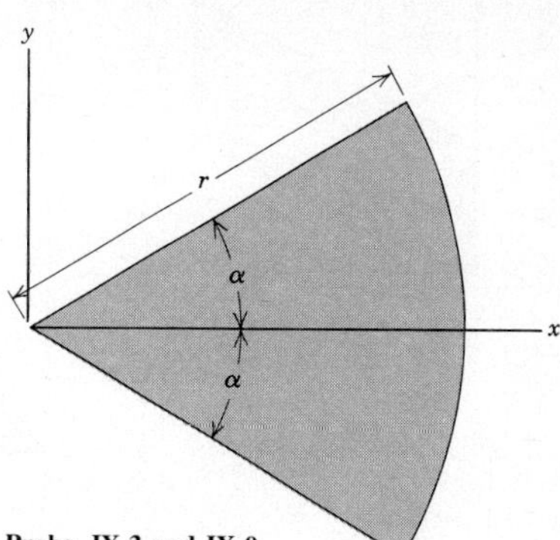

Probs. IX.3 and IX.9

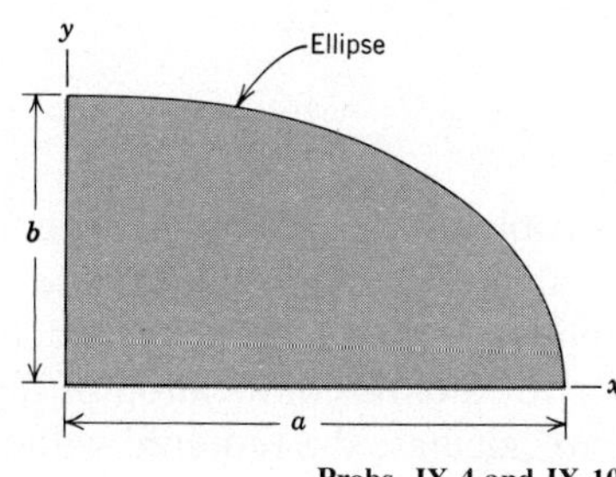

Probs. IX.4 and IX.10

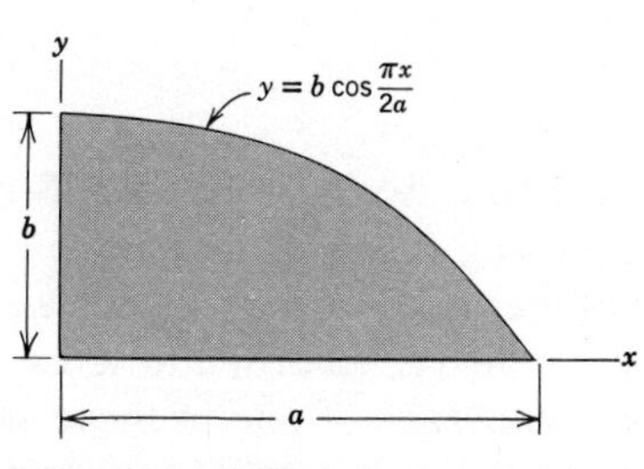

Probs. IX.5 and IX.11

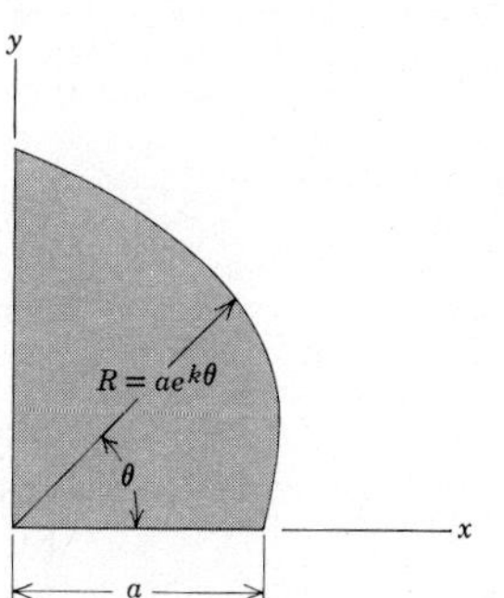

Probs. IX.6 and IX.12

IX.7–IX.12 By integration, determine the polar area moment of inertia J_z and the product of inertia I_{xy} for the shaded area shown.

IX.13–IX.18 Determine the radius of gyration k_x of the shaded area shown.

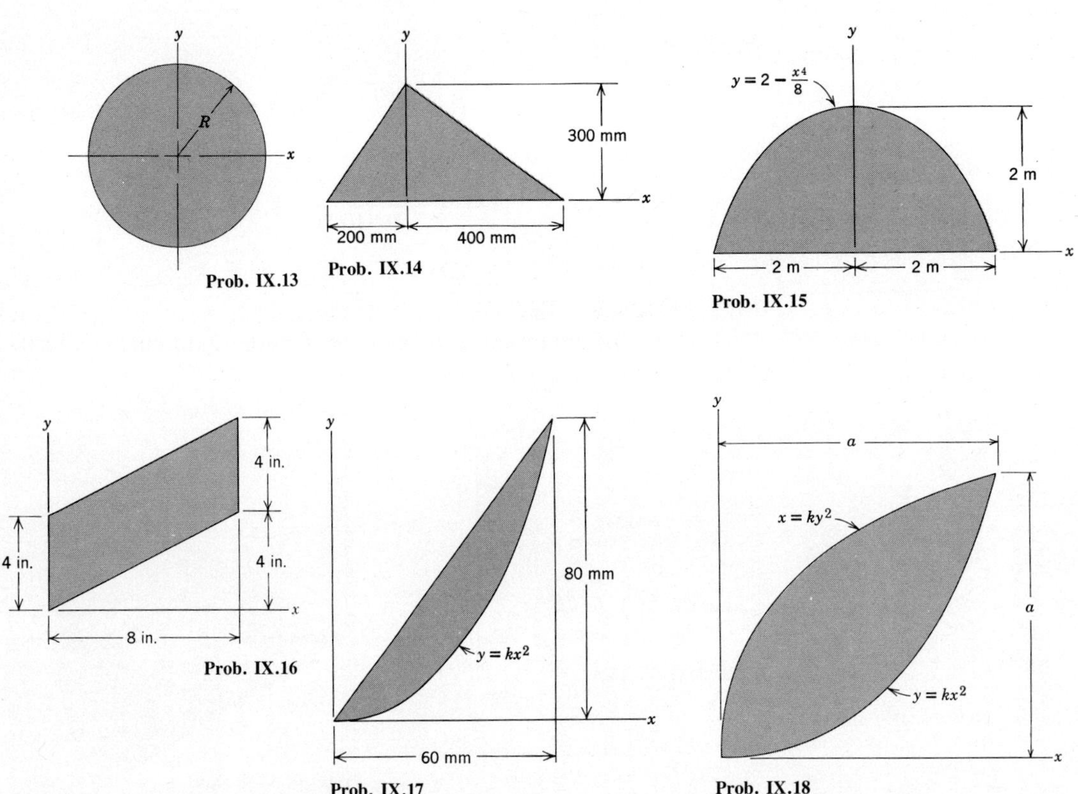

C. PARALLEL AXIS THEOREMS

In the preceding examples and homework problems the area moments and product of inertia of various shapes were evaluated for a specific choice of location of the xyz coordinate axes. Suppose that we now wanted to know the properties of one of these shapes for a different location of the coordinate system. It would be preferable to calculate the required values by means of algebraic formulas, rather than by evaluating a new set of integrals. In this section we shall study the relationships between the moments and products of inertia for two mutually parallel coordinate systems.

Consider the two coordinate systems xyz and $\hat{x}\hat{y}\hat{z}$ used to describe the

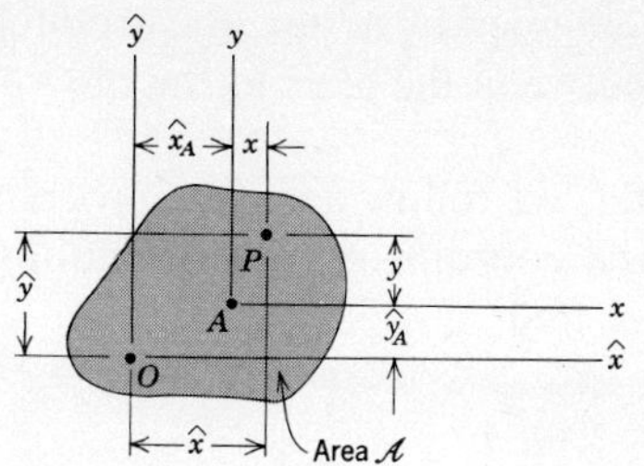

Figure 7

arbitrary shape in Figure 7. As shown in the figure, the coordinates of the origin A of the xyz coordinate system with respect to the $\hat{x}\hat{y}\hat{z}$ system are $(\hat{x}_A, \hat{y}_A, 0)$. Note that the distance between the x and $\hat{x}$ axis is $|\hat{y}_A|$ and the distance between the y and $\hat{y}$ axes is $|\hat{x}_A|$, and that the signs for these quantities are chosen in accord with the quadrant of the $\hat{x}\hat{y}\hat{z}$ system in which the origin A is located. From Figure 7, we see that the coordinates of an arbitrary point P are related by

$$x = \hat{x} - \hat{x}_A \qquad y = \hat{y} - \hat{y}_A \tag{8}$$

Let us first focus attention on determining I_x. From its definition and equations (8), we find that

$$\begin{aligned} I_x &= \int_{\mathcal{A}} y^2\, d\mathcal{A} = \int_{\mathcal{A}} (\hat{y} - \hat{y}_A)^2\, d\,\mathcal{A} \\ &= \int_{\mathcal{A}} \hat{y}^2\, d\mathcal{A} - 2\hat{y}_A \int_{\mathcal{A}} \hat{y}\, d\mathcal{A} + (\hat{y}_A)^2 \int_{\mathcal{A}} d\mathcal{A} \end{aligned}$$

The last of the integrals on the right side of the preceding equation is readily identified to give the area $\mathcal{A}$, and the first integral is seen to be the definition of $I_{\hat{x}}$. Recall that the integral in the second term is the first moment of area with respect to the $\hat{y}$ coordinate. Hence, *by choosing the origin O to be the centroid of the area, we can make this integral vanish.*

It is not difficult to show that similar steps for I_y and J_z yield the following set of *parallel axis theorems* for converting the moments of inertia for one coordinate system into those of the other.

$$\boxed{\begin{aligned} I_x &= I_{\hat{x}} + \mathcal{A}\,(\hat{x}_A)^2 \\ I_y &= I_{\hat{y}} + \mathcal{A}\,(\hat{y}_A)^2 \\ J_z &= J_{\hat{z}} + \mathcal{A}\,[(\hat{x}_A)^2 + (\hat{y}_A)^2] \end{aligned}} \tag{9}$$

Two important facts to bear in mind when applying equations (9) are that they are valid only if the origin of the $\hat{x}\hat{y}\hat{z}$ coordinate system coincides with the centroid of the area, and that $(\hat{x}_A, \hat{y}_A)$ are the $\hat{x}$, $\hat{y}$ coordinates of the origin A of the xyz system.

The interpretation of equations (9) are fairly direct. As we noted earlier, $|\hat{y}_A|$ and $|\hat{x}_A|$ represent the distances between the parallel pairs of x and y axes, respectively. Similarly, it may be seen by applying the Pythagorean theorem to Figure 7 that $[(\hat{x}_A)^2 + (\hat{y}_A)^2]^{1/2}$ is the distance between the parallel pair of z axes. Thus, as we transfer from an axis passing through the centroid of the area to another parallel axis, the moment of inertia about this new axis is obtained by adding the product of the area and the square of the distance between the axes to the old value of the moment of inertia.

With respect to the product of inertia, we follow a similar procedure to the one just presented. Substituting the coordinate transformations of equation (8) into the definition of I_{xy}, we find

$$I_{xy} = \int_{\mathcal{A}} xy\, d\mathcal{A} = \int_{\mathcal{A}} (\hat{x} - \hat{x}_A)(\hat{y} - \hat{y}_A)\, d\mathcal{A}$$

$$= \int_{\mathcal{A}} \hat{x}\hat{y}\, d\mathcal{A} - \hat{x}_A \int_{\mathcal{A}} \hat{y}\, d\mathcal{A} - \hat{y}_A \int_{\mathcal{A}} \hat{x}\, d\mathcal{A} + \hat{x}_A\hat{y}_A \int_{\mathcal{A}} d\mathcal{A}$$

The first integral is the definition of $I_{\hat{x}\hat{y}}$, whereas the second and third integrals are zero because the $\hat{x}$ and $\hat{y}$ axes pass through the centroid of the area. Hence, the parallel axis theorem for the product of inertia is

$$\boxed{I_{xy} = I_{\hat{x}\hat{y}} + \mathcal{A}\,\hat{x}_A\,\hat{y}_A} \tag{10}$$

Notice that in equations (9) the sign of $\hat{x}_A$ and $\hat{y}_A$ are not significant, because those parallel axis theorems depend only on the distance between corresponding axes. However, particular attention must be paid to the sign of $\hat{x}_A$ and $\hat{y}_A$ when applying equation (10). Again we remind you that $(\hat{x}_A, \hat{y}_A, 0)$ *are the coordinates of the origin A of the xyz system with respect to the $\hat{x}\hat{y}\hat{z}$ system whose origin is the centroid of the area.*

EXAMPLE 3

The area moments and product of inertia of the parabolic sector with respect to the *xyz* coordinate system were determined in Example 1. Determine these parameters for the $x'y'z'$ system shown.

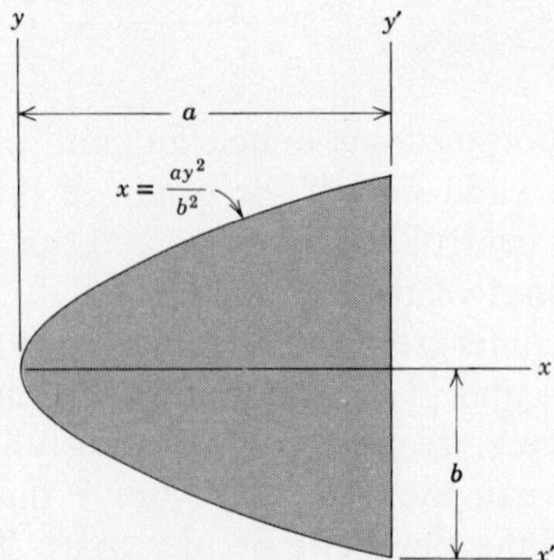

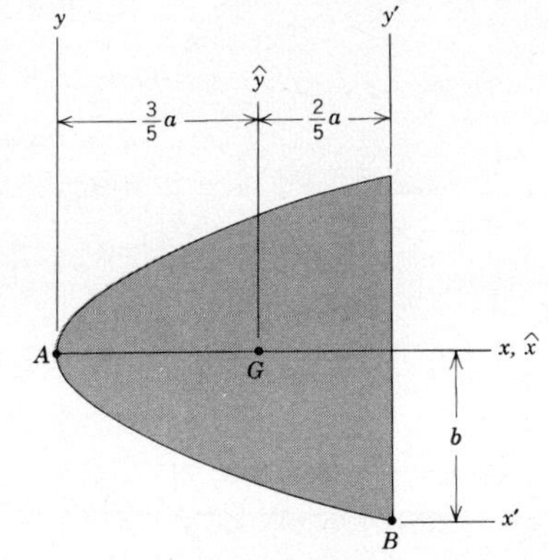

Solution

A common error is solving problems such as this is to try to transfer directly between the xyz and $x'y'z'$ coordinate systems, thereby applying the parallel axis theorems only once. This procedure is wrong because it overlooks the fact that the parallel axis theorems as we derived them are valid only when transforming between a centroidal and a noncentroidal coordinate system. Therefore, to solve the problem we must first locate the centroid of the parabolic sector. We will also need its area.

The properties of a semiparabolic section, such as that on either side of the x axis, are tabulated in Appendix B. In view of the symmetry of a parabola, we determine the centroid G to be situated as shown to the left. The area $\mathcal{A}$ is also given in Appendix B, so

$$\mathcal{A} = \tfrac{4}{3}ab$$

As noted earlier, we need a parallel $\hat{x}\hat{y}\hat{z}$ coordinate system whose origin is at the centroid. This is shown in the sketch, where the origins of the given coordinate systems have been labeled as points A and B. With respect to the $\hat{x}\hat{y}\hat{z}$ coordinate system, the coordinates of point A are $(-3a/5, 0, 0)$ and those of point B are $(2a/5, -b, 0)$.

We next determine the values of $I_{\hat{x}}$, $I_{\hat{y}}$, and $I_{\hat{x}\hat{y}}$ by substituting into the parallel axis theorems. Using the results of Example 1, this gives

$$I_{\hat{x}} = I_x - \mathcal{A}(\hat{y}_A)^2 = \tfrac{4}{15}ab^3 - (\tfrac{4}{3}ab)(0) = \tfrac{4}{15}ab^3$$

$$I_{\hat{y}} = I_y - \mathcal{A}(\hat{x}_A)^2 = \tfrac{4}{7}a^3b - (\tfrac{4}{3}ab)(-\tfrac{3}{5}a)^2 = \tfrac{16}{175}a^3b$$

$$I_{\hat{x}\hat{y}} = I_{xy} - \mathcal{A}\,\hat{x}_A\,\hat{y}_A = 0 - (\tfrac{4}{3}ab)(-\tfrac{3}{5}a)(0) = 0$$

Note that it is not necessary to apply the parallel axis theorem for $J_{\hat{z}}$ because we will be able to use equation (3) to determine $J_{z'}$.

We now apply the parallel axis theorems again in order to transfer from the $\hat{x}\hat{y}\hat{z}$ system to the $x'y'z'$ system. This yields

$$I_{x'} = I_{\hat{x}} + \mathcal{A}(\hat{y}_B)^2 = \tfrac{4}{15}ab^3 + (\tfrac{4}{3}ab)(-b)^2$$

$$= \tfrac{8}{5}ab^3$$

$$I_{y'} = I_{\hat{y}} + \mathcal{A}(x_B)^2 = \tfrac{16}{175}a^3b + (\tfrac{4}{3}ab)(\tfrac{2}{5}a)^2$$

$$= \tfrac{32}{105}a^3b$$

$$I_{x'y'} = I_{\hat{x}\hat{y}} + \mathcal{A}\,\hat{x}_B\,\hat{y}_B = 0 + (\tfrac{4}{3}ab)(\tfrac{2}{5}a)(-b)$$

$$= -\tfrac{8}{15}a^2b^2$$

We then find

$$J_{z'} = I_{x'} + I_{y'} = \tfrac{8}{105}ab(21b^2 + 4a^2)$$

HOMEWORK PROBLEMS

IX.19–IX.26 Determine the moment of inertia I_x, the polar moment of inertia J_z, and the product of inertia I_{xy} for the shaded area and corresponding coordinate system shown. The properties in Appendix B may be employed where necessary.

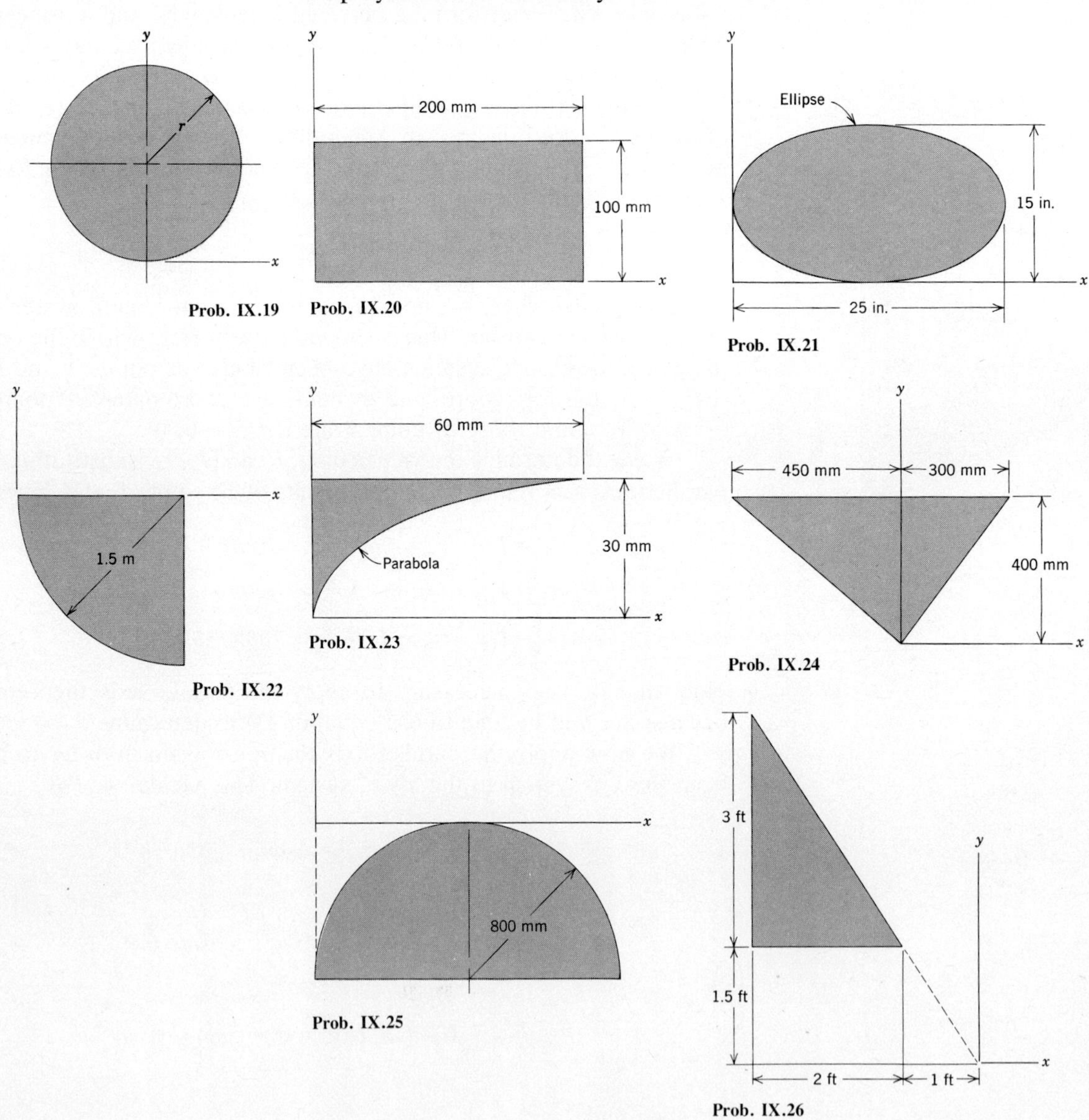

Prob. IX.19 Prob. IX.20 Prob. IX.21 Prob. IX.22 Prob. IX.23 Prob. IX.24 Prob. IX.25 Prob. IX.26

IX.27 Point C is the centroid of the shaded area. It is known that for the xyz coordinate system shown, $I_x = 1230 \text{ in.}^4$ and $J_z = 2580 \text{ in.}^4$, whereas for the centroidal $\hat{x}\hat{y}\hat{z}$ coordinate system $J_{\hat{z}} = 1830 \text{ in.}^4$. Determine (a) the area $\mathcal{A}$, (b) the values of $I_{\hat{x}}$ and $I_{\hat{y}}$.

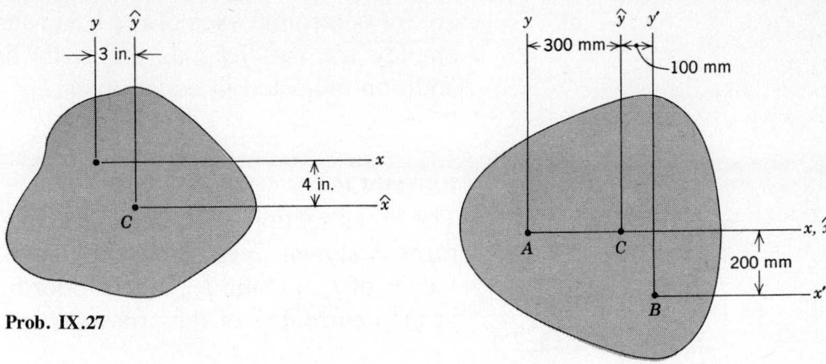

Prob. IX.27

Prob. IX.28

IX.28 Point C is the centroid of the shaded area shown. Because of symmetry it is known that the centroidal $\hat{x}\hat{y}\hat{z}$ axes are principal axes for which the products of inertia with respect to these axes is zero. It is also known that for the other coordinate systems shown, $I_y = 8.50(10^{-2}) \text{ m}^4$, $I_{x'} = 3.25(10^{-2}) \text{ m}^4$, and $I_{y'} = 6.50(10^{-2}) \text{ m}^4$. Determine (a) the area $\mathcal{A}$, (b) the moment of inertia $I_{\hat{x}}$, (c) the products of inertia I_{xy} and $I_{x'y'}$.

D. COMPOSITE AREAS

It is common practice to tabulate the properties of various basic shapes. Appendix B gives the area moments and product of inertia for axes having an origin at the centroid. We may then determine the properties of a planar area whose shape does not appear in Appendix B by considering it to be a composite area.

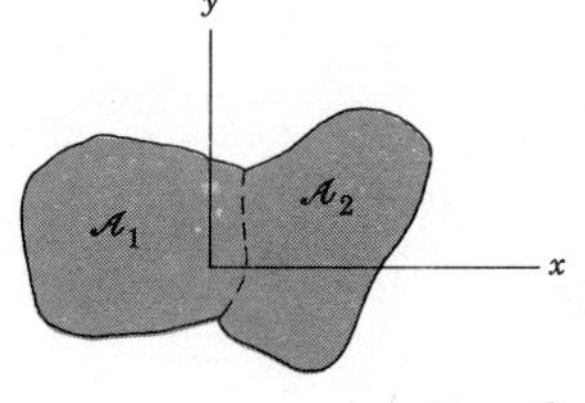

Figure 8

Figure 8 depicts an area that is formed from two basic shapes having areas $\mathcal{A}_1$ and $\mathcal{A}_2$. In the definitions of the inertia properties of an area, integrals extending over the total area $\mathcal{A}$ may be expressed as the sum of integrals extending over the individual areas $\mathcal{A}_1$ and $\mathcal{A}_2$. The following identities are then found.

$$\boxed{\begin{aligned} I_x &= (I_x)_1 + (I_x)_2 \\ I_y &= (I_y)_1 + (I_y)_2 \\ J_z &= (J_z)_1 + (J_z)_2 \\ I_{xy} &= (I_{xy})_1 + (I_{xy})_2 \end{aligned}} \tag{11}$$

In other words the area moments and product of inertia are additive.

An aspect of equations (11) that can lead to errors is that these properties are additive only when they have been expressed with respect to the same coordinate system. Thus, noting that the properties in Appendix B are for centroidal axes of the basic shapes, it will usually be necessary to employ the parallel axis theorems before carrying out the appropriate addition indicated in equations (11).

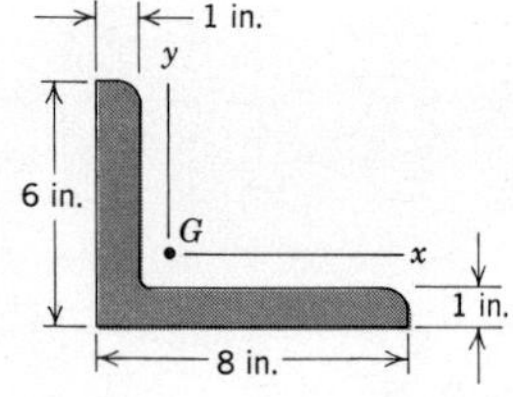

EXAMPLE 4

The cross section of an angle iron used in the construction of steel structures is shown. Neglecting the fillets (rounded-off edges), determine the values of I_x, I_y, and I_{xy} for the coordinate system shown having its origin at the centroid G of the cross section.

Solution

Let us consider the cross section to be a composite formed from the rectangular shape of each leg. First, we must determine the location of the centroid G. To do this we define a convenient $x'y'z'$ coordinate system parallel to the given system. The component rectangles, as well as the x' and y' axes, are illustrated in the sketch to the left.

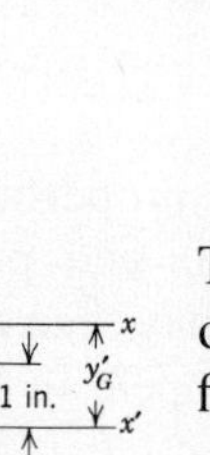

The areas are

$$\mathcal{A}_1 = 7(1) = 7 \text{ in.}^2 \qquad \mathcal{A}_2 = 6(1) = 6 \text{ in.}^2$$

$$\mathcal{A} = \mathcal{A}_1 + \mathcal{A}_2 = 13 \text{ in.}^2$$

The centroid of each rectangle is at its middle, so the (x', y') coordinates of centroid G_1 are (4.5, 0.5) and those of centroid G_2 are (0.5, 3.0). The first moments of area with respect to the x' and y' coordinates then give

$$13(x'_G) = 7(4.5) + 6(0.5)$$

$$x'_G = 2.654 \text{ in.}$$

$$13(y'_G) = 7(0.5) + 6(3.0)$$

$$y'_G = 1.654 \text{ in.}$$

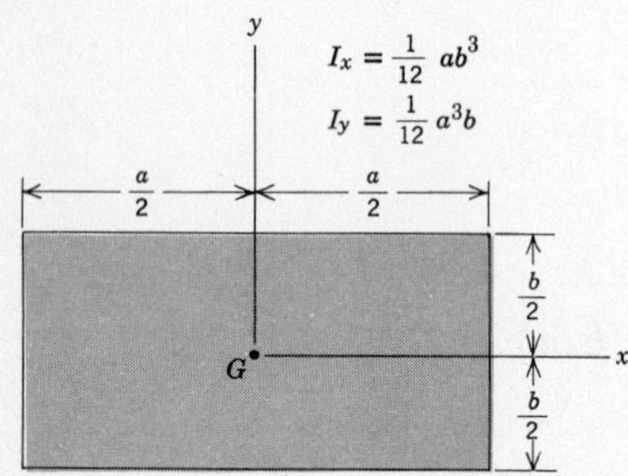

Now that we have located point G, we may determine the moments and product of inertia of the two basic shapes with respect to the centroidal xyz coordinate system. We refer to Appendix B to determine the inertia properties of each shape with respect to its own centroidal coordinate system. The properties of a rectangle are repeated in the adjacent sketch. Note that $I_{xy} = 0$ because of symmetry.

In this problem there are two rectangles, so we will need centroidal axes $\hat{x}_i\hat{y}_i\hat{z}_i$ $(i = 1, 2)$ for each, as shown in the sketch on the next page.

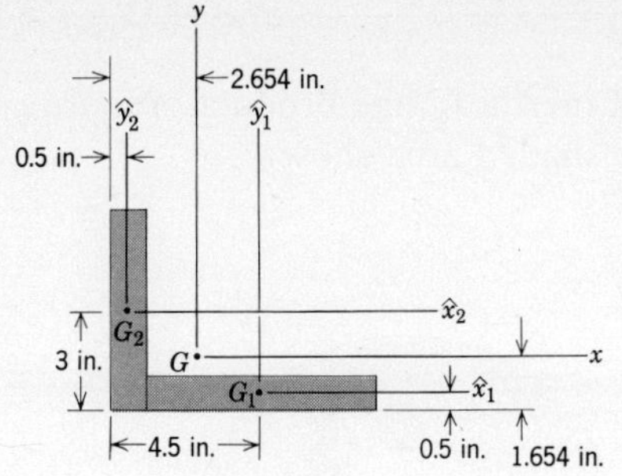

Considering rectangle 1 first, we see that in comparison with the rectangle reproduced on the previous page from Appendix B, $a = 7$ in., $b = 1$ in. In order to employ the parallel axis theorems to transfer from the $\hat{x}_1\hat{y}_1\hat{z}_1$ system to the xyz system, we need the $(\hat{x}_1, \hat{y}_1)$ coordinates of point G. These values are $(-4.50 + 2.654, 1.654 - 0.50)$ inches. Thus, using the parallel axis theorems, we have

$$(I_x)_1 = (I_{\hat{x}_1})_1 + \mathscr{A}_1(1.654 - 0.50)^2 = \tfrac{1}{12}(7)(1)^3 + 7(1.332)$$

$$= 9.91 \text{ in.}^4$$

$$(I_y)_1 = (I_{\hat{y}_1})_1 + \mathscr{A}_1(-4.50 + 2.654)^2 = \tfrac{1}{12}(1)(7)^3 + 7(3.408)$$

$$= 52.44 \text{ in.}^4$$

$$(I_{xy})_1 = (I_{\hat{x}_1\hat{y}_1})_1 + \mathscr{A}_1(-4.50 + 2.654)(1.654 - 0.50)$$

$$= 0 + 7(-2.130) = -\ 14.91 \text{ in.}^4$$

For rectangle 2, we see that $a = 1$ in. and $b = 6$ in. The $(\hat{x}_2, \hat{y}_2)$ coordinates of point G are $(2.654 - 0.5, -3.0 + 1.654)$ inches, so the parallel axis theorems give

$$(I_x)_2 = (I_{\hat{x}_2})_2 + \mathscr{A}_2(-3.0 + 1.654)^2 = \tfrac{1}{12}(1)(6)^3 + 6(1.812)$$

$$= 28.87 \text{ in.}^4$$

$$(I_y)_2 = (I_{\hat{y}_2})_2 + \mathscr{A}_2(2.654 - 0.50)^2 = \tfrac{1}{12}(6)(1)^3 + 6(4.640)$$

$$= 28.34 \text{ in.}^4$$

$$(I_{xy})_2 = (I_{\hat{x}_2\hat{y}_2}) + \mathscr{A}_2(2.654 - 0.50)(-3.0 + 1.654)$$

$$= 0 + 6(-2.899) = -17.40 \text{ in.}^4$$

We now add corresponding inertia properties to obtain the results for the composite cross section. This yields

$$I_x = (I_x)_1 + (I_x)_2 = 38.8 \text{ in.}^4$$

$$I_y = (I_y)_1 + (I_y)_2 = 80.8 \text{ in.}^4$$

$$I_{xy} = (I_{xy})_1 + (I_{xy})_2 = -32.3 \text{ in.}^4$$

Notice that it was not necessary to calculate the polar moments of inertia of the individual rectangles, because we may use equation (3) to write

$$J_z = I_x + I_y = 119.6 \text{ in.}^4$$

HOMEWORK PROBLEMS

IX.29–IX.32 Determine the moment of inertia I_x, the product of inertia I_{xy}, and the radius of gyration k_x of the shaded area shown.

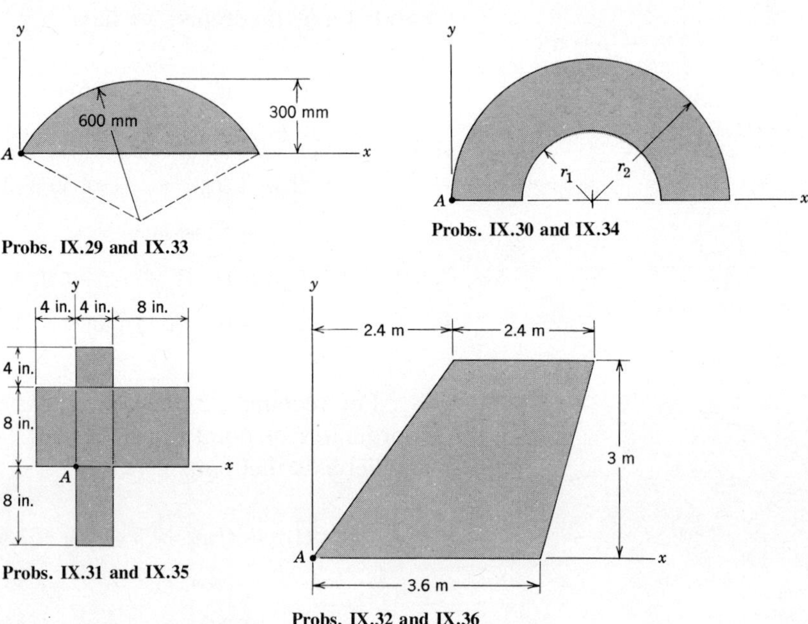

Probs. IX.29 and IX.33

Probs. IX.30 and IX.34

Probs. IX.31 and IX.35

Probs. IX.32 and IX.36

IX.33–IX.36 Determine the polar moment of inertia J_z and the radius of gyration k_z for the shaded area shown.

IX.37–IX.43 Point G is the centroid of the area shown. Determine the location of this point and the moments and product of inertia of the area with respect to the xyz coordinate system having origin G.

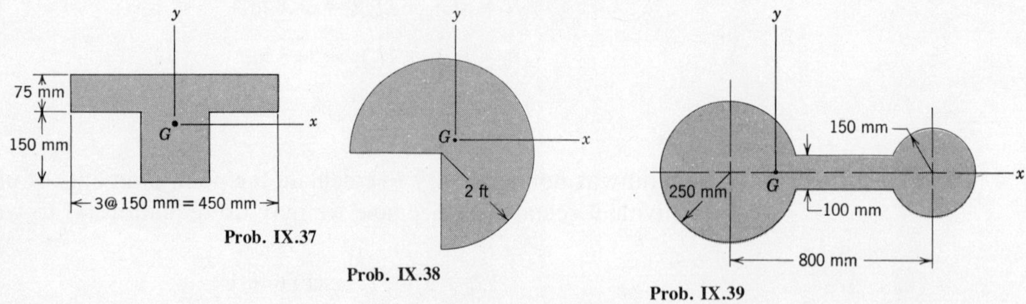

Prob. IX.37

Prob. IX.38

Prob. IX.39

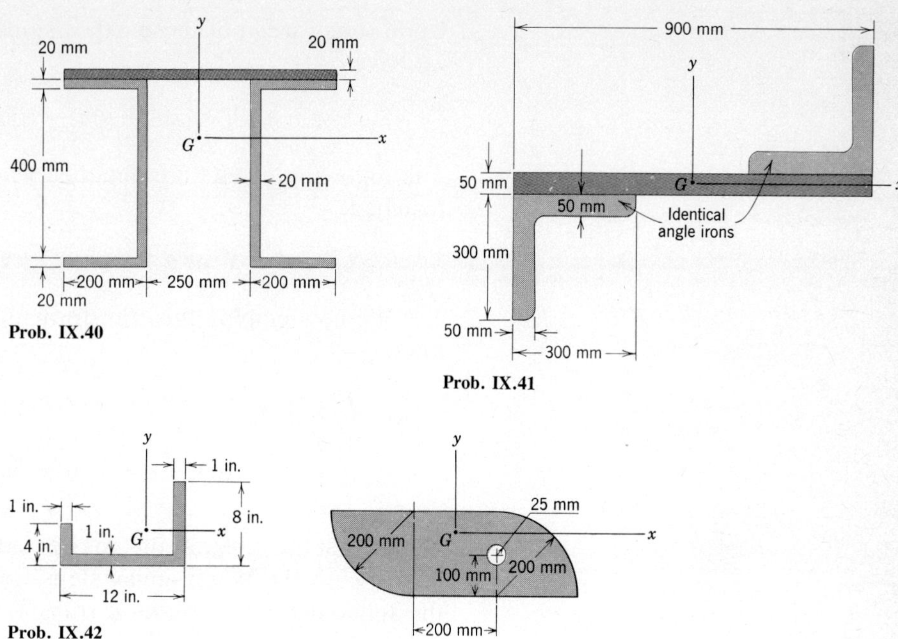

Prob. IX.40

Prob. IX.41

Prob. IX.42

Prob. IX.43

E. ROTATION TRANSFORMATION — MOHR'S CIRCLE

The parallel axis theorems enable us to determine the moments and product of inertia corresponding to one coordinate system in terms of the properties for another parallel coordinate system. In what follows we shall determine the algebraic relationship between the inertia properties for two coordinate systems having identical origins, but different orientations. A typical situation is shown in Figure 9, where the xyz system is the one for which the moments and product of inertia are known. To proceed with the derivation we need expressions for the (x', y') coordinates in terms of the (x, y) values. These could be obtained by using trigonometry to measure various lengths in Figure 9.

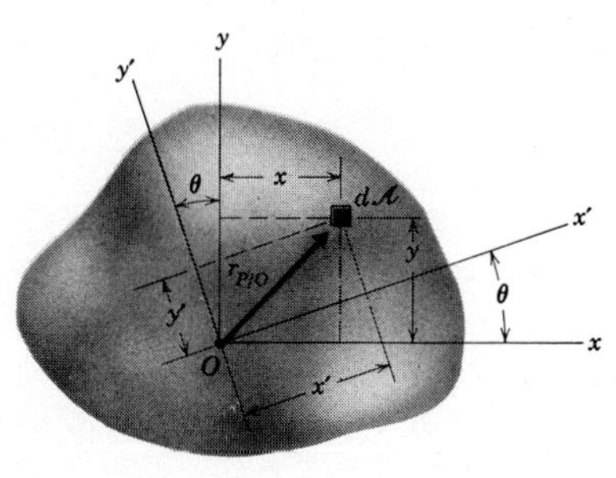

Figure 9

An alternative method employs the fact that the position vector $\bar{r}_{P/O}$ may be described in terms of either coordinate system. Thus, we write

$$\bar{r}_{P/O} = x'\bar{i}' + y'\bar{j}' = x\bar{i} + y\bar{j}$$

Expressing the unit vectors $\bar{i}$ and $\bar{j}$ in terms of their components with respect to the $x'y'z'$ system, we find

$$\bar{i} = (\cos\theta)\,\bar{i}' - (\sin\theta)\,\bar{j}'$$

$$\bar{j} = (\sin\theta)\,\bar{i}' + (\cos\theta)\,\bar{j}'$$

Upon substitution of these expressions into the earlier equation for $\bar{r}_{P/O}$, we have

$$\bar{r}_{P/O} = x'\bar{i}' + y'\bar{j}' = x[(\cos\theta)\,\bar{i}' - (\sin\theta)\,\bar{j}'] + y[(\sin\theta)\,\bar{i}' + (\cos\theta)\,\bar{j}']$$

The following results are obtained when corresponding components are matched.

$$x' = x\cos\theta + y\sin\theta \qquad y' = -x\sin\theta + y\cos\theta \tag{12}$$

We now employ this transformation in the expression for $I_{x'}$, as follows.

$$\begin{aligned} I_{x'} &\equiv \int_{\mathcal{A}} (y')^2\, d\mathcal{A} = \int_{\mathcal{A}} (-x\sin\theta + y\cos\theta)^2\, d\mathcal{A} \\ &= \cos^2\theta \int_{\mathcal{A}} y^2\, d\mathcal{A} - 2\cos\theta\sin\theta \int_{\mathcal{A}} xy\, d\mathcal{A} + \sin^2\theta \int_{\mathcal{A}} x^2\, d\mathcal{A} \end{aligned}$$

Notice that the integrals on the right side of this expression are I_x, I_{xy}, and I_y, respectively. When similar steps are followed for $I_{y'}$ and $I_{x'y'}$, we obtain the following set of *rotation transformations* for the area moments of inertia.

$$\boxed{\begin{aligned} I_{x'} &= I_x\cos^2\theta + I_y\sin^2\theta - 2\sin\theta\cos\theta\, I_{xy} \\ I_{y'} &= I_x\sin^2\theta + I_y\cos^2\theta + 2\sin\theta\cos\theta\, I_{xy} \\ I_{x'y'} &= I_{xy}(\cos^2\theta - \sin^2\theta) + (I_x - I_y)\sin\theta\cos\theta \end{aligned}} \tag{13}$$

There is no need to consider $J_{z'}$; the coincidence of the z and z' axes means that $J_{z'} = J_z$. This is confirmed by the foregoing expressions, which show that

$$J_{z'} = I_{x'} + I_{y'} \equiv I_x + I_y$$

This also shows that the sum of the I_x and I_y is a constant regardless of the angular orientation of the coordinate axes.

In your later studies you will encounter transformations similar to these, for example, in mechanics of materials for stress and strain and in dynamics for mass moments of inertia.

Equations (13) are easy enough to employ. Nevertheless, by performing a few simple manipulations we may obtain a pictorial representation of the transformation that makes it unnecessary to remember formulas. To do this, we employ the following trigonometric identities.

$$\cos^2\theta = \tfrac{1}{2}(1 + \cos 2\theta) \qquad \sin^2\theta = \tfrac{1}{2}(1 - \cos 2\theta) \qquad \sin\theta\cos\theta = \tfrac{1}{2}\sin 2\theta$$

Substitution of the foregoing into equations (13) then yields

$$I_{x'} = \tfrac{1}{2}(I_x + I_y) + \tfrac{1}{2}(I_x - I_y)\cos 2\theta - I_{xy}\sin 2\theta \tag{14a}$$

$$I_{y'} = \tfrac{1}{2}(I_x + I_y) - \tfrac{1}{2}(I_x - I_y)\cos 2\theta + I_{xy}\sin 2\theta \tag{14b}$$

$$I_{x'y'} = I_{xy}\cos 2\theta + \tfrac{1}{2}(I_x - I_y)\sin 2\theta \tag{14c}$$

We next transpose the terms $\frac{1}{2}(I_x + I_y)$ from the right side to the left in equations (14a) and (14b). Then, squaring either of equations (14a) or (14b), and adding the result to the square of equation (14c), we obtain the following expression.

$$\left[I_{p'} - \left(\frac{I_x + I_y}{2}\right)\right]^2 + (I_{x'y'})^2 = \left(\frac{I_x - I_y}{2}\right)^2 + (I_{xy})^2 \tag{15}$$

where the symbol p' represents either x' or y', depending on which equation was squared.

The significance of this expression becomes apparent when we recall that I_x, I_y, and I_{xy} are specific known values. Therefore, equation (15) can be regarded as a relationship between $I_{x'}$ and $I_{x'y'}$, or alternatively, between $I_{y'}$ and $I_{x'y'}$. In terms of a graph with $I_{p'}$ plotted along the abscissa and $I_{x'y'}$ plotted along the ordinate, equation (15) represents a circle of radius R that is centered on the abscissa at a distance d from the origin, where

$$R = \left[\left(\frac{I_x - I_y}{2}\right)^2 + (I_{xy})^2\right]^{1/2} \qquad d = \tfrac{1}{2}(I_x + I_y) \tag{16}$$

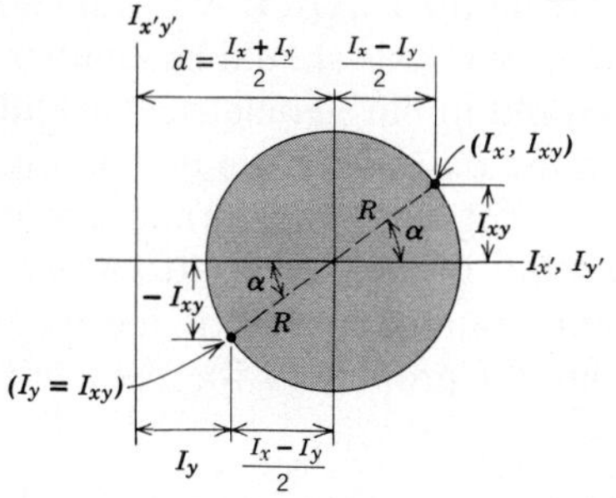

Figure 10

The parameters R and d are readily determined from a geometric construction by plotting the points (I_x, I_{xy}) and $(I_y, -I_{xy})$ and then connecting these points by a straight line. This construction and the resulting circle are illustrated in Figure 10.

When $\theta = 0°$, the x and x' axes in Figure 9 are coincident, as are the y and y' axes. Hence, the points we have plotted in Figure 10 may be considered to represent $I_{x'}$, $I_{y'}$, and $I_{x'y'}$ when $\theta = 0$. The question we must now consider is which points on the circle correspond to the values of $I_{x'}$ and $I_{x'y'}$, or $I_{y'}$ and $I_{x'y'}$, for a nonzero value of θ. To answer this question, we need the angle α. Referring to Figure 10, we see that this angle satisfies

$$R\cos\alpha = \tfrac{1}{2}(I_x - I_y) \qquad R\sin\alpha = I_{xy} \tag{17}$$

Substituting the foregoing expressions and the definition of the distance d into the basic relationships of equations (14), and making use of the trigonometric identities for the sine and cosine of the sum of two angles, we find that

$$I_{x'} = d + R(\cos\alpha\cos 2\theta - \sin\alpha\sin 2\theta) \equiv d + R\cos(\alpha + 2\theta)$$

$$I_{y'} = d - R(\cos\alpha\cos 2\theta - \sin\alpha\sin 2\theta) \equiv d - R\cos(\alpha + 2\theta)$$

$$I_{x'y'} = R(\sin\alpha\cos 2\theta + \cos\alpha\sin 2\theta) \equiv R\sin(\alpha + 2\theta)$$

As illustrated in Figure 11, these relations show that the point $(I_{x'}, I_{x'y'})$ is

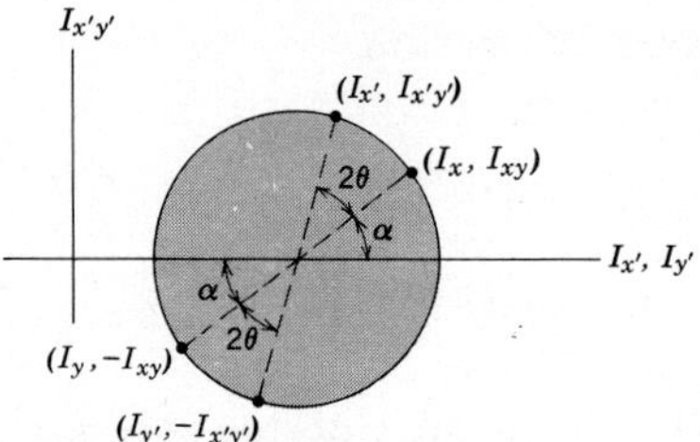

Figure 11

obtained by rotating the radial line from the point (I_x, I_{xy}) counterclockwise through an angle 2θ. A similar statement applies to the point $(I_{y'}, -I_{x'y'})$. It is significant that in this construction the angle 2θ is positive in the counterclockwise sense, for that is the same sense as the positive rotation from the x and y axes to the x' and y' axes originally depicted in Figure 9.

A question that bothers some students about this derivation is, "Why do we plot the point $(I_{y'}, -I_{x'y'})$, and not the point $(I_{y'}, I_{x'y'})$?" To answer this question it is only necessary to note that the x' and y' axes are separated by 90°. Thus, if we give the x' axis an additional rotation of 90° beyond its initial orientation at an angle θ from the x axis, it will coincide with the original y' axis. Hence, $I_{x'}$ for the new x' axis should be identical to $I_{y'}$ for the old y' axis. This result is assured in our geometrical sketch because angles of rotation there are twice the physical angle of rotation. This pictorial method for representing the rotation transformations of equations (13) is known as *Mohr's circle* after the engineer Otto Mohr.

In summation, the Mohr's circle for area moments of inertia is employed as follows to determine the inertia properties for the $x'y'z'$ coordinate system.

1 Using the abscissa to plot moments of inertia and the ordinate for the product of inertia, plot the two points (I_x, I_{xy}) and $(I_y, -I_{xy})$.
2 Connect the two points plotted in the previous step. The length of this line is the diameter of the Mohr's circle, and the point of intersection of this line with the abscissa is the center of the circle. Sketch the corresponding circle.

3 Draw a diameter in the Mohr's circle that is at an angle 2θ from the line in step 2, in the same sense as the angle of rotation θ from the x axis to the x' axis.
4 The point of intersection of the diameter just drawn with the Mohr's circle, adjacent to the point (I_x, I_{xy}) in the sense of the rotation from the x axis to the x' axis, has the coordinates $(I_{x'}, I_{x'y'})$. The point of intersection adjacent to the point $(I_y, -I_{xy})$ in the sense of the rotation is the point $(I_{y'}, -I_{x'y'})$.

Note that when constructing Mohr's circle, the points should be plotted consistently with the sign of I_{xy}. Also, there is no restriction that I_x be larger than I_y, even though this was the case in the derivation.

We will not utilize Mohr's circle to obtain a graphical solution. Rather, Mohr's circle will be employed as a device that makes it unnecessary to remember equations (13).

An important aspect of the rotation transformation is exhibited directly by Mohr's circle. Specifically, referring to Figure 11, notice that the two points where the circle intersects the abscissa correspond to orientations of the $x'y'z'$ system for which $I_{x'y'} = 0$. A coordinate system for which the product of inertia is zero is called a set of *principal axes,* and the corresponding moments of inertia are the *principal* values. Earlier, in discussing shapes having an axis of symmetry, we saw that any coordinate system containing this axis is a set of principal axes. We now see that all planar figures have a set of principal axes for any choice of the origin.

The angle of rotation $\tilde{\theta}$ from the xyz axes having known properties to the principal axes is easily determined from Figure 10 or Figure 11. Recalling that the angle of rotation is one-half the angle in Mohr's circle, and considering counterclockwise rotations to be positive rotations, we see that $2\tilde{\theta} = -\alpha$. (The alternative $2\tilde{\theta} = 180° - \alpha$ gives the same coordinate system as the foregoing, except that the coordinate axes are labeled differently.) Using equations (17), we then have

$$\tilde{\theta} = -\tfrac{1}{2}\tan^{-1}\left(\frac{2I_{xy}}{I_x - I_y}\right) \tag{18}$$

In practice, there is no need to remember equation (18), for the magnitude and sense of the angle of rotation $\tilde{\theta}$ are easily determined after the Mohr's circle has been constructed.

Additional features of the rotation transformation are apparent from the Mohr's circle in Figure 11. We see that the principal moments of inertia are the maximum and minimum values of I_x and I_y that can be obtained in a rotation of the coordinate axes. Also, the axes for which the product of inertia is a maximum are at an angle of 45° ($2\theta = 90°$) from the principal axes.

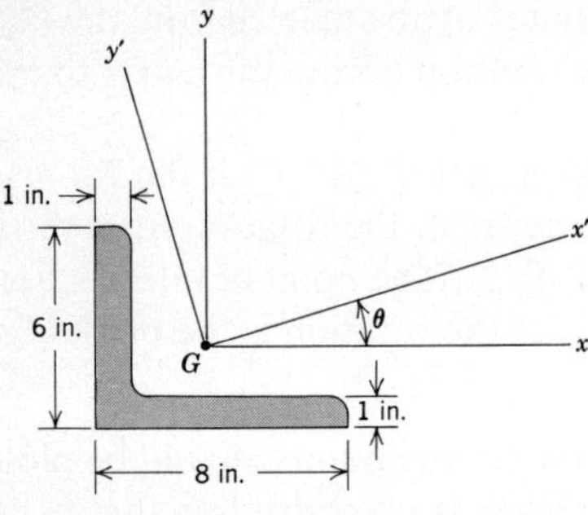

EXAMPLE 5

The moments and product of inertia of the cross section of the angle iron shown with respect to the centroidal *xyz* coordinate system were determined in Example 4. Determine (a) the values of $I_{x'}$, $I_{y'}$, and $I_{x'y'}$ when $\theta = 40°$, (b) the angle $\bar{\theta}$ for principal axes and the corresponding principal values of the moments of inertia.

Solution

From Example 4 we know that $I_x = 38.8$ in.^{4}, $I_y = 80.8$ in.^{4}, $I_{xy} = -32.3$ in.^{4}. We plot the points (I_x, I_{xy}) and $(I_{y'} -I_{xy})$, and label them as points *A* and *B*, respectively, for future reference. We then connect points *A* and *B* by a straight line and draw a circle having line *AB* as a diameter. This is shown in the following sketch.

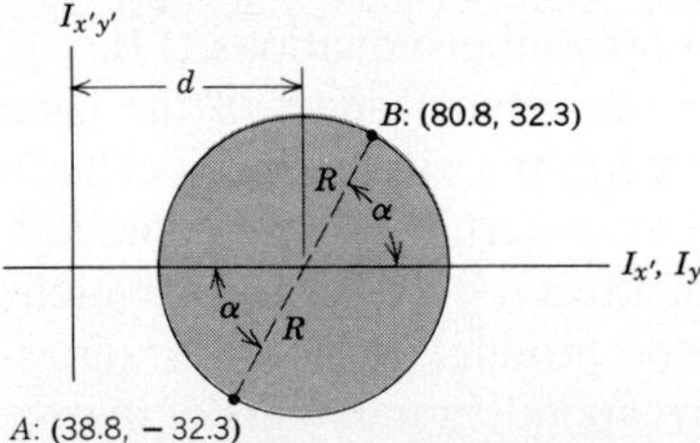

In order to locate other points on the Mohr's circle, we need the distance *d*, the radius *R*, and the angle α. From the sketch we see that

$$d = \tfrac{1}{2}(80.8 + 38.8) = 59.8$$

$$R = [(80.8 - 59.8)^2 + (32.3)^2]^{1/2} = 38.5$$

$$\alpha = \tan^{-1}\left(\frac{38.5}{80.8 - 59.8}\right) = 61.41°$$

For part (a) we are interested in the properties with respect to the $x'y'z'$ axis when $\theta = 40°$ counterclockwise, so we draw a diameter that is rotated from line *AB* by $2\theta = 80°$. We label this diameter as *CD*. In part (b) we want the principal axes, which are obtained by rotating line *AB* clockwise by $2\bar{\theta} = \alpha = 61.41°$, thus resulting in the diameter *EF* coincident with the abscissa. These diameters are illustrated in the sketch to the left.

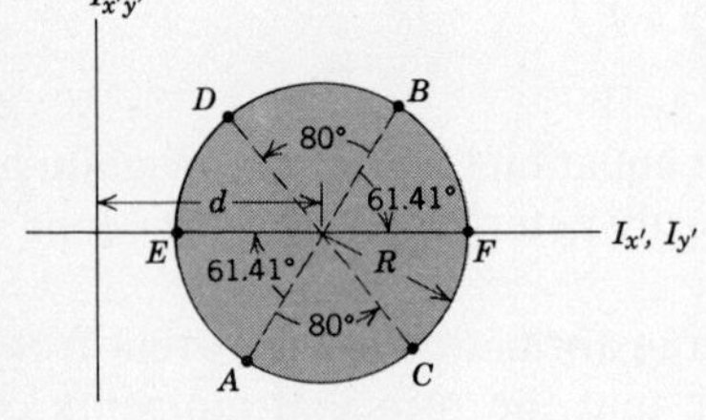

Solving part (a) first, as a result of the rotation point *A* goes to point *C*, so the coordinates of this point are $(I_{x'}, I_{x'y'})$. Similarly, point *B* goes to point *D*, so the corresponding coordinates are $(I_{y'}, -I_{x'y'})$. Using trigonometry to evaluate the distances, we then have

$$I_{x'} = d + R\cos(180° - 61.41° - 80°) = 89.9 \text{ in.}^4$$

$$I_{y'} = d - R\cos(180° - 61.41° - 80°) = 29.7 \text{ in.}^4$$

$$I_{x'y'} = -R\sin(180° - 61.41° - 80°) = -24.01 \text{ in.}^4$$

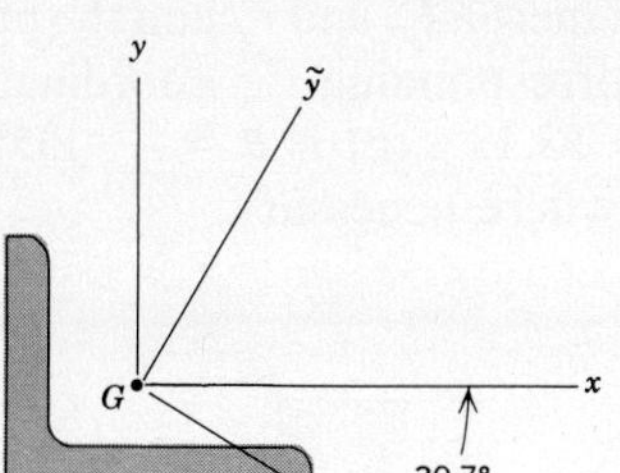

For part (b), point A goes to point E and point B goes to point F. Hence, point E has the coordinates $(I_{\tilde{x}}, 0)$ and point F has the coordinates $(I_{\tilde{y}}, 0)$, where $I_{\tilde{x}}$ and $I_{\tilde{y}}$ are the principal moments of inertia. Thus, for principal axes we have

$$\theta = \tilde{\theta} = \tfrac{1}{2}(61.41°) = 30.7° \text{ clockwise}$$

$$I_{\tilde{x}} = d - R = 21.3 \text{ in.}^4$$

$$I_{\tilde{y}} = d + R = 98.3 \text{ in.}^4$$

The resulting principal axes are shown in the adjacent sketch. Notice that their orientation could not be guessed on the basis of the geometry of the cross section.

HOMEWORK PROBLEMS

IX.44–IX.47 Determine the moments of inertia $I_{x'}$ and $I_{y'}$ and the product of inertia $I_{x'y'}$ for the shaded area and corresponding centroidal $x'y'z'$ axes (a) if $\theta = 36.87°$, (b) if $\theta = 135°$, (c) if $\theta = -30°$. The properties in Appendix B may be used where necessary.

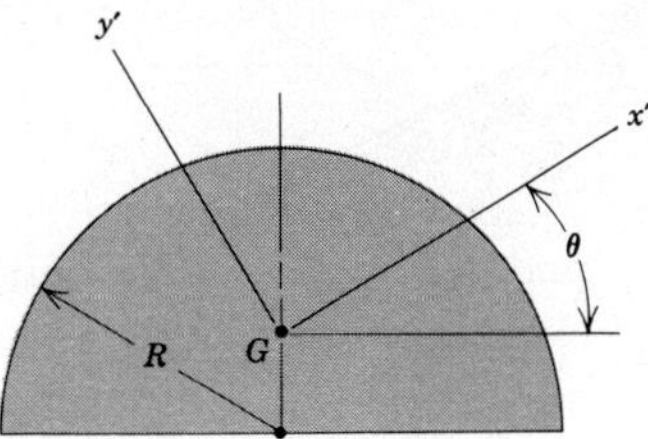

Prob. IX.44

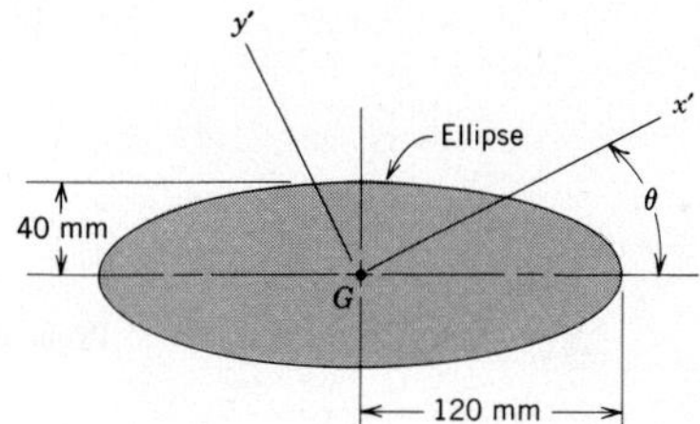

Prob. IX.45

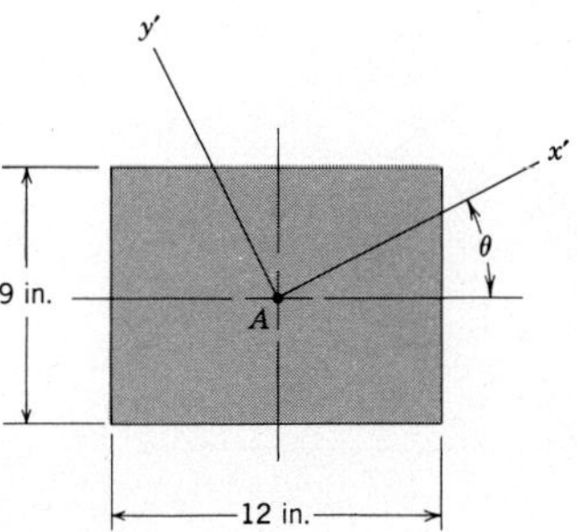

Prob. IX.46

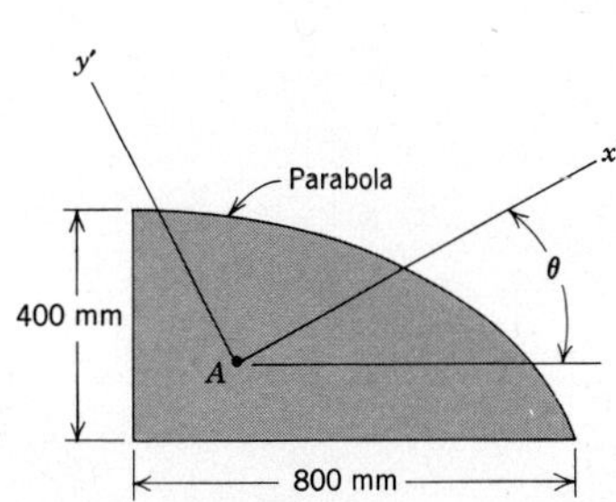

Prob. IX.47

IX.48–IX.51 Determine the moments of inertia $I_{x'}$ and $I_{y'}$ and the product of inertia $I_{x'y'}$ for the shaded area and corresponding xyz coordinate system shown (a) if $\theta = 60°$, (b) if $\theta = -53.13°$, (c) if $\theta = -135°$. The properties in Appendix B may be used where necessary.

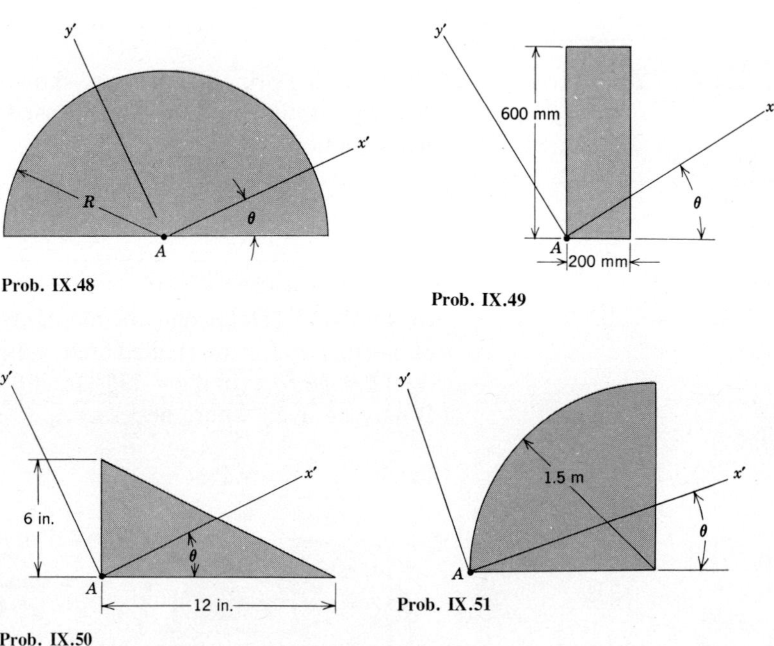

Prob. IX.48

Prob. IX.49

Prob. IX.50

Prob. IX.51

IX.52–IX.55 Determine the moments of inertia I_x I_y and the product of inertia I_{xy} for the composite shape with respect to the coordinate axes shown.

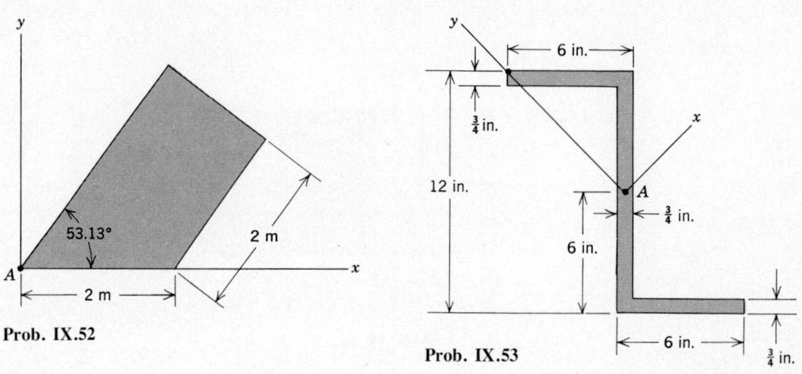

Prob. IX.52

Prob. IX.53

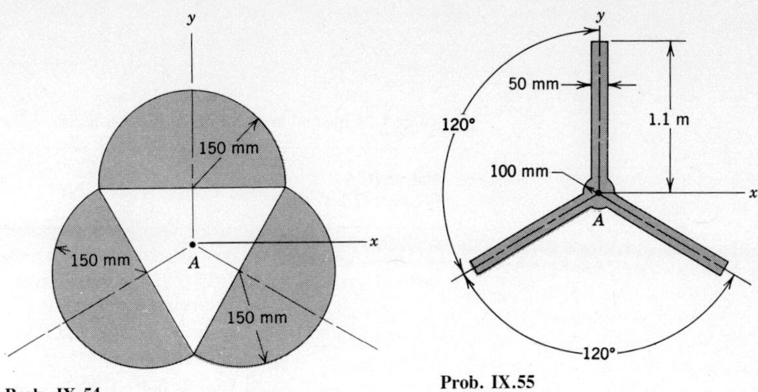

Prob. IX.54

Prob. IX.55

IX.56–IX.61 For the shape in the problem listed below, determine the orientation of the principal axes having origin A, and also determine the corresponding principal moments of inertia.

Homework Problem	**IX.56**	**IX.57**	**IX.58**	**IX.59**	**IX.60**	**IX.61**
Shape in Problem	**IX.46**	**IX.47**	**IX.49**	**IX.50**	**IX.52**	**IX.53**

IX.62–IX.67 For the shape in the problem listed below, determine the orientation of the axes having origin A for which the product of inertia has its maximum value. Also determine the moments and product of inertia for this coordinate system.

Homework Problem	**IX.62**	**IX.63**	**IX.64**	**IX.65**	**IX.66**	**IX.67**
Shape in Problem	**IX.46**	**IX.47**	**IX.49**	**IX.50**	**IX.52**	**IX.53**

IX.68 Letting $\bar{x}\bar{y}\bar{z}$ denote a set of principal axes for an arbitrary area and xyz be another set of axes having the same origin, prove that

$$I_{xy} = \sqrt{I_x I_y - I_{\bar{x}} I_{\bar{y}}}$$

APPENDIX A
SI UNITS

Table 1 Conversion factors from British units

PHYSICAL QUANTITY	U.S.-BRITISH UNIT	= SI EQUIVALENT
		BASIC UNITS
Length	1 foot (ft)	= $3.048(10^{-1})$ meter (m)*
	1 inch (in.)	= $2.54(10^{-2})$ meter (m)*
	1 mile (U.S. statute)	= $1.6093(10^{3})$ meter (m)
Mass	1 slug ($\text{lb-s}^2/\text{ft}$)	= $1.4594(10)$ kilogram (kg)
	1 pound mass (lbm)	= $4.5359(10^{-1})$ kilogram (kg)
		DERIVED UNITS
Acceleration	1 foot/second^2 (ft/s^2)	= $3.048(10^{-1})$ meter/second^2 (m/s^2)*
	1 inch/second^2 (in./s^2)	= $2.54(10^{-2})$ meter/second^2 (m/s^2)*
Area	1 foot^2 (ft^2)	= $9.2903(10^{-2})$ meter^2 (m^2)
	1 inch^2 (in.^2)	= $6.4516(10^{-2})$ meter^2 (m^2)*
Density	1 slug/foot^3 ($\text{lb-s}^2/\text{ft}^4$)	= $5.1537(10^{2})$ kilogram/meter^3 (kg/m^3)
	1 pound mass/foot^3 (lbm/ft^3)	= $1.6018(10)$ kilogram/meter^3 (kg/m^3)
Energy and Work	1 foot-pound (ft-lb)	= 1.3558 joules (J)
	1 kilowatt-hour (kW-hr)	= $3.60(10^{6})$ joules (J)*
	1 British thermal unit (Btu)	= $1.0551(10^{3})$ joules (J)
	(1 joule ≡ 1 meter-newton)	
Force	1 pound (lb)	= 4.4482 newtons (N)
	1 kip (1000 lb)	= $4.4482(10^{3})$ newtons (N)
	(1 newton ≡ 1 kilogram-meter/second^2)	
Power	1 foot-pound/second (ft-lb-s)	= 1.3558 watt (W)
	1 horsepower (hp)	= $7.4570(10^{2})$ watt (W)
	(1 watt ≡ 1 joule/second)	
Pressure and Stress	1 pound/foot^2 (lb/ft^2)	= $4.7880(10)$ pascal (Pa)
	1 pound/inch^2 (lb/in.^2)	= $6.8948(10^{3})$ pascal (Pa)
	1 atmosphere (standard, 14.7 lb/in.^2)	= $1.0133(10^{5})$ pascal (Pa)
	(1 pascal ≡ 1 newton/m^2)	
Speed	1 foot/second (ft/s)	= $3.048(10^{-1})$ meter/second (m/s)*
	1 mile/hr	= $4.4704(10^{-1})$ meter/second (m/s)
	1 mile/hr	= 1.6093 kilometer/hr (km/hr)
Volume	1 foot^3 (ft^3)	= $2.8317(10^{-2})$ meter^3 (m^3)
	1 inch^3 (in.^3)	= $1.6387(10^{-5})$ meter^3 (m^3)
	1 gallon (U.S. liquid)	= $3.7854(10^{-3})$ meter^3 (m^3)

*Denotes an exact factor.

Table 2 Conversion factors from "old" metric units

PHYSICAL QUANTITY	"OLD" METRIC UNIT	= SI EQUIVALENT
Energy	1 erg	= 1.00(10^{-7}) joule (J)*
Force	1 dyne	= 1.00(10^{-5}) newton (N)*
Length	1 angstrom	= 1.00(10^{-10}) meter (m)*
	1 micron	= 1.00(10^{-6}) meter (m)*
Pressure	1 bar	= 1.00(10^{5}) pascal (Pa)*
Volume	1 liter	= 1.00(10^{-3}) meter^3 (m^3)*

*Denotes an exact factor

Table 3 SI prefixes

PREFIX	SYMBOL	FACTOR BY WHICH UNIT IS MULTIPLIED
tera*	T	10^{12}
giga*	G	10^{9}
mega*	M	10^{6}
kilo*	k	10^{3}
hecto	h	10^{2}
deka	da	10
deci	d	10^{-1}
centi	c	10^{-2}
milli*	m	10^{-3}
micro*	μ	10^{-6}
nano*	n	10^{-9}
pico*	p	10^{-12}
femto*	f	10^{-15}
atto*	a	10^{-18}

*Denotes preferred prefixes.

APPENDIX B AREAS, CENTROIDS, AND AREA MOMENTS OF INERTIA

SHAPE	AREA	AREA MOMENT OF INERTIA FOR CENTROIDAL AXES ($J_{\hat{z}} = I_{\hat{x}} + I_{\hat{y}}$)
Rectangle	$\mathcal{A} = ab$	$I_{\hat{x}} = \frac{1}{12}ab^3$ $I_{\hat{y}} = \frac{1}{12}a^3b$
Scalene triangle	$\mathcal{A} = \frac{1}{2}ab$	$I_{\hat{x}} = \frac{1}{36}a^3b$ $I_{\hat{y}} = \frac{1}{36}ab(b^2 + c^2 - bc)$ $I_{\hat{x}\hat{y}} = \frac{1}{72}a^2b(2c - b)$
Circle	$\mathcal{A} = \pi r^2$	$I_{\hat{x}} = I_{\hat{y}} = \frac{1}{4}\pi r^4$

Rectangle

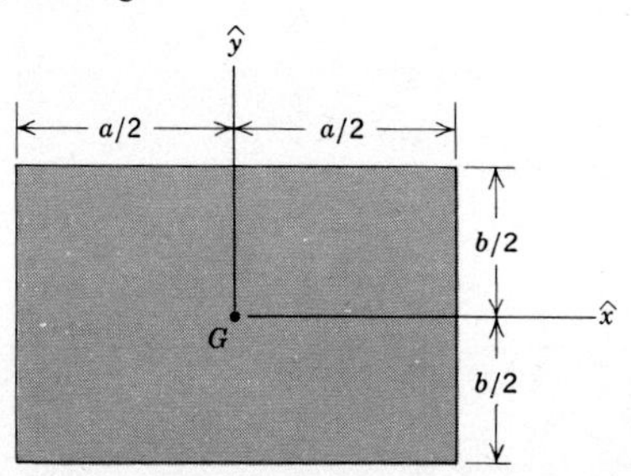

Scalene triangle

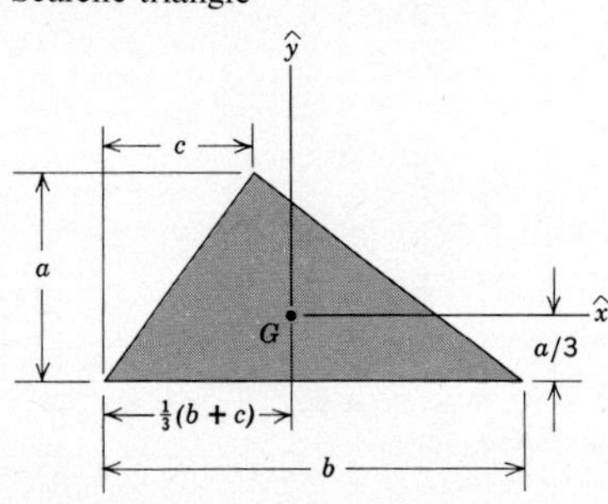

Circle

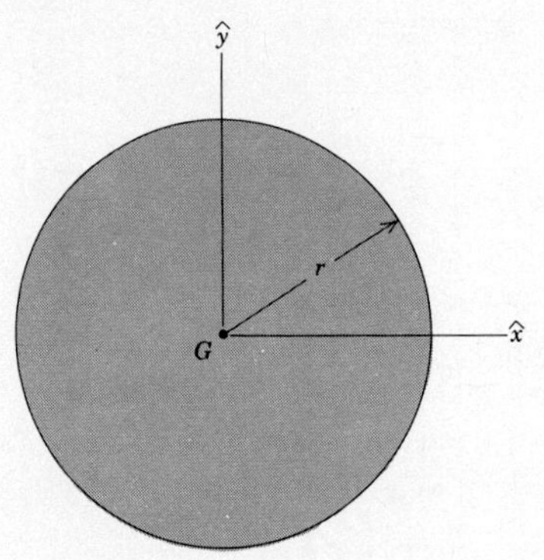

Quarter circle

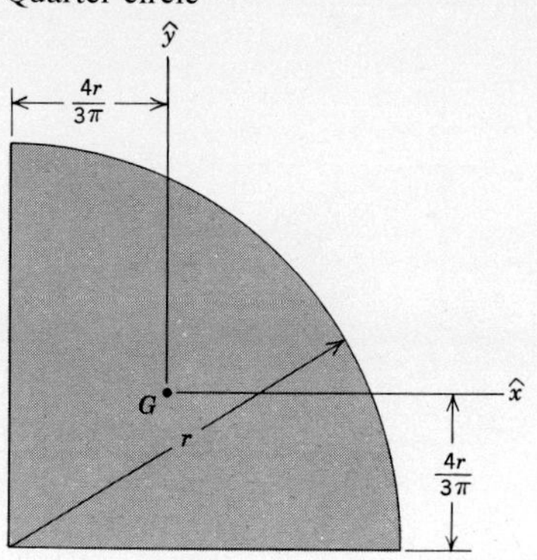

$\mathcal{A} = \frac{1}{4}\pi r^2$

$$I_{\hat{x}} = I_{\hat{y}} = \left(\frac{9\pi^2 - 64}{144\pi}\right)r^4$$

$$I_{\hat{x}\hat{y}} = \left(\frac{9\pi - 32}{72\pi}\right)r^4$$

Circular sector

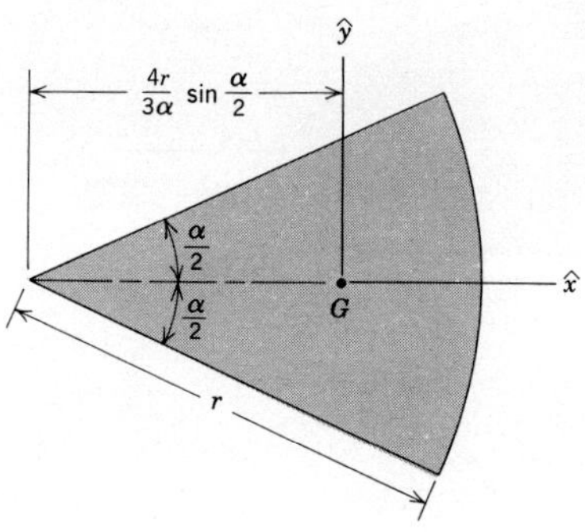

$\mathcal{A} = \frac{1}{2}\alpha r^2$

$$I_{\hat{x}} = \frac{1}{8}(\alpha - \sin\alpha)r^4$$

$$I_{\hat{y}} = \left[\frac{\alpha + \sin\alpha}{8} - \frac{4}{9\alpha}(1 - \cos\alpha)\right]r^4$$

Circular arc

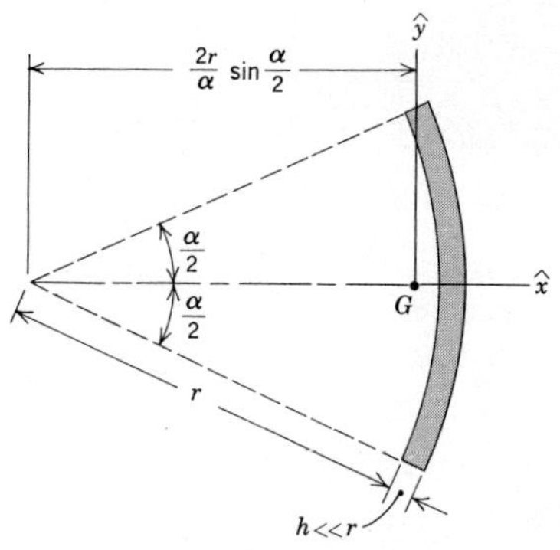

$\mathcal{A} = \alpha r h$

$$I_{\hat{x}} = \frac{1}{2}(\alpha - \sin\alpha)r^3 h$$

$$I_{\hat{y}} = \left[\frac{\alpha + \sin\alpha}{2} - \frac{2}{\alpha}(1 - \cos\alpha)\right]r^3 h$$

Ellipse

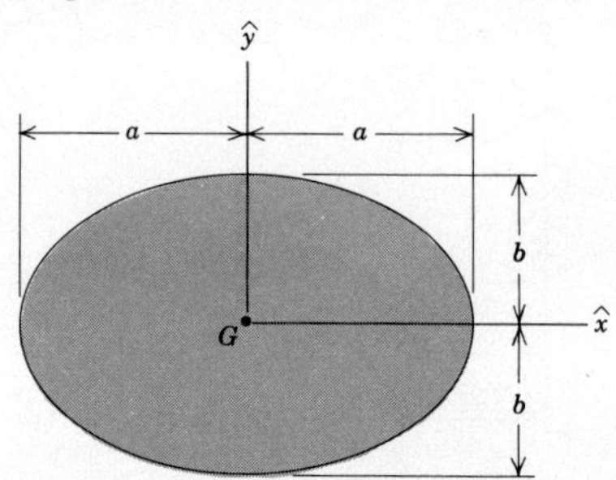

$\mathcal{A} = \pi ab$

$I_{\hat{x}} = \frac{\pi}{4}ab^3$

$I_{\hat{y}} = \frac{\pi}{4}a^3b$

Quarter ellipse

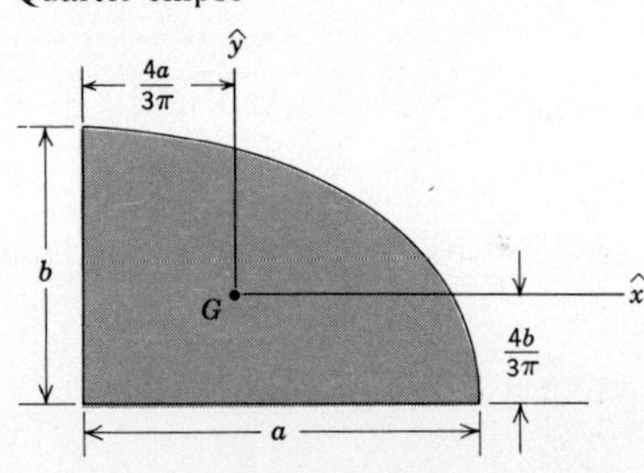

$\mathcal{A} = \frac{1}{4}\pi ab$

$I_{\hat{x}} = \left(\frac{9\pi^2 - 64}{144\pi}\right)ab^3$

$I_{\hat{y}} = \left(\frac{9\pi^2 - 64}{144\pi}\right)a^3b$

$I_{\hat{x}\hat{y}} = \left(\frac{9\pi - 32}{72\pi}\right)a^2b^2$

Parabolic section

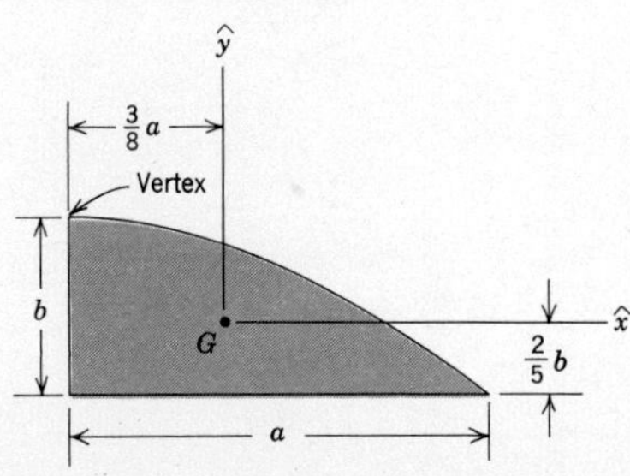

$\mathcal{A} = \frac{2}{3}ab$

$I_{\hat{x}} = \frac{8}{175}ab^3$

$I_{\hat{y}} = \frac{19}{480}a^3b$

$I_{\hat{x}\hat{y}} = -\frac{1}{60}a^2b^2$

Parabolic spandrel

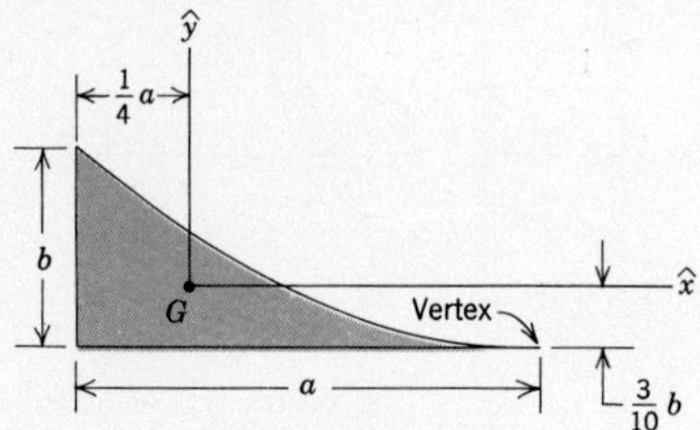

$\mathcal{A} = \frac{1}{3}ab$

$I_{\hat{x}} = \frac{19}{1050}ab^3$

$I_{\hat{y}} = \frac{1}{80}a^3b$

$I_{\hat{x}\hat{y}} = -\frac{1}{120}a^2b^2$

APPENDIX C VOLUMES AND CENTROIDS

SHAPE	VOLUME
Hemisphere	
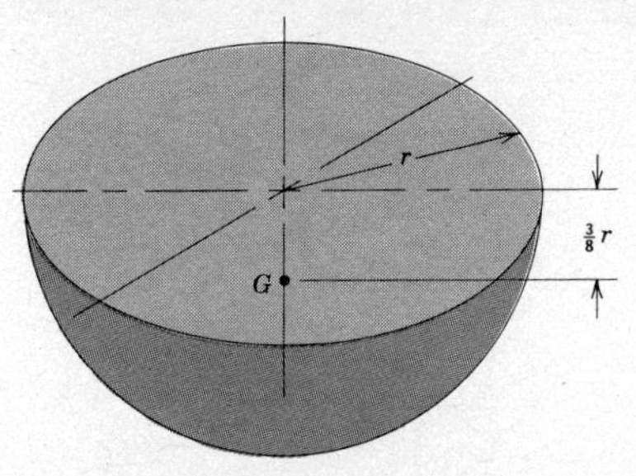	$V = \frac{2}{3}\pi r^3$
Hemispherical shell	
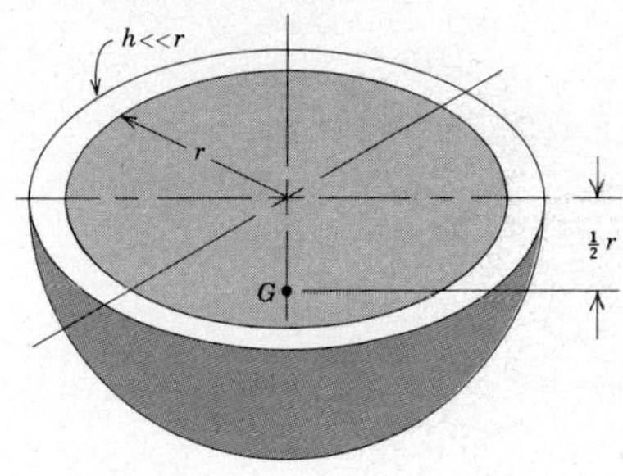	$V = 2\pi r^2 h$
Semicylinder	
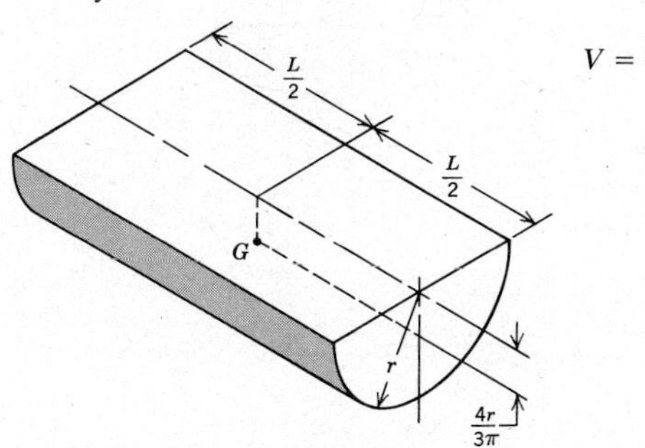	$V = \frac{1}{2}\pi r^2 L$

Semicylindrical shell

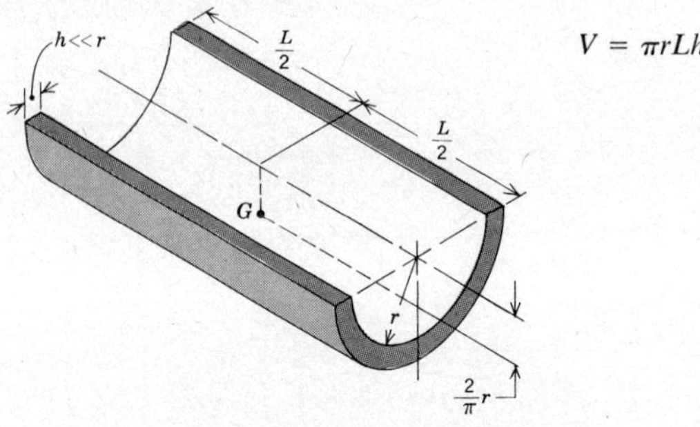

$V = \pi r L h$

Semicone

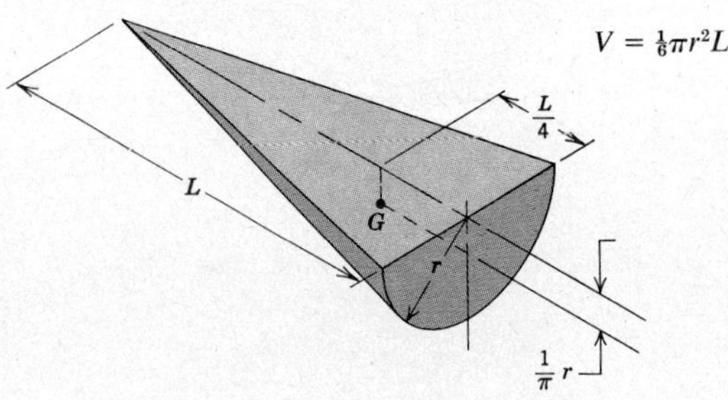

$V = \frac{1}{6}\pi r^2 L$

Semiconical shell

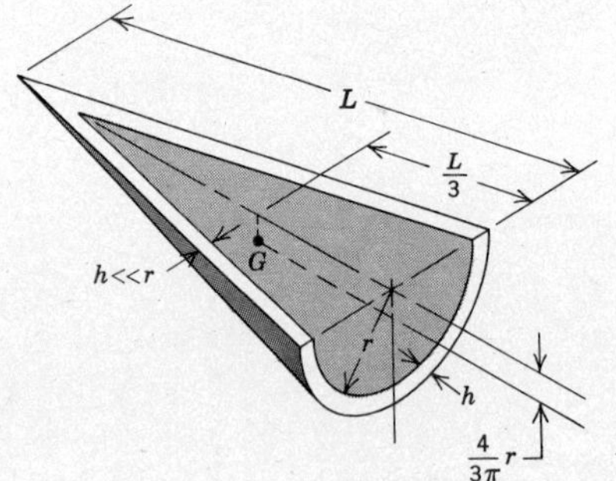

$V = \frac{\pi}{2} r h (r^2 + L^2)^{1/2}$

Orthogonal tetrahedron

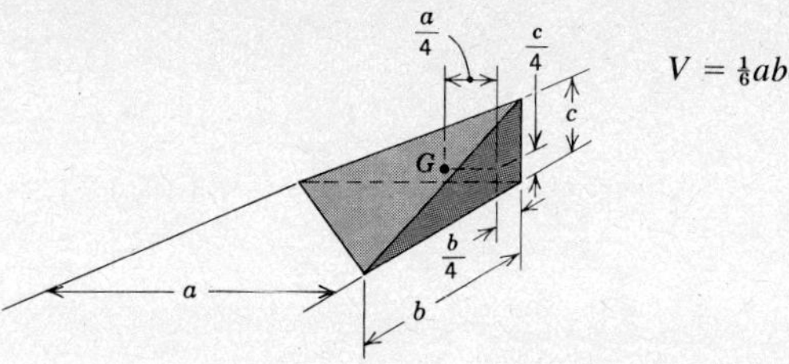

$V = \frac{1}{6}abc$

Triangular prism

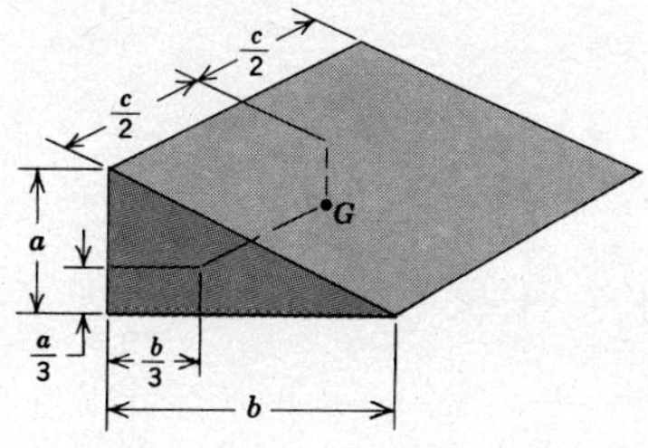

$V = \frac{1}{2}abc$

APPENDIX D
DENSITIES

MATERIAL	SPECIFIC GRAVITY (ρ/ρ_{water})*
Aluminum	2.69
Concrete (av.)	2.40
Copper	8.91
Earth (av. wet)	1.76
(av. dry)	1.28
Glass	2.60
Iron (cast)	7.21
Lead	11.38
Mercury	13.57
Oil (av.)	0.90
Steel	7.84
Water (fresh, liquid)	1.00
(ice)	0.90
Wood (soft pine)	0.48
(hard oak)	0.80

*$\rho_{water} = 1.00(10^3)$ kg/m^3.
$\gamma_{water} = \rho_{water}g = 62.4$ lb/ft^3.

ANSWERS TO ODD-NUMBERED PROBLEMS

I.1 437 N at 11.49° above lt horiz

I.3 (a) 21.28 at 25.59° E of N
(b) 9.23 at 5° N of W
(c) 21.75 at 32.24° N of W

I.5 36.22 at 39.59° N of E

I.7 35.64 at 23.46° E of S

I.9 (a) 1 in.^{2} = 6.45(10^{-4}) m^2
(b) 1 ft^3 = 2.83(10^{-2}) m^3
(c) 1 oz = 0.278 N
(d) 1 lb/in.^{2} = 6.90(10^3) N/m^2
(e) 1 lb/ft^2 = 47.9 N/m^2
(f) 1 ft/s = 0.305 m/s
(g) 1 in./hr = 7.06(10^{-6}) m/s
(h) 1 mile/hr^2 = 1.242(10^{-4}) m/s^2

I.11 1 touchdown = 91.4 m
1 dash = 9 seconds
1 golf ball = 0.450 N

I.13 k = 150 lb/in. = 26.27 kN/m

I.15 L^2/T

I.17 (a) 1 kg − m^2/s^2
(b) 1 kg − m^2/s^3
(c) 1 J/s

II.1 F_X = 50.7 N
F_Y = 108.8 N

II.3 $\theta = 63.43°$, $F_X = 107.3$, $F_Y = 53.7$ N; or
$\theta = -116.57°$, $F_X = -107.3$, $F_Y = -53.7$ N

II.5 (a) 45.40°
(b) 300 lb (dn)
(c) $F_{\|}$ = 351 lb (uphill), $F_{\perp}$ = 356 lb (dn)

II.7 (a) F_{hor} = 80 lb (rt), F_{vert} = 60 lb (up)
(b) F_{hor} = 80 lb (lt), F_{vert} = 60 lb (dn)

II.9 (a) $400\,\bar{I} + 600\,\bar{J} + 200\,\bar{K}$ m
(b) $\alpha = 57.69°$, $\beta = 36.70°$
($\bar{I}$ south, $\bar{J}$ east)

II.11 $\bar{F} = F(0.8365\,\bar{I} + 0.50\,\bar{J} + 0.2241\,\bar{K})$
$\alpha = 33.23°$, $\beta = 60°$, $\gamma = 77.05°$

II.13 $\bar{F} = 60\,\bar{I} - 40\,\bar{J} + 120\,\bar{K}$
$\alpha = 64.62°$, $\beta = 106.60°$, $\gamma = 31.00°$

II.15 $\bar{F} = -7.16\,\bar{I} - 3.58\,\bar{J}$ kN
$\bar{e}_{C/D} = -0.8944\,\bar{I} - 0.4472\,\bar{J}$

II.17 $\bar{e}_{B/A} = \bar{K}$
$\bar{e}_{C/A} = -0.9487\,\bar{I} + 0.3167\,\bar{K}$
$\bar{e}_{D/A} = -0.5455\,\bar{I} - 0.8182\,\bar{J} + 0.1818\,\bar{K}$

II.19 244 N

II.21 39.8 N at 59.02° below rt horiz

II.23 5.04 kN at 57.52° below rt horiz

II.25 (a) F = 4.70 tons, $\alpha = 90°$
(b) F = 1.71 tons, $\alpha = 180°$

II.27 (a) 335 N
(b) 900 N

II.29 −54.71° or −145.29°

II.31 F_2 = 5 kN, F_4 = 3 kN

II.33 (a) F = 2.25 kN, $\beta = 90°$
(b) F = 3 kN, $\beta = 90°$

II.35 $-1.846\,\bar{I} + 2.26\,\bar{J} - 3.02\,\bar{K}$ kN
($\bar{I} = \bar{e}_{C/B}$, $\bar{K}$ up)

II.37 $-49.3\,\bar{J} - 186.9\,\bar{K}$
($\bar{J} = \bar{e}_{D/C}$, $\bar{K}$ up)

II.39 F_{AC} = 1.719 kN
F_{AD} = 1.613 kN
R = 0.957 kN

II.41 F_{AB} = 72N, F_{AC} = 664N

II.43 T_{AD} = 6.76 kN, T_{BD} = 8.65 kN, T_{CD} = 5.53 kN

II.45 −4.11(10^4) N^2

II.47 (a) 75.64°
(b) 100.66°
(c) 90°

II.49 39.88°

II.51 $-0.5196\,\bar{I} + 0.30\,\bar{J} + 0.8\,\bar{K}$

II.53 (a) $F_t = 5R/(h^2 + R^2)^{1/2}$ kN
$F_s = 5h/(h^2 + R^2)^{1/2}$ kN
(b) $\bar{F}_t = 5R(0.3420h\,\bar{I} + 0.9397h\,\bar{J} + R\,\bar{K})/(h^2 + R^2)$ kN
$\bar{F}_s = 5h(-0.3420R\,\bar{I} - 0.9397R\,\bar{J} + h\,\bar{K})/(h^2 + R^2)$ kN

II.55 (a) F_t = 2.56 kN
F_s = 4.29 kN
(b) $\bar{F}_t = 1.55\,\bar{I} + 1.55\,\bar{J} + 1.31\,\bar{K}$ kN
$\bar{F}_s = -1.55\,\bar{I} - 1.55\,\bar{J} + 3.69\,\bar{K}$ kN

II.57 W_2 = 20 lb, W_3 = 40 lb

II.59 F = 70.3 N, $\beta = 58.83°$

II.61 N_{lt} = 0.335 N
N_{rt} = 0.490 N

II.63 N_{lt} = 0.389 N
N_{rt} = 1.074 N

II.65 $\cos 2\theta - 2(kR/mg)(\sin 2\theta - \sin\theta) = 0$

II.67 (a) $mg/(k_1 + k_2)$
(b) $k_1 + k_2$

II.69 0.1011 m

II.71 $m = 160y/(y^2 + 0.01)^{1/2}$ kg

II.73 0.384 W

II.75 $l_1 = 0.228$ m
$l_2 = 0.150$ m

II.77 1125 lb

II.79 $T_{lt} = 628$ N
$T_{rt} = 439$ N
$T_{rear} = 844$ N

II.81 0.1719 m

II.83 $T_{lt} = 550$ lb
$T_{rt} = 599$ lb

II.85 $N_{groove} = 167$ N
$N_{block} = 1035$ N

III.1 (a) $-2\,\bar{I} - 16\,\bar{J} + 6\,\bar{K}$
(b) $-4\,\bar{I} + 8\,\bar{J} - 3\,\bar{K}$
(c) $-4\,\bar{I} - 2\,\bar{J} + 8\,\bar{K}$

III.3 (a) $22\,\bar{I} + 7\,\bar{J} - 2\,\bar{K}$
(b) $\bar{I} - 2\,\bar{J} - \bar{K}$

III.7 $Y_B = -1.70$ m, $Z_C = -0.0125$ m

III.9 $X_B = 2.1 - 2Z_C$ m
$\bar{e}_n = [0.4Z_C\,\bar{I} + (-2Z_C^2 + 2.4Z_C - 0.42)\,\bar{J} + (-0.8Z_C + 0.84)\,\bar{K}]/(4Z_C^4 - 9.6Z_C^3 + 2.48Z_C^2 - 3.36Z_C + 0.882)^{1/2}$

III.11 $M_E = 4620$ lb-in. (ccw)

III.13 $M_O = 12.99$ N-m (ccw)

III.15 $M_B = 20.49$ N-m (ccw)

III.17 $\theta = 53.13°$
$M_C = 3.75$ kN-m (cw)

III.19 $\phi = 90°$
$M_A = 250$ lb-ft (cw)

III.21 (a) 5.12°
(b) 95.12°

III.23 (a) $-4.8\,\bar{I} - 8.31\,\bar{J} + 16.63\,\bar{K}$ N-m
(b) 16.63 N-m
(c) $r_Z = 0.1386$ m
$|\bar{r}| = 0.160$ m
($\bar{J}$ rt, $\bar{K}$ up)

III.25 28.3 kip-in.

III.27 (a) $87.8\,\bar{J} + 202.8\,\bar{K}$ N-m
(b) 131.8 N-m

III.29 $M_{AB} = 0.360$ kN-m

III.31 (a) 8000 lb-ft

III.33 $F_A = 25.1$ N @ 26.57° cw from upward vert
$F_B = 25.1$ N @ 26.57° cw from downward vert

III.35 247 N-m (ccw)

III.37 $M = 15.81$ kN-m
$\alpha = 71.57°$
$\beta = 161.57°$
$\gamma = 90°$

III.39 $F = 7.071$ kN
$\alpha = 74.05°$
$\bar{M} = -0.8\,\bar{J} + 0.6\,\bar{K}$ kN-m
($\bar{J}$ rt, $\bar{K}$ up)

III.41 (a) & (b) $\bar{R} = 50\,\bar{I} - 86.6\,\bar{J}$ N-m
(a) $\bar{M}_A = -80\,\bar{K}$ N-m
(b) $\bar{M}_O = -85\,\bar{K}$ N-m
($\bar{I}$ rt, $\bar{J}$ up)

III.43 (a), (b) & (c) $\bar{R} = 1.04\,\bar{I} + 3.86\,\bar{J}$ kN
(a) $\bar{M}_A = -0.630\,\bar{K}$ kN-m
(b) $\bar{M}_B = 0.091\,\bar{K}$ kN-m
(c) $\bar{M}_C = 0.721\,\bar{K}$ kN-m

III.45 $\bar{R} = -10\,\bar{I} + 14.64\,\bar{J}$ N
$\bar{M}_O = -6\,\bar{K}$ N-m
($\bar{I}$ rt, $\bar{J}$ up)

III.47 $F = 267$ N
$\bar{R} = 154\,\bar{I} - 455\,\bar{J}$ N
($\bar{I}$ rt, $\bar{J}$ up)

III.49 $T = 172.8$ lb
$\bar{R} = -72\,\bar{I} - 522\,\bar{J}$ lb
($\bar{I}$ rt, $\bar{J}$ up)

III.51 $F_A = 1.0$ kips (up)
$F_B = 3.0$ kips (dn)

III.53 $F_A = -4.48\,\bar{J}$ kN
$F_B = 4.10\,\bar{I} - 8.39\,\bar{J}$ kn
($\bar{I}$ rt, $\bar{J}$ up)

III.55 (a) & (b) $\bar{R} = -40\,\bar{I}$ lb
(a) $\bar{M}_A = 170\,\bar{J} + 1449\,\bar{K}$ lb-in
(b) $\bar{M}_B = 170\,\bar{J} + 849\,\bar{K}$ lb-in
($\bar{J} = \bar{e}_{B/A}$, $\bar{K}$ up)

III.57 (a) $\bar{R} = -120\,\bar{K}$ N
$\bar{M}_A = 4.50\,\bar{K}$ N-m
(b) $M_{AD} = 4.50$ N-m
($\bar{I} = \bar{e}_{B/D}$, $\bar{K}$ up)

III.59 (a) & (b) $\bar{R} = -0.6\,\bar{I} + 2.5\,\bar{J} + 0.9\,\bar{K}$ kN
(a) $\bar{M}_A = -0.78\,\bar{K}$ kN-m
(b) $\bar{M}_B = -1.50\,\bar{I} - 0.36\,\bar{J} - 0.78\,\bar{K}$ kN-m
($\bar{J}$ rt, $\bar{K}$ up)

III.61 $\bar{F} = 70.7\,\bar{I} - 10.7\,\bar{K}$ N
$\bar{M} = 28.6\,\bar{I} + 70.7\,\bar{J} - 141.4\,\bar{K}$ N-m
($\bar{I} = \bar{e}_{B/A}$, $\bar{J} = \bar{e}_{C/B}$)

III.63 (a) $F = 3400$ lb
(b) $\bar{R} = -1200\,\bar{I} - 6600\,\bar{J}$ lb

$\bar{M}_A = -6000\,\bar{J} + 216{,}000\,\bar{K}$ lb-ft
($\bar{I} = \bar{e}_{A/B}$, $\bar{J} = \bar{e}_{C/B}$)

III.65 $R = 90$ kN (dn) @ 0.722 m from lt end

III.67 $R = 424$ N, 0.66 m above B @ 45° cw from $\bar{e}_{A/B}$

III.69 21.7 lb, 21.3 in. rt of C @ 64.03° cw from $\bar{e}_{D/C}$

III.71 $\bar{R} = 0.18\,\bar{I} - 2.04\,\bar{J}$ kN, 3.33 m above A
($\bar{I}$ rt, $\bar{J}$ up)

III.73 $R = 213.6$ kN, 69.44° cw from $\bar{e}_{B/A}$, 11.30 m rt of A

III.75 3.6 in. lt and 5.4 in. below the centroid

III.77 $\bar{R} = -61\,\bar{K}$ kN @ $1.72\,\bar{I} + 1.23\,\bar{J}$ m from near lt corner
($\bar{I}$ rt, $\bar{K}$ up)

III.79 $\bar{R} = -150\,\bar{K}$ N @ $0.373\,\bar{I} + 0.149\,\bar{J}$ m

III.81 40 N dn, along lt edge 0.125 m from far corner, or
40 N up, along rt edge 0.125 m from near corner

III.99 $\bar{F}_A = -0.10\,\bar{I} + 12.05\,\bar{J}$ kN
($\bar{I}$ rt, $\bar{J}$ up)

III.101 0.497 kN-m

III.103 For 20° dn: $N_{rt} = 5.23$ kN, $N_{lt} = 6.75$ kN
For no slope: $N_{rt} = 6.12$ kN, $N_{lt} = 6.63$ kN

III.105 $m_B/m_A = (3\cos\theta + \sin\theta) / (3\cos\theta - \sin\theta)$

III.107 $T_1 = 6.83$ kN
$T_2 = 11.32$ kN
$\alpha = 83.66°$

III.109 40 lb (lt), bears on rt surface

III.111 $\bar{F}_A = 224\,\bar{I}$ N
$\bar{F}_B = -224\,\bar{I} + 736\,\bar{J}$ N
($\bar{I}$ rt, $\bar{J}$ up)

III.113 3000 lb (lt)

III.115 235 kips

III.117 832 N (T)

III.119 $F_{lt} = 0.5$ kN (up)
$F_{rt} = 0.5$ kN (dn)

III.121 $F_{lt} = 500$ lb (up)
$M_{lt} = 2000$ lb-ft (ccw)

III.123 $\bar{F}_{lt} = -0.54\,\bar{I} + 7.07\,\bar{J}$ kN
$\bar{F}_{rt} = 2.54\,\bar{I} + 4.39\,\bar{J}$ kN
($\bar{I}$ rt, $\bar{J}$ up)

III.125 $\bar{F}_A = -158.6\,\bar{I} + 80\,\bar{J}$ kN
$\bar{F}_B = 118.6\,\bar{I}$ kn
($\bar{I}$ rt, $\bar{J}$ up)

III.127 (a) $A = 37.5$ kN (T), $B = 50$ kN (C), $C = 7.5$ kN (T)
(b) $A = 31.3$ kN (T), $B = 42.5$ kN (C),
$C = 6.3$ kN (T)

III.129 4.02 kN

III.131 $\theta - \dfrac{mgd}{kR}\sin\theta = 0$

III.133 $m_A/m_B = 2\,[\cos\alpha\cos\beta/\cos(\alpha + \beta)] - 1$

III.135 $\theta = \sin^{-1}[(2d/L)^{1/3}]$, for $d < L/2$

III.137 (a) complete constraint: 1,2,3,5,6,9,10,11
partial constraint: 4
improper constraint: 7,8,12
(b) determinate: 1,2,6,10
indeterminate: 3,5,9,11

III.139 0.5 m

III.141 0

III.143 143.13°

III.145 $\bar{F}_A = -100\,\bar{I} - 200\,\bar{J} + 150\,\bar{K}$ N
$\bar{M}_A = 100\,\bar{I} - 72\,\bar{J} + 156\,\bar{K}$ N-m
($\bar{I} = \bar{e}_{D/C}$, $\bar{J} = \bar{e}_{B/A}$)

III.147 $m = 35.7$ kg
$R = 0.0756$ m
$\theta = 160.89°$

III.149 44.1 N

III.151 $T_{BD} = 160.1$ N
$T_{BE} = 98.1$ N

III.153 $T_{CE} = 15.81$ kips
$T_{BD} = 16.59$ kips
$T_{CF} = 7.91$ kips

III.155 $\bar{F}_A = -50\,\bar{I} - 33.3\,\bar{J}$ N
$\bar{F}_B = -66.7\,\bar{J} - 16.7\,\bar{K}$ N
$\bar{F}_C = 16.7\,\bar{K}$ N
($\bar{J}$ rt, $\bar{K}$ up)

III.157 $T_{CD} = 3000$ lb
$F_B = 0$

III.159 $\bar{N}_D = 4.25\,\bar{I} + 7.35\,\bar{J}$ N
$\bar{F}_A = -4.25\,\bar{I} - 7.45\,\bar{J} + 26.28\,\bar{K}$ N
$\bar{F}_B = 0.10\,\bar{J} + 12.95\,\bar{K}$ N
($\bar{I} = \bar{e}_{A/B}$, $\bar{K}$ up)

III.161 49.0 N *(lt)*

IV.1 $F_{AB} = 8$ kN (T)
$F_{AC} = 6$ kN (T)
$F_{BC} = 10$ kN (C)

IV.3 $F_{1-2} = 205$ lb (C)
$F_{1-3} = 220$ lb (C)
$F_{2-3} = 564$ lb (C)

IV.5 $F_{AB} = 0$
$F_{AD} = 6$ kN (C)
$F_{BC} = 5$ kN (T)
$F_{BD} = 3.61$ kN (T)
$F_{CD} = 4$ kN (C)

IV.7 $F_{AB} = 8$ kN (T)
$F_{AC} = 11.31$ kN (C)
$F_{AF} = 8$ kN (T)
$F_{BC} = 12$ kN (C)
$F_{CD} = 12$ kN (C)
$F_{CE} = 11.31$ kN (C)
$F_{CF} = 16$ kN (T)
$F_{DE} = 0$
$F_{EF} = 8$ kN (T)

IV.9 $F_{AB} = F_{CE} = F_{CF} = 0$
$F_{AG} = 29.6$ kN (C)
$F_{BC} = 11.5$ kN (C)
$F_{BF} = 23.1$ kN (T)
$F_{CD} = 11.5$ kN (C)
$F_{DE} = 10.5$ kN (C)
$F_{EF} = 10.5$ kN (C)
$F_{FG} = 29.6$ kN (C)

IV.11 $F_{1-2} = 22.5$ kN (C)
$F_{1-8} = 16.88$ kN (C)
$F_{2-3} = 3.75$ kN (C)
$F_{2-6} = 12.5$ kN (C)
$F_{2-7} = 5$ kN (C)
$F_{2-8} = 18.75$ kN (T)
$F_{3-4} = 3.75$ kN (C)
$F_{3-6} = 0$
$F_{4-5} = 5$ kN (C)
$F_{5-6} = 0$
$F_{6-7} = 11.25$ kN (T)
$F_{7-8} = 11.25$ kN (T)

IV.13 $F_{AB} = F_{DE} = 260$ lb (T)
$F_{AI} = F_{EF} = 520$ lb (C)
$F_{BC} = F_{CD} = 673$ lb (T)
$F_{BH} = F_{DG} = 312$ lb (C)
$F_{BI} = F_{DF} = 520$ lb (T)
$F_{CG} = F_{CH} = 104$ lb (T)
$F_{GH} = 667$ lb (C)
$F_{FG} = F_{HI} = 460$ lb (C)

IV.15 $F_{AB} = F_{AC} = F_{AD} = F_{AE} = F_{BF} = F_{CF} = F_{DF} = F_{EF} =$ 1.382 kN (T)
$F_{BC} = F_{BE} = F_{CD} = F_{DE} = 0.833$ kN (C)

IV.17 $F_{AB} = 760$ kN (T)
$F_{AC} = 13.0$ kN (C)
$F_{AD} = F_{BD} = 72.1$ kN (C)
$F_{BC} = 65.0$ kN (C)
$F_{CD} = 78.0$ kN (T)
$F_{DE} = 101.8$ kN (T)

IV.19 $F_{AB} = 6.15$ kips (T)
$F_{AC} = 1.04$ kips (C)
$F_{AD} = 9.22$ kips (C)
$F_{BC} = 0.67$ kips (C)
$F_{BD} = 0.90$ kips (C)
$F_{CD} = 6.04$ kips (T)

IV.21 $F_{1-2} = 207$ lb (C)
$F_{1-3} = 1172$ lb (T)
$F_{1-4} = 325$ lb (T)
$F_{2-3} = 75$ lb (T)
$F_{2-4} = 207$ lb (C)
$F_{2-5} = 400$ lb (T)
$F_{3-4} = 1586$ lb (C)

IV.23 $F_{AD} = 180$ kN (T)
$F_{BD} = 39.0$ kN (C)
$F_{BE} = 36.0$ kN (C)
$F_{CD} = 58.5$ kN (C)
$F_{CE} = 10.1$ kN (C)
$F_{CF} = 45.0$ kN (C)
$F_{DE} = 15.0$ kN (T)
$F_{DF} = 22.5$ kN (T)
$F_{DG} = 90.0$ kN (T)
$F_{EF} = 4.5$ kN (C)
$F_{EG} = 39.0$ kN (C)
$F_{EH} = 9.0$ kN (C)
$F_{FG} = 58.5$ kN (C)
$F_{FH} = 10.1$ kN (T)
$F_{FI} = 0$
$F_{GH} = 15.0$ kN (T)
$F_{GI} = 22.5$ kN (T)
$F_{HI} = 13.5$ kN (C)

IV.25 $F_{3-4} = 7.5$ kips (T)
$F_{3-6} = 20.0$ kips (T)

IV.27 $F_{CD} = 32$ kN (C)
$F_{DG} = 22.6$ kN (T)

IV.29 $F_{2-9} = 0$
$F_{8-9} = 40.6$ kips (C)

IV.31 31.2 kN (T)

IV.33 $F_{2-7} = 5.66$ kN (T)
$F_{3-7} = 6.0$ kN (C)
$F_{6-7} = 13.42$ kN (C)

IV.35 $F_{4-12} = 0$
$F_{5-12} = 166.7$ N (C)

IV.37 $F_{4-9} = 1.08$ kN (C)
$F_{9-11} = 6.15$ kN (T)

IV.39 $F_{1-2} = 1667$ lb (C)
$F_{1-3} = 4800$ lb (T)
$F_{4-12} = 2692$ lb (C)
$F_{6-10} = 2115$ lb (T)

IV.41 $d = 5$ m: $F_{CD} = 67.1$ kN (C) & $F_{FG} = 35$ kN (T)
$d = 2.5$ m: no

IV.43 $F_{IJ} = 50.0$ kips (C)
$F_{IO} = 6.25$ kips (T)

IV.45 $F_{BF} = 0$
$F_{CG} = 21.3$ kN (T)

IV.47 $F_{DE} = 15.0$ kN (T)
$F_{DF} = 22.5$ kN (T)
$F_{DG} = 90.0$ kN (T)

IV.49 $F_3 = 15.56$ kN (C)
$F_4 = 13.33$ kN (C)

IV.51 $\bar{F}_A = -211\,\bar{I} - 70\,\bar{J}$ lb
$\bar{F}_C = 211\,\bar{I} + 470\,\bar{J}$ lb
($\bar{I}$ rt, $\bar{J}$ up)

IV.53 $F_{bolt} = 4P$
$F_A = 5P$ (dn on AC)

IV.55 20 N-m (ccw)

IV.57 500 lb-ft (ccw)

IV.59 $k = 418$ N/m
$\bar{F}_A = 5.7\,\bar{I} + 78.5\,\bar{J}$ N
($\bar{I}$ rt, $\bar{J}$ up)

IV.61 $F_A = 1$ kN (up)
$F_B = 5$ kN (up)
$M_B = 1.6$ kN-m (cw)

IV.63 $F_A = 2$ kN (up)
$M_A = 0.8$ kN-m (ccw)
$F_B = 4$ kN (up)
$M_B = 1.2$ kN-m (cw)

IV.65 member AC: $\bar{F}_A = 362\,\bar{I} - 43\,\bar{J}$ N
$\bar{F}_B = -696\,\bar{I} + 187\,\bar{J}$ N
$\bar{F}_C = 335\,\bar{I} - 230\,\bar{J}$ N
member CE: $\bar{F}_C = -335\,\bar{I} + 230\,\bar{J}$ N
$\bar{F}_D = -696\,\bar{I} - 387\,\bar{J}$ N
$\bar{F}_E = -362\,\bar{I} + 243\,\bar{J}$ N
($\bar{I}$ rt, $\bar{J}$ up)

IV.67 bar AB: $\bar{F}_A = 45\,\bar{I} + 38.7\,\bar{J}$ kN
$\bar{F}_B = -60\,\bar{I} + 6.3\,\bar{J}$ kN
bar BC: $\bar{F}_B = 60\,\bar{I} - 6.3\,\bar{J}$ kN
$\bar{F}_C = -75\,\bar{I} + 51.3\,\bar{J}$ kN
($\bar{I}$ rt, $\bar{J}$ up)

IV.69 $\bar{F}_A = -28.7\,\bar{I} + 100.3\,\bar{J}$ kN
$\bar{F}_D = 28.7\,\bar{I} + 49.7\,\bar{J}$ kN
($\bar{I}$ rt, $\bar{J}$ up)

IV.71 $\bar{F}_A = -268\,\bar{I} + 3000\,\bar{J}$ lb
$\bar{F}_E = 268\,\bar{I} + 5000\,\bar{J}$ lb
($\bar{I}$ rt, $\bar{J}$ up)

IV.73 (a) bar AB: $\bar{F}_A = -0.289W\,\bar{I} + 0.5\,W\,\bar{J}$
$\bar{F}_B = 0.289W\,\bar{I} + 0.5W\,\bar{J}$
bar AC: $\bar{F}_A = 0.289W\,\bar{I} + W\,\bar{J}$
$\bar{F}_C = -0.289W\,\bar{I}$
bar BC: $\bar{F}_B = -0.289W\,\bar{I} + W\,\bar{J}$
$\bar{F}_C = 0.289W\,\bar{I}$
(b) bar AB: $\bar{F}_A = -\bar{F}_B = -0.289W\,\bar{I}$
bar AC: $\bar{F}_A = -\bar{F}_C = 0.289W\,\bar{I} + 0.5W\,\bar{J}$
bar BC: $\bar{F}_B = -\bar{F}_C = -0.289W\,\bar{I} + 0.5W\,\bar{J}$
($\bar{I}$ rt, $\bar{J}$ up)

IV.75 $M_C = 3750$ N-m (cw)

IV.77 $M_2 = 2(1 + r_2/r_1)M_1$ (cw)
$M_3 = (1 + 2r_2/r_1)M_1$ (ccw)

IV.79 $\bar{F}_A = 2.67\,\bar{I} + 1.60\,\bar{J}$ kips
$\bar{F}_F = -2.67\,\bar{I} + 3.40\,\bar{J}$ kips
($\bar{I}$ rt, $\bar{J}$ up)

IV.81 $F_A = 4$ kN (up)
$F_B = 4$ kN (dn)

IV.83 67.5 N

IV.85 $\bar{F}_{plate} = -145.9\,\bar{I}$ lb
$\bar{F}_A = -173.8\,\bar{J}$ lb
$\bar{F}_E = -192.3\,\bar{I} + 192.3\,\bar{J}$ lb
($\bar{I}$ rt, $\bar{J}$ up)

IV.87 $\bar{F}_A = 8\,\bar{I} - 21.6\,\bar{J}$ kN
$\bar{F}_C = -8\,\bar{I} + 25.6\,\bar{J}$ kN
$\bar{F}_D = -1.01\,\bar{I} - 0.08\,\bar{J}$ kN
$\bar{F}_F = 1.01\,\bar{I} - 3.92\,\bar{J}$ kN
($\bar{I}$ rt, $\bar{J}$ up)

IV.89 $1.5\,P\tan\theta$ (rt)

IV.91 $F_{sp} = 27.26$ kN (C)
$\bar{F}_A = -3.12\,\bar{I} - 1.25\,\bar{J}$ kN
$\bar{F}_B = 10.18\,\bar{I} - 21.08\,\bar{J}$ kN
($\bar{I}$ rt, $\bar{J}$ up)

V.1 $\frac{a}{3}$ forward of rear face, $\frac{b}{3}$ above bottom,
$\frac{h}{2}$ lt of rt face

V.3 $(\frac{2}{3}h, 0, 0)$

V.5 $(\frac{3}{8}a, 0, 0)$

V.7 $\left(\frac{5}{16}h, -\frac{a}{4}, 0\right)$

V.9 $0.3619a$ on ctr line from flat face

V.11 $(0.807a, 0.540b, 0.2421c)$

V.13 $\left(\frac{4}{3\pi}a, \frac{4}{3\pi}b\right)$

V.15 $(\frac{4}{5}a, \frac{2}{7}b)$

V.17 $\left[\dfrac{(n-1)^2a}{2n\left(n-1+\ln\frac{1}{n}\right)}, \dfrac{(n-1)^2b}{2n\left(n-1+\ln\frac{1}{n}\right)}\right]$

V.19 $(0.574a, 0.410a)$

V.21 $\frac{3}{4}h$ from apex on ctr line

V.23 $\dfrac{h}{5}\left[\dfrac{3(a^2+4h^2)^{3/2}}{(a^2+4h^2)^{3/2}-a^3} - \dfrac{a^2}{2h^2}\right]$ from apex on ctr line

V.25 0.1611 m dn from top ctr

V.27 $\dfrac{4}{3\pi}\dfrac{r_2^2 + r_1r_2 + r_1^2}{r_1 + r_2}$ up from bot ctr

V.29 7.25 in. lt, 3.25 in. up from rt tip

V.31 0.289 m rt, 0.296 m up from lower lt corner

V.33 0.01045 ft lt, 0.00523 ft up from ctr

V.35 0.125 m dn from top ctr

V.37 0.779 m from lt tip on ctr line

V.39 11.592 in. rt, 3.147 in. up from lower lt corner

V.41 0.367 in. up from ctr

V.43 1.5°

V.45 8.36° or 38.36°

V.47 4.25 m from lt apex on ctr line

V.49 (a) $m = 2.85$ kg, G is 22.2 mm up from bot ctr
(b) $m = 3.81$ kg, G is 21.0 mm up, 297 mm lt from bot rt ctr

V.51 $m = 8.58$ kg, 163.1 mm from apex on ctr line

V.53 $R = 2.4$ kN @ 0.667 m rt of A
$R_A = 1.40$ kN (up)
$R_B = 1.00$ kN (up)

V.55 couple $= \frac{11}{6} wL^2$ N-m (cw)
$R_A = \frac{11}{12} wL$ N (dn)
$R_B = \frac{11}{12} wL$ N (up)

V.57 $R = 2w_0L/\pi$ @ $L(1-2/\pi)$ rt of A
$R_A = 4w_0L/\pi^2$ (up)
$R_B = (2w_0L/\pi)(1-2/\pi)$ (up)

V.59 $R_A = 4.33$ kips (up)
$R_B = 0.67$ kips (up)

V.61 $R_A = 10.92$ kN (up)
$R_B = 15.48$ kN (up)

V.63 (a) $q_1 = 2q_2$, $R_A = \frac{1}{2} q_2L$ (dn)
(b) $q_1 = q_2$, $M_A = q_2L^2/6$ (ccw)

V.67 $\bar{R}_A = 1.571\ qr\ \bar{K}$
$\bar{M}_A = qr^2(\bar{I} + 0.571\ \bar{J})$
($\bar{J}$ rt, $\bar{K}$ up)

V.69 $\bar{R}_A = q_0r\ \bar{J}$
$\bar{M}_A = q_0r^2(0.5\ \bar{I} + 0.7854\ \bar{K})$
($\bar{I}$ rt, $\bar{J}$ up)

V.71 $\bar{R}_A = 3.93\ \bar{I}$ kN
$\bar{M}_A = 9.82\ \bar{J}$ kN-m
($\bar{J}$ rt, $\bar{K}$ up)

V.73 $\bar{R}_A = 2.50\ \bar{I}$ kN
$\bar{M}_A = 6.67\ \bar{J} - 0.83\ \bar{K}$ kN-m
($\bar{J}$ rt, $\bar{K}$ up)

V.75 $\bar{R}_A = 3.82\ \bar{I}$ kN
$\bar{M}_A = 9.54\ \bar{J} + 0.01\ \bar{K}$ kN-m
($\bar{J}$ rt, $\bar{K}$ up)

V.77 $R = 178.5$ kN (dn) @ 9.07 m from lt edge along ctr line

V.79 $R = 48.7$ kN (dn) @ 1 m rt, 1.5 m forward from pt of max p

V.81 $\bar{R} = -0.25p_0ab\ \bar{K}$ @ $2a/3\ \bar{I} + 2b/3\ \bar{J}$

V.83 $R_{front} = 280.8$ lb @ 0.5 ft up from bot on ctr line
$R_{side} = 140.4$ lb @ 0.5 ft up from bot on ctr line
$R_{bot} = 748.8$ lb @ ctr of bot

V.85 P = 8356 kN (⊥ wall) @ 8 m depth

V.87 (a) 71.3 kN
(b) 159.5 kN

V.89 (a) 71.9 kN
(b) 194.1 kN

V.91 398 lb

V.93 7.27 kN

V.95 $R = 59.96$ kN @ 11.11° below lt horiz, intersecting ctr

V.97 $\bar{R} = -499\ \bar{I} + 1082\ \bar{J}$ lb @ $1.016\ \bar{I} + 0.254\ \bar{J}$ ft

V.99 1.961 kN

V.101 2.43 kN

V.103 (a) 5611 lb
(b) 23,587 lb

V.105 $a = 0.5656R$

V.107 1: 13.15 kN @ 25.9 mm below ctr
2: 18.59 kN @ ctr
3: 18.59 kN @ 73.2 mm below ctr

V.109 (a) vertex up: $R = \frac{1}{6}\rho g\ ab(3h + 2a)$ @ $\frac{a}{2}\,\frac{2h + a}{3h + 2a}$ above A
vertex dn: $R = \frac{1}{6}\rho g\ ab(3h + a)$ @ $\frac{a}{2}\left(\frac{4h + a}{3h + a}\right)$ above A
(b) vertex up: $\bar{R}_A = \frac{1}{6}\rho g\ ab(3h + 2a)\ \bar{k}$
$\bar{M}_A = -\frac{1}{12}\rho g\ a^2b\ (2h + a)\ \bar{\imath}$
vertex dn: $\bar{R}_A = \frac{1}{6}\rho g\ ab\ (3h + a)\ \bar{k}$
$\bar{M}_A = -\frac{1}{12}\rho g\ a^2b\ (4h + a)\ \bar{\imath}$
($\bar{\imath}$ rt, $\bar{\jmath}$ dn)

V.111 243 N

V.113 $T_A = T_B = 77.1$ lb
$T_C = 154.1$ lb

VI.1 (a) $R_a = 0$
$|R_s| = 7.5$ kN
$|M_b| = 0.75$ kN-m
(b) $R_a = 0$
$|R_s| = 2.5$ kN
$|M_b| = 2$ kN-m

VI.3 (a) $R_a = 3$ kips
$|R_s| = 10$ kips
$|M_B| = 21$ kip-ft

(b) $R_a = 3$ kips
$|R_s| = 6$ kips
$|M_b| = 5$ kip-ft

VI.5 (a) $R_a = 0$
$|R_s| = 1.85$ kN
$|M_b| = 2.35$ kN-m
(b) $R_a = 4.8$ kN
$|R_s| = 4.0$ kN
$|M_b| = 1.6$ kN-m

VI.7 (a) $R_a = -1.73$ kips
$|R_s| = 2.63$ kips
$|M_b| = 97.0$ kip-in.
(b) $R_a = 0$
$|R_s| = 1.63$ kips
$|M_b| = 8.15$ kip-in.

VI.9 (a) $R_a = -10.33$ kN
$|R_s| = 4.10$ kN
$|M_b| = 0.459$ kN-m
(b) $R_a = 1.518$ kN
$|R_s| = 3.83$ kN
$|M_b| = 1.290$ kN-m

VI.11 (a) $R_a = 0$
$|R_s| = 2.57$ kips
$|M_b| = 15.4$ kip-ft
(b) $R_a = 0$
$|R_s| = 6$ kips
$|M_b| = 42.0$ kip-ft

VI.13 (a) $R_a = 0.333q_0R$
$|R_s| = 0.083q_0R$
$|M_b| = 0.0417q_0R^2$
(b) $R_a = -0.455q_0R$
$|R_s| = 0.12q_0R$
$|M_b| = 0.1725q_0R^2$

VI.15 $R_a = 4$ kN
$|R_s| = 5$ kN
$|M_b|$ 2.72 kN-m
$|M_t| = 0.45$ kN-m

VI.17 $R_a = 0.707\ F$
$|R_s| = 1.225\ F$
$|M_b| = FR$
$|M_t| = 0.293\ FR$

VI.19 $|V| = P$
$|M|_{max} = PL$ @ lt end

VI.21 $|V|_{max} = 3.2$ kN @ lt end
$|M|_{max} = 5.12$ kN-m @ 3.2 m from lt end

VI.23 $|V|_{max} = \frac{1}{4}\ w_0L$ @ midpt
$|M|_{max} = \frac{1}{6}\ w_0L^2$ @ lt end

VI.25 $|V|_{max} = 12.5$ lb on $8 < x < 16$ in. from lt end
$|M|_{max} = 120$ lb-in. on $16 < x < 24$ in. from lt end

VI.27 $|V|_{max} = 32$ kN @ lt end
$|M|_{max} = 192$ kN-m @ 8 m from lt end

VI.29 $|V|_{max} = 10$ kn @ midpt
$|M|_{max} = 10$ kN-m @ midpt

VI.31 $|V|_{max} = 600$ N @ 2 m from lt end
$|M|_{max} = 133.3$ N-m @ 1 m and 2 m from lt end

VI.33 $|V|_{max} = 100$ N on $0 < x < 0.1$ m from rt end
$|M|_{max} = 10$ N-m @ 0.1 m from rt end

VI.35 $|V|_{max} = 12$ kN on $0 < x < 2$ m from lt end
$|M|_{max} = 24$ kN-m on $2 < x < 6$ m from lt end

VI.37 $|V|_{max} = 20.7$ kN on $0 < x < 0.4$ m from lt end
$|M|_{max} = 14.67$ kN-m @ 1 m from lt end

VI.39 $|V|_{max} = 12.5$ lb on $8 < x < 16$ in. from lt end
$|M|_{max} = 120$ lb-in on $16 < x < 24$ in. from lt end

VI.41 $|V|_{max} = \frac{1}{3}w_0L$ @ lt end
$|M|_{max} = 0.0642w_0L^2$ @ $0.423L$ from lt end

VI.43 $|V|_{max} = 32$ kN @ lt end
$|M|_{max} = 192$ kN-m @ 8 m from lt end

VI.45 $|V|_{max} = 21.3$ kips @ rt end
$|M|_{max} = 15.80$ kip-ft @ 1.723 ft from lt end

VI.47 $|V|_{max} = 30$ kN @ rt end
$|M|_{max} = 60$ kN @ rt end

VI.49 (a) $q = 5$ kips/ft (dn), on $0 < x < 5$ ft
$F_1 = 10$ kips (up), @ $x = 0$
$F_2 = 5$ kips (dn), @ $x = 3$ ft
$F_3 = 25$ kips (up), @ $x = 5$ ft
$F_4 = 5$ kips (dn), @ $x = 9$ ft
(b) 9 ft
(c) $|M|_{max} = 22.5$ kip-ft @ $x = 5$ ft

VI.51 $|V|_{max} = 2650$ N @ 15 m from lt end
$|M|_{max} = 8640$ N-m @ 7.84 m from lt end

VI.53 (a) 14.40 lb
(b) 50 in.

VI.55 (a) 3.83 m
(b) $T_{AB} = 3.54$ kN
(c) $\theta_{AB} = 62.45°$

VI.57 (a) $T_{DE} = 1.18$ kN
(b) 15.49 m

VI.59 (a) $T_{min} = 1.226(10^5)$ kN
$T_{max} = 1.320(10^5)$ kN
(b) 1059 m

VI.61 $\dfrac{L_2}{h_2} = \left[\dfrac{L_1^2}{L_2^2}\left(\dfrac{L_1^2}{h_1^2} + 16\right) - 16\right]^{1/2}$

VI.63 $h = 75$ ft
$T_A = T_B = 1.061(10^5)$ lb

VI.65 (a) $T_A = 231$ kN
(b) 106.09 m

VI.67 $h = 0.333$ m
$s = 56.30$ m

VI.69 (a) 131.7 kips
(b) 24.93 ft
(c) 2.41 ft

VI.71 (a) $y = \frac{1}{9}\frac{W}{T_O L^2} x^2(x + 3L)$
(b) $\frac{4}{9}\frac{WL}{T_O}$
(c) $(T_O^2 + W^2)^{1/2}$

VI.73 $s = 202.23$ ft
$h = 12.99$ ft

VI.75 $0.203\ w_0 d$

VI.77 (a) $0.1468\ w_0 d$
(b) $0.1288\ w_0 d$

VI.79 (a) 45.46 ft
(b) $F_{hor} = 59.6$ lb
$F_{vert} = 50$ lb

VI.81 0.1408

VI.83 0.66 m or 74.08 m

VII.1 147.1 N

VII.3 7.01 N

VII.5 2.83 kg

VII.7 3.93 N (slipping betw A & B)

VII.9 All slide together

VII.11 0.884

VII.13 0

VII.15 (a) 0.484
(b) $\bar{e} = -0.659\ \bar{I} + 0.706\ \bar{J} - 0.257\ \bar{K}$

VII.17 8.53°

VII.19 $74 < P < 368$ N

VII.21 287 N-m

VII.23 $L = 0.5\left(\frac{d}{\mu} - h\right)$

VII.25 (a) 1.439 kN (lt)
(b) 0.746

VII.27 0.311

VII.29 0.570 kN

VII.31 210 N

VII.33 Yes

VII.35 8.83 kN (up)

VII.37 (a) 43.7 lb (lt)
(b) 1.017

VII.39 (a) 550 N
(b) 1450 N

VII.41 $\mu_A = 0.420$
$\mu_C = 0.141$
$\mu_D = 0.598$

VII.43 34.23°

VII.45 25.02°

VII.47 (a) $1.5 < d < 2$ m
(b) $2 < d < 3$ m

VII.49 $0 < h < 3.57$ m

VII.51 (a) & (b) 235 N
(c) 136 N

VII.53 (a) 2.67 m
(b) & (c) 36.8 N

VII.55 (a) 4.0 m
(b) 24.5 N
(c) 49.0 N

VII.57 No sliding: $\beta < \tan^{-1}\mu$
No tipping: $\beta < \tan^{-1}\frac{b}{h}$

VII.59 1.197 kN

VII.61 0.245 kN

VII.63 2.20 kips

VII.65 (a) 2.69 kN
(b) 0.0145
(c) Yes

VII.67 (a) $F_B = 1.880$ kN
(b) $F_A = 0$, $F_C = 2$ kN (rt)

VII.69 219 lb

VII.71 1.842 N-m

VII.73 1.047 N-m

VII.75 110 N

VII.77 238 lb-in

VII.79 273 N

VII.81 8.47 N-m

VII.83 (a) 31.0 lb
(b) 149.0 lb

VII.85 $0.4 < T < 1.071(10^5)$ N

VII.87 $T = 68.9$ N

VII.89 (a) 39.4 N-m
(b) 33.7 N-m

VII.91 $M_A = 70.2$ N-m
$M_B = 281$ N-m
Slipping at A first

VII.93 $M_A = 92.1$ lb-in (cw)
$M_B = 184.2$ lb-in (ccw)

VIII.1 $(mgr_2 - kr_1^2\,\theta)\,\delta\theta$

VIII.3 $[-kl^2(1 - \cos\theta)\sin\theta/\cos^2\theta]\delta\theta$

VIII.5 $[M_A - FL(4 \sin^2 \theta + 1)/2]\delta\theta$

VIII.7 $-\dfrac{a}{(4b^2 - a^2)^{1/2}}(F_2 + 2F_3)\delta a$

VIII.9 $\theta = \sin^{-1}\left[\left(\dfrac{2a}{L}\right)^{1/3}\right]$

VIII.11 26.0 lb (rt)

VIII.13 $M_A = 450\dfrac{\sin\theta\,(1-\cos\theta)}{\cos^3\theta}$ N-m

VIII.15 232 lb

VIII.17 26.1 kN/m

VIII.19 $P = 150\left(1 - 6\tan\dfrac{\theta}{2}\right)$ N

VIII.21 1.6 kN-m (cw)

VIII.23 5.11 kN (up)

VIII.25 3.45 kN-m (cw)

VIII.27 $m_B/m_A = 0.5 \sin\alpha/\sin\beta$

VIII.29 1204 N (rt)

VIII.31 (a) 1158 lb-in (cw)
(b) 442 lb-in (cw)

VIII.33 $m = \dfrac{2kL}{g}\cot^3\theta\left[1 + \left(\dfrac{T_0}{kL} - 1\right)\sin\theta\right]$

VIII.35 $M_A = 450\,\dfrac{\sin\theta\,(1-\cos\theta)}{\cos^3\theta}$ N-m

VIII.37 $m = (0.8\,kl/g)\,(2\sin\theta - 1)$

VIII.41 3520 lb (lt)

VIII.43 $F = 780\tan\theta\,[1 - 10(1252 - 1248\cos\theta)^{-1/2}]$ lb

VIII.45 (a) & (b) $\theta = 90°$ is unstable

For $m < kl/g$, $\theta = \sin^{-1}\left(1 - \dfrac{mg}{kl}\right)$ is stable

For $m > kl/g$, $\theta = 0$ is stable
(θ is angle of elev of bar AB)

VIII.47 (a) $\sin c\theta = -\sin\theta$
(b) $\theta = 0$ is unstable
$\theta = 180°$ is stable
(c) $\theta = 0°$ is unstable
$\theta = 180°$ is unstable
$\theta = 120°$ is stable

VIII.49 If $k < 1.531\,mg$, $\theta = 0$ is stable
If $k > 1.531\,mg$, $\theta = 0$ is unstable
and $\theta = \cos^{-1}(1.531\,mg/k)$ is stable

VIII.51 $k = mg\,a/L^2$ is neutral equil

VIII.53 $K > 0.5\,mg/b^2$

VIII.55 $P = c^2\,kl$

IX.1 $I_x = ab^3/12$
$I_y = a^3b/12$

IX.3 $I_x = r^4\,(2\alpha - \sin 2\alpha)/8$
$I_y = r^4\,(2\alpha + \sin 2\alpha)/8$

IX.5 $I_x = 4ab^3/(9\pi)$
$I_y = (2/\pi^3)\,(\pi^2 - 8)a^3b$

IX.7 $J_z = ab(a^2 + b^2)/12$
$I_{xy} = a^2b^2/24$

IX.9 $J_z = R^4\alpha/2,\ I_{xy} = 0$

IX.11 $J_z = 2ab[9(\pi^2 - 8)a^2 + 2\pi^2b^2]/(9\pi^2)$
$I_{xy} = (\pi^2 - 4)a^2b^2/(8\pi^2)$

IX.13 $R/2$

IX.15 1.046 m

IX.17 37.0 mm

IX.19 $I_x = (5\pi/4)r^4$
$J_z = (3\pi/2)r^4$
$I_{xy} = 0$

IX.21 $I_x = 2.071(10^4)$ in.^{4}
$J_z = 7.823(10^4)$ in.^{4}
$I_{xy} = 2.761\,(10^4)$ in.^{4}

IX.23 $I_x = 2.94(10^{-7})$ m^4
$J_z = 5.10(10^{-7})$ m^4
$I_{xy} = 2.16(10^{-7})$ m^4

IX.25 $I_x = 0.2581$ m^4
$J_z = 1.0624$ m^4
$I_{xy} = -0.3703$ m^4

IX.27 (a) $A = 30$ in.^{2}
(b) $I_{\hat{x}} = 750$ in.^{4}
$I_{\hat{y}} = 1080$ in.^{4}

IX.29 $I_x = 6.29(10^{-3})$ m^4
$I_{xy} = 1.413(10^{-2})$ m^4
$k_x = 0.1686$ m

IX.31 $I_x = 5035$ in.^{4}
$I_{xy} = 2112$ in.^{4}
$k_x = 5.349$ in.

IX.33 $J_z = 7.89(10^{-2})$ m^4
$K_z = 0.573$ m

IX.35 $J_z = 10{,}070$ in.^{4}
$k_z = 7.56$ in.

IX.37 G is 142.5 mm above lower edge on ctr line
$I_x = 2.29(10^{-4})$ m^4
$I_y = 6.12(10^{-4})$ m^4
$I_{xy} = 0$
$J_z = 8.41(10^{-4})$ m^4

IX.39 G is 242.8 mm rt of ctr of lt circle
$I_x = 3.50(10^{-3})$ m^4
$I_y = 3.92(10^{-2})$ m^4
$I_{xy} = 0$
$J_z = 4.27(10^{-2})$ m^4

IX.41 G is at midpt
$I_x = 1.221(10^{-3})$ m^4
$I_y = 1.048(10^{-2})$ m^4
$I_{xy} = 2.58(10^{-3})$ m^4
$J_z = 1.170(10^{-2})$ m^4

IX.43 G is 285 mm rt & 100 mm below the upper lt corner
$I_x = 3.23(10^{-4})$ m^4
$I_y = 2.44(10^{-3})$ m^4
$I_{xy} = -2.28(10^{-4})$ m^4
$J_z = 2.76(10^{-3})$ m^4

IX.45 (a) $I_{x'} = 2.34(10^{-5})$ m^4
$I_{y'} = 3.69(10^{-5})$ m^4
$I_{x'y'} = -2.32(10^{-5})$ m^4
(b) $I_{x'} = I_{y'} = 3.02(10^{-5})$ m^4
$I_{x'y'} = 2.41(10^{-5})$ m^4
(c) $I_{x'} = 1.81(10^{-5})$ m^4
$I_{y'} = 4.22(10^{-5})$ m^4
$I_{x'y'} = 2.09(10^{-5})$ m^4

IX.47 (a) $I_{x'} = 6.06(10^{-3})$ m^4
$I_{y'} = 4.39(10^{-3})$ m^4
$I_{x'y'} = -3.25(10^{-3})$ m^4
(b) $I_{x'} = 3.52(10^{-3})$ m^4
$I_{y'} = 6.93(10^{-3})$ m^4
$I_{x'y'} = 2.88(10^{-3})$ m^4
(c) $I_{x'} = 2.30(10^{-3})$ m^4
$I_{y'} = 8.14(10^{-3})$ m^4
$I_{x'y'} = 1.64(10^{-3})$ m^4

IX.49 (a) $I_{x'} = 1.682(10^{-3})$ m^4
$I_{y'} = 1.4318(10^{-2})$ m^4
$I_{x'y'} = 3.74(10^{-3})$ m^4
(b) $I_{x'} = 9.66(10^{-3})$ m^4
$I_{y'} = 6.34(10^{-3})$ m^4
$I_{x'y'} = -7.15(10^{-3})$ m^4
(c) $I_{x'} = 4.40(10^{-3})$ m^4
$I_{y'} = 1.16(10^{-3})$ m^4
$I_{x'y'} = 6.40(10^{-3})$ m^4

IX.51 (a) $I_{x'} = 0.531$ m^4
$I_{y'} = 2.058$ m^4
$I_{x'y'} = -0.788$ m^4
(b) $I_{x'} = 2.391$ m^4
$I_{y'} = 0.198$ m^4
$I_{x'y'} = -0.007$ m^4
(c) $I_{x'} = 0.240$ m^4
$I_{y'} = 2.349$ m^4
$I_{x'y'} = -0.301$ m^4

IX.53 $I_x = 356.4$ in.^{4}
$I_y = 90.6$ in.^{4}
$I_{xy} = 134.1$ in.^{4}

IX.55 $I_x = I_y = 0.0334$ m^4
$I_{xy} = 0$

IX.57 $\tilde{\theta} = 15.32°$ (cw)
$I_{\tilde{x}} = 1.874(10^{-3})$ m^4
$I_{\tilde{y}} = 8.574(10^{-3})$ m^4

IX.59 $\tilde{\theta} = 20.82°$ (ccw)
$I_{\tilde{x}} = 106$ in.^{4}
$I_{\tilde{y}} = 974$ in.^{4}

IX.61 $\tilde{x}$ is 22.37° ccw from rt horiz
$I_{\tilde{x}} = 412$ in.^{4}
$I_{\tilde{y}} = 34.7$ in.^{4}

IX.63 $\theta = 29.69°$ (ccw)
$I_{x'} = I_{y'} = 5.22(10^{-3})$ m^4
$I_{x'y'} = -3.35(10^{-3})$ m^4

IX.65 $\theta = 24.19°$ (cw)
$I_{x'} = I_{y'} = 540$ in.^{4}
$I_{x'y'} = 434$ in.^{4}

IX.67 x' is 22.63° cw from rt horiz
$I_{x'} = I_{y'} = 224$ in.^{4}
$I_{x'y'} = -189$ in.^{4}

INDEX

M

N

P

R

S

CONVERSION FACTORS

ALPHABETICAL LIST OF UNITS
U.S.-BRITISH UNITS TO SI UNITS

To convert from	to	Multiply by
atmosphere	pascal (Pa)	1.013 25 (10^5)
bar	pascal (Pa)	1.000 000 (10^5)*
British thermal unit	joule (J)	1.055 056 (10^3)
Btu/hour	watt (W)	2.930 711 (10^{-1})
centimeter of mercury (0°C)	pascal (Pa)	1.333 22 (10^3)
dyne	newton (N)	1.000 000 (10^{-5})*
dyne-centimeter	newton-meter (N-m)	1.000 000 (10^{-7})*
dyne/centimeter^2	pascal (Pa)	1.000 000 (10^{-1})*
erg	joule (J)	1.000 000 (10^{-7})*
erg/second	watt (W)	1.000 000 (10^{-7})*
foot	meter (m)	3.048 000 (10^{-1})*
foot^2	meter^2 (m^2)	9.290 304 (10^{-2})*
foot^3	meter^3 (m^3)	2.831 685 (10^{-2})
foot/second	meter/second (m/s)	3.048 000 (10^{-1})*
foot-pound	joule (J)	1.355 818
foot-pound/second	watt (W)	1.355 818
foot/second^2	meter/second^2 (m/s^2)	3.048 000 (10^{-1})*
gallon (U.S. liquid)	meter^3 (m^3)	3.785 412 (10^{-3})
gram/centimeter^3	kilogram/meter^3 (kg/m^3)	1.000 000 (10^3)*
horsepower (550 ft-lb/s)	watt (W)	7.456 999 (10^2)
inch	meter (m)	2.540 000 (10^{-2})*
inch^2	meter^2 (m^2)	6.451 600 (10^{-4})*

*Denotes an exact quantity.